W9-ASB-927

Readings in the Philosophy of Social Science

Readings in the Philosophy of Social Science

edited by Michael Martin and Lee C. McIntyre

A Bradford Book
The MIT Press
Cambridge, Massachusetts
London, England

Third printing, 1996

© 1994 Massachusetts Institute of Technology

All rights reserved. No part of this book may be reproduced in any form by any electronic or mechanical means (including photocopying, recording, or information storage and retrieval) without permission in writing from the publisher.

This book was set in Palatino by Asco Trade Typesetting Ltd., Hong Kong and was printed and bound in the United States of America.

Library of Congress Cataloging-in-Publication Data

Readings in the philosophy of social science / edited by Michael Martin and
Lee C. McIntyre.
p. cm.
Includes bibliographical references and index.
"A Bradford book."
ISBN 0-262-13296-6 (hc). — ISBN 0-262-63151-2 (pbk)
1. Social sciences—Philosophy. I. Martin, Michael, 1932 Feb. 3– II. McIntyre, Lee C.
H61.R33 1994
300'.1—dc20 93-43185
CIP

To Carl G. Hempel

Contents

Acknowledgments

Many people have aided us in editing this anthology. At The MIT Press, Betty Stanton and Teri Mendelsohn were always there when we needed them, cheerfully answering our questions and doing everything good editors do.

In the crucial selection process—deciding which papers and sections of books to include—we were advised by many friends and colleagues. Alexander Rosenberg, Daniel Little, Harold Kincaid, Merrilee Salmon, Alison Wylie, Owen Flanagan, Brian Fay, Jane Roland Martin, David Braybrooke, Jeremy Shearmur, Ian Jarvie, Valdir Ramalho, Dagfinn Føllesdal, Josh Cohen, Sarah Patterson, and Leon Goldstein made important suggestions, and we are grateful for their help. We wish we had been able to incorporate all of their ideas. As it is, we take full responsibility for the ultimate decision.

We especially thank Robert Cohen, Richard Boyd, and J. D. Trout, who gave us invaluable editorial advice and support. Special thanks also go to Ed Freedman for the enormous job of proofreading and indexing, and to Barbara Burg for assistance with the bibliography.

Finally, we thank our wives, Jane Roland Martin and Josephine Hernández. But for their moral support and understanding, this anthology would never have been completed.

Introduction

After a period of relative dormancy, the philosophy of social science is beginning to reemerge as a vibrant and exciting field of philosophical inquiry. This book is a manifestation of this emergence, as well as an attempt to further it. Anthologies such as May Brodbeck's *Readings in the Philosophy of the Social Sciences* and Leonard Krimerman's *The Nature and Scope of Social Science,* which almost defined the field through the 1970s, are now long out of print and out of date.[1] When they appeared, they encompassed the best work in the field, including both classical and (what were then) contemporary pieces; consequently, they framed the debate for a generation of scholarship, along the terms set out by their selections. Since then, work in the philosophy of social science has continued to move forward, fueled by new scholarship on issues such as the debate about interpretation, reduction, and problems in the special sciences, as well as continued focus on problems of longstanding concern in the field, such as the nature of explanation, prediction, objectivity, and functionalism.

Our aim in this anthology is to provide a comprehensive collection of both classical and contemporary pieces in order to bring the resources available in the field up to date. Indeed, since the early 1980s, there has been a resurgence of interest in the philosophy of social science, spawned by increasing methodological self-consciousness among practitioners of social science and recognition of the importance of closer ties between philosophical inquiry and ongoing work in the social sciences, as well as renewed interest in traditional philosophical debates about the nature of the relationship between the natural and social sciences. This revival of interest led to three introductory textbooks in the field: David Braybrooke's *Philosophy of Social Science,* Alexander Rosenberg's *Philosophy of Social Science,* and Daniel Little's *Varieties of Social Explanation.*[2] There has not, however, appeared anything like the wealth of supporting anthologic resources that were available in the late 1960s. Our hope is that this book will fill the gap.

Over the intervening years, however, there has been one constant: the dispute between the naturalist and the antinaturalist in the philosophy of social science. These two poles still define much of the debate in the field, even while their positions have been modified and made way for new contenders. Still, any resource for scholarship or teaching in the field must begin with an examination of the terms of this debate.

Naturalism, Antinaturalism, Pluralism, and Critical Social Science

Traditionally, there have been two basic approaches to the study of social phenomena, both of them well represented in this anthology: the "naturalist" and the "antinaturalist."[3] Briefly stated, the naturalist maintains that the social sciences should approach the study of social phenomena in the same way that the natural sciences have approached the study of natural phenomena—that the social sciences should have as their goals prediction and nomological explanation. Advocates of this approach have admitted that the search for social scientific laws may be more difficult than in the natural sciences (due to certain allegedly unique obstacles faced in the study of human behavior) and that the laws produced may be more general, statistical, or less well supported than they are in natural science. Nonetheless, they have denied that such differences represent any fundamental difference in kind between the problems generated by the underlying subject matter or the explanatory goals across natural and social science.

Antinaturalists deny what the naturalists assert. They hold that the social sciences should not approach the study of social phenomena in the same way as the natural sciences approach the study of nature because of basic differences in the subject matter at hand and differences in what we want to know about the phenomena in question. In particular, they maintain the social sciences should use the method of *Verstehen* in their attempt to understand social phenomena from the point of view of the social agent instead of focusing exclusively on the causal context. Advocates of this view claim that there are no laws in the social sciences, and even if there were, they would be impractical or irrelevant, given what it is that we seek to understand about social phenomena. Thus, by this view, there is not just a difference in degree concerning the methodology appropriate to the study of natural and social science but an unbridgeable gulf in the terms of investigation.

Although naturalism is no longer a popular view (and many contemporary philosophers of social science believe it is not a viable position), this controversy has persisted, albeit sometimes under different labels. For example, instead of appealing to *Verstehen* in describing their preferred approach, antinaturalists today speak about the interpretation of meaning or of hermeneutical understanding.[4] The thrust of the underlying hostility to nomological explanation, however, remains, and the wide gap between naturalism and antinaturalism has persisted.

To be sure, some philosophers of social science, taking a pluralistic stance, maintain that naturalism and antinaturalism are compatible or even complementary.[5] The two approaches are compatible, the pluralist argues, because social scientists can simultaneously pursue naturalist *and* antinaturalist research programs—for example, by attempting to explain the phenomenon of cargo cults by laws and by trying to interpret their meaning from the point of view of the natives. Of course, there are problems with this approach. One might argue that each social scientist is trying to answer a different question. Moreover, the individuals involved may not be doing the same kinds of research. Still, it could be argued that the definition of social science is wide enough to incorporate both research programs and could tolerate one anthropologist's asking why cargo cults developed under specific social and economic conditions and not under others, while in the process appealing to underlying laws, while another could be asking what the meaning of the cult might be for its participants, employing hermeneutical techniques. Moreover, the pluralist might also maintain that naturalism and antinaturalism are complementary because each program illuminates a different aspect of human action, both important to complete understanding. Such an approach lies at the foundation of David Braybrooke's text, and is explicitly represented in this anthology in chapter 2 by Brian Fay and J. Donald Moon.

Another approach has also been articulated recently, and it too is represented in this book. On this view the social sciences should expose the hidden biases and ideologies behind social thinking. The advocates of "critical social science" maintain that the deep unconscious prejudices we hold about class, race, and gender influence our research[6] and that the social sciences should raise them to consciousness so that we can escape their influence and become liberated. Feminist scholars, for instance, argue that the social sciences display a deep bias against women and have attempted to bring this to the surface in social research. Unlike the naturalist and antinaturalist approaches (where, despite their differences about what it is that we want explained, there is general agreement that the goal of social science is explanation), critical social science has as its direct goal human emancipation: liberation from prejudice, ignorance, and actual oppression.

Some scholars have argued that critical social science is actually dependent on naturalistic or antinaturalistic methodologies.[7] For example, it might be argued that in order to show that a particular social practice has a gender bias, inquiry into the meaning of the practice for its female practitioners, as well as its causal effect on women, would be indispensable. But even if the methodology needed to do critical social science is rooted in naturalism or antinaturalism, the uses to which these are put in critical social science are quite different from the goals of naturalism or antinaturalism. Thus, it is misleading to say that critical social science is reducible to naturalism or antinaturalism, just as it is to say that medicine is reducible to biology.

The Value of Philosophy of Social Science

We have outlined four basic approaches to the study of social phenomena. One might now ask, Why is the philosophy of social science worth studying at all? The answer, in short, is that the social sciences have as their subject some of the most fascinating practices in which humans engage and fulfill our desire to understand them. Yet many of us, while knowledgeable about the details of individual social sciences, have a confused or at least incomplete view of their purposes, methodologies, and implications. Philosophical study of the social sciences can help to eliminate such confusion by presenting a more balanced picture of the social sciences and by providing illumination of the epistemological context within which social science presents its discoveries. For example, many people assume without question that sociology will be reduced to psychology one day. Others are equally confident that such a reduction is impossible or inappropriate. As some of the chapters in this book indicate, those on both sides of the argument take too simple a view of the issue by attempting to settle it on largely a priori grounds. Understanding what reduction involves, therefore, will help to clarify the discussion, and, since reduction plays a crucial role in social scientific theorizing, it will illuminate the social sciences more generally.

Studying the philosophy of social science, therefore, not only provides the conceptual clarity that is needed to understand fundamental concepts used in social scientific theorizing, but it also can provide the tools to defend social science against its critics. For example, detractors of the social sciences have long maintained that these fields are inferior to the natural sciences and suffer from unique conceptual and epistemological problems. It has been argued that the social sciences are plagued by problems of reflexivity, complexity, the inability to perform controlled experiments, and subjective bias. These problems, it is alleged, uniquely implicate factors found in the study of social phenomena. However, as some of the chapters in this book make clear, the thesis of the inferiority of the social sciences is more of an open question than is usually supposed.

In addition, the philosophy of social science can help us come to grips with the inherent biases in social research and thus help us to improve our understanding of them and the policy implications that grow out of them. As several chapters in this book show, the objectivity of social scientific research is not something that can be taken for granted, since it is unclear in which instances, and in what way, it prevents understanding. And as some of the following chapters show, it may be possible to reduce such bias by appropriate research methods.

Different Concepts of the Philosophy of Social Science

What kind of inquiry is undertaken in the philosophy of social science? There have been several different conceptions of it.

One traditional conception is that philosophers should attempt to unify the various social sciences under some general perspective. One way to accomplish this would be for the philosopher to provide some classificatory scheme in which all of the major methods of the social sciences could be understood and related.[8] Another would be to construct a grand explanatory theory in which a variety of phenomena could be explained and understood.

There is no doubt that these two conceptions of the philosophy of social science have great historical importance and viability even today. Far too often it is assumed that the social sciences use a single method of investigation, and a classificatory scheme would be helpful in the recognition of this error. Furthermore, the attempt to unify a wide variety of social science data under one theoretical framework, if successful, would give a deeper understanding of social reality and would provide powerful theories that are perhaps analogous to those found in the physical sciences.

These views, however, are not without limitations. In some classificatory schemes, the social sciences as currently practiced are accepted in too uncritical a spirit. The philosopher of the social sciences should not just bring the social sciences under some unifying conception but should also critically evaluate their practices. This view of the philosophy of social science underestimates the normative value of philosophical analysis. Philosophers need to cast their nets not only widely but deeply. There is much that needs clarifying in the social sciences, and this takes detailed analysis rather than mere classification. For example, social scientists sometimes claim that a theory enables us to explain social phenomena. But the term *explain* is often used ambiguously in the literature. Part of the job of the philosopher of social science is to unravel such ambiguity and to provide needed clarification—to put social scientific inquiry on a more solid epistemological footing.

The conception of the philosophy of social science as a super-theory builder and synthesizer is not without problems. First, it is not clear that philosophers have any special training or talent for this sort of empirical theory building. Certainly in contemporary times the major attempts at this have been made by theoretically inclined social scientists rather than by philosophers. Second, it is not at all obvious that such a unifying theory is possible in the social sciences as currently practiced. Indeed, whether a unifying theory will be possible in the social sciences at any time has been questioned. Moreover, it is a source of some debate whether such a goal would be consistent with explanatory desiderata in the individual social sciences. Certainly the problems with this type of unification give one pause in viewing this as the goal of the philosophy of social science. This does not mean, however, that unifying theories do not provide valuable insights and points of departure for social inquiry. Indeed, some of the chapters included here show how these insights may be utilized.

Another conception of the philosophy of social science has the philosopher analyzing the concepts and methods of the social sciences. According to this view, which is well represented in current scholarship, the philosopher dissects the social sciences in order to reveal ambiguities and to expose gaps in argumentation. No doubt this approach makes a valuable contribution to social inquiry, and the special training of philosophers in logic and analysis makes them well suited to the job. However, this approach too has critics. First, exclusive concern with analysis can blind the philosopher to a more comprehensive picture of the social sciences. It is easy for philosophical analysis to become an end in itself instead of a tool that can be used for understanding and critical evaluation. Second, some philosophers have supposed that the analysis of concepts and methods of the social sciences can proceed entirely in an a priori fashion. However, the analysis of social scientific concepts and methods requires rather intimate knowledge of social scientific research, both substantive and methodological. Indeed, such insight has informed some of the deepest understanding of social scientific investigation of the past decade. Third, philosophical analysis may wrongly be used as a justification for the status quo, since the analysis of a concept is often taken as justification of it, even when it is not.

A third conception of the philosophy of the social sciences views the philosopher as a social critic, evaluating the social uses and abuses of the social sciences and specifying the responsibilities of social scientists by suggesting moral guidelines for social scientific research. Closely related to this conception of the philosophy of social science is one in which the philosopher exposes the prejudices and biases of particular social sciences. As we have seen, critical social science advances a view that is sympathetic with this conception of the philosophy of social science. But it is important to realize that not all philosophers who concern themselves with the ethical and social dimensions of social research would necessarily identify themselves as critical theorists. In any case, concerned as they are with ethical problems, philosophers can make useful contributions to this sort of social criticism of the practices of social science. Recent criticisms of social science by feminist scholars, some of them contained in this book, are examples of this approach.

Such a view of the philosophy of social science is surely not complete if it excludes from its task either the synthetic or analytic job of philosophy. These two tasks are not only compatible with social criticism; they complement it. As critical social scientists might well benefit from the insights of naturalistic and antinaturalistic social science, it may well be the case that a unified picture of the social sciences is needed in order to have a good grasp of the social scientist's responsibility or of individual implicit biases. Furthermore, philosophical analysis of certain key ideas in social science can be helpful in any critique of the uses of these findings.

A fourth conception of the philosophy of social science would have the philosopher constructing theories of social scientific growth in order to answer questions as to how social scientific knowledge best develops. Although this is an approach that has gained prominence in recent philosophy of science, its proponents argue that it was advocated by classical philosophers of science such as Mill and Bacon. Historical origins aside, although it has several advocates in the philosophy of natural science, this view is less well represented in the philosophy of social science.[9] Although no chapters represent this position in its pure form, several speak indirectly to the issue of how the social sciences best develop.

Philosophers can and surely should make contributions to theories of social scientific growth. Yet once again this view, when taken by itself, has its critics. It is not, however, incompatible with other conceptions of the philosophy of social science but, rather, complements them. In order to construct theories of social scientific growth, for instance, it may be useful to have a comprehensive view of the methodology of the social sciences.

Is Philosophy of Social Science Necessary?

The four conceptions of the philosophy of social science are important, but each one by itself is limited. This does not mean that every philosopher of social science needs to practice in accordance with all four. Some may specialize in building grand explanatory theories or classificatory schemes, some may favor constructing theories of scientific growth, others may specialize in the philosophical analysis of social scientific concepts and methods, and still others may concentrate on the social criticism of social scientific theories and their use in defining policy. Nevertheless, the work of a person engaged in one type of enterprise can be useful for someone engaged in another.

But the question now arises as to whether the philosophy of social science is really necessary at all. Could not all of the things that philosophers of social science do be done by social scientists? After all, theoretically minded social scientists have attempted to provide a synthetic picture of social scientific methodology, they have clarified and analyzed their concepts and methods, they have raised questions about the responsibility of social scientists, and they have criticized the uses of social scientific information. Social scientific information and research are essential to evaluating theories of social scientific growth, for instance, and it seems to be social scientists who are best able to make such evaluations. Given this, what special contribution can philosophers make?

In some ways, there is no sharp difference between the job of the social scientist and the philosopher of social science. This does not suggest that there is no difference at all, or that the philosophy of social science fulfills no useful function. Although they may share a subject matter, the social scientist and the philosopher have different skills and focus on different aspects of the issues at hand. The philosopher, for example, is usually concerned primarily with one of the four functions outlined and indeed is usually a specialist in one or more of them. But the typical social scientist approaches these four tasks, if at all, not as a primary function but as a preliminary to the empirical task at hand. As a consequence, philosophy of social science is needed, given the philosopher's unique perspective and training in the analysis of ideas. This said, it is also important to acknowledge that good philosophy of social science must be informed by the realization that if this work is to be useful to practicing social scientists, it must be tied to ongoing work in the field. Thus, one may come to appreciate the unique contribution of philosophy in thinking about social scientific inquiry, even while respecting the fact that, without social scientists, there would be nothing to think about.

Organization and Selection of Chapters

Every selection and organization of readings in an anthology indicates the editors' vision of the subject matter; this one is no exception. We have, however, made a deliberate effort to compile a collection that is balanced enough to be useful for both scholarship and teaching irrespective of one's ideological commitments and that is comprehensive enough to be used either as a companion to one of the excellent textbooks now available or as the sole resource. We have tried not to duplicate material that is readily available in specialized anthologies, on topics such as rationality, relativism, and interpretation, but only to provide representative selections of work in these areas.[10] In contrast, we have strived to include pieces that are of relevance to practitioners of social science, based on special problems in their disciplines, in Part VIII of this work. All other topics of more traditional philosophical concern have also been covered, throughout the other sections.

The readings in Part I set the stage by defending the social sciences from the charge of inferiority and provide an overview of the necessary components of a philosophy of social science that is either nomological, interpretive, or critical. Part II, on explanation, prediction, and laws, presents selections that both defend and criticize the naturalist's goal of nomological social scientific explanation. In Part III, we take up the issues of interpretation and meaning, whose study has been offered as an alternative to nomological social science, including material that both defends and criticizes an interpretive approach to social scientific understanding. The problem of rationality has been crucial in advancing such interpretive explanations but is also an important issue in its own right and is explored in part IV. Part V, on functional explanation, contains chapters analyzing another important type of social scientific explanation. Part VI, on reductionism, individualism, and holism, confronts one of the most difficult sets of issues within the philosophy of social science: whether social science should deal with wholes, such as cultures and institutions, or with individuals, and with whether there is or should be a reductive relationship between these two levels of understanding. The chapters in Part VII reveal that much of social science is based on bias and ideology and consider whether and how social science can be objective. Finally, in Part VIII, we consider problems of the special sciences: economics, psychology, history, and archaeology, in which some of the more general debates considered in this anthology are given even more particular focus.

Notes

1. May Brodbeck, *Readings in the Philosophy of the Social Sciences* (London: Macmillan, 1968), Leonard Krimerman, *The Nature and Scope of Social Science* (New York: Appleton-Century-Crofts, 1969).
2. David Braybrooke, *Philosophy of Social Science* (Englewood-Cliffs, N.J.: Prentice-Hall, 1987); Alexander Rosenberg, *Philosophy of Social Science* (Boulder, Colo.: Westview Press, 1988); Daniel Little, *Varieties of Social Explanation* (Boulder, Colo.: Westview Press, 1991).
3. For recent attempts to characterize these two approaches, see Braybrooke, *Philosophy*, chap. 1, and Little, *Varieties*, chap. 11. Maurice Natanson, *Philosophy of Social Science: A Reader* (New York: Random House, 1963).
4. See Paul Rabinow and William Sullivan, *Interpretive Social Science: A Second Look* (Berkeley: University of California Press, 1987).
5. See Braybrooke, *Philosophy* and Little, *Varieties.*
6. See Braybrooke, *Philosophy*, chaps. 1, 4. See also Brian Fay, *Critical Social Science* (Ithaca: Cornell University Press, 1987).
7. Braybrooke, *Philosophy*, p. 69.

8. Paul Diesing attempts to do this in *Patterns of Discovery in the Social Sciences* (Chicago: Aldine-Atherton, 1971). Some of Diesing's views are evaluated in Michael Martin, "The Philosopher of Social Science as Participant Observer", in *Social Science and Philosophical Analysis* (Washington, D.C.: University Press of America, 1978), pp. 422–433.
9. See Imre Lakatos and Alan Musgrave, *Criticism and the Growth of Knowledge* (Cambridge: Cambridge University Press, 1970).
10. Such as, Bryan Wilson (ed.), *Rationality* (Oxford: Blackwell, 1970); Martin Hollis and Steven Lukes (eds.), *Rationality and Relativism* (Cambridge: MIT Press, 1982); Paul Rabinow and William Sullivan (eds.), *Interpretive Social Science: A Second Look* (Berkeley: University of California Press, 1987).

Sources and Acknowledgments

Fritz Machlup
"Are the Social Sciences Really Inferior?"
From *Southern Economic Journal*, vol. 17 (1961), pp. 173–184, by permission of the journal.

Brian Fay and J. Donald Moon
"What Would an Adequate Philosophy of Social Science Look Like?"
From *Philosophy of Social Science*, vol. 7 (1977), pp. 209–227, by permission of the authors and journal.

Carl G. Hempel
"The Function of General Laws in History"
From *Journal of Philosophy*, vol. 39 (1942), pp. 35–48, by permission of the author and journal.

F. A. Hayek
"The Theory of Complex Phenomena"
From *Studies in Philosophy, Politics and Economics* (Chicago: The University of Chicago Press, 1967), pp. 22–42. Copyright © 1967 by the University of Chicago. Reprinted with the permission of the publisher and Stephen Kresge. Permission to reprint in the United Kingdom and British Empire, excluding Canada, granted by Routledge and Kegan Paul.

Michael Scriven
"A Possible Distinction between Traditional Scientific Disciplines and the Study of Human Behavior"
From H. Feigl and M. Scriven (eds.), *Minnesota Studies in the Philosophy of Science*, vol. 1 (1956), pp. 330–339, published by University of Minnesota Press. Copyright © 1956 by the University of Minnesota. Reprinted with the permission of the author and publisher.

Donald Davidson
"Psychology as Philosophy"
From *Philosophy of Psychology*, edited by S. C. Brown (London: Macmillan Press, 1974), pp. 41–52. Reprinted by permission of The Royal Institute of Philosophy and the author.

Brian Fay
"General Laws and Explaining Human Behavior"
From Daniel R. Sabia and Jerald Wallulis (eds.), *Changing Social Science* (Albany: SUNY Press, 1983), pp. 103–128, by permission of the author and the State University of New York Press. Copyright © 1983 by State University of New York.

Harold Kincaid
"Defending Laws in the Social Sciences"
From *Philosophy of Social Science*, vol. 20 (1990), pp. 56–83, by permission of the author and journal.

Lee McIntyre
"'Complexity' and Social Scientific Laws"
From *Synthese*, vol. 97 (1993). Copyright © 1993 Kluwer Academic Publishers. Reprinted, with slight modifications to the original, by permission of Kluwer Academic Publishers and the author.

George Romanos
"Reflexive Predictions"
From *Philosophy of Science*, vol. 40 (1973), pp. 97–109, by permission of the author and the Philosophy of Science Association.

R. G. Collingwood
"Human Nature and Human History"
pp. 205–217 From *The Idea of History* by R. G. Collingwood (1946). Reprinted by permission of Oxford University Press.

William Dray
"The Rationale of Action"
pp. 118–131 from *Laws and Explanation in History* by William Dray (1957). Reprinted by permission of Oxford University Press and the author.

Charles Taylor
"Interpretation and the Sciences of Man"
From *Review of Metaphysics*, vol. 25 (1971), pp. 3–51, by permission of the author and publisher.

Clifford Geertz
"Thick Description: Toward an Interpretive Theory of Culture"
Excerpt from *The Interpretation of Cultures* by Clifford Geertz. Copyright © 1983 by Clifford Geertz. Reprinted by permission of Basic Books, a division of HarperCollins Publishers, and the author.

Dagfinn Føllesdal
"Hermeneutics and the Hypothetico-Deductive Method"
From *Dialectica*, vol. 33 (1979), pp. 319–336, by permission of the author and publisher.

Jane Roland Martin
"Another Look at the Doctrine of *Verstehen*"
From *British Journal for the Philosophy of Science*, vol. 20 (1969), pp. 53–67, by permission of the author and publisher.

Michael Martin
"Taylor on Interpretation and the Sciences of Man"
This chapter appears for the first time in this volume.

Steven Lukes
"Some Problems about Rationality"
From *Archives Européennes de Sociologie*, vol. 8 (1967), pp. 247–264, reprinted with the permission of the author and publisher.

Dagfinn Føllesdal
"The Status of Rationality Assumptions in Interpretation and the Explanation of Action"
From *Dialectica*, vol. 36 (1982), pp. 301–316, by permission of the author and publisher.

Jon Elster
"The Nature and Scope of Rational-Choice Explanation"
From E. LePore and B. McLaughlin (eds.), *Actions and Events: Perspectives on Donald Davidson* (Oxford: Blackwell Publishers, 1985), pp. 60–72, by permission of the author and publisher.

David Henderson
"The Principle of Charity and the Problem of Irrationality (Translation and the Problem of Irrationality)"
From *Synthese*, vol. 73 (1987), pp. 225–252. Copyright © 1987 Kluwer Academic Publishers. Reprinted by permission of Kluwer Academic Publishers and the author.

Carl G. Hempel
"The Logic of Functional Analysis"
From *Symposium on Sociological Theory* by Llewellyn Gross. Copyright © 1959 by Harper and Row Publishers, Inc. Reprinted by permission of HarperCollins Publishers and the author.

R. P. Dore
"Function and Cause"
From *American Sociological Review*, vol. 16 (1961), pp. 843–853, by permission of the author.

G. A. Cohen
"Functional Explanation in Marxism"
From Cohen, G. A., *Karl Marx's Theory of History: A Defense* (Princeton: Princeton University Press, 1978), pp. 278–296. Copyright © 1978 by Princeton University Press. Reprinted with the permission of Princeton University Press and the author.

Jon Elster
"Functional Explanation"
From *Explaining Technical Change* (Cambridge: Cambridge University Press, 1983), pp. 55–68, 241–243. Copyright © Cambridge University Press and Universitetsforlaget 1983. Reprinted with the permission of Cambridge University Press and the author.

Harold Kincaid
"Assessing Functional Explanation in the Social Sciences"
From A. Fine, M. Forbes, and L. Wessels (eds.), *PSA 1990*, vol. I, pp. 341–354, by permission of the Philosophy of Science Association and the author.

Emile Durkheim
"Social Facts"
Reprinted with the permission of The Free Press, a Division of MacMillan, Inc. from *The Rules of Sociological Method* by Emile Durkheim, translated by Sarah A. Solovay and John H. Mueller. Edited by George E. G. Catlin. Copyright © 1938 by George E. G. Catlin; copyright renewed 1966 by Sarah Solovay, John H. Mueller, and George E. G. Catlin.

J. W. N. Watkins
"Historical Explanations in the Social Sciences"
From *British Journal for the Philosophy of Science*, vol. 8 (1957), pp. 104–117, by permission of the author and publisher.

Steven Lukes
"Methodological Individualism Reconsidered"
From *British Journal of Sociology*, vol. 19 (1968), pp. 119–129, by permission of Routledge Publishers, the journal, and the author.

Richard Miller
"Methodological Individualism and Social Explanation"
From *Philosophy of Science*, vol. 45 (1978), pp. 387–414, by permission of the author and the Philosophy of Science Association.

Daniel Little
"Microfoundations of Marxism"
This chapter appears for the first time in this volume.

Harold Kincaid
"Reduction, Explanation and Individualism"
From *Philosophy of Science*, vol. 53 (1986), pp. 492–513, by permission of the author and the Philosophy of Science Association.

Alan Nelson
"Social Science and the Mental"
Selection taken from *Midwest Studies in Philosophy, Volume XV, Philosophy of Human Sciences*, edited by P. French, T. Uehling, and H. Wettstein. © 1990 by University of Notre Dame Press. Reprinted by permission of the publisher and author.

Max Weber
"'Objectivity' in Social Science and Social Policy"
Reprinted with the permission of The Free Press, a Division of Macmillan, Inc. from *The Methodology of the Social Sciences* by Max Weber, translated and edited by Edward A. Shils and Henry A. Finch. Copyright © 1949 by the Free Press; copyright renewed 1977 by Edward A. Shils.

Charles Taylor
"Neutrality in Political Science"
From *Philosophy, Politics and Society,* 3d series, ed. P. Laslett and W. G. Runciman (Oxford: Blackwell, 1967), pp. 25–57, by permission of the author.

Ernest Nagel
"The Value-Oriented Bias of Social Inquiry"
From Ernest Nagel, *The Structure of Science,* 1979, Hackett Publishing Co., Inc., Indianapolis, IN, and Cambridge, MA. Reprinted with permission.

Michael Martin
"The Philosophical Importance of the Rosenthal Effect"
From *Journal for the Theory of Social Behavior,* vol. 7 (1977), pp. 81–96, by permission of the author and journal.

Naomi Weisstein
"Psychology Constructs the Female"
From *Social Education,* vol. 35 (1971), pp. 362–373, by permission of the author.

Alison Wylie
"Reasoning about Ourselves: Feminist Methodology in the Social Sciences"
From E. Harvey and K. Okruhik (eds.), *Women and Reason* (1992). Copyright © 1992 by The University of Michigan Press. Reprinted with the permission of the publisher and author.

Donald Comstock
"A Method of Critical Research"
From E. Bredo and W. Feinberg (eds.), *Knowledge and Values in Social and Educational Research* (Philadelphia: Temple University Press, 1982), pp. 370–390. Reprinted with the permission of the publisher and author.

Milton Friedman
"The Methodology of Positive Economics"
From *Essays in Positive Economics* (Chicago: University of Chicago Press, 1953), pp. 3–23, 39–43. Copyright © 1953 by University of Chicago. Reprinted with the permission of the publisher and author.

Alexander Rosenberg
"If Economics Isn't Science, What Is It?"
From *Philosophical Forum,* vol. 14 (1983), pp. 296–314, by permission of the author and journal.

Donald Davidson
"Actions, Reasons and Causes"
© Donald Davidson 1963. Reprinted from *Essays on Actions and Events* by Donald Davidson (1980) by permission of Oxford University Press, and the author.

Jerry Fodor
"Special Sciences, or the Disunity of Science as a Working Hypothesis"
From *Synthese,* vol. 28 (1974), pp. 97–115. Copyright © 1974 Kluwer Academic Publishers. Reprinted by permission of Kluwer Academic Publishers and the author.

Paul Roth
"Narrative Explanations: The Case of History"
Copyright © 1988 Wesleyan University. Reprinted with permission from *History and Theory,* 27 (1988), 1–13, and from the author.

Louis O. Mink
"The Autonomy of Historical Understanding"
Copyright © 1965 Wesleyan University. Reprinted with permission from *History and Theory,* 5 (1965), 24–47.

Merrilee Salmon
"On the Possibility of Lawful Explanation in Archaeology"
From *Critica,* vol. 22 (1990), pp. 87–112. Reprinted by permission of *Critica,* Instituto de Investigaciones Filosoficas, Universidad Nacional Autonoma de Mexico. Reprinted by permission of the publisher and author.

Alison Wylie
"Evidential Constraints: Pragmatic Objectivism in Archaeology."
This chapter appears for the first time in this volume.

PART I

Introduction

Introduction to Part I

It is fitting to begin with two chapters that set the stage for much of what follows and pose in their titles two crucial questions about the social sciences: Fritz Machlup's "Are the Social Sciences Really Inferior?" and Brian Fay and J. Donald Moon's "What Would an Adequate Philosophy of Social Science Look Like?"

The social sciences have often been compared invidiously to the natural sciences, with some critics going so far as to claim that economics, sociology, anthropology, and the like are not truly sciences. Considering the question of whether the social sciences are inferior to the natural sciences in relation to nine "grounds of comparison," Machlup holds that with respect to some grounds (e.g., the invariability of observation), the social sciences are inferior, while with respect to others (e.g., objectivity), they are not. Machlup's final score card suggests that the social sciences may well be more scientifically respectable than their critics would allow. Furthermore, he claims that even if they are inferior in some respects, little follows from this admission. In particular, it does not follow that we should stop studying them or spending money on social science research.

Defenders of naturalism in the social sciences might argue that even Machlup's strong defense of the status of the social sciences can be strengthened. He concludes that the grounds in relation to which the social sciences are inferior cannot be remedied. Thus, he maintains, for example, that social phenomena are more complex than natural phenomena because there are more relevant factors to consider and that this is something that cannot be changed but has "to be grasped, accepted, and taken into account." Machlup seems to assume here that one phenomenon is more complex than another independently of the way it is described and conceptualized. But is this so? Is there any a priori reason to suppose that descriptions and conceptualizations of social phenomena in terms of new theories might not make them as complex or simple as natural phenomena? For their part, antinaturalists would point out that Machlup assumes the appropriateness of comparing the social and natural sciences: he assumes that the same standards are relevant to both fields and that the two should be judged according to the same terms. But antinaturalists would maintain that the goals and methodologies of the social sciences are fundamentally different from those of the natural sciences and that comparisons are not appropriate.

Is there any way to reconcile the conflict between naturalism and antinaturalism? In chapter 2, Fay and Moon present a pluralistic approach to the social sciences, one in which naturalistic, antinaturalistic, and critical elements complement one another. In so doing they indirectly raise fundamental questions about the social sciences.

Fay and Moon maintain that neither the naturalistic nor the interpretive position—what we have called the antinaturalistic position—can answer three questions that any adequate philosophy of the social sciences must answer: What is the relation between interpretation and explanation? What is the nature of scientific theory? What is the role of critique? Interpreting human action is essential to social science, they argue, for descriptions in terms of human action can neither be reduced to behavior nor eliminated. In this respect, the interpretive position seems right and naturalism wrong. In their view, however, this does not mean that social science consists only in the interpretation of human action, as the interpretive position claims. Maintaining that social scientists give causal explanations of actions and of many other phenomena, they conclude that neither position provides an adequate account of the relation between interpretation and explanation. Nor, according to these authors, does either position give an adequate account of social scientific theory. The interpretive position neglects the topic entirely, they say, and the naturalistic position wrongly assimilates social scientific theories to naturalistic ones. Further, Fay and Moon hold that neither position gives an account of the role of critique. Whereas the interpretive position is unable to transcend the limited and perhaps false and irrational perspective of social actors or to show that their outlook is mistaken, the naturalistic position is unable to assess the rationality of their beliefs since it is not concerned with the rationality of belief systems.

How does Machlup's comparison relate to Fay and Moon's approach? If Fay and Moon are correct, there are some dimensions—the interpretive and critical—of the social sciences that Machlup's approach does not elucidate and perhaps even obscures. For these dimensions, comparisons other than to the natural sciences would be appropriate. For example, the interpretive aspect of the social sciences might be illuminated by comparing it to literary interpretations, while light might be shed on the critical dimension by relating it to the insights and awareness gained through psychoanalysis. Nevertheless, the usefulness of drawing these other comparisons does not gainsay Machlup's argument so long as it is simply construed as maintaining that one important aspect of the social sciences compares more favorably with the natural sciences than is usually believed.

Chapter 1

Are the Social Sciences Really Inferior?

Fritz Machlup

If we ask whether the "social sciences" are "really inferior," let us first make sure that we understand each part of the question.

"Inferior" to what? Of course to the natural sciences. "Inferior" in what respect? It will be our main task to examine all the "respects," all the scores on which such inferiority has been alleged. I shall enumerate them presently.

The adverb *"really"* which qualifies the adjective "inferior" refers to allegations made by some scientists, scholars, and laymen. But it refers also to the "inferiority complex" which I have noted among many social scientists. A few years ago I wrote an essay entitled "The Inferiority Complex of the Social Sciences."[1] In that essay I said that "an inferiority complex may or may not be justified by some 'objective' standards," and I went on to discuss the consequences which "the *feeling* of inferiority"—conscious or subconscious—has for the behavior of the social scientists who are suffering from it. I did not then discuss whether the complex has an objective basis, that is, whether the social sciences are "really" inferior. This is our question to-day.

The subject noun would call for a long disquisition. What is meant by *"social sciences,"* what is included, what is not included? Are they the same as what others have referred to as the "moral sciences," the "Geisteswissenschaften," the "cultural sciences," the "behavioral sciences"? Is Geography, or that part of it that is called "Human Geography," a social science? Is History a social science—or perhaps even *the* social science *par excellence,* as some philosophers have contended? We shall not spend time on this business of defining and classifying. A few remarks may later be necessary in connection with some points of methodology, but by and large we shall not bother here with a definition of "social sciences" and with drawing boundary lines around them.

The Grounds of Comparison

The social sciences and the natural sciences are compared and contrasted on many scores, and the discussions are often quite unsystematic. If we try to review them systematically, we shall encounter a good deal of overlap and unavoidable duplication. Nonetheless, it will help if we enumerate in advance some of the grounds of comparison most often mentioned, grounds on which the social sciences are judged to come out "second best":

1. Invariability of observations
2. Objectivity of observations and explanations
3. Verifiability of hypotheses
4. Exactness of findings
5. Measurability of phenomena

6. Constancy of numerical relationships
7. Predictability of future events
8. Distance from everyday experience
9. Standards of admission and requirements

We shall examine all these comparisons.

Invariability of Observations

The idea is that you cannot have much of a science unless things recur, unless phenomena repeat themselves. In nature we find many factors and conditions "invariant." Do we in society? Are not conditions in society changing all the time, and so fast that most events are unique, each quite different from anything that has happened before? Or can one rely on the saying that "history repeats itself" with sufficient invariance to permit generalizations about social events?

There is a great deal of truth, and important truth, in this comparison. Some philosophers were so impressed with the invariance of nature and the variability of social phenomena that they used this difference as the criterion in the definitions of natural and cultural sciences. Following Windelband's distinction between generalizing ("nomothetic") and individualizing ("ideographic") propositions, the German philosopher Heinrich Rickert distinguished between the generalizing sciences of nature and the individualizing sciences of cultural phenomena; and by individualizing sciences he meant historical sciences.[2] In order to be right, he redefined both "nature" and "history" by stating that reality is "nature" if we deal with it in terms of the *general* but becomes "history" if we deal with it in terms of the *unique*. To him, geology was largely history, and economics, most similar to physics, was a natural science. This implies a rejection of the contention that all fields which are normally called social sciences suffer from a lack of invariance; indeed, economics is here considered so much a matter of immutable laws of nature that it is handed over to the natural sciences.

This is not satisfactory, nor does it dispose of the main issue that natural phenomena provide *more* invariance than social phenomena. The main difference lies probably in the number of factors that must be taken into account in explanations and predictions of natural and social events. Only a small number of reproducible facts will normally be involved in a physical explanation or prediction. A much larger number of facts, some of them probably unique historical events, will be found relevant in an explanation or prediction of economic or other social events. This is true, and methodological devices will not do away with the difference. But it is, of course, only a difference in degree.

The physicist Robert Oppenheimer once raised the question whether, if the universe is *a unique* phenomenon, we may assume that *universal* or *general* propositions can be formulated about it. Economists of the Historical School insisted on treating each "stage" or phase of economic society as a completely unique one, not permitting the formulation of universal propositions. Yet, in the physical world, phenomena are not quite so homogeneous as many have liked to think; and in the social world, phenomena are not quite so heterogeneous as many have been afraid they are. (If they were, we could not even have generalized concepts of social events and words naming them.) In any case, where reality seems to show a bewildering number of variations, we construct an ideal world of abstract models in which we create enough homogeneity to permit us to apply reason and deduce the implied consequences of assumed constellations. This artificial homogenization of types of phenomena is carried out in natural and social sciences alike.

There is thus no difference in invariance in the sequences of events in nature and in society as long as we theorize about them—because in the abstract models homogeneity is assumed. There is only a difference of degree in the variability of phenomena of nature and society if we talk about the real world—as long as heterogeneity is not reduced by means of deliberate "controls." There is a third world, between the abstract world of theory and the real unmanipulated world, namely, the artificial world of the experimental laboratory. In this world there is less variability than in the real world and more than in the model world. But this third world does not exist in most of the social sciences (nor in all natural sciences). We shall see later that the mistake is often made of comparing the artificial laboratory world of manipulated nature with the real world of unmanipulated society.

We conclude on this point of comparative invariance, that there is indeed a difference between natural and social sciences, and that the difference—apart from the possibility of laboratory experiments—lies chiefly in the number of relevant factors, and hence of possible combinations, to be taken into account for explaining or predicting events occurring in the real world.

Objectivity of Observations and Explanations

The idea behind a comparison between the "objectivity" of observations and explorations in the natural and social sciences may be conveyed by an imaginary quotation: "Science must be objective and not affected by value judgments; but the social sciences are inherently concerned with values and, hence, they lack the disinterested objectivity of science." True? Frightfully muddled. The trouble is that the problem of "subjective value," which is at the very root of the social sciences, is quite delicate and has in fact confused many, including some fine scholars.

To remove confusion one must separate the different meanings of "value" and the different ways in which they relate to the social sciences, particularly economics. I have distinguished eleven different kinds of value-reference in economics, but have enough sense to spare you this exhibition of my pedagogic dissecting zeal. But we cannot dispense entirely with the problem and overlook the danger of confusion. Thus, I offer you a bargain and shall reduce my distinctions from eleven to four. I am asking you to keep apart the following four meanings in which value judgment may come into our present discussion: (1) The analyst's judgment may be biased for one reason or another, perhaps because his views of the social "Good" or his personal pecuniary interests in the practical use of his findings interfere with the proper scientific detachment. (2) Some normative issues may be connected with the problem under investigation, perhaps ethical judgments which may color some of the investigator's incidental pronouncements—obiter dicta—without however causing a bias in his reported findings of his research. (3) The interest in solving the problems under investigation is surely affected by values since, after all, the investigator selects his problems because he believes that their solution would be of value. (4) The investigator in the social sciences has to explain his observations as results of human actions which can be interpreted only with reference to motives and purposes of the actors, that is, to values entertained by them.

With regard to the first of these possibilities, some authorities have held that the social sciences may more easily succumb to temptation and may show obvious biases.

The philosopher Morris Cohen, for example, spoke of "the subjective difficulty of maintaining scientific detachment in the study of human affairs. Few human beings can calmly and with equal fairness consider both sides of a question such as socialism, free love, or birth-control."[3] This is quite true, but one should not forget similar difficulties in the natural sciences. Remember the difficulties which, in deference to religious values, biologists had in discussions of evolution and, going further back, the troubles of astronomers in discussions of the heliocentric theory and of geologists in discussions of the age of the earth. Let us also recall that only 25 years ago [1936] German mathematicians and physicists rejected "Jewish" theorems and theories, including physical relativity, under the pressure of nationalistic values, and only ten years ago [1951] Russian biologists stuck to a mutation theory which was evidently affected by political values. I do not know whether one cannot detect in our own period here in the United States an association between political views and scientific answers to the question of the genetic dangers from fallout and from other nuclear testing.

Apart from political bias, there have been cases of real cheating in science. Think of physical anthropology and its faked Piltdown Man. That the possibility of deception is not entirely beyond the pale of experimental scientists can be gathered from a splendid piece of fiction, a recent novel, *The Affair*, by C. P. Snow, the well-known Cambridge don.

Having said all this about the possibility of bias existing in the presentation of evidence and findings in the natural sciences, we should hasten to admit that not a few economists, especially when concerned with current problems and the interpretation of recent history, are given to "lying with statistics." It is hardly a coincidence if labor economists choose one base year and business economists choose another base year when they compare wage increases and price increases; or if for their computations of growth rates expert witnesses for different political parties choose different statistical series and different base years. This does not indicate that the social sciences are in this respect "superior" or "inferior" to the natural sciences. Think of physicists, chemists, medical scientists, psychiatrists, etc., appearing as expert witnesses in court litigation to testify in support of their clients' cases. In these instances the scientists are in the role of analyzing concrete individual events, of interpreting recent history. If there is a difference at all between the natural and social sciences in this respect, it may be that economists these days have more opportunities to present biased findings than their colleagues in the physical sciences. But even this may not be so. I may underestimate the opportunities of scientists and engineers to submit expert testimonies with paid-for bias.

The second way in which value judgments may affect the investigator does not involve any bias in his findings or his reports on his findings. But ethical judgments may be so closely connected with his problems that he may feel impelled to make evaluative pronouncements on the normative issues in question. For example, scientists may have strong views about vivisection, sterilization, abortion, hydrogen bombs, biological warfare, etc., and may express these views in connection with their scientific work. Likewise, social scientists may have strong views about the right to privacy, free enterprise, free markets, equality of income, old-age pensions, socialized medicine, segregation, education, etc., and they may express these views in connection with the results of their research. Let us repeat that this need not imply that their findings are biased. There is no difference on this score between the natural and the social sciences. The research and its results may be closely connected with values of all sorts, and value judgments may be expressed, and yet the objectivity of the research and of the reports on the findings need not be impaired.

The third way value judgments affect research is in the selection of the project, in the choice of the subject for investigation. This is unavoidable and the only question is what kinds of value and whose values are paramount. If research is financed by foundations or by the government, the values may be those which the chief investigator believes are held by the agencies or committees that pass on the allocation of funds. If the research is not aided by outside funds, the project may be chosen on the basis of what the investigator believes to be "social values," that is, he chooses a project that may yield solutions to problems supposed to be important for society. Society wants to know how to cure cancer, how to prevent hay fever, how to eliminate mosquitoes, how to get rid of crab grass and weeds, how to restrain juvenile delinquency, how to reduce illegitimacy and other accidents, how to increase employment, to raise real wages, to aid farmers, to avoid price inflation, and so on, and so forth. These examples suggest that the value component in the project selection is the same in the natural and in the social sciences. There are instances, thank God, in which the investigator selects his project out of sheer intellectual curiosity and does not give "two hoots" about the social importance of his findings. Still, to satisfy curiosity is a value too, and indeed a very potent one. We must not fail to mention the case of the graduate student who lacks imagination as well as intellectual curiosity and undertakes a project just because it is the only one he can think of, though neither he nor anybody else finds it interesting, let alone important. We may accept this case as the exception to the rule. Such exceptions probably are equally rare in the natural and the social sciences.

Now we come to the one real difference, the fourth of our value-references. Social phenomena are defined as results of human action, and all human action is defined as motivated action. Hence, social phenomena are explained only if they are attributed to definite types of action which are "understood" in terms of the values motivating those who decide and act. This concern with values—not values which the investigator entertains but values he understands to be effective in guiding the actions which bring about the events he studies—is the crucial difference between the social sciences and the natural sciences. To explain the motion of molecules, the fusion or fission of atoms, the paths of celestial bodies, the growth or mutation of organic matter, etc., the scientist will not ask why the molecules want to move about, why atoms decide to merge or to split, why Venus has chosen her particular orbit, why certain cells are anxious to divide. The social scientist, however, is not doing his job unless he explains changes in the circulation of money by going back to the decisions of the spenders and hoarders, explains company mergers by the goals that may have persuaded managements and boards of corporate bodies to take such actions, explains the location of industries by calculations of such things as transportation costs and wage differentials, and economic growth by propensities to save, to invest, to innovate, to procreate or prevent procreation, and so on. My social-science examples were all from economics, but I might just as well have taken examples from sociology, cultural anthropology, political science, etc., to show that explanation in the social sciences regularly requires the interpretation of phenomena in terms of idealized motivations of the idealized persons whose idealized actions bring forth the phenomena under investigation.

An example may further elucidate the difference between the explanatory principles in nonhuman nature and human society. A rock does not say to us: "I am a beast,"[4] nor does it say: "I came here because I did not like it up there near the glaciers, where I used to live: here I like it fine, especially this nice view of the valley." We do not inquire into value judgments of rocks. But we must not fail to take account of valuations of humans; social phenomena must be explained as the results of motivated human actions.

The greatest authorities on the methodology of the social sciences have referred to this fundamental postulate as the requirement of "subjective interpretation," and all such interpretation of "subjective meanings" implies references to values motivating actions. This has of course nothing to do with value judgments impairing the "scientific objectivity" of the investigators or affecting them in any way that would make their findings suspect. Whether the postulate of subjective interpretation which *differentiates* the social sciences from the natural sciences should be held to make them either "inferior" or "superior" is a matter of taste.

Verifiability of Hypotheses

It is said that verification is not easy to come by in the social sciences, while it is the chief business of the investigator in the natural sciences. This is true, though many do not fully understand what is involved and, consequently, are apt to exaggerate the difference.

One should distinguish between what a British philosopher has recently called "high-level hypotheses" and "low-level generalizations."[5] The former are postulated and can never be *directly* verified; a single high-level hypothesis cannot even be *indirectly* verified, because from one hypothesis standing alone nothing follows. Only a *whole system* of hypotheses can be tested by deducing from some set of general postulates and some set of specific assumptions the logical consequences, and comparing these with records of observations regarded as the approximate empirical counterparts of the specific assumptions and specific consequences.[6] This holds for both the natural and the social sciences. (There is no need for *direct* tests of the fundamental postulates in physics—such as the laws of conservation of energy, of angular momentum, of motion—or of the fundamental postulates in economics—such as the laws of maximizing utility and profits.)

While entire theoretical systems and the low-level generalizations derived from them are tested in the natural sciences, there exist at any one time many unverified hypotheses. This holds especially with regard to theories of creation and evolution in such fields as biology, geology, and cosmogony; for example (if my reading is correct), of the theory of the expanding universe, the dust-cloud hypothesis of the formation of stars and planets, of the low-temperature or high-temperature theories of the formation of the earth, of the various (conflicting) theories of granitization, etc. In other words, where the natural sciences deal with nonreproducible occurrences and with sequences for which controlled experiments cannot be devised, they have to work with hypotheses which remain untested for a long time, perhaps forever.

In the social sciences, low-level generalizations about recurring events are being tested all the time. Unfortunately, often several conflicting hypotheses are consistent with the observed facts and there are no crucial experiments to eliminate some of the hypotheses. But every one of us could name dozens of propositions that have been disconfirmed, and this means that the verification process has done what it is supposed to do. The impossibility of controlled experiments and the relatively large number of relevant variables are the chief obstacles to more efficient verification in the social sciences. This is not an inefficiency on the part of our investigators, but it lies in the nature of things.

Exactness of Findings
Those who claim that the social sciences are "less exact" than the natural sciences often have a very incomplete knowledge of either of them, and a rather hazy idea of the meaning of "exactness." Some mean by exactness measurability. This we shall discuss under a separate heading. Others mean accuracy and success in predicting future events, which is something different. Others mean reducibility to mathematical language. The meaning of exactness best founded in intellectual history is the possibility of constructing a theoretical system of idealized models containing abstract constructs of variables and of relations between variables, from which most or all propositions concerning particular connections can be deduced. Such systems do not exist in several of the natural sciences—for example, in several areas of biology—while they do exist in at least one of the social sciences: economics.

We cannot foretell the development of any discipline. We cannot say now whether there will soon or ever be a "unified theory" of political science, or whether the piecemeal generalizations which sociology has yielded thus far can be integrated into one comprehensive theoretical system. In any case, the quality of "exactness," if this is what is meant by it, cannot be attributed to all the natural sciences nor denied to all the social sciences.

Measurability of Phenomena
If the availability of numerical data were in and of itself an advantage in scientific investigation, economics would be on the top of all sciences. Economics is the only field in which the raw data of experience are already in numerical form. In other fields the analyst must first quantify and measure before he can obtain data in numerical form. The physicist must weigh and count and must invent and build instruments from which numbers can be read, numbers standing for certain relations pertaining to essentially nonnumerical observations. Information which first appears only in some such form as "relatively" large, heavy, hot, fast, is later transformed into numerical data by means of measuring devices such as rods, scales, thermometers, speedometers. The economist can begin with numbers. What he observes are prices and sums of moneys. He can start out with numerical data given to him without the use of measuring devices.

The compilation of masses of data calls for resources which only large organizations, frequently only the government, can muster. This, in my opinion, is unfortunate because it implies that the availability of numerical data is associated with the extent of government intervention in economic affairs, and there is therefore an inverse relation between economic information and individual freedom.

Numbers, moreover, are not all that is needed. To be useful, the numbers must fit the concepts used in theoretical propositions or in comprehensive theoretical systems. This is rarely the case with regard to the raw data of economics, and thus the economic analyst still has the problem of obtaining comparable figures by transforming his raw data into adjusted and corrected ones, acceptable as the operational counterparts of the abstract constructs in his theoretical models. His success in this respect has been commendable, but very far short of what is needed; it cannot compare with the success of the physicist in developing measurement techniques yielding numerical data that can serve as operational counterparts of constructs in the models of theoretical physics.

Physics, however, does not stand for all natural sciences, nor economics for all social sciences. There are several fields, in both natural and social sciences, where quantification of relevant factors has not been achieved and may never be achieved. If Lord Kelvin's phrase, "Science is Measurement," were taken seriously, science might miss some of the most important problem. There is no way of judging whether nonquantifiable factors are more prevalent in nature or in society. The common reference to the "hard" facts of nature and the "soft" facts with which the student of society has to deal seems to imply a judgment about measurability. "Hard" things can be firmly gripped and measured, "soft" things cannot. There may be something to this. The facts of nature are perceived with our "senses," the facts of society are interpreted in terms of the "sense" they make in a motivational analysis. However, this contrast is not quite to the point, because the "sensory" experience of the natural scientist refers to the *data,* while the "sense" interpretation by the social scientist of the ideal-typical inner experience of the members of society refers to basic *postulates* and intervening variables.

The conclusion, that we cannot be sure about the prevalence of nonquantifiable factors in natural and social sciences, still holds.

Constancy of Numerical Relationships

On this score there can be no doubt that some of the natural sciences have got something which none of the social sciences has got: "constants," unchanging numbers expressing unchanging relationships between measurable quantities.

The discipline with the largest number of constants is, of course, physics. Examples are the velocity of light ($c = 2.99776 \times 10^{10}$ cm/sec), Planck's constant for the smallest increment of spin or angular momentum ($h = 6.624 \times 10^{-27}$ erg sec), the gravitation constant ($G = 6.6 \times 10^{-8}$ dyne cm^2 gram^{-2}), the Coulomb constant ($e = 4.8025 \times 10^{-10}$ units), proton mass ($M = 1.672 \times 10^{-24}$ gram), the ratio of proton mass to electron mass ($M/m = 1836.13$), the fine-structure constant ($\alpha^{-1} = 137.0371$). Some of these constants are postulated (conventional), others (the last two) are empirical, but this makes no difference for our purposes. Max Planck contended, the postulated "universal constants" were not just "invented for reasons of practical convenience, but have forced themselves upon us irresistibly because of the agreement between the results of all relevant measurements."[7]

I know of no numerical constant in any of the social sciences. In economics we have been computing certain ratios which, however, are found to vary relatively widely with time and place. The annual income-velocity of circulation of money, the marginal propensities to consume, to save, to import, the elasticities of demand for various goods, the savings ratios, capital-output ratios, growth rates—none of these has remained constant over time or is the same for different countries. They all have varied, some by several hundred percent of the lowest value. Of course, one has found "limits" of these variations, but what does this mean in comparison with the virtually immutable physical constants? When it was noticed that the ratio between labor income and national income in some countries has varied by "only" ten percent over some twenty years, some economists were so perplexed that they spoke of the "constancy" of the relative shares. (They hardly realized that the 10 percent variation in that ratio was the same as about a 25 percent variation in the ratio between labor income and nonlabor income.) That the income velocity of circulation of money has rarely risen above 3 or fallen below 1 is surely interesting, but this is anything but a "constant." That the marginal propensity to consume cannot in the long run be above 1 is rather obvious, but in the short run it may vary between .7 and 1.2 or even more. That saving ratios (to national income) have never been above 15 percent in any country regardless of the economic system (communistic or capitalistic, regulated or essentially free) is a very important fact; but saving ratios have been known to be next to zero, or even negative, and the variations from time to time and country to country are very large indeed.

Sociologists and actuaries have reported some "relatively stable" ratios—accident rates, birth rates, crime rates, etc.—but the "stability" is only relative to the extreme variability of other numerical ratios. Indeed, most of these ratios are subject to "human engineering," to governmental policies designed to change them, and hence they are not even thought of as constants.

The verdict is confirmed: while there are important numerical constants in the natural sciences, there are none in the social sciences.

Predictability of Future Events

Before we try to compare the success which natural and social sciences have had in correctly predicting future events, a few important distinctions should be made. We must distinguish hypothetical or conditional predictions from unconditional predictions or forecasts. And among the former we must distinguish those where all the stated conditions can be controlled, those where all the stated conditions can be either controlled or unambiguously ascertained before the event, and finally those where some of the stated conditions can neither be controlled nor ascertained early enough (if at all). A conditional prediction of the third kind is such an "iffy" statement that it may be of no use unless one can know with confidence that it would be highly improbable for these problematic conditions (uncontrollable and not ascertainable before the event) to interfere with the prediction. A different kind of distinction concerns the numerical definiteness of the prediction: one may predict that a certain magnitude (1) will change, (2) will increase, (3) will increase by at least so-and-so much, (4) will increase within definite limits, or (5) will increase by a definite amount. Similarly, the prediction may be more or less definite with respect to the time within which it is supposed to come true. A prediction without any time specification is worthless.

Some people are inclined to believe that the natural sciences can beat the social sciences on any count, in unconditional predictions as well as in conditional predictions fully specified as to definite conditions, exact degree and time of fulfilment. But what they have in mind are the laboratory experiments of the natural sciences, in which predictions have proved so eminently successful; and then they look at the poor record social scientists have had in predicting future events in the social world which they observe but cannot control. This comparison is unfair and unreasonable. The artificial laboratory world in which the experimenter tries to control all conditions as best as he can is different from the real world of nature. If a comparison is made, it must be between predictions of events in the real natural world and in the real social world.

Even for the real world, we should distinguish between predictions of events which we try to bring about by design and predictions of events in which we have no part at all. The teams of physicists and engineers who have been designing and developing machines and apparatuses are not very successful in predicting their performance when the design is still new. The record of predictions of the paths of moon shots and space missiles has been rather spotty. The so-called bugs that have to be worked out in any new contraption are nothing but predictions gone wrong. After a while predictions become more reliable. The same is true, however, with predictions concerning the performance of organized social institutions. For example, if I take an envelope, put a certain address on it and a certain postage stamp, and deposit it in a certain box on the street, I can predict that after three or four days it will be delivered at a certain house thousands of miles away. This prediction and any number of similar predictions will prove correct with a remarkably high frequency. And you don't have to be a social scientist to make such successful predictions about an organized social machinery, just as you don't have to be a natural scientist to predict the result of your pushing the electric-light switch or of similar manipulations of a well-tried mechanical or electrical apparatus.

There are more misses and fewer hits with regard to predictions of completely unmanipulated and unorganized reality. Meteorologists have a hard time forecasting the weather for the next 24 hours or two or three days. There are too many variables involved and it is too difficult to obtain complete information about some of them. Economists are only slightly better in forecasting employment and income, exports and tax revenues for the next six months or for a year or two. Economists, moreover, have better excuses for their failures because of unpredictable "interferences" by governmental agencies or power groups which may even be influenced by the forecasts of the economists and may operate to defeat their predictions. On the other hand, some of the predictions may be self-fulfilling in that people, learning of the predictions, act in ways which bring about the predicted events. One might say that economists ought to be able to include the "psychological" effects of their communications among the variables of their models and take full account of these influences. There are, however, too many variables, personal and political, involved to make it possible to allow for all effects which anticipations, and anticipations of anticipations, may have upon the end results. To give an example of a simple self-defeating prediction from another social science: traffic experts regularly forecast the number of automobile accidents and fatalities that are going to occur over holiday weekends, and at the same time they hope that their forecasts will influence drivers to be more careful and thus to turn the forecasts into exaggerated fears.

We must not be too sanguine about the success of social scientists in making either unconditional forecasts or conditional predictions. Let us admit that we are not good in the business of prophecy and let us be modest in our claims about our ability to predict. After all, it is not our stupidity which hampers us, but chiefly our lack of information, and when one has to make do with bad guesses in lieu of information the success cannot be great. But there is a significant difference between the natural sciences and the social sciences in this respect: Experts in the natural sciences usually do not try to do what they know they cannot do; and nobody expects them to do it. They would never undertake to predict the number of fatalities in a train wreck that might happen under certain conditions during the next year. They do not even predict next year's explosions and epidemics, floods and mountain slides, earthquakes and water pollution. Social scientists, for some strange reason, are expected to foretell the future and they feel badly if they fail.

Distance from Everyday Experience

Science is, almost by definition, what the layman cannot understand. Science is knowledge accessible only to superior minds with great effort. What everybody can know cannot be science.

A layman could not undertake to read and grasp a professional article in physics or chemistry or biophysics. He would hardly be able to pronounce many of the words and he might not have the faintest idea of what the article was all about. Needless to say, it would be out of the question for a layman to pose as an expert in a natural science. On the other hand, a layman might read articles in descriptive economics, sociology, anthropology, social psychology. Although in all these fields technical jargon is used which he could not really understand, he might *think* that he knows the sense of the words and grasps the meanings of the sentences; he might even be inclined to poke fun at some of the stuff. He believes he is—from his own experience and from his reading of newspapers and popular magazines—familiar with the subject matter of the social sciences. In consequence, he has little respect for the analyses which the social scientists present.

The fact that social scientists use less Latin and Greek words and less mathematics than their colleagues in the natural science departments and, instead, use everyday words in special, and often quite technical, meanings may have something to do with the attitude of the layman. The sentences of the sociologist, for example, make little sense if the borrowed words are understood in their nontechnical, everyday meaning. But if the layman is told of the special meanings that have been bestowed upon his words, he gets angry or condescendingly amused.

But we must not exaggerate this business of language and professional jargon because the problem really lies deeper. The natural sciences talk about nuclei, isotopes, galaxies, benzoids, drosophilas, chromosomes, dodecahedrons, Pleistocene fossils, and the layman marvels that anyone really cares. The social sciences, however—and the layman usually finds this out—talk about—him. While he never identifies himself with a positron, a pneumococcus, a coenzyme, or a digital computer, he does identify himself with many of the ideal types presented by the social scientist, and he finds that the likeness is poor and the analysis "consequently" wrong.

The fact that the social sciences deal with man in his relations with fellow man brings them so close to man's own everyday experience that he cannot see the analysis of this experience as something above and beyond him. Hence he is suspicious of the analysts and disappointed in what he supposes to be a portrait of him.

Standards of Admission and Requirements

High-school physics is taken chiefly by the students with the highest IQs. At college the students majoring in physics, and again at graduate school the students of physics, are reported to have on the average higher IQs than those in other fields. This gives physics and physicists a special prestige in schools and universities, and this prestige carries over to all natural sciences and puts them somehow above the social sciences. This is rather odd, since the average quality of students in different departments depends chiefly on departmental policies, which may vary from institution to institution. The preeminence of physics is rather general because of the requirement of calculus. In those universities in which the economics department requires calculus, the students of economics rank as high as the students of physics in intelligence, achievement, and prestige.

The lumping of all natural sciences for comparisons of student quality and admission standards is particularly unreasonable in view of the fact that at many colleges some of the natural science departments, such as biology and geology, attract a rather poor average quality of student. (This is not so in biology at universities with many applicants for a premedical curriculum.) The lumping of all social sciences in this respect is equally wrong, since the differences in admission standards and graduation requirements among departments, say between economics, history, and sociology, may be very great. Many sociology departments have been notorious for their role as refuge for mentally underprivileged undergraduates. Given the propensity to overgeneralize, it is no wonder then that the social sciences are being regarded as the poor relations of the natural sciences and as disciplines for which students who cannot qualify for *the* sciences are still good enough.

Since I am addressing economists, and since economics departments, at least at some of the better colleges and universities, are maintaining standards as high as physics and mathematics departments, it would be unfair to level exhortations at my present audience. But perhaps we should try to convince our colleagues in all social science departments of the disservice they are doing to their fields and to the social sciences at large by admitting and keeping inferior students as majors. Even if some of us think that one can study social sciences without knowing higher mathematics, we should insist on making calculus and mathematical statistics absolute requirements—as a device for keeping away the weakest students.

Despite my protest against improper generalizations, I must admit that averages may be indicative of something or other, and that the average IQ of the students in the natural science departments is higher than that of the students in the social science department.[8] No field can be better than the men who work in it. On this score, therefore, the natural sciences would be superior to the social sciences.

The Score Card

We may now summarize the tallies on the nine scores.

1. With respect to the invariability or recurrence of observations, we found that the greater number of variables—of relevant factors—in the social sciences makes for more variation, for less recurrence of exactly the same sequences of events.

2. With respect to the objectivity of observations and explanations, we distinguished several ways in which references to values and value judgments enter scientific activity. Whereas the social sciences have a requirement of "subjective interpretation of value-motivated actions" which does not exist in the natural sciences, this does not affect the proper "scientific objectivity" of the social scientist.

3. With respect to the verifiability of hypotheses, we found that the impossibility of controlled experiments combined with the larger number of relevant variables does make verification in the social sciences more difficult than in most of the natural sciences.

4. With respect to the exactness of the findings, we decided to mean by it the existence of a theoretical system from which most propositions concerning particular connections can be deduced. Exactness in this sense exists in physics and in economics, but much less so in other natural and other social sciences.

5. With respect to the measurability of phenomena, we saw an important difference between the availability of an ample supply of numerical data and the availability of such numerical data as can be used as good counterparts of the constructs in theoretical models. On this score, physics is clearly ahead of all other disciplines. It is doubtful that this can be said about the natural sciences in general relative to the social sciences in general.

6. With respect to the constancy of numerical relationships, we entertained no doubt concerning the existence of constants, postulated or empirical, in physics and in other natural sciences, whereas no numerical constants can be found in the study of society.

7. With respect to the predictability of future events, we ruled out comparisons between the laboratory world of some of the natural sciences and the unmanipulated real world studied by the social sciences. Comparing only the comparable, the real worlds—and excepting the special case of astronomy—we found no essential differences in the predictability of natural and social phenomena.

8. With respect to the distance of scientific from everyday experience, we saw that in linguistic expression as well as in their main concerns the social sciences are so much closer to prescientific language and thought that they do not command the respect that is accorded to the natural sciences.

9. With respect to the standards of admission and requirements, we found that they are on the average lower in the social than in the natural sciences.

The last of these scores relates to the current practice of colleges and universities, not to the character of the disciplines. The point before the last, though connected with the character of the social sciences, relates only to the popular appreciation of these disciplines; it does not aid in answering the question whether the social sciences are "really" inferior. Thus the last two scores will not be considered relevant to our question. This leaves seven scores to consider. On four of the seven no real differences could be established. But on the other three scores, on "Invariance," "Verifiability," and "Numerical Constants," we found the social sciences to be inferior to the natural sciences.

The Implications of Inferiority

What does it mean if one thing is called "inferior" to another with regard to a particular "quality"? If this "quality" is something that is highly valued in any object, and if the absence of this "quality" is seriously missed regardless of other qualities present, then, but only then, does the noted "inferiority" have any evaluative implications. In order to show that "inferiority" sometimes means very little, I shall present here several statements about differences in particular qualities.

> "Champagne is inferior to rubbing alcohol in alcoholic content."
> "Beef steak is inferior to strawberry Jello in sweetness."
> "A violin is inferior to a violoncello in physical weight."
> "Chamber music is inferior to band music in loudness."
> "Hamlet is inferior to Joe Palooka in appeal to children."
> "Sandpaper is inferior to velvet in smoothness."
> "Psychiatry is inferior to surgery in ability to effect quick cures."
> "Biology is inferior to physics in internal consistency."

It all depends on what you want. Each member in a pair of things is inferior to the other in some respect. In some instances it may be precisely this inferiority that makes the thing desirable. (Sandpaper is wanted *because* of its inferior smoothness.) In other instances the inferiority in a particular respect may be a matter of indifference. (The violin's inferiority in physical weight neither adds to nor detracts from its relative value.) Again in other instances the particular inferiority may be regrettable, but nothing can be done about it and the thing in question may be wanted nonetheless. (We need psychiatry, however much we regret that in general it cannot effect quick cures; and we need biology, no matter how little internal consistency has been attained in its theoretical systems.)

We have stated that the social sciences are inferior to the natural sciences in some respects, for example, in verifiability. This is regrettable. If propositions cannot be readily tested, this calls for more judgment, more patience, more ingenuity. But does it mean much else?

The Crucial Question: "So What?"

What is the pragmatic meaning of the statement in question? If I learn, for example, that drug E is inferior to drug P as a cure for hay fever, this means that, if I want such a cure, I shall not buy drug E. If I am told Mr. A is inferior to Mr. B as an automobile mechanic, I shall avoid using Mr. A when my car needs repair. If I find textbook K inferior to textbook S in accuracy, organization, as well as exposition, I shall not adopt textbook K. In every one of these examples, the statement that one thing is inferior to another makes pragmatic sense. The point is that all these pairs are *alternatives* between which a choice is to be made.

Are the natural sciences and the social sciences alternatives between which we have to choose? If they were, a claim that the social sciences are "inferior" could have the following meanings:

1. We should not study the social sciences.
2. We should not spend money on teaching and research in the social sciences.
3. We should not permit gifted persons to study social sciences and should steer them toward superior pursuits.
4. We should not respect scholars who so imprudently chose to be social scientists.

If one realizes that none of these things could possibly be meant, that every one of these meanings would be preposterous, and that the social sciences and the natural sciences can by no means be regarded as alternatives but, instead, that both are needed and neither can be dispensed with, he can give the inferiority statement perhaps one other meaning:

5. We should do something to improve the social sciences and remedy their defects.

This last interpretation would make sense if the differences which are presented as grounds for the supposed inferiority were "defects" that can be remedied. But they are not. That there are more variety and change in social phenomena; that, because of the large number of relevant variables and the impossibility of controlled experiments, hypotheses in the social sciences cannot be easily verified; and that no numerical constants can be detected in the social world—these are not defects to be remedied but fundamental properties to be grasped, accepted, and taken into account. Because of these properties research and analysis in the social sciences hold greater complexities and difficulties. If you wish, you may take this to be a greater challenge, rather than a deterrent. To be sure, difficulty and complexity alone are not sufficient reasons for studying certain problems. But the problems presented by the social world are certainly not unimportant. If they are also difficult to tackle, they ought to attract ample resources and the best minds. Today they are getting neither. The social sciences are "really inferior" regarding the place they are accorded by society and the priorities with which financial and human resources are allocated. This inferiority is curable.

Notes

1. Published in *On Freedom and Free Enterprise: Essays in Honor of Ludwig von Mises*, ed. Mary Sennholz, pp. 161–172.
2. H. Rickert, *Die Grenzen der naturwissenschaftlichen Begriffsbildung*.
3. M. Cohen, *Reason and Nature*, p. 348.
4. H. Kelsen, *Allgemeine Staatslehre*, p. 129. Quoted with illuminating comments in A. Schutz, *Der sinnhafte Aufbau der sozialen Welt*.
5. R. B. Braithwaite, *Scientific Explanation: A Study of the Function of Theory, Probability and Law in Science*.
6. F. Machlup, "The Problem of Verification in Economics," *Southern Economic Journal*, XXII, 1955, 1–21.
7. M. Planck, *Scientific Autobiography and Other Papers*, p. 173.
8. The average IQ of students receiving bachelor's degrees was, according to a 1954 study, 121 in the biological sciences, and 122 in economics, 127 in the physical sciences, and 119 in business. See D. Wolfe, *America's Resources of Specialized Talent: The Report of the Commission on Human Resources and Advanced Training*, pp. 319–322.

Chapter 2

What Would an Adequate Philosophy of Social Science Look Like?

Brian Fay and J. Donald Moon

I

During the last twenty years an enormous literature has grown up around the question, what is the nature of social science? Two positions have dominated these discussions, the "naturalist" view which holds that social science involves no essential differences from the natural sciences, and the "humanist" view which holds that social life cannot adequately be studied "scientifically." Whole models of social science have been propounded that argue for one position and view the other as an incompatible alternative.[1] Given such a vigorous tradition of discourse, it may seem odd that anyone would now ask the question, what would an adequate philosophy of social science look like? Unfortunately, however, neither naturalism nor humanism is capable of answering the three questions which the idea of a science of behavior raises. These questions are: first, What is the *relationship between interpretation and explanation in social science*? second, What is the *nature of social scientific theory*? and third, What is *the role of critique*?

In this essay we will show why these three questions must be answered by any compelling account of social science, and why humanism and naturalism are unable to answer them. The first question will be taken up in section II, the second in section III, and the third in section IV. By showing that the dualism which dominates current philosophical thinking makes it impossible to answer these questions adequately, we will point to the need for a new synthesis in the philosophy of social science, one that transcends the antimony of humanism and natrualism.

II

One way of beginning to talk about the nature of social phenomena is to invoke the now familiar prima facia distinction between human action, on the one hand, and mere bodily movements on the other—between raising one's arm and one's arm rising, to use the time-worn example. According to this distinction, actions differ from mere movements in that they are intentional and rule-governed: they are performed in order to achieve a particular purpose, and in conformity to some rules. These purposes and rules constitute what we shall call the "semantic dimension" of human behavior[2]—its symbolic or expressive aspect. An action, then, is not simply a physical occurrence, but has a certain intentional content which specifies what sort of an action it is, and which can be grasped only in terms of the system of meanings in which the action is performed. A given movement counts as a vote, a signal, a salute or an attempt to reach something, only against the background of a set of applicable rules and conventions, and the purposes of the actor involved.

To the prima facie face that human actions are intentional events in the sense that their identity is a function of their content—what they express or the states of affairs they refer to—and, consequently, that they are characterized by invoking the rules and intentions which define them to be what they are, there are three possible responses. The first of these is to accept this prima facie fact and to try to construct a science of intentional objects in terms of it; this it the intentionalist response.[3] The second is to attempt an analysis of the concepts "intention," "meaning" and "action" in purely observational (usually behavioral) terms, so that one can use these concepts in one's science but in a purified form; this is the tack of the definitional behaviorist.[4] The third response is to accept that one cannot capture the meaning of intentional concepts without reference to mental states such as beliefs and institutional norms such as rules, and so conclude that these concepts are radically defective for scientific purposes; it therefore seeks to develop a science of behavior without using these concepts at all. This is the position of the eliminative materialists who ultimately wish to confine their accounts of language and other social behavior to a purely extensional terminology.[5]

The important thing to realize about the third response is that it requires a radically different approach from anything remotely resembling what is understood to be social science as it is practiced today. Broadly speaking, social scientists seek to offer accounts of events described in terms of their significance; thus, they want to understand why it is that a certain group is *dancing* (and not why the feet of its members are *twitching* in a manner describable in purely spatiotemporal terms), or *voting* (and not why the arms of certain bodies are rising), and they characterize the speech of people in terms of its content rather than in terms of its purely phonetic qualities. However, if the approach of the eliminative materialists came to dominate the science of behavior, it would become a sort of mechanics or neurophysiology whose explanatory concepts would be drawn from the natural sciences.

Of course, this observation does not in itself show that this approach is incoherent, or that it cannot be realized. Just as natural science abandoned intentional concepts—a strategy that was unthinkable to many at the time—so the sciences of human behavior might also be transformed in this way. The question of what might be called the conceptual solvency of a "natural science of man," as well as the problems that would arise in attempting to implement this program, are exceedingly interesting ones. However, in some sense they lie outside the boundaries of our inquiry, just because we are trying to offer an account of the many forms which *social* science now takes. It is for this reason that we feel justified in setting this position to one side.

Nor does the second response seem to be adequate. It appears to be the case that any attempt to translate intentional concepts, which involve reference to such things as rules and beliefs, into dispositional terms, which specify a set of dispositions to engage in overt movements under particular stimulus-conditions, is bound to fail. No matter how one tries to construe these concepts, it is ultimately necessary to employ another intentional concept in order to explicate its meaning.[6]

Take, for example, the statement, "Jones asked the cashier to deposit the money into his account"; the concepts "cashier," "deposited" and "money" are all prima facie intentional, in that their meanings involve certain rules (a cashier is a person who has a certain role to play in a specific institution, with certain duties and orders to follow), beliefs (in order to deposit the money, the cashier must believe that Jones has an account), and desires (Jones must what to put his money into his account in order for it to be said that he deposited it). Now, to take one of these intentional concepts, a behaviorist might argue that the beliefs involved in making a deposit can be explicated in purely dispositional terms. A line he might take is this: when a certain sound is made in the cashier's presence ("Do you think that Jones has an account?"), he will produce another ("yes"). But this construal is adequate only if the person understands the question, and understanding is an intentional state. The behaviorist, of course, may then try to give a nonintentionalist account of understanding a question—for example, that a person may be said to understand a question if he or she is able to answer it correctly most of the time. But this account also involves an intentional object, since an answer is correct only in terms of certain rules indicating what is appropriate and what is not. And so the discussion will proceed, until gradually it will become clear that what is wrong is not simply this particular attempt to reduce intentional concepts to nonintentional ones, but that there is something in principle wrong with the whole definitional behaviorist program.

Thus, we are left with the intentionalist response to the prima facie meaningful character of human actions, mental events and social institutions. The question then arises, what implications does this have for social science? The most obvious task which an intentionalist perspective imposes on the study of human action is the need for interpretation. In order to study human behavior as meaningful performances, we must grasp the meanings expressed in speech and action, and this requires that we understand the system of concepts, rules, conventions and beliefs which give such behavior its meaning. This is the doctrine of understanding, or *verstehen*, which figures as a prominent methodological principle in the humanist account of social science. It marks an essential methodological difference between the human sciences and the study of nature, expressing itself most clearly in the principles of concept-formation appropriate to each. Briefly, concept-formation in the natural sciences is governed by two related sets of considerations—those of theory, and those of measurement. We require that concepts be developed which permit the formation of testable laws and theories, and other issues—e.g., those deriving from ordinary language—may simply be set aside. But in the human sciences there is another set of considerations as well: the concepts we use to describe and explain human activity must be drawn from the social life that is being studied, and not from the observer's theories, at least in the first instance. Because the very identity of a particular action depends upon its meanings for the social actors, the concepts we use to describe it must capture this meaning.

Another way of putting this point is to say that concepts bear a fundamentally different relationship to social phenomena from that which they bear to natural phenomena. In the social sciences, concepts partially constitute the reality we study, while in the latter case they merely serve in describing and explaining it. As Winch has argued, something can be an "order" only if the *social actors* involved have the concept of an order, and such related concepts as obedience, authority, etc.; but the natural event of lightning is the same whether it is conceptualized as an expression of Zeus's anger, or as an atmospheric electrical discharge: its identity is not a function of its meaning or intentional content.[7]

The interpretation of the meanings of actions, practices and cultural objects is an extremely difficult and complicated enterprise. The basic reason for this is that, as Wittgenstein has shown, the meaning of something depends upon the role which it has in the system of which it is a part. To understand a particular action, we must grasp the beliefs and intentions which motivated it, and this further requires that we know the social contexts of practices and institutions which specify what the action in question "counts as," what sort of an action it is. To return to our check depositing example: in order for the social scientist to know what the overt movements he observes actually mean, i.e., to understand what action is being performed, it is necessary that he have an understanding of the beliefs, desires and values of the particular people involved. But in order to understand these, he must know the vocabulary in terms of which they are expressed, and this, in turn, will require that he know the social rules and conventions which specify what a certain movement or object will count as. Moreover, in order to grasp these particular rules, he will also have to know the set of institutional practices (in this case, those of banking) of which they are a part, and how these are related to other practices of the society (in this case, the institutions of a money economy).

Nor can our scientist stop here. For the conventions of a social group, as Taylor has convincingly argued,[8] presuppose a set of fundamental conceptualizations or basic assumptions regarding man, nature and society. These basic conceptualizations might be called the "constitutive meanings of a form of life," for they are the basic ideas or notions in terms of which the meanings of specific practices and schemes of activity must be analyzed. For example, the social practice of banking can only occur given the shared constitutive meanings of (say) some conception of property, some notion of being a unit with a particular identity, some idea of exchange value. An adequate account of the practices of a particular society, by setting out the basic ideas and conceptualizations which underlie these practices, will show how various aspects of the social order are related to each other, and how (or the extent to which) the social order constitutes a coherent whole.

The need for such a high level of interpretation may be missed if one focuses one's attention only on studies of one's own culture by other members of it. For in these situations, the scientists do not have to make explicit their interpretive scheme in order to identify and characterize the class of actions and institutions in which they are interested. They, as well as their readers, already know what banks are and what depositing funds means. However, this point should not be pushed too far, because the sorts of implicit self-understandings which we have as practitioners are generally going to be inadequate for the tasks of social science. This is the reason why some of the very best work in social science will partially consist in explicating the sets of shared rules and constitutive meanings which underlie quite ordinary, everyday practices. (Here we are thinking of such works as Beer's *Modern British Politics*, Douglas's *The Social Meanings of Suicide*, Cicourel's *The Social Organization of Juvenile Justice*, and Goffman's *Asylums*.)[9]

Impressed by the elegance and penetration of interpretive theories, humanist philosophers of social science have assumed or argued that interpretation is all there is. They have gone from the correct observation that social theories must be interpretive, to the incorrect conclusion that they can *only* be interpretive. For social phenomena do not consist in abstract structures of meanings which can be set forth and analyzed, but they consist in actions (and other events) which actually occur in particular places at particular times. And, while we cannot even approach our subject without understanding what these actions *mean*, such understanding does not, by itself, constitute an explanation of why they *occur*. To know, e.g., what someone said, and what it means, is not to know why he or she said it.

Accounts of why something happened are commonly said to be causal explanations, for they explain why it occurred by setting out what led it to happen. In the case of actions, e.g., we explain why an agent does something by pointing out the motives, or purposes which led him or her to do it. Thus, for example, Weber explained the type of behavior typical of capitalists in the sixteenth and seventeenth centuries by citing the set of religious beliefs and desires which caused certain sorts of Protestants to act in this manner.

One of the principal tenets of humanism over the last twenty years [1957–1977] has been that beliefs, purposes, values, desires, and so forth—reasons, for short—cannot be causes, and that therefore there is no real "explanation" in social science but only a further form of interpretation in which the scientist tries to uncover the rationale or warrant for the actions in question.[10] But such arguments are now generally recognized to have been inadequate because, while reasons cannot be causes (the *are* utterly different sorts of things), the having of reasons, the believing in reasons, the giving of reasons, etc., are all psychological events and, as such, nothing prevents them from figuring in causal explanations.[11] (We say this though we are aware that in order to actually detail the nature of these causes one needs to develop a philosophy of mental events which will do justice to their peculiar qualities, e.g., their having an intentional content, and their very close relationship to overt behavior. Unfortunately, this sort of philosophical analysis has been strangely omitted in most discussions of action theory and its relationship to the philosophy of social science, even by those who advocate a causalist position.)[12]

Moreover, social scientists are interested in explaining a great many phenomena other than actions. They want to explain why it is that people have certain beliefs and values (as in the sociology of knowledge); to account for patterns of unintended consequences of actions; to discover why a social structure arose in the first place, and why it continues to exist despite a changing membership; and so forth. In these, and in all the other questions in which a social scientist is interested, the form of explanation is causal. For in each of them what is required as an explanation is the identification of the necessary and/or sufficient condition or events which produced the phenomena in question.

We will return to the question of causal explanations in social science in a moment when we come to discuss theory in social science, but already enough has been said to demonstrate that social science is an explanatory enterprise as well as an interpretive one. And so the questions which immediately arise are: What is the relationship between interpretation and explanation in social science? How does one influence or restrain the other? How do the criteria for a good interpretation fit with the criteria for a good explanation? These questions arise just because social science is the systematic scientific study of intentional phenomena. Because humanists have failed to appreciate the explanatory task of social sciences (i.e., they have failed to see in what way these disciplines are scientific), and because naturalists have misunderstood the crucial role which interpretation plays in the social sciences (i.e., they have given insufficient or misleading analysis of what it means for a phenomenon to be intentional), both of them have neglected such questions. This is one reason why the current traditions in the analytical philosophy of social science are not only inadequate but, given their terms of reference, incapable of getting onto the right track.[13]

III

The dichotomy between humanist and naturalist also makes it impossible to answer the second question which is critical for a science of behavior: What is the nature of social-scientific theory? For many writers in the humanist tradition, particularly as represented in recent analytical philosophy, the question scarcely seems to exist; one can look in vain in the work of Louch, Winch, Taylor or von Wright—to mention the most important humanist statements of the last fifteen years—for even a mention of social-scientific theories, let alone a discussion of them.

The reason for this is not hard to find. From the humanist perspective, there is neither a need for theories nor a place for them in the study of society. (At least this is true if we understand "theory" to refer to systematic, unified explanations of a diverse range of social phenomena.) There is no place for theories in the humanist position because its cardinal point is that social science is simply interpretive: it seeks to provide us with an understanding of the meanings of particular actions or practices of a given society. As we have already shown, such understanding may require that we grasp the worldview of the society or culture in question, and elaborate and sophisticated intellectual structures may be necessary to do so.

But an account of a society's worldview, or its intersubjective or constitutive meanings, is not a theory which explains why the society has the institutions it has, or why certain processes of social change occur, or why it is characterized by certain regularities, or why people of a certain sort perform particular kinds of actions. To explain such phenomena we need theories that are, broadly speaking, causal, and the fixation of the humanist tradition with the meaningful dimension of human action has prevented it from developing an account of this kind of social-scientific theory.

This failure to give an account of explanatory theory has proved a particular embarrassment to those espousing the humanist case because it has meant that they have failed to deal with just those aspects of social-scientific work which are of paramount importance to many of its practitioners, and which constitute some of its most conspicuous successes. The clearest example of this is Keynesian economic theory; but all the social sciences possess theories of one sort or another. Thus, kinship theory in anthropology, exchange theory in sociology, the theory of transformational grammar in linguistics, modernization theory in political science and cognitive dissonance theory in psychology are all examples of the theoretical dimension operative in modern social science. Although humanism is popular among analytical philosophers, naturalism is still the dominant position among social scientists; one of the reasons for this is that the antitheoretical stance of the humanist model has made it appear patently deficient and even irrelevant to those actually engaged in doing substantive social scientific work.

Social science must be theoretical because on of its aims is to give causal explanations of events, and even singular causal explanations require some sort of general law or laws. To say that an event, *x*, causes another event, *y*, is to say (speaking very roughly) that *x*'s occurrence is a necessary and/or sufficient condition for the occurrence of *y*. The idea that the occurrence of one event is a condition for the occurrence of the other distinguishes causal statements of the form, "Under *C*, *x* caused *y*," from mere statements of conjunction of the form, "Under *C*, *x* occurred and then *y* occurred." But this is to say that when we give causal explanations we are implicitly asserting that, whenever an *x*-type event occurs under conditions *C*, a *y*-type event will also occur, which is to say that causal explanations ultimately rest on general laws.

This does not mean that we must actually be able to state a law in order to offer a valid causal explanation, for we may have good reasons for believing that two events are causally related even though we cannot provide the appropriate covering law. Indeed, it may even be the case that we will not be able to state the covering law until we redescribe the events in question in the language of some theory.[14] Thus, we may be warranted in explaining the decrease in the mass of a piece of wood by its having been burned, even though we cannot state the general law upon which this explanation rests, and even though we would have to redescribe this event in terms of the theory of oxidation before we could do so. In these cases we must justify our claim by presenting reasons to believe that there is a causal law operating here. Such reasons will consist of reports of other instances in which the two events are conjoined, together with evidence that the relationship is actually a causal one. Such evidence could include our ability to manipulate the putatively causal variables so as to bring about or suppress the effects in question, and/or a specification of the causal mechanisms by which one event produces the other.

In explaining the occurrence of one event or condition in terms of another, it is not sufficient merely to offer a generalization reporting the covariance of these two events. Rather, what we require is a general statement that is lawlike in the sense that it explains its instances. Take, for example, a social scientist trying to explain why it is that in Western Europe support for totalitarian parties is inversely related to education. In the first instance, he many attempt to explain this finding with the observation that, in Western Europe, less educated people tend to have authoritarian personalities, and with the generalization that people with authoritarian personalities support authoritarian political movements. Here he tries to offer an explanation by showing that the phenomenon in question is an instance of a deeper, generally recurring pattern by embedding descriptions of it in higher-level generalizations. However, there is something problematic about this putative explanation, and that is the status of the generalizations it contains. For it immediately leads us to ask what it is about people with authoritarian personalities which leads them to support antidemocratic parties. Is it just a coincidence that they do, or is their behavior somehow necessitated by their having the kind of personality they have? If we could change a person's personality, would his or her political preferences change as well? In short, unless the general statement is not simply an empirical generalization, but what we have called a nomic generalization, so that it can support contrary-to-fact and subjunctive conditionals, then this account is not a genuine explanation. A generalization, we might say, cannot serve as the required basis for making a causal explanation unless it can explain its instances, and it cannot explain its instances unless one can give an account of *why* the generalization holds.

It is precisely at this point that theories are required, for it is in terms of theories that such an account can be forthcoming. (This is the reason why it is said that causal generalizations must be theory-impregnated.) Theories provide a systematic account of a diverse set of phenomena by showing that the events in question all result from the operation of a few basic principles. A theory goes beyond particular generalizations by showing why the generalizations hold, and it does this by specifying the basic entities which constitute the phenomena to be explained, and their modes of interaction, from which the observed generalizations can be inferred. Thus, a theory not only provides unity and coherence to a field of inquiry, but it also gives us the grounds required for asserting subjunctive conditionals, or the reasons for believing that these generalizations are, in some sense, necessary. Thus, we are inevitably led from the need to explain particular occurrences to the need for social theories.

Moreover, as a social science attempts to become more rigorously scientific, it will naturally attempt to organize and structure its various particular causal explanations and the relatively specific nomic generalizations upon which they rest by systematically interrelating them, and by subjecting them to experimental and other empirical verification. In this process, the self-conscious development of "large-scale" theory is absolutely crucial, and it is for this reason that the sciences of behavior have developed the social theories of extremely wide scope and power which we have already mentioned.

However, if the humanists fail to provide an account of social theories, we fare little better at the hands of the naturalists. Of course, the naturalists spend a great deal of time talking about scientific theories, but they analyze scientific theories in general, and give little attention to the specific problems of *social* theories. For the naturalists, the human sciences and the natural sciences share the same methodology, and so there is no need to discuss social theories apart from physical theories: what can be said of the latter applies a fortiori to the former. And since theories in the physical sciences are far more elaborate and developed than theories in the social sciences, discussion of the nature of theories is usually focused on physical theories. Moreover, when social theories are discussed in the naturalist tradition, it is often to set forth their deficiencies in terms of the naturalist ideal, rather than to analyze them in their own right.[15]

This is not a happy situation, however, because theories in the social sciences are needed to explain phenomena which are different from those in the natural world—they are intentional—and therefore we cannot assume that they will have the same structure as, or be similar in all important respects to, theories in the natural sciences. Indeed, there are at least three ways that it can be seen that theory-construction in the social sciences faces problems quite unlike theory-construction in the natural sciences.

In the first place, because intentional actions are rule-governed, they have an irreducibly normative character. Speech acts, e.g., are performed in accordance with linguistic rules, and so they can be assessed as correct or incorrect. Similarly, instrumental actions can be assessed as more or less rational depending on the extent to which they are likely to realize their intended aims. Because of this normative character of action, a distinction can be made between the *competence* of an actor and his or her *actual performance*. An actor's competence is his or her mastery of the rules (or norms of rationality) which apply to a particular area of activity; performance, on the other hand, refers to the person's actual behavior, which is determined not only by his or her competence, but also by such other factors as fatigue, inattention, misperception, learning failure and the like.

Now, corresponding to this distinction between competence and performance is a distinction between two types of theory. A theory of competence is designed to explain the competence of an actor, or, more likely, the competence of an idealized actor who is perfectly rational, or has perfectly mastered the relevant rules. A theory of performance, on the other hand, while perhaps making use of, or presupposing, a theory of competence, is designed to explain what a person actually does, and so it would encompass all of the causal factors which bear upon behavior. In his theory of transformational grammar, for example, Chomsky attempts to set forth the basic rules or principles which generate all grammatically well-formed sentences of a language, and only such sentences. An adequate theory, then, would model the (idealized) native speaker's *mastery* of his or her language, his or her (potential) capacity to recognize well-formed utterances.[16] Similarly, modern economics is based upon a theory of choice which sets forth the rules which must be followed by an ideally rational actor in different kinds of choice situations.

In both linguistics and economics there is considerable controversy over the role of these theories of competence in accounting for actual performance, since it is not obvious what relevance such idealized accounts could have for explaining the behavior of any particular actor. But what is crucial for our purposes is the failure of naturalists to recognize this problem at all. Because they take social theories to have the same structure as physical theories, they are not even able to ask what the relationship is between theories which model competence and causal theories which explain overt behavior.

A second problem for theory construction that is unique to the human sciences is the relationship between the concepts and principles which the scientist uses to account for social phenomena, and those which inform the actions and beliefs of social actors. As we have already shown, because social phenomena are intentional, their very identity depends upon the concepts and self-understandings of social actors, and so in order to explain social behavior social scientists are constrained to use the actors' framework. If social scientists wish to go beyond these self-understandings by introducing concepts and principles which may be at variance with them, they face the problem of relating these new principles to those employed by the actors themselves. Failure to make this relationship would result in the scientists' failing to capture the phenomena they wish to explain, since the events in question would slip through the conceptual net the scientists had constructed.

Nor is this the problem that natural scientists face in giving empirical content to theoretical terms and principles by means of correspondence rules or bridge principles, since the concepts which the actors employ are no more "empirical" or "observable" than the concepts of the scientist: in terms of the distinction between theoretical and observational terms, concepts such as "belief" or "decision" are as "theoretical" as concepts such as "social structure" or "national income." Moreover, the problem of theoretical interpretation in the social sciences may not simply be one of developing bridge principles which specify, in part, what it is that theoretical terms refer to, or how statements employing theoretical terms can be tested. Rather, it may be a matter of establishing that different behaviors have the same or similar meanings, as when aspects of American Halloween customs are shown to be similar in meaning to the myths and rituals of the Kayapo, a people living in the Amazon basin.[17] Once again, because social theories are theories of intentional objects, they pose problems for analysis which cannot be grasped merely from an understanding of theories of physical things.

A third problem for theory construction which the naturalists have failed to discuss is the nature of paradigms or research programs in the social sciences. For a long time, of course, this was due to the dominance of positivism within the philosophy of science, and so it reflected a failure to discuss the conceptual presuppositions of scientific or theoretical work within all branches of empirical science. In the past decade or so this deficiency has been corrected, particularly with the seminal work of Kuhn and Lakatos. Lakatos's account of science in terms of the concept of a research program is of particular importance, for the touchstone of his work is not the history of actual scientific developments, as it was for Kuhn, but the logic of science itself. A research program, according to Lakatos,[18] sets out the fundamental conceptual framework or conceptualization of the phenomena we wish to explain, and the rules in accordance with which theoretical innovations or developments will be made. We require such rules, Lakatos argues, because we must have criteria which can be used to recognize adjustments to a theory which are essentially ad hoc, or unconnected with the rest of the theory. Theoretical developments which are ad hoc must then be rejected, for they do not represent genuine scientific progress.

Given that we recognize the need for research programs for theory construction in the social sciences, the question immediately arises whether the intentional nature of social phenomena constrains what can count as an adequate research program. For to identify a phenomenon as intentional is to identify it as something which was brought about for some reason: it is part of what we mean by "intentional" that it was done for a reason or purpose. And so describing something in intentional terms is implicitly to make an explanatory claim. If this argument is correct, it suggests that an adequate explanation of a social phenomenon would have to include, or be based upon, an account of the reasons or motivations which led to the behavior which brought about the phenomenon in question. If this is the case, then research programs in the social sciences would have to include a conception of human needs, purposes, rationality, etc., in terms of which these motivational accounts could be constructed. Research programs, such as certain versions of systems theory, which dispensed entirely with the motivations and orientations of social actors, could be dismissed as inadequate to explain intentional phenomena.[19]

The purpose of this discussion of research programs in social science is not to show that they must have some particular form. Rather, it is to point to yet another problem that is distinctive to theory construction in the social sciences, and which the naturalist tradition in the philosophy of social science cannot address, let alone solve. Until we can transcend the sterile antinomy between naturalist and humanist in the philosophy of social science, we will be completely unable to provide an adequate account of the nature of social theory.

IV

To this point we have argued that the self-understandings which people have play a causal role in bringing about the behavior in which they engage. Now this fact has often served as the basis for humanist philosophers and social scientists to make the further claim that explanations of social behavior consist solely of reconstructions of these self-understandings. Since actions are events that occur because they are warranted by the beliefs and desires of the actor, the task of explaining them is thought to consist of laying out the structure of reasons which justifies them. According to the humanist model, social science grasps the intelligibility of a particular form of behavior by making explicit the conceptual links that, it is hypothesized, implicitly exist between various sorts of activities, institutions and psychological states like beliefs and desires. A good interpretation, then, is one which demonstrates the coherence which an initially unintelligible act, rule or belief has in terms of the whole of which it is a part.

Humanists often draw two important conclusions from this construal of social science. The first is that the social scientist must assume that the beliefs, practices and actions which he encounters are congruent with one another insofar as they are explicable. The second is that, since it is the conceptual linkages between the actors' beliefs, actions and practices which he must uncover, the explanations which the social scientist puts forward must employ essentially the same concepts which an ideal, fully informed and articulate participant would give.[20] Both of these conclusions support a view of social life which takes it to be, by definition, rational at some level and understandable in its own terms.

Unfortunately, such a view is woefully inadequate just because it ignores crucial elements of social experience which are obviously present in social life, and which are often studied by social scientists. These include cases in which people's self-understandings are at variance with their actual situation and behavior, or in which a specific belief and action system is incompatible with other norms of the culture, or in which there are endemic conflicts as the result of conflicts in social-structural principles.[21]

In short, people may systematically misunderstand their own motives, wants, values and actions, as well as the nature of their social order, and—given what we have said about the constitutive role of self-understandings in social life—these misunderstandings may underlie and sustain particular forms of social interaction. In these situations, the actors' ideas may mask social reality as much as reveal it, and so the social scientist cannot confine himself to explicating the way in which the actors' concepts and self-understandings form a coherent system. In order to understand these cases, the social scientist must recognize how the actors' self-understandings are incoherent, and he must show what consequences these incoherencies have.

Concrete examples of social phenomena which cannot be understood in their own terms include the idea of nobility in feudal society, and the witch craze of early modern Europe. The concept of "nobility," as Gellner has pointed out,[22] was used to legitimate rulership in feudal society: one is entitled to rule because one is "noble" or virtuous. But, at the same time, a person was a member of the ruling class, or a "noble," simply by virtue of birth, not personal merit. Thus, the concept of nobility is at best equivocal, if not thoroughly incoherent, and the failure to notice and correct this incoherency is a misunderstanding or confusion that is a condition of the feudal form of political domination.

The European witch craze is also a social phenomenon that is, in many significant respects, irrational. As Trevor-Roper has argued,[23] the belief in "witches" was not necessarily irrational in the intellectual context of the time, but the belief in witches did not cause the witch craze. What was distinctive about the witch craze, and what requires explanation going beyond the self-understandings of the actors involved, are such factors as the ferocity of the persecutions, the sudden and dramatic increase in the number of putative witches who were discovered and condemned, the geographic and social patterns of persecution, and the widespread use of torture. By focusing only on the concepts available to the actors involved, we could not explain these phenomena adequately: it would certainly not do to say that the cause of the witch craze was the fact that the number of witches has dramatically increased! Moreover, in doing so we would also fail to set the witch craze in the context of the social tensions of the time, and we would fail to see how it involved a process of scapegoating that served to deflect social discontent. By focusing only on witchcraft in terms of the system of beliefs and values of which it was a part, we would miss much that is essential to the social reality of the witch craze.

Irrational social phenomena, unfortunately, are quite common. Consider, e.g., sociologists' and psychologists' attempts to uncover the "real meaning" of neurotic behavior (like compulsive handwashing), of violent prejudicial behavior toward minority groups, of recurring self-destructive patterns of social interaction, and so forth. Moreover, in situations such as these, the particular form of irrational behavior may not be just an isolated feature of a person's life, but may instead be systematically related to a wide range of different emotions, beliefs and actions. The very basis of a person's life—the terms in which he talks about himself in his most lucid and reflective moments, and the fears, aspirations, beliefs, passions, and values which he ascribes to himself at these times—may be fundamentally mistaken, and, as a result, he may be unable to adequately explain his behavior to himself or others. Worse than this, as a result of such misunderstanding he may pursue ends he cannot achieve, and the goals he does reach may not be satisfying. Such frustration may lead him to intensify his efforts, and so to perpetuate his misery. And just as it is possible for a person to be systematically mistaken, so whole forms of life may be based upon such self-misunderstandings, or what might be called "false consciousness." This is the picture of life that is painted for us by Rousseau, Hegel, Marx, and, more recently, by Freud, Brown, Habermas, Becker, and a host of others.

The social scientist attempts to explain such irrational phenomena by treating the actors' beliefs and desires as ciphers for something else that constitutes the actors' actual reason for acting, or the real need which they are trying to fill. Thus, according to Rousseau, people desire wealth, but what they really want is social distinction, and money is an expression of social distinction in certain societies.[24] Similarly, according to Marx, people engage in religious practices because they desire to be complete and whole human beings, and they believe that God will provide that fulfillment; but God is really nothing more than a picture of themselves fully actualized, and what would really satisfy them is to develop and exercise their productive capacities in forms of cooperative, social labor.[25] Finally, to offer a third example, Becker argues that people pursue sexual romance and contact, because sex is a cipher for everlasting life, and what they really what is to overcome the fear of their own death.[26]

Such accounts of human motivation and behavior immediately lead to the question, How is it possible for people to be so ignorant and confused about their own needs and motives, thereby leading them to engage in destructive and frustrating activities? To answer this question we must have an account of what causes people to mistake some purpose or object (wealth, God, sex), for what they really want (social distinction, happiness, eternal life), and how these delusions are maintained. Freud's notions of sublimation and repression, and Marx's notions of alienation and ideology, are examples of concepts created in order to explain the process by which an activity acquires symbolic import and with it causal power, and how this process itself is hidden from the agent's view.

Thus, systematic misunderstandings of the meanings of one's activities, reinforced by repressive mechanisms, can result in irrational behavior whose upshot is social conflict and the experience of frustration. And this is the case just because human behavior is intentional in the sense of being undertaken on the basis of the ideas, desires and perceptions of those who perform it. But in these situations the traditional humanist goal of understanding intentional phenomena by grasping the coherence which exists among their meanings must be replaced by the need to critique these phenomena. Or better, the only way to understand such a social situation is to engage in a critique in which one lays bare the ways in which the ideas people have of themselves mask the social reality which their behavior creates, and in which one tries to demonstrate that the coherence of the relevant behavior occurs at a level so deep that it is beyond the capacity of the actors to appreciate it given the conceptual and emotional responses open to them. In doing this, the social scientist will undoubtedly have to make use of concepts and conceptual distinctions which in a basic way go beyond those operative in the social life which is being studied. It is in this way that the humanist model will be transcended.

Of course, the naturalist model will be of no help in this matter either. For though the naturalists have always been insistent that social theorists need not be confined to the categories of thought of the people they are analyzing, there is nothing in the natural sciences comparable to assessing the rationality of a particular belief system, institution or system of actions, and deciding on a certain type of explanation depending on this assessment. Only of an intentional phenomenon can one ask: Are the factors which support it mistaken? Could it have been undertaken out of ignorance? What role does deception play in its continuation? And so forth.

The humanist cannot appreciate the role of critique in social science because he artificially confines himself to interpreting the meanings which various aspects of a social life are supposed to have by grasping the coherence which he thinks exists between these aspects understood in their own terms. By so confining himself, he not only ensures that he will fail to see the conflict, irrationality and mechanisms of repression operative in all social orders, but also deprives himself of the means necessary to understand these phenomena, namely, a categorial scheme which allows him to speak about the relevant social order in terms radically opposed to that of the participants. The naturalist, on the other hand, cannot give an account of critique because, by neglecting the particular features of intentional phenomena, he cannot appreciate the crucial role which rationality plays in social life, or its assessment plays in social science. These inadequacies of both the humanist model and the naturalist model in elucidating the role of critique in social theory give a third reason why the dualist approach of humanism versus naturalism must be overcome if a satisfactory philosophy of social science is to be forthcoming.

V

In this essay we have not tried to set out a philosophical account of social science, but to show that neither of the two prevailing account is adequate. An adequate philosophy of social science must be capable of answering the three question we have discussed: first, What is the relationship between interpretation and explanation? second, What is the nature of social scientific theory? and third, What is the role of critique in social science? Broadly speaking, these questions arise because of the conjunction of two important features of social science. In the first place, these sciences are *social,* which is to say that the phenomena they study are intentional phenomena, and so must be identified in terms of their meanings. Secondly, these sciences are *sciences,* in the sense that they try to develop systematic theories to explain the underlying causal interconnections among phenomena of a widely divergent sort. Because they each fasten on only one of these features, humanism and naturalism fail to provide an adequate account of social science.

This does not mean, however, that we reject both of these traditions of thought entirely. On the contrary, these philosophical metatheories are partial realizations of the task of giving an account of social science. What is wrong with them is not that they are false, but that they are one-sided. Indeed, as we have suggested throughout our analysis, these two positions can be reformulated in such a way as to render them compatible, and their insights complementary. By showing just where humanism and naturalism are inadequate, we hope to have contributed to the construction of the framework for a new synthesis in the philosophy of social science.

Notes

The essay was written while the authors were receiving a Joint Research Grant from the National Endowment for the Humanities (RO-22106-75-139). The authors wish to acknowledge the importance of the free time which this grant made possible, and to publicly thank the National Endowment. The argument presented here does not necessarily represent the view of NEH. An earlier version of this essay was read at Williams College.

1. See Maurice Roche, *Phenomenology, Language and the Social Sciences*, London 1973, and G. H. von Wright, *Explanation and Understanding*, Ithaca, N.Y. 1971, for recent examples of this opposition between these two models of social science.
2. Following von Wright, p. 6.
3. The intentionalist response may be adopted by a broad spectrum of what otherwise might be strange bedfellows, including humanists (such as phenomenologists) and naturalists who are not at the same time explanatory reductionists (such as those who adopt a functionalist theory of mind). Moreover there are some who think that a simple intentionalist response is inadequate, because the science of man will ultimately be a hybrid science which employs both intentional and extensional terminology (much as computer science today employs both the language of programming and the language of electronics); for this, see Daniel Dennett, "Intentional Systems," *Journal of Philosophy*, 68, 1971. Of course, in this case the problems associated with an intentionalist analysis would still remain.
4. The classical statement of such a positions is G. Ryle, *The Concept of Mind*, New York 1949. Behaviorist psychology is based on this response; see B. F. Skinner, *Verbal Behavior*, New York 1957. In *The Nature of Cultural Things*, New York 1964, Marvin Harris advocates such a program for anthropology.
5. See W. V. O. Quine, *Word and Object*, Cambridge, Mass. 1960. Actually, the third response is more varied that it might at first appear. For, on the one hand, it characterizes those who think that the task of a science of man is to discover the contingent identities between those states and events now characterized in intentional terms and these same states and events designated in purely physical terms, such that a mapping of one terminology into the other via their supposed common extension, followed by the replacement of the intentionalist vocabulary by the physicalist one, would characterize the development of human sciences. (This is the view of the Identity theorists; see D. Armstrong, *A Materialist Theory of Mind*, London 1968.) And, on the other hand, it also characterizes those who believe that the intentionalist idiom should be abandoned altogether in the construction of a science of man. See P. K. Feyerabend, "Mental Events and the Brain," *Journal of Philosophy*, 60, 1963.
6. See Roderick Chisholm, *Perceiving*, Ithaca, N.Y. 1957. ch. 11.
7. Peter Winch, *The Idea of a Social Science*, London 1958, p. 125.
8. Charles Taylor "Interpretation and the Sciences of Man," *Review of Metaphysics*, 25, 1971.
9. It is in anthropology that the need for interpretation is most obvious, because the anthropologist does not have an "insider's" implicit understanding of the society he studies, and so he must develop an explicit scheme of the whole in his work. It is, thus, no accident that it is in anthropology that the interpretive enterprise is most highly developed.
10. See, e.g., A. I. Melden, *Free Action*, London 1961, passim.
11. The classic statement of this position is Donald Davidson, "Actions, Reasons, and Causes," *Journal of Philosophy*, 60, 1963.
12. The exception to this is Arthur Danto, *Analytical Philosophy of Action*, Cambridge 1973, and especially the writings of those who propound a functionalistic theory of mind, such as J. Fodor, *Psychological Explanation*, New York 1968; Daniel Dennett, *Content and Consciousness*, London 1969; and the essays by Hilary Putnam in his *Mind, Language and Reality*, Cambridge 1975.

13. This was, of course, Max Weber's position, especially in his "Critical Studies in the Logic of the Cultural Science," and in this respect we feel a return to Weber would be a progressive step in the philosophy of social science. We say this even though we do not agree with his answers to our questions.
14. See Donald Davidson, "Causal Relations," *Journal of Philosophy*, 64, 1967, and his "Mental Events," in L. Foster and J. Swanson (eds.), *Experience and Theory*, Amherst, Mass. 1970. In the latter essay Davidson argues that the causal relationships which hold between the having of reasons and actions can only be stated in a nonintentionalist vocabulary.
15. Richard Rudner, e.g., in his discussion of social theory in his *Philosophy of Social Science*, Englewood Cliffs, New Jersey 1966, first gives an account of physical theories, and then discusses various theoretical formulations found in the social sciences, including typologies and analytical conceptual schemata, pointing out how these fail to meet the criteria for genuine theories.
16. Noam Chomsky discusses the relevance of competence theories for linguistics in *Aspects of a Theory of Syntax*. Cambridge, Mass 1965, part I.
17. See Victory Turner, *The Ritual Process*, Chicago 1967, pp. 172ff.
18. Imre Lakatos, "Falsification and the Methodology of Scientific Research Programmes," in Lakatos and Musgrave (eds.), *Criticism and the Growth of Knowledge*, Cambridge 1970.
19. For further discussion of the use of the idea of a "research program" in explicating the structure of social science theories, see J. Donald Moon, "The Logic of political Inquiry," in Fred I. Greeenstein and Nelson W. Polsby (eds.), *The Handbook of Political Science*, vol. 1, Reading, Mass. 1975, pp. 192ff.
20. R. G. Collingwood, *The Idea of History*, New York 1945, pp. 308ff., and pp. 282ff.
21. For examples of studies investigating each of these three types of situations, see V. Aubert, *Sociology of the Law*, Oslo 1964; Chalmers Johnson, *Revolutionary Change*, Boston 1966; and Victor Turner, *Schism and Continuity in an African Society*, Manchester 1957, respectively.
22. E. Gellner, "Concepts and Society," reprinted in his *Cause and Meaning in the Social Sciences*, London 1973, pp. 18–46.
23. See H. R. Trevor-Roper, *The European Witch-Craze*, New York 1969, and Alasdair MacIntyre, "Rationality and the Explanation of Action," in his *Against the Self-Images of the Age*, New York 1971.
24. *Discourse on the Origins of Inequality*, ed. Masters, New York 1964, pp. 265–66.
25. See "A Contribution to the Critique of Hegel's 'Philosophy of Right,' Introduction," in Karl Marx, *Critique of Hegel's "Philosophy of Right,"* ed. Joseph O'Malley, Cambridge 1970, pp. 129ff.
26. See *The Denial of Death*, New York 1973, pp. 160–70.

PART II

Explanation, Prediction, and Laws

Introduction to Part II

We now come to grips with one of the most important issues in the philosophy of social science: What sort of explanation of social action is appropriate? And what role, if any, do laws and predictions have in them?

Traditionally, the thesis that laws are important to the explanation of human behavior has been informed by an analogy between the natural and social sciences. It seems appropriate, therefore, that the first selection in part II is by the person who initiated the debate on the role of laws in natural and social science, Carl G. Hempel.

In "The Function of General Laws in History," Hempel argues that, as is the case in the explanation of natural scientific phenomena, social scientific explanation looks to an account based on an understanding of the inevitability of the relationship between causal forces and behavioral outcome; we want to show that the action we are trying to explain was "no accident," and that it could have been predicted if sufficient information had been known at the time of the event. Thus, the logical parity between explanation and prediction is upheld, by way of the view that any appropriate scientific account of the phenomenon will be one that embeds it in the system of influences leading up to it.

The authors of the next four chapters, however, mount a powerful dissent from this conception of the adequacy of nomological explanation in the social sciences. They question whether the subject matter of social science is relevantly similar to that examined in the natural sciences (and presents difficulties unique to the exploration of social action) and also whether it is fair to say that we even want to have social phenomena explained in terms of their governing laws, if any such indeed exist. That is, critics of Hempel's deductive-nomological model of explanation have held that social scientific laws are impossible, impractical, or irrelevant and therefore do not or cannot help in the task of explaining human action.

In chapter 4, the first of these dissents, "The Theory of Complex Phenomena," F. A. Hayek presents one of the most popular criticisms of the attempt to use laws in the explanation of human action: the phenomena are far too complex. Hayek's goal is to show that this complexity would undercut the possibility of social scientific laws, due to this essential difference in the subject matter at hand.

Critics have held, however, that this view is marred by the failure to recognize that the phenomena *as such* are never complex, but only *as described*, and that it is the phenomena as described that we hope to capture in scientific explanations. Such a sweeping attempt as Hayek's to dismiss the efficacy of social scientific laws on a priori grounds, therefore, may run into problems. In an attempt to ameliorate such difficulties, Michael Scriven, in "A Possible Distinction between Traditional Scientific Disciplines and the Study of Human Behavior" (chapter 5), has picked up on Hayek's central theme, the problem of complexity. Instead of trying to show that it would make all attempts to provide laws impossible, Scriven's goal is to show that given the level of description that we desire when engaging in social inquiry, the phenomena are captured in a way that renders them complex (even if only derivatively so) and would therefore make social scientific laws impractical.

It is feared, however, that such an ambitious attempt itself runs into difficulties, given the fact that it ignores the possibility of redescriptions of social phenomena, in terms that would allow for a simplification of our understanding of human affairs, and codification into a lawlike account. But would this then be social science? Aren't the phenomena as captured at certain levels of description just constituitive of the subject matter of social science?

These issues are taken up by Donald Davidson in "Psychology as Philosophy" (chapter 6), in which he upholds the idea that although we may believe that there are genuine laws underlying human behavior, these are not, and never will be, laws of social science. Social science, Davidson argues, is based on a certain descriptive framework of social action and will not support nomological explanations. Redescriptions would be just to change the subject matter. Thus, Davidson advocates a position of abandoning the search for social scientific laws while recognizing that it is legitimate to continue doing non-nomological social science, and even to continue looking for the laws that underlie this phenomenon at other levels of description, as long as one resists the temptation to unify explanatory accounts across different levels of inquiry.

Consideration of these issues reaches its highest stage in the argument against social scientific laws in Brian Fay's "General Laws and Explaining Human Behavior" (chapter 7). Fay here speaks not of the complexity of social phenomena but of a subtle and closely related problem (originally outlined by Karl Popper), concerning the "openness" of human systems: our logical inability to predict the novel thoughts of human beings and their inevitable influence on human action. Fay's view is rooted in the same idea as Scriven and Hayek's—that there is a distinction in kind between the subject matter of natural and social science—but he presents a degree of sophistication that allows him to recognize the efficacy of laws, the power of redescription, the importance of avoiding parochialism in definitions of the subject matter, and the poverty of a priori claims concerning the differences between the phenomena in natural and social science. In the end, however, Fay advocates the idea that such laws, even if they could be produced in social science, would not be explanatory, given what it is that we seek to know about human affairs. That is, despite the sophistication of his critique, he ends up advocating a position close to Scriven's (and Davidson's) in which the redescription of the subject matter of social science is taken inevitably to lead only to a truistic understanding of human affairs.

Harold Kincaid and Lee McIntyre, in the next two chapters, however, defend the power of nomological social science against its critics. Kincaid, in "Defending Laws in the Social Sciences," stakes out a view that such criticisms have largely overlooked the fact that the methodological situations in natural and social science are similar and that if the argument outlined above were cogent, they would probably succeed in preventing laws in most of natural science. Lee McIntyre, in "Complexity and Social Scientific Laws" (chapter 9), argues that the idea that theory-based redescriptions of the phenomena in other terms, in the search for laws, will inevitably be truistic and will result in inquiry that is other than "social scientific" is a repercussion of a naive understanding of the constraints put on lawlike explanation. This general theme is developed by way of specific examination of the argument from complexity and the failure to compare the methodological situations facing natural and social science on appropriate grounds. McIntyre also develops a general defense against the criticisms regarding redescription in the social sciences through his claim that it is used throughout the natural sciences, as well as his claim that all such "descriptivist" criticisms—like the four outlined above—are rooted in an overly narrow conception of the requirements of scientific explanation, either natural or social.

Finally, George Romanos considers the interesting issue of "Reflexive Predictions" (chapter 10) and explores the suitability of the analogy between prediction and explanation in the social sciences. As human beings with a reflexive self-consciousness that allows us to learn about a prediction governing our behavior, aren't we free to defy it? Romanos reviews this debate, which goes to the very heart of the claim that the subject matter of social science is somehow distinctive. This issue is taken up again later in this book in an examination of Romanos's argument in Michael Martin's "The Philosophical Importance of the Rosenthal Effect" (chapter 37).

Chapter 3

The Function of General Laws in History

Carl G. Hempel

1

It is a rather widely held opinion that history, in contradistinction to the so-called physical sciences, is concerned with the description of particular events of the past rather than with the search for general laws which might govern those events. As a characterization of the type of problem in which some historians are mainly interested, this view probably cannot be denied; as a statement of the theoretical function of general laws in scientific historical research, it is certainly unacceptable. The following considerations are an attempt to substantiate this point by showing in some detail that general laws have quite analogous functions in history and in the natural sciences, that they form an indispensable instrument of historical research, and that they even constitute the common basis of various procedures which are often considered as characteristic of the social in contradistinction to the natural sciences.

By a general law, we shall here understand a statement of universal conditional form which is capable of being confirmed or disconfirmed by suitable empirical findings. The term "law" suggests the idea that the statement in question is actually well confirmed by the relevant evidence available; as this qualification is, in many cases, irrelevant for our purpose, we shall frequently use the term "hypothesis of universal form" or briefly "universal hypothesis" instead of "general law," and state the condition of satisfactory confirmation separately, if necessary. In the context of this chapter, a universal hypothesis may be assumed to assert a regularity of the following type: In every case where an event of a specified kind *C* occurs at a certain place and time, an event of a specified kind *E* will occur at a place and time which is related in a specified manner to the place and time of the occurrence of the first event. (The symbols *C* and *E* have been chosen to suggest the terms "cause" and "effect," which are often, though by no means always, applied to events related by a law of the above kind.)

2

2.1 The main function of general laws in the natural sciences is to connect events in patterns which are usually referred to as *explanation* and *prediction*.

The explanation of the occurrence of an event of some specific kind *E* at a certain place and time consists, as it is usually expressed, in indicating the causes or determining factors of *E*. Now the assertion that a set of events—say, of the kinds $C_1, C_2, \ldots, C_n$— have caused the event to be explained, amounts to the statement that, according to certain general laws, a set of events of the kinds mentioned is regularly accompanied by an event of kind *E*. Thus, the scientific explanation of the event in question consists of

1. a set of statements asserting the occurrence of certain events $C_1, \ldots, C_n$ at certain times and places,
2. a set of universal hypotheses, such that
a. the statements of both groups are reasonably well confirmed by empirical evidence,
b. from the two groups of statements the sentence asserting the occurrence of event E can be logically deduced.

In a physical explanation, group 1 would describe the initial and boundary conditions for the occurrence of the final event; generally, we shall say that group 1 states the *determining conditions* for the event to be explained, while group 2 contains the general laws on which the explanation is based; they imply the statement that whenever events of the kind described in the first group occur, an event of the kind to be explained will take place. As an illustration, let the event to be explained consist in the cracking of an automobile radiator during a cold night. The sentences of group 1 may state the following initial and boundary conditions: The car was left in the street all night. Its radiator, which consists of iron, was completely filled with water, and the lid was screwed on tightly. The temperature during the night dropped from 39°F. in the evening to 25°F. in the morning; the air pressure was normal. The bursting pressure of the radiator material is so and so much. Group 2 would contain empirical laws such as the following: Below 32°F., under normal atmospheric pressure, water freezes. Below 39.2°F., the pressure of a mass of water increases with decreasing temperature, if the volume remains constant or decreases; when the water freezes, the pressure again increases. Finally, this group would have to include a quantitative law concerning the change of pressure of water as a function of its temperature and volume.

From statements of these two kinds, the conclusion that the radiator cracked during the night can be deduced by logical reasoning; an explanation of the considered event has been established.

2.2 It is important to bear in mind that the symbols E, C, C_1, C_2, and so forth, which were used above, stand for kinds or properties of events, not for what is sometimes called individual events. For the object of description and explanation in every branch of empirical science is always the occurrence of an event of a certain *kind* (such as a drop in temperature by 14°F., an eclipse of the moon, a cell division, an earthquake, an increase in employment, a political assassination) at a given place and time, or in a given empirical object (such as the radiator of a certain car, the planetary system, a specified historical personality, etc.) at a certain time.

What is sometimes called the *complete description* of an individual event (such as the earthquake of San Francisco in 1906 or the assassination of Julius Caesar) would require a statement of all the properties exhibited by the spatial region or the individual object involved, for the period of time occupied by the event in question. Such a task can never be completely accomplished.

A fortiori, it is impossible to give a *complete explanation* of an individual event in the sense of accounting for *all* its characteristics by means of universal hypotheses, although the explanation of what happened at a specified place and time may gradually be made more and more specific and comprehensive.

But there is no difference, in this respect, between history and the natural sciences: both can give an account of their subject matter only in terms of general concepts, and history can "grasp the unique individuality" of its objects of study no more and no less than can physics or chemistry.

3

The following points result more or less directly from the above study of scientific explanation and are of special importance for the questions here to be discussed.

3.1. A set of events can be said to have caused the event to be explained only if general laws can be indicated which connect "causes" and "effect" in the manner characterized above.

3.2. No matter whether the cause-effect terminology is used or not, a scientific explanation has been achieved only if empirical laws of the kind mentioned in section 2.1 have been applied.[1]

3.3. The use of universal empirical hypotheses as explanatory principles distinguishes genuine from pseudo-explanation, such as, say, the attempt to account for certain features of organic behavior by reference to an entelechy, for whose functioning no laws are offered, or the explanation of the achievements of a given person in terms of his "mission in history," his "predestined fate," or similar notions. Accounts of this type are based on metaphors rather than laws; they convey pictorial and emotional appeals instead of insight into factual connections; they substitute vague analogies and intuitive "plausibility" for deduction from testable statements and are therefore unacceptable as scientific explanations.

Any explanation of scientific character is amenable to objective checks; these include

a. an empirical test of the sentences which state the determining conditions;
b. an empirical test of the universal hypotheses on which the explanation rests;
c. an investigation of whether the explanation is logically conclusive in the sense that the sentence describing the events to be explained follows from the statements of groups 1 and 2.

4

The function of general laws in *scientific prediction* can now be stated very briefly. Quite generally, prediction in empirical science consists in deriving a statement about a certain future event (for example, the relative position of the planets to the sun, at a future date) from (1) statements describing certain known (past or present) conditions (for example, the positions and momenta of the planets at a past or present moment) and (2) suitable general laws (for example, the laws of celestial mechanics). Thus, the logical structure of a scientific prediction is the same as that of a scientific explanation, which has been described in section 2.1. In particular, prediction no less than explanation throughout empirical science involves reference to universal empirical hypotheses.

The customary distinction between explanation and prediction rests mainly on a pragmatic difference between the two: While in the case of an explanation, the final event is known to have happened, and its determining conditions have to be sought, the situation is reversed in the case of a prediction: here, the initial conditions are given, and their "effect"—which, in the typical case, has not yet taken place—is to be determined.

In view of the structural equality of explanation and prediction, it may be said that an explanation as characterized in section 2.1 is not complete unless it might as well have functioned as a prediction: If the final event can be derived from the initial conditions and universal hypotheses stated in the explanation, then it might as well have been predicted, before it actually happened, on the basis of a knowledge of the initial conditions and the general laws. Thus, for example, those initial conditions and general laws which the astronomer would adduce in explanation of a certain eclipse of the sun are such that they might also have served as a sufficient basis for a forecast of the eclipse before it took place.

However, only rarely, if ever, are explanations stated so completely as to exhibit this predictive character (which the test referred to under (c) in 3.3 would serve to reveal). Quite commonly, the explanation offered for the occurrence of an event is incomplete. Thus, we may hear the explanation that a barn burned down "because" a burning cigarette was dropped in the hay, or that a certain political movement has spectacular success "because" it takes advantage of widespread racial prejudices. Similarly, in the case of the broken radiator, the customary way of formulating an explanation would be restricted to pointing out that the car was left in the cold, and the radiator was filled with water. In explanatory statements like these, the general laws which confer upon the stated conditions the character of "causes" or "determining factors" are completely omitted (sometimes, perhaps, as a "matter of course"), and, furthermore, the enumeration of the determining conditions of group 1 is incomplete; this is illustrated by the preceding examples, but also by the earlier analysis of the broken radiator case: as a closer examination would reveal, even that much more detailed statement of determining conditions and universal hypotheses would require amplification in order to serve as a sufficient basis for the deduction of the conclusion that the radiator broke during the night.

In some instances, the incompleteness of a given explanation may be considered as inessential. Thus, e.g., we may feel that the explanation referred to in the last example could be made complete if we so desired; for we have reasons to assume that we know the kind of determining conditions and of general laws which are relevant in this context.

Very frequently, however, we encounter "explanations" whose incompleteness cannot simply be dismissed as inessential. The methodological consequences of this situation will be discussed later (especially in 5.3 and 5.4).

5

5.1 The preceding considerations apply to *explanation in history* as well as in any other branch of empirical science. Historical explanation, too, aims at showing that the event in question was not "a matter of chance," but was to be expected in view of certain antecedent or simultaneous conditions. The expectation referred to is not prophecy or divination, but rational scientific anticipation which rests on the assumption of general laws.

If this view is correct, it would seem strange that while most historians do suggest explanations of historical events, many of them deny the possibility of resorting to any general laws in history. It is possible, however, to account for this situation by a closer study of explanation in history, as may become clear in the course of the following analysis.

5.2 In some cases, the universal hypotheses underlying a historical explanation are rather explicitly stated, as is illustrated by the italicized passages in the following attempt to explain the tendency of government agencies to perpetuate themselves and to expand:

> As the activities of the government are enlarged, more people develop a vested interest in the continuation and expansion of governmental functions. *People who have jobs do not like to lose them; those who are habituated to certain skills do not welcome change; those who have become accustomed to the exercise of a certain kind of power do not like to relinquish their control—if anything, they want to develop greater power and correspondingly greater prestige....*
>
> Thus, government offices and bureaus, once created, in turn institute drives, not only to fortify themselves against assault, but to enlarge the scope of their operations.[2]

Most explanations offered in history or sociology, fail to include an explicit statement of the general regularities they presuppose; and there seem to be at least two reasons which account for this:

First, the universal hypotheses in question frequently relate to individual or social psychology, which somehow is supposed to be familiar to everybody through his everyday experience; thus, they are tacitly taken for granted. This is a situation quite similar to that characterized in section 4.

Second, it would often be very difficult to formulate the underlying assumptions explicitly with sufficient precision and at the same time in such a way that they are in agreement with all the relevant empirical evidence available. It is highly instructive, in examining the adequacy of a suggested explanation, to attempt a reconstruction of the universal hypotheses on which it rests. Particularly, such terms as "hence," "therefore," "consequently," "because," "naturally," and "obviously" are often indicative of the tacit presupposition of some general law: they are used to tie up the initial conditions with the event to be explained; but that the latter was "naturally" to be expected as a "consequence" of the stated conditions follows only if suitable general laws are presupposed. Consider, for example, the statement that the dust bowl farmers migrated to California "because" continual drought and sandstorms made their existence increasingly precarious, and because California seemed to them to offer so much better living conditions. This explanation rests on some such universal hypothesis as that populations will tend to migrate to regions which offer better living conditions. But it would obviously be difficult accurately to state this hypothesis in the form of a general law which is reasonably well confirmed by all the relevant evidence available. Similarly, if a particular revolution is explained by reference to the growing discontent, on the part of a large part of the population, with certain prevailing conditions, it is clear that a general regularity is assumed in this explanation, but we are hardly in a position to state just what extent and what specific form the discontent has to assume, and what the environmental conditions have to be, to bring about a revolution. Analogous remarks apply to all historical explanations in terms of class struggle, economic or geographic conditions, vested interests of certain groups, tendency to conspicuous consumption, etc.: all of them rest on the assumption of universal hypotheses[3] which connect certain characteristics of individual or group life with others; but in many cases, the content of the hypotheses which are tacitly assumed in a given explanation can be reconstructed only quite approximately.

5.3 It might be argued that the phenomena covered by the type of explanation just mentioned are of a statistical character, and that therefore only probability hypotheses need to be assumed in their explanation, so that the question as to the "underlying general laws" would be based on a false premise. And indeed, it seems possible and justifiable to construe certain explanations offered in history as based on the assumption of probability hypotheses rather than of general "deterministic" laws, i.e., laws in the form of universal conditionals. This claim may be extended to many of the explanations offered in other fields of empirical science as well. Thus, e.g., if Tommy comes down with the measles two weeks after his brother, and if he has not been in the company of other persons having the measles, we accept the explanation that he caught the disease from his brother. Now, there is a general hypothesis underlying this explanation; but it can hardly be said to be a general law to the effect that any person who has not had the measles before will get it without fail if he stays in the company of somebody else who has the measles; that contagion will occur can be asserted only with high probability.

Many an explanation offered in history seems to admit of an analysis of this kind: if fully and explicitly formulated, it would state certain initial conditions, and certain probability hypotheses,[4] such that the occurrence of the event to be explained is made highly probable by the initial conditions in view of the probability hypotheses. But no matter whether explanations in history be construed as causal or as probabilistic, it remains true that in general the initial conditions and especially the universal hypotheses involved are not clearly indicated, and cannot unambiguously be supplemented. (In the case of probability hypotheses, for example, the probability values involved will at best be known quite roughly.)

5.4 What the explanatory analyses of historical events offer is, then, in most cases not an explanation in one of the senses indicated above, but something that might be called an *explanation sketch*. Such a sketch consists of a more or less vague indication of the laws and initial conditions considered as relevant, and it needs "filling out" in order to turn into a full-fledged explanation. This filling out requires further empirical research, for which the sketch suggests the direction. (Explanation sketches are common also outside history; many explanations in psychoanalysis, for instance, illustrate this point.)

Obviously, an explanation sketch does not admit of an empirical test to the same extent as does a complete explanation; and yet, there is a difference between a scientifically acceptable explanation sketch and a pseudo-explanation (or a pseudo-explanation sketch). A scientifically acceptable explanation sketch needs to be filled out by more specific statements; but it points in the direction where these statements are to be found; and concrete research may tend to confirm or to infirm those indications; i.e., it may show that the kind of initial conditions suggested are actually relevant; or it may reveal that factors of a quite different nature have to be taken into account in order to arrive at a satisfactory explanation.

The filling-out process required by an explanation sketch will in general effect a gradual increase in the precision of the formulations involved; but at any stage of this process, those formulations will have some empirical import: it will be possible to indicate, at least roughly, what kind of evidence would be relevant in testing them, and what findings would tend to confirm them. In the case of nonempirical explanations or explanation sketches, on the other hand—say, by reference to the historical destiny of a certain race, or to a principle of historical justice—the use of empirically meaningless terms makes it impossible even roughly to indicate the type of investigation that would have a bearing upon those formulations, and that might lead to evidence either confirming or infirming the suggested explanation.

5.5 In trying to appraise the soundness of a given explanation, one will first have to attempt to reconstruct as completely as possible the argument constituting the explanation or the explanation sketch. In particular, it is important to realize what the underlying explanatory hypotheses are, and to appraise their scope and empirical foundation. A resuscitation of the assumptions buried under the gravestones "hence," "therefore," "because," and the like will often reveal that the explanation offered is poorly founded or downright unacceptable. In many cases this procedure will bring to the light the fallacy of claiming that a large number of details of an event have been explained when, even on a very liberal interpretation, only some broad characteristics of it have been accounted for. Thus, for example, the geographic or economic conditions under which a group lives may account for certain general features of, say, its art or its moral codes; but to grant this does not mean that the artistic achievements of the group or its system of morals has thus been explained in detail; for this would imply that from a description of the prevalent geographic or economic conditions alone, a detailed account of certain aspects of the cultural life of the group can be deduced by means of specifiable general laws.

A related error consists in singling out one of several important groups of factors which would have to be stated in the initial conditions, and then claiming that the phenomenon in question is "determined" by that one group of factors and thus can be explained in terms of it.

Occasionally, the adherents of some particular school of explanation or interpretation in history will adduce, as evidence in favor of their approach, a successful historical prediction which was made by a representative of their school. But though the predictive success of a theory is certainly relevant evidence of its soundness, it is important to make sure that the successful prediction is in fact obtainable by means of the theory in question. It happens sometimes that the prediction is actually an ingenious guess which may have been influenced by the theoretical outlook of its author, but which cannot be arrived at by means of his theory alone. Thus, an adherent of a quite metaphysical "theory" of history may have a sound feeling for historical developments and may be able to make correct predictions, which he will even couch in the terminology of his theory, though they could not have been attained by means of it. To guard against such pseudo-confirming cases would be one of the functions of test c in section 3.3.

6

We have tried to show that in history no less than in any other branch of empirical inquiry, scientific explanation can be achieved only by means of suitable general hypotheses, or by theories, which are bodies of systematically related hypotheses. This thesis is clearly in contrast with the familiar view that genuine explanation in history is obtained by a method which characteristically distinguishes the social from the natural sciences, namely, *the method of empathic understanding*: The historian, we are told, imagines himself in the place of the persons involved in the events which he wants to explain; he tries to realize as completely as possible the circumstances under which they acted and the motives which influenced their actions; and by this imaginary self-identification with his heroes. he arrives at an understanding and thus at an adequate explanation of the events with which he is concerned.

This method of empathy is, no doubt, frequently applied by laymen and by experts in history. But it does not in itself constitute an explanation; it rather is essentially a heuristic device; its function is to suggest psychological hypotheses which might serve as explanatory principles in the case under consideration. Stated in crude terms, the idea underlying this function is the following: The historian tries to realize how he himself would act under the given conditions, and under the particular motivations of his heroes; he tentatively generalizes his findings into a general rule and uses the latter as an explanatory principle in accounting for the actions of the persons involved. Now, this procedure may sometimes prove heuristically helpful; but it does not guarantee the soundness of the historical explanation to which it leads. The latter rather depends upon the factual correctness of the generalizations which the method of understanding may have suggested.

Nor is the use of this method indispensable for historical explanation. A historian may, for example, be incapable of feeling himself into the role of a paranoiac historic personality, and yet he may well be able to explain certain of his actions by reference to the principles of abnormal psychology. Thus, whether the historian is or is not in a position to identify himself with his historical hero is irrelevant for the correctness of his explanation; what counts is the soundness of the general hypotheses involved, no matter whether they were suggested by empathy or by a strictly behavioristic procedure. Much of the appeal of the "method of understanding" seems to be due to the fact that it tends to present the phenomena in question as somehow "plausible" or "natural" to us;[5] this is often done by means of persuasive metaphors. But the kind of "understanding" thus conveyed must clearly be separated from scientific understanding. In history as anywhere else in empirical science, the explanation of a phenomenon consists in subsuming it under general empirical laws; and the criterion of its soundness is not whether it appeals to our imagination, whether it is presented in terms of suggestive analogies or is otherwise made to appear plausible—all this may occur in pseudo-explanations as well—but exclusively whether it rests on empirically well confirmed assumptions concerning initial conditions and general laws.

7

7.1 So far, we have discussed the importance of general laws for explanation and prediction, and for so-called understanding in history. Let us now survey more briefly some other procedures in historical research which involve the assumption of universal hypotheses.

Closely related to explanation and understanding is the so-called *interpretation of historical phenomena* in terms of some particular approach or theory. The interpretations which are actually offered in history consist either in subsuming the phenomena in question under a scientific explanation or explanation sketch; or in an attempt to subsume them under some general idea which is not amenable to any empirical test. In the former case, interpretation clearly is explanation by means of universal hypotheses; in the latter, it amounts to a pseudo-explanation which may have emotive appeal and evoke vivid pictorial associations, but which does not further our theoretical understanding of the phenomena under consideration.

7.2 Analogous remarks apply to the procedure of ascertaining the "*meaning*" of given historical events; its scientific import consists in determining what other events are relevantly connected with the event in question, be it as "causes," or as "effects";

and the statement of the relevant connections assumes, again, the form of explanations or explanation sketches which involve universal hypotheses; this will be seen more clearly in the next subsection.

7.3 In the historical explanation of some social institutions great emphasis is laid upon an analysis of the *development* of the institution up to the stage under consideration. Critics of this approach have objected that a mere description of this kind is not a genuine explanation. This argument may be given a slightly different form in terms of the preceding reflections: An account of the development of an institution is obviously not simply a description of *all* the events which temporally preceded it; only those events are meant to be included which are "*relevant*" to the formation of that institution. And whether an event is relevant to that development is not a matter of evaluative opinion, but an objective question depending upon what is sometimes called a causal analysis of the rise of that institution.[6] Now, the causal analysis of an event establishes an explanation for it, and since this requires reference to general hypotheses, so do assumptions about relevance, and, consequently, so does the adequate analysis of the historical development of an institution.

7.4 Similarly, the use of the notions of *determination* and *dependence* in the empirical sciences, including history, involves reference to general laws.[7] Thus, e.g., we may say that the pressure of a gas depends upon its temperature and volume, or that temperature and volume determine the pressure, in virtue of Boyle's law. But unless the underlying laws are stated explicitly, the assertion of a relation of dependence or of determination between certain magnitudes or characteristics amounts at best to claiming that they are connected by some unspecified empirical law; and that is a very meager assertion indeed: If, for example, we know only that there is some empirical law connecting two metrical magnitudes (such as length and temperature of a metal bar), we cannot even be sure that a change of one of the two will be accompanied by a change of the other (for the law may connect the same value of the "dependent" or "determined" magnitude with different values of the other), but only that with any specific value of one of the variables, there will always be associated one and the same value of the other; and this is obviously much less than most authors mean to assert when they speak of determination or dependence in historical analysis.

Therefore, the sweeping assertion that economic (or geographic, or any other kind of) conditions "determine" the development and change of all other aspects of human society has explanatory value only insofar as it can be substantiated by explicit laws which state just what kind of change in human culture will regularly follow upon specific changes in the economic (geographic, etc.) conditions. Only the establishment of specific laws can fill the general thesis with scientific content, make it amenable to empirical tests, and confer upon it an explanatory function. The elaboration of such laws with as much precision as possible seems clearly to be the direction in which progress in scientific explanation and understanding has to be sought.

8

The considerations developed in this chapter are entirely neutral with respect to the problem of "*specifically historical laws*": they do not presuppose a particular way of distinguishing historical from sociological and other laws, nor do they imply or deny the assumption that empirical laws can be found which are historical in some specific sense, and which are well confirmed by empirical evidence.

But it may be worth mentioning here that those universal hypotheses to which historians explicitly or tacitly refer in offering explanations, predictions, interpretations, judgments of relevance, etc., are taken from *various* fields of scientific research, insofar as they are not prescientific generalizations of everyday experiences. Many of the universal hypotheses underlying historical explanation, for instance, would commonly be classified as psychological, economical, sociological, and partly perhaps as historical laws; in addition, historical research has frequently to resort to general laws established in physics, chemistry, and biology. Thus, e.g. the explanation of the defeat of an army by reference to lack of food, adverse weather conditions, disease, and the like, is based on a—usually tacit—assumption of such laws. The use of tree rings in dating events in history rests on the application of certain biological regularities. Various methods of testing the authenticity of documents, paintings, coins, etc., make use of physical and chemical theories.

The last two examples illustrate another point which is relevant in this context: Even if a historian should propose to restrict his research to a "pure description" of the past, without any attempt at offering explanations or statements about relevance and determination, he would continually have to make use of general laws. For the object of his studies would be the past—forever inaccessible to his direct examination. He would have to establish his knowledge by indirect methods: by the use of universal hypotheses which connect his present data with those past events. This fact has been obscured partly because some of the regularities involved are so familiar that they are not considered worth mentioning at all; and partly because of the habit of relegating the various hypotheses and theories which are used to ascertain knowledge about past events, to the "auxiliary sciences" of history. Quite probably, some of the historians who tend to minimize, if not to deny, the importance of general laws for history, are prompted by the feeling that only "genuinely historical laws" would be of interest for history. But once it is realized that the discovery of historical laws (in some specified sense of this very vague notion) would not make history methodologically autonomous and independent of the other branches of scientific research, it would seem that the problem of the existence of historical laws ought to lose some of its importance.

The remarks made in this section are but special illustrations of two broader principles of the theory of science: first, the separation of "pure description" and "hypothetical generalization and theory-construction" in empirical science is unwarranted; in the building of scientific knowledge the two are inseparably linked. And, second, it is similarly unwarranted and futile to attempt the demarcation of sharp boundary lines between the different fields of scientific research, and an autonomous development of each of the fields. The necessity, in historical inquiry, to make extensive use of universal hypotheses of which at least the overwhelming majority come from fields of research traditionally distinguished from history is just one of the aspects of what may be called the methodological unity of empirical science.

Notes

1. Maurice Mandelbaum, in his generally very clarifying analysis of relevance and causation in history (*The Problem of Historical Knowledge*, New York, 1938, Chs. 7, 8) seems to hold that there is a difference between the "causal analysis" or "causal explanation" of an event and the establishment of scientific laws governing it in the sense stated above. He argues that "scientific laws can only be formulated on the basis of causal analysis," but that "they are not substitutes for full causal explanations" (p. 238). For the reasons outlined above, this distinction does not appear to be justifiable: every "causal explanation" is an "explanation by scientific laws"; for in no other way than by reference to empirical laws can the assertion of a causal connection between events be scientifically substantiated.

2. Donald W. McConnell et al., *Economic Behavior*; New York, 1939; pp. 894–95. (Italics supplied.)
3. What is sometimes misleadingly called an explanation by means of a certain *concept* is, in empirical science, actually an explanation in terms of *universal hypotheses* containing that concept. "Explanations" involving concepts which do not function in empirically testable hypotheses—such as "entelechy" in biology, "historic destination of a race" or "self-unfolding of absolute reason" in history—are mere metaphors without cognitive content.
4. E. Zilsel, in a stimulating paper on "Physics and the Problem of Historico-Sociological Laws" (*Philosophy of Science*, Vol. 8, 1941, pp. 567–79), suggests that all specifically historical laws are of a statistical character similar to that of the "macro-laws" of physics. The above remarks, however, are not restricted to specifically historical laws since explanation in history rests to a large extent on nonhistorical laws (cf. section 8 of this chapter).
5. For a criticism of this kind of plausibility, cf. Zilsel, pp. 577–78, and sections 7 and 8 in the same author's "Problems of Empiricism," in *International Encyclopedia of Unified Science*, Vol. II, 8 (Chicago: University of Chicago Press, 1941).
6. See the detailed and clear exposition of this point in M. Mandelbaum's book, chapters 6–8.
7. According to Mandelbaum, history, in contradistinction to the physical sciences, consists "not in the formulation of laws of which the particular case is an instance, but in the description of the events in their actual determining relationships to each other; in seeing events as the products and producers of change" (pp. 13–14). This is, in effect, a conception whose untenability has been pointed out already by Hume, namely, that a careful examination of two specific events alone, without any reference to similar cases and to general regularities, can reveal that one of the events produces or determines the other. This thesis does not only run counter to the scientific meaning of the concept of determination which clearly rests on that of general law, but it even fails to provide any objective criteria which would be indicative of the intended relationship of determination or production. Thus, to speak of empirical determination independently of any reference to general laws is to use a metaphor without cognitive content.

Chapter 4
The Theory of Complex Phenomena

F. A. Hayek

Pattern Recognition and Pattern Prediction

Man has been impelled to scientific inquiry by wonder and by need. Of these wonder has been incomparably more fertile. There are good reasons for this. Where we wonder we have already a question to ask. But however urgently we may want to find our way in what appears just chaotic, so long as we do not know what to look for, even the most attentive and persistent observation of the bare facts is not likely to make them more intelligible. Intimate acquaintance with the facts is certainly important; but systematic observation can start only after problems have arisen. Until we have definite questions to ask we cannot employ our intellect; and questions presuppose that we have formed some provisional hypothesis or theory about the events.[1]

Questions will arise at first only after our senses have discerned some recurring pattern or order in the events. It is a recognition of some regularity (or recurring pattern, or order), of some similar feature in otherwise different circumstances, which makes us wonder and ask "why?"[2] Our minds are so made that when we notice such regularity in diversity we suspect the presence of the same agent and become curious to detect it. It is to this trait of our minds that we owe whatever understanding and mastery of our environment we have achieved.

Many such regularities of nature are recognized "intuitively" by our senses. We see and hear patterns as much as individual events without having to resort to intellectual operations. In many instances these patterns are of course so much part of the environment which we take for granted that they do not cause questions. But where our senses show us new patterns, this causes surprise and questioning. To such curiosity we owe the beginning of science.

Marvelous, however, as the intuitive capacity of our senses for pattern recognition is, it is still limited.[3] Only certain kinds of regular arrangements (not necessarily the simplest) obtrude themselves on our senses. Many of the patterns of nature we can discover only *after* they have been constructed by our mind. The systematic construction of such new patterns is the business of mathematics.[4] The role which geometry plays in this respect with regard to some visual patterns is merely the most familiar instance of this. The great strength of mathematics is that it enables us to describe abstract patterns which cannot be perceived by our senses, and to state the common properties of hierarchies or classes of patterns of a highly abstract character. Every algebraic equation or set of such equations defines in this sense a class of patterns, with the individual manifestation of this kind of pattern being particularized as we substitute definite values for the variables.

It is probably the capacity of our senses spontaneously to recognize certain kinds of patterns that has led to the erroneous belief that if we look only long enough, or at a sufficient number of instances of natural events, a pattern will always reveal itself. That this often is so means merely that in those cases the theorizing has been done already by our senses. Where, however, we have to deal with patterns for the development of which there has been no biological reason, we shall first have to invent the pattern before we can discover its presence in the phenomena—or before we shall be able to test its applicability to what we observe. A theory will always define only a kind (or class) of patterns, and the particular manifestation of the pattern to be expected will depend on the particular manifestation of the pattern to be expected will depend on the particular circumstances (the "initial and marginal conditions" to which, for the purposes of this chapter, we shall refer as "data"). How much in fact we shall be able to predict will depend on how many of those data we can ascertain.

The description of the pattern which the theory provides is commonly regarded merely as a tool which will enable us to predict the particular manifestations of the pattern that will appear in specific circumstances. But the prediction that in certain general conditions a pattern of a certain kind will appear is also a significant (and falsifiable) prediction. If I tell somebody that if he goes to my study he will find there a rug with a pattern made up of diamonds and meanders, he will have no difficulty in deciding "whether that prediction was verified or falsified by the result,"[5] even though I have said nothing about the arrangement, size, color, etc., of the elements from which the pattern of the rug is formed.

The distinction between a prediction of the appearance of a pattern of a certain class and a prediction of the appearance of a particular instance of this class is sometimes important even in the physical sciences. The mineralogist who states that the crystals of a certain mineral are hexagonal, or the astronomer who assumes that the course of a celestial body in the field of gravity of another will correspond to one of the conic sections, makes significant predictions which can be refuted. But in general the physical sciences tend to assume that it will in principle always be possible to specify their predictions to any degree desired.[6] The distinction assumes, however, much greater importance when we turn from the relatively simple phenomena with which the natural sciences deal, to the more complex phenomena of life, of mind, and of society, where such specifications may not always be possible.[7]

Degrees of Complexity

The distinction between simplicity and complexity raises considerable philosophical difficulties when applied to statements. But there seems to exist a fairly easy and adequate way to measure the degree of complexity of different kinds of abstract patterns. The minimum number of elements of which an instance of the pattern must consist in order to exhibit all the characteristic attributes of the class of patterns in question appears to provide an unambiguous criterion.

It has occasionally been questioned whether the phenomena of life, of mind, and of society are really more complex than those of the physical world.[8] This seems to be largely due to a confusion between the degree of complexity characteristic of a peculiar *kind* of phenomenon and the degree of complexity to which, by a combination of elements, any kind of phenomenon can be built up. Of course, in this manner physical phenomena may achieve any degree of complexity. Yet when we consider the question from the angle of the minimum number of distinct variables a formula or model must possess in order to reproduce the characteristic patterns of structures of different fields (or to exhibit the general laws which these structures obey), the increasing complexity as we proceed from the inanimate to the ("more highly organized") animate and social phenomena becomes fairly obvious.

It is, indeed, surprising how simple in these terms, i.e., in terms of the number of distinct variables, appear all the laws of physics, and particularly of mechanics, when we look through a collection of formulas expressing them.[9] On the other hand, even such relatively simple constituents of biological phenomena as feedback (or cybernetic) systems, in which a certain combination of physical structures produces an overall structure possessing distinct characteristic properties, require for their description something much more elaborate than anything describing the general laws of mechanics. In fact, when we ask ourselves by what criteria we single out certain phenomena as "mechanical" or "physical," we shall probably find that these laws are simple in the sense defined. Nonphysical phenomena are more complex because we call physical what can be described by relatively simple formulas.

The "emergence" of "new" patterns as a result of the increase in the number of elements between which simple relations exist means that this larger structure as a whole will possess certain general or abstract features which will recur independently of the particular values of the individual data, so long as the general structure (as described, e.g., by an algebraic equation) is preserved.[10] Such "wholes," defined in terms of certain general properties of their structure, will constitute distinctive objects of explanation for a theory, even though such a theory may be merely a particular way of fitting together statements about the relations between the individual elements.

It is somewhat misleading to approach this task mainly from the angle of whether such structures are "open" or "closed" systems. There are, strictly speaking, no closed systems within the universe. All we can ask is whether in the particular instance the points of contact through which the rest of the universe acts upon the system we try to single out (and which for the theory become the data) are few or many. These data, or variables, which determine the particular form which the pattern described by the theory will assume in the given circumstances, will be more numerous in the case of complex wholes and much more difficult to ascertain and control than in the case of simple phenomena.

What we single out as wholes, or where we draw the "partition boundary,"[11] will be determined by the consideration whether we can thus isolate recurrent patterns of coherent structures of a distinct kind which we do in fact encounter in the world in which we live. Many complex patterns which are conceivable and might recur we shall not find it worthwhile to construct. Whether it will be useful to elaborate and study a pattern of a particular kind will depend on whether the structure it describes is persistent or merely accidental. The coherent structures in which we are mainly interested are those in which a complex pattern has produced properties which make self-maintaining the structure showing it.

Pattern Prediction with Incomplete Data

The multiplicity of even the minimum of distinct elements required to produce (and therefore also of the minimum number of data required to explain) a complex phenomenon of a certain kind creates problems which dominate the disciplines concerned with such phenomena and gives them an appearance very different from that of those concerned with simpler phenomena. The chief difficulty in the former becomes one of in fact ascertaining all the data determining a particular manifestation of the phenomenon in question, a difficulty which is often insurmountable in practice and sometimes even an absolute one.[12] Those mainly concerned with simple phenomena are often inclined to think that where this is the case a theory is useless and that scientific procedure demands that we should find a theory of sufficient simplicity to enable us to derive from it predictions of particular events. To them the theory, the knowledge of the pattern, is merely a tool whose usefulness depends entirely on our capacity to translate it into a representation of the circumstances producing a particular event. Of the theories of simple phenomena this is largely true.[13]

There is, however, no justification for the belief that it must always be possible to discover such simple regularities and that physics is more advanced because it has succeeded in doing this while other sciences have not yet done so. It is rather the other way round: physics has succeeded because it deals with phenomena which, in our sense, are simple. But a simple theory of phenomena which are in their nature complex (or one which, if that expression be preferred, has to deal with more highly organized phenomena) is probably merely of necessity false—at least without a specified *ceteris paribus* assumption, after the full statement of which the theory would no longer be simple.

We are, however, interested not only in individual events, and it is also not only predictions of individual events which can be empirically tested. We are equally interested in the recurrence of abstract patterns as such; and the prediction that a pattern of a certain kind will appear in defined circumstances is a falsifiable (and therefore empirical) statement. Knowledge of the conditions in which a pattern of a certain kind will appear, and of what depends on its preservation, may be of great practical importance. The circumstances or conditions in which the pattern described by the theory will appear are defined by the range of values which may be inserted for the variables of the formula. All we need to know in order to make such a theory applicable to a situation is, therefore, that the data possess certain general properties (or belong to the class defined by the scope of the variables). Beyond this we need to know nothing about their individual attributes so long as we are content to derive merely the sort of pattern that will appear and not its particular manifestation.

Such a theory destined to remain "algebraic,"[14] because we are in fact unable to substitute particular values for the variables, ceases then to be a mere tool and becomes the final result of our theoretical efforts. Such a theory will, of course, in Popper's terms,[15] be one of small empirical content, because it enables us to predict or explain only certain general features of a situation which may be compatible with a great many particular circumstances. It will perhaps enable us to make only what M. Scriven has called "hypothetical predictions,"[16] i.e., predictions dependent on yet unknown future events; in any case the range of phenomena compatible with it will be wide and the possibility of falsifying it correspondingly small. But as in many fields this will be for the present, or perhaps forever, all the theoretical knowledge we can achieve, it will nevertheless extend the range of the possible advance of scientific knowledge.

The advance of science will thus have to proceed in two different directions: while it is certainly desirable to make our theories as falsifiable as possible, we must also push forward into fields where, as we advance, the degree of falsifiability necessarily decreases. This is the price we have to pay for an advance into the field of complex phenomena.

Statistics Impotent to Deal with Pattern Complexity

Before we further illustrate the use of those mere "explanations of the principle"[17] provided by "algebraic" theories which describe only the general character of higher-level generalities, and before we consider the important conclusions which follow from the insight into the boundaries of possible knowledge which our distinction provides, it is necessary to turn aside and consider the method which is often, but erroneously, believed to give us access to the understanding of complex phenomena: statistics. Because statistics is designed to deal with large numbers it is often thought that the difficulty arising from the large number of elements of which complex structures consist can be overcome by recourse to statistical techniques.

Statistics, however, deals with the problem of large numbers essentially by eliminating complexity and deliberately treating the individual elements which it counts as if they were not systematically connected. It avoids the problem of complexity by substituting for the information on the individual elements information on the frequency with which their different properties occur in classes of such elements, and it deliberately disregards the fact that the relative position of the different elements in a structure may matter. In other words, it proceeds on the assumption that information on the numerical frequencies of the different elements of a collective is enough to explain the phenomena and that no information is required on the manner in which the elements are related. The statistical method is therefore of use only where we either deliberately ignore, or are ignorant of, the relations between the individual elements with different attributes, i.e., where we ignore or are ignorant of any structure into which they are organized. Statistics in such situations enables us to regain simplicity and to make the task manageable by substituting a single attribute for a the unascertainable individual attributes in the collective. It is, however, for this reason irrelevant to the solution of problems in which it is the relations between individual elements with different attributes which matters.

Statistics might assist us where we had information about many complex structures of the same kind, that is, where the complex phenomena and not the elements of which they consist could be made the elements of the statistical collective. It may provide us, e.g., with information on the relative frequency with which particular properties of the complex structures, say of the members of a species of organisms, occur together; but it presupposes that we have an independent criterion for identifying structures of the kind in question. Where we have such statistics about the properties of many individuals belonging to a class of animals, or languages, or economic systems, this may indeed be scientifically significant information.[18]

How little statistics can contribute, however, even in such cases, to the explanation of complex phenomena is clearly seen if we imagine that computers were natural objects which we found in sufficiently large numbers and whose behavior we wanted to predict. It is clear that we should never succeed in this unless we possessed the mathematical knowledge built into the computers, that is, unless we knew the theory determining their structure. No amount of statistical information on the correlation between input and output would get us any nearer our aim. Yet the efforts which are currently made on a large scale with regard to the much more complex structures which we call organisms are of the same kind. The belief that it must be possible in this manner to discover by observation regularities in the relations between input and output without the possession of an appropriate theory in this case appears even more futile and naive than it would be in the case of the computers.[19]

While statistics can successfully deal with complex phenomena where these are the elements of the population on which we have information, it can tell us nothing about the structure of these elements. It treats them, in the fashionable phrase, as "black boxes" which are presumed to be of the same kind but about whose identifying characteristics it has nothing to say. Nobody would probably seriously contend that statistics can elucidate even the comparatively not very complex structures of organic molecules, and few would argue that it can help us to explain the functioning of organisms. Yet when it comes to accounting for the functioning of social structures, that belief is widely held. It is here of course largely the product of a misconception about what the aim of a theory of social phenomena is, which is another story.

The Theory of Evolution as an Instance of Pattern Prediction

Probably the best illustration of a theory of complex phenomena which is of great value, although it describes merely a general pattern whose detail we can never fill in, is the Darwinian theory of evolution by natural selection. It is significant that this theory has always been something of a stumbling block for the dominant conception of scientific method. It certainly does not fit the orthodox criteria of "prediction and control" as the hallmarks of scientific method.[20] Yet it cannot be denied that it has become the successful foundation of a great part of modern biology.

Before we examine its character we must clear out of the way a widely held misconception as to its content. It is often represented as if it consisted of an assertion about the succession of particular species of organisms which gradually changed into each other. This, however, is not the theory of evolution but an application of the theory to the particular events which took place on earth during the last two billion years or so.[21] Most of the misapplications of evolutionary theory (particularly in anthropology and the other social sciences) and its various abuses (e.g., in ethics) are due to this erroneous interpretation of its content.

The theory of evolution by natural selection describes a kind of process (or mechanism) which is independent of the particular circumstances in which it has taken place on earth, which is equally applicable to a course of events in very different circumstances, and which might result in the production of an entirely different set of organisms. The basic conception of the theory is exceedingly simple and it is only in its application to the concrete circumstances that its extraordinary fertility and the range of phenomena for which it can account manifests itself.[22] The basic proposition which has this far-reaching implication is that a mechanism of reduplication with transmittable variations and competitive selection of those which prove to have a better chance of survival will in the course of time produce a great variety of structures adapted to continuous adjustment to the environment and to each other. The validity of this general proposition is not dependent on the truth of the particular applications which were first made of it: if, for example, it should have turned out that, in spite of their structural similarity, man and ape were not joint descendants from a comparatively near common ancestor but were the product of two convergent strands starting from ancestors which differed much more from each other (such as is true of the externally very similar types of marsupial and placental carnivores), this would not have refuted Darwin's general theory of evolution but only the manner of its application to the particular case.

The theory as such, as is true of all theories, describes merely a range of possibilities. In doing this it excludes other conceivable courses of events and thus can be falsified. Its empirical content consists in what it forbids.[23] If a sequence of events should be observed which cannot be fitted into its pattern, such as, e.g., that horses suddenly should begin to give birth to young with wings, or that the cutting off of a hind paw in successive generations of dogs should result in dogs being born without that hind paw, we should regard the theory as refuted.[24]

The range of what is permitted by the theory is undeniably wide. Yet one could also argue that it is only the limitation of our imagination which prevents us from being more aware of how much greater is the range of the prohibited—how infinite is the variety of conceivable forms of organisms which, thanks to the theory of evolution, we know will not in the foreseeable future appear on earth. Common sense may have told us before not to expect anything widely different from what we already knew. But exactly what kinds of variations are within the range of possibility and what kinds are not, only the theory of evolution can tell us. Though we may not be able to write down an exhaustive list of the possibilities, any specific question we shall, in principle, be able to answer.

For our present purposes we may disregard the fact that in one respect the theory of evolution is still incomplete because we still know only little about the mechanism of mutation. But let us assume that we knew precisely the circumstances in which (or at least the probability that in given conditions) a particular mutation will appear, and that we similarly knew also the precise advantages which any such mutation would in any particular kind of environment confer upon an individual of a specific constitution. This would not enable us to explain why the existing species or organisms have the particular structures which they possess, or to predict what new forms will spring from them.

The reason for this is the actual impossibility of ascertaining the particular circumstances which, in the course of two billion years, have decided the emergence of the existing forms, or even those which, during the next few hundred years, will determine the selection of the types which will survive. Even if we tried to apply our explanatory scheme to a single species consisting of a known number of individuals each of which we were able to observe, and assuming that we were able to ascertain and record every single relevant fact, their sheer number would be such that we should never be able to manipulate them, i.e., to insert these data into the appropriate blanks of our theoretical formula and then to solve the "statement equations" thus determined.[25]

What we have said about the theory of evolution applies to most of the rest of biology. The theoretical understanding of the growth and functioning of organisms can only in the rarest of instances be turned into specific predictions of what will happen in a particular case, because we can hardly ever ascertain all the facts which will contribute to determine the outcome. Hence, "prediction and control, usually regarded as essential criteria of science, are less reliable in biology."[26] It deals with pattern-building forces, the knowledge of which is useful for creating conditions favorable to the production of certain kinds of results, while it will only in comparatively few cases be possible to control all the relevant circumstances.

Theories of Social Structures

It should not be difficult now to recognize the similar limitations applying to theoretical explanations of the phenomena of mind and society. One of the chief results so far achieved by theoretical work in these fields seems to me to be the demonstration that here individual events regularly depend on so many concrete circumstances that we shall never in fact be in a position to ascertain them all; and that in consequence not only the ideal of prediction and control must largely remain beyond our reach, but also the hope remain illusory that we can discover by observation regular connections between the individual events. The very insight which theory provides, for example, that almost any event in the course of a man's life may have some effect on almost any of his future actions, makes it impossible that we translate our theoretical knowledge into predictions of specific events. There is no justification for the dogmatic belief that such translation must be possible if a science of these subjects is to be achieved, and that workers in these sciences have merely not yet succeeded in what physics has done, namely to discover simple relations between a few observables. If the theories which we have yet achieved tell us anything, it is that no such simple regularities are to be expected.

I will not consider here the fact that in the case of mind attempting to explain the detail of the working of another mind of the same order of complexity, there seems to exist, in addition to the merely "practical" yet nevertheless unsurmountable obstacles, also an absolute impossibility: because the conception of a mind fully explaining itself involves a logical contradiction. This I have discussed elsewhere.[27] It is not relevant here because the practical limits determined by the impossibility of ascertaining all the relevant data lie so far inside the logical limits that the latter have little relevance to what in fact we can do.

In the field of social phenomena only economics and linguistics[28] seem to have succeeded in building up a coherent body of theory. I shall confine myself here to illustrating the general thesis with reference to economic theory, though most of what I have to say would appear to apply equally to linguistic theory.

Schumpeter well described the task of economic theory when he wrote that "the economic life of a non-socialist society consists of millions of relations or flows between individual firms and households. We can establish certain theorems about them, but we can never observe them all."[29] To this must be added that most of the phenomena in which we are interested, such as competition, could not occur at all unless the number of distinct elements involved were fairly large, and that the overall pattern that will form itself is determined by the significantly different behavior of the different individuals so that the obstacle of obtaining the relevant data cannot be overcome by treating them as members of a statistical collective.

For this reason economic theory is confined to describing kinds of patterns which will appear if certain general conditions are satisfied, but can rarely if ever derive from this knowledge any predictions of specific phenomena. This is seen most clearly if we consider those systems of simultaneous equations which since Léon Walras have been widely used to represent the general relations between the prices and the quantities of all commodities bought and sold. They are so framed that *if* we were able to fill in all the blanks, i.e., *if* we knew all the parameters of these equations, we could calculate the prices and quantities of all the commodities. But, as at least the founders of this theory clearly understood, its purpose is not "to arrive at a numerical calculation of prices," because it would be "absurd" to assume that we can ascertain all the data.[30]

The prediction of the formation of this general kind of pattern rests on certain very general factual assumptions (such as that most people engage in trade in order to earn an income, that they prefer a larger income to a smaller one, that they are not prevented from entering whatever trade they wish, etc.—assumptions which determine the scope of the variables but not their particular values); it is, however, not dependent on the knowledge of the more particular circumstances which we would have to know in order to be able to predict prices or quantities of particular commodities. No economist has yet succeeded in making a fortune by buying or selling commodities on the basis of his scientific prediction of future prices (even though some may have done so by selling such predictions).

To the physicist it often seems puzzling why the economist should bother to formulate those equations although admittedly he sees no chance of determining the numerical values of the parameters which would enable him to derive from them the values of the individual magnitudes. Even many economists seem loath to admit that those systems of equations are not a step toward specific predictions of individual events but the final results of their theoretical efforts, a description merely of the general character of the order we shall find under specifiable conditions which, however, can never be translated into a prediction of its particular manifestations.

Predictions of a pattern are nevertheless both testable and valuable. Since the theory tells us under which general conditions a pattern of this sort will form itself, it will enable us to create such conditions and to observe whether a pattern of the kind predicted will appear. And since the theory tells us that this pattern ensures a maximization of output in a certain sense, it also enables us to create the general conditions which will assure such a maximization, though we are ignorant of many of the particular circumstances which will determine the pattern that will appear.

It is not really surprising that the explanation of merely a sort of pattern may be highly significant in the field of complex phenomena but of little interest in the field of simple phenomena, such as those of mechanics. The fact is that in studies of complex phenomena the general patterns are all that is characteristic of those persistent wholes which are the main object of our interest, because a number of enduring structures have this general pattern in common and nothing else.[31]

The Ambiguity of the Claims of Determinism

The insight that we will sometimes be able to say that data of a certain class (or of certain classes) will bring about a pattern of a certain kind, but will not be able to ascertain the attributes of the individual elements which decide which particular form the pattern will assume, has consequences of considerable importance. It means, in the first instance, that when we assert that we know how something is determined, this statement is ambiguous. It may mean that we merely know what class of circumstances determines a certain kind of phenomena, without being able to specify the particular circumstances which decide which member of the predicted class of patterns will appear; or it may mean that we can also explain the latter. Thus we can reasonably claim that a certain phenomenon is determined by known natural forces and at the same time admit that we do not know precisely how it has been produced. Nor is the claim invalidated that we can explain the principle on which a certain mechanism operates if it is pointed out that we cannot say precisely what it will do at a particular place and time. From the fact that we do know that a phenomenon is determined by certain kinds of circumstances it does not follow that we must be able to know even in one particular instance all the circumstances which have determined all its attributes.

There may well be valid and more grave philosophical objections to the claim that science can demonstrate a universal determinism; but for all practical purposes the limits created by the impossibility of ascertaining all the particular data required to derive detailed conclusions from our theories are probably much narrower. Even if the assertion of a universal determinism were meaningful, scarcely any of the conclusions usually derived from it would therefore follow. In the first of the two senses we have distinguished we may, for instance, well be able to establish that every single action of a human being is the necessary result of the inherited structure of his body (particularly of its nervous system) and of all the external influences which have acted upon it since birth. We might even be able to go further and assert that if the most important of these factors were in a particular case very much the same as with most other individuals, a particular class of influences will have a certain kind of effect. But this would be an empirical generalization based on a *ceteris paribus* assumption which we could not verify in the particular instance. The chief fact would continue to be, in spite of our knowledge of the principle on which the human mind works, that we should not be able to state the full set of particular facts which brought it about that the individual did a particular thing at a particular time. The individual personality would remain for us as much a unique and unaccountable phenomenon which we might hope to influence in a desirable direction by such empirically developed practices as praise and blame, but whose specific actions we could generally not predict or control, because we could not obtain the information on all the particular facts which determined it.

The Ambiguity of Relativism

The same sort of misconception underlies the conclusions derived from the various kinds of "relativism." In most instances these relativistic positions on questions of history, culture, or ethics are derived from the erroneous interpretations of the theory of evolution which we have already considered. But the basic conclusion that the whole of our civilization and all human values are the result of a long process of evolution in the course of which values, as the aims of human activity appeared, continue to change, seems inescapable in the light of our present knowledge. We are probably also entitled to conclude that our present values exist only as the elements of a particular cultural tradition and are significant only for some more or less long phase of evolution—whether this phase includes some of our prehuman ancestors or is confined to certain periods of human civilization. We have no more ground to ascribe to them eternal existence than to the human race itself. There is thus one possible sense in which we may legitimately regard human values as relative and speak of the probability of their further evolution.

But it is a far cry from this general insight to the claims of the ethical, cultural, or historical relativists or of evolutionary ethics. To put it crudely: while we know that all those values are relative to something, we do not know to what they are relative. We may be able to indicate the general class of circumstances which have made them what they are, but we do not know the particular conditions to which the values we hold are due, or what our values would be if those circumstances had been different. Most of the illegitimate conclusions are the result of the erroneous interpretation of the theory of evolution as the empirical establishment of a trend. Once we recognize that it gives us no more than a scheme of explanation which might be sufficient to explain particular phenomena *if* we knew all the facts which have operated in the course of history, it becomes evident that the claims of the various kinds of relativism (and of evolutionary ethics) are unfounded. Though we may meaningfully say that our values are determined by a class of circumstances definable in general terms, so long as we cannot state which particular circumstances have produced the existing values, or what our values would be under any specific set of other circumstances, no significant conclusions follow from the assertion.

It deserves brief notice in passing how radically opposed are the practical conclusions which are derived from the same evolutionary approach according as it is assumed that we can or cannot in fact know enough about the circumstances to derive specific conclusions from our theory. While the assumption of a sufficient knowledge of the concrete facts generally produces a sort of intellectual hubris which deludes itself that reason can judge all values, the insight into the impossibility of such full knowledge induces an attitude of humility and reverence toward that experience of mankind as a whole that has been precipitated in the values and institutions of existing society.

A few observations ought to be added here about the obvious significance of our conclusions for assessing the various kinds of "reductionism." In the sense of the first of the distinctions which we have repeatedly made—in the sense of general description—the assertion that biological or mental phenomena are "nothing but" certain complexes of physical events, or that they are certain classes of structures of such events, these claims are probably defensible. But in the second sense—specific prediction—which alone would justify the more ambitious claims made for reductionism, they are completely unjustified. A full reduction would be achieved only if we were able to substitute for a description of events in biological or mental terms a description in physical terms which included an exhaustive enumeration of all the physical circumstances which constitute a necesary and sufficient condition of the biological or mental phenomena in question. In fact such attempts always consist—and can consist only—in the illustrative enumeration of classes of events, usually with an added "etc.," which might produce the phenomenon in question. Such "etc.-reductions" are not reductions which enable us to dispense with the biological or mental entities, or to substitute for them a statement of physical events, but are mere explanations of the general character of the kind of order or pattern whose specific manifestations we know only through our concrete experience of them.[32]

The Importance of Our Ignorance

Perhaps it is only natural that in the exuberance generated by the successful advances of science the circumstances which limit our factual knowledge, and the consequent boundaries imposed upon the applicability of theoretical knowledge, have been rather disregarded. It is high time, however, that we take our ignorance more seriously. As Popper and others have pointed out, "the more we learn about the world, and the deeper our learning, the more conscious, specific, and articulate will be our knowledge of what we do not know, our knowledge of our ignorance."[33] We have indeed in many fields learned enough to know that we cannot know all that we would have to know for a full explanation of the phenomena.

These boundaries may not be absolute. Though we may never know as much about certain complex phenomena as we can know about simple phenomena, we may partly pierce the boundary by deliberately cultivating a technique which aims at more limited objectives—the explanation not of individual events but merely of the appearance of certain patterns or orders. Whether we call these mere explanations of the principle or mere pattern predictions or higher-level theories does not matter. Once we explicitly recognize that the understanding of the general mechanism which produces patterns of a certain kind is not merely a tool for specific predictions but important in its own right, and that it may provide important guides to action (or sometimes indications of the desirability of no action), we may indeed find that this limited knowledge is most valuable.

What we must get rid of is the naive superstition that the world must be so organized that it is possible by direct observation to discover simple regularities between all phenomena and that this is a necessary presupposition for the application of the scientific method. What we have by now discovered about the organization of many complex structures should be sufficient to teach us that there is no reason to expect this, and that if we want to get ahead in these fields our aims will have to be somewhat different from what they are in the fields of simple phenomena.

A Postscript on the Role of "Laws" in the Theory of Complex Phenomena[34]

Perhaps it deserves to be added that the preceding considerations throw some doubt on the widely held view that the aim of theoretical science is to establish "laws"—at least if the word "law" is used as commonly understood. Most people would probably accept some such definition of "law" as that "a scientific law is the rule by which two phenomena are connected with each other according to the principle of causality, that is to say, as cause and effect."[35] And no less an authority than Max Planck is reported to have insisted that a true scientific law must be expressible in a single equation.[36]

Now the statement that a certain structure can assume only one of the (still infinite) number of states defined by a system of many simultaneous equations is still a perfectly good scientific (theorectical and falsifiable) statement.[37] We might still call, of course, such a statement a "law," if we so wish (though some people might rightly feel that this would do violence to language); but the adoption of such a terminology would be likely to make us neglectful of an important distinction: for to say that such a statement describes, like an ordinary law, a relation between cause and effect would be highly misleading. It would seem, therefore, that the conception of law in the usual sense has little application to the theory of complex phenomena, and that therefore also the description of scientific theories as "nomologic" or "nomothetic" (or by the German term *Gesetzeswissenschaften*) is appropriate only to those two-variable or perhaps three-variable problems to which the theory of simple phenomena can be reduced, but not to the theory of phenomena which appear only above a certain level of complexity. If we assume that all the other parameters of such a system of equations describing a complex structure are constant, we can of course still call the dependence of one of the latter on the other a "law" and describe a change in the one as "the cause" and the change in the other as "the effect." But such a "law" would be valid only for one particular set of values of all the other parameters and would change with every change in any one of them. This would evidently not be a very useful conception of a "law," and the only generally valid statement about the regularities of the structure in question is the whole set of simultaneous equations from which, if the values of the parameters are continuously variable, an infinite number of particular laws, showing the dependence of one variable upon another, could be derived.

In this sense we may well have achieved a very elaborate and quite useful theory about some kind of complex phenomenon and yet have to admit that we do not know of a single law, in the ordinary sense of the word, which this kind of phenomenon obeys. I believe this to be in a great measure true of social phenomena: though we possess theories of social structures, I rather doubt whether we know of any "laws" which social phenomena obey. It would then appear that the search for the discovery of laws is not an appropriate hallmark of scientific procedure but merely a characteristic of the theories of simple phenomena as we have defined these earlier; and that in the field of complex phenomena the term "law" as well as the concepts of cause and effect are not applicable without such modification as to deprive them of their ordinary meaning.

In some respect the prevalent stress on "laws", i.e., on the discovery of regularities in two-variable relations, is probably a result of inductivism, because only such simple co-variation of two magnitudes is likely to strike the senses before an explicit theory or hypothesis has been formed. In the case of more complex phenomena it is more obvious that we must have our theory first before we can ascertain whether the things do in fact behave according to this theory. It would probably have saved much confusion if theoretical science had not in this manner come to be identified with the search for laws in the sense of a simple dependence of one magnitude upon another. It would have prevented such misconception as that, e.g., the biological theory of evolution proposed some definite "law of evolution" such as a law of the necessary sequence of certain stages or forms. It has of course done nothing of the kind and all attempts to do this rest on a misunderstanding of Darwin's great achievement. And the prejudice that in order to be scientific one must produce laws may yet prove to be one of the most harmful of methodological conceptions. It may have been useful to some extent for the reason given by Popper, that "simple statements ... are to be prized more highly"[38] in all fields where simple statements are significant. But it seems to me that there will always be fields where it can be shown that all such simple statements must be false and where in consequence also the prejudice in favor of "laws" must be harmful.

Notes

1. See already Aristotle, *Metaphysics*, I, II 9, 9826b (Loeb ed., p. 13): "It is through wonder that men now begin and originally began to philosophize ... it is obvious that they pursued science for the sake of knowledge, and not for any practical utility", also Adam Smith, "The Principles which Lead and Direct Philosophical Inquiries, as Illustrated by the History of Astronomy," in *Essays*, London, 1869, p. 340: "Wonder, therefore, and not any expectation of advantage from its discoveries, is the first principle which prompts mankind to the study of philosophy, that science which pretends to lay open the concealed connections that unite the various appearances of nature; and they pursue this study for its own sake, as an original pleasure or good in itself, without regarding its tendency to procure them the means of many other pleasures." Is there really any evidence for the now popular contrary view that, e.g., "hunger in the Nile Valley led to the development of geometry" (as Gardner Murphy in the *Handbook of Social Psyobalogy*, ed. by Gardner Lindzey, 1954, Vol. 11, p. 616, tells us)? Surely the fact that the discovery of geometry turned out to be useful does not prove that it was discovered because of its usefulness. On the fact that economics has in some degree been an exception to the general rule and has suffered by being guided more by need than by detached curiosity, see my lecture on "The Trend of Economic Thinking" in *Economica*, 1933.
2. See K. R. Popper, *The Poverty of Historicism*, London, 1957, p. 121: "Science ... cannot start with observations, or with the 'collection of data', as some students of method believe. Before we can collect data, our interest in *data of a certain kind* must be aroused: the *problem* always comes first." Also in his *The Logic of Scientific Discovery*, London, 1959, p. 59: "observation is always *observation in the light of theories*."

3. Although in some respects the capacity of our senses for pattern recognition clearly also exceeds the capacity of our mind for specifying these patterns. The question of the extent to which this capacity of our senses is the result of another kind of (presensory) experience is another matter. See, on this and on the general point that all perception involves a theory or hypothesis, my book *The Sensory Order*, London and Chicago, 1952, esp. para. 7.37. Cf. also the remarkable thought expressed by Adam Ferguson (and probably derived from George Berkeley) in *The History of Civil Society*, London, 1767, p. 39, that "the inferences of thought are sometimes not to be distinguished from the perception of sense"; as well as H. von Helmholtz's theory of the "unconscious inference" involved in most perceptions. For a recent revival of these ideas see N. R. Hanson, *Patterns of Discovery*, Cambridge University Press, 1958, esp. p. 19, and the views on the role of "hypotheses" in perception as developed in recent "cognition theory" by J. S. Bruner, L. Postman and others.
4. Cf. G. H. Hardy, *Mathematician's Apology*, Cambridge University Press, 1941, p. 24: "A mathematician, like a painter or poet, is a maker of patterns."
5. Charles Dickens, *David Copperfield*, p. 1.
6. Though it may be permissible to doubt whether it is in fact possible to predict, e.g., the precise pattern which the vibrations of an airplane will at a particular moment produce in the standing wave on the surface of the coffee in my cup.
7. Cf. Michael Scriven, "A Possible Distinction between Traditional Scientific Disciplines and the Study of Human Behavior," *Minnesota Studies in the Philosophy of Science*, I, 1956, p. 332: "The difference between the scientific study of behavior and that of physical phenomena is thus partly due to the relatively greater complexity of the simplest phenomena we are concerned to account for in a behavioral theory."
8. Ernest Nagel, *The Structure of Science*, New York, 1961, p. 505: "though social phenomena may indeed be complex, it is by no means certain that they are in general more complex than physical and biological phenomena." See, however, Johann von Neumann, "The General and Logical Theory of Automata," *Cerebral Mechanism in Bebavior*. The Hixon Symposium, New York, 1951, p. 24: "we are dealing here with parts of logic with which we have practically no experience. The order of complexity is out of all proportion to anything we have ever known." It may be useful to give here a few illustrations of the orders of magnitude with which biology and neurology have to deal. While the total number of electrons in the universe has been estimated at 10^{79} and the number of electrons and protons at 10^{100}, there are in chromosomes with 1,000 locations [genes] with 10 allelomorphs 10^{1000} possible combinations; and the number of possible proteins is estimated at 10^{2700} (L. von Bertalanffy, *Problems of Life*, New York, 1952, p. 103). C. Judson Herrick (*Brains of Rats and Men*, New York), suggests that "during a few minutes of intense cortical activity the number of interneuronic connections actually made (counting also those that are actuated more than once in different associational patterns) may well be as great as the total number of atoms in the solar system" (i.e., 10^{56}); and Ralph W. Gerard (*Scientific American*, September 1953, p. 118) has estimated that in the course of seventy years a man may accumulate 15×10^{12} units of information ("bits"), which is more than 1,000 times larger than the number of nerve cells. The further complications which social relations superimpose upon this are, of course, relatively insignificant. But the point is that if we wanted to "reduce" social phenomena to physical events, they would constitute an additional complication, superimposed upon that of the physiological processes determining mental events.
9. Cf. Warren Weaver, "A Quarter Century in the Natural Science," *The Rockefeller Foundation Annual Report*, 1958, Chapter I, "Science and Complexity," which, when writing this, I knew only in the abbreviated version which appeared in the *American Scientist*, XXXVI, 1948.
10. Lloyd Morgan's conception of "emergence" derives, via G. H. Lewes (*Problems of Life and Mind*, 1st series, Vol. II, problem V, Ch. III, section headed "Resultants and Emergents," American ed., Boston, 1891, p. 368), from John Stuart Mill's distinction of the "heteropathic" laws of chemistry and other complex phenomena from the ordinary "composition of causes" in mechanics, etc. See his *System of Logic*, London, 1843, Bk. III, Ch. 6, in Vol. I, p. 431 of the first edition, and C. Lloyd Morgan, *The Emergence of Novelty*, London, 1933, p. 12.
11. Lewis White Beck, "The 'Natural Science Ideal' in the Social Sciences," *Scientific Monthly*, LXVIII, June 1949, p. 388.
12. Cf. F. A. Hayek, *The Sensory Order*, paras, 8.66–8.86.
13. Cf. Ernest Nagel, "Problems of Concept and Theory Formation in the Social Sciences," in *Science, Language and Human Rights* (American Philosophical Association, Eastern Division, Vol. I), University of Pennsylvania Press, 1952, p. 620: "In many cases we are ignorant of the appropriate initial and

boundary conditions, and cannot make precise forecasts even though available theory is adequate for that purpose."

14. The useful term "algebraic theories" was suggested to me by J. W. N. Watkins.
15. K. R. Popper, *The Logic of Scientific Discovery*, London, 1959, p. 113.
16. M. Scriven, "Explanation and Prediction in Evolutionary Theory," *Science*, August 28, 1959, p. 478 and cf. K. R. Popper, "Prediction and Prophecy in the Social Sciences" (1949), reprinted in his *Conjectures and Refutations*, London, 1963, especially pp. 339 et seq.
17. Cf. F. A. Hayek, "Degrees of Explanation," *British Journal for the Philosophy of Science*, VI, No. 23, 1955.
18. See F. A. Hayek, *The Counter-Revolution of Science*, Glencoe, III., 1952, pp. 60–63.
19. Cf. J. G. Taylor, "Experimental Design: A Cloak for Intellectual Sterility," *British Journal of Psychology*, 49, 1958, esp. pp. 107–8.
20. Cf., e.g., Stephen Toulmin, *Foresight and Prediction*, London, 1961, p. 24: "No scientist has ever used this theory to foretell the coming into existence of creatures of a novel species, still less verified his forecast."
21. Even Professor Popper seems to imply this interpretation when he writes (*Poverty of Historicism*, p. 107) that "the evolutionary hypothesis is not a universal law of nature but a particular (or, more precisely, singular) historical statement about the ancestry of a number of terrestrial plants and animals." If this means that the essence of the theory of evolution is the assertion that particular species had common ancestors, or that the similarity of structure always means a common ancestry (which was the hypothesis from which the theory of evolution was derived), this is emphatically not the main content of the present theory of evolution. There is, incidentally, some contradiction between Popper's treatment of the concept of "mammals" as a universal (*Logic*, p. 65) and the denial that the evolutionary hypothesis describes a universal law of nature. The same process might have produced mammals on other planets.
22. Charles Darwin himself well knew, as he once wrote to Lyell, that "all the labour consists in the application of the theory" (quoted by C. C. Gillispie, *The Edge of Objectivity*, Princeton, 1960, p. 314).
23. K. R. Popper, *Logic*, p. 41.
24. Cf. Morton Beckner, *The Biological Way of Thought*, Columbia University Press, 1954, p. 241.
25. K. R. Popper, *Logic*, p. 73.
26. Ralph S. Lillie, "Some Aspects of Theoretical Biology," *Philosophy of Science*, XV, 2, 1948, p. 119.
27. See *The Sensory Order*, 8.66–8.86, also *The Counter-Revolution of Science*, Glencoe, I, 22 1952, p. 48.
28. See particularly Noam Chomsky, *Syntactic Structures*, 's Gravenhage, 1957, who characteristically seems to succeed in building up such a theory after frankly abandoning the striving after an inductivist "discovery procedure" and substituting for it the search after an "evaluation procedure" which enables him to eliminate false theories of grammars and where these grammars may be arrived at "by intuition, guess-work, all sorts of partial methodological hints, reliance on past experience, etc." (p. 56).
29. J. A. Schumpeter, *History of Economic Analysis*, Oxford University Press, 1954, p. 241.
30. V. Pareto, *Manuel d'économie politique*, 2d ed., Paris, 1927, pp. 223–4.
31. A characteristic instance of the misunderstanding of this point (quoted by E. Nagel, l.c., p. 61) occurs in Charles A. Beard, *The Nature of the Social Sciences*, New York, 1934, p. 29, where it is contended that if a science of society "were a true science, like that of astronomy, it would enable us to predict the essential movements of human affairs for the immediate and the indefinite future, to give pictures of society in the year 2000 or the year 2500 just as astronomers can map the appearances of the heavens at fixed points of time in the future."
32. Cf. My *Counter-Revolution of Science*, pp. 48 et seq., and William Craig, "Replacement of Auxiliary Expressions," *Philosophical Review*, 65, 1956.
33. K. R. Popper, "On the Sources of Knowledge and Ignorance," *Proceedings of the British Academy*, 46, 1960, p. 69. See also Warren Weaver, "A Scientist Ponders Faith," *Saturday Review*, January 3, 1959: "Is science really gaining in its assault on the totality of the unsolved? As science learns one answer, it is characteristically true that it also learns several new questions. It is as though science were working in a great forest of ignorance, making an ever larger circular clearing within which, not to insist on the pun, things are clear.... But, as that circle becomes larger and larger, the circumference of contact with ignorance also gets longer and longer. Science learns more and more. But there is an ultimate sense in which it does not gain; for the volume of the appreciated but not understood keeps getting larger. We keep, in science, getting a more and more sophisticated view of our ignorance."

34. This last section was not contained in the original version.
35. The particular wording which I happened to come across while drafting this is taken from H. Kelsen, "The Natural Law Doctrine Before the Tribunal of Science" (1949), reprinted in *What Is Justice?* University of California Press, 1960, p. 139. It seems to express well a widely held view.
36. Sir Karl Popper comments on this that it seems extremely doubtful whether any *single* one of Maxwell's equations could be said to express anything of real significance if we knew none of the others; in fact, it seems that the repeated occurrence of the symbols in the various equations is needed to secure that these symbols have the intended meanings.
37. Cf. K. R. Popper, *Logic of Scientific Discovery*, sec. 17, p. 73: "Even if the system of equations does not suffice for a unique solution, it does not allow every conceivable combination of values to be substituted for the 'unknowns' (variables). Rather, the system of equations characterizes certain combinations of values or value systems as admissible, and others as inadmissible; it distinguishes the class of admissible value systems from the class of inadmissible value systems." Note also the application of this in the following passages to "statement equations."
38. Ibid., p. 142.

Chapter 5

A Possible Distinction between Traditional Scientific Disciplines and the Study of Human Behavior

Michael Scriven

I wish to discuss what I believe to be a difference between subjects such as physics, chemistry, and astronomy on the one hand, and economics, anthropology, sociology, and psychology on the other. These groups of subjects have often been referred to as the physical (or natural) sciences and the social (or behavioral) sciences respectively. Many people, feeling that there is an essential difference between the groups, have wished to mark it by withholding the term "science" from the description of the latter group. Not wishing to prejudge the issue, I have formulated the title of this chapter in such a way as to avoid the disputed terminology.

Let me begin by pointing out some respects in which it will not be possible to draw a categorical distinction. First, in nearly all studies of the world from bird-watching to biochemistry, there is a descriptive, nonexplanatory aspect. This is exemplified in astronomy when we give the constituents of the planetary atmospheres or the approximate diameter of a spiral nebula, and in economics when we give a breakdown of the Swedish taxation revenue or the approximate value of the British national debt. It is certainly true that there are important connections between the units and descriptive terms used in giving such data and the accepted theories in the field. But even if it were granted that economic theories are scientifically disreputable, one could not deny the existence of an exact though perhaps incomplete descriptive language in which the data of an economic science can be formulated. Now, in some subjects this description forms that most important part of the material, while in others prediction and explanation are admittedly more important; but this distinction does not coincide with the one we seek. For astronomy (as opposed to celestial mechanics or cosmology) is mainly a descriptive science, and so is primitive anthropology, while thermodynamics and welfare economics are not.

Second, I cannot see any way in which one could establish the claim that human behavior is in principle undetermined and, hence, distinguish its study from that of the presumably determined inanimate world. No matter in how many respects two human state-descriptions are the same, if the ensuing behavior differs, we shall regard that as evidence that somewhere in the individual or genetic histories or in the current circumstances there must be a difference in the value of a parameter.[1] I regard this as an empirical (though meta-theoretical) belief because I can conceive circumstances which might lead us to abandon it, but not in a world that resembles the one we inhabit.

Third, I see no other reason to doubt that precise explanations and predictions are available or possible in the study of some behavioral phenomena, e.g., in the psychological field. I have in mind two types of behavior here: on the one hand, behavior under restraints or compulsions; and, on the other, choice-behavior with one heavily weighted alternative. It is not hard to predict or explain what is in one sense the behavior of a man chained hand and foot to a rotating wheel, or that of a claustrophobe when he sees an exit from a confined space, or that of a man needing an operation to save his life who is offered the money by a wealthy philanthropist.

These predictions or explanations are not based on anything which is referred to as a psychological theory, but merely on our observations and "commonsense" inference. One might wish to say that there is an implicit theory in our classifications and inference in these cases. If so, then my point can be rephrased to state that in the cases in which we feel most certain most often of our behavioral predictions and explanations, they are based on the theory implicit in common sense and not on theories developed by professional psychologists. Some obvious qualifications must be made; but I think most of them fall under the heading of special types of behavior which had not previously been observed in sufficient detail or with sufficient frequency to permit generalization, e.g., neurosis, overlearning, parapsychology. Of course, the souls of all good psychologists will revolt at this claim. But if justified, it is not a reflection on their past achievements or present ability. It is a description of their difficulties. The difference between the scientific study of behavior and that of physical phenomena is thus partly due to the relatively greater complexity of the *simplest phenomena we are concerned to account for* in a behavioral theory. There are, I think, three related explanations of this difficulty.

1. The basic generalizations are more complex, in the sense that more standing conditions must be specified for a functional relationship of comparable simplicity, and consequently more variables must be measured in obtaining the basic data to which the basic generalizations refer.
2. The useful concepts, i.e., those occurring in observation-statements and theory, include many from physics and mathematics as a proper subset.
3. The ordinary procedures for explaining behavior that are embedded in our everyday language contain a considerable proportion of the low-level laws (albeit imprecisely formulated)[2] obtained simply as a result of long experience; thus, some of the cream has been skimmed from the subject in a way not possible in, e.g., spectrochemistry.

There are two important consequences of the preceding propositions.

A. Students of human behavior have to theoretically run before they can theoretically walk; i.e., they have to look for higher-order theories in order to get a lead on the variables to isolate for sound generalizations. This is not unique to, but is more common in, this area.
B. Practical problems of prediction, or explanation at *any level* (i.e., including theory building) are more like to be *insoluble* in the study of behavior.

Now, it is the latter point that I wish to recommend as a salutary point of distinction between the two groups of studies. It requires some expansion.

Must there be a theory or theories that explain the phenomena (meaning basic observational data and generalizations) in a given field? Naturally we do our very best to find one, and in many fields we have come very close to success or have completely succeeded in the attempt. Naturally, too, we can never absolutely prove that there is not one. And it is true that the longer we wait, the more material (or the more *reliable* material) we shall have on which to build one. But is there any reason—other than our past experience in some field—to believe that the proliferation of evidence makes a theory any more possible? There is an obvious sense in which an increase in data may make the job of finding an adequate explanation more difficult. Now, a scientific theory is typically a system of propositions which organizes the evidence internally and in relation to other propositions of the system which concern certain (possibly hypothetical) entities or states; so that we can see it as a consistent and connected whole, where the connection consists of *explanation* (not necessarily deduction) in the direction

> (propositions about [possibly hypothetical] entities or states)
> to (propositions describing the phenomena)

and of *inference* (not necessarily induction in the narrow sense) in the other direction.

In what sense *must* there be such relations among a given set of statements and such inferable or actual entities or conditions? If our notion of what constitutes a satisfactory explanation were completely a priori, then it would indeed be surprising if we were always able to find explanations; and if that notion were so accommodating that any fresh observation-statement was acceptable as a new postulate (when not immediately explicable by some other accepted statement), then we should never be at a loss for an explanation. Compare the question: In what sense must there be a formula which gives us all the prime numbers? If the primes are like clues in a numerical crossword puzzle, we are quite right to assume there is a solution; if they are like the number of squares on a chessboard that have not been occupied prior to the nth move of a game, we are mistaken in thinking there is a solution—though, of course, we can establish certain limits. And the primes present a problem that is in certain ways very like each of these. It is not possible to conclude that there *must* be a formula. Is it possible even to say that, if it is true, there *must* be a proof of Goldbach's hypothesis (that any even number can be expressed in at least one way as the sum of two primes)? I do not think that even a convinced formalist impressed by proof-theoretic successes would be able to establish such a claim—though we may be justified in thinking it probably correct. How, then, could we support the view that there *must* be an explanation of, or a theory about, the phenomena in a given field?

In the field of the empirical disciplines, the concept of explanation is not, it seems to me, as precise as that of a prime-number function or a number-theoretic proof. If it were, then the analogy would be fairly close, and we could only assert, "There must always be an explanation," as an empirically well-founded slogan. But it is also true that we occasionally admit new types of explanation, which makes the slogan more like a tautology, and that we impose certain restrictions on what constitute acceptable postulates and rules of inference, which makes the slogan more empirical. With these simple introductory remarks about the inevitability of explanations (and we can substitute "predictions" or "theory" throughout), let us return to consideration of the special thesis that new explanations will be less readily found in the field of behavior than they have been in the development of the first-born sciences (and may sometimes never be found). I wish to distinguish two arguments for this thesis. The first point I regard as moderately important, certainly true, and fairly obvious: it is that the student of behavior is, in general, faced with problems of explanation that are very much more difficult than those that faced the early physicists. This is apparent in the practical field for one reason already cited: viz., we already have commonsensical and well-supported explanations of nearly all the easy cases, and we are therefore left with the problems we haven't been able to solve exactly by common sense: How do pilots judge height? How can we predict a student's examination performances? What lighting suits a machine assembly line best? What causes amnesia? Even here, we do have rough, untested, unquantified ideas about the answers. These problems can be approached in a perfectly scientific way and valuable answers produced, prior to the formalization of the theory of normal perception, problem-performance, etc. But even if it is thought possible to produce a precise general theory of basic behavior, it would be wrong to conclude that exact predictions and faultless explanations will be possible in the field of practical problems; for they are now very rarely achieved in this area by any sciences. The practical problems of physics today are engineering problems, meteorological problems aerodynamical problems; and to these there are not often exact solutions, but only compromises and approximations. How far will a missile of this shape travel with this propellant? We cannot tell accurately from experiments with scale models because we cannot scale down the size of air molecules or the critical mass of the propellant. Even a full-scale test does not yield exactly repeatable results, and there are no precise general formulas for air resistance. But we can give quite good answers or, as in fact we have done, produce radio-controls that circumvent the problem of prediction. In general, the psychologist (or economist, etc.) has to work with a larger number of critical variables and cannot run full-scale repeatable tests. So he is not even as successful as scientists in those fields; but my main point is that he has to deal with the hardest type of problem that they ever face.

Some theoretical psychologist will want to reply by saying that what remains to be done in psychology is the job that Galileo did—the mathematicizing of its basic theory. Given that, he might say, we shall be able to improve practical predictions until they are comparable with those made by physicists. However, the importance of my point rests on the fact that in the area of fundamental research the number of critical variables is absolutely crucial; or to put it another way, the degree of actualized approximation to the ideal case is important. Galileo wished to measure the rate at which bodies naturally fell. His ingenuity lay in transposing the problem to that of measuring the rate at which spheres roll down an inclined plane. His good fortune lay in the fact that the loss of energy due to rolling friction, elastic absorption, and air resistance was negligible; and that a comparatively simple law roughly relates any two of the remaining three variables (distance traveled along the plane [or vertically], the time taken, and the velocity attained). Furthermore, he had measuring instruments which were sufficiently sensitive to variations in the most recalcitrant variable (a waterclock reading to one-tenth of a "pulse-beat") to reveal the law, while not so sensitive that they would yield the progressively greater inaccuracy as the absolute size of any variable increased (due to the mounting energy losses). Galileo was indeed fortunate; no less so were Boyle, Charles, Gay-Lussac, Van der Waals, and the others who discovered the gas laws: no less so Kepler whose very favorite law about the planets, relating their orbits to the regular-solids, has *not* survived the test of more accurate measurements. Students of behavior are not so fortunate, and it would be misguided of them to labor at the task of proving otherwise.

I am not making the absurd claim that there will be no progress, but simply the claim that simple laws will very rarely be found even under the most idealized laboratory conditions. This is a claim based on the empirical evidence, not on any a priori necessity. If the evidence was simply that we have so far failed to discover any simple laws, then it would seem rash to claim that none will be found. I believe, however, that the evidence also suggests an explanation for the lack of such successes in terms of several factors, the most important of which has been discussed above: viz., the multiplicity of critical variables *in the simplest interesting cases*. I wish also to combine the basic fact of relative lack of success with what I shall call the finitude of the set of usable hypotheses to provide another, substantially weaker, reason. It seems to me that in the field of individual behavior, to restrict the area sharply, the talk of "another Newton" is inappropriate. I would venture to say that it is extremely improbable that anything remotely corresponding to the simplicity and importance of the concept of universal gravitation can possibly be found in the field of psychology. The apparent exception is Freud. His importance is, to me, tempered by the recognition that his work was not only nonquantitative but in other respects imprecise, even as now reformulated (cf. evolutionary theory, which does not require statistical genetics for precision), and applies most successfully to *abnormal* psychology and is irrelevant to, or only partly relevant to, e.g., perception and learning.[3] (Now of course, gravitation did not explain electromagnetic phenomena; I wish only to point out that important restrictions do exist and that they exclude most of the area from which our fund of common observation stems). I am skeptical that a basic concept or set of concepts that will provide a new and fundamental insight into *ordinary* human behavior is discoverable. Man might have been a simple creature, his behavior governed by the stars or by enlightened self-interest, but he is not.

If the arguments that I have previously given convince us that a new conceptual scheme or theory in terms of some behavioral construct is unlikely to prove revelatory, it may be on the grounds that we must look for our revelations in the brain, as Dalton sought the explanations of overt chemical behavior in the atom. But the comparison is unsound. Dalton was faced with a large number of precise laws of chemical combination, and had precise data about the substances which thus interacted. Suppose that he had observed only the results of experiments with highly stable, highly heterogeneous mixtures labeled perhaps A, B, or C. His results would be inexplicable in terms of any simple theory. It is this situation which faces the student of behavior—and for him there can be no reduction of the macroscopic data to simple invariable regularities, whose existence could perhaps be explained by reference to a theory of the microstructure. Why not? Because the fundamental experimental element is the human being, or his responses, enormously complex in structure and function and reared in an enormously complex environment. There can be no *practical* sense in which this element can be reduced to simpler ones. There is, of course, a *theoretical* sense in which we can analyze an individual's motivation in terms, say, of primary and acquired drives or his purchasing in terms of consumption, savings, or production. But these analyses are neither precise nor productive of simple laws, so we are not empirically justified in claiming that micro-concepts will be found that will yield simple laws. Here I do not intend "simple" to be very restrictive: the laws of radiation include fourth powers, those of electromagnetism vector differentials, those of elasticity tensor quantities, yet nothing as simple as these can be expected in behavior. This is not to deny that, under certain specified circumstances, the behavior of an organism can be precisely predicted, e.g., in a Skinner box. This I earlier said was even possible with humans under nonexperimental conditions. I am denying that this possibility is more than an unimportant (though *perhaps* necessary) condition for developing a manageable *general* theory of behavior which will usefully predict the aspects of ordinary behavior that we need predictions about. The meteorologist is in the same position; he knows that under some circumstances he can predict rain with tremendous reliability, and he can always do quite well at predicting barometric pressure, and he can almost always see in retrospect how to explain his errors, i.e., can give a general account of weather. But the practical definition of his problems makes the long-range prediction of precipitation in all cases immensely important, and predicting morning rain in London for the next test match with Australia with the 51 percent accuracy an insurance company would be interested in is just a pipe dream, whereas in astronomy it's child's play. The practical problems in psychology are often worse than the meteorologist's.

The statistics of mathematical economics, perhaps the most highly formalized area in the study of behavior, is particularly interesting in this respect. It is quite true that if men always chose rationally, were equally well informed, and had identical needs and desires, precise laws in economics could probably be found. And it is true that rationality, informedness, and requirements are dominating factors in the actual situation. It does not follow from this that a theory which describes the action of an agent or economic group in the ideal case is in the least valuable as a practical aid. It must also be proved that the variations from the ideal produce effects that are small compared with the total effects in the ideal case—or if the effects of the variations are large, they must themselves be predictable. I think that overlooking this point has led many mathematical economists to believe that, because they have a theory of the ideal case, they are therefore in as satisfactory a position as that of the theoretical chemist trying to predict the behavior of a mixture, granted a knowledge of the behavior of the ingredients: just a problem of measurement and calculation. Though it is true that scientific laws typically refer to an ideal case, and it is true that in the practical field they very often do not yield precise predictions, it does not follow that a theory which deals with an idealization of some or even all the factors involved in a behavior situation is properly called a theory of behavior. One might say that the essence of the success of the natural sciences is the possibility of finding simple laws referring to ideal cases that are or can be realized in empirical cases to an indefinitely high degree of approximation.

It is, I think, very important to distinguish two theses about the future of the behavioral sciences. First, there is the thesis that it will be possible to improve predictions and explanations indefinitely. Second, there is the thesis that it will be possible to improve predictions and explanations in *every case, and to an indefinitely high degree* of approximation. The first thesis I agree with; the second I think false for two reasons:

1. It is typical of problems in the field of behavior that we have limited access to data (extreme case: precise prediction of individual behavior may prove possible *given* certain data about values of neurological variables. But it would be dogmatic to insist that there *must* be a way to determine these that does not involve long study and perhaps surgery on that individual; therefore, the problem of prediction given only limited observational access will probably never be precisely solved).
2. There is a great deal of difference between indefinite improvement in a field, i.e., continual progress with some aspects of some problems, and the attainment of indefinitely high approximations in solving a given aspect of a given problem. Science has not advanced by solving all problems but often by abandoning them; we never solved the problem of finding how the stars affected our lives; we never found a real philosopher's stone; we never found an elixir of life, or the vital essence, or the language of animals. Why then must there be laws in the social sciences that will enable us *in every case* to predict or explain? May we not here also come to see that the search is for something we may wish for but cannot expect?

There is a second reason for my main conclusion and this one I regard as extremely important, not at all obvious, and only probably right. This is the existence of non-deductive explanations, which are the central type in the field of behavior, and it is connected with the requirements of universality and repeatability of effects. I shall discuss it elsewhere. The conclusion I hope then to establish more clearly, but which I have here tried to support, is not in general form a very exciting one, for we have all realized that in some sense psychological laws and theories have been harder to find than were the early physical ones. Its merit, if any, lies in the stress on the particular respects in which this is true and the reasons for these particular differences.

Notes

1. Making some allowance for the amplification of quantum effects does not affect this as a methodological principle but would require modification to it as a descriptive account.
2. As Skinner has said, "If a reasonable order [in human behavior] was not discoverable, we could scarcely be effective in dealing with human affairs." B. F. Skinner, *Science and Homan Behavior* (New York: Macmillan, 1953).
3. The theory of dream interpretation and the parapraxes in Freud are, I would think it clear, too simple to be correct except (1) occasionally in normal people or (2) generally in some abnormal people. The vagueness of the postulates in, e.g., the theory of slips does not make it any less true that they are simple in the sense that they exclude motor errors.

Chapter 6

Psychology as Philosophy

Donald Davidson

Not all human motion is behavior. Each of us . . . is moving eastward at about 700 miles an hour, carried by the diurnal rotation of the earth, but this is not a fact about our behavior. When I cross my legs, the raised foot bobs gently with the beat of my heart, but I do not move my foot. Behavior consists in things we do, whether by intention or not, but where there is behavior, intention is relevant. In the case of actions, the relevance may be expressed this way: an event is an action if and only if it can be described in a way that makes it intentional. For example, a man may stamp on a hat, believing it is the hat of his rival when it is really his own. Then stamping on his own hat is an act of his, and part of his behavior, though he did not do it intentionally. As observers we often describe the actions of others in ways that would not occur to them. This does not mean that the concept of intention has been left behind, however, for happenings cease to be actions or behavior only when there is no way of describing them in terms of intention.

These remarks merely graze a large subject, the relation between action and behavior on the one hand, and intention on the other. I suggest that even though intentional action, at least from the point of view of description, is by no means all the behavior there is, intention is conceptually central; the rest is understood and defined in terms of intention. If this is true, then any considerations which show that the intentional has traits that segregate it conceptually from other families of concepts (particularly physical concepts) will apply *mutatis mutandis* to behavior generally. If the claim is mistaken, then the following considerations apply to psychology only to the extent that psychology employs the concepts of intention, belief, desire, hope, and other attitudes directed (as one says) upon propositions.

Can intentional human behavior be explained and predicted in the same way other phenomena are? On the one hand, human acts are clearly part of the order of nature, causing and being caused by events outside ourselves. On the other hand, there are good arguments against the view that thought, desire and voluntary action can be brought under deterministic laws, as physical phenomena can. An adequate theory of behavior must do justice to both these insights and show how, contrary to appearance, they can be reconciled. By evaluating the arguments against the possibility of deterministic laws of behavior, we can test the claims of psychology to be a science like others (some others).

When the world impinges on a person, or he moves to modify his environment, the interactions can be recorded and codified in ways that have been refined by the social sciences and common sense. But what emerge are not the strict quantitative laws embedded in sophisticated theory that we confidently expect in physics, but irreducibly statistical correlations that resist, and resist in principle, improvement without limit. What lies behind our inability to discover deterministic psychophysical laws is this. When we attribute a belief, a desire, a goal, an intention or a meaning to an agent, we necessarily operate within a system of concepts in part determined by the structure of beliefs and desires of the agent himself. Short of changing the subject, we cannot escape this feature of the psychological; but this feature has no counterpart in the world of physics.

The nomological irreducibility of the psychological means, if I am right, that the social sciences cannot be expected to develop in ways exactly parallel to the physical sciences, nor can we expect ever to be able to explain and predict human behavior with the kind of precision that is possible in principle for physical phenomena. This does not mean there are any events that are in themselves undetermined or unpredictable; it is only events as described in the vocabulary of thought and action that resist incorporation into a closed deterministic system. These same events, described in appropriate physical terms, are as amenable to prediction and explanation as any.

I shall not argue here for this version of monism, but it may be worth indicating how the parts of the thesis support one another. Take as a first premise that psychological events such as perceivings, rememberings, the acquisition and loss of knowledge, and intentional actions are directly or indirectly caused by, and the causes of, physical events. The second premise is that when events are related as cause and effect, then there exists a closed and deterministic system of laws into which these events, when appropriately described, fit. (I ignore as irrelevant the possibility that microphysics may be irreducibly probabilistic.) The third premise, for which I shall be giving reasons, is that there are no precise psychophysical laws. The three premises, taken together, imply monism. For psychological events clearly cannot constitute a closed system; much happens that is not psychological, and affects the psychological. But if psychological events are causally related to physical events, there must, by premise 2, be laws that cover them. By premise 3, the laws are not psychophysical, so they must be purely physical laws. This means that the psychological events are describable, taken one by one, in physical terms, that is, they are physical events. Perhaps it will be agreed that this position deserves to be called *anomalous monism*: *monism*, because it holds that psychological events are physical events; *anomalous*, because it insists that events do not fall under strict laws when described in psychological terms.

My general strategy for trying to show that there are no strict psycholophysical laws depends, first, on emphasizing the holistic character of the cognitive field. Any effort at increasing the accuracy and power of a theory of behavior forces us to bring more and more of the whole system of the agent's beliefs and motives directly into account. But in inferring this system from the evidence, we necessarily impose conditions of coherence, rationality, and consistency. These conditions have no echo in physical theory, which is why we can look for no more than rough correlations between psychological and physical phenomena.

Consider our commonsense scheme for describing and explaining actions. The part of this scheme that I have in mind depends on that fact that we can explain why someone acted as he did by mentioning a desire, value, purpose, goal, or aim the person had, and a belief connecting the desire with the action to be explained. So, for example, we may explain why Achilles returned to the battle by saying he wished to avenge the death of Patroclus. (Given this much, we do not need to mention that he believed that by returning to the battle he could avenge the death of Patroclus.) This style of explanation has many variants. We may adumbrate explanation simply by expanding the description of the action. "He is returning to battle with the intention of avenging the death of Patroclus." Or we may more simply redescribe: "Why is he putting on his armor?" "He is getting ready to avenge Patroclus' death." Even the answer, "He just wanted to" falls into the pattern. If given in explanation of why Sam played the piano at midnight, it implies that he wanted to make true a certain proposition, that Sam play the piano at midnight, and he believed that by acting as he did, he would make it true.

A desire and a belief of the right sort may explain an action, but not necessarily. A man might have good reasons for killing his father, and he might do it, and yet the reasons not be his reasons in doing it (think of Oedipus). So when we offer the fact of the desire and belief in explanation, we imply not only that the agent had the desire and belief, but that they were *efficacious* in producing the action. Here we must say, I think, that causality is involved, i.e., that the desire and belief were causal conditions of the action. Even this is not sufficient, however. For suppose, contrary to the legend, that Oedipus, for some dark oedipal reason, was hurrying along the road intent on killing his father, and, finding a surly old man blocking his way, killed him so he could (as he thought) get on with the main job. Then not only did Oedipus want to kill his father, and actually kill him, but his desire caused him to kill his father. Yet we could not say that in killing the old man he intentionally killed his father, nor that his reason in killing the old man was to kill his father.

Can we somehow give conditions that are not only necessary, but also sufficient, for an action to be intentional, using only such concepts as those of belief, desire and cause? I think not. The reason, very sketchily stated, is this. For a desire and a belief to explain an action in the right way, they must cause it in the right way, perhaps through a chain or process of reasoning that meets standards of rationality. I do not see how the right sort of causal process can be distinguished without, among other things, giving an account of how a decision is reached in the light of conflicting evidence and conflicting desires. I doubt whether it is possible to provide such an account at all, but certainly it cannot be done without using notions like evidence, or good reasons for believing, and these notions outrun those with which we began.

What prevents us from giving necessary and sufficient conditions for acting on a reason also prevents us from giving serious laws connecting reasons and actions. To see this, suppose we had the sufficient conditions. Then we could say: whenever a man has such-and-such beliefs and desires, and such-and-such further conditions are satisfied, he will act in such-and-such a way. There are no serious laws of this kind. By a serious law, I mean more than a statistical generalization (the statistical laws of physics are serious because they give sharply fixed probabilities, which spring from the nature of the theory); it must be a law that, while it may have provisos limiting its application, allows us to determine in advance whether or not the conditions of application are satisfied. It is an error to compare a truism like, "If a man wants to eat an acorn omelette, then he generally will if the opportunity exists and no other desire overrides" with a law that says how fast a body will fall in a vacuum. It is an error, because in the latter case, but not the former, we can tell in advance whether the condition holds, and we know what allowance to make if it doesn't. What is needed in the case of action, if we are to predict on the basis of desires and beliefs, is a quantitative calculus that brings all relevant beliefs and desires into the picture. There is no hope of refining the simple pattern of explanation on the basis of reasons into such a calculus.

Two ideas are built into the concept of acting on a reason (and hence, the concept of behavior generally): the idea of cause and the idea of rationality. A reason is a rational cause. One way rationality is built in is transparent: the cause must be a belief and a desire in the light of which the action is reasonable. But rationality also enters more subtly, since the way desire and belief work to cause the action must meet further, and unspecified, conditions. The advantage of this mode of explanation is clear: we can explain behavior without having to know too much about how it was caused. And the cost is appropriate: we cannot turn this mode of explanation into something more like science.

Explanation by reasons avoids coping with the complexity of causal factors by singling out one, something it is able to do by omitting to provide, within the theory, a clear test of when the antecedent conditions hold. The simplest way of trying to improve matters is to substitute for desires and beliefs more directly observable events that may be assumed to cause them, such as flashing lights, punishments and rewards, deprivations, or spoken commands and instructions. But perhaps it is now obvious to almost everyone that a theory of action inspired by this idea has no chance of explaining complex behavior unless it succeeds in inferring or constructing the pattern of thoughts and emotions of the agent.

The best, though by no means the only, evidence for desires and beliefs is action, and this suggests the possibility of a theory that deals directly with the relations between actions, and treats wants and thoughts as theoretical constructs. A sophisticated theory along these lines was proposed by Frank Ramsey.[1] Ramsey was primarily interested in providing a foundation in behavior for the idea that a person accords one or another degree of credence to a proposition. Ramsey was able to show that if the pattern of an individual's preferences or choices among an unlimited set of alternatives meets certain conditions, then that individual can be taken to be acting so as to maximize expected utility, that is, he acts as if he assigns values to the outcomes on an interval scale, judges the plausibility of the truth of propositions on a ratio scale, and chooses the alternative with the highest computed expected yield.

Ramsey's theory suggests an experimental procedure for disengaging the roles of subjective probability (or degree of belief) and subjective value in choice behavior. Clearly, if it may be assumed that an agent judges probabilities in accord with frequencies or so-called objective probabilities, it is easy to compute from his choices among gambles what his values are; and similarly one can compute his degree of belief in various propositions if one can assume that his values are, say, linear in money. But neither assumption seems justified in advance of evidence, and since choices are the resultant of both factors, how can either factor be derived from choices until the other is known? Here, in effect, is Ramsey's solution: we can tell that a man judges an event as likely to happen as not if he doesn't care whether an attractive or an unattractive outcome is tied to it, if he is indifferent, say, between these two options:

	Option 1	*Option 2*
If it rains you get:	$1,000	a kick
If it doesn't rain:	a kick	$1,000

Using this event with a subjective probability of one-half, it is possible to scale values generally, and using these values, to scale probabilities.

In many ways, this theory takes a long step toward scientific respectability. It gives up trying to explain actions one at a time by appeal to something more basic, and instead postulates a pattern in behavior from which beliefs and attitudes can be inferred. This simultaneously removes the need for establishing the existence of beliefs and attitudes apart from behavior, and takes into systematic account (as a construct) the whole relevant network of cognitive and motivational factors. The theory assigns numbers to measure degrees of belief and desire, as is essential if it is to be adequate to prediction, and yet it does this on the basis of purely qualitative evidence (preferences or choices between pairs of alternatives). Can we accept such a theory of decision as a scientific theory of behavior on a par with a physical theory?

Well, first we must notice that a theory like Ramsey's has no predictive power at all unless it is assumed that beliefs and values do not change over time. The theory merely puts restrictions on a temporal cross-section of an agent's dispositions to choose. If we try experimentally to test the theory, we run into the difficulty that the testing procedure disturbs the pattern we wish to examine. After spending several years testing variants of Ramsey's theory on human subjects, I tried the following experiment (with Merrill Carlsmith). Subjects made all possible pairwise choices within a small field of alternatives, and in a series of subsequent sessions, were offered the same set of options over and over. The alternatives were complex enough to mask the fact of repetition, so that subjects could not remember their previous choices, and payoffs were deferred to the end of the experiment so that there was no normal learning or conditioning. The choices for each session and each subject were then examined for inconsistencies—cases where someone had chosen *a* over *b*, *b* over *c*, and *c* over *a*. It was found that as time went on, people became steadily more consistent; intransitivities were gradually eliminated; after six sessions, all subjects were close to being perfectly consistent. This was enough to show that a static theory like Ramsey's could not, even under the most carefully controlled conditions, yield accurate predictions: merely making choices (with no reward or feedback) alters future choices. There was also an entirely unexpected result. If the choices of an individual over all trials were combined, on the assumption that his "real" preference was for the alternative of a pair he chose most often, then there were almost no inconsistencies at all. Apparently, from the start there were underlying and consistent values which were better and better realized in choice. I found it impossible to construct a formal theory that could explain this, and gave up my career as an experimental psychologist.

Before drawing a moral from this experiment, let me return to Ramsey's ingenious method for abstracting subjective values and probabilities simultaneously from choice behavior. Application of the theory depends, it will be remembered, on finding a proposition with a certain property: it must be such that the subject does not care whether its truth or its falsity is tied to the more attractive of two outcomes. In the context of theory, it is clear that this means, *any* two outcomes. So, if the theory is to operate at all, if it is to be used to measure degrees of belief and the relative force of desire, it is first necessary that there be a proposition of the required sort. Apparently, this is an empirical question; yet the claim that the theory is true is then a very sweeping empirical claim. If it is ever correct, according to the theory, to say that for a given person a certain event has some specific subjective probability, it must be the case that a detailed and powerful theory is true concerning the pattern of that person's choice behavior. And if it is ever reasonable to assert, for example, that one event has a higher subjective probability than another for a given person, then there must be good reason to believe that a very strong theory is true rather than false.

From a formal point of view, the situation is analogous to fundamental measurement in physics, say of length, temperature, or mass. The assignment of numbers to measure any of these assumes that a very tight set of conditions holds. And I think that we can treat the cases as parallel in the following respect. Just as the satisfaction of the conditions for measuring length or mass may be viewed as constitutive of the range of application of the sciences that employ these measures, so the satisfaction of conditions of consistency and rational coherence may be viewed as constitutive of the range of applications of such concepts as those of belief, desire, intention and action. It is not easy to describe in convincing detail an experiment that would persuade us that the transitivity of the relation of *heavier than* had failed. Though the case is not as extreme, I do not think we can clearly say what should convince us that a man at a given time (without change of mind) preferred *a* to *b*, *b* to *c*, and *c* to *a*. The reason for our difficulty is that we cannot make good sense of an attribution of preference except against a background of coherent attitudes.

The significance of the experiment I described a page or so back is that it demonstrates how easy it is to interpret choice behavior so as to give it a consistent and rational pattern. When we learn that apparent inconsistency fades with repetition but no learning, we are apt to count the inconsistency as merely apparent. When we learn that frequency of choice may be taken as evidence for an underlying consistent disposition, we may decide to write off what seem to be inconsistent choices as failures of perception or execution. My point is not merely that the data are open to more than one interpretation, though this is obviously true. My point is that if we are intelligibly to attribute attitudes and beliefs, or usefully to describe motions as behavior, then we are committed to finding, in the pattern of behavior, belief, and desire, a large degree of rationality and consistency.

A final consideration may help to reinforce this claim. In the experiments I have been describing, it is common to offer the subject choices verbally, and for him to respond by saying what he chooses. We assume that the subject is choosing between the alternatives described by the experimenter, i.e. that the words used by subject and experimenter have the same interpretation. A more satisfying theory would drop the assumption by incorporating in decision theory a theory of communication. This is not a peripheral issue, because except in the case of the most primitive beliefs and desires, establishing the correctness of an attribution of belief or desire involves much the same problems as showing that we have understood the words of another. Suppose I offer a person an apple and a pear. He points to the apple, and I record that he has chosen the apple. By describing his action in this way, I imply that he intended to point to the apple, and that by pointing he intended to indicate his choice. I also imply that he believed he was choosing an apple. In attributing beliefs we can make very fine distinctions, as fine as our language provides. Not only is there a difference between his believing he is choosing an apple and his believing he is choosing a pear. There is even a difference between his believing he is choosing the best apple in the box and his believing he is choosing the largest apple, and this can happen when the largest is the best.

All the distinctions available in our language are used in the attribution of belief (and desire and intention); this is perhaps obvious from the fact that we can attribute a belief by putting any declarative sentence after the words. "He believes that." There is every reason to hold, then, that establishing the correctness of an attribution of belief is no easier than interpreting a man's speech. But I think we can go further, and say that the problems are identical. Beliefs cannot be ascertained in general without command of a man's language; and we cannot master a man's language without knowing much of what he believes. Unless someone could talk with him, it would not be possible to know that a man believed Fermat's last theorem to be true, or that he believed Napoleon had all the qualities of a great general.

The reason we cannot understand what a man means by what he says without knowing a good deal about his beliefs is this. In order to interpret verbal behavior, we must be able to tell when a speaker holds a sentence he speaks to be true. But sentences are held to be true partly because of what is believed, and partly because of what the speaker means by his words. The problem of interpretation therefore is the problem of abstracting simultaneously the roles of belief and meaning from the pattern of sentences to which a speaker subscribes over time. The situation is like that in decision theory: just as we cannot infer beliefs from choices without also inferring desires, so we cannot decide what a man means by what he says without at the same time constructing a theory about what he believes.

In the case of language, the basic strategy must be to assume that by and large a speaker we do not yet understand is consistent and correct in his beliefs—according to our own standards, of course. Following this strategy makes it possible to pair up sentences the speaker utters with sentences of our own that we hold true under like circumstances. When this is done systematically, the result is a method of translation. Once the project is under way, it is possible, and indeed necessary, to allow some slack for error or difference of opinion. But we cannot make sense of error until we have established a base of agreement.

The interpretation of verbal behavior thus shows the salient features of the explanation of behavior generally: we cannot profitably take the parts one by one (the words and sentences), for it is only in the context of the system (language) that their role can be specified. When we turn to the task of interpreting the pattern, we notice the need to find it in accord, within limits, with standards of rationality. In the case of language, this is apparent, because understanding it is *translating* it into our own system of concepts. But in fact the case is no different with beliefs, desires, and actions.

The constitutive force in the realm of behavior derives from the need to view others, nearly enough, as like ourselves. As long as it is behavior and not something else we want to explain and describe, we must warp the evidence to fit this frame. Physical concepts have different constitutive elements. Standing ready, as we must, to adjust psychological terms to one set of standards and physical terms to another, we know that we cannot insist on a sharp and lawlike connection between them. Since psychological phenomena do not constitute a closed system, this amounts to saying they are not, even in theory, amenable to precise prediction or subsumption under deterministic laws. The limit thus placed on the social sciences is set not by nature, but by us when we decide to view men as rational agents with goals and purposes, and as subject to moral evaluation.

Comments and Replies

This essay was delivered (in slightly altered form) at a conference on the Philosophy of Psychology organized by the Royal Institute of Philosophy and held at the

University of Kent at Canterbury in September, 1971. The proceedings were published in *Philosophy of Psychology*, edited by Stuart Brown, and contained comments on the remarks by the chairman of the session, Professor Richard Peters, and questions asked by Mr. Robin Attfield, Professor Les Holborrow, and Professor Robert Solomon. The author's replies follow.

Reply to Peters

The conclusion defended in this essay is a familiar one, and is shared by many philosophers and, probably, psychologists. The position might be put this way: the study of human action, motives, desires, beliefs, memory, and learning, at least so far as these are logically tied to the so-called propositional attitudes, cannot employ the same methods as, or be reduced to, the more precise physical sciences. Many would agree, too, that we cannot expect to find strict psychophysical laws. If there is anything new in what I say on this topic, it is in the details of the reasons I give for saying that generalizations that combine psychological and physical predicates are not lawlike in the strong sense that wholly physical laws can be. What apparently arouses the most doubt and opposition is my attempt to combine the view that psychological concepts have an autonomy relative to the physical with a monistic ontology and a causal analysis of action.

I thought, then, that my *conclusions* (in contrast, perhaps, to my arguments) concerning the nature of cognitive psychology as a science were neither new nor apt to excite much debate. But hearing Professor Peters's generous and sensitive remarks ... made me realize I had given the impression that I was making some sort of attack on psychology generally, or at least on its right to be called a science. That is certainly not what I intended, but I do see how things I wrote could bear that interpretation. So here I will briefly try to set matters straight.

First, let me reemphasize the fact that my arguments are limited in application to branches of psychology that make essential reference to "propositional attitudes" such as belief, desire, and memory, or use concepts logically tied to these, such as perception, learning, and action. (Some of these concepts may not always show intensionality, and in such cases are also exempt.)

Second, I made much of the fact that psychophysical generalizations must be treated as irreducibly statistical in character, in contrast to sciences where in principle exceptions can be taken care of by refinements couched in a homogeneous vocabulary. This is not a *reproach* to psychology, nor does it mean its predictions and explanations are in fact less precise than those of many other sciences. I assume that in application meteorology and geology, for example, are far less precise than much work on perception. The point is not the actual degree of looseness in psychology, but what guarantees that it can't be eliminated, namely the conceptually hermaphroditic character of its generalizations.

Third, I argued that the part of psychology with which I was concerned cannot be, or be incorporated in, a closed science. This is due to the irreducibility of psychological concepts, and to the fact that psychological events and states often have causes that have no natural psychological descriptions. I do not want to say that analogous remarks may not hold for some other sciences, for example, biology. But I do not know how to show that the concepts of biology are nomologically irreducible to the concepts of physics. What sets apart certain psychological concepts—their intentionality—does not apply to the concepts of biology.

"Science" being the honorific word it is in some quarters, it would be meretricious to summarize these points by saying that psychology (the part with which we are concerned) is not a science; the conclusion is rather that psychology is set off from other sciences in an important and interesting way. The argument against the existence of strict psychophysical laws provides the key to psychology's uniqueness: the argument led from the necessarily holistic character of interpretations of propositional attitudes to the recognition of an irreducibly normative element in all attributions of attitude. In the formulation of hypotheses and the reading of evidence, there is no way psychology can avoid consideration of the nature of rationality, of coherence and consistency. At one end of the spectrum, logic and rational decision theory are psychological theories from which the obviously empirical has been drained. At the other end, there is some form of behaviorism better imagined than described from which all taint of the normative has been subtracted. Psychology, if it deals with propositional attitudes, hovers in between. This branch of the subject cannot be divorced from such questions as what constitutes a good argument, a valid inference, a rational plan, or a good reason for acting. These questions also belong to the traditional concerns of philosophy, which is my excuse for my title.

Reply to Attfield

Mr. Attfield believes that what I have called anomalous monism is inconsistent; in his view, psychological events cannot be identical with physical events while being nomologically unrelated to them. His argument, if I follow it correctly, has this form:

> *Premise 1.* Some psychological event, say Attfield's perceiving of a fly at *t*, is identical with a certain physical event, say the neurological change in Attfield at *t* that is *P*.
> *Premise 2.* Since there is a causal law that connects physical events with their causes, there is a causal law that connects the change in Attfield at *t* that is *P* with some physical event that caused it (say the change at $t - 1$ that is *P*′).
> *Conclusion.* There is a (psychophysical) causal law that connects Attfield's perceiving of a fly at *t* with its cause, the change at $t - 1$ that is *P*′.

Attfield is right that anomalous monism is committed to the premises, and that it rejects the conclusion. But I would urge that the conclusion does not follow from the premises. Attfield's argument has the same basic semantic structure as an argument that would infer that Jones believes Scott wrote *Waverley* from the facts that Scott is the author of *Waverley* and Jones believes the author of *Waverley* wrote *Waverley*. This argument fails because after psychological verbs like "believes" normal substitutivity of co-referring singular terms breaks down. The same must be said about the positions occupied by *a* and *b* in, "There is a law that connects events *a* and *b*." We cannot conclude from the fact that there is a causal law that connects events *a* and *b*, and the fact that $a = c$, that there is a causal law that connects events *c* and *b*. The reason is that laws (and nomological explanations) do not deal directly (i.e., extensionally) with events, but with events as described in one way or another.

The point may be made without reference to nonextensional contexts. Suppose there is an event that is uniquely characterized both by, "The event that is *F* at *t*" and, "The event that is *G* at *t*," and that there is a law, "Every event that is *H* is followed a minute later by exactly one event that is *F*." It does not follow that there is a law that says, "Every event that is *H* is followed a minute later by exactly one event that is *G*." Here *H* and *F* may be thought of as physical predicates, and *G* as a psychological predicate.

Reply to Holborrow
Where Mr. Attfield thinks that the identity of psychological and physical events entails the existence of psychophysical laws, Mr. Holborrow is worried that unless psychophysical laws are introduced, the dualism of psychological and physical concepts will entail a dualism of autonomous "causal systems." "At the psychological level," he writes, "the event which we describe as an action is caused by a reason. At the level of physical phenomena, the same event differently described is caused by an entirely different set of factors." But causal relations, in my view, hold between events however described. So one and the same event, whether described as an action or as a physical event, will have the *same* causes, whether these are described as reasons or as physical states or events. There is no dualism of "causal factors," "causal systems," or "types of causation." Nor is there any reason to suppose, from the dualism of descriptions, that there are two kinds of law. If *a* caused *b*, then *some* descriptions of *a* and *b* instantiate a strict causal law. But the law is never, if I am right, a psychophysical law, nor can it be purely psychological (since the mental does not constitute a closed system).

Reply to Solomon
We cannot make good sense of the idea that there are seriously different total conceptual schemes, or frames of reference, or that there may be radically "incommensurate" languages (to use Whorf's word). So I have argued elsewhere[2] and Professor Solomon understandably wonders how I can reconcile my rejection of conceptual relativity with the claim that psychological concepts are nomologically and otherwise irreducible to physical concepts. He does make me regret saying that in interpreting the verbal behavior of others we must translate into our own scheme of concepts. Of course interpretation is essentially translation, and so there is no avoiding that. But if translation succeeds, we have shown there is no need to speak of two conceptual schemes, while if translation fails, there is no ground for speaking of two. If I am right then, there never can be a situation in which we can intelligibly compare or contrast divergent schemes, and in that case we do better not to say that there is one scheme, as if we understood what it would be like for there to be more.

A conceptual scheme, in the context of the above remarks, is supposed to correspond to a whole language; nothing, at any rate, can be left out that is needed to make sense of the rest. A concept, or set of concepts, may be irreducible and yet be essential to making sense of some, or all, of the rest. Indeed if a concept *C* is essential to making sense of some other concept *C′* in a set of concepts *S*, *C* must be primitive (indefinable) relative to the resources of *S*.

Psychological concepts, I have been arguing, cannot be reduced, even nomologically, to others. But they are essential to our understanding of the rest. We cannot conceive a language without psychological terms or expressions—there would be no way to translate it into our own language. Of course there could be a part or fragment of a language that lacked psychological expressions, provided there was a (complete) language in which to incorporate or explain the fragment. It makes sense to speak of irreducible or semi-autonomous systems of concepts, or schemes of description and explanation, but only as these are less than the whole of what is available for understanding and communication.

Notes

1. "Truth and Probability." Ramsey's theory, in a less interesting form, was later, and independently, rediscovered by von Neumann and Morgenstern, and is sometimes called a theory of decision under uncertainty, or simply decision theory, by economists and psychologists.
2. In "On the Very Idea of a Conceptual Scheme."

Chapter 7

General Laws and Explaining Human Behavior

Brian Fay

I

In this chapter I argue for three major theses that are often thought to be antithetical to one another. The three theses are: first, that explanations of human behavior in terms of its reasons (beliefs, desires, motives, goals) rest upon general laws because such explanations are causal in nature; second, that it is extremely unlikely that these general laws are statable in the intentionalist vocabulary of the social sciences; and third, that the social sciences must be genuinely theoretical if they are to be at all viable.

My purpose in this is, in the first instance, to present a model of the role general laws and what I shall call causal generalizations have in the explanation of human behavior. But there is a second, larger purpose as well, namely, to outline a picture of social-scientific theory. Briefly stated, I hope to show that, although general laws properly so called will not emerge from the social sciences, a certain sort of genuinely theoretical science of human behavior is still possible. The "sort" I have in mind is what is sometimes called Critical Theory. The chapter is ultimately concerned to show, therefore, that a proper understanding of the nature and basis of the causal explanations of human behavior leads to a critical metatheoretic conception of social-scientific theory.

II

There is a widespread belief among certain (mostly post-Wittgensteinian) philosophers in what I shall call the "singularity thesis of human action."[1] According to this thesis, reason-explanations can account for human action without invoking or presupposing any general law; in the words of Hart and Honore, this thesis consists of the claim that, "The statement that a given person acted for a given reason does not require for its defense generalizations asserting connections between types of events."[2] This thesis, if it were true, would have profound consequences for any science of action, for it would mean that that explanations in this science would be particularistic, and it would mean that such a science could not be genuinely theoretical.[3] Instead, the social sciences involved in explaining action would be confined to elaborating the character of the particular reasoning process that results in the performance of certain historically located events: They would be backward looking, ad hoc, and ideographic.

I wish to begin by examining what I take to be the two most compelling arguments for this belief, and to show why I think they are mistaken. These two arguments might be called the "logical-connection argument" and the "essential-nature argument." My aim in doing this is to demonstrate, in an indirect manner, that reason-explanations do rest on general laws.

One major support for the singularity thesis is the claim that the connection between that which explains an action and the action itself is a logical one, and that therefore this connection is both intuitively clear and qualitatively different from the relationship which exists between events which figure in causal explanations.

Thus, William Dray maintains that reason-explanations invoke principles of action (as opposed to empirical laws) to explain human behavior, and that the relationship between principles and their outcomes is not essentially one of a recurring pattern, but is rather one in which the outcome (in this case the action) is intrinsically (logically, conceptually) connected with the principle itself. Reasons give the grounds for which the action is a consequent, and, since the relationship between ground and consequent is logical rather than empirical, he argues that reason-explanations do not require general statements linking a kind of reason with a kind of action.[4]

To take a concrete case. Alasdair MacIntyre, in his well-known (though now self-repudiated) article, "A Mistake about Causality in Social Science,"[5] analyzed Max Weber's explanation of the rise of capitalist behavior in terms of certain theological beliefs of Protestants. He concluded that, just because the connection between these beliefs and actions is a conceptual one, Weber's tactic of supporting his thesis by embedding it in large-scale historical generalizations was irrelevant; as MacIntyre wrote:

> The use of Mill's methods is entirely out of place; we do not need to juggle with causal alternatives. India and China do not strengthen and could not have weakened his case about Europe. For it is not a question of whether there is a purely contingent relationship between isolable phenomena. And so constant conjunction is neither here or there.[6]

Now, in this argument I wish to support Weber against MacIntyre, for I want to maintain that explanations of particular kinds of action in terms of particular sorts of reasons do in fact rest on at least implicit general laws.

The crucial mistake in the logical-connection argument lies in its account of the way reason-explanations account for an action. According to it, explaining an action involves specifying the reasons that rationalize it, i.e., that show it to be the appropriate thing for the agent to have done, given his situation. However, this account is inadequate because it does not distinguish between those beliefs and desires that are a reason for the behavior but that did not cause it to happen ("a" reason for doing it) from those which in fact were responsible for its occurrence ("the" reason for doing it).

The distinction that is relevant here is between acting *and* having a reason and acting *because* of that reason. In the former case an agent may have a reason for his behavior, and it may therefore have been a rational and justifiable thing to have done. But unless the having of this reason was the cause of the agent's acting as he did, the reason does not explain the act, i.e., it does not show that the act occurred because the agent had the specified reason.

Broadly speaking, reason-explanations succeed in explaining when they show that it was because the agent thought that the act was the appropriate way to achieve his ends that he acted as he did. In other words, it isn't the reason that explains the act, but rather the agent's *having* this reason, and this having caused him to act in the way he did, that explains it. In another article, I have described this by saying that we explain the behavior in question in discovering the agent's practical reasoning processes that brought it about.[7]

If explaining an act by means of a reason-explanation is knowing the reasoning process which caused it, then I think it can be demonstrated that such explanations implicitly rest on general laws. Such a demonstration is a simple matter if one adopts a broadly Humean construal of causality. It is much more complex if one subscribes to a broadly realist construal of causal explanation. I will discuss each of these in turn.

On a broadly Humean construal of causality, a claim that *x* causes *y* involves the claim, among others, that *x* is regularly related to *y*. Thus, to employ Mill's account of the regularity thesis, to say that *x* causes *y* is to say that *x* is a sufficient condition of *y*'s occurring, and/or that it is a necessary condition as well. But relations of necessary and sufficient conditionship obviously rest on general laws. For any claim that *x* causes *y* involves the assumption that whenever *x* occurs *y* will occur, ceteris paribus (otherwise *x* cannot be a sufficient condition of *y*), or that whenever *x* does not occur *y* will not occur, ceteris paribus (otherwise *x* cannot be a necessary condition of *y*), or both.

Given a realist construal, the matter is somewhat different.[8] According to the realist, providing a causal explanation of the form "*x* causes *y*" is to relate *x* and *y* by means of an actual mechanism which, in suitable conditions captured by "*x*," generates the observed outcomes described as "*y*." (The realist doesn't intend anything specifically mechanical by the term "mechanism." Thus, a practical reasoning process could itself be such a mechanism. Indeed, my account of reason-explanations as causal in form is broadly realist in that, on my reading, the reasons of the actor are connected to his behavior by means of real psychological reasoning processes.) In the realist view, causal explanations may start with an observed regularity between *x* and *y*, but this regularity is only evidence that a causal relationship actually exists. Only when the underlying mechanism that has certain natural powers is discovered can a genuine causal explanation be said to have been given. It follows from this, according to the realist, that causal explanations do not therefore rest on general laws.

But does this follow? I think not.[9] The reason why is that the realist's account of causal explanation surreptitiously smuggles Humean regularities back in on another level. The realist invokes a basic mechanism as a way of explaining why a particular sort of event will occur in certain circumstances. This mechanism is meant to have a particular nature such that, subject to conditions of an appropriate kind, it will perform in a specific manner. But all of this presupposes that there is a regular operation of the mechanism, that under certain circumstances the mechanism will act in a predictable manner. If this were not the case, then invoking the mechanism would not be genuinely explanatory, because then one would need to know why the mechanism worked as it did to produce the effect *y* in the case at hand. It is only because the notion of a causal mechanism carries with it the backing of a general law that such a mechanism can be a relevant part of a causal explanation.

To this the realist has an answer. He will claim that knowledge of the underlying mechanism does not consist of general laws that supposedly govern its operation; rather, such knowledge consists of knowing the "nature" or "essence" of the mechanism in question, and this includes knowing the powers that it has. This response is thus a variant of the second major argument which supports the singularity thesis, namely, the essential-nature argument.

I will turn almost immediately to the essential-nature argument, but before I do so I want to point out that no matter how one interprets either the Humean or the realist construal of causal explanation, both accounts agree as to the relevance of generalizations in the assessment of causal explanations. In the case of the Humean, this is so because causal explanations just are generalizations of a certain sort. In the case of the realist, this is so because generalizations indicate that deeper causal mechanisms are at work. Thus, insofar as reason-explanations are a type of causal explanation, generalizations linking particular beliefs and desires with particular actions will be relevant in determining the worth of the explanation at hand. This is directly contrary to the singularity thesis.

Thus, to return to the example of Weber, it may well be the case that even though the matrix of beliefs, desires, values, and so on associated with Calvinism is an initially appealing explanation because it rationalizes capitalist behavior, and because it invokes elements in the experience and thought of a group of people that appear to be crucial motivational factors in their lives, it is indeed false. For it may well be the case either that many of the particular Protestants who did possess the relevant beliefs and desires nevertheless did not in fact act in the way that Weber thought they did, and/or that many people acted in a capitalist manner who were not in fact Protestants. If this turned out to be so, then Weber would have had to reject his interpretation as the explanation of capitalist behavior because he would now be in possession of evidence which indicated that "a" reason for the puzzling behavior was not, upon empirical investigation, "the" reason why people behaved as they did.[10]

So far I have argued that reason-explanations are causal in nature, and that on at least one construal (the Humean) of causal explanation this means that reason-explanations rest on general laws. I have also argued that another construal of causal explanation (the realist), though apparently not nomological, actually is so at a deeper level. Insofar as these arguments are right, the singularity thesis cannot be correct. However, I did allow that the realist could salvage his case by invoking the essential-nature argument, and I said earlier that this argument was itself one of the most important supports of the singularity thesis. I must, therefore, consider this argument.

The essential-nature argument amounts to the claim that good explanations are those which ultimately rest on an account of the nature of the basic entities involved. Knowing that the essence of an entity is to act in a certain manner means that the operation of this entity does not require further explanation in terms of some general law under which one could subsume its fundamental dispositions. Thus, to turn to the case of a practical reasoning process, the essential-nature argument says that, because it is part of the very nature of such a process to result in an action, the relationship between coming to have a reason and the action which this event explains is immediately apprehensible without recourse to any generalizations. We don't need to see the occurrence of these two events as instances of some generally recurring pattern in order for the occurrence of one to explain the occurrence of the other once we know what a practical reasoning process is. Thus, for example, when one is told that a person crossed the street because he inferred from his belief that this was the only way to buy cigarettes, and from his desire to have a smoke now, that he ought to cross the street, this explanation appears satisfying in itself: One doesn't need any further information to the effect that the person was of type x, and type x people engaged in type y reasoning act in z type way, because one can understand the relationship between the reasoning process and the act immediately.

Now it *is* a fact that in explaining an action we grasp a connection between a singular explanans and a singular explanandum such that we do not feel the need to subsume them under some general statement. But this is not because reason-explanations are not causal in form (as the logical-connection argument would have it), or because the causal powers of the mechanism involved in producing the act involve some sort of "natural necessity" that does not admit or require further explanation in terms of a general law (as the realist would have it). *The reason is that reasoning processes are partially defined in functional terms.* One doesn't feel the need for general laws when explaining act *z* by the agent's desire for *y* and his belief that his doing *z* is the best means for achieving *y*, any more than we feel the need for a general law when we explain why a person feels relaxed by the fact that he ingested a tranquilizer. In neither case do we feel the further need for a general statement which supports the particular causal explanation, because the disposition to produce certain sorts of outcomes is built right into the concept "desire" and "practical reasoning process" just as it is built into the concept "tranquilizer." In both sorts of cases, discovering that the cause of an event is another event or object which is identified as one which characteristically produces events of the first type is (psychologically) satisfying in itself as an explanation.

Indeed, it is because reasoning processes (and poisons and tranquilizers) are entities specified at least partially in functional terms that scientists are generally interested in those cases in which they don't operate, i.e., when their causal force is defeated by external circumstances. In these cases, what is sought is some law which states that when a particular set of (necessary) conditions is not present, then the expected causal force of the entity will be inoperative. In other words, with functionally defined causes the general laws which are usually sought are those which explain their *breakdown* rather than their operation, just because in "normal" circumstances the entity's being of a certain kind is a sufficient condition for the occurrence of a certain event.

But it certainly does not follow from this that in the normal cases in which the appropriate effect is forthcoming the relationship between the explanans event and the explanandum event does not presuppose a general law to the effect that under normal circumstances these kinds of events are related in a specified way. On the contrary, as I argued above when I showed that the realist smuggles Humean regularities into his account of causality, it presupposes just this sort of statement. For to characterize entities functionally is partly to characterize them in terms of the general causal outcomes that they will produce.

Moreover, it is a deep mistake to think with the realists that explanations in terms of "basic nature" and "causal power" are as deep as science can go. For one of the ways that science progresses is specifying in more detailed and sharply defined terms under just what circumstances these sorts of events are related. Furthermore, in this process of articulating the general laws which govern the relations between functionally defined entities and certain events, it is quite often the case—contrary to Charles Taylor's thesis of asymmetrical explanation[11]—that the scientist will try to (causally) explain just why it is that a certain entity ordinarily produces the effects that it does.

Thus, while a relaxed condition is a natural and in some sense privileged outcome of a person's taking a tranquilizer—in Taylor's words, there is a certain "bent or pressure of events towards a certain consummation," an outcome which "does not come about by 'accident' but is somehow part of its 'essential nature'"[12]—the neurophysiologist will certainly try to explain why this is the case by investigating the neurochemical processes through which the tranquilizer causes this result. And in the same way, a psychologist may well try to explain why it is that people of a certain sort who engage in particular reasoning processes will ordinarily or "naturally" behave in a certain manner, by referring, for example, to their schedules of reinforcement or to the development of their mental capacities. The reason for this *further* level of explanation is not only the commonplace one that science seeks to include phenomena in a wider and wider range explainable by a smaller and smaller number of principles, but as well the more pertinent one that explanations in terms of functionally characterized entities tend toward the vacuous (á la the explanation of Molière's opium in terms of its dormitive powers).

Thus, though it is often the case that we feel satisfied with an explanation of an action which specifies the particular desires and beliefs which brought it about, and that we feel it is unnecessary to invoke a general law in virtue of which the particulars of our explanation would be seen to be instances of a generally recurring pattern, these psychological facts should not blind us to the logic of the situation. For these psychological facts are rooted in the peculiar feature of our characterizations of the mental events which cause actions, namely, that they are partially functional characterizations; and functionally characterized events are so characterizable just because we believe that they regularly produce certain outcomes, and thus that some general law involving their description is in the offing. This is why singular explanations which invoke functionally characterized events do not seem to require subsumption—the general law is implicitly brought into the situation in the very meaning of the description of the particular causal event. Moreover, it should not be forgotten that although we often do not seek to elaborate these general laws, social scientists must try to discover the larger causal patterns in virtue of which these conditions hold.[13]

If what I have argued so far is correct, then two conclusions can be drawn. The first is that reason-explanations account for actions by seeing them as the causal outcome of certain mental events, namely, practical reasoning processes. The second is that because they are causal and because causal explanations are essentially nomological, reason-explanations necessarily rest on general laws, at least implicity.

However, there is a glaring fact regarding the explanations of action that seems to conflict sharply with my whole analysis of the nomological foundation of reason-explanations. The fact is that we *do* presently have reason-explanations for all sorts of actions, but we do *not* have available to us any general laws properly so-called which link the having of certain reasons with the performing of certain actions—indeed, we are far more certain of singular causal connections than we are of any putative law governing the cases in which we assert their existence. (Thus, for example, Weber's causal generalizations are not general laws.[14]) This fact lends credence to the singularity thesis, and it seems to undermine the nomological thesis that I am supporting.

However, the drawing of such a conclusion because of this fact would be a result of a deep misunderstanding of the nature of the nomological thesis.[15] For this thesis does not consist of the claim that for every particular causal explanation there is ready at hand a general law under which it can be subsumed; indeed, the thesis does *not* entail even that it be known what form the relevant general law would take if it were statable. All that the nomological thesis asserts is that there is *a* general law under which the events invoked in a causal explanation fall.

There are three important ways in which it can be seen that this is so. In the first place, there are many cases in which the claim that "*x* causes *y*," or even that "*x*'s cause *y*'s" is true, and yet the general laws under which such claims are subsumed involve no use of *x* and *y* at all. In such cases the events which we initially described as *x* and *y* are redescribed by means of *a* and *b*, and only then are they linkable by means of a general law. It is quite consistent with the nomological thesis that the general law which figures in a causal explanation be formulated in terms quite unlike those used to assert a particular causal connection or even a particular type of causal connection.

Indeed, it is normally the case that scientists have had to redescribe events which they believed (correctly) to be causally related in order to be able to formulate the general laws which govern them. Thus, for example, it was necessary to redescribe the event type originally described as "the production of warmth" as "the increase in molecular motion" in order to generate the causal laws governing heat.

In this discussion, an extremely important, if a relatively obvious, point to remember is that phenomena as such are never explained, *but only phenomena as described in some way*. And it is also important to remember that there may be any number of different descriptions of the same phenomenon. By keeping in mind these two considerations, one can easily see how events that are described in one set of terms, and related to one another by means of these terms, may well be redescribed in conceptually quite dissimilar terms from those employed in the original description, and, as a result, only then be able to be seen as part of a generally recurring pattern of events.

Of course, in order for this to happen there must be a specific kind of relationship between the terms describing the events in question. In the first place, there must either be an equivalence in their extension, or at least the extension of the first term must be a subset of the extension of the second, redescribing term. In the second place, the redescribing terms must figure in a more comprehensive theoretical scheme that allows one to understand why the event as initially described could have the causal power ascribed to it, and that gives one the capability of articulating more general and more precise formulations of the causal relationships involved.

I should mention in passing that the sorts of considerations I have been discussing are particularly apt in the context of the philosophy of social science, since there seems to be a widespread belief among a number of its practitioners from quite divergent perspectives in what I call "the doctrine of *superficial generalization*." This doctrine holds that if one claims one event is the cause of another event, one is thereby claiming that the law upon which this explanation rests will consist of the very same terms as used in the particular descriptions of these singular events. Hart and Honore appear to be holders of this doctrine when they write: "To make such a singular causal statement is therefore to claim that the events which it relates are instances of such a universal connection between types of events."[16] And an instance of this doctrine can be found in Hempel's famous article, "The Function of General Laws in History"; there, in trying to demonstrate that the explanation of particular historical events requires a covering law, he writes: "Now the assertion that a set of events ... have caused the event to be explained, amounts to the statement that, according to certain general laws, a set of events *of the kinds mentioned* is regularly accompanied by an event (of the kind for which an explanation is sought)." And then he says by way of example: "Consider, for example, the statement that the Dust Bowl farmers migrated to California 'because' continual drought and sandstorms made their existence increasingly precarious, and because California seemed to them to offer so much better living conditions. *This explanation rests on some such universal hypothesis as that populations will tend to migrate to regions which offer better living conditions*"[17]

Here the law which Hempel adduces is simply a more general version of the singular explanatory statement itself.[18]

This doctrine has often been responsible for objections to the nomological thesis on the grounds that social scientists are quite often willing to accept a singular explanatory statement which asserts a causal relationship between two events, and yet to deny the truth of any putative law or causal generalization formulated by using the same terms found in the singular explanation.[19] Thus, no matter how Hempel formulates his "universal hypothesis," it seems extremely implausible that we would be willing to accept it, even though his particular causal explanation seems evidently to be true. And this wedge between the particular and the general is supposed to demonstrate that the nomological thesis is false.

But it shows no such thing. For this sort of objection is rooted in the mistaken assumption that the nomological thesis consists of the claim that the laws which a particular explanation instantiates will be formulated in the same sorts of terms as those to be found in the descriptions of the particular case. In fact, however, the nomological thesis only asserts that there must be a covering law in order for a singular causal statement to be true; and it is quite in keeping with this that the actual laws that do cover these instances will be formulated in terms other than those found in the particular explanation.

The second important way in which it can be seen that a holder of the nomological thesis is not committed to the truth of any available general law, even though he is willing to assert the truth of some singular causal explanations, is to see that it is perfectly consistent with this thesis that there not be available a law under which a true particular causal statement is subsumable. For, as I have already had occasion to mention, all that the thesis maintains is that there is a law; but it does not follow from this that this law be currently known. Thus, for example, it is perfectly consistent for someone to claim that smoking causes lung cancer (under certain circumstances), to believe that such a claim rests on a general law which links together the two events now described as "smoking" and "the development of lung cancer," and yet to admit that as of this moment no such law exists. Indeed, it is probably the case that a majority of those causal ascriptions which both ordinary people and natural scientists currently make are not supported by fully adequate general laws. And thus it is no argument against the relevance of the nomological thesis for human behavior that although we are willing to believe a whole array of causal statements linking motives, beliefs, desires, and values with actions, we cannot provide a genuine general law under which they are subsumable.

Of course (as the example of smoking clearly shows), it does not follow from this that generalized statements and a whole range of empirical evidence are not therefore relevant to our making singular causal ascriptions. On the contrary, in order to provide an adequate causal explanation, we must have good evidence for believing that full-fledged causal laws which cover the relevant events actually exist. I will take this up again in section IV; at this point I just wish to forestall a possible misinterpretation to the effect that, since the nomological thesis apparently does not require that there actually exists a formulated general law for every (true) causal explanation, it allows us to dispense with the need for generalizations of any sort.

The third important way in which it can be seen that the nomological thesis is consistent with the glaring fact that we presently do have reason-type causal explanations (some of which are undoubtedly true), but that we do not have available to us any general laws properly so-called, is really only a product of the first two ways. This is that it is certainly possible for someone to be a proponent of the nomological thesis and at the same time believe that the general laws under which his singular causal explanations are subsumable will not be formulated in the same terms as those found in the causal explanations he presently gives.

Thus, for example, a historian may assert that soil erosion and the decline of agricultural production in a particular area are causally related, and he may assert this even though he also believes that it is impossible to formulate the laws which link these events by using such a gross term as "soil erosion." In fact, he may even believe that the event which he now describes as soil erosion will have to be redescribed in terms of a radically different sort before the appropriate law could be forthcoming—for example, it may be that he will have to introduce quite determinate physical concepts drawn from chemistry, such as the relative amounts of nitrogen.

To borrow a distinction from Davidson,[20] the generalizations we have may be either *homonomic* or *heteronomic*. Homonomic generalizations are those whose positive instances give use reason to believe that the form and vocabulary of the finished law will be of the same type as the generalizations themselves; heteronomic generalizations are those which lead us to believe that the precise law at work can only be stated by switching to a different vocabulary altogether. Thus, in terms of our example of soil erosion, this causal explanation involves a heteronomic generalization.

The important point about heteronomic generalizations is that they allow for the possibility of one's developing a whole range of causal stories without at the same time committing oneself to the belief that the general laws which underlie these stories will be formulated in the same terms that one is currently employing. We need not wait on the actual development of the relevant scientific theories before offering causal explanations of the events around us. And we may offer these explanations even though we might well expect that ultimately the laws which govern the phenomena involved will be expressed in a radically different terminology from what we curently employ.

With this understanding of the nomological thesis—and particularly the possibility of heteronomic generalizations—I wish to turn to the social sciences that explain human behavior, with an eye toward understanding the nature of the causal explanations which they can and do offer. In particular, I want to examine whether it isn't the case that the causal generalizations found in these social sciences aren't heteronomic. For if this is the case, it will point to a deep difference between the sciences of intentional action and the sciences of nature.

III

In this section I hope to demonstrate that there is a good reason to believe that the laws which underlie the causal processes of mental events that bring about actions will not be forthcoming at the level of discourse that social scientists use to describe and explain actions, namely, intentional discourse. I want to show that the generalizations they employ possess features which make them unusable in highly deterministic theories, and which make them incapable of being indefinitely refined so that they might become so usable; that is, that they are heteronomic generalizations. In the social sciences there are genuine causal explanations rooted in genuine causal generalizations about how certain kinds of people think and act in certain sorts of circumstances; but these generalizations are *not* genuine laws, nor is it at all probable that they ever will be purified into general laws properly so-called.

In order to see why this is the case it is crucial to understand that the identity of intentional objects and events described as such is partially a function of the propositional attitudes which they embody. Another way of putting this is that intentional objects are what they are partially in virtue of their content, i.e., in virtue of the ideas they instantiate. Thus, for example, an arm extended from the window of a turning car is an act of signaling partly because of the beliefs and desires which it expresses. This means that the very identity of human actions, institutions, and psychological states is partially determined by the conceptual distinctions on which they rest. We might call these conceptual distinctions the "constitutive meanings" of an intentional object.

As an example of this from social science, take the nature of the political realm. What politics is in a given social setting (i.e., what the nature and function of government is; what political power is; what political relationships consist of; and so on) depends to a large extent upon the ideas which the actors themselves hold, at least implicitly. Thus, the nature of political behavior can be a profoundly different thing depending on whether one is referring to the political activity in an African tribe, or the ancient *polis*, or Elizabethan England, or twentieth-century America; and the reason for this is that the political realm in each of these societies is rooted in fundamentally different constitutive meanings. This is, of course, a fact well recognized by social scientists; and a book like Samuel Beer's *Modern British Politics* is an exceptionally good one in showing how the periods in the political life of modern Britain are in some sense discontinuous partially because each of them has been structured around different sets of beliefs about the nature of the political.

The situation is similar for the mental phenomena which figure in the explanations of social behavior. Mental states and events are representational states and, as such, are what they are by virtue of what they are about, i.e., their contents. Thus, one belief is distinguished from another by specifying the content of each belief—the belief that it is cold outside is distinguished from the belief that it is warm outside by indicating *what* a person is committed to by virtue of having either of these beliefs. And the same is true for desires, motives, perceptions, and the like.

Now, an extremely important fact is that the self-understandings which constitute social and psychological objects and events are inherently historical because they are subject to the constant change resulting from the various conceptual innovations which a group's members introduce and come to accept.[21] These conceptual innovations assume a bewildering number of types and forms, and they may arise from any number of sources both internal and external to a given social group. (Of course it is true that rates of conceptual change may vary widely, that there are some kinds of societies which are more resistant to such changes than others, and that there are even some that institutionalize means by which such changes can be prevented. But *all* societies, even the most closed and isolated, are subject to the developments of thought occasioned by changes in what must be done to survive and prosper, by the shifting of relationships within the social whole, by contact with foreign groups, and by the widespread tendency of humans to ask further and further questions about their world.)

Thus, to return to our example of the political realm, starting in the seventeenth century in Western Europe, people gradually came to understand themselves and others as the possessors of individual rights. Now, this new self-understanding marks the occurrence of a real conceptual innovation that not only changed the way people

—both theorists and laymen alike—talked *about* their political relationships and institutions, it also altered the very nature *of* these relationships and institutions as well. For example, governments had limits placed on their activity which they never had before, and they had duties to perform—such as defending the civil liberties of their citizens—that were essentially new. Such changes are themselves only moments in a continually evolving historical process in which the ideas that form the social space of people's lives alter and shift and combine in novel ways.

In fact, the kinds of historical changes that I am discussing are not merely accidental ones in human life (in the way in which a change in the average height of humans is accidental). For human beings are self-educable creatures capable of transforming the social and natural settings in which they live, and themselves in the process. That is, it is one of the distinguishing features of humans that they reflect on their experience and, within a certain range, alter the forms of this experience as a result of this reflection. Human life is essentially historical, not because changes in how it is lived have occurred, but because parts of these changes have been authored by the participants themselves in this historical process. I shall return to this point at the end of the chapter, for it will serve as the foundation of my remarks on the critical theoretical character of social science.

The crucial point in all of this for my purposes lies in the pragmatic epistemic unpredictability of these sorts of conceptual innovations.[22] In a very well-known argument, Maurice Cranston has shown that it is logically impossible for anyone to make predictions about the occurrence of conceptual innovations,[23] and though this is *not* the argument I wish to make here (for reasons which will become clear in a moment), a brief examination of it will be useful in order to bring out what would be involved in making predictions about the future course of human beliefs.

Stated simply, Cranston's argument is that in order to predict an invention (whether conceptual or mechanical is immaterial) one would already have to be in possession of it; but if this is the case, then one could not be said to predict its novel appearance at some time in the future. The example he employs is the corkscrew, which he supposes to have been invented in 1650. In order to be able to predict at $1650 - n$ that in 1650 the corkscrew would be invented, the predictor would have to know at $1650 - n$ what a corkscrew is; but if this were the case, then he himself would be the inventor of this gadget, not the poor fellow in 1650. Trying to predict the discovery of an invention puts the predictor in the self-contradictory position of predicting at time $t - n$ the invention at t of a device that he himself had known of at $t - n$!

The same situation would exist in the case of a predictor trying to forecast in 1890 that Albert Einstein would originate the Special Theory of Relativity in 1905. For in order to be able to do so, the predictor would have to know at least roughly the contents of the Special Theory, and this means that Einstein could not have been the discoverer of the theory. Once again, the predictor would be in the logical bind of predicting the creation in the future of something already in existence.

However, while this argument makes a clever logical point, it really is of limited interest. For there is nothing in the argument which makes it a logical impossibility for someone to predict that at a future date a particular object will be fashioned or a particular theory will be formulated. All that Cranston's argument rules out is the possibility of predicting the discovery of a *novel* or *original* theory or invention; it is against he possibility of predicting an event described as the *first* of its kind that his argument is telling. Even if a predictor knew the Special Theory of Relativity in 1890, for all of Cranston's argument there would be nothing contradictory in his predicting that Einstein would produce such a theory in 1905.

Nevertheless, the main thrust of this argument is useful because it demonstrates what would be involved in predicting the future course of human thought. For what the argument does reveal is that in order to accomplish such a task one would have to be able oneself to make all the creative leaps that will recur later at some specified time. In order to have predicted in 1850 that the General Theory of Employment, Interest and Money would be proposed by Keynes in 1935, a scientist would have already had to have formulated the rough outlines of the theory himself.

And while there is nothing about such an event which make it a priori impossible, from a pragmatic point of view such a Herculean effort is extremely unlikely. This is especially so for innovations which themselves depend on a whole range of other innovations, or for innovations which involve fundamental alterations in the basic theories and principles which underlie the broad mass of our knowledge. To predict in 1600 the emergence of Marx's social theory would require that the predictor be in possession of concepts which themselves depended on developments in philosophy ("dialectic," "alienation"), economics ("capital," "commodity"), sociology ("civil society," "bureaucracy"), and a whole range of other areas of thought, and so it would be necessary for him to elaborate these conceptual distinctions and intellectual strategies in order to predict for any distance into the future a social theory of the complexity of Marx's.

Moreover, the unlikeliness of such a situation is not based solely on the limits of the human mind; there is another reason such a situation is almost unthinkable. It is that, as Popper has forcefully pointed out,[24] successful prediction is only possible when one is dealing with a closed system, that is, with a system which is protected from external influences that would tend to upset the regular interaction of members in the system. All scientific predictions taken the form, "if *C*, then *E*, in situation *X*, ceteris paribus; but *C* in situation *X*, ceteris paribus; therefore *E*"; and they are applicable to real situations only when the ceteris paribus clause has been satisfied, which is to say, when no untoward event occurs to upset the relationship between *C* and *E*.

But the collection of individuals who comprise a given social group (say, all atomic physicists, or all the members of the Ndembu tribe) is a most unlikely candidate to be a closed system. The possible influences on the minds of people are practically innumerable, and the amount and intensity of interaction between such collections so great, that the idea that a human group might be isolated enough so that a scientific prediction about its conceptual developments might be forthcoming sounds like a mad millenarian dream.

The difficulties involved in predicting conceptual developments are enormous; indeed, from a practical standpoint, it may be taken as a given that such predictions are so unlikely as to be almost certainly not realizable. And this means that even if it were the case that the course of human thought is law governed in some fully deterministic way there is a *pragmatic* epistemic unpredictability about the development of human thought.

Moreover, it should be clear from what I said about the constitutive role that concepts play in human actions and institutions that this unpredictability is not confined to the history of human *thought*; human actions and institutions are also unpredictable insofar as they change as a result of people's coming to think of themselves, each other, and the natural world in novel ways. Thus, as the lessons of Keynesian theory came to be appreciated first by the leaders of government and industry, and then in an attenuated form by the populace at large, new demands on the government were made and were seen to be justifiable, and a whole new class of laws, regulations, institutions, and practices emerged. As a result, the very nature of the relationship between the government and its citizens changed.

The conclusion to be drawn from all of this is that the objects of social science are open-ended in a practically unpredictable way. Social institutions and practices, as well as the beliefs and desires of the members of particular social groups, are continually in a state of flux and evolution which will always appear to be indeterminate to those who wish to study them. To understand what this actually means, it might be useful to draw an analogy suggested by Alasdair MacIntyre[25] to some imagined geologist's attempting to study rocks which changed their shapes, sizes, colors, and chemical compositions in a manner which (even though lawful) always eluded his predictive capacity. Retrospectively he would be able to understand why it was that a class of rocks assumed the form that it did; but prospectively he would be unable to know what form it will take: The objects of his research would be constantly changing in surprising ways. Now, this imagined situation of the geologist is like the real situation of the social scientist interested in explaining intentional behavior, just because all such behavior is what it is by virtue of its place in a social community, and because the life history of social communities is constantly changing in unpredictable ways.

What does all of this have to do with the heteronomic character of the generalizations in a science of action? Just this. In order to frame general laws properly so-called, it is necessary that one use concepts which refer to objects which are in some sort of steady state, or which change in some regular way which is apprehensible. The reason for this is that general laws are universal well-confirmed empirical hypotheses which state that under situations *X*, if *a* then *b*, ceteris paribus; but if there is no way of describing an object or event so that it can be seen to be part of a regularly recurring sequence of events, then there is no way that one can know whether one's hypotheses are either universal or well confirmed enough in order to be accorded the title of "general law."

It is as if the objects in the world will not stay still enough, or evolve in a predictable enough fashion, so that one can pick them out as part of a genuine natural pattern. To return to the example of the geologist and his rocks, if the objects to which his terms "rocks of type *a*" and "rocks of type *b*" refer are forever shifting in unpredictable ways, there is no way that the geologist could frame a general law using the concepts "rocks of type *a*" and "rocks of type *b*"; this is because there would simply be no way to know whether the purported general law which resulted by using such concepts held or not, and, if it did, over what range of phenomena. In other words, the basic problem about general laws in the social sciences of action is one of confirmation: In order to have general laws one must be able to predict outcomes as the result of the presence of a certain factor; but it is extremely unlikely that such predictions of social and psychological phenomena will occur, just because the concepts which partially make these phenomena what they are are subject to unpredictable change, and so the phenomena themselves are unlikely to be enough like the original to provide confirmatory instances.

Take, for example, the hypothetical case of a social scientist in 1800 who is trying to frame a general law about the political life of tribal societies, and yet who does not possess—and could not possibly possess, given the limits of his ability to predict conceptual changes—the concepts of "imperialism" and "socialism." No generalization that he formulates using the terms he has available to him could ever become a general law properly so-called because the very nature of tribal politics in the nineteenth and twentieth centuries would be so deeply altered by the development and spread of these notions: His subject matter would change on him, and so his carefully wrought generalizations could never be tested and applied over a wide variety of situations so that they could become more than generalizations.

This hypothetical social scientist is in the same situation as was Marx in trying to formulate the iron laws of capitalist economy—say, the inevitable and immense pauperization of the working class. Such an attempt was doomed to fail just because capitalist social systems are constantly evolving: For instance, Marx would never have known of the theoretical innovations that would be made by Keynes, innovations which would irrevocably alter the set of institutions, practices, and beliefs about which Marx was trying to theorize. Such unpredictable change inevitably made Marx's putative general law just that, a *putative* general law.

Nor can the social scientist avoid this situation by arguing that his generalizations can be transformed into lawlike statements by recognizing their inherently statistical character. For the statistical generalizations found in the social sciences are not likely to become genuine statistical laws for exactly the same reason that causal generalizations in social science are not, namely, because to become so they must be confirmed in a wide variety of instances, and such confirmation is subject to the same sorts of difficulty I have been discussing. (Actually, the logic of confirmation for statistical laws is exactly the same as that for causal laws, except that in the latter case one is concerned with the occurrence of individual events, whereas in the former one is concerned with the occurrence of sets of events.)

It should be noted before proceeding that there is nothing in what I have said which would be incompatible with the social world being as deterministic as one pleases (deterministic in the sense of being fully law governed)—or being indeterministic, for that matter. All that is required in order for my argument to work is that social phenomena be unpredictable because they are constituted by the self-understandings of the relevant actors, self-understandings which themselves change in unforetellable ways.

Nor should it be concluded from this that no general statements at all are possible in the social sciences of action. This would be the case if the sorts of changes I have been discussing occurred extremely fast and appeared to be totally random (such a situation would be analogous to our geologist's confronting a world in which the rocks in it changed in irregular ways every month or so). But fortunately this is not the situation in which we find ourselves. The structure of the social world is *relatively* stable, and its changes are usually confined to some roughly definable area; indeed, if this were not the case, it would completely undermine the possibility of sustained social interaction, and hence the possibility of there being some sort of genuine social order at all. There is a kind of regularity and continuity which must be present if there is to be social life; and it is on these facts that the generalizations that we do find in the social sciences rest.

Thus, for example, the sociology of knowledge tries to provide us with a general understanding of the relationship between certain sorts of social structures and certain sorts of belief systems. Moreover, it also tries to provide some understanding of the sorts of intellectual innovations that are likely to find acceptance in societies of a certain description. In so doing, it gives us some understanding of the range of ideas that are likely to gain a foothold in a social order and thereby alter it. And on the basis of such knowledge, one can make certain genuinely testable generalizations about the rate and kind of change in various social systems.

The question is not, therefore, whether generalizations are possible—in the first place, they must be if there is going to be a social order at all, and, in the second, social scientists have already given us a whole slew of them. The question instead is whether we can expect these generalizations to be purified and rigorously stated so that they may thereby become genuine general laws. To this question, because of the constitutive role of concepts and beliefs in making human social life what it is, and because of the practical unpredictability of the development of these concepts and beliefs, we have every reason to believe that the answer will be "no." For these generalizations refer to what I hope to have shown to be epistemically anomalous phenomena, and such phenomena are not the sort requisite for formulations of a genuinely nomological character.

IV

My discussion about the possibility of general statements in social science leads directly into a discussion of the nature and role of causal theories in explaining actions, assuming it is the case that the causal generalizations in social science are indeed heteronomic. It is necessary to include this discussion because only then will it be clear what we can expect to be the nature of those sciences that try to give systematic causal explanations of human behavior. I want to conclude by emphasizing that although I believe that general laws properly so-called will not emerge from the social sciences—and in this respect they differ from the natural sciences—I do not think it follows from this that a genuinely theoretical science of human behavior is impossible.[26] On the contrary, I believe that until the essentially theoretical character of social scientific explanation is appreciated, no account of the nature of the sciences of behavior will be adequate.

Toward the end of section II, I mentioned that all causal explanations require the existence of causal generalizations; this is true in the natural as well as the social sciences. The reason for this is that in order to justifiably claim that a particular (kind of) event causes another (kind of) event, one must have good reason to believe that the two events are not merely accidentally related, or even the joint outcomes of a third unknown (kind of) event. That is, one must have evidence that in certain circumstances the first (sort of) event actually is a necessary and/or sufficient condition for the other (sort).

The evidence called for here is of two kinds. In the first place, because the explanation rests, at least implicitly, on a general claim (that, ceteris paribus, the first [type of] event is enough to produce the other, and/or that the second [type] cannot occur without the first having occurred), the relevant evidence will involve generalizations which report other instances in which the two events are conjoined. In this regard, the more *unlike* the circumstances in which the relation is observed, the stronger the evidence that it is indeed the (type of) event in question, and not some other one in the environment, that is the cause.

But—and this leads to the second sort of evidence—generalizations of covariance alone cannot provide enough weight to support an imputation of a causal relationship. The two events might be causally unrelated to one another, and yet if they were the common effect of another, but unknown event, one would still have a generalization of covariance. This shows that something stronger than this sort of generalization is required. What is needed besides is a generalization which might be said to explain its instances, in the sense that implicit in it is an account of *why* the relationship between the events is indeed one of necessary and/or sufficient conditionship. It is only when one is in possession of this sort of generalization that one can with any degree of confidence make the contrary-to-fact and subjunctive conditionals that one must be able to make in order to claim that in certain circumstances one event is a sufficient and/or necessary condition of the other.

It is precisely at this point that theories are required, for it is from theories that such an account derives. Theories provide a systematic explanation of a diverse set of phenomena by showing that the events in question all result from the operation of a few basic principles. A theory goes beyond generalizations by showing why the generalizations hold, and it does this by specifying the basic entities which constitute the phenomena to be explained, and their modes of interaction, on the basis of which the observed generalizations can be inferred. One might say—with an acknowledgment to the realist theory of science, which I discussed earlier—that theories provide us with knowledge of the causal mechanisms that relate the events in question, and that are the means by which one event brings about the other. It is knowledge of these mechanisms (understood in the broadest sense, such that the having of motives or beliefs might figure in such mechanisms) that supplements mere empirical generalizations and enables them to be evidence that a genuinely causal relationship is involved. It is thus that we are inevitably led from the desire to explain causally a particular (sort of) occurrence to the need for theories.

Of course, as science develops, these theories become much more rigorous and explicit. Moreover, it will naturally attempt to organize and structure the various causal explanations and causal generalizations upon which they rest by systematically interrelating them, and by subjecting them to experimental and other empirical verification. In this process the self-conscious development of "large-scale" theories is absolutely essential.

It is a very odd fact that the most sophisticated statements in the analytic tradition about the nature of social science have consistently failed to mention its essentially theoretical character[27]—a fact which has made much in this tradition appear irrelevant to many of its practitioners. I say an odd fact because even a cursory glance reveals the kind of theorizing I am discussing. In neoclassical economic theory, in structural-functional theory in anthropology, in exchange theory in sociology, in structuralist theories of cognitive development in psychology—in these and countless other cases the fact that causal explanations require the development and articulation of large-scale theories is evident.

Of course, such theories have never lived up to the aspirations of Hobbes or Comte; that is, they have never approached the universal scope and precision of theories in the natural sciences. And if what I have argued in this chapter is correct they never will. The causal generalizations that figure in the theories of the social sciences are heteronomic (principally because their objects, human behavior and society, are intentional and historical entities), and consequently these theories are limited. These limits manifest themselves in a number of ways: in the precision of the terminology in these sciences; in the sharpness with which a scientist will be able to specify the conditions in which his theories hold; in the range of application in both space and time of such theories; and in their predictive power, and therefore their testability.

Moreover, there is one last limit on the causal theories of social science which deserves particular mention because of its bearing on the question of the sort of theory that social and psychological phenomena call for. The limit I have in mind is that many of the causal generalizations in a social theory will be restricted to a given cultural context. The reason for this limitation is the constitutive role beliefs play in human life. If, as I have argued, social and psychological phenomena are what they are, and therefore have the causal relations they have, partly because of the beliefs of the actors involved, then these actors coming to have a radically different set of beliefs will likely mean a whole new set of relations among them, and this will consequently require a different set of causal generalizations to explain their social and psychological experience. Thus the causal theories a social scientist develops are likely to be more or less confined to particular cultures or types of culture.

This is an extremely important point, and not just for theoretical reasons alone. For if one interprets the causal generalizations produced by social science as if they were general laws which applied over a whole range of cultures, or even if one thinks that social science is capable of producing such laws, this may have the terribly unfortunate political repercussion of stifling political change. Let me explain, and in the process reveal why I think the theories in social science ought sometimes to be what has been called "critical."[28]

The causal generalizations in social science are about essentially conventional activity just because that activity is partly constituted by the beliefs of those involved. However, if one takes these generalizations to be actual or possible laws, one may be unwittingly reifying the particular conventions one is observing, i.e., treating them as if they were nature-like necessities such that the particular way a group of people interacts is taken to be the way it *must* interact. The reason why a commitment to general laws properly so-called inevitably leads to this reification is the generality involved in such laws: If the generalizations one discovers are indeed (potential) general laws, then what might at first appear to be a local or idiosyncratic practice must be seen as an instance of something that is in the nature of things, and thus as something not alterable.

However, from the perspective of this chapter, in which I have argued that causal theories in social science are limited in scope to a particular culture, or perhaps to particular sorts of culture, this reification is a form of ideological distortion. For in such cases the social scientist is illicitly transforming the generalizations which account for one particular way of doing things into purported general laws which supposedly govern human life as such. The effect of this concealed ideological transformation can be particularly oppressive, for it can reinforce the social actors' acceptance of a status quo which may be deeply frustrating to them. It can do this both by giving them reason to believe that their social life must be as it is, and by failing to provide them with an analysis of their situation which might help them to change it (and so falsify the causal generalizations which now characterize their behavior!).

I have claimed that ultimately the nature of social-scientific theory (with its heteronomic generalizations, limitations in scope and specificity, and restriction to particular [sorts of] cultural setting) is shaped by the essential historicity of the objects it seeks to explain. And earlier in the chapter I suggested that this historicity is itself rooted in the capacity for self-transformation characteristic of human beings. This suggests that any conception of social science which fails to take historicity into proper account will be defective. It will be theoretically defective because it will fail to appreciate the special character of causal theory in social science, and it will be practically defective in the oppression that it can cause in the way I have just indicated.

It is on just this feature of historicity that critical theory focuses. According to it, social-scientific theories not only must self-consciously recognize that they are limited because they are about creatures capable of self-transformation, but they must make this feature play an essential role in their construction. That is, critical theory insists that social science ought to be a means by which such transformation is fostered.

How can social theory do this? By assuming a particular form, namely, one that isolates in the lives of a group of people those causal conditions that depend for their power on the ignorance of those people as to the nature of their collective existence, and that are frustrating them. The intention here is to enlighten this group of people about these causal conditions and the ways in which they are oppressive, so that, *being enlightened, these people might change these conditions and so transform their lives* (and, coincidentally, transcend the original theory). Examples of critical theory are Marx's theory of capitalism and Freud's theory of neurosis.

A critical metatheoretical understanding of social science grows quite naturally out of the account of explanation, cause, action, law, generalization, and theory that I have given in this chapter (although obviously it is not entailed by my account). The reason why it does is that both the heteronomic character of causal generalizations in social science and the idea of a critical social theory derive from the same special feature of human beings, namely, what I have called their historicity. It is because humans learn about themselves and their world that they are instrumental in transforming themselves and their relations, thereby defeating the causal generalizations which a social scientist might have used to describe their lives. This is why such generalizations are heteronomic. But it is this very same capacity to be enlightened by these theories about the world and to alter their social arrangements partly on the basis of this enlightenment that makes humans fit subjects for a critical social science.

Moreover, while to a theory of social science that seeks to model it on the natural sciences the heteronomic character of social scientific causal generalizations is a pronounced liability, to a critical social scientist this heteronomy is a virtue. Heteronomy is a virtue for critical theory because it means that humans are capable of self-reflection and self-transformation, and it is just these that a critical social science is meant to foster. Indeed, a critical social scientist actually desires to see his causal generalizations made otiose by a group of actors who, having learned them, alter the way they live. He desires this because it means that he has been successful as a theorist in helping to alter the social world which he is studying.

A consideration of the nature of causal generalizations in explaining human behavior has led me into a critical theoretic conception of social theory. This, it seems to me, is no accident. In the first place, such a conception is based on the belief that social science must be theoretical, but also on the self-conscious recognition of the heteronomic character of social scientific theories. Moreover, such a conception sees humans as natural creatures in a natural world of cause and effect, and thus as fit subjects for science; but it also sees humans as capable of a kind of initiative which distinguishes them from other natural creatures, and thus it argues that social science must be of a novel form. It is critical theory which understands the basis for the heteronomy of social science (namely, the historicity of human beings) and which builds this understanding into its account of social scientific theory.

Notes

1. See A. R. Louch, *Explanation and Human Action* (Berkeley: University of California Press, 1966), passim; R. Peters, *The Concept of Motivation* (New York: Humanities Press, 1958), ch. 1; D. Hamlyn, "Behavior," in *Philosophy*, 28 (1953); H. L. A. Hart and A. M. Honore, *Causation and the Law* (Oxford: Oxford University Press, 1959), esp. pp. 48–55; and William Dray, *Laws and Explanation in History* (Oxford: Clarendon Press, 1957), ch. 5.
2. Hart and Honore, p. 21.
3. This is the central thesis of Louch's book, for example.
4. Dray, ch. 5, part 4. Also cf. Hart and Honore, p. 52.
5. In *Philosophy, Politics and Society* (Second Series), ed. Peter Laslett and W. G. Runciman (Oxford: Blackwell, 1962), pp. 48–70.
6. Ibid., p. 55.
7. I have defended this distinction in detail in my "Practical Reasoning, Rationality, and the Explanation of Action," *Journal for the Theory of Social Behavior*, 8 (1978), pp. 77–101.
8. The Realist position has been developed by Rom Harre. See, for example, Harre and Peter Secord, *The Explanation of Social Behavior* (Oxford: Blackwell, 1972). See also Russell Keat and John Urrey, *Social Theory as Science* (London: Routledge and Kegan Paul, 1975).
9. I learned a great deal on this matter from David Miller of Warwick University. See Miller's review of Rom Harre's *The Principles of Scientific Thinking* (London: Macmillan, 1970), entitled "Back to Aristotle?" in the *British Journal of the Philosophy of Science*, 23 (1972), pp. 69–78.
10. This is just the sort of debate that developed over Weber's thesis. See, for example, the essays by Robertson, Samuelson, and Hansen, in *Protestantism, Capitalism, and Social Science*, ed. Robert Green (Lexington, Mass.: Heath, 1973).
11. *The Explanation of Behavior* (London: Routledge and Kegan Paul, 1964), pp. 21–25. There Taylor claims that the fact "that the system achieves this (normal) result-condition neither calls for nor admits of explanation" (p. 22).
12. Ibid., p. 24.
13. In the case of practical reasoning processes, I take it that this is just what developmental psychologists do; for example, see Jean Piaget, *The Origins of Intelligence in Children* (New York: Norton, 1963), particularly Part II.
14. Of course, in order for this to be so one must be able to distinguish between *causal generalizations* and *causal laws* properly so-called. For this chapter, general laws properly so-called are universal well-confirmed empirical hypotheses of conditional form capable of supporting counterfactuals. They state that under certain specified boundary conditions every case in which events of the type *N* occur an event of the type *E* will occur, ceteris paribus. Causal generalizations, on the other hand, although of essentially the same form as that of general laws, are much more rough-and-ready: they are not universal or well confirmed; their boundary conditions are not well articulated; and their capacity to support counterfactuals is limited to a quantifiably unspecified range of events.
15. This part is a direct borrowing from Donald Davidson's "Causal Relations" in the *Journal of Philosophy*, LXIV (Novermber 9, 1967), pp. 691–703.
16. Hart and Honore, pp. 13–14.
17. Reprinted in Carl G. Hempel, *Aspects of Scientific Explanation* (New York: Free Press, 1965), p. 232 and p. 236 (italics mine).
18. In the first half of his article, "Theory in History" (*Philosophy of Science*, 34 (1967), pp. 23–40), Leon Goldstein gives other instances of this doctrine to be found in the work of the philosophers Patrick Gardner (in *The Nature of Historical Explanation*, pp. 57–89), Ernest Nagel (in "Determination in History," *Philosophy and Phenomenological Research*, 20 (1967), p. 307), and Karl Popper (in *The Open Society*, p. 448).
19. See, for example, Louch, chs. 1, 2.

20. See Donald Davidson, "Mental Events," reprinted in L. Foster and J. Swanson, *Experience and Theory* (Amherst: University of Massachusetts Press, 1970), pp. 79–101. See also William Alston, "Do Actions Have Causes?" in the *Proceedings of the Seventh Inter-American Congress of Philosophy* (Quebec: Laval university Press, 1970), pp. 256–276.
21. One might be tempted to interpret this as an excessively idealist account of social life because it seems to make ideas the crucial factor in social change. Such an interpretation would be a mistake, however. As "materialist" a theory of social change as one likes is compatible with what I say here, provided that this theory includes the assertion that social practices and psychological states are partly constituted by the self-understandings of the actors involved. (Marxism is such a materialist theory, for instance.)
22. The argument in this section was suggested by Karl R. Popper, *The Poverty of Historicism* (London: Routledge and Kegan Paul, 1957); and by Louis Mink, "Philosophical Analysis and Historical Understanding," *Review of Metaphysics*, XXI (June 1968), pp. 667–698.
23. Maurice Cranston, *Freedom* (London: Longmans, 1953), p. 118.
24. In *The Poverty of Historicism*, passim.
25. See Alasdair MacIntyre, "Predictability and Explanation in Social Science," *Philosophic Exchange*, 1 (Summer 1972), pp. 5–13. In chapter 8 of *After Virtue* (Notre Dame: University of Notre Dame Press, 1981) MacIntyre develops an argument quite like the one I am making in this chapter.
26. This is in contrast to the views of MacIntyre, for example, who, after having shown that the particulars characteristically studied by social scientists are not predictable, claims to have shown thereby "that the aspiration to construct theories of scientific or a quasi-scientific sort in this area *must* fail" ("Predictability and Explanation in the Social Sciences," *Philosophic Exchange*, 1 (Summer 1972), p. 12). He makes the same claim in "Ideology, Social Science, and Revolution," *Comparative Politics*, 5 (April 1973), p. 336.
27. For example: Georg Henrik von Wright, *Explanation and Understanding* (Ithaca: Cornell University Press, 1972); A. R. Louch; Peter Winch, *The Idea of a Social Science* (London: Routledge and Kegan Paul, 1958).
28. See Jürgen Habermas, *Knowledge and Human Interests*, trans. Jeremy J. Shapiro (Boston: Beacon Press, 1972), chs. 10–12, and *Theory and Practice*, trans. John Viertel (Boston: Beacon Press, 1973), essay 7; and Brian Fay, *Social Theory and Political Practice* (London: Allen and Unwin, 1976), esp. ch. 5.

Chapter 8

Defending Laws in the Social Sciences

Harold Kincaid

Physics has long been the paradigm of a science—both to scientists and to philosophers. Quantitative by nature, universal in scope, and apparently indefinitely refinable, physics has set the standard for scientific practice. Measured by that paradigm, the social sciences have generally fared poorly. Scientists (of the hard variety) and philosophers alike have doubted that the social sciences have produced any real explanations, with the prime complaint being that social scientists have failed to produce any well-confirmed laws. Taking this alleged fact at face value, philosophers have rushed to explain why the social sciences are doomed to inexactitude. However, they have not proceeded by carefully examining the empirical work in the social sciences and arguing that it fails to meet standards of scientific adequacy. Instead, philosophers have given various more or less a priori reasons to explain why social laws were impossible all along.

Physics worship was part and parcel of the positivist tradition. No doubt physics occupies a special place among the sciences. But the positivists in part misunderstood what science was about—and what was special about physics. If physicalism is true, then physics is indeed unique among the sciences—but there is little reason to think physics shows there is a logic of confirmation, a single scientific method, a sharp distinction between laws and accidental generalizations, or that all adequate sciences are ultimately reducible to physics. Nonetheless, while few adhere to the positivist account of science, such positivist assumptions still distort how philosophers think about the social sciences. Laws in the social sciences are rejected for reasons that cannot stand up to any sophisticated account of science—or so I shall argue.

Thus my main concern in this chapter is to defend laws—both their possibility and their reality—in the social sciences. No doubt social explanation is not exhausted by subsumption under laws—just as explanation in general involves both more and less than laws.[1] Nonetheless, laws do play an important role in explanation. Singular causal explanations, functional explanations, explanations via classification and grouping (explaining "what"), and explanations in terms of underlying mechanisms may well presuppose or implicitly invoke laws. And even if one denies—contra what seem to me powerful arguments[2]—that laws are involved in or entailed by such explanations, there can be little doubt that laws can greatly strengthen these explanations. Showing that an alleged particular cause instantiates a law, that some mechanism always underlies some macroprocess, that some set of classifications are related in a lawlike fashion, and so on, gives those explanations a depth they would not have otherwise. Furthermore, no explanation is completely compelling unless the explanatory statements are well confirmed; confirmation generally comes from repeatable manipulation of data under controlled conditions, and laws both result from and help make possible that process. Laws, explanation, and confirmation go hand in hand. Thus if one is inclined to believe, as I am, that some parts of the social sciences *explain*, then it is natural and perhaps necessary to defend social laws.

Section I of this chapter defends social laws against three common criticisms: that social laws are impossible because (1) social kinds have multiple physical realizations, (2) social theories are not "closed," and (3) there can be no laws relating belief, desire, and action. Another popular argument rejects social laws because they are often about social entities rather than individuals and/or are teleological in nature. Section II spells out and rejects these criticisms. More serious problems for social laws result from the fact that those laws are usually qualified ceteris paribus. Section III argues that such qualifications are not an inherent problem and spells out some criteria for determining when ceteris paribus laws are confirmed and explanatory. Finally, section IV makes a prima facie case for the claim that some current social science has produced laws by (1) examining some well-established work in economics and (2) arguing that at least some parts of the social sciences produce laws, evidence, and explanations quite similar in form to those of evolutionary biology and ecology.[3]

I

Philosophers have had no trouble finding reasons to deny that social laws are possible. Some think that social scientists cannot produce laws because the basic kinds or predicates of the social sciences have no determinate physical definition—they are "multiply realizable." Others think that the social sciences will never produce real laws because they are not "closed"—the social sciences, unlike physics, do not describe all the causes operative in their domain. Finally, some philosophers have argued that all social laws depend on explaining human behavior in terms of belief and desire. Such folk psychological explanations, however, cannot give us laws of behavior. So, social laws are impossible. In this section, I sketch these arguments and show that all are unsound.

John Searle recently argued against social laws on the grounds that social kinds have multiple realizations; his argument is a variant of earlier attacks on psychological laws. According to Searle," the defining principle of ... social phenomena set no physical limits on what can count as the physical realization."[4] "Money," for example, is social in nature—its definition refers to its social function, not its physical attributes. As a result, nearly anything can serve as money. Most or all social kinds are in the same boat: They have indefinitely many diverse, physical realizations. How does this fact argue against social laws? It "means that there can't be any systematic connections between the physical and the social ... [and] therefore there can't be any matching ... of the sort that would be necessary to make strict laws of the social sciences possible."[5] Similar arguments have been offered by Churchland and Rosenberg to show that folk psychology is bereft of laws.[6]

Let us tighten up Searle's argument a bit. He seems to be saying something like this:

1. Social kinds have indefinitely many physical realizations.
2. When a kind has indefinitely many physical realizations, it has no systematic connection to the physical.
3. If a kind is not systematically connected to the physical, it cannot support genuine laws.
4. Thus social kinds cannot support genuine laws.

This premise seems to be Searle's argument. How does it fare?

This argument is, so far as I can see, either invalid or unsound, depending on how we read "systematic connection." If we take systematic connection to refer to a lawlike relation between social and physical predicates, then premise 3 makes a highly implausible claim: that the social sciences cannot produce laws unless they are reducible to physics. That claim is implausible because, among other things, it threatens to rule out laws in the natural sciences as well. For example, the fundamental predicate in population genetics and evolutionary theory—that is, fitness—seems clearly to have no determinate physical definition. Fitness supervenes on physical properties but is not reducible to them[7] Similarly, program states of computers can be brought about by indefinitely many machine states—and thus are at most supervenient on them. Even molecular biology faces such problems: Essential predicates like "antibody," "signals," "receptors," and so on all have open-ended physical realizations.[8] So, if we read "systematic connection" as type-type connectability, then lawlike statements in biology, computer science, and elsewhere are ruled out. I assume this result is unacceptable.

We can, of course, read systematic connection as the looser relation described by supervenience (which means in this case, "once all the physical facts are set, so are the social facts"). But then premise 2 is clearly false—for supervenience does not rule out multiple realizations. Thus this argument either has a false premise—premise 2 or 3 depending on how you read "systematic connection"—or is invalid. In the end, Searle makes an unwarranted reductionist assumption—that no physical definition means no definition at all.

A second common argument against laws in the social sciences turns on the fact that the social realm is not "closed."[9] Laws by nature must be universal. But if the social constitutes an open realm subject to outside forces—physical or biological events, for example—then social theory will remain forever incomplete and forever without true laws. Laws seem precluded.

This argument can be taken in two ways: as arguing (1) from a fact about social systems as objects in the world, or (2) from an alleged fact about social theories: they are not closed in the sense that they do not cover all the forces or causes effective in their domain. Neither rendering produces a sound argument.

Social systems—a given institution, society, or even world system—are obviously not closed systems. They depend in important ways on both physical and biological factors. This fact alone cannot entail that social laws are impossible—unless we are prepared to grant that even physics produces no laws. Every physical system short of the entire universe is influenced by outside causes. So merely describing open systems cannot preclude laws. If it does, then the only laws of physics are those that describe the totality of the universe, an unacceptable conclusion, I assume.

Perhaps the crucial issue concerns not the open or closed nature of actual systems but rather a theory's ability to handle those outside factors. A closed theory is complete: It can describe and explain in its own terms all the forces acting in its domain. So, the argument runs, forces affecting open physical systems can be fully handled within physics itself. In the social sciences, however, outside factors are not social in nature—and thus cannot be handled by social theory. Consequently, alleged social laws are bound to be incomplete and thus not laws.

This argument fails for several reasons. On some reasonable assumptions, this argument proves too much—that is, that no physical laws are possible, either. We know that biological, psychological, and social events influence the physical universe. If, however, our biological, psychological, and social theories are even in part irreducible, then there is little prospect that exceptions to physical laws can be handled in physical terms.

Biological factors, for example, will interfere in physical processes, thus creating apparent exceptions to physical laws. Although such factors are in the end composed of physical entities, that does not solve the problem. For merely listing all known biological factors under their particular physical description will not show physical laws completely refinable in their own terms. Assuming that biological kinds are not captured in physics—and this is what irreducibility means—then physics will have no systematic way to identify and incorporate biological factors. Physics will be in a certain sense incomplete as well—and thus without laws. I again assume that this conclusion is unacceptable.

The open nature of social science carries little weight for another reason. Why does a real or strict law have to invoke language only from one theory? The above argument assumes that it must, but that assumption seems quite unwarranted. Cellular biology invokes chemical facts, and evolutionary theory does the same with physical facts about the environment. Why should that undercut laws if the law identifies kinds and relates them in whatever manner laws require? This problem becomes particularly acute when we realize that individual laws in physics may invoke only a subset of the total physical language. When other *physical* forces interfere, that subset obviously will not have the vocabulary to handle this more complicated situation. Are those laws only apparent because they are not refinable in their own terms? Here it just seems silly to make lawfulness turn on some prior notion of the "right" vocabulary. The real issue is whether potential exceptions can be handled in a systematic way—not whether their descriptions fall into one vocabulary or another. Thus there might well be social laws even if they are not refinable in a purely social vocabulary.

A final argument which I want to briefly discuss in this section rests on the alleged nature of psychological explanation. Davidson argued that there can be no laws relating mental states like belief, desire, and behavior, at least when those states are described in mentalistic terms. Philosophers—for example, Rosenberg—as well as social scientists—for example, Porpora[10]—have seen Davidson's conclusion as a powerful argument against social laws (and as an explanation of why laws are so scarce in the social sciences). Since the social sciences seem wedded to belief-desire explanations, social laws seem ruled out by the nature of the mental.

While the difficulties of folk psychological concepts may partially help explain why the social sciences progress slowly, those difficulties are little reason to think that social laws in general are impossible or unlikely for several reasons:

1. Belief-desire psychological theories are not the last word in behavioral theory: Variants of behaviorism, cognitive psychology, and so on might serve to better explain human behavior while avoiding the belief-desire framework altogether—so alleged obstacles to belief-desire theories is a shaky basis for ruling on what the social sciences can and cannot do.
2. It has been persuasively argued by Horgan and Woodward,[11] among others, that attacks on the folk psychological explanation often pick on the weakest version of belief-desire psychology, that they rely on unwarranted reductionist assumptions about intertheoretic relations, and that suitably refined and improved folk concepts may be part of well-confirmed psychological explanations.

Furthermore, objections to folk psychology often turn on either the open nature of the psychological theory or their use of ceteris paribus escape clauses. The former obstacle has already been answered here; the latter problem, I shall argue in section III, is no inherent obstacle to successful explanation.

3. Most important, much social science proceeds at the macro-level, describing large-scale social behavior and structure based on aggregate data. As such, it is unaffected by the failure of specific theories of *individual* behavior.

Thus this last argument for the impossibility of social laws fares no better than its predecessors. Nonetheless, these arguments were in part motivated by a concern I have yet to address: the desire to explain the noticeable lack of progress in the social sciences. That concern will be taken up indirectly in section III and directly in section IV. Before that, however, it is necessary to defend a controversial assumption used in my previous arguments, that is, that purely macro-level social laws can be adequate.

II

Led by Jon Elster, an increasing number of social scientists are rejecting macro laws in the social sciences on grounds other than those discussed earlier. Elster is representative: He holds that macro-level laws are inadequate so long as we have not specified the underlying mechanism that makes them possible.[12] That mechanism, he holds, must be specified in terms of individual behavior. Purely macro-level laws are thus inadequate.

In what sense are such laws inadequate unless the underlying mechanisms are specified? Sometimes, critics talk as if purely social laws are simply *nonexplanatory*: Explanation is had only when the relevant mechanism is identified. Sometimes, the question seems to be whether macro-level laws without mechanisms can really be *confirmed*. Let me consider first the claim that purely macro-level laws cannot explain.

This version of the argument is prima facie quite unreasonable. On either of the two most influential accounts of explanation—the covering law model of Hempel and the question model of van Fraassen and others[13]—well-confirmed macro-level laws seem to explain. Hempel's hypothetical deductive account does not restrict explanation to laws at some given level. Events subsumed under well-confirmed laws are thereby explained. If there are confirmed laws relating macro-level variables, then they would serve to explain. Similarly, if I ask why the output of a given industry has expanded and request an answer citing known causal regularities, then social laws also explain on the van Fraassen model. Social laws answer questions and thus explain.

A more promising tack would be to question whether any purely social laws can be confirmed in the first place. Rather than making mechanisms necessary for explanation, critics might argue, as does Elster,[14] that no macro-level law is confirmed until the mechanism bringing it about is identified. Mechanisms are thus identified with intermediate causes. Successful confirmation requires that we control for spurious correlations, and we can do that only when we have identified the immediate cause—so the argument runs.

If successful, this argument has ramifications that reach far beyond a debate about social laws. Similar issues come up in public policy debates. Acid rain, for example, is thought by many to destroy forests and lakes. To fend off costly laws mandating pollution controls, the power industry and its friends (some quite respectable) argue that the connection between acid rain and biological destruction is not well established.

The reason given is that biologists have not identified the relevant mechanism. At issue is a general question of some import.

To evaluate this attack on social laws, it will be helpful to subdivide our question. Social science often appeals to teleological causes: Some social practice or institution exists because of its usefulness. Since laws of this sort may present special difficulties, I shall first consider the above argument as it applies to ordinary, nonteleological laws in the social sciences. Teleological laws will then be considered in turn.

The claim under scrutiny is that macro-level social laws are unconfirmed so long as no mechanism involving individuals is identified. This is so because no alleged causal law is confirmed until we control for all other possible causal factors. Mechanisms are intermediate causes and may show the alleged connection to be spurious, that is, a mere by-product of the true cause. Thus we must identify mechanisms before we can confirm social laws.

As an argument against *all* macro-level social laws, this reasoning fails for three reasons:

1. The argument appeals to certain alleged facts about confirming causal laws. However, *causal* laws do not exhaust explanation in the social sciences. Many suggested laws of social behavior are functional laws—they describe changes in one variable as the function of another. Of course, this procedure is not peculiar to the social sciences; it is part and parcel of explanation in physics. However, when social laws describe functional relations between variables, spurious causation is no threat since no causal claim is being made. Thus the above argument leaves many purported social laws untouched.
2. The argument under consideration assumes that the relevant mechanism is an intervening causal variable. That is not always the case. Sometimes, a law might describe how one social whole causes another. In that instance, the relevant mechanism—what makes the connection possible—is the relations between individuals that constitute the wholes in question. Take, for example, this purported law of Marxist theory: In every class society, the oppressed class overthrows the ruling class or both classes are destroyed. The causal connection between the oppressed and the ruling class would not be some intermediate causal factor between them. Rather what brings this laws about are the actions of individuals—those constituting the oppressed class. When one whole acts on another, so do its collected parts. It makes no sense to describe the latter as an *intervening* variable.
3. Causes, I assume, must begin before or simultaneous with their effects. In some cases, this fact allows us to rule out spurious causes even without knowing the precise intervening mechanism. If we can (1) identify the list of possible intermediate causes and (2) know the moment of their onset, we can avoid the problem of spurious causation. Possible intermediate causes that begin after the purported initial cause cannot be the cause of the latter—thus spurious causation is not a threat. For example, imagine that our causal law asserts that a rise in the rate of profit in an industry causes an increase in output. The standard mechanism for such a law would be an increase in new investors—an event which could occur after the rise in profits. Thus, while it may often be that intermediate causes are simultaneous with the relevant initial cause, I see no reason to think that that is always the case. Intermediate causes do not show that no social law is confirmed until we identify the precise intermediate cause.

Apparently, we do not need to identify the relevant mechanism to confirm all social laws. Nonetheless, the argument of Elster and others might hold for a significant range of cases. Many social laws are causal laws. Many involve mechanisms whose onset is for all practical purposes simultaneous with the onset of the initial cause. And many social laws are brought about by mechanisms that really are intermediate causes. Are laws in these cases unconfirmed so long as we cannot identify the relevant mechanism?

I think not. While there are, of course, notorious difficulties in analyzing the notions of causes, laws, and confirmation, in practice, we have a fairly good idea of how to make causal inferences: We look for the appropriate correlations among variables after controlling for other possible causal factors. As simple-minded as this observation is, it nonetheless shows why identifying mechanisms is not essential to confirmation. Let me explain how.

To confirm that A causes C, I must find a significant correlation between them while controlling for factors $B_1, \ldots, B_n$ which might be the real cause of C and/or of the correlation between A and C. If I suspect that there is an intermediate cause between A and C, it must be controlled for. However, nothing requires me to identify the precise mechanism. Rather, I need only decide, based on my background theory, on a list of possible mechanisms. So long as I control for them all, the causal law can be confirmed without knowledge of the specific mechanism. In fact, I need not even list the possible mechanisms. If I can cite some general feature, which any mechanism must have, then that feature can be controlled for—thus allowing confirmation even without precise knowledge of the possible mechanisms.

Consider the following example from biology that illustrates the above point. Evidence for the theory of evolution goes far beyond anecdotes about moths adapting due to pollution or mosquitoes and roaches to DDT. Biologists have developed a variety of methods to test a fundamental evolutionary tenet: that natural selection causes changes in trait or gene frequencies.[15] For example, field studies can establish that certain traits are correlated with certain environments. If the other relevant factors are controlled for (e.g., variation due to age differences, variability in phenotypic expression, and so on), then such information provides good evidence for a causal claim. Nonetheless, these results are silent on the actual mechanism. Environmental factors may cause changes in trait frequencies in at least two ways: by directly picking out the trait in question or by picking out some other associated trait. To use Sober's terminology, we can establish the causal relation between selection and the existence of a trait without knowing whether selection of or selection for was the relevant mechanism.[16] Furthermore, even when we know that the trait in question was specifically chosen by selection, we still frequently do not know the *genetic* mechanism involved—and may not know for the indefinite future. Yet so long as we know that the underlying mechanism is genetic in nature, we can use standard methods to show that a trait was selected for.[17] In short, we can have good evidence for causal claims without knowing the intervening mechanism.

Of course, precisely identifying the actual mechanism does confirm a causal law. But while this method may be sufficient, it is not necessary—at least when it comes to imperfect knowers such as ourselves. In fact, if we insisted that mechanisms always be cited, we might never confirm any causal law. For most mechanisms B that connect A and C, we can always ask what allows B to play its causal role. If certain individual behavior brings about some social law, that behavior is not the final mechanism. Deeper accounts can be given in terms of neurophysiology and then chemistry and then particle physics. So identifying the mechanism is an open-ended process that admits of degrees. Demanding a complete specification would not only rule out social laws, it would eliminate much of current science.

Is there something special about the social sciences that make them different in this regard? Is there something unique to the social sciences that makes an exact identification of the intervening mechanisms an absolute requirement for confirmation? The difference cannot simply be the macro-level nature of social laws—the gas laws, the laws of population genetics, ecology, and so on are likewise macro in nature. Is it perhaps that the relevant difference is the teleological nature of (many) social laws?

We thus should consider next whether citing mechanisms is essential for confirming teleological laws. I think the points made earlier concerning ordinary causal laws hold equally for the confirmation of teleological laws. Before I can argue for that claim, it is necessary to look at just what such laws assert.

Much of the current debate over these issues has focused on teleological or functional laws as defended by G. A. Cohen.[18] Cohen has sought to clarify Marxist statements such as "the relations of production exist because they develop the forces of production." So "teleological laws" in this context assert that some entity exists because of its beneficial or useful effect. Claims of this sort are rife in the social sciences. According to Cohen, such explanations are confirmed when we confirm the appropriate "consequence law" of the form "if X would be useful, then X comes to exist." Cohen believes such laws can be confirmed without knowing the relevant mechanisms. Elster and others disagree.

Unfortunately, the debate has been somewhat obscured, I think, by Cohen's account. Consequence laws are not sufficient for teleological laws.[19] They are not sufficient because we could have a well-confirmed consequence law and yet know that the relevant teleological law was false. The consequence law only requires a correlation between usefulness and occurrence. Such correlations, however, do not establish a causal connection—the correlation could be the result of some third factor. Until we rule out a third factor, the consequence law is no guarantee. Biologists, for example, who inferred selection for a trait from its usefulness would be doing poor science. For example, pleiotropy and gene linkage make correlations between usefulness and existence weak evidence for causal connections.

What then is involved in confirming a teleological claim? Nothing more is required, I think, than the usual procedures for establishing causal claims. To show that X exists because it is useful we must (1) specify the relevant causal effect of X that constitutes its usefulness, and (2) show that effect plays a positive causal role in X's continued or expanded existence. Establishing the latter requires that we test the relation between X's usefulness and its existence *while controlling for other possible causes*. For example, we identify a widespread social practice that may exist because of its usefulness. After establishing precisely what positive effect it has, we then try to show that the practice persists or spreads because of those effects. Is the practice correlated with some other fact that might be the real causal factor? Does the practice sometimes arise with the same effects and yet does not persist? Does the practice sometimes exist without being useful and yet persist? If so, then its usefulness is probably not the cause of its persistence.

Given that teleological laws are confirmed much like ordinary causal laws, we have no reason to think that they cause special problems. Just as citing precise mechanisms is unnecessary for confirming causal laws, the same applies for teleological laws. In both cases, we can control for spurious causation without specifying the precise mechanism or even identifying all possible mechanisms. Our confidence is, of course, increased by more precisely specifying the relevant mechanisms. But in both cases, the macro law can be relatively well confirmed without doing so.

III

So far, I have defended the *possibility* of social laws. Skeptics, however, are still probably not convinced. After all, merely showing that social laws are possible is a small achievement. It certainly does not show that there really are social laws or that there ever will be. And a close look at the social sciences, so the skeptical reply runs, reveals that what social scientists take to be laws are nothing of the sort. The accepted wisdom seems to be that the social sciences obviously have not produced real laws. The real puzzle is why they have so obviously failed (cf. Rosenberg).[20]

Skeptics cite three facts to explain their insistence that current social science has not produced laws: (1) the alleged laws in the social sciences lack the requisite generality, (2) alleged laws in the social sciences are really only accidental generalizations, not laws (Rosenberg, for example, makes both these criticisms)[21], and (3) the purported laws are either deduced from obviously false assumptions or have unspecified (and/or unspecifiable) ceteris paribus clauses. What social scientists call laws do not meet the necessary requirements, requirements that the laws of physics do pass.

In my defense of social laws against these criticisms, I must first sketch a reply to the generality objection and to the claim that social laws are only accidental generalizations. In my view, these objections were shown to be misguided long ago by Goodman, among others.[22] Nonetheless, these objections are worth discussing briefly, if only because philosophers continue to think that reference to particulars and being a mere generalization are sufficient grounds for rejecting social laws. My main concern, however, is the problems raised by ceteris paribus clauses; this difficulty is potentially much more serious. Against this objection, I argue that (1) ceteris paribus qualifications are not in principle an obstacle to laws, as a look at physics will reveal; (2) there are some relatively straightforward criteria to determine when ceteris paribus laws are confirmed and explanatory, and (3) there are laws in the social sciences that meet those criteria. Ceteris paribus clauses are neither principled nor insurmountable practical obstacles to social laws.

Of course, social scientists have produced few if any completely general laws. The results of neoclassical economics hold only for market economies. Macroeconomic generalizations obviously do not apply to primitive societies. Claims about bureaucracy do not hold for just any hierarchical organization. But is this really a difficulty? I do not think so. Being appropriately general is a matter of degree, not an all-or-nothing property. Newton's laws do not apply to numbers, temperature is determined by mean kinetic energy only in gases, and so on. Furthermore, appeal to logical form alone will not separate the really universal from the rest, for we can always transform references to specific individuals into a universally quantified statement. While some laws are intuitively more universal than others, complete universality is an elusive goal—even for the physical sciences. Kepler's laws, for example, refer to a particular: our solar system.

Does this mean that the universality requirement ought to be given up? Not necessarily. Relative to disciplines and theories, we can perhaps distinguish those statements that are universal from those that are not. The fundamental processes described by molecular biology—translation, transcription, protein synthesis, and the like—only hold for entities governed by DNA. Nonetheless, these accounts are universal once we set molecular biology as our domain of reference—unlike any statement that refers to specific species or cell kind in describing basic processes. So, while social laws may be specific to various domains, that still leaves room for universality, properly understood.

A second complaint against social laws is that they are merely generalizations and thus lack the "nomic force" necessary for real laws. Such a complaint presupposes that there is some sharp distinction between "real laws" and those lawlike statements that are merely accidental generalizations. While we can no doubt describe clear instances of necessary laws and accidental correlations, it is doubtful that the difference is one of kind rather than of degree. Goodman argued thirty years ago that the distinction between laws and accidental generalizations was best captured by their degree of entrenchment and projectability; those are, of course, not all-or-nothing criteria. Similarly, Skyrms recently analyzed lawfulness in terms of what he called "resilience" (defined roughly in terms of statistical invariance)—one again a notion that makes the law/accidental generalization distinction a relative one.[23]

What does this mean then for social laws? It implies that criticizing social laws as mere accidental generalizations is misguided. All laws are accidental generalizations after a fashion. To show that social laws are really inadequate, one would have to establish case by case that alleged social laws are so low on the entrenchment or resilience scale that they really should not be called scientific laws at all. Needless to say, that arduous task has not been undertaken by critics of the social sciences.[24] (And I will present below some evidence to suggest that task could not be carried out.)

Now, let us look at the ceteris paribus objection. Certainly, social laws are generally implicitly qualified, often in unspecified and unspecifiable ways. However, there is no principled difference here between the social and physical sciences. As Cartwright and others have persuasively argued,[25] most theoretical laws of physics are either false or implicitly qualified with ceteris paribus (or, more accurately, ceteris absentus) clauses. They describe relations between variables that would hold if no other factors intervened. The force between two bodies, for example, varies inversely with the square of their distance only if magnetic forces and so on are not present. Real events typically involve multiple laws and the factors they describe. Getting an explanation of the individual event in all its particularity requires some way to compose the joint factors, some way of tying the counterfactual law to reality. Typically, however, physics provides no automatic and precise way to deal with such messy complexity. Rather, we have at best numerous rules of thumb and somewhat ad hoc and piecemeal principles for conjoining multiple ceteris paribus laws with reality. Ceteris paribus clauses are prevalent and uneliminable in physics.

Cartwright often makes this picture sound paradoxical: Theoretical laws explain only when they are false and cannot explain when they are true. But her point is a perfectly coherent one about models and reality. Theoretical laws hold only ceteris paribus—they describe the way things *would* be if certain simplified conditions held. By describing the counterfactual situation, theoretical laws isolate basic processes, forces, and so on. Counteracting and complicating factors are removed; the basic nature of the force in question is then revealed.

Thus physics essentially involves ceteris paribus laws—both with unspecified and unactualized qualifications. Unless we are prepared to jettison large parts of physics, we cannot reject social laws simply because they hold only ceteris paribus. Nonetheless, this "physics does it, too" response would be much more compelling if we could say exactly how it is that ceteris paribus laws explain, and are confirmed, and show that those conditions are met by at least some alleged social laws. I thus turn to these tasks.

Ceteris paribus laws present puzzles and complexities. They describe connections that hold under conditions that may never obtain—they describe what things *would* be like if certain conditions held. Sometimes, we cannot even fully specify those conditions. So, how can we ever confirm ceteris paribus laws? That is one question we need to answer.

Answering that question is only a first step. Even if we have good evidence for a ceteris paribus law, we still need to say how and when a law about a counterfactual situation explains the real world. Ceteris paribus laws describe what would be the case. How can they then explain what is the case? And once we say how that is possible, we still need some criteria for judging when ceteris paribus laws explain a given event and when they do not. Since many qualified laws generally do not strictly apply to the facts to be explained—otherwise, they would not be ceteris paribus laws—how do we tell genuinely explanatory laws from irrelevant ones?

These are important problems, for they point to inadequacies in the account of laws and explanation employed here. Without some way to say how and when ceteris paribus laws explain, Cartwright's account remains incomplete and threatens to make explanation a quite subjective affair. So, resolving these puzzles is important both to the project of this chapter and to defending the general approach to explanation and laws that it presupposes.

Let us consider first *how* it is that ceteris paribus laws can explain. Imagine that we have a well-confirmed law about what *would* happen if the ceteris paribus clause did hold (so there is no paradox about explaining with false laws as the problem is sometimes put).[26] How can we explain an actual state of affairs using such a law? We cannot really subsume the event to be explained under the ceteris paribus law for the conditions described by the law do not strictly hold. So how can the law explain?

One reasonable answer goes something like this: Ceteris paribus laws generally describe forces, relations, and so on as they would operate in isolation. When things are not ceteris paribus, the laws in question still apply. But they now describe tendencies—partial elements of a complex situation. Tendency statements describe an element—call it a force, process, or whatever—which would produce a certain result in the presence of some larger set of conditions. When those conditions are not met, the element in question is only a tendency. Thus ceteris paribus laws and tendencies go hand in hand—and that seems reasonable enough. The law of gravitation is implicitly qualified with a ceteris paribus clause. While the law itself holds only counter-factually, it nonetheless says something about reality. Attraction in proportion to squared distance is a tendency for any two bodies. Citing the law explains because it identifies that tendency—even if the conditions cited by the law in fact never strictly obtain.

So, we can see how laws about what does not obtain can nonetheless explain what does obtain—they cite factors, aspects, and tendencies of a complex situation. Given that we can at least make sense of how ceteris paribus laws explain, we must next (1) say how we can confirm ceteris paribus laws in the first place—since they are often about conditions that do not exist—and (2) give criteria for telling when a particular ceteris paribus law explains a particular event and when it is irrelevant. The two questions are intimately related, I think, and answering one goes hand in hand with answering the other. Standard scientific methods do help confirm ceteris paribus laws and they tell us when those laws explain. While I cannot defend these claims in detail here, I can sketch a variety of testing practices that lend credence to ceteris paribus laws:[27]

1. We can sometimes show that in some narrow range of cases the ceteris paribus are satisfied. Rational economic man is an idealization, but sometime consumers do act on well-ordered preferences and maximize. We can then confirm the law directly.
2. We can sometimes show that although other things are not equal, it make little difference: The law holds for the large part.
3. When a ceteris paribus law fails to hold for reality, we can nonetheless explain away its failure. Sometimes, the counteracting factors can be cited and relevant laws invoked, giving us at least an approximate prediction of their combined effect. Other times, the interfering factors may be unique and fall under no law, yet we can reasonably explain away their influence.
4. Sometimes, we can provide inductive evidence for a ceteris paribus law by showing that as conditions approach those required by the ceteris paribus clause, the law becomes more predictively accurate.
5. Unspecified ceteris paribus laws may nonetheless predict striking or novel facts. Even if we do now know counteracting factors, the law may make some predictions which are borne out and which are unexplained by alternative hypotheses.
6. A ceteris paribus law may provide a single explanation for some set of diverse phenomena, even though we cannot fully specify the ceteris paribus clause or explain away all counteracting factors.

All the above methods provide evidence for the claim that if other things were equal, the law would hold. Methods 2 through 6 also give us reason to believe that the alleged law applied to or explains real phenomena even though it makes false assumptions. Finally, these methods are at least part of what is required to show that a lawlike statement is entrenched or resilient (i.e., to show that, relatively speaking, it is not an accidental generalization).

Obviously, there is no fixed algorithm for using these procedures. Nonetheless, they have proved to be powerful methods for establishing laws and showing that those laws describe tendencies which explain the data at issue. Of course, exactly when ceteris paribus laws are supported and explanatory is a judgment call. But that is the nature of science.

Let me emphasize one important factor involved in making such judgments. Laws in the social sciences, I have been arguing naturally involve abstractions from real world complexity, abstractions which serve to isolate important forces or tendencies. However, which forces are important or need to be explained is in part a reflection of our explanatory interests. The explanatory power of any ceteris paribus law will thus depend on what we are trying to explain. Purported laws that abstract from the very features of interest obviously fare poorly in explaining the real world.

For example, neoclassical economic theory seems sterile to critics, both from the left (e.g., Marxists) and the right (e.g., Austrians). While its detractors often point to the ceteris paribus nature of neoclassical theory, that is not, I suggest, the real sticking point. Rather, Marxists and Austrians have different explanatory interests than do mainstream economists. Someone interested in explaining economic crises, the origin of surplus or profit, or the role of class power in an economy—as do Marxists—is bound to find neoclassical results nonexplanatory, for they focus on equilibrium conditions where markets clear, profits are zero, and each producer and consumer has no influence over price. So, evaluating ceteris paribus laws requires prior agreement over just what features in a complex situation ought to be explained. Different goals will produce different evaluations.

Are the methods described earlier available to the social sciences? If the arguments of sections I and II succeed, then a priori or conceptual answers to this question are unpromising. Armchair theorizing about what science can and cannot do must be eschewed in favor of a close look at real empirical work in the social sciences. No doubt, some or much social science is empirically weak and does not make convincing use of the methods outlined earlier. Nonetheless, it is my belief that no such blanket claim can be supported. Parts of the social sciences are relatively well confirmed; standard methods for evaluating ceteris paribus laws have been successfully applied. Establishing this judgment is a huge task which obviously cannot be undertaken here. Section IV, however, provides some reasons to think that this task can be carried out.

IV

In this final section, I want to make at least a prima facie case that the social laws are a reality: There are sections of the social sciences that have produced confirmed laws according to more or less standard scientific procedures. My argument will take both direct and indirect routes. The more direct argument will cite two fundamental laws in economics that are relatively well confirmed via the processes outlined in the last section. Proceeding more indirectly, I shall then argue that some parts of the social sciences are roughly in the same boat—when it comes to kinds of laws, qualifications, explanations, and level of confirmation—as is much good work in the biological sciences. The case for social laws should thus be strengthened by a kind of "merit by association."

Economics is arguably the most developed of the social sciences. Within economic theory, laws of market behavior are probably most thoroughly developed. They are also a fundamental part of divergent economic theories: Marxists and Austrians share with defenders of neoclassical theory, I would argue, a commitment to certain laws relating supply and demand. Albeit qualified with ceteris paribus clauses, these laws have been steadily confirmed by the routes listed above. They have a strong claim to scientific respectability.

Consider the following elementary laws about market behavior:

1. A rise in the price of a good will result in a decrease in the quantity demanded.
2. A decline in the supply of a good will result in a rise in price.

These laws, I would claim, are well confirmed.

While the above laws are not *explicitly* qualified with a ceteris paribus clause, clearly they must be. They hold only assuming that other things (e.g., nuclear war, the explosion of the sun, the heat death of the universe, and so on) do not interfere! More serious, these laws are derived from a standard set of assumptions—perfect information, transitivity of preferences, and so on—that frequently do not hold. How should we treat these assumptions? At least two approaches suggest themselves, which are commensurate with the argument of this article:

1. We can take the assumptions as specifying the mechanism that brings about these macro-level laws. The laws describe the behavior of aggregates; the assumptions specify the mechanisms that realize them. Proceeding this way, these assumptions need not be established to confirm the two laws because (a) in line with the previous arguments of this chapter, the mechanism must not be identified to confirm, and (b) alternative mechanisms may be possible if these assumptions are false (e.g., by substituting selection mechanisms for rational preference mechanisms to explain profit maximization).
2. On the other hand, we can take these assumptions as implicit ceteris paribus clauses. Price and quantity demanded, for example, are related, assuming that preferences are transitive and so on.

Although I think both approaches are reasonable, I shall focus on the second in order to illustrate more clearly the claim that some ceteris paribus laws in the social sciences are confirmed.

To show that the above laws are explanatory, we need to establish that they are confirmed—that if the designated ceteris paribus clauses held, the laws would be true—and that they explain actual economic operations. Both tasks are possible. Because these laws follow almost[28] deductively from the respective ceteris paribus clauses, there is little doubt that they describe the way that things would be. Showing that they explain actual economic reality is also possible, although less obvious. A horde of empirical evidence establishes that these laws meet the necessary requirements for explanation described earlier:

1. Studies based on questionnaires, controlled bargaining situations, and animal experimentation suggest that one crucial ceteris paribus clause is often met. Preferences are frequently transitive.[29]
2. There is always some reason to believe that even when preferences are not orderly as the ceteris paribus clauses imply, little deviation from the above laws will result. As the work of Becker suggests, disorderly preferences do not necessarily undercut supply and demand theory if the other components of market context are fixed.[30]
3. We have evidence that as the ceteris paribus clauses on these laws are made more realistic, the laws hold with greater accuracy. For example, the standard approach takes the quantity demanded as a function of the preferences of *households*. Of course, households do not strictly have preferences—their members do. Recent work has refined this assumption, using models that take household preferences to be the outcome of a bargaining situation between members. Tests against data indicate that this refinement leads to greater empirical accuracy.[31]
4. These laws have both been fruitfully applied to a diverse range of phenomena and have made striking predictions. The literature on these topics is enormous, but two obvious examples are the work of Becker using these laws, among others, to explain various aspects of crime, education,[32] and so on, and the work of Buchanan and others in explaining political processes.[33]

So, it seems that these basic social laws are in the same boat as are many relatively well-confirmed laws in the other special sciences. Without doubt, they apply to a certain range of phenomena. In many other cases, however, these laws do not strictly hold, but they nonetheless explain, because we have some idea of the counteracting influences and because they unify a diverse set of events. Of course, these laws depend on abstractions and idealizations—noneconomic influences, for example, are entirely ignored. Nonetheless, we remain convinced that the connection among supply, demand, and price tells us something about reality, despite these complications. The evidence cited earlier helps justify that belief.

Two laws of supply and demand are of course not much. Are they simply rare jewels in the morass of bad social science? A careful look at other empirical work in the social sciences suggests not. In what follows I first discuss some central results in biology in order to draw a general characterization of what laws, explanation, and confirmation in that domain look like, and then argue that some empirical work in the social sciences proceeds in much the same way. The upshot should be a prima facie stronger case that social laws are a reality.

Both evolutionary biology and ecology rely primarily on field data. While laboratory or field manipulation is sometimes possible, much evidence comes from field observations of relative abundance of species or individuals, rates of change in abiotic factors, relative survivor rates for one phenotype or another, and so on. Such data are manipulated using standard statistical techniques searching for connections while holding other explanatory causes constant.

From such data, ecology and evolutionary biology produce some lawlike claims. Typically, those claims describe a connection between two phenomena that holds only ceteris paribus. And just as typically, we do not know all the likely intervening factors that might counteract the law. Rather, we can identify the primary counteracting causes; sometimes, we can specify their precise influence and how they interact, but sometimes we cannot. similarly, it is often the case that "major" causes and limiting factors are all that we can describe with confidence.

The laws produced are often functional in nature—they relate changes in one variable as a function of another, without necessarily citing the causal basis for that relation. Laws in ecology and evolution are also frequently functional in the other sense: They describe the functions which various items have and sometimes go on to attribute these items' existence to that function. Finally, the vast majority of lawlike claims in these sciences are relativized to specific domains—a type of ecological community, environment, species, and so on.

Presumably, this characterization rings true for those familiar with empirical work in evolutionary theory and ecology. Let me mention some obvious examples that illustrate these features:

> 1. A standard kind of ecological law—laws describing community succession—is specific to communities. Different laws of development hold for different kinds of communities.
> 2. Ecology describes laws of succession. Such laws are known to have exceptions; among the counteracting factors are radical environmental change, although other unknown factors may be involved. Similarly, another prime ecological law—organisms with the same ecologies cannot coexist in the same environment—may fail to hold when the environment is sufficiently variable, when population

sizes at equilibrium are equal, and when selective forces are weak; the complexities of real competitive situations make it unlikely that these are the only possible counteracting factors.
3. Evolutionary theory is at its clearest in idealized situations involving a few factors of known magnitude operating within populations. However, once factors such as group selection, internal constraints, and tendencies to speciate, as well as the factors making up environments and fitness (e.g., competition, predation, and patchiness, and costs of mate search and territory defense, respectively) are factored in, then there is no mechanical way to combine these factors to answer every relevant question. Similarly, density dependence, time lags, competition, predation, and environmental fluctuation have relatively well-understood effects in simplified situations; putting them together is a piecemeal and ad hoc process.
4. Evolutionary theory relates trait, organism, and community types as a function of environment type. A major set of ecological laws describe community structure, nutrient cycling, community energetics, and level of homeostasis as a function of successional level.
5. Evolutionary theory is, of course, teleological in that it explains existence of traits via their useful effects. Ecology likewise explains niche structure and trophic organization as due to their adaptive value in given environments.

This general picture of laws, explanation, and confirmation also applies to at least some empirical work in the social sciences. I want to support this claim by looking at some of the best empirical work in the social sciences: explanations of cultural evolution and ecological adaptation in small-scale societies.

Small-scale societies are ideal candidates for explanation for several reasons. Small-scale societies are generally less complex than modern social structures. They are likewise more isolated from outside cultural factors while at the same time more directly tied to their ecological environment. So, small-scale societies present cases where the number of variables are relatively reduced and provide us with a rich set of data for testing hypotheses.

Through both individual studies and statistical analysis of many different small-scale societies, anthropologists and economists have been able to reach a number of relatively well-confirmed conclusions about cultural evolution, economic development, and environmental influences which are given below.

As small-scale societies evolve in complexity, they develop greater concentrations of wealth, political authority becomes more concentrated, family structure tends toward monogamy, kinship ties decline in importance, the frequency of external war increases, and so on.[34] Similarly, as the level of economic development increases, so does the extent of market exchange, private property, and the rental of land and capital.[35] Both sets of claims have been established by standard procedures: Multiple independent measures of complexity and development are used without altering results, and connections between complexity/development and associated changes hold after other potential explanatory variables have been controlled for.

Of course, the above claims must be qualified ceteris paribus, as is obvious from the fact that the relevant correlations are high but not perfect. Among the complicating causes are family structure and women's role in subsistence. The extent of market exchange and also private property is in part influenced by whether monogamy or polygyny is prevalent and by the relative level of participation of women in the productive process. Once again, these connections have been established via standard scientific practices controlling for other variables and for measurement errors.

As one might expect, small-scale societies tend to develop in response to the ecological situation. Ancestral rituals provide mechanisms that keep husbandry and agriculture within the carrying capacity of the local environment.[36] Matrilocal residence is sometimes an adaptive response to intercommunal competition—warfare, in particular.[37]

This body of work exhibits much the same status and structure as the ecological and evolutionary explanations cited earlier. Field observations manipulated by standard statistical techniques are the prime source of evidence. From such evidence, lawlike claims are made, but those claims are clearly qualified. Major interfering factors are known, although their precise composite effects often are not. Many important laws describe basic patterns or functional roles; most hold only for specific domains. Of course, the work cited here can be challenged. But it does present a prima facie case that some social laws may be relatively well confirmed and explanatory—for at least this part of the social sciences produces kinds of evidence and explanation closely akin to those found in large parts of biology.

In closing, it is worth taking up one last skeptical reaction to the case for social laws—an inductive argument based on the poor success of the social sciences. If social laws are indeed possible, why have the social sciences been so notoriously unsuccessful in producing anything like well-confirmed social theories? What explains the lackluster progress of social theory? And if the social sciences are roughly in the same boat as the biological sciences, why do we know so much more about biological processes than social ones?

While these questions really call for their own article-length discussion, generalized answers will have to suffice here.

First, the question, "Why have the social sciences been so unsuccessful at developing well-confirmed theory?" is in part based on a false presupposition—if the thrust of this chapter is correct. Critics have failed to see progress because they looked at the social sciences through a distorted conception of what science is. While the social sciences have not produced exceptionless, completely general laws, neither have many other respectable sciences—evolutionary biology and ecology, for example. But such laws are not required for good science and explanation. Thus, if the arguments of this chapter succeed, it is just wrong to say that the social sciences have largely failed. No doubt, some parts have. However, as I argued above, significant sections of social research—like economic anthropology and parts of microeconomics—have produced relatively well-confirmed explanations. The judgment that the social sciences have been more or less entirely unsuccessful relies on an untenable account of science.

The enormous complexity of social systems is often cited to explain slow progress in the social sciences. While that response is unconvincing without some measure of complexity, I think it is nonetheless on the right track. Social phenomena are more complex in this respect. We frequently lack the ability—either for technical or ethical reasons—to use controlled experimentation to isolate single causes in complex social systems. Of course, controlled experimentation is not the only way to confirm laws, and once again, the social sciences are in much the same boat in this respect as are parts of ecology and evolutionary biology. In both cases, progress is slowed because we much rely entirely on nonexperimental data. As Glymour and others[38] recently pointed out, statistical data can support many different causal interpretations. But once again, this is not a problem peculiar to the social sciences nor is it insurmountable (cf. Glymour).

Relying on purely observational data causes problems in yet another way. Because phenomena cannot be created at will in the laboratory, social scientists face a problem of limited sample size. Important social phenomena may reoccur infrequently, and in particular, they may seldom reoccur in precisely the same context. Of course, there are statistical methods for dealing with this situation—a situation that is again not special to the social sciences (for example, particular kinds of ecosystems are likewise limited). Nonetheless, this fact obviously helps to explain why progress comes faster in the laboratory sciences, and why some parts of biology—cell biology, physiology, and so on—are indeed much more successful than the social sciences and why biological sciences, like ecology, look much more like the social sciences.

Values play a role in social research that makes progress both more difficult and harder to recognize when it comes. Whatever your position on value neutrality in social explanation, it is clear that values are inevitably involved in decisions about what questions to pursue. While this is true for any science, it has far greater implications in the social sciences. Few theoretical questions in physics or even in most parts of biology have any direct bearing on interest groups in society. Exactly the opposite holds in the social sciences, where different political, economic, sexual, and racial allegiances entail differing views about what questions ought to be pursued in social research. As a result, divergent research agendas are inevitable; debates are often at cross-purposes, and the corresponding slow progress typical of "preparadigm" science plagues the social sciences in ways that the physical and biological sciences can avoid.

Finally, there are contingent, historical reasons why progress in the social sciences has been harder. For example, grand theorizing has dominated the social sciences—at the expense of more narrow empirical work—since their inception. Furthermore, the social sciences have typically attracted those individuals who are less mathematically and quantitatively inclined. Levels of funding for social science research pale in comparison to those available for research in physical and biological sciences. All these trends have arguably slowed empirical progress. Much more could be said, I believe, along these lines.

Thus the social sciences have progressed slowly—but that fact can be explained without appealing to special, insurmountable obstacles to a science of human society and without undervaluing the real progress that parts of the social sciences have made.

Summary

The crux of my overall argument is this: Once we give up the positivist assumptions that there is some simple property that separates science from nonscience, laws from exceptionless generalizations, and so on, then a number of consequences follow for our thinking about laws in the social sciences. For one, the science/nonscience distinction is a continuum, and judging where some domain stands on that continuum can seldom be done short of a full-fledged analysis of the methods, evidence, and kind of explanation employed in that science. As a result, simple conceptual arguments, like those of Searle and Davidson, against a real social science are likely to fail. Furthermore, any account of confirmation and laws will have to, in part, generalize from real scientific practice. But once we see what confirmation and laws in the special sciences—and physics as well, if Cartwright is correct—look like, then at least parts of the social sciences will not seem significantly less scientific than important parts of biology, for example. To argue otherwise, philosophers will have to dirty their hands with real empirical data. Gone are the days when philosophers can declare on simple methodological or conceptual grounds that entire domains of painstaking research are unscientific.

Notes

Work on this chapter was supported by a grant from the American Council of Learned Societies. George Graham, Alexander Rosenberg, Marthe Chandler, and Scott Arnold made helpful comments on earlier drafts.

1. Cf. B. van Fraassen, *The scientific image* (New York: Oxford University Press, 1980), Chap. 5.
2. For example, Donald Davidson, "Causal relations," *Journal of Philosophy* 64 (1967): 691–703.
3. I do not pretend to cover all popular objections to social laws; in particular, some traditional criticisms flowing from the *verstehen* tradition concerning interpretation and objectivity are beyond the scope of this chapter.
4. J. Searle, *Minds, brains and behavior* (Cambridge, MA: Harvard University Press, 1984), 78.
5. Ibid., 79.
6. P. Churchland, *Scientific realism and the plasticity of mind* (Cambridge: Cambridge University Press, 1979), 113; A. Rosenberg, *Sociobiology and the pre-emption of the social sciences* (Baltimore: Johns Hopkins University Press, 1980), 107.
7. See A. Rosenberg, "The supervenience of biological concepts," *Philosophy of Science* 45 (1978): 368–86.
8. See my "Molecular biology and the unity of science," *Philosophy of Science*, 57 (1990): 575–93.
9. Donald Davidson has argued in this fashion. See "Mental events," in *Experience and theory* edited by L. Foster and J. Swanson (Amherst: University of Massachusetts Press, 1970).
10. D. Porpora, "On the prospects for a nomothetic theory of social structure," *Journal for the Theory of Social Behavior* 13 (1983): 243–64.
11. T. Horgan and J. Woodward, "Folk psychology is here to stay," *Philosophical Review* 94 (1985): 197–226.
12. J. Elster, *Making sense of Marx* (Cambridge: Cambridge University Press, 1985), 5–8.
13. van Fraassen, *The scientific image*; P. Achinstein, *The nature of explanation* (New York: Oxford University Press, 1980).
14. Elster, *Making sense of Marx*.
15. J. Endler, *Natural selection in the wild* (Princeton, NJ: Princeton University Press, 1986).
16. Elliot Sober, *The nature of selection* (Cambridge: MIT Press, 1985), Chap. 3.
17. See A. Grafen, "Natural selection, kin selection and group selection," in J. Krebs and N. Davies, *Behavioral ecology: An evolutionary approach* (Sunderland: Sinauer Associates, 1984), 65.
18. G. A. Cohen, *Karl Marx's theory of history* (Princeton, NJ: Princeton University Press, 1978), Chap. IX.
19. It is not clear to me that they are, strictly speaking, necessary either. Locomotion probably exists among organisms because of its contribution to fitness, despite the fact that the corresponding consequence law is false—trees would be better off if they could move to the sun, but structural factors have prevented that. So while it is not true that if locomotion would be good for organisms, it thus comes to exist.
20. Rosenberg, "Surveillance," and *Sociobiology and the pre-emption of the social sciences*.
21. Rosenberg, "Surveillance." But to be fair, it should be noted that Rosenberg does so as part of a larger and more subtle argument that cannot be discussed here. I should also note that my views here are in fact much closer to Rosenberg's earlier position in *Microeconomic laws* (Baltimore: Johns Hopkins University Press, 1976).
22. Nelson Goodman, *Fact, fiction and forecast* (Indianapolis, IN: Bobbs-Merrill, 1965).
23. Brian Skyrms, *Causal necessity* (New Haven, CT: Yale University Press, 1980).
24. It should be noted that this same point can be made even if one rejects a Humean *account* of laws—for even on a realist, necessitarian account of laws, our *epistemological* judgments about which generalizations are real laws will rely on something like projectability or resilience. See Fred Wilson, *Laws and other worlds* (Dordrecht: Reidel, 1986), Chap. 2.
25. Nancy Cartwright, *How the laws of physics lie* (New York: Oxford University Press, 1983).

26. D. Hausman, *Capital, profits and prices* (New York: Columbia University Press, 1981), Chap. 7.
27. Criteria similar to some of those listed here are also discussed in D. Hausman, *Capital, profits and prices*, Chap. 7.
28. I say almost because the deduction follows only assuming that extraeconomic events (e.g., the death of the solar system) do not intervene.
29. A. Weinstein, "Transitivity of preference," *Journal of Political Economy* 76 (1968): 307–11; R. Battalio et al.," A test of consumer demand theory using observations of individual consumer preferences," *Western Economic Journal* 11 (1973): 415–21.
30. G. Becker, *The economic approach to human behavior* (Chicago: University of Chicago Press, 1976).
31. M. McElroy, and M. Horney, "Nash-bargained household decisions: Toward a generalization of the theory of demand," *International Economic Review* 22 (1981): 333–48.
32. G. Becker, *The economic approach.*
33. J. Buchanan, *The limits of liberty* (Chicago: University of Chicago Press, 1975).
34. For a review of the data, see David Levinson and Martin Malone, *Toward explaining human culture* (New Haven, CT: HRAF Press, 1980), Chap. 2.
35. Frederick Pryor, *The origins of the economy* (New York: Academic Press, 1977).
36. Roy Rappaport, *Pigs for the ancestors* (New Haven, CT: Yale University Press, 1984).
37. William Divale, "Migration, external warfare, and matrilocal residence," *Behavior Science Research* 9 (1974): 75–133.
38. Clark Glymour, R. Scheines, P. Spirtes, and K. Kelly, *Discovering causal structures* (New York: Academic Press, 1987).

Chapter 9
Complexity and Social Scientific Laws
Lee C. McIntyre

Laws in social science, if we had them, would contain many more variables than those in physics. Yet we berate the social scientist for not being able to do what even the natural scientist cannot do. The multiplicity and complexity of factors in social phenomena impose limitations upon what we can reasonably expect to achieve. These limitations are only a practical, though perhaps practically insuperable, difficulty and we simply do the best we can.
May Brodbeck, "On the Philosophy of the Social Sciences"

There are many arguments purporting to show that social scientific laws are either impossible or impractical. Most often, these claims are made in terms of the intractability of the subject matter with which social scientists must contend (Kincaid 1988; Rosenberg 1980, 1988).[1] It is widely believed by many critics of social scientific laws that there is a difference "in kind" between the subject matter of natural and of social science, which precludes the use of laws in the explanation of human behavior. It is argued that, while natural scientific phenomena are well isolated, stationary, recurrent, and simple, human phenomena are, on the contrary, interactive, variable, singular, and complex (Popper 1965). Given such differences, it is thought to be a hopeless task to attempt to discover social scientific laws.

But are such arguments persuasive? Do they serve to demonstrate that it is impossible or impractical to have laws in the social sciences? It is my thesis that they do not. In this chapter, I will examine one of the most influential arguments for the claim that there is a difference in the subject matter of the social and the natural sciences sufficient to guarantee the impossibility or impracticality of laws in the former—the argument from complexity.[2]

Complexity claims are ubiquitous in the philosophy of social science. In almost every account of why it is that social scientists have not yet been able to find any laws of human behavior, one reason offered is that the subject matter is so complex that it precludes their discovery. In social science, it is held, we face a subject matter that is constantly changing, due to the continual realignment of the relationship between the critical influential variables that underlie human behavior, as a result of human consciousness. As humans, we affect and are affected by one another; our decisions feed off one another in a reciprocal way. Consequently, it is argued that human systems will never remain stationary long enough to determine the exact relationship between the variables. Once we think we have it, the system has had time to change. But, in order to come up with laws, we *would* need some type of regularity or stability in the phenomena under investigation. So, if human systems are prohibitively complex, we shall not be able to discover such patterns. There will always be too many critical influential variables behind human action, and the interaction between them will be too intricate and fluid, for us to ever be able to capture them in a comprehensive way. As a result, we shall be prevented from finding social scientific laws.

But what is the status of this claim? Is it a contention that there is a fundamental difference in the subject matter of natural versus social science *as such*—that human phenomena are somehow *inherently* complex, and therefore that social scientific laws are unavailable in principle (Hayek 1967)? Surely there are problems with this interpretation.

For one thing, this "naive" interpretation of the argument from complexity is not at all sensitive to the idea that social phenomena are not complex *as such,* but only *as described and defined* at a given level of inquiry. Surely the complexity of human phenomena, as all other phenomena, depends crucially on the level of description and investigation we are using, and the subject matter is thereby shaped by the nature of our engagement with it. We should remember here that a subject matter does not arise fully formed out of the phenomena—questions do not magically jump out begging to be asked. Instead, the way that we have described the phenomena, and the things that we are interested in knowing about them, shapes and frames the methods and modes of explanation that we shall be able to use in our investigation, by first framing the subject matter itself.

A subject matter is defined by the question that we ask about the phenomena we see. So, if we choose to describe them by using natural kinds and categorizations that do not reveal the nature of the underlying regularities, laws may elude us. But, at a different level of inquiry, they may not be so difficult to find. That is, one cannot say that at *any* level the phenomena are too complex for us to find laws, for complexity *just is* a function of the level of inquiry that we are using.

It is a mistake, then, to argue that there is a fundamental difference in subject matter between natural and social science, on the grounds of complexity, for we have been facing a possible confusion about what a "subject matter" is. It is not merely an ontological issue. We cannot treat a subject matter as if it marks off a natural kind in the world, and we come upon it fully formed. Instead, both the phenomena in the world and the level of our interest in and engagement with them contribute to what becomes our "subject matter." At the ontological level, the subject matter is just "matter"; it is only when we begin to ask questions about it that a "subject" comes forward. That is, the phenomena per se, are only phenomena. And it is only at the level of description, explanation, and inquiry which a particular type of investigation warrants that any distinctions begin to arise. A subject matter is not so much discovered, then, as it is *defined* by the questions that we ask about the phenomena we are interested in having explained—it is shaped by our descriptions.

So we see that the complexity of human phenomena is *not* inherent, but *derivative,* in that it is dependent on the nature of our interests. The possibility of finding social scientific laws, therefore, is a function of the level of our inquiry. Moreover, complexity at any one level does not guarantee complexity at all others. The complexity of the subject matter is a continuum, and we choose our place on it by the level of description and investigation we undertake. Thus, the absence of laws at one level of investigation cannot be used to rule out their possibility at all others; the possibility of inquiry at different levels holds out the prospect of future social scientific laws.

But it is now important to consider a somewhat stronger and more sophisticated interpretation of the argument from complexity. For someone might object, in response to the above account, that the prospect of investigation at different levels of inquiry is not adequate to the task before us in social science. It can be argued that in no area of inquiry do we really have a "choice" about the level at which we are going to inquire. Certain phenomena just seem interesting to us and, whether we like it or not, we *just are interested*, sometimes, in investigating phenomena at a level of inquiry at which we are enjoined from discovering laws. So it is not as if we may simply choose to describe the phenomena at another level of inquiry, where laws may be possible. That would be to let the tail wag the dog.

It is easy to see that such a position is more sophisticated than the simple claim that complexity is inherent in the phenomena, and somehow absolute. The "naive" view held that, at *any* level of investigation, the phenomena will be too complex for us to discover laws. We saw this view come to grief on the idea that we cannot infer, based on the difficulty of obtaining laws at one level, that they will be impossible at all others. But the sophisticated view goes beyond this, and holds that, *at* the level at which we are interested in having human phenomena explained to us, the phenomena are too complex to yield laws. At precisely *this* level of inquiry laws seem precluded.

We see here the subtlety of this view, in that it takes into account the idea that complexity is a function of the level of description and investigation we are using, and admits that complexity is derivative. But, this view holds, at precisely the level at which we want to frame our inquiry, complexity forbids laws. This account does not, therefore, attempt to rule out the possibility of truistic laws, pursued at other levels of investigation. But what it does rule out is the possibility of laws that are *explanatory*. Why? Because it argues that, at any level of inquiry at which laws are possible, the phenomena are described so simply that laws at that level would not be explanatory, and, at a level at which the phenomena are engaged in such a way that laws would be explanatory, things are too complex for us to discover them. In short, the phenomena are complex at precisely the level at which laws, if possible, would be explanatory.

The most developed version of this argument has been offered by its chief advocate, Michael Scriven. Scriven rounds out the above argument by making a case for the idea that the situation faced by the early physicists was comparatively easy, compared with that faced by today's social scientists, and that people like Galileo were lucky that they were not facing a field of such intractable complexity that there were no laws to discover. In the social sciences, Scriven claims, we face a task far more difficult than that in physics. He accounts for this by arguing that in social science all of the the easy laws have been skimmed off by long nonformal reflection on human affairs and poaching by other fields (such as literature). Social science, in contrast to a popular myth, is not young at all, but is among the oldest areas of inquiry. Consequently, after millennia of observation into human affairs, social scientists are left with all of the most difficult problems; they are stranded with a subject matter that is complex, because we demand explanation at a level of description and investigation where we have no good reason to suppose that we shall be able to find laws. Scriven admits that the subject matter of social science is "derivatively" complex—that it depends on the level of our description of it—but, given our interest in a particular level of inquiry, we are left with the dilemma of choosing between nonexplanatory regularities pursued at a truistic level of description and the absence of laws in the face of complexity. As a result, the social scientist faces a hopeless task. Scriven writes:

> The difference between the scientific study of human behavior and that of physical phenomena is thus partly due to the relatively greater complexity of the *simplest phenomena we are concerned to account for* in a behavioral theory. (1956, p. 332)

Such observations have led Scriven to claim it is merely good fortune that in the natural sciences we happen to be mainly interested in a level of investigation that does not fatally frustrate the search for laws; but, in the social sciences we have not been so fortunate. Yet given our interests, there is little that we can do about it. Scriven is willing to accept the fact that the complexity of certain subjects may well be a matter of an inefficacious level of description—of bad luck—but, *given* the level of our interest in social scientific explanation, laws will continue to elude us. He sums up the situation aptly:

> One is inclined to respond to comments of this kind by saying that this is surely no different from the situation in physics. We know what happens to falling bodies in a vacuum, but, when it comes to the way bodies behave when we drop them in air, we are not able to say very precisely what they will do. And, when it comes to the question of how a particular leaf falls from a particular tree on a particular autumn day, we are almost helpless. This is true, but nobody feels that it is very important to be able to predict the behavior of a leaf. If this were the kind of crucial problem in physics then it would be the case that physics would always be a subject of a very unsatisfactory kind. (1964, p. 171)

The unfortunate situation for social science, however, is that we *have* been concerned to know how a "particular leafs falls from a particular tree."[3]

But does this interpretation of the argument from complexity fare any better in its attempt to demonstrate the unattainability of social scientific laws? I maintain that it does not. We saw that the naive view foundered because it overlooked the important notion of investigating at different levels of inquiry. The view that human phenomena are too complex, at any level of description, for there to be laws belies the fact that at some levels of investigation we do have fairly accurate generalizations available. Of course, the sophisticated view objects here and tells us that such generalizations are not genuine laws. But the sophisticated view fails for a similar reason—for it is no more reputable to claim that laws will always fail at a *given* level than it is to say that laws will always fail at *all* levels. For who is to say what is possible, even at a particular level of description? In principle, it is always possible that, even using the descriptive terms desired, we could nevertheless come up with a nomological relationship between the phenomena. Perhaps one might think that this is unlikely, but it is important to remember that there is nothing in the argument from complexity that rules it out. The sophisticated complexity argument does not show that laws are unavailable at even the most detailed levels of description.

Indeed, what guarantee do we have that the dividing line between simple and complex phenomena will always remain where it is today? What is simple and what is complex changes over time, and is dependent not just on the level of our inquiry but also on the concepts and theoretical tools available to us at any given time. Has the division between simple and complex phenomena always been as it is now? Will it always be so? Rollo Handy has observed that "physical phenomena that now seem simple probably struck the original investigators as bafflingly complex" (1964, p. 112).

To think otherwise, to maintain that what is today regarded as the dividing line between simplicity and complexity will always be so, is a position that others have been forced to abandon in the face of scientific advancement. The extension of the set of complex phenomena, even at one level of inquiry, is not absolute, but is relative to scientific progress.

Moreover, why couldn't one suppose that it is open to us to use different descriptive terms and theories to capture the phenomena we are interested in having explained? After all, there should be no single privileged vocabulary or theory that is thought to capture all facets of human action. Indeed, it is important to remember that the subject matter of social science is most basically *human behavior*, not that behavior as captured by any particular theory or description of it. So, just as we earlier realized that the phenomena are not *complex* as such but only as described, we may now realize that laws do not *explain* the phenomena as such but only as described at some given level of inquiry. But why are we not free to change our descriptions, and thereby to allow for the possibility of nomological explanation? Why think that the subject matter of social science is fixed by a single level of interest, at which laws are unavailable? Surely there are more descriptive categorizations than have been previously discovered in our social scientific inquiry. Thus the prospect of redescription holds out the possibility of social scientific laws.

Yet, at this point, there arises another objection. For what guarantee do we have that the descriptive terms and categorizations that may end up supporting laws will be the ones that we find explanatory or that capture the phenomena we are concerned to know about? Many have objected to the possibility of redescription in the social sciences by pointing out that in social science we just find certain descriptive terms "constituitive" of the subject matter, and therefore seek explanation in terms of them. To redescribe, it is argued, is to change the subject matter, so that we are really no longer doing social science at all. Indeed, the foundation of Scriven's earlier claim about the danger of truism in the search for social scientific laws seems rooted in just such a "descriptivist" vision—that there is a single level of description of the phenomena of social science that is legitimate and that captures what it is that we want to know about it. And, if there are no laws linking these terms, then there are really no laws of social science.[4]

However, despite the subtlety of this revitalized account of the argument from complexity, it still falls prey to one overarching and recurring error. For Scriven confuses the thesis that there can be no laws of social phenomena at one level of description with the thesis that there can be no social scientific laws *at all* (using any other descriptions). Why does he do this? Because he believes that there is only one level of description of social phenomena that is legitimate, and that does not inevitably lead to truism.

Now, of course Scriven is right to guard against truism. He correctly recognizes that it is just not adequate to come up with trivial laws, based on some truistic redescription of social phenomena into unfamiliar categories, as if, in concert with the underdeterminist thesis that there are an infinite number of ways of connecting the same data, any regularity would do. Surely this would defeat the very purpose of explanation. Thus, Scriven is right to believe that wholesale redescription of social phenomena may lead to truism and to the degeneration of the search for social scientific laws into mere naive correlation., Moreover, Scriven also seems right to point out that we just seem to be interested in social phenomena at a particular level of description, where it is unfortunately the case that complexity may be present.

But where Scriven fails is in pushing the implications of these two theses too far—for be seems to believe that *any* redescription will *inevitably* lead to truism, an assumption which, if true, would preempt almost all of natural scientific, as well as social scientific, practice. Scriven has not allowed for the idea that redescription may indeed be useful in the search for laws *without* leading to truism; he has failed to recognize that our interest in having social phenomena explained at a certain level does not preempt nontruistic inquiry at other levels (where our goal may be to link the redescription back up to the original characterization), or the use of different natural kinds even *at* the original level of inquiry.[5]

Here Scriven has confused the worthy goal of explaining human phenomena at a level at which we find them puzzling with the spurious goal of engaging them using only a particularly narrow set of theories, descriptions, and vocabulary. He treats it as if once we have chosen the level of our interest, we have also determined all of the "natural kinds" that govern that level as well. But he has not realized that even *at* one level of interest there may be many different possible descriptions, categorizations, theories, distinctions, and vocabularies—some of which are more amenable to nomological inquiry than others. Thus, Scriven is not allowing the categorization of social phenomena to evolve along with our theories about them, and has insisted that, if they are to be explained at all, they must be captured in terms of a vocabulary and a theoretical framework that we have antecedently deemed to be legitimate—as if we had already discovered all of the proper natural kinds governing social phenomena, and the job of social science was merely to find the connections between them. But science, either natural or social, does not work like this; even *at* one level of description, there is more than one way of proceeding. We must not be so narrow in our definition of "social science."

In criticizing Scriven's account, it is therefore important to realize that even in natural science, we could not produce lawlike explanations under the conditions he imposes on social science. For one level of interest does *not* necessarily afford only a single descriptive vocabulary or set of natural kinds, and redescription does *not* necessarily lead to engagement at a truistic level. Throughout natural science, we allow our theories, descriptions, and categorization into kinds to evolve in response to our search for regularity. We assume that laws exist, and formulate our theories in an attempt to uncover them. Consequently, we may find ourselves investigating natural phenomena in terms that are, at first glance, far removed from the terms of our original query. But this does not necessarily mean that our inquiry is now truistic!

What is required for scientific explanation is that the theories and laws reflected in our inquiry *link up with* the real-world questions that we ask about natural phenomena. There must be a legitimate theoretical relationship between the explanations that we give in terms of "wavelength" and "the spectrum," and the everyday questions we ask about "light" and "color." But scientific inquiry does not require that this relationship be so close that we are bound by a single set of descriptive terms and distinctions. Instead, science grows by allowing our theories and categorizations of the familiar to expand, as we look for the regularities that underlie them, and to link up with our commonsense questions in novel and illuminating ways. But we must learn that this linkage does not involve refusing to allow science to make its own distinctions, develop its own theoretical framework, or even redirect the focus of our inquiry. We now realize that it would be improper to handicap social science by insisting that there is only one proper mode of description of social phenomena (or that use of new theories makes the subject matter something other than social scientific), and that if we fail to find lawlike connections then there is no role for laws in the explanation of human behavior.

Consequently, Scriven may be right that we demand explanation of human behavior that connects with the phenomena at a level at which we are interested in having them explained, but he is wrong to think that social inquiry must therefore be carried out only in terms of the particular vocabulary that has heretofore framed our interest. Imagine our dilemma if we insisted that the job of science was only to find connections between natural kinds that seemed a priori plausible. Under such circumstances, it would not be surprising is we failed to find laws, even in physics (Kincaid 1988).

Indeed, it is important to realize that natural science proceeds in its search for laws precisely by using redescription. In natural science the search for laws is primary. When faced with the dilemma of choosing between a descriptive vocabulary that seems sacred (but is not producing laws) and the challenge of producing a new categorization of nature that may yet reveal lawlike connections, science has relentlessly (although sometimes over great intervals) chosen the latter course. In natural science, it has been recognized that redescription, even when it means giving up a favored mode of inquiry, is valuable for its potential in revealing more basic insights into the regularity that stands behind surface diversity.

In the history of chemistry, as an example, we can see that this is true. In the transition from theories about "phlogiston" to those concerning "oxygen" in the explanation of combustion, one quickly recognizes the value of redescription. If one had insisted, in the face of growing anomalies, that "phlogiston" adequately captured a true natural kind, and that we should give up on lawlike explanation if we could not turn up a law that used it as a referent, it is easy to see the poverty of the explanations that might result. Similarly, one might imagine the state of physiology if the "four humors" had continued to be regarded as the only legitimate descriptive terms for nomological investigation. The point here is that we just recognize the "descriptivist" view to be wrong in its supposition that explanation can only be given—and inquiry can only be carried out—at a single level of description, based on a privileged categorization of the phenomena. For if this path were followed, science would produce laws only in exceptional cases.

It is unfortunate, however, that many people have misunderstood the foundation of the success of natural science in discovering laws and have attributed it to the efficacy of such factors as its "simple" subject matter. This has contributed to a highly idealized view of natural scientific practice, however (based on a mythical conception of the subject matter faced and the criteria needed for the formulation of laws), which it is doubtful natural science has ever met. Worse, this picture is then often used in the rejection of the possibility of nomological inquiry in the social sciences, based on their unfavorable comparison with the conditions allegedly met in natural science.

But this is to overlook those instances in which the natural sciences have faced genuine complexity in their subject matter. Certainly at high levels of description, such as those routinely faced in meteorology or evolutionary biology, we face what may seem to be well nigh intractable complexity. And yet this does not deter us in the search for laws.[6] Such instances, however, are often obscured, and are thought of as special cases as compared to the more paradigmatic enterprise of large-scale physics.

Indeed, the most salient examples of the success of nomological explanation in natural science are usually drawn form those cases in which the phenomena are examined at a level of description that is simple enough that the laws discovered tap some intuitive understanding we may already have had about the phenomena in question. The most often cited instances of the success of laws in science—the prediction of a new planet the time of an eclipse, or the movement of the tides—all deal with the phenomena at a level at which we are already used to thinking about them, and merely provides the missing link, properly embedded in a theory of course, that connects them. But the important point here is that the theory seldom changes our conception of what the phenomena *are*. To a large extent, the "natural kinds" that serve as the referents of the laws, in these cases, match the natural kinds we already use to think about the phenomena. The theory we have about them may change, but our taxonomy does not.

But this, I maintain, is a misleading picture of how science works and gives rise to false expectations about the role that laws might play in the explanation of scientific phenomena. Indeed it tempts us to overlook the very foundation of the success of natural science in its search for laws—which is flexibility in the development of new theories and redescriptions of familiar phenomena—and instead to attribute that success to features like simplicity, which merely *facilitate* the search for credible theories. The search for laws proceeds not because of the good fortune afforded us by examining an "easy" subject matter, however, but by learning to use tools like redescription in overcoming the difficulties provided by the subject matter we have chosen. Thus the idealized view persuades us to underestimate the extent to which science can deal precisely with those cases in which we *cannot* find simple and straightforward relationships between an antecedently sanctioned set of variables, instead of reinterpreting the phenomena in order to tell us what these most basically are. We are encouraged by this idealized view both to underestimate science and to expect too much of it. But it is important to realize that science has a role not only in those cases in which we are dealing with "simple," "repeatable," "human-sized" phenomena, but also those which are "complex," "variable," and "derivative." Although the simple cases may be the most salient, we should not overgeneralize them and ignore what science is most basically about.

Mary Williams, in a brilliant essay entitled "Similarities and Differences between Evolutionary Theory and the Theories of Physics"(1980) has addressed this issue indirectly, by way of an analogy between social science and evolutionary biology.[7] She points out that we are often misled to think that we cannot have laws—that science will not be able to succeed—in those cases in which we are dealing with phenomena that are apparently unlike those in the most salient examples of successful physics. She points out that there indeed *are* laws underlying the relationships between phenomena that are expressed at a level of description at which complexity inhibits their formulation (in physics and elsewhere). But, we just do not always realize this because the nomological relationship is obscured by the "size" of the theories we use to examine them. So, we are led into the false belief that the practical barriers generated by the subject matter, for instance, prevent the formulation of laws. But what does she mean by the "size" of a theory? The notion is a subtle one.

Williams uses the notion of "human-sized" and "non-human-sized" theories to make the point that there are some theories in science that deal with the phenomena at a level of description at which we are not used to thinking of them. This is meant to be a point about the nature of our perception (or conception) of the phenomena, which gives us insight into the intuitions that we may make about scientific phenomena. In short, her point is this: there are some scientific theories that deal with phenomena at a level of description at which we are used to thinking about them, and where we find it easy to understand the nature of the variables being used—the "natural kinds" we encounter perceptually are matched up with those used in the scientific theory. However, Williams notes, there are also other cases in which we are unable to identify the referents of any purported laws (i.e., the primitive terms), because they are so heavily embedded within a theory that defines the very nature of the phenomena. An example of the former might be the planets moving around the sun or balls rolling down an inclined plane (Williams, 1980, p. 389). An example of the latter might be "charm" or "color" in quantum theory. However, it is important to realize, as Williams points out, that such a distinction is "clearly not independent of background conditions and theories" (ibid., pp. 389–390). The planetary model of the atom, for instance, though not an object of direct perception, is nevertheless "human-sized" since it is intuitively familiar to us *conceptually*, and thus may not obscure the nomological referents. That is, whether a theory is "human-sized" has to do neither with the actual physical size of the referents nor with their perceptual status, but instead with the extent to which we find them intuitively familiar enough to conceive of them without the aid of a scientific theory.

Williams defines this notion as follows:

> I will hereinafter refer to phenomena which our ordinary intuition regards as individuals or individual events as "human-sized"; their physical size may differ by many orders of magnitude, but if the untrained intuition sees them as a whole, in a single gestalt, their size, psychologically, is "human-sized." (1973, p. 533)

Thus, Williams has drawn a distinction between two importantly different kinds of scientific theories.

But what use is it? Williams uses this distinction to explain why it is that in some cases we may fail to recognize an underlying nomological relationship even if there are no fundamental, or even practically insuperable, barriers to their discovery. The reason is that we often mistakenly assume that there is no role for nomological explanation in those cases in which the entities or processes to be explained do not deal with phenomena that are "human-sized." Why? Precisely because many critics of laws, like Scriven, have the *wrong picture of science*; consequently, they have been led to suppose that we can only have laws that connect those referents that we are used to dealing with, and mistakenly believe that there is no nomological relationship that could be legitimately explanatory outside of this. If the nomological referents are not "human-sized," she explains, we often jump to the conclusion that these are beyond our ability to comprehend—that barriers like complexity, for instance, would render the discovery of social scientific laws either impossible or impractical.[8]

But, of course, this criticism is precisely in concert with the one that we offered earlier against Scriven. In Williams's terminology, Scriven is saying that we cannot have nomological explanations connecting "non-human-sized" natural kinds. But why couldn't we? In our criticism of Scriven we saw that it was illicit to think that science could only succeed in finding nomological relationships between those referents that we antecedently deemed legitimate—a view we called "descriptivism." With Williams, we now understand that science may also succeed in formulating causal relationships by redefining what the entities are that underlie the phenomena we would like to explain. So, we may be able to find laws even when the phenomena are complex at the level at which we are used to conceiving of them, since it is not required that science proceed using only these descriptive terms.

Thus, as Williams points out, we often practice science precisely by *redescribing* the phenomena in terms that may *not* be familiar to us or similar to those in which we frame our original protoscientific questions about the phenomena. That is, we often seek explanation by connecting newly defined "theoretical" entities—new natural kinds—that may be nothing more than redescriptions of the phenomena we observe in everyday life, conceived of in a new way. And yet we *do* permit science the liberty of cutting nature at different perceptual "joints" than we do in our everyday experience. Why? Because we recognize that science *is* robust enough to deal with complex phenomena, and indeed it is capable of formulating lawlike explanations about them, despite the barriers generated by the subject matter as described at any given level of inquiry. And, as Williams explains, we use scientific theories to codify these redescriptions, in an attempt to capture the underlying relationships that may elude us at the surface level. We allow our scientific investigation to evolve a new vocabulary, perhaps unfamiliar to common sense, in order to redefine perceptually familiar phenomena in novel ways, and to systematize these redescriptions into sometimes highly derivative theories *precisely because* we understand science to be about something different from what the critics of social scientific laws have understood it to be.

The import of all this for the philosophy of social science is that there may well be more continuity between the subject matters, and methods of explanation, in natural and social science than one might have thought. If one throws out the unrealistic picture of natural scientific practice, motivated by the highly idealized view of the requirements for the formulation of scientific laws, we may begin to see certain similarities in the barriers facing natural and social sciences. And, methodological continuity in dealing with them would therefore appear to be more reasonable.

On this subject, by way of analogy with the natural science which is thought to have the most in common with social science—evolutionary biology—Mary Williams has written:

> I think the most important lesson to be learned is that our understanding of the structure of a science is likely to be seriously obscured by the failure to properly identify the referents of the laws (i.e., the relevant variables—or, in more formal language, the primitive terms). The extent to which this has obscured the structure of laws, predictions, and explanations, etc., in evolutionary theory for more than 100 years after the basic theory was understood (at an intuitive level) is astonishing. My feeling is that the same will be true in the social sciences (although the lesson learned from the experience in evolutionary theory should shorten the time needed to recognize the relevant variables). I suspect that it will not be possible to recognize them until social science has had its Darwin—I just don't think you can recognize such highly theoretical entities until the theory is available; Newtonian entities are visible in everyday experience, and can be recognized without a strong theory, but entities which are not literally visible are probably not recognizable until they are embedded in theory.[9]

Thus, although she is skeptical about the extent to which it may be recognized, Williams concurs that there is an underlying continuity between the situations facing natural and social science.

And, even if it does not turn out to be the kind of continuity hoped for by the positivists, in their desire for a unified science brought about by making the social sciences live up to the rigorous standards set by natural science, it may well be that we have learned an important methodological lesson for the social sciences by understanding that now perhaps even natural science does not live up to the idealized standards that many have supposed. Perhaps natural science faces problems in its subject matter that are more like what we have come to think of as the exclusive domain of social science. The lesson to be derived from this, however, is not to despair about the suddenly unbecoming status of natural science (and attempt to show that it could not produce laws either), but instead to find renewed grounds for optimism about the possibility and practicality of finding laws in social science, despite the barriers provided by factors like complexity.

Indeed, there are practical barriers to the achievement of laws brought about by complexity, in both the natural and the social sciences. But the extent to which these impinge on our ability to find laws probably does *not*, contrary to earlier critical views, split neatly along disciplinary lines. And, it is important to remember that the problem of figuring out how to ameliorate such difficulties is a perfectly appropriate job for science itself. It is a matter of individual empirical judgment, on a case-by-case basis, to determine those instances in which the barriers have so far proven to be so great as to suggest that we pursue alternative modes of explanation, in addition to the search for laws. But, even while doing this, we must not be so anxious to abandon the search for social scientific laws, and take refuge in such convenient doctrines as "descriptivism." In fact, descriptivism (from which the argument from complexity stems), stands as a barrier to the recognition of laws in both the natural and the social sciences, insofar as it tempts us to turn our focus away from the phenomena, and toward one particular preferred vocabulary for capturing them. We may gain insight here into the poverty of *all* purported arguments against the availability of social scientific laws, based on the alleged "intractability" of its subject matter. For a subject matter should not presuppose any particular theory or preferred description for capturing the phenomena to be explained.

Thus, I hope to have shown that it would be rash to give up on the nomological ideal as a useful structure for the explanation of social scientific phenomena, in light of the failure of the argument from complexity in showing that social scientific laws are unavailable, and based on the strength of the analogy between the methodological situations faced by natural and by social science. The nomological ideal has proven robust in the structuring of explanations, even in those cases in which we may ultimately face barriers to the immediate achievement of lawlike accounts. We may remain hopeful, therefore, about the prospects for a nomological social science, even while recognizing that the search for such laws cannot proceed without facing such problems as complexity—for in this way, as in others, social science reveals itself to be just like natural science.

Notes

I would like to thank Richard Adelstein, Brian Fay, Harold Kincaid, Daniel Little, Michael Martin, Peter Railton, Alex Rosenberg, Merrilee Salmon, Lawrence Sklar, Hal Varian, and Mary Williams for their many helpful comments on earlier versions of this chapter.

1. Arguments against social scientific laws are roughly divisible into claims about the intractability of the subject matter and claims about the irrelevance of laws to what we want explained about human behavior. I will focus here on the former.

2. Other arguments rooted in the alleged "intractability" of the subject matter of social science are the argument from openness (Popper 1961, 1965), and other various claims about the effects of "singularity," "uniqueness," "historicism," and the inability to perform controlled experiments on the subject matter of social science.
3. The leaf-falling analogy is popular in the philosophy of social science. Brodbeck writes: "The situation in social science ... is no different from that in physics. The physicist may know all of the principles involved yet be quite at a loss to predict, say, how many leaves will blow off a tree in the next storm. The poignant difference is, of course, that in social matters we desperately want explanation in detail, while in physical changes we are frequently indifferent" (1962, p. 47).
4. Of course, even this objection does not answer the earlier claim that one cannot rule out the possibility of discovering laws, even *given* a privileged vocabulary.
5. That is, one might say that there are two different kinds of redescription: one that is truistic (because it is based on naive correlation), and one that is grounded in a scientific theory (and is therefore not truistic). Perhaps there is no easy way to demarcate between these two in principle (just as philosophers of science have had a hard time providing a logical distinction between a naive correlation and a law), but *in practice* we may recognize a genuine distinction between truistic and nontruistic redescription, and should be wary of lumping them together.
6. In my dissertation (McIntyre 1991), I have explored an extended example from evolutionary biology (Dollo's Law) in order to show that it is indeed practical to formulate laws even in the face of complexity. One might also refer to Mary Williams's excellent "debunking" of complexity and other intractability arguments against laws in evolutionary biology (1970, 1973, 1980).
7. It is interesting to note that Michael Scriven has also drawn a link between social science and evolutionary biology. Scriven (1959a, 1969) has argued that complexity serves as a barrier to laws in evolutionary biology as well as social science. Yet, if Williams (1970, 1973, 1980) is right, then complexity is *not* an absolute, or even a practically insuperable, barrier to laws in evolutionary biology. Consequently, one might wonder whether the argument from complexity is less credible as a prohibition against laws in social science. Thus Williams maintains that evolutionary biology and social science *are* analogous, but in a radically different sense than Scriven; Williams contends that evolutionary biology and social science both face complexity, and yet, after suitable theory development, both are fully capable of supporting nomological explanations.
8. Williams herself deplores this tendency, and has written several landmark articles in which she upholds the nomological status of certain regularities in evolutionary biology (Williams 1970, 1973, 1980). See also Rosenberg (1985).
9. Mary Williams, personal communication, 17 October 1989.

References

Bernstein, R. 1976. *The Restructuring of Social and Political Theory*. New York: Harcourt Brace Jovanovich.

Brodbeck, M. 1962. "On the Philosophy of the Social Sciences." In E. C. Harwood (ed.), *Reconstruction of Economics*, pp. 39–58. 3d ed. Great Barrington, Mass.: American Institute for Economic Research.

Cohen, M. 1959. *Reason and Nature: An Essay on the Meaning of Scientific Method*. Glencoe, Ill.: Free Press.

Davidson, D. 1980. *Essays on Actions and Events*. Oxford: Oxford University Press.

Fay, B. 1983. "General Laws and Explaining Human Behavior." In D. Sabia, Jr. and J. Wallulis (eds.), *Changing Social Science*, pp. 103–128. Albany, N.Y.: SUNY Press.

Goodman, N. 1955. *Fact, Fiction, and Forecast*. Cambridge: Harvard University Press.

Handy, R. 1964. *Methodology of the Behavioral Sciences: Problems and Controversies*. Springfield, Ill.: Charles C. Thomas.

Hayek, F. 1967. "The Theory of Complex Phenomena." In *Studies in Philosophy, Politics and Economics*, pp. 22–42. Chicago: University of Chicago Press.

Hayek, F. 1979. *The Counter-Revolution of Science: Studies on the Abuse of Reason*. 2d edition. Indianapolis: Liberty Press.

Hempel, C. 1965. *Aspects of Scientific Explanation and Other Essays in the Philosophy of Science*. New York: Free Press.

Kincaid, H. 1988. "Confirmation, Complexity and Social Laws." In A. Fine and J. Leplin (eds.), *PSA 1988*, 2: 299–307. East Lansing, Mich.: The Philosophy of Science Association.

Kincaid, H. 1990. "Defending Laws in the Social Sciences." *Philosophy of the Social Sciences* 20, no. 1: 56–83.

McIntyre, L. 1991. "Problems in the Philosophy of Social Science: Towards a Defense of Nomological Explanation in the Social Sciences." Ph.D. dissertation, University of Michigan.
Nagel, E. 1961. *The Structure of Science: Problems in the Logic of Scientific Explanation.* New York: Harcourt, Brace, and World.
Popper, K. 1961. *The Poverty of Historicism.* New York: Harper Torchbooks.
Popper, K. 1965. "Prediction and Prophecy in the Social Sciences." In K. Popper (ed.), *Conjectures and Refutations: The Growth of Scientific Knowledge.* New York: Harper Torchbooks.
Rosenberg, A. 1980, *Sociobiology and the Preemption of Social Science.* Baltimore: Johns Hopkins University Press.
Rosenberg, A. 1985. *The Structure of Biological Science.* Cambridge: Cambridge University Press.
Rosenberg, A. 1988. *Philosophy of Social Science.* Boulder, Colo.: Westview Press.
Salmon, M. 1989. "Explanation in the Social Sciences." In P. Kitcher and W. Salmon (eds.), *Scientific Explanation,* pp. 384–409. Minnesota Studies in the Philosophy of Science, vol. 13. Minneapolis: University of Minnesota Press.
Scriven, M. 1956. "A Possible Distinction between Traditional Scientific Disciplines and the Study of Human Behavior." *The Foundations of Science and the Concepts of Psychology and Psychoanalysis,* pp. 330–339. Minnesota Studies in the Philosophy of Science, vol. 1. Minnesota: University of Minnesota Press.
Scriven, M. 1959a. "Explanation and Prediction in Evolutionary Theory." *Science* 130, no. 3374: 477–482.
Scriven, M. 1959b. "Truisms as the Grounds for Historical Explanations." In P. Gardiner (ed.), *Theories of History,* pp. 443–471. Glencoe, Ill.: Free Press.
Scriven, M. 1964. "Views of Human Nature." In T. Wann (ed.), *Behaviorism and Phenomenology: Contrasting Bases for Modern Psychology,* pp. 163–190. Chicago: University of Chicago Press.
Scriven, M. 1969, "Explanation in the Biological Sciences." *Journal of the History of Biology* 2: 187–198.
Williams, M. 1970. "Deducing the Consequences of Evolution: A Mathematical Model." *Journal of Theoretical Biology* 29: 343–385.
Williams, M. 1973. "Falsifiable Predictions of Evolutionary Theory." *Philosophy of Science* 40: 518–537.
Williams, M. 1980. "Similarities and Differences between Evolutionary Theory and the Theories of Physics." In P. Asquith and R. Giere (eds.), *PSA 1980,* 2: 385–396. East Lansing, Mich.: The Philosophy of Science Association.
Wootton, B. 1950. *Testament for Social Science: An Essay in the Application of Scientific Method to Human Problems.* New York: W. W. Norton.

Chapter 10
Reflexive Predictions
George D. Romanos

I

The problem of reflexive predictions is a familiar one for social scientists. The most well-known examples of such predictions probably are those which occur in political science where the publication or dissemination of predictions of future political phenomena, such as voting behavior, is frequently assessed as the very cause of these predictions' coming out true or false. Perhaps the single, most notorious instance of such a prediction, even if not the most clear cut, was the prediction of the outcome of the 1948 presidential election. Here the widespread publication of predictions that Dewey would achieve a landslide victory is commonly supposed to have been one of the most important factors leading to his ultimate defeat.

Still, self-frustrating and self-fulfilling predictions, prophecies, and forecasts are hardly restricted to this one discipline, and they are generally conceded to be likely to occur just as frequently, if somewhat less spectacularly, throughout the entire range of social sciences. What is not so generally conceded is their ultimate scientific importance and whether or not it is conceivable that they may occur altogether outside of a social science setting, i.e. in the purely "physical" contexts of the natural sciences. The problem of determining once and for all just what sort of problems reflexive predictions do in fact pose for scientists and whether or not their occurrence is unique to the social sciences can be attributed in large part to the general confusion over just what a reflexive prediction really is and just how it operates in helping to bring about its own ultimate truth or falsity.

Roger Buck in his paper "Reflexive Predictions" [1] undertakes to do three things: first, to clarify the notion of reflexive predictions; second, to assess their methodological import for the social sciences; and third, to offer "indirect" support for the thesis that reflexive predictions are restricted to the social sciences and mark a "philosophically interesting difference" between these and the natural sciences. Buck claims to achieve this last objective through a supposed refutation of a counterexample proposed by Adolf Grünbaum to the so-called unique-to-human-affairs thesis, but Buck's argument here leans heavily on the conception of prediction reflexivity he sets forth at the start of his paper.

I would like to show first that even granting the uniqueness of reflexive predictions to the social sciences, the degree of philosophic import which this appears to entail is largely illusory. In so doing I will examine only the most conventional and well-recognized sorts of reflexive predictions—those commonly encountered in the social sciences and more or less notorious for the difficulties they have in fact presented to social scientists in the past. In my discussion of the nature of the problems presented by such predictions I will assume Buck's own conception of a reflexive prediction, which I take to be an essentially fair description of the "conventional" examples.

I will not explicitly consider Buck's conception until later on in my chapter where I intend to attack it insofar as it purports to restrict and define the class of reflexive predictions, and in so doing stacks the cards heavily in favor of the "uniqueness" thesis. This criticism, however, will occur only after I examine the particular argument Buck offers against Grünbaum's proposed counterexample.

In concluding, I will offer an alternative characterization of what is involved in a prediction's being reflexive, as a replacement for the one put forth by Buck. This will be carried out in a somewhat broad, schematic manner, but I think it will both illustrate the essentially simple nature of the problem of reflexive predictions and make it intuitively more clear why such phenomena themselves fail to present any really "philosophically interesting" issues.

II

The phenomena of reflexive predictions have frequently been thought to present serious methodological obstacles for the social scientist, obstacles in theory testing "impossible" to overcome. This has to do with the way the very publishing or even "making" of certain kinds of "scientific" predictions—what Buck calls the "dissemination status"—is supposed sometimes to be enough, in itself, to cause the prediction to come out true or false, e.g. in the case of predictions of bank failures and election results. Such reflexive predictions are classed by Buck into two groups, the "self-fulfilling" and the "self-frustrating." The "nature" of these predictions is thought to deny a scientist the conditions necessary to conduct a genuine test of the theory from which the prediction has originally been derived.

In section III of his paper Buck concludes that the occurrence of reflexive predictions in the social sciences entails no overwhelming methodological problems. He gives essentially two key reasons for this. First, Buck claims that any given prediction is open to empirical investigation of a sort which will be able to determine whether or not it is indeed reflexive. Second, Buck points out that the scientist may at any time keep the prediction private or restrict its dissemination in various ways, whenever it is deemed necessary in order to guard against the self-fulfilling or self-frustrating effects of its reflexivity.

I do not disagree with this argument or its conclusion. However, I do feel it is not absolutely clear that Buck indeed makes the important point he might. Buck appears to be arguing simply that reflexive predictions need not present insuperable obstacles to the social scientist, that there exist ways of dealing with this problem. What Buck does not do is stress the implication here, that there is no logical connection between a prediction's being reflexive and its being a problem for the scientist in the process of theory testing, and therefore that the problem presented, to begin with, is only *technical* and not *methodological* at all.

By keeping the scientific considerations involved in testing a hypothesis distinct from, say, ethical or pragmatic considerations, which may determine whether it is, or ought to be, publicly disseminated, we may clarify two correspondingly different uses of the term "prediction."

When a scientist makes a determination of the truth value of a prediction, in the process of testing the theory from which it has been derived, the predicted state of affairs or event need have no intrinsic interest or value for him (or anyone else) whatsoever. All that concerns him is that if the prediction comes out true or false, generally speaking, there is accordingly increased or lessened confidence in the theory itself. Thus a prediction employed primarily for the purpose of testing a theory is a prediction which has an essentially *scientific* motivation and purpose behind it, and a determination of its truth value awaits the outcome of future events.

Another sort of prediction, which may or may not be based on "theories" (interpreted loosely) of varying degrees of popular and professional acceptance, is the prediction that is taken to be genuinely informative or predictive, i.e. as conveying actual information. Examples possessing such an informative character are predictions by fortune-tellers, government economic predictions, weather forecasts, astronomical predictions, and many others. These sorts of predictions purport to be simply true, or "probably-true" statements about the future. The states of affairs or events which they predict generally possess an intrinsic interest or social importance which is altogether separate from any scientific purpose (e.g., confirmatory function) which they may or may not simultaneously serve. This second kind of prediction has a conspicuously informative character and purpose.

It is entirely possible that a scientist might derive a prediction from a theory which could be even "crucially" important for determining the acceptability of that theory, and yet at the same time the prediction might be one that, if publicly disseminated, would excite great and widespread interest among other persons because of the value placed on the information about the future which the prediction is taken by them to convey. At such times the scientist may be under great pressure of various kinds to divulge or "publish" the prediction. It should be noted that no "scientific" prediction need have any such social value or importance any more than there need be scientific interest surrounding a prediction purporting to be informative or "predictive." Still it is entirely possible for scientific motivation and purpose and nonscientific interest to converge in the case of certain predictions.

It is a special case of this convergence that presents serious and perplexing obstacles to the scientist attempting to test a theory. In such an instance the "nonscientific" import of the prediction is such that whether or not the prediction is publicly disseminated happens to be causally related to the truth value it will have. It is in these cases, therefore, that scientific considerations vie with nonscientific considerations over whether or not the prediction ought to be published. For the purposes of science there is simply no a priori reason why any genuine hypothesis (prediction) must, or must not, be publicly disseminated at any given time. The reflexivity of such a prediction has no *logical* consequences with respect to testing or validating the theory from which the prediction has been derived. Buck's reasoning certainly implies as much, but he continues to refer to the problems caused by reflexive predictions as "methodological."

The reason I find this point worth making is that although Buck holds that the problems of prediction reflexivity can easily be overcome, as long as he maintains that the problems are of a *methodological* nature and peculiar to the social sciences he has a prima facie argument for a *methodological* (and as a result "philosophically interesting") distinction between the natural and the social sciences. What I want to show is that even allowing (as I do not) that reflexive predictions are peculiar to the social sciences, Buck's own comments make it clear that any distinction between the two branches of science so rooted must be purely *technical*.

It must be made clear that the particular reflexive predictions discussed here are simply the ones which have as a matter of fact caused the problems and generated the controversy. It should not be supposed that I assume these and closely related ones (i.e., those which would behave reflexively if disseminated, or others, disseminated, but not "scientifically" utilized, whose dissemination nonetheless contributes to the outcome of the very events about which they are predictions) to exhaust the entire class of reflexive predictions. In a moment I will try to show how logically and practically speaking, "public dissemination" is only one *means* by which predictions may come to exhibit reflexivity. As a result the notion of a prediction's conveying important or valuable information will be seen, strictly speaking, as relevant only to a certain subset of reflexive predictions, those which happen to exhibit their reflexive nature through "public dissemination." And therefore it will follow that it is only this same subset of reflexive predictions for which we will deem Buck's phrase "dissemination status" as appropriate, i.e. as designating the relevant kind of causal factor that brings about the "reflexive behavior" of such predictions.

III

Adolf Grünbaum does not think that instances of reflexive predictions are *necessarily* restricted to the realm of the social sciences. To make his point he cites the example of a computer's predicting that a missile will miss its target. This prediction is passed on to the missile in such a way that the missile alters its course and hits its target, thereby apparently disconfirming the computer's prediction ([4], pp. 239–240). In describing this case, Grünbaum appears to be relying heavily on an analogy between the missile's behavior and the human behavior supposedly exemplified during "ordinary" occurrences of reflexive predictions. Whether or not Grünbaum believes such an analogy to be necessary in order to validate his example is not clear. It is clear, however, that Buck thinks it is necessary and takes Grünbaum to think likewise. Buck therefore undertakes to show that the missile's behavior is not analogous to the appropriate human behavior and that it can never be so construed.

Buck insists that in order for Grünbaum's counterexample to stand up it must be plausible to view the missile as "acting on the belief" that the computer's prediction is true. Buck quickly points out that two alternative extreme responses to the statement of this requirement both beg the question; i.e. either, that it is simply impossible for machines to believe or to "act-on-beliefs," since only humans can do that; or, "diametrically opposed" to this view, the so-called reductivist answer, that the phenomenon of belief can be easily reduced to physical or mechanical principles ([1], p. 367). Buck himself chooses to respond in a different vein. He wishes to allow the possibility of analogy between machine behavior and that of humans, but he does not intend to let such an analogy extend to the point where it will be considered proper to say of a machine either that it "believes" or that it is acting on a belief.

Buck attempts to justify this position by pointing out the wording of Grünbaum's counterexample, in which it is said that the missile receives the prediction from the computer "in the form of a set of instructions" ([1], p. 367). Buck indicates this as an example of how it is proper to speak of a machine's *acting-on-orders* but not of it *acting-on-beliefs*. That is, Buck claims that the appropriate analogy between machine behavior and human behavior extends only up to the level of receiving and acting upon orders or instructions, and not to the point of forming beliefs and acting upon them. Buck feels he has made a correct distinction between the behavior of men and that of machines insofar as he has indicated the distinction between the imperative mood in which instructions are issued and the indicative mood in which statements appropriate to being believed or disbelieved are expressed.

Now certainly it may be allowed as a cognitive fact that it is quite possible for a person to understand and obey orders without necessarily having any sort of knowledge or belief concerning the truth or falsity of a given prediction, but this in no way explains why "taking orders" is so clearly an appropriate analogue for describing machine behavior while "receiving information" and "forming beliefs" is not. It is not at all clear why it should be so correct to think of a machine understanding an order but not of it believing a prediction.

While I may grant that "taking orders" represents a somewhat more common mode of talking about machine behavior, I nowhere see the hard and fast distinction Buck is trying to draw. Machines ideally do not respond to either the imperative or the indicative mood. Buck fails to explain why marking this grammatical distinction between the imperative and indicative moods, or the epistemological one between understanding (orders *or* statements) and believing is helpful in drawing the distinction between man and machines or human and nonhuman behavior. That all these distinctions exist is granted, but that the latter ones are significantly related to the former two is not at all obvious. That they are so related is almost totally unargued for by Buck except for a single erroneous claim about actual speech usage ([1], p. 445).

IV

Buck and Grünbaum both attribute the original thesis that reflexive predictions are peculiar to the social sciences to Robert Merton in his book, *Social Theory and Social Structure*. Here Merton points specifically to the absence of "self-stultifying" or self-frustrating predictions in the natural domain saying, "that a meteorologist's prediction of continued rainfall has until now not perversely led to the occurrence of a drought" ([5], p. 122) and that "predictions of the return of Halley's comet do not influence its orbit" ([5], p. 181). It was essentially in response to these and similar remarks by Merton that Grünbaum originally offered his counterexample, and this point is stressed by Grünbaum in his reply to Buck's criticism ([3], p. 370).

Grünbaum does not contest the general validity and relevance of Buck's distinction between acting-on-orders and acting-on-beliefs, and he even endorses the decision to regard it as unreasonable to speak of a machine's acting-on-beliefs, but reasonable to speak of one acting-on-orders (as long as this in itself does not presuppose some sort of ability to "believe" on the part of the machine) ([3], p. 371). Grünbaum's reply is simply that he originally proposed the case of the missile as a counterexample to Merton's position only, as stated in his book, and that no claim was made therein by Merton that a "self-stultifying" prediction need exhibit its reflexive character only through being believed and thus acted upon. Furthermore Grünbaum points out that it is the definite implication of the "physical examples" Merton employs that he would require no more of, say, a self-frustrating meteorological prediction of continued rainfall than just that it lead to something like the occurrence of a drought ([3], p. 320). Grünbaum cites as his final justification Merton's own concession (albeit qualified) to his (Grünbaum's) counterexample in the revised edition of his (Merton's) book ([6], p. 129n.).

One may well grant that Grünbaum has given a legitimate counterexample to Merton's thesis, but this ignores the more important question. At this juncture we want to know not simply whether the counterexample satisfies Merton's conditions but rather if its admitted failure to meet those set by Buck rules out its counting as a plausible instance of a reflexive prediction. Grünbaum is more or less silent on this question as to whose standards, Buck's or Merton's, we are to regard as more correct. This is even more unfortunate since Buck never intended to refer to Merton as an authority, and even explicitly rejects Merton's later concession to Grünbaum as entirely unwarranted ([1], p. 444).

Accordingly, in his rejoinder to Grünbaum, Buck justifiably points out Grünbaum's failure to deal with the main point at issue, and reiterates his position that what in fact really qualifies a prediction as self-frustrating (or just reflexive) is the "causal mechanisms [beliefs, and actions predicated on such beliefs] which mediate between the event which is the issuance of that prediction and the nonoccurrence of the event predicted" ([2], p. 373). This contention remains the basis for Buck's claim that in order to construct the analogy required to demonstrate that reflexive predictions may conceivably occur in the natural sciences, it is necessary first to postulate a machine whose behavior "differs strikingly" from Grünbaum's, i.e. a machine of which it will not be improper to say that it is acting-on-a-belief.

The question now is clearly reduced to: "Why is it necessary to postulate such a machine in order to construct the needed analogy?" And what this comes down to is simply the conception of reflexive predictions with which Buck is working. In section II of his paper Buck attempts to clarify the concept of "reflexive prediction" and he there sets forth the following four conditions to be taken together as both necessary and sufficient for determining whether or not a given prediction is reflexive:

> (1) Its truth-value would have been different had its dissemination status been different,
> (2) The dissemination status it actually had was causally necessary for the social actors involved to hold relevant and causally efficacious beliefs,
> (3)The prediction was, or if disseminated, would have been believed and acted upon, and finally
> (4) Something about the dissemination status or its causal consequences was abnormal, or at the very least unexpected by the predictor, by whoever calls it reflexive, or by those to whose attention its reflexive character is called. ([1], pp. 361–362)

Now Buck has previously stated in this section that the "dissemination status" of a reflexive prediction (i.e., whether or not the prediction is published or revealed to certian "social actors") "must be *a* causal factor relative to what it [the prediction] predicts" ([1], p. 361). Now (2), (3), and to a somewhat less obvious extent (4), serve the purpose of establishing this condition as necessary for a reflexive prediction while simultaneously making sure that the class of reflexive predictions is strictly limited to just those predictions whose "dissemination status" plays such a causal role only insofar as it has something to do with the holding of "causally efficacious beliefs" by certain "social actors" who are said to "act on" them.[1] In other words Buck is here stating as part of his formal criteria for reflexive predictions that they be amenable to the acting-on-beliefs mode of interpretation.

How does Buck justify this restriction? Why does he consider it *necessary*? The only basis for this which he mentions in his entire paper is simply his own observation that "at least in all the standard examples in the literature, the causal efficacy of such dissemination must be mediated by the formation of beliefs on the part of the various actors on the social scene, and by their behaving in such a way that can be reasonably described as 'acting on' those beliefs" ([1], p. 360).

My criticism here is simply that by Buck's own formulation of "reflexivity" for predictions there is little or no way to distinguish "standard" examples from "non-standard" examples (like Grünbaum's) except by virtue of considering whether or not they are appropriate to the acting-on-beliefs model. Yet Buck here attempts to justify the necessity that the acting-on-belief model be applicable by appealing to the same notion of a "standard" case of a reflexive prediction. It is clear that the reasoning by which Buck here restricts the class of "standard" (interpreted as "legitimate" or "proper") is hopelessly circular.

To interpret his use of "standard examples" in some weaker sense, such as "*most examples*," "the *average* example" or "the *most common* examples," would not suffice to warrant inclusion of (2), (3), and (4) in their present form as necessary conditions for a prediction's being reflexive. What I am suggesting now is that Buck's insistence that we must necessarily be able plausibly to explain the reflexive behavior of a prediction according to the acting-on-beliefs model, if such a prediction is to be allowed as genuinely reflexive, is entirely unwarranted. Moreover, it seems mistakenly to stem from the mere fact that in *most* of the cases (or perhaps all) where the pseudo methodological problems of reflexive predictions have happened to occur, the reflexive behavior of such predictions has been commonly explicable according to the acting-on-beliefs model.

If we waive the acting-on-beliefs restrictions, we are left, for consideration, something like the following more or less general condition as necessary and sufficient which a prediction must meet if it is to be correctly classed as reflexive:

> *R* The dissemination status of the prediction must be a causal factor relative to the prediction's coming out true or false.

This condition I take to capture sufficiently what is important from Buck's list after all talk about believing and acting-on-beliefs is eliminated.

Essentially I have only one reservation concerning the adequacy of the above condition, and that reservation has to do with the use of the term "dissemination status." According to Buck's use the dissemination status of a prediction is said to be "public" or "private," and "published" or "unpublished." The use of such language, if not taken in a somewhat strained metaphorical sense, may, as previously noted, seem to imply the "standard" sort of reflexive predictions. I should like presently to propose an altogether different way of speaking in this situation. For now, however, it will be enough if talk about the dissemination status of reflexive predictions is *not* taken as appropriate only in those circumstances where the acting-on-beliefs mode of explanation is conventionally understood. With this qualification in mind I shall tentatively accept *R* as both a necessary and sufficient condition for determining a reflexive prediction.

To return to the question of the aptness of Grünbaum's counterexample, it is now evident that if there is any "question begging" going on here it is on Buck's part. His introduction of talk, about believing and acting-on-beliefs, into such a central role in the discussion must now appear thoroughly unjustified; and his demand that we bring our reflexive prediction analogues from the natural sciences up to his standards appears as quite arbitrary. That *belief*, or even *understanding*, is involved in the case of certain reflexive predictions is simply *a hypothesis* about the cause of the reflexivity exhibited by those predictions, a hypothesis, in Buck's own words, about the nature "of the causal mechanisms which mediate between the event which is the issuance of that prediction and the non-occurrence [or occurrence] of the event predicted" ([2], p. 373).

Buck has, therefore, failed in his attack on Grünbaum's counterexample in two respects. First, even granting his thesis that the acting-on-beliefs model must be applicable to examples of genuine reflexive predictions, he has failed to show why "acting-on-orders" is appropriate for describing some machine behavior while "acting-on-beliefs" is not. Second, and more important, it has been shown that Buck's demand, that Grünbaum's missile do anything like believe or act-on-beliefs, is without foundation from the start.

Grünbaum certainly attempts to show an essential similarity between the occurrence of a certain phenomenon in natural science and the occurrence of others (reflexive predictions) in the social sciences. However, the success of this endeavor does not depend upon his demonstrating that thinking, understanding, believing or acting can occur within a purely "physical" (or "natural") context. What Grünbaum must, and does, show is only that the "issuance" of the computer's prediction in his example is *somehow* responsible for bringing about conditions under which the prediction itself turns out false, just as in the more conventional cases of self-frustrating predictions in the social sciences. Any further analogy, between the behavior of Grünbaum's missile and some sort of human behavior, is irrelevant and unnecessary for the purpose of establishing the validity of his counterexample.

V

The essential thing, of course, in both Grünbaum's example and the more conventional ones from the social sciences, is that in each case the prediction possesses something loosely described as a "self-fulfilling property" or "self-frustrating property" in virtue of which the prediction is said to be reflexive. A prediction possessing this reflexive character is more or less likely to turn out true or false *on its own accord*, independently, as it were, of the conditions indicated as causally relevant by any theory from which it may have been originally derived. I have criticized Buck's attempt to clarify the issue, and now I will attempt to formulate somewhat more specifically just what is involved in a prediction's turning out reflexive.

Central to my analysis of the problem is the replacement of Buck's notion of a prediction's "dissemination status" with a different one which, I will attempt to show, operates more effectively in essentially the same role. The substitute notion I will propose is that of a *formulation/dissemination style*, and before I can show that it is an improvement I must first try to explain just what it is.

Every prediction, in order to be *made* at all, must be formulated in some way. By this I mean that there must be something like a certain sequence of sounds or inscriptions which is said to convey or express the prediction. Even predictions made only "in the mind" may be said to have some manner of formulation insofar as they would admit of being uttered or written down. Prediction formulations needn't be viewed as possible only in natural languages which can be spoken or written in the conventional manner, but may be constructed out of such things as electric impulses, bodily movements, puffs of smoke, or anything else which may be interpreted as expressing a prediction. Thus every prediction will have a formal or "syntactical" property, which I will call a *formulation style* (F-style), corresponding to the manner of the original and paradigm production of some well-formed formula subject to interpretation as expressing that prediction. A complete description of a prediction's formulation style will therefore be possible in the appropriate empirical or "physical" terms.

Now any prediction will be considered subject to possible dissemination, and many predictions are in some way actually disseminated. Here we must carefully avoid any implications "dissemination" may seem to involve of a notion of prediction reflexivity based ultimately on Buck's acting-on-beliefs model. We will speak therefore of a prediction's *dissemination style* (D-style) in purely neutral language, meaning only the manner of reproduction and/or transmission of a given prediction *in a certain formulation style*. Any prediction undergoing such reproduction and/or transmission we will regard as disseminated. In this way the transmission of the computer prediction to the missile in Grünbaum's example shall be considered as genuine an instance of prediction dissemination as the publishing in a newspaper of an election prediction. Thus any disseminated prediction will have in addition to its formulation style, a dissemination style, likewise describable in suitable empirical terms.

Now every prediction made must have at least one F-style and may have a number of them. Any prediction may have one or more D-styles.[2] Now, for any prediction, a full description of one of its D-styles conjoined with a full description of the appropriate F-style will constitute a complete description of one of that prediction's *formulation/dissemination styles* (*F/D-style*). A prediction may have many F/D-styles. For any prediction which is not disseminated, and therefore has no D-style as such, each of its F-styles will constitute a distinct F/D-style as well. Similarly for F-styles of disseminated predictions which, however, do not appear in such dissemination. As the separation of the D-style from its F-style may prove unfeasible other than in a purely logical sense it will be sufficient to stipulate that every prediction must have at least one F/D-style which may be distinctively described according to its manner of formulation, reproduction and transmission in at least one instance.

I now propose to substitute "formulation/dissemination style" for "dissemination status" in *R* and the result may be taken as stating, finally, the necessary and sufficient condition for a prediction's being reflexive:

> R_1 The formulation/dissemination style of the prediction must be a causal factor relative to the prediction's coming out true or false.

The "F/D-style" is an improvement over "dissemination status," first of all, because it does not carry the aforementioned undesireable connotations of the latter expression and thus is not automatically inimical to cases in the natural sciences. A second important reason, not unrelated to the first, is that inasmuch as the dissemination status of a prediction admits only of being either (relatively) private or (relatively) public, the concept is inadequate to cover all possible causally relevant features which may render a prediction reflexive. This limitation of "dissemination status" yields unsatisfactory consequences even in contexts for which Buck primarily intended it. For example, imagine a prediction of bank failure published only in Japanese among an English-speaking population. A satisfactory account of even such simple cases thus obviously takes us beyond the notion of a prediction's dissemination status as merely public or private. The concept of an F/D-style will facilitate a more detailed and complete approach in the attempt to identify and assess the causal factors at work in any given case of a reflexive prediction. Most important it will be seen to cover a far wider range of cases.

Realizing that though "dissemination status" suffers from a narrowness and limited applicability, it may, as a concept, nonetheless be granted an intuitive palatability in those situations for which it initially seemed more or less sufficient. To alleviate the somewhat awkward nature of my corresponding notion of an F/D-style I will attempt to elaborate on just how I suppose the F/D-style of a prediction operates in the manner mentioned in R_1. I will give a brief sketch of what goes on in a typical case of a reflexive prediction, illustrating clearly, I hope, how the F/D-style, regardless of its individual nature in any particular instance, plays its causal role.

I would like to consider first a case of a self-fulfilling prediction. Let us suppose that we are scientists of some kind and we wish to test some theory (hypothesis) or other which we will refer to as T_1. From T_1 (together with, we may suppose, certain empirical assumptions) we derive a certain prediction, P, that an event, e_1 will occur. Generally speaking, if e_1 fails to occur we will regard this as tending to disconfirm our theory (T_1) while if e_1 does occur we shall regard this as tending to confirm our theory (T_1). Now we have described above the way in which any prediction may be said to possess at least one F/D-style. Let us suppose that our prediction P has just one such F/D-style and that its *occurrence in this F/D-style* is an event which we will refer to as e_2. Now I suggest that P will turn out to be a self-fulfilling (reflexive) prediction just in case there is some other, well accepted or likely theory, T_2, given certain conditions C (which obtain), e_2 is sufficient to bring about the occurrence of e_1 (i.e., the event originally predicted by P).

The analogous case for a self-frustrating prediction would be for the theory in the T_2 position to causally relate e_2 with the nonoccurrence of e_1 (the event predicted by P), paralleling the way it (e_2) was related to the occurrence of e_1 in the self-fulfilling example.

Now for us, as scientists interested in testing our theory (T_1), we shall not know, in the self-fulfilling instance, whether to count the occurrence of e_1 (i.e., P's turning out true) as confirmation of T_1 or merely to attribute it to the explanation provided by T_2. In the self-frustrating example we cannot take the nonoccurrence of e_1 (i.e., P's turning out false) as a falsifying instance of our original theory, as long as it seems at least as likely that this turn of events is adequately explained by the theory now in the T_2 position.

It is simple to see how to relate this explanation to the conventional reflexive predictions of the social sciences, such as forecasts of bank failures or election outcomes. In these cases the original economic or political theory from which the forecast or prediction is derived occupies the T_1 position. The theory which relates the written or auditory verbal strings and their dissemination to human behavior (perhaps through talk of "acting-on-beliefs") and the outcome of future events (which of course the prediction in question happens also to be "about") occupies the T_2 position.

In a case like that described in Grünbaum's counterexample the relevance is perhaps even easier to see. The (physical) theory according to which the computer derives its original prediction that the missile will miss its target is in the T_1 position. The theory which causally relates the transmission and reception of certain electromechanical "messages" to the subsequent behavior of the missile in flight is in the T_2 position.

Now in this chapter, due to our initial concern over the so-called methodological problems presented by reflexive predictions, we have been concerned primarily with the occurrence of reflexive predictions in situations where there is an earnest attempt to validate or test the theory in what we now refer to as the T_1 position. However, apart from these situations there is altogether no necessity that a prediction, in order to be reflexive, be originally derived from any "scientific" theory at all. A prediction simply "made up" or gotten from sheer "intuition" is as likely to behave reflexively as any other.

It is well known that reflexive predictions are frequently "manufactured" for the sole purpose of manipulating the outcome of future events. In these situations the prediction is produced *in a certain F/D-style* because (according to a theory (T_2)) it will effect (or prevent) the occurrence of the predicted event. Well-known examples of this kind are the predictions issued during political campaigns with a hope of producing what is commonly called a "bandwagon effect."

In the case of Grünbaum's missile example, on the other hand, we have a sort of mixed situation. There is a theory in the T_1 position, but it is not undergoing a test at all. Rather, both it and the theory in the T_2 position are being employed to make sure the missile hits its target. The F/D-style of the prediction of a missed target (obtained from T_1) is purposely selected for its causal efficacy (according to T_2) in preventing what the prediction predicts.

The requirement that the theory in the T_2 position always be something like credible or likely true is simply to ensure that we do not overstretch the sense of saying (as we did in R_1) that in any given instance a reflexive prediction's F/D-style is causally related to the truth value it (the prediction) will have. This point may be better illustrated by saying if the only theory (in the T_2 position), whereby the occurrence of the prediction is related to its eventually coming out true or false, is extremely unlikely, then it will be as unjustified to call the prediction reflexive as it is incredible to say its occurrence (in a given F/D-style) is causally related to the truth value it will have.

It should be noted that, according to this view, reflexivity is a property which, properly speaking, cannot be possessed simply by a prediction as such, but only by a prediction *in a certain F/D-style*. Thus it is not really legitimate to speak of a prediction's being reflexive in itself, but only a certain *instance* or *occurrence* of it. This I assume to hold equally true regardless of whether prediction reflexivity is construed along my lines or along the narrower ones set by Buck.

This then is the view which I feel captures all that is essential to the notion of a reflexive prediction, as well as generally comporting with the facts. I think it makes clearer just why reflexive predictions are in principle so easily avoided in the process of theory testing; for it is manifestly apparent that there is no a priori reason why we should have to produce any prediction in one F/D-style rather than another. Also, this account allows for the possibility that a prediction may behave reflexively not only as a result of its being clearly "disseminated" in some way, but just in virtue of its simply being made ("formulated") at all.

Most important, the view outlined here does not restrict in any way the range of possible explanations which may be advanced in order to account for the reflexive behavior of a given prediction. It allows for the existence of reflexive predictions in the natural sciences without any implications whatsoever of the "crude reductionism" that Buck claimed would be necessary. Buck assumes one causal account of the reflexive behavior of some reflexive predictions and then takes it to be, a priori, true of all reflexive predictions. Here, in contrast, Buck's acting-on-beliefs mode of explanation has been seen to be just one possible alternative to, one other candidate for, the T_2 position. The most common examples of reflexive predictions may well be those from the social sciences; and the explanation of their behavior in terms of certain relevant "social actors acting-on" certain relevant beliefs may well prove generally the most satisfactory. Yet, this is no reason to suppose that the very concept of "reflexive prediction" entails the truth of just such statements.

Notes

1. (4), however, is also concerned to ensure that anything which is to be correctly attributed a causal role in bringing about a given effect must, all things being equal, be considered "abnormal" or "unexpected." This point, even if undisputed, properly relates rather to the notion of "cause" or "*the* cause" and as such is of only derivative importance here where it is superfluous and out of place.
2. The problem, should it arise at all, of ascertaining the identity of a prediction through different F- and D-styles, is considered essentially a problem of deciding on acceptable or suitable translations.

References

[1] Buck, R. C. "Reflexive Predictions." *Philosophy of Science* 30 (1963): 359–369.
[2] Buck, R. C. "Rejoinder to Grünbaum." *Philosophy of Science* 30 (1963): 373–374.
[3] Grünbaum, A. "Comments on Professor Roger Buck's Paper 'Reflexive Predictions.'" *Philosophy of Science* 30 (1963): 370–372.
[4] Grünbaum, A. "Historical Determinism, Social Activism, and Predictions in the Social Sciences." *British Journal for the Philosophy of Science* 7 (1956).
[5] Merton, R. K. *Social Theory and Social Structure.* (1st edn.) Glencoe: Free Press, 1949.
[6] Merton, R. K. *Social Theory and Social Structure.* (Revised edn.) Glencoe: Free Press, 1957.

PART III

Interpretation and Meaning

Introduction to Part III

According to naturalism, the primary goals of the social sciences are the explanation and prediction of social phenomena by means of laws. In part II, criticisms as well as defenses of this position were considered. As an alternative, antinaturalists have maintained that the goal of the social sciences should be the understanding of meaning and that its method should be interpretation.

Interpretation as an approach to social and historical phenomena has a complex history that can only be sketched here.[1] Hermeneutics—a method of interpreting meaning—was developed centuries ago in the context of literary[2] and religious exegesis[3] in order to try to understand authoritative texts in which meaning was disputed or hidden. Although it is indirectly influenced by this older hermeneutic tradition, the modern interpretive approach to the social sciences and history can be more directly traced back to the late nineteenth- and early twentieth-century reaction of German intellectuals such as Wilhelm Dilthey and Heinrich Richert to positivistic approaches to social and historical phenomena.[4] Scholars also cite two other influences on the development of this tradition. Phenomenology, especially through the work of sociologist Alfred Schutz and ethnomethodologists such as Harold Garfinkel, has had an impact on the contemporary interpretive approach. In addition, the later philosophy of Wittgenstein as represented by Peter Winch, especially in his widely read book, *The Idea of a Social Science*, has provided an interpretivist criticism of naturalism that has greatly influenced some British and American analytic philosophers.[5] Despite Winch's contribution, however, the interpretive approach has had its greatest effect on continental philosophy. Indeed, such well-known contemporary continental philosophers as Paul Ricoeur, Hans-Georg Gadamer, Karl-Otto Apel, and Jürgen Habermas have been influenced by this tradition to varying degrees.

Historical origins aside, the interpretive approach makes three basic claims.[6] It holds that social phenomena—social practices, institutions, behavior—are intrinsically meaningful and that their meanings are constituted by the meanings that social actors give to them. Moreover, it claims that social phenomena can be understood only by unraveling the meanings that constitute them, a process that involves understanding the social phenomena from the actor's point of view. Finally, it insists that causal explanations, inductive generalizations, and predictions have little or no importance in the social sciences.

The criticisms of this interpretive program are many. Critics argue that despite what interpretivists say, their method is not distinct from those used in the natural sciences. They maintain that although some interpretivists hold that the criteria of what constitutes a valid interpretation are subjective and arbitrary, objective standards of validity are possible. Contending that the interpretive approach has too limited a view of what the social sciences can do, they say that there is no reason why social science cannot causally explain a social phenomenon and also interpret it in terms of meaningful categories. And finally, they claim that the interpretive approach takes too limited a view of what constitutes understanding. There are many ways of understanding social phenomena besides adopting the categories of the social actors, critics note.

In this part, we provide selections by four advocates of the interpretive tradition—R. G. Collingwood, William Dray, Charles Taylor, and Clifford Gertz—and three critics—Dagfinn Føllesdal, Jane Roland Martin, and Michael Martin.

In a selection from *The Idea of History* (chapter 11), Collingwood maintains that it is a serious mistake to construe the science of human nature on the model of the natural sciences, that is, to regard human beings as "mere phenomena" that one observes from "the outside." We must understand human nature historically from "the inside," he says. By this he means that to understand human nature we must know people's desires, beliefs, and thoughts. Collingwood maintains that one discerns a person's thoughts by rethinking them in one's own mind. Although he does not rule out causal concepts in history, he argues that "cause" has different meanings in history and in the natural sciences. The cause of an event for a historian, Collingwood says, is a thought in the mind of the actor "by whose agency the event came about: and this is not something other than the event, it is the inside of the event itself."

In a selection from *Laws and Explanation in History*, a book that dominated the field of philosophy of history for many years, Dray, in chapter 12, builds on Collingwood's idealistic insights concerning rethinking historical events. Developing these insights in the context of the Oxford analytic philosophy of the late 1950s, Dray maintains that explanation by laws does not help us understand human action in the sense proper to the subject matter. Although he believes that the idealists were mistaken and confused to suppose that historians must actually rethink or reexperience historical events, Dray argues that they had an important point to make. He proceeds to give the point a logical basis by means of his analysis of historical explanations. According to Dray, in history one can understand a human action only by giving a rational explanation of it. A rational explanation is one that takes the actor's point of view and attempts either to reconstruct the agent's actual rationale or to construct a possible rationale for the action. As Dray sees it, such an explanation shows that the action was the appropriate one to take under the circumstances.

Taylor's "Interpretation and the Sciences of Man" (chapter 13) is one of the most widely cited essays on interpretation to appear in English. Drawing on the analogy of the interpretation of a literary text, Taylor argues that the aim of the social sciences is to provide an interpretation of the social meanings connected with the practices and institutions of particular societies rather than to furnish causal explanations and predictions. He holds that the social meanings revealed by interpretive procedures are non-subjective and that one can contrast these with the subjective meanings appealed to, for example, by mainstream political scientists. According to Taylor, there is no objective way of validating such interpretations, for the validation of an interpretation is a circular procedure (the hermeneutic circle), and disagreements about interpretations ultimately rest on conflicting intuitions. He further maintains that exact prediction in the social sciences is impossible.

Geertz, perhaps the best-known advocate of the interpretive approach among contemporary social scientists,[7] assumes in "Thick Description: Toward an Interpretive Theory of Culture" (chapter 14), that one of the primary aims of anthropology is to understand culture. In this respect his views accord with the tradition of American cultural anthropology. However, in contrast to the traditional concept of culture, the one that Geertz assumes is semiotic. Culture, he says, is constituted by "webs of significance" that human beings have "spun." To discover these webs of meaning, anthropologists must do a kind of ethnography that involves providing what Geertz, borrowing from Gilbert Ryle, calls thick descriptions, that is, a detailed description of the complex web of social relations that constitute a culture. An example of the kind of cultural interpretation he has in mind is found in his well-known essay, "Deep Play: Notes on the Balinese Cockfight," which contains a description of the sport of cock-fighting in Bali that reveals its relation to deep elements of the Balinese culture, concept of self, and worldview.[8]

What is the connection between the hypothetico-deductive method that is widely used in the natural sciences and the hermeneutic method used by interpretivists? In answering this question, Føllesdal argues in "Hermeneutics and the Hypothetico-Deductive Method" (chapter 15) that the hermeneutic method is simply the hypothetico-deductive method applied to materials that are "meaningful," such as the system of beliefs and values of human beings. Using five different interpretations of the role of the stranger in lbsen's *Peer Gynt*, Føllesdal shows how one can judge these by deducing consequences from them and checking them against the data—for example, the text and the biography of the author. Føllesdal concludes that despite what some interpretivists say, there is no fundamental difference between the natural and the social sciences.

When it first appeared, Dray's account of human action was discussed and criticized primarily as a theory of explanation. In "Another Look at the Doctrine of *Verstehen*" (chapter 16), Jane Martin evaluates it as a theory of understanding. Distinguishing between internal and external understanding, she argues that one can understand something in many different ways. In external understanding, a given thing is treated as a whole, and understanding involves relating it to something apart from it; in internal understanding, a thing is taken in isolation, and understanding consists in relating parts or aspects of it. Both external and internal understanding, Martin argues, can take many different forms depending on how the thing understood is described, for example, as a war, a civil war, a revolution. Dray has arbitrarily restricted understanding, she maintains, to one type of internal understanding. Even if one construes his claim to apply only to historical contexts, she continues, there are many ways to understand actions.

Michael Martin in "Taylor on Interpretation and the Sciences of Man" (chapter 17), points to serious problems in the interpretive approach as it is presented in Taylor's essay. He maintains that the scope of the social sciences, as Taylor conceives of them, excludes important questions with which social science can and should be concerned. Even if one limits one's concerns to the interpretation of meaning, Martin argues, causality is relevant since interpreting the meaning of a social practice often involves understanding the causal connections between aspects of the practice. Moreover, he maintains that, despite what Taylor says, objective interpretations are possible. Even if there is a hermeneutic circle, this does not exclude evidence from either falsifying or supporting interpretations by means of probabilistic reasoning nor does it rule out placing objective constraints on interpretations. In addition, Taylor's theory is restrictive in still another way: it would keep social scientists from using theoretical concepts that go beyond the understanding of social actors.

Notes

1. For a detailed account, see Josef Bleicher, *Contemporary Hermeneutics* (London: Routledge and Kegan Paul, 1980), chap. 1.
2. According to Bleicher, "Literary interpretation had its origin in the Greek educational system where it assisted in the interpretation and criticism of Homer and other poets." Ibid., p. 11.
3. According to Bleicher, "Rabbis had rules for the interpretation of the Talmud and the Misraschim. Dithey himself refers to Philo, an Alexandrian, to indicate the origin of canons for the interpretation of allegories.... Biblical interpretation reached its first major formulation in the course and after-effects of the Reformation with Matthias Flacius. " Ibid.
4. See ibid., chap. 1.
5. See Fred R. Dallmayr and Thomas A. McCarthy (eds.), *Understanding and Social Inquiry* (Note Dame, Ind.: University of Notre Dame Press, 1977).
6. Cf. Daniel Little, *Varieties of Social Explanation* (Boulder, Colo.: Westview, 1991), p. 233.
7. Not all social scientists have embraced the interpretive program. For example, some anthropologists have argued that there are serious problems with Geertz's brand of interpretivism. See Paul Shankman, "The Thick and the Thin: On the Interpretive Program of Clifford Geertz," *Current Anthropology* 25 (1984): 261–270.
8. Cifford Geertz, "Deep Play: Notes on the Balinese Cockfight," *Interpretation of Cultures* (New York: Basic Books, 1973), pp. 412–453.

Chapter 11

Human Nature and Human History

R. G. Collingwood

The Science of Human Nature

Man, who desires to know everything, desires to know himself. Nor is he only one (even if, to himself, perhaps the most interesting) among the things he desires to know. Without some knowledge of himself, his knowledge of other things is imperfect: for to know something without knowing that one knows it is only a half-knowing, and to know that one knows is to know oneself. Self-knowledge is desirable and important to man, not only for its own sake, but as a condition without which no other knowledge can be critically justified and securely based.

Self-knowledge, here, means not knowledge of man's bodily nature, his anatomy and physiology; nor even a knowledge of his mind, so far as that consists of feeling, sensation, and emotion; but a knowledge of his knowing faculties, his thought or understanding or reason. How is such knowledge to be attained? It seems an easy matter until we think seriously about it; and then it seems so difficult that we are tempted to think it impossible. Some have even reinforced this temptation by argument, urging that the mind, whose business it is to know other things, has for that very reason no power of knowing itself. But this is open sophistry: first you say what the mind's nature is, and then you say that because it has this nature no one can know that it has it. Actually, the argument is a counsel of despair, based on recognizing that a certain attempted method of studying the mind has broken down, and on failure to envisage the possibility of any other.

It seems a fair enough proposal that, in setting out to understand the nature of our own mind, we should proceed in the same way as when we try to understand the world about us. In studying the world of nature, we begin by getting acquainted with the particular things and particular events that exist and go on there; then we proceed to understand them, by seeing how they fall into general types and how these general types are interrelated. These interrelations we call laws of nature; and it is by ascertaining such laws that we understand the things and events to which they apply. The same method, it might seem, is applicable to the problem of understanding mind. Let us begin by observing, as carefully as possible, the ways in which our own minds and those of others behave under given circumstances; then, having become acquainted with these facts of the mental world, let us try to establish the laws which govern them.

Here is a proposal for a "science of human nature" whose principles and methods are conceived on the analogy of those used in the natural sciences. It is an old proposal, put forward especially in the seventeenth and eighteenth centuries, when the principles and methods of natural science had been lately perfected and were being triumphantly applied to the investigation of the physical world. When Locke undertook his inquiry into that faculty of understanding which "sets Man above the rest of sensible Beings, and gives him all the Advantage and Dominion which he has over them," the novelty of his project lay not in his desire for a knowledge of the human mind, but in his attempt to gain it by methods analogous to those of natural science: the collection of observed facts and their arrangement in classificatory schemes. His own description of his method as a "historical, plain Method" is perhaps ambiguous; but his follower Hume was at pains to make it clear that the method to be followed by the science of human nature was identical with the method of physical science as he conceived it: its "only solid foundation," he wrote, "must be laid on experience and observation." Reid, in his *Inquiry into the Human Mind*, was if possible even more explicit. "All that we know of the body, is owing to anatomical dissection and observation, and it must be by an anatomy of the mind that we can discover its powers and principles." And from these pioneers the whole English and Scottish tradition of a "philosophy of the human mind" was derived.

Even Kant did not take an essentially different view. He certainly claimed that his own study of the understanding was something more than empirical; it was to be a demonstrative science; but then he held the same view concerning the science of nature; for that also, according to him, has in it an a priori or demonstrative element, and is not based merely on experience.

It is evident that such a science of human nature, if it could attain even a tolerable approximation to the truth, could hope for results of extreme importance. As applied to the problems of moral and political life, for example, its results would certainly be no less spectacular than were the results of seventeenth-century physics when applied to the mechanical arts in the eighteenth century. This was fully realized by its promoters. Locke thought that by its means he could "prevail with the busy Mind of Man, to be more cautious in meddling with things exceeding its Comprehension; to stop, when it is at the utmost of its Tether; and to sit down in a quiet Ignorance of those Things, which, upon Examination, are found to be beyond the reach of our Capacities." At the same time, he was convinced that the powers of our understanding are sufficient for our needs "in this state," and can give us all the knowledge we require for "the comfortable provision for this life, and the way that leads to a better." "If [he concludes] we can find out those Measures, whereby a Rational creature, put in the state which Man is in this World, may and ought to govern his Opinions and Actions depending thereon, we need not be troubled that some other things escape our knowledge."

Hume is even bolder. "'Tis evident," he writes, "that all the sciences have a relation, more or less, to human nature ... since they lie under the cognizance of men, and are judged of by their powers and faculties. 'Tis impossible to tell what changes and improvements we might make in these sciences were we thoroughly acquainted with the extent and force of human understanding." And in sciences directly concerned with human nature, like morals and politics, his hopes of a beneficent revolution are proportionately higher. "In pretending, therefore, to explain the principles of human nature, we in effect propose a complete system of the sciences, built on a foundation almost entirely new, and the only one upon which they can stand with any security." Kant, for all his habitual caution, claimed no less when he said that his new science would put an end to all the debates of the philosophical schools, and make it possible to solve all the problems of metaphysics at once and forever.

It need not imply any underestimate of what these men actually achieved if we admit that these hopes were in the main unfulfilled, and that the science of human nature, from Locke to the present day, has failed to solve the problem of understanding what understanding is, and thus giving the human mind knowledge of itself. It was not through any lack of sympathy with its objects that so judicious a critic as John Grote found himself obliged to treat the "philosophy of the human mind" as a blind alley out of which it was the duty of thought to escape.

What was the reason for this failure? Some might say that it was because the undertaking was in principle a mistake: mind cannot know itself. This objection we have already considered. Others, notably the representatives of psychology, would say that the science of these thinkers was not sufficiently scientific: psychology was still in its infancy. But if we ask these same men to produce here and now the practical results for which those early students hoped, they excuse themselves by saying that psychology is still in its infancy. Here I think they wrong themselves and their own science. Claiming for it a sphere which it cannot effectively occupy, they belittle the work it has done and is doing in its proper field. What that field is, I shall suggest in the sequel.

There remains a third explanation: that the "science of human nature" broke down because its method was distorted by the analogy of the natural sciences. This I believe to be the right one.

It was no doubt inevitable that in the seventeenth and eighteenth centuries, dominated as they were by the new birth of physical science, the eternal problem of self-knowledge should take shape as the problem of constructing a science of human nature. To anyone reviewing the field of human research, it was evident that physics stood out as a type of inquiry which had discovered the right method of investigating its proper object, and it was right that the experiment should be made of extending this method to every kind of problem. But since then a great change has come over the intellectual atmosphere of our civilization. The dominant factor in this change has not been the development of other natural sciences like chemistry and biology, or the transformation of physics itself since more began to be known about electricity, or the progressive application of all these new ideas to manufacture and industry, important though these have been; for in principle they have done nothing that might not have been foreseen as implicit in seventeenth-century physics itself. The really new element in the thought of today as compared with that of three centuries ago is the rise of history. It is true that the same Cartesian spirit which did so much for physics was already laying the foundations of critical method in history before the seventeenth century was out;[1] but the modern conception of history as a study at once critical and constructive, whose field is the human past in its entirety, and whose method is the reconstruction of that past from documents written and unwritten, critically analyzed and interpreted, was not established until the nineteenth, and is even yet not fully worked out in all its implications. Thus history occupies in the world of today a position analogous to that occupied by physics in the time of Locke: it is recognized as a special and autonomous form of thought, lately established, whose possibilities have not yet been completely explored. And just as in the seventeenth and eighteenth centuries there were materialists, who argued from the success of physics in its own sphere that all reality was physical, so among ourselves the success of history has led some people to suggest that its methods are applicable to all the problems of knowledge, in other words, that all reality is historical.

This I believe to be an error. I think that those who assert it are making a mistake of the same kind which the materialists made in the seventeenth century. But I believe, and in this essay I shall try to show, that there is at least one important element of truth in what they say. The thesis which I shall maintain is that the science of human nature was a false attempt—falsified by the analogy of natural science—to understand the mind itself, and that, whereas the right way of investigating nature is by the methods called scientific, the right way of investigating mind is by the methods of history. I shall contend that the work which was to be done by the science of human nature is actually done, and can only be done, by history: that history is what the science of human nature professed to be, and that Locke was right when he said (however little he understood what he was saying) that the right method for such an inquiry is the historical, plain method.

The Field of Historical Thought[2]

I must begin by attempting to delimit the proper sphere of historical knowledge as against those who, maintaining the historicity of all things, would resolve all knowledge into historical knowledge. Their argument runs in some such way as this.

The methods of historical research have, no doubt, been developed in application to the history of human affairs: but is that the limit of their applicability? They have already before now undergone important extensions: for example, at one time historians had worked out their methods of critical interpretation only as applied to written sources containing narrative material, and it was a new thing when they learned to apply them to the unwritten data provided by archaeology. Might not a similar but even more revolutionary extension sweep into the historian's net the entire world of nature? In other words, are not natural processes really historical processes, and is not the being of nature an historical being?

Since the time of Heraclitus and Plato, it has been a commonplace that things natural, no less than things human, are in constant change, and that the entire world of nature is a world of "process" or "becoming." But this is not what is meant by the historicity of things; for change and history are not at all the same. According to this old-established conception, the specific forms of natural things constitute a changeless repertory of fixed types, and the process of nature is a process by which instances of these forms (or quasi-instances of them, things approximating to the embodiment of them) come into existence and pass out of it again. Now in human affairs, as historical research had clearly demonstrated by the eighteenth century, there is no such fixed repertory of specific forms. Here, the process of becoming was already by that time recognized as involving not only the instances or quasi-instances of the forms, but the forms themselves. The political philosophy of Plato and Aristotle teaches in effect that city-states come and go, but the idea of the city-state remains for ever as the one social and political form toward whose realization human intellect, so far as it is really intelligent, strives. According to modern ideas, the city-state itself is as transitory a thing as Miletus or Sybaris. It is not an eternal ideal, it was merely the political ideal of the ancient Greeks. Other civilizations have had before them other political ideals, and human history shows a change not only in the individual cases in which these ideals are realized or partially realized, but in the ideals themselves. Specific types of human organization, the city-state, the feudal system, representative government, capitalistic industry, are characteristic of certain historical ages.

At first, this transience of specific forms was imagined to be a peculiarity of human life. When Hegel said that nature has no history, he meant that whereas the specific forms of human organization change as time goes on, the forms of natural organization do not. There is, he grants, a distinction of higher and lower in the specific forms of nature, and the higher forms are a development out of the lower; but this development is only a logical one, not a temporal, and in time all the "strata" of nature exist simultaneously.[3] But this view of nature has been overthrown by the doctrine of evolution. Biology has decided that living organisms are not divided into kinds each permanently distinct from the rest, but have developed their present specific forms through a process of evolution in time. Nor is this conception limited to the field of biology. It appeared simultaneously, the two applications being closely connected through the study of fossils, in geology. Today even the stars are divided into kinds which can be described as older and younger; and the specific forms of matter, no longer conceived in the Daltonian manner, as elements eternally distinct like the living species of pre-Darwinian biology, are regarded as subject to a similar change, so that the chemical constitution of our present world is only a phase in a process leading from a very different past to a very different future.

This evolutionary conception of nature, whose implications have been impressively worked out by philosophers like M. Bergson, Mr. Alexander, and Mr. Whitehead, might seem at first sight to have abolished the difference between natural process and historical process, and to have resolved nature into history. And if a further step in the same resolution were needed, it might seem to be provided by Mr. Whitehead's doctrine that the very possession of its attributes by a natural thing takes time. Just as Aristotle argued that a man cannot be happy at an instant, but that the possession of happiness takes a lifetime, so Mr. Whitehead argues that to be an atom of hydrogen takes time—the necessary for establishing the peculiar rhythm of movements which distinguishes it from other atoms—so that there is no such thing as "nature at an instant."

These modern views of nature do, no doubt, "take time seriously." But just as history is not the same thing as change, so it is not the same thing as "timefulness," whether that means evolution or an existence which takes time. Such views have certainly narrowed the gulf between nature and history of which early nineteenth-century thinkers were so conscious; they have made it impossible to state the distinction any longer in the way in which Hegel stated it; but in order to decide whether the gulf has been really closed and the distinction annulled, we must turn to the conception of history and see whether it coincides in essentials with this modern conception of nature.

If we put this question to the ordinary historian, he will answer it in the negative. According to him, all history properly so called is the history of human affairs. His special technique, depending as it does on the interpretation of documents in which human beings of the past have expressed or betrayed their thoughts, cannot be applied just as it stands to the study of natural processes; and the more this technique is elaborated in its details, the further it is from being so applicable. There is a certain analogy between the archaeologist's interpretation of a stratified site and the geologist's interpretation of rock-horizons with their associated fossils; but the difference is no less clear than the similarity. The archaeologist's use of his stratified relics depends on his conceiving them as artifacts serving human purposes and thus expressing a particular way in which men have thought about their own life; and from his point of view the paleontologist, arranging his fossils in a time-series, is not working as a historian, but only as a scientist thinking in a way which can at most be described as quasi-historical.

Upholders of the doctrine under examination would say that here the historian is making an arbitrary distinction between things that are really the same, and that his conception of history is an unphilosophically narrow one, restricted by the imperfect development of his technique, very much as some historians, because their equipment was inadequate to studying the history of art or science or economic life, have mistakenly restricted the field of historical thought to the history of politics. The question must therefore be raised. Why do historians habitually identify history with the history of human affairs? In order to answer this question, it is not enough to consider the characteristics of historical method as it actually exists, for the question at issue is whether, as it actually exists, it covers the whole field which properly belongs to it. We must ask what is the general nature of the problems which this method is designed to solve. When we have done so, it will appear that the special problem of the historian is one which does not arise in the case of natural science.

The historian, investigating any event in the past, makes a distinction between what may be called the outside and the inside of an event. By the outside of the event I mean everything belonging to it which can be described in terms of bodies and their movements: the passage of Caesar, accompanied by certain men, across a river called the Rubicon at one date, or the spilling of his blood on the floor of the senate-house at another. By the inside of the event I mean that in it which can only be described in terms of thought: Caesar's defiance of Republican law, or the clash of constitutional policy between himself and his assassins. The historian is never concerned with either of these to the exclusion of the other. He is investigating not mere events (where by a mere event I mean one which has only an outside and no inside) but actions, and an action is the unity of the outside and inside of an event. He is interested in the crossing of the Rubicon only in its relation to Republican law, and in the spilling of Caesar's blood only in its relation to a constitutional conflict. His work may begin by discovering the outside of an event, but it can never end there; he must always remember that the event was an action, and that his main task is to think himself into this action, to discern the thought of its agent.

In the case of nature, this distinction between the outside and the inside of an event does not arise. The events of nature are mere events, not the acts of agents whose thought the scientist endeavors to trace. It is true that the scientist, like the historian, has to go beyond the mere discovery of events; but the direction in which he moves is very different. Instead of conceiving the event as an action and attempting to rediscover the thought of its agent, penetrating from the outside of the event to its inside, the scientist goes beyond the event, observes its relation to others, and thus brings it under a general formula or law of nature. To the scientist, nature is always and merely a "phenomenon," not in the sense of being defective in reality, but in the sense of being a spectacle presented to his intelligent observation; whereas the events of history are never mere phenomena, never mere spectacles for contemplation, but things which the historian looks, not at, but through, to discern the thought within them.

In thus penetrating to the inside of events and detecting the thought which they express, the historian is doing something which the scientist need not and cannot do. In this way the task of the historian is more complex than that of the scientist. In another way it is simpler: the historian need not and cannot (without ceasing to be a historian) emulate the scientist in searching for the causes or laws of events. For science, the event is discovered by perceiving it, and the further search for its cause is conducted by assigning it to its class and determining the relation between that class and others. For history, the object to be discovered is not the mere event, but the thought expressed in it. To discover that thought is already to understand it. After the historian has ascertained the facts, there is no further process of inquiring into their causes. When he knows what happened, he already knows why it happened.

This does not mean that words like "cause" are necessarily out of place in reference to history; it only means that they are used there in a special sense. When a scientist asks, "Why did that piece of litmus paper turn pink?" he means, "On what kinds of occasions do pieces of litmus paper turn pink?" When a historian asks, "Why did Brutus stab Caesar?" he means, "What did Brutus think, which made him decide to stab Caesar?" The cause of the event, for him, means the thought in the mind of the person by whose agency the event came about: and this is not something other than the event, it is the inside of the event itself.

The processes of nature can therefore be properly described as sequences of mere events, but those of history cannot. They are not processes of mere events but processes of actions, which have an inner side, consisting of processes of thought, and what the historian is looking for is these processes of thought. All history is the history of thought.

But how does the historian discern the thoughts which he is trying to discover? There is only one way in which it can be done: by rethinking them in his own mind. The historian of philosophy, reading Plato, is trying to know what Plato thought when he expressed himself in certain words. The only way in which he can do this is by thinking it for himself. This, in fact, is what we mean when we speak of "understanding" the words. So the historian of politics or warfare, presented with an account of certain actions done by Julius Caesar, tries to understand these actions, that is, to discover what thoughts in Caesar's mind determined him to do them. This implies envisaging for himself the situation in which Caesar stood, and thinking for himself what Caesar thought about the situation and the possible ways of dealing with it. The history of thought, and therefore all history, is the reenactment of past thought in the historian's own mind.

This reenactment is only accomplished, in the case of Plato and Caesar respectively, so far as the historian brings to bear on the problem all the powers of his own mind and all his knowledge of philosophy and politics. It is not a passive surrender to the spell of another's mind; it is a labor of active and therefore critical thinking. The historian not only reenacts past thought, he reenacts it in the context of his own knowledge and therefore, in reenacting it, criticizes it, forms his own judgment of its value, corrects whatever errors he can discern in it. This criticism of the thought whose history he traces is not something secondary to tracing the history of it. It is an indispensable condition of the historical knowledge itself. Nothing could be a completer error concerning the history of thought than to suppose that the historian as such merely ascertains "what so-and-so thought," leaving it to someone else to decide "whether it was true." All thinking is critical thinking; the thought which reenacts past thoughts, therefore, criticizes them in reenacting them.

It is now clear why historians habitually restrict the field of historical knowledge to human affairs. A natural process is a process of events, a historical process is a process of thoughts. Man is regarded as the only subject of historical process, because man is regarded as the only animal that thinks, or thinks enough, and clearly enough, to render his actions the expressions of his thoughts. The belief that man is the only animal that thinks at all is no doubt a superstition; but the belief that man thinks more, and more continuously and effectively, than any other animal, and is the only animal whose conduct is to any great extent determined by thought instead of by mere impulse and appetite, is probably well enough founded to justify the historian's rule of thumb.

It does not follow that all human actions are subject matter for history; and indeed historians are agreed that they are not. But when they are asked how the distinction is to be made between historical and nonhistorical human actions, they are somewhat at a loss how to reply. From our present point of view we can offer an answer: so far as man's conduct is determined by what may be called his animal nature, his impulses and appetites, it is nonhistorical; the process of those activities is a natural process. Thus, the historian is not interested in the fact that men eat and sleep and make love and thus satisfy their natural appetites; but he is interested in the social customs which they create by their thought as a framework within which these appetites find satisfaction in ways sanctioned by convention and morality.

Consequently, although the conception of evolution has revolutionized our idea of nature by substituting for the old conception of natural process as a change within the limits of a fixed system of specific forms the new conception of that process as involving a change in these forms themselves, it has by no means identified the idea of natural process with that of historical process; and the fashion, current not long ago, of using the word "evolution" in a historical context, and talking of the evolution of parliament or the like, though natural in an age when the science of nature was regarded as the only true form of knowledge, and when other forms of knowledge, in order to justify their existence, felt bound to assimilate themselves to that model, was the result of confused thinking and a source of further confusions.

There is only one hypothesis on which natural processes could be regarded as ultimately historical in character: namely, that these processes are in reality processes of action determined by a thought which is their own inner side. This would imply that natural events are expressions of thoughts, whether the thoughts of God, or of angelic or demonic finite intelligences, or of minds somewhat like our own inhabiting the organic and inorganic bodies of nature as our minds inhabit our bodies. Setting aside mere flights of metaphysical fancy, such a hypothesis could claim our serious attention only if it led to a better understanding of the natural world. In fact, however, the scientist can reasonably say of it, "Je n'ai pas eu besoin de cette hypothèse," and the theologian will recoil from any suggestion that God's action in the natural world resembles the action of a finite human mind under the conditions of historical life. This at least is certain: that, so far as our scientific and historical knowledge goes, the processes of events which constitute the world of nature are altogether different in kind from the processes of thought which constitute the world of history.

Notes

1. "Historical criticism was born in the seventeenth century from the same intellectual movement as the philosophy of Descartes." E. Bréhier, in *Philosophy and History: Essays presented to Ernst Cassirer* (Oxford, 1936), p. 160.

2. In the argument of this section I owe much to Mr. Alexander's admirable essay on "The Historicity of Things," in the volume on *Philosophy and History* already quoted. If I seem to be controverting his main thesis, that is not because I disagree with his argument or any part of it, but only because I mean more than he does by the word "historicity." For him, to say that the world is "a world of events" is to say that "the world and everything in it is historical." For me, the two things are not at all the same.
3. *Naturphilosophie: Einleitung. System der Philosophie*, sec. 249, *Zusatz* (*Werks*, Glockner's edition, vol. ix, p. 59).

Chapter 12
The Rationale of Actions
William Dray

Historical Understanding as "Empathetic"

... [In this chapter, I] direct attention to ... the kind of explanation historians generally give of the *actions* of those individuals who are important enough to be mentioned in the course of historical narrative. It will be my thesis ... that the explanation of individual human behavior as it is usually given in history has features which make the covering law model peculiarly inept.

What I ... wish to say may be regarded as an attempt to rehabilitate to some extent a second traditional doctrine of idealist philosophers of history which Gardiner has attacked at length: the view that the objects of historical study are fundamentally different from those, for example, of the natural sciences, because they are the actions of beings like ourselves; and that even if (for the sake of argument) we allow that natural events may be explained by subsuming them under empirical laws, it would still be true that this procedure is inappropriate in history. Sometimes such a view will be supported by the belief that human actions—at any rate the ones we call "free"—do not fall under law at all. Sometimes it will be alleged only that even if they do fall under law, discovery of the law would still not enable us to understand them in the sense proper to this special subject matter. It is the second of these claims which I especially want to consider here.

The doctrine is commonly expressed with the aid of a characteristic set of terms. To understand a human action, it will be said, it is necessary for the inquirer somehow to discover its "thought-side"; it is not sufficient merely to know the pattern of overt behavior. The historian must *penetrate* behind appearances, achieve *insight* into the situation, *identify* himself sympathetically with the protagonist, *project* himself imaginatively into his situation. He must *revive, reenact, rethink, reexperience* the hopes, fears, plans, desires, views, intentions, of those he seeks to understand. To explain action in terms of covering law would be to achieve, at most, an external kind of understanding. The historian, by the very nature of his self-imposed task, seeks to do more than this.

It is worth noticing that historians themselves, and not just professional philosophers of history, often describe their task in these terms. Professor Butterfield is representative of a large group of his professional colleagues when he insists that "the only understanding we ever reach in history is but a refinement, more or less subtle and sensitive, of the difficult—and sometimes deceptive—process of imagining oneself in another person's place." And elsewhere in *History and Human Relations*, he writes:

> Our traditional historical writing ... has refused to be satisfied with any merely causal or stand-offish attitude towards the personalities of the past. It does not treat them as mere things, or just measure such features of them as the scientist might measure; and it does not content itself with merely reporting about them

> in the way an external observer would do. It insists that the story cannot be told correctly unless we see the personalities from the inside, feeling with them as an actor might feel the part he is playing—thinking their thoughts over again and sitting in the position not of the observer but of the doer of the action. If it is argued that this is impossible—as indeed it is—not merely does it still remain the thing to aspire to, but in any case the historian must put himself in the place of the historical personage, must feel his predicament, must think as though he were that man. Without this art not only is it impossible to tell the story correctly but it is impossible to interpret the very documents on which the reconstruction depends. Traditional historical writing emphasizes the importance of sympathetic imagination for the purpose of getting inside human beings. We may even say that this is part of the science of history for it produces communicable results—the insight of one historian may be ratified by scholars in general, who then give currency to the interpretation that is produced.[1]

Among covering law logicians there is an "official" answer to philosophers or historians who talk in this way about the peculiarities of "historical understanding." The answer is that although there is something right about it, the element of truth in such an account is not a point of logic; it is a mixture of psychological description and methodological precept. As a psychological description of the historian's state of mind when he succeeds in explaining the action of one of his characters, the notion of "empathy" or "imaginative understanding," as it is often called, will be allowed some merit—although it will be represented as involving us all too easily in the philosophical error of thinking that merely having certain experiences, or thinking certain thoughts similar to those of the historical agents, itself constitutes understanding or explaining. Similarly, as a suggestion as to how to go about discovering what the agent's motives were, the "empathy" theory will be admitted to have a certain methodological point—although the reservation will be made that the principle involved often leads the investigator astray. Professor Hempel puts the position succinctly in the following passage:

> The historian, we are told, imagines himself in the place of the persons involved in the events which he wants to explain; he tries to realize as completely as possible the circumstances under which they acted, and the motives which influenced their actions; and by this imaginary self-identification with his heroes, he arrives at an understanding and thus at an adequate explanation of the events with which he is concerned.
>
> This method of empathy is, no doubt, frequently applied by laymen and by experts in history. But it does not in itself constitute an explanation; it rather is essentially a heuristic device; its function is to suggest certain psychological hypotheses which might serve as explanatory principles in the case under consideration. Stated in crude terms, the idea underlying this function is the following: the historian tries to realize how he himself would act under the given conditions, and under the particular motivations of his heroes; he tentatively generalizes his findings into a general rule and uses the latter as an explanatory principle in accounting for the actions of the persons involved. Now, this procedure may sometimes prove heuristically helpful; but its use does not guarantee the soundness of the historical explanation to which it leads. The latter rather depends upon the factual correctness of the empirical generalizations which the method of understanding may have suggested.

> Nor is the use of this method indispensable for historical explanation. A historian may, for example, be incapable of feeling himself into the role of a paranoiac historic personality, and yet be able to explain certain of his actions; notably by reference to the principles of abnormal psychology. Thus whether the historian is or is not in a position to identify himself with his historical hero, is irrelevant for the correctness of his explanation; what counts, is the soundness of the general hypotheses involved, no matter whether they were suggested by empathy, or by a strictly behavioristic procedure.[2]

Now I do not wish to deny that there is any value at all in this sort of objection. But I think it important to show that the argument does not cut as deeply as covering law theorists commonly assume. For in recognizing the mixture of psychological and methodological elements in many statements of the idealist position, and in denying that these amount to an analysis of logical structure, these theorists fail to notice what it is about explanations of human actions in history which make the idealists want to say what they do—albeit in a quasi-psychological and quasi-methodological way. And what is left out, I wish to maintain, should properly be taken into account in a *logical* analysis of explanation as it is given in history. I shall argue that idealist theory partially, and perhaps defectively, formulates a certain pragmatic criterion operating in explanations of action given by historians, and that when this is ignored, we are quite properly puzzled as to why certain alleged explanations, which meet the covering law requirements, would be dismissed by historians as unsatisfactory—perhaps even as "no explanation at all."

The discussion to follow may be regarded in part as an attempt to "make sense" of what Collingwood, in particular, has to say about historical understanding—and I make no apology for this. But although some reference will be made to dicta of his, I shall not offer any close textual discussion of his account. I shall try, rather, to bring out independently, by reference to examples, features which covering law theory seems to me to miss, going on thereafter to discuss likely misunderstandings of, and objections to, the logical point which appears to emerge out of such an examination.

Explaining and Justifying Actions

The following extract from G. M. Trevelyan's *The English Revolution* is typical of a wide range of explanations of individual actions to be found in ordinary historical writing. In the course of an account of the invasion of England by William of Orange, Trevelyan asks: "Why did Louis make the greatest mistake of his life in withdrawing military pressure from Holland in the summer of 1688?" His answer is:

> He was vexed with James, who unwisely chose this moment of all, to refuse the help and advice of his French patron, upon whose friendship he had based his whole policy. But Louis was not entirely passion's slave. No doubt he felt irritation with James, but he also calculated that, even if William landed in England, there would be civil war and long troubles, as always in that factious island. Meanwhile, he could conquer Europe at leisure. "For twenty years," says Lord Acton, "it had been his desire to neutralize England by internal broils, and he was glad to have the Dutch out of the way (in England) while he dealt a blow at the Emperor Leopold (in Germany)." He thought "it was impossible that the conflict between James and William should not yield him an opportunity." This

> calculation was not as absurd as it looks after the event. It was only defeated by the unexpected solidity of a new type of Revolution.[3]

What Trevelyan here makes quite explicit is that, when we ask for the explanation of an action, what we very often want is a reconstruction of the agent's *calculation* of means to be adopted toward his chosen end in the light of the circumstances in which he found himself. To explain the action we need to know what considerations convinced him that he should act as he did.

But the notion of discovering the agent's calculation, it must be admitted, takes us no more than one preliminary step toward a satisfactory analysis of such explanations; and it may in itself be misleading. It must not be assumed, for instance, that the agent "calculated" in the sense of deriving by strict deductive reasoning the practical conclusion he drew—i.e., that the various considerations are elements in a calculus. Indeed, Trevelyan's explanation provides an obvious example to the contrary. Nor should we assume that the explanatory calculation must have been recited in propositional form, either aloud or silently—a notion which one might be forgiven for extracting out of Collingwood's discussion of the way thought must be reenacted by historians in order to understand intelligent, purposive actions. Not all high-grade actions are performed deliberately in the sense that they are undertaken with a plan consciously preformulated.

Indeed, it is tempting to say that in such cases there is *no* calculation to be *re*constructed by the historian. But such an admission need not affect the main point; for insofar as we say an action is purposive at all, no matter at what level of conscious deliberation, there is a calculation which could be constructed for it; the one the agent would have gone through if he had had time, if he had not seen what to do in a flash, if he had been called upon to account for what he did after the event, etc. And it is by eliciting some such calculation that we explain the action. It might be added that if the agent is to understand his *own* actions, i.e., after the event, he may have to do so by constructing a calculation in exactly the same way, although at the time he recited no propositions to himself. No doubt there are special dangers involved in such construction after the fact. But although we may have to examine very critically any particular example, the point is that when we do consider ourselves justified in accepting an explanation of an individual action, it will most often assume the general *form* of an agent's calculation.

Since the calculation gives what we should normally call the agent's *reasons* for acting as he did, I shall refer hereafter to this broad class of explanations as "rational." It should be clear that this use of the expression "rational explanation" is a narrower one than is often found in philosophical and semi-philosophical literature. It is sometimes said, for instance, that all science, all systematic inquiry, seeks a rational explanation for what is observed, where all that is meant is an explanation which takes account of all the facts considered puzzling, and which does not violate, say, the canons of coherence and induction. I intend something much more restricted than this: an explanation which displays the *rationale* of what was done.

The goal of such explanation is to show that what was done was the thing to have done for the reasons given, rather than merely the thing that is done on such occasions, perhaps in accordance with certain laws (loose or otherwise). The phrase "thing to have done" betrays a crucially important feature of explanations in terms of agent calculations—a feature quite different from any we have noticed so far. For the infinitive "to do" here functions as a value term. I wish to claim therefore that there is an element of *appraisal* of what was done in such explanations; that what we want to know when we ask to have the action explained is in what way it was *appropriate*. In the ordinary course of affairs, a demand for explanation is often recognized to be at the same time a challenge to the agent to produce either justification or excuse for what was done. In history, too, I want to argue, it will often be found impossible to bring out the point of what is offered as explanation unless the overlapping of these notions, when it is human actions we are interested in, is explicitly recognized.

Once again, however, I must be on guard against overstating the point; for I do not wish to imply that anything that is explained on the rational model is thereby certified *without qualification* as the right, or proper, or intelligent thing to have done. In saying that the explanation must exhibit what was done as appropriate or justified it is always necessary to add the philosopher's proviso: "in a sense."

The sense in question may be clarified if we note a scale along which rational explanations can be ranged. The scale falls away from the simple case in which we can say: "I find his action perfectly intelligible; he did exactly as I should have done." It is a small step from such a case to one where we can understand an action when we see that it is what we should agree was the thing to do in view of the agent's peculiar circumstances. In such a case the explanation would consist of an account of these circumstances; they are the missing data which permit the construction of a calculation certifying the action as appropriate. Sometimes, of course, the agent is found to have been mistaken about the facts—including (as Trevelyan's example of Louis XIV shows) his views about what the results of certain lines of action will be. The agent is thus mistaken about the nature of his circumstances; yet his action can still be explained in the rational way so long as by bringing his erroneous beliefs to bear, the calculation can be satisfactorily constructed. It may also be necessary, at times, to take note explicitly of the agent's purposes, which may be quite different from the ones which the investigator would have had in the same circumstances, or even in the circumstances the agent envisaged. And the calculation may also have to take into account certain peculiar principles of the agent; for the action is rationally explained if it is in accordance with the agent's principles—no matter what we think of these.

There are thus gradations of rational explanation, depending on the amount of "foreign" data which the investigator must bring in to complete the calculation: beliefs, purposes, principles, etc., of the agent which are different from those we might have assumed in absence of evidence to the contrary. Rational explanation may be regarded as an attempt to reach a kind of logical equilibrium at which point an action is *matched* with a calculation. A demand for explanation arises when the equilibrium is upset—when from the "considerations" obvious to the investigator it is impossible to see the point of what was done. The function of the historian's explanatory story will in many cases be to sketch in the corrections to these "obvious" considerations which require to be made if the reader is to be able to say: "Now I understand what he was about."

In the light of this account, it should be clear how restricted is the sense in which a rational explanation, as I use the term here, must show that what was done was the appropriate or right thing to have done. It is not necessary for the historian to show that the agent had reason for what he did; it is sufficient for explanation to show that he had reasons. But the element of appraisal remains in that what the historian declares to have been the agent's reasons must really *be* reasons (from the agent's point of view). To record what the agent *said* his reasons were would not be enough to provide a rational explanation unless the cogency of such reported reasons could be appreciated by the historian, when any peculiar beliefs, purposes, or principles of the agent were taken into account. Reported reasons, if they are to be explanatory in the rational way, must be *good* reasons at least in the sense that *if* the situation had been as the agent envisaged it (whether or not we, from our point of vantage, concur in his view of it), then what was done would have been the thing to have done. The historian must be able to "work" the agent's calculation.

The Point of the "Identification" Metaphor

If my account of rational explanation is correct, what should we say about the view that historical understanding is "empathetic"? It seems to me that our being able to range rational explanations along a scale in the way described above gives a real point to the "projection" metaphors used by empathy theorists. Perhaps it is because the scale has been either ignored or misunderstood that what such theorists have said has been so easily written off as obvious but uninteresting, or as interesting but dangerous.

Covering law logicians commonly speak of empathy as a "methodological dodge." And it might, I suppose, be claimed that if an old, practiced historian were to say to a novice: "You will never understand the way medieval knights behaved unless you drop your twentieth century prejudices and try to see things from their point of view," he *may* be telling the novice how to get on with his job, and thus be making a point which might be called "methodological." But I cannot believe that what the old hand offers his young colleague is (in Hempel's words) "a heuristic device" whose function is "to suggest certain psychological hypotheses which might serve as explanatory principles in the case under consideration." As Hempel goes on to explain, by this he means that the historian, since he lacks empirically tested psychological laws which fit, say, the behavior of medieval knights, must do something about repairing the deficiency if he is ever to give an explanation of knightly activities; for according to the covering law theory there is no explanation without empirical laws. Clearly the historian, especially the novice, is in no position to work over the whole field himself in search of the required laws. So, according to Hempel, he takes a short cut; he imagines himself in the knight's position, asks himself what *he* would have done, generalizes the answer as an empirical law covering knights (i.e., from a single imaginary case), and in this way satisfies the logical requirements of the model.

Hempel warns us, of course, that the use of the "device" does not "guarantee the soundness of the historical explanation to which it leads," which depends rather "upon the factual correctness of the empirical generalizations which the method of understanding may have suggested." That is, we may presume, further empirical confirmation of the generalization must come in before we can regard the explanation as anything more than an inspired guess. In Hempel's terminology, the generalization is only a "hypothesis" until it has received the sort of empirical confirmation and testing that any respectable scientific law must undergo, losing in the process he marks of its Athena-like origin.

In the light of what was said in the previous section, it should be clear how misleading this is as an account of "empathetic understanding." No doubt there *is* a methodological side to the doctrine; and it might be formulated in some such way as: "Only by putting yourself in the agent's position can you *find out* why he did what he did." Here the suggestion is admittedly that by an imaginative technique we shall discover some *new information*—the agent's motives or reasons for acting. When Collingwood says that historical understanding consists of penetrating to the thought-side of actions—discovering the thought and nothing further—the temptation to interpret this in the methodological way is understandably strong. But there is another way in which the doctrine can be formulated: "Only by putting yourself in the agent's position can you *understand* why he did what he did." The point of the "projection" metaphor is, in this case, more plausibly interpreted as a logical one. Its function is not to remind us of *how we come to know* certain facts, but to formulate, however tentatively, certain *conditions which must be satisfied* before a historian is prepared to say: "Now I have the explanation."

To dismiss "empathy" as a mere "methodological dodge" is to assume, falsely, that all there is to notice when rational explanations are given is a second-rate method of obtaining the same sort of result as can be obtained more reliably by direct attempts to subsume what is to be explained under an empirical covering law. But, as I have tried to show, at least part of what is meant by talking about the "need to project," etc., is not achievable at all by the method recommended by covering law theorists. To accept Hempel's argument against "empathy" is to obliterate a distinction between explanation types: a distinction between representing something as the thing generally done, and representing it as the appropriate thing to have done. Thus, when Hempel, after the passage quoted, goes on to say: "The kind of understanding thus conveyed must be clearly separated from scientific understanding," I have no objection to make, provided that by "scientific understanding" is meant "knowing to fall under an empirical law." But Hempel's account of the alternative is quite unsatisfactory. For "empathetic understanding," interpreted as "rational explanation," is *not* a matter of "presenting the phenomena in question as somehow 'plausible' or 'natural' to us ... by means of attractively worded metaphors."

No doubt the widespread resistance to admitting the need to cite anything more than antecedent conditions and a general law in explaining actions owes something to the air of mystery surrounding the language in which "empathy" theory is often framed: "projection," "identification," "imagination," "insight," 'intuition," etc. Such words arouse the suspicion that, if the conditions of the covering law theory are not met, it will be necessary to claim that the historian's explanation somehow goes beyond the limits of empirical inquiry into the realm of the unverifiable. As Gardiner puts it, historians often seem to be credited with "an additional power of knowing which allows them to 'penetrate into' the minds of the subjects of their study and take, as it were, psychological X-ray photographs."[4] And in the bulletin of the American Social Science Research Council..., historians are warned against a view of "historical understanding" supposed to be "achieved not by introducing general laws or relevant antecedent events, but by an act of 'intuition,' 'imaginative identification,' 'empathy' or 'valuation' which makes the historical occurrence plausible or intelligible," and whose adequacy is determined by "a self-certifying insight."[5] To allow the legitimacy of empathy appears to many of its opponents as the granting of a license to eke out scanty evidence with imaginative filler.

It is therefore worth my denying explicitly that what I have called rational explanation is in any damaging sense beyond empirical inquiry. As I have pointed out already, it has an inductive, empirical side, for we build up to explanatory equilibrium *from the evidence*. To get inside Disraeli's shoes the historian does not simply ask himself: "What would I have done?"; he reads Disraeli's dispatches, his letters, his speeches, etc.—and not with the purpose of discovering antecedent conditions falling under some empirically validated law, but rather in the hope of appreciating the problem as Disraeli saw it. The attempt to provide rational explanation is thus—if you like the term—"scientific" explanation in a broad sense; there is no question of the investigator letting his imagination run riot. Indeed, many "empathy" theorists have expressly guarded against such a misinterpretation of their views. To Butterfield, for instance, historical understanding is not a deliberate commission of the sin of anachronism; it is a "process of emptying oneself in order to catch the outlook and feelings of men not like-minded with oneself."[6]

It is true, of course, that the *direction* of inquiry in the explanation of actions is generally from what the inquirer presumes the relevant agent calculation to be—using his own, or his society's conception of rational purposes and principles—to what he discovers to be the peculiar data of the historical agent: a direction suggested by the scale already indicated. In view of this, Butterfield's admonition to "empty ourselves" is a little sweeping. In achieving rational explanation of an action we do project—but we project from our own point of view. In each case, the inclusion of "foreign" data in the calculation requires positive evidence that the agent was *not* like-minded with us. The historian does not build up to explanatory equilibrium form scratch. But this is far from admitting the covering law objection that the whole direction of the inquiry amounts to a vicious methodology. The procedure is self-corrective.

There is thus no reason to think that what I am calling "rational" explanations are put forward as self-evidently true, as some philosophers who talk of "insight" may seem to imply. Collingwood has sometimes been thought to provide justification for those who attack empathy theory on this account—e.g., when he represents the understanding of an action as an immediate leap to the discovery of its "inside," without the aid of any general laws, and (it may appear) without the use of any inductive reasoning at all.[7] But it is always possible that a mistake has been made in the inductive reasoning which provided the factual information for the calculation. It is always possible that further data may come in which will upset the logical equilibrium—perhaps evidence that the agent did not know something which it was at first thought he did. The ability of the historian to go through what he takes to be a relevant calculation does not guarantee the correctness of the explanation given; correct *form* is never a guarantee of correct *content*. But this is nothing more than the normal hazard of any empirical inquiry.

Notes

1. Pp. 145–46. See also pp. 116–17.
2. C. G. Hempel, "The Function of General Laws in History," in *Readings in Philosophical Analysis*, H. Feigl and W. Sellars, eds. (New York, 1949) p. 467. A similar argument is used by Crawford (R. M. Crawford, "History as Science," *Historical Studies, Australia and New Zealand*, 1947, p. 157); Peters (R. S. Peters, "Motives and Causes," *Proceedings of the Aristotelian Society*, supp. vol. 1952, p. 143); Gardiner (P. L. Gardiner, *The Nature of Historical Explanations*, Oxford, 1952, p. 129); and A. Danto ("Mere Chronicles and History Proper," *Journal of Philosophy*, 1953, p. 176).
3. *The English Revolution* (London, 1938), pp. 105–6.
4. Op. cit., p. 128.
5. *Bulletin No. 54*, p. 128.
6. Op. cit., p. 146.
7. E.g. "When [the historian] knows what happened, he already knows why it happened" (*The Idea of History*, p. 214).

Chapter 13

Interpretation and the Sciences of Man

Charles Taylor

I

i

Is there a sense in which interpretation is essential to explanation in the sciences of man? The view that it is, that there is an unavoidably "hermeneutical" component in the sciences of man, goes back to Dilthey. But recently the question has come again to the fore, for instance, in the work of Gadamer,[1] in Ricoeur's interpretation of Freud,[2] and in the writings of Habermas.[3]

Interpretation, in the sense relevant to hermeneutics, is an attempt to make clear, to make sense of an object of study. This object must, therefore, be a text, or a text-analogue, which in some way is confused, incomplete, cloudy, seemingly contradictory—in one way or another unclear. The interpretation aims to bring to light an underlying coherence or sense.

This means that any science which can be called "hermeneutical," even in an extended sense, must be dealing with one or another of the confusingly interrelated forms of meaning. Let us try to see a little more clearly what this involves.

We need, first, an object or a field of objects, about which we can speak in terms of coherence or its absence, or making sense or nonsense.

Second, we need to be able to make a distinction, even if only a relative one, between the sense of coherence made, and its embodiment in a particular field of carriers or signifiers. For otherwise, the task of making clear what is fragmentary or confused would be radically impossible. No sense could be given to this idea. We have to be able to make for our interpretations claims of the order: the meaning confusedly present in this text or text-analogue is clearly expressed here. The meaning, in other words, is one which admits of more than one expression, and in this sense a distinction must be possible between meaning and expression.

The point of the above qualification, that this distinction may be only relative, is that there are cases where no clear, unambiguous, nonarbitrary line can be drawn between what is said and its expression. It can be plausibly argued (I think convincingly, although there is no space to go into it here) that this is the normal and fundamental condition of meaningful expression, that exact synonymy, or equivalence of meaning, is a rare and localized achievement of specialized languages or uses of civilization. But this, if true (and I think it is), does not do away with the distinction between meaning and expression. Even if there is an important sense in which a meaning reexpressed in a new medium cannot be declared identical, this by no means entails that we can give no sense to the project of expressing a meaning in a new way. It does of course raise an interesting and difficult question about what can be meant by expressing it in a clearer way: what is the "it" which is clarified if equivalence is denied? I hope to return to this in examining interpretation in the sciences of man.

Hence the object of a science of interpretation must be describable in terms of sense and nonsense, coherence and its absence; and must admit of a distinction between meaning and its expression.

There is also a third condition it must meet. We can speak of sense or coherence, and of their different embodiments, in connection with such phenomena as gestalts, or patterns in rock formations, or snow crystals, where the notion of expression has no real warrant. What is lacking here is the notion of a subject for whom these meanings are. Without such a subject, the choice of criteria of sameness and difference, the choice among the different forms of coherence which can be identified in a given pattern, among the different conceptual fields in which it can be seen, is arbitrary.

In a text or text-analogue, on the other hand, we are trying to make explicit the meaning expressed, and this means expressed by or for a subject or subjects. The notion of expression refers us to that of a subject. The identification of the subject is by no means necessarily unproblematical, as we shall see further on; it may be one of the most difficult problems, an area in which prevailing epistemological prejudice may blind us to the nature of our object of study. I think this has been the case, as I will show below. And moreover, the identification of a subject does not assure us of a clear and absolute distinction between meaning and expression as we saw above. But any such distinction, even a relative one, is without any anchor at all, is totally arbitrary, without appeal to a subject.

The object of a science of interpretation must thus have sense, distinguishable from its expression, which is for or by a subject.

ii

Before going on to see in what way, if any, these conditions are realized in the sciences of man, I think it would be useful to set out more clearly what rides on this question, why it matters whether we think of the sciences of man as hermeneutical, what the issue is at stake here.

The issue here is at root an epistemological one. But it is inextricable from an ontological one, and hence, cannot but be relevant to our notions of science and of the proper conduct of inquiry. We might say that it is an ontological issue which has been argued ever since the seventeenth century in terms of epistemological considerations which have appeared to some to be unanswerable.

The case could be put in these terms: what are the criteria of judgment in a hermeneutical science? A successful interpretation is one which makes clear the meaning originally present in a confused, fragmentary, cloudy form. But how does one know that this interpretation is correct? Presumably because it makes sense of the original text: what is strange, mystifying, puzzling, contradictory is no longer so, is accounted for. The interpretation appeals throughout to our understanding of the "language" of expression, which understanding allows us to see that this expression is puzzling, that it is in contradiction to that other, etc., and that these difficulties are cleared up when the meaning is expressed in a new way.

But this appeal to our understanding seems to be crucially inadequate. What if someone does not "see" the adequacy of our interpretation, does not accept our reading? We try to show him how it makes sense of the original non- or partial sense. But for him to follow us he must read the original language as we do, he must recognize these expressions as puzzling in a certain way, and hence be looking for a solution to our problem. If he does not, what can we do? The answer, it would seem, can only be more of the same. We have to show him through the reading of other expressions why this expression must be read in the way we propose. But success here requires that he follow us in these other readings, and so on, it would seem, potentially forever. We cannot escape an ultimate appeal to a common understanding of the expressions, of the "language" involved. This is one way of trying to express what has been called the "hermeneutical circle." What we are trying to establish is a certain reading of text or expressions, and what we appeal to as our grounds for this reading can only be other readings. The circle can also be put in terms of part-whole relations: we are trying to establish a reading for the whole text, and for this we appeal to readings of its partial expressions; and yet because we are dealing with meaning, with making sense, where expressions only make sense or not in relation to others, the readings of partial expressions depend on those of others, and ultimately of the whole.

Put in forensic terms, as we started to do above, we can only convince an interlocutor if at some point he shares our understanding of the language concerned. If he does not, there is no further step to take in rational argument; we can try to awaken these intuitions in him or we can simply give up; argument will advance us no further. But of course the forensic predicament can be transferred into my own judging: if I am this ill equipped to convince a stubborn interlocutor, how can I convince myself? how can I be sure? Maybe my intuitions are wrong or distorted, maybe I am locked into a circle of illusion.

Now one, and perhaps the only sane response to this would be to say that such uncertainty is an ineradicable part of our epistemological predicament. That even to characterize it as "uncertainty" is to adopt an absurdly severe criterion of "certainty," which deprives the concept of any sensible use. But this has not been the only or even the main response of our philosophical tradition. And it is another response which has had an important and far-reaching effect on the sciences of man. The demand has been for a level of certainty which can only be attained by breaking beyond the circle.

There are two ways in which this breakout has been envisaged. The first might be called the "rationalist" one and could be thought to reach a culmination in Hegel. It does not involve a negation of intuition, or of our understanding of meaning, but rather aspires to attainment of an understanding of such clarity that it would carry with it the certainty of the undeniable. In Hegel's case, for instance, our full understanding of the whole in "thought" carries with it a grasp of its inner necessity, such that we see how it could not be otherwise. No higher grade of certainty is conceivable. For this aspiration the word "breakout" is badly chosen; the aim is rather to bring understanding to an inner clarity which is absolute.

The other way, which we can call "empiricist," is a genuine attempt to go beyond the circle of our own interpretations, to get beyond subjectivity. The attempt is to reconstruct knowledge in such a way that there is no need to make final appeal to readings or judgments which cannot be checked further. That is why the basic building block of knowledge on this view is the impression, or sense datum, a unit of information which is not the deliverance of a judgment, which has by definition no element in it of reading or interpretation, which is a brute datum. The highest ambition would be to build our knowledge from such building blocks by judgments which could be anchored in a certainty beyond subjective intuition. This is what underlies the attraction of the notion of the association of ideas, or if the same procedure is viewed as a method, induction. If the original acquisition of the units of information is not the fruit of judgment or interpretation, then the constatation that two such elements occur together need not either be the fruit of interpretation, of a reading or intuition which cannot be checked. For if the occurrence of a single element is a brute datum, then so is the co-occurrence of two such elements. The path to true knowledge would then repose crucially on the correct recording of such co-occurrences.

This is what lies behind an ideal of verification which is central to an important tradition in the philosophy of science, whose main contemporary protagonists are the logical empiricists. Verification must be grounded ultimately in the acquisition of brute data. By "brute data," I mean here and throughout data whose validity cannot be questioned by offering another interpretation or reading, data whose credibility cannot be founded or undermined by further reasoning.[4] If such a difference of interpretation can arise over given data, then it must be possible to structure the argument so as to distinguish the basic, brute data from the inferences made on the basis of them.

The inferences themselves, of course, to be valid must similarly be beyond the challenge of a rival interpretation. Here the logical empiricists added to the armory of traditional empiricism which set great store by the method of induction, the whole domain of logical and mathematical inference which had been central to the rationalist position (with Leibniz at least, although not with Hegel), and which offered another brand of unquestionable certainty.

Of course, mathematical inference and empirical verification were combined in such a way that two theories or more could be verified within the same domain of facts. But this was a consequence to which logical empiricism was willing to accommodate itself. As for the surplus meaning in a theory which could not be rigorously coordinated with brute data, it was considered to be quite outside the logic of verification.

As a theory of perception, this epistemology gave rise to all sorts of problems, not least of which was the perpetual threat of skepticism and solipsism inseparable from a conception of the basic data of knowledge as brute data, beyond investigation. As a theory of perception, however, it seems largely a thing of the past, in spite of a surprising recrudescence in the Anglo-Saxon world in the thirties and forties. But there is no doubt that it goes marching on, among other places, as a theory of how the human mind and human knowledge actually function.

In a sense, the contemporary period has seen a better, more rigorous statement of what this epistemology is about in the form of computer-influenced theories of intelligence. These try to model intelligence as consisting of operations on machine-recognizable input which could themselves be matched by programs which could be run on machines. The machine criterion provides us with our assurance against an appeal to intuition or interpretations which cannot be understood by fully explicit procedures operating on brute data—the input.[5]

The progress of natural science has lent great credibility to this epistemology, since it can be plausibly reconstructed on this model, as has been done, for instance, by the logical empiricists. And of course the temptation has been overwhelming to reconstruct the sciences of man on the same model; or rather to launch them in lines of inquiry that fit this paradigm, since they are constantly said to be in their "infancy." Psychology, where an earlier vogue of behaviorism is being replaced by a boom of computer-based models, is far from the only case.

The form this epistemological bias—one might say obsession—takes is different for different sciences. Later I should like to look at a particular case, the study of politics, where the issue can be followed out. But in general, the empiricist orientation must be hostile to a conduct of inquiry which is based on interpretation, and which encounters the hermeneutical circle as this was characterized above. This cannot meet the requirements of intersubjective, nonarbitrary verification which it considers essential to science. And along with the epistemological stance goes the ontological belief that reality must be susceptible to understanding and explanation by science so understood. From this follows a certain set of notions of what the sciences of man must be.

On the other hand, many, including myself, would like to argue that these notions about the sciences of man are sterile, that we cannot come to understand important dimensions of human life within the bounds set by this epistemological orientation. This dispute is of course familiar to all in at least some of its ramifications. What I want to claim is that the issue can be fruitfully posed in terms of the notion of interpretation as I began to outline it above.

I think this way of putting the question is useful because it allows us at once to bring to the surface the powerful epistemological beliefs that underlie the orthodox view of the sciences of man in our academy, and to make explicit the notion of our epistemological predicament implicit in the opposing thesis. This is in fact rather more way-out and shocking to the tradition of scientific thought than is often admitted or realized by the opponents of narrow scientism. It may not strengthen the case of the opposition to bring out fully what is involved in a hermeneutical science as far as convincing waverers is concerned, but a gain in clarity is surely worth a thinning of the ranks—at least in philosophy.

iii

Before going on to look at the case of political science, it might be worth asking another question: why should we even pose the question whether the sciences of man are hermeneutical? What gives us the idea in the first place that men and their actions constitute an object or a series of objects which meet the conditions outlined above?

The answer is that on the phenomenological level or that of ordinary speech (and the two converge for the purposes of this argument) a certain notion of meaning has an essential place in the characterization of human behavior. This is the sense in which we speak of a situation, an action, a demand, a prospect having a certain meaning for a person.

Now it is frequently thought that "meaning" is used here in a sense that is a kind of illegitimate extension from the notion of linguistic meaning. Whether it can be considered an extension is another matter; it certainly differs from linguistic meaning. But it would be very hard to argue that it is an illegitimate use of the term.

When we speak of the "meaning" of a given predicament, we are using a concept which has the following articulation. (1) Meaning is for a subject: it is not the meaning of the situation *in vacuo*, but its meaning for a subject, a specific subject, a group of subjects, or perhaps what its meaning is for the human subject as such (even though particular humans might be reproached with not admitting or realizing this). (2) Meaning is of something; that is, we can distinguish between a given element—situation, action, or whatever—and its meaning. But this is not to say that they are physically separable. Rather we are dealing with two descriptions of the element, in one of which it is characterized in terms of its meaning for the subject. But the relations between the two descriptions are not symmetrical. For, on the one hand, the description in terms of meaning cannot be unless descriptions of the other kind apply as well; or put differently, there can be no meaning without a substrate. But on the other hand, it may be that the same meaning may be borne by another substrate—e.g., a situation with the same meaning may be realized in different physical conditions. There is a necessary role for a potentially substitutable substrate; or all meanings are of something.

And (3) things only have meaning in a field, that is, in relation to the meanings of other things. This means that there is no such thing as a single, unrelated meaningful element; and it means that changes in the other meanings in the field can involve changes in the given element. Meanings cannot be identified except in relation to others, and in this way resemble words. The meaning of a word depends, for instance, on those words with which it contrasts, on those that define its place in the language (e.g., those defining "determinable" dimensions, like color, shape), on those that define the activity or "language game" it figures in (describing, invoking, establishing communion), and so on. The relations between meanings in this sense are like those between concepts in a semantic field.

Just as our color concepts are given their meaning by the field of contrast they set up together, so that the introduction of new concepts will alter the boundaries of others, so the various meanings that a subordinate's demeanor can have for us, as deferential, respectful, cringing, mildly mocking, ironical, insolent, provoking, downright rude, are established by a field of contrast; and as with finer discrimination on our part, or a more sophisticated culture, new possibilities are born, so other terms of this range are altered. And as the meaning of our terms "red," "blue," "green" is fixed by the definition of a field of contrast through the determinable term "color," so all these alternative demeanors are only available in a society which has, among other types, hierarchical relations of power and command. And corresponding to the underlying language game of designating colored objects is the set of social practices which sustain these hierarchical structures and are fulfilled in them.

Meaning in this sense—let us call it experiential meaning—thus is for a subject, of something, in a field. This distinguishes it from linguistic meaning which has a four- and not a three-dimensional structure. Linguistic meaning is for subjects and in a field, but it is the meaning of signifiers and it is about a world of referents. Once we are clear about the likenesses and differences there should be little doubt that the term "meaning" is not a misnomer, the product of an illegitimate extension into this context of experience and behavior.

There is thus a quite legitimate notion of meaning which we use when we speak of the meaning of a situation for an agent. And that this concept has a place is integral to our ordinary consciousness and hence speech about our actions. Our actions are ordinarily characterized by the purpose sought and explained by desires, feelings, emotions. But the language by which we describe our goals, feelings, desires is also a definition of the meaning things have for us. The vocabulary defining meaning—words like "terrifying," "attractive"—is linked with that describing feeling—"fear," "desire"—and that describing goals—"safety," "possession."

Moreover, our understanding of these terms moves inescapably in a hermeneutical circle. An emotion term like "shame," for instance, essentially refers us to a certain kind of situation, the "shameful," or "humiliating," and a certain mode of response, that of hiding oneself, of covering up, or else "wiping out" the blot. That is, it is essential to this feeling's identification as shame that it be related to this situation and give rise to this type of disposition. But this situation in its turn can only be identified in relation to the feelings it provokes; and the disposition is to a goal that can similarly not be understood without reference to the feelings experienced: the "hiding" in question is one which will cover up my shame; it is not the same as hiding from an armed pursuer; we can only understand what is meant by "hiding" here if we understand what kind of feeling and situation is being talked about. We have to be within the circle.

An emotion term like "shame" can only be explained by reference to other concepts which in turn cannot be understood without reference to shame. To understand these concepts we have to be in on a certain experience, we have to understand a certain language, not just of words, but also a certain language of mutual action and communication, by which we blame, exhort, admire, esteem each other. In the end we are in on this because we grow up in the ambit of certain common meanings. But we can often experience what it is like to be on the outside when we encounter the feeling, action, and experiential meaning language of another civilization. Here there is no translation, no way of explaining in other, more accessible concepts. We can only catch on by getting somehow into their way of life, if only in imagination. Thus if we look at human behavior as action done out of a background of desire, feeling, emotion, then we are looking at a reality which must be characterized in terms of meaning. But does this mean that it can be the object of a hermeneutical science as this was outlined above?

There are, to remind ourselves, three characteristics that the object of a science of interpretation has: it must have sense or coherence; this must be distinguishable from its expression; and this sense must be for a subject.

Now insofar as we are talking about behavior as action, hence in terms of meaning, the category of sense or coherence must apply to it. This is not to say that all behavior must "make sense," if we mean by this be rational, avoid contradiction, confusion of purpose, and the like. Plainly a great deal of our action falls short of this goal. But in another sense, even contradictory, irrational action is "made sense of," when we understand why it was engaged in. We make sense of action when there is a coherence between the actions of the agent and the meaning of his situation for him. We find his action puzzling until we find such a coherence. It may not be bad to repeat that this coherence in no way implies that the action is rational: the meaning of a situation for an agent may be full of confusion and contradiction, but the adequate depiction of this contradiction makes sense of it.

Making sense in this way through coherence of meaning and action, the meanings of action and situation, cannot but move in a hermeneutical circle. Our conviction that the account makes sense is contingent on our reading of action and situation. But these readings cannot be explained or justified except by reference to other such readings, and their relation to the whole. If an interlocutor does not understand this kind of reading, or will not accept it as valid, there is nowhere else the argument can go. Ultimately, a good explanation is one which makes sense of the behavior; but then to appreciate a good explanation one has to agree on what makes good sense; what makes good sense is a function of one's readings; and these in turn are based on the kind of sense one understands.

But how about the second characteristic, that sense should be distinguishable from its embodiment? This is necessary for a science of interpretation because interpretation lays a claim to make a confused meaning clearer; hence there must be some sense in which the "same" meaning is expressed, but differently.

This immediately raises a difficulty. In talking of experiential meaning above, I mentioned that we can distinguish between a given element and its meaning, between meaning and substrate. This carried the claim that a given meaning *may* be realized in another substrate. But does this mean that we can *always* embody the same meaning in another situation? Perhaps there are some situations, standing before death, for instance, which have a meaning which cannot be embodied otherwise.

But fortunately this difficult question is irrelevant for our purposes. For here we have a case in which the analogy between text and behavior implicit in the notion of a hermeneutical science of man only applies with important modifications. The text is replaced in the interpretation by another text, one which is clearer. The text-analogue of behavior is not replaced by another such text-analogue. When this happens we have revolutionary theater, or terroristic acts designed to make propaganda of the deed, in which the hidden relations of a society are supposedly shown up in a dramatic confrontation. But this is not scientific understanding, even though it may perhaps be based on such understanding, or claim to be.

But in science the text-analogue is replaced by a text, an account. Which might prompt the question, how we can even begin to talk of interpretation here, of expressing the same meaning more clearly, when we have two such utterly different terms of comparison, a text and a tract of behavior? Is the whole thing not just a bad pun?

This question leads us to open up another aspect of experiential meaning which we abstracted from earlier. Experiential meanings are defined in fields of contrast, as words are in semantic fields.

But what was not mentioned above is that these two kinds of definition are not independent of each other. The range of human desires, feelings, emotions, and hence meanings is bound up with the level and type of culture, which in turn is inseparable from the distinctions and categories marked by the language people speak. The field of meanings in which a given situation can find its place is bound up with the semantic field of the terms characterizing these meanings and the related feelings, desires, predicaments.

But the relationship involved here is not a simple one. There are two simple types of models of relation which could be offered here, but both are inadequate. We could think of the feeling vocabulary as simply describing preexisting feelings, as marking distinctions that would be there without them. But this is not adequate, because we often experience in ourselves or others how achieving, say, a more sophisticated vocabulary of the emotions makes our emotional life more sophisticated and not just our descriptions of it. Reading a good, powerful novel may give me the picture of an emotion which I had not previously been aware of. But we cannot draw a neat line between an increased ability to identify and an altered ability to feel emotions which this enables.

The other simple inadequate model of the relationship is to jump from the above to the conclusion that thinking makes it so. But this clearly won't do either, since not just any new definition can be forced on us, nor can we force it on ourselves; and some that we do gladly take up can be judged inauthentic, in bad faith, or just wrong-headed by others. These judgments may be wrong, but they are not in principle illicit. Rather we make an effort to be lucid about ourselves and our feelings, and admire a man who achieves this.

Thus, neither the simple correspondence view is correct, nor the view that thinking makes it so. But both have prima facie warrant. There is such a thing as self-lucidity, which points us to a correspondence view; but the achievement of such lucidity means moral change, that is, it changes the object known. At the same time, error about oneself is not just an absence of correspondence; it is also in some form inauthenticity, bad faith, self-delusion, repression of one's human feelings, or something of the kind;

it is a matter of the quality of what is felt just as much as what is known about this, just as self-knowledge is.

If this is so, then we have to think of man as a self-interpreting animal. He is necessarily so, for there is no such thing as the structure of meanings for him independently of his interpretation of them; one is woven into the other. but then the text of our interpretation is not that heterogeneous from what is interpreted, for what is interpreted is itself an interpretation: a self-interpretation which is embedded in a stream of action. It is an interpretation of experiential meaning which contributes to the constitution of this meaning. Or to put it in another way, that of which we are trying to find the coherence is itself partly constituted by self-interpretation.

Our aim is to replace this confused, incomplete, partly erroneous self-interpretation by a correct one. And in doing this we look not only to the self-interpretation but to the stream of behavior in which it is set, just as in interpreting a historical document we have to place it in the stream of events which it relates to. But of course the analogy is not exact, for here we are interpreting the interpretation and the stream of behavior in which it is set together, and not just one or the other.

There is thus no utter heterogeneity of interpretation to what it is about; rather there is a slide in the notion of interpretation. Already to be a living agent is to experience one's situation in terms of certain meanings, and this in a sense can be thought of as a sort of proto-"interpretation." This is in turn interpreted and shaped by the language in which the agent lives these meanings. This whole is then at a third level interpreted by the explanation we proffer of his actions.

In this way the second condition of a hermeneutical science is met. But this account poses in a new light the question mentioned at the beginning whether the interpretation can ever express the same meaning as the interpreted. And in this case, there is clearly a way in which the two will not be congruent. For if the explanation is really clearer than the lived interpretation, then it will be such that it would alter in some way the behavior if it came to be internalized by the agent as his self-interpretation. In this way a hermeneutical science that achieves its goal, that is, attains greater clarity than the immediate understanding of agent or observer, must offer us an interpretation that is in this way crucially out of phase with the explicandum.

Thus human behavior seen as action of agents who desire and are moved, who have goals and aspirations, necessarily offers a purchase for descriptions in terms of meaning—what I have called "experiential meaning." The norm of explanation which it posits is one that "makes sense" of the behavior, that shows a coherence of meaning. This "making sense of" is the proffering of an interpretation, and we have seen that what is interpreted meets the conditions of a science of interpretation: first, that we can speak of its sense or coherence; and second, that this sense can be expressed in another form, so that we can speak of the interpretation as giving clearer expression to what is only implicit in the explicandum. The third condition, that this sense be for a subject, is obviously met in this case, although who this subject is, is by no means an unproblematical question as we shall see later on.

This should be enough to show that there is a good prima facie case to the effect that men and their actions are amenable to explanation of a hermeneutical kind. There is therefore some reason to raise the issue and challenge the epistemological orientation that would rule interpretation out of the sciences of man. A great deal more must be said to bring out what is involved in the hermeneutical sciences of man. But before getting on to this, it might help to clarify the issue with a couple of examples drawn from a specific field, that of politics.

II

i

In politics, too, the goal of a verifiable science has led to the concentration on features that can supposedly be identified in abstraction from our understanding or not understanding experiential meaning. These—let us call them brute data identifications—are what supposedly enable us to break out from the hermeneutical circle and found our science foursquare on a verification procedure which meets the requirements of the empiricist tradition.

But in politics the search for such brute data has not gone to the lengths which it has in psychology, where the object of science has been thought of by many as behavior *qua* "colorless movement," or as machine-recognizable properties. The tendency in politics has been to stop with something less basic, but—so it is thought—the identification of which cannot be challenged by the offering of another interpretation or reading of the data concerned. This is what is referred to as "behavior" in the rhetoric of political scientists, but it has not the rock-bottom quality of its psychological homonym.

Political behavior includes what we would ordinarily call actions, but ones that are supposedly brute-data-identifiable. How can this be so? Well, actions are usually described by the purpose or end state realized. But the purposes of some actions can be specified in what might be thought to be brute data terms; some actions, for instance, have physical end states, like getting the car in the garage or climbing the mountain. Others have end states which are closely tied by institutional rules to some unmistakable physical movement; thus, when I raise my hand in the meeting at the appropriate time, I am voting for the motion. The only questions we can raise about the corresponding actions, given such movements or the realization of such end states, are whether the agent was aware of what he was doing, was acting as against simply emitting reflex behavior, knew the institutional significance of his movement, and so forth. Any worries on this score generally turn out to be pretty artificial in the contexts political scientists are concerned with; and where they do arise they can be checked by relatively simple devices, for example, asking the subject: did you mean to vote for the motion?

Hence it would appear that there are actions which can be identified beyond fear of interpretative dispute; and this is what gives the foundation for the category of "political behavior." Thus, there are some acts of obvious political relevance which can be specified as such in physical terms, such as killing, sending tanks into the streets, seizing people and confining them to cells; and there is an immense range of others that can be specified from physical acts by institutional rules, such as voting, for instance. These can be the object of a science of politics which can hope to meet the stringent requirements of verification. The latter class particularly has provided matter for study in recent decades—most notably in the case of voting studies.

But of course a science of politics confined to such acts would be much too narrow. For on another level these actions also have meaning for the agents which is not exhausted in the brute data descriptions, and which is often crucial to understanding why they were done. Thus in voting for the motion I am also saving the honor of my party, or defending the value of free speech, or vindicating public morality, or saving civilization from breakdown. It is in such terms that the agents talk about the motivation of much of their political action, and it is difficult to conceive a science of politics which does not come to grips with it.

Behavioral political science comes to grips with it by taking the meanings involved in action as facts about the agent, his beliefs, his affective reactions, his "values," as the term is frequently used. For it can be thought verifiable in the brute-data sense that men will agree to subscribe or not to a certain form of words (expressing a belief, say); or express a positive or negative reaction to certain events, or symbols; or agree or not with the proposition that some act is right or wrong. We can thus get at meanings as just another form of brute data by the techniques of the opinion survey and content analysis.

An immediate objection springs to mind. If we are trying to deal with the meanings which inform political action, then surely interpretive acumen is unavoidable. Let us say we are trying to understand the goals and values of certain group, or grasp their vision of the polity; we might try to probe this by a questionnaire asking them whether they assent to a number of propositions, which are meant to express different goals, evaluations, beliefs. But how did we design the questionnaire? How did we pick these propositions? Here we relied on our understanding of the goals, values, vision involved. But then this understanding can be challenged, and hence the significance of our results questioned. Perhaps the finding of our study, the compiling of proportions of assent and dissent to these propositions, is irrelevant, is without significance for understanding the agents or the polity concerned. This kind of attack is frequently made by critics of mainstream political science, or for that matter social science in general.

To this the proponents of this mainstream reply with a standard move of logical empiricism: distinguishing the process of discovery from the logic of verification. Of course it is our understanding of these meanings which enables us to draw up the questionnaire which will test people's attitudes in respect to them. And of course interpretive dispute about these meanings is potentially endless; there are no brute data at this level, every affirmation can be challenged by a rival interpretation. But this has nothing to do with verifiable science. What is firmly verified is the set of correlations between, say, the assent to certain propositions and certain behavior. We discover, for instance, that people who are active politically (defined by participation in a certain set of institutions) are more likely to consent to certain sets of propositions supposedly expressing the values underlying the system.[6] This finding is a firmly verified correlation no matter what one thinks of the reasoning, or simply hunches, that went into designing the research which established it. Political science as a body of knowledge is made up of such correlations; it does not give a truth value to the background reasoning or hunch. A good interpretive nose may be useful in hitting on the right correlations to test, but science is never called on to arbitrate the disputes between interpretations.

Thus in addition to those overt acts which can be defined physically or institutionally, the category of political behavior can include assent or dissent to verbal formulas, or the occurrence or not of verbal formulas in speech, or expressions of approval or rejection of certain events or measures as observed in institutionally defined behavior (for instance, turning out for a demonstration).

Now there are a number of objections which can be made to this notion of political behavior; one might question in all sorts of ways how interpretation free it is in fact. But I should like to question it from another angle. One of the basic characteristics of this kind of social science is that it reconstructs reality in line with certain categorical principles. These allow for an intersubjective social reality which is made up of brute data, identifiable acts and structures, certain institutions, procedures, actions. It allows for beliefs, affective reactions, evaluations as the psychological properties of individuals. And it allows for correlations, for example, between these two orders of reality: that certain beliefs go along with certain acts, certain values with certain institutions, and so forth.

To put it another way, what is objectively (intersubjectively) real is brute-data-identifiable. This is what social reality *is*. Social reality described in terms of its meaning for the actors, such that disputes could arise about interpretation that could not be settled by brute data (e.g., are people rioting to get a hearing, or are they rioting to redress humiliation, out of blind anger, because they recover a sense of dignity in insurrection?), is given subjective reality; that is, there are certain beliefs, affective reactions, evaluations which individuals make or have about or in relation to social reality. These beliefs or reactions can have an effect on this reality; and the fact that such a belief is held is a fact of objective social reality. But the social reality which is the object of these attitudes, beliefs, reactions can only be made up of brute data. Thus any description of reality in terms of meanings which is open to interpretive question is only allowed into this scientific discourse if it is placed, as it were, in quotes and attributed to individuals as their opinion, belief, attitude. That this opinion, belief, and so forth is held is thought of as a brute datum, since it is redefined as the respondent's giving a certain answer to the questionnaire.

This aspect of social reality which concerns its meanings for the agents has been taken up in a number of ways, but recently it has been spoken of in terms of political culture. Now the way this is defined and studied illustrates clearly the categorical principles above. For instance, political culture is referred to by Almond and Powell as the "psychological dimension of the political system."[7] Further on they state: "Political culture is the pattern of individual attitudes and orientations towards politics among the members of a political system. It is the subjective realm which underlies and gives meaning to political actions." The authors then go on to distinguish three different kinds of orientations, cognitive (knowledge and beliefs), affective (feelings), and evaluative (judgments and opinions).

From the point of view of empiricist epistemology, this set of categorical principles leaves nothing out. Both reality and the meanings it has for actors are coped with. But what it in fact cannot allow for are intersubjective meanings; that is, it cannot allow for the validity of descriptions of social reality in terms of meanings, hence not as brute data, which are not in quotation marks and attributed as opinion, attitude, and so forth to individual(s). Now it is this exclusion that I should like to challenge in the name of another set of categorical principles, inspired by a quite other epistemology.

ii

We spoke earlier about the brute-data identification of acts by means of institutional rules. Thus, putting a cross beside someone's name on a slip of paper and putting this in a box counts in the right context as voting for that person; leaving the room, saying or writing a certain form of words, counts as breaking off the negotiations; writing one's name on a piece of paper counts as signing the petition, and so forth. But what is worth looking at is what underlies this set of identifications. These identifications are the application of a language of social life, a language which marks distinctions among different possible social acts, relations, structures. But what underlies this language?

Let us take the example of breaking off negotiations above. The language of our society recognizes states or actions like the following: entering into negotiation, breaking off negotiations, offering to negotiate, negotiating in good (bad) faith, concluding negotiations, making a new offer. In other more jargon-infested language, the semantic "space" of this range of social activity is carved up in a certain way, by a certain set of distinctions which our vocabulary marks; and the shape and nature of these distinctions is the nature of our language in this area. These distinctions are applied in our society with more or less formalism in different contexts.

But of course this is not true of every society. Our whole notion of negotiation is bound up, for instance, with the distinct identity and autonomy of the parties, with the willed nature of their relations; it is a very contractual notion. But other societies have no such conception. It is reported about the traditional Japanese village that the foundation of its social life was a powerful form of consensus, which put a high premium on unanimous decision.[8] Such a consensus would be considered shattered if two clearly articulated parties were to separate out, pursuing opposed aims and attempting either to vote down the opposition or push it into a settlement on the most favorable possible terms for themselves. Discussion there must be, and some kind of adjustment of differences. But our idea of bargaining, with the assumption of distinct autonomous parties in willed relationship, has no place there; nor does a series of distinctions, like entering into and leaving negotiation, or bargaining in good faith (sc. with the genuine intention of seeking agreement).

Now the difference between our society and one of the kind just described could not be well expressed if we said we have a vocabulary to describe negotiation which they lack. We might say, for instance, that we have a vocabulary to describe the heavens which they lack, namely, that of Newtonian mechanics; for here we assume that they live under the same heavens as we do, only understand it differently. But it is not true that they have the same kind of bargaining as we do. The word, or whatever word of their language we translate as "bargaining," must have an entirely different gloss, which is marked by the distinctions their vocabulary allows in contrast to those marked by ours. But this different gloss is not just a difference of vocabulary, but also one of social reality.

But this still may be misleading as a way of putting the difference. For it might imply that there is a social reality which can be discovered in each society and which might exist quite independently of the vocabulary of that society, or indeed of any vocabulary, as the heavens would exist whether men theorized about them or not. And this is not the case; the realities here are practices; and these cannot be identified in abstraction from the language we use to describe them, or invoke them, or carry them out. That the practice of negotiation allows us to distinguish bargaining in good or bad faith, or entering into or breaking off negotiations, presupposes that our acts and situation have a certain description for us, for example, that we are distinct parties entering into willed relations. But they cannot have these descriptions for us unless this is somehow expressed in our vocabulary of this practice; if not in our descriptions of the practices (for we may as yet be unconscious of some of the important distinctions) in the appropriate language for carrying them on. (Thus, the language marking a distinction between public and private acts or contexts may exist even where these terms or their equivalents are not part of this language; for the distinction will be marked by the different language which is appropriate in one context and the other, be it perhaps a difference of style, or dialect, even though the distinction is not designated by specific descriptive expressions.)

The situation we have here is one in which the vocabulary of a given social dimension is grounded in the shape of social practice in this dimension; that is, the vocabulary would not make sense, could not be applied sensibly, where this range of practices did not prevail. And yet this range of practices could not exist without the prevalence of this or some related vocabulary. There is no simple one-way dependence here. We can speak of mutual dependence if we like, but really what this points up is the artificiality of the distinction between social reality and the language of description of that social reality. The language is constitutive of the reality, is essential to its being the kind of reality it is. To separate the two and distinguish them as we quite rightly distinguish the heavens from our theories about them is forever to miss the point.

This type of relation has been recently explored, for example, by John Searle, with his concept of a constitutive rule. As Searle points out,[9] we are normally induced to think of rules as applying to behavior which could be available to us whether or not the rule existed. Some rules are like this, they are regulative like commandments: Don't take the goods of another. But there are other rules, for example, that governing the Queen's move in chess, which are not so separable. If one suspends these rules, or imagines a state in which they have not yet been introduced, then the whole range of behavior in question, in this case, chess playing, would not be. There would still, of course, be the activity of pushing a wood piece around on an eight-by-eight-inch board made of squares; but this is not chess any longer. Rules of this kind are constitutive rules. By contrast again, there are other rules of chess, such as that one say "j'adoube" when one touches a piece without intending to play it, which are clearly regulative.[10]

I am suggesting that this notion of the constitutive be extended beyond the domain of rule-governed behavior. That is why I suggest the vaguer word "practice." Even in an area where there are no clearly defined rules, there are distinctions between different sorts of behavior such that one sort is considered the appropriate form for one action or context, the other for another action or context; for example, doing or saying certain things amounts to breaking off negotiations; doing or saying other things amounts to making a new offer. But just as there are constitutive rules, that is, rules such that the behavior they govern could not exist without them, and which are in this sense inseparable from that behavior, so I am suggesting that there are constitutive distinctions, constitutive ranges of language which are similarly inseparable, in that certain practices are not without them.

We can reverse this relationship and say that all the institutions and practices by which we live are constituted by certain distinctions and hence a certain language which is thus essential to them. We can take voting, a practice which is central to large numbers of institutions in a democratic society. What is essential to the practice of voting is that some decision or verdict be delivered (a man elected, a measure passed), through some criterion of preponderance (simple majority, two-thirds majority, or whatever) out of a set of microchoices (the votes of the citizens, MPs, delegates). If there is not some such significance attached to our behavior, no amount of marking and counting pieces of paper, raising hands, walking out into lobbies amounts to voting. From this it follows that the institution of voting must be such that certain distinctions have application: for example, that between someone being elected, or a measure passed, and their failing of election, or passage; that between a valid vote and an invalid one which in turn requires a distinction between a real choice and one which is forced or counterfeited. For no matter how far we move from the Rousseauian notion that each man decide in full autonomy, the very institution of the vote requires that in some sense the enfranchised choose. For there to be voting in a sense recognizably like ours, there must be a distinction in men's self-interpretations between autonomy and forced choice.

This is to say that an activity of marking and counting papers has to bear intentional descriptions which fall within a certain range before we can agree to call it voting, just as the intercourse of two men or teams has to bear descriptions of a certain range before we will call it negotiation. Or in other words, that some practice is voting or negotiation has to do in part with the vocabulary established in a society as appropriate for engaging in it or describing it.

Hence implicit in these practices is a certain vision of the agent and his relation to others and to society. We saw in connection with negotiation in our society that it requires a picture of the parties as in some sense autonomous, and as entering into willed relations. And this picture carries with it certain implicit norms, such as that of good faith mentioned above, or a norm of rationality, that agreement correspond to one's goals as far as attainable, or the norm of continued freedom of action as far as attainable. These practices require that one's actions and relations be seen in the light of this picture and the accompanying norms, good faith, autonomy, and rationality. But men do not see themselves in this way in all societies, nor do they understand these norms in all societies. The experience of autonomy as we know it, the sense of rational action and the satisfactions thereof, are unavailable to them. The meaning of these terms is opaque to them because they have a different structure of experiential meaning open to them.

We can think of the difference between our society and the simplified version of the traditional Japanese village as consisting in this, that the range of meaning open to the members of the two societies is very different. But what we are dealing with here is not subjective meaning which can fit into the categorical grid of behavioral political science, but rather intersubjective meanings. It is not just that all or most people in our society have a given set of ideas in their heads and subscribe to a given set of goals. The meanings and norms implicit in these practices are not just in the minds of the actors but are out there in the practices themselves, practices which cannot be conceived as a set of individual actions, but which are essentially modes of social relation, of mutual action.

The actors may have all sorts of beliefs and attitudes which may be rightly thought of as their individual beliefs and attitudes, even if others share them; they may subscribe to certain policy goals or certain forms of theory about the polity, or feel resentment at certain things, and so on. They bring these with them into their negotiations, and strive to satisfy them. But what they do not bring into the negotiations is the set of ideas and norms constitutive of negotiation themselves. These must be the common property of the society before there can be any question of anyone entering into negotiation or not. Hence they are not subjective meanings, the property of one or some individuals, but rather intersubjective meanings, which are constitutive of the social matrix in which individuals find themselves and act.

The intersubjective meanings which are the background to social action are often treated by political scientists under the heading "consensus." By this is meant convergence of beliefs on certain basic matters, or of attitude. But the two are not the same. Whether there is consensus or not, the condition of there being either one or the other is a certain set of common terms of reference. A society in which this was lacking would not be a society in the normal sense of the term, but several. Perhaps some multiracial or multitribal states approach this limit. Some multinational states are bedeviled by consistent cross-purposes, for example, Canada. But consensus as a convergence of beliefs or values is not the opposite of this kind of fundamental diversity. Rather the opposite of diversity is a high degree of intersubjective meanings. And this can go along with profound cleavage. Indeed, intersubjective meanings are a condition of a certain kind of very profound cleavage, such as was visible in the Reformation, or the American Civil War, or splits in left-wing parties, where the dispute is at fever pitch just because both sides can fully understand the other.

In other words, convergence of belief or attitude or its absence presupposes a common language in which these beliefs can be formulated, and in which these formulations can be opposed. Much of this common language in any society is rooted in its institutions and practices; it is constitutive of these institutions and practices. It is part of the intersubjective meanings. To put the point another way, apart from the question of how much people's beliefs converge is the question of how much they have a common language of social and political reality in which these beliefs are expressed. This second question cannot be reduced to the first; intersubjective meaning is not a matter of converging beliefs or values. When we speak of consensus we speak of beliefs and values which could be the property of a single person, or many, or all; but intersubjective meanings could not be the property of a single person because they are rooted in social practice.

We can perhaps see this if we envisage the situation in which the ideas and norms underlying a practice are the property of single individuals. This is what happens when single individuals from one society interiorize the notions and values of another, for example, children in missionary schools. Here we have a totally different situation. We *are* really talking now about subjective beliefs and attitudes. The ideas are abstract, they are mere social "ideals." Whereas in the original society, these ideas and norms are rooted in their social relations, and are that on the basis of which they can formulate opinions and ideals.

We can see this in connection with the example we have been using all along, that of negotiations. The vision of a society based on negotiation is coming in for heavy attack by a growing segment of modern youth, as are the attendant norms of rationality and the definition of autonomy. This is a dramatic failure of "consensus." But this cleavage takes place in the ambit of this intersubjective meaning, the social practice of negotiation as it is lived in our society. The rejection would not have the bitter quality it has if what is rejected were not understood in common, because it is part of a social practice which we find hard to avoid, so pervasive is it in our society. At the same time there is a reaching out for other forms which have still the "abstract" quality of ideals which are subjective in this sense, that is, not rooted in practice; which is what makes the rebellion look so "unreal" to outsiders, and so irrational.

iii

Intersubjective meanings, ways of experiencing action in society which are expressed in the language and descriptions constitutive of institutions and practices, do not fit into the categorical grid of mainstream political science. This allows only for an intersubjective reality that is brute-data-identifiable. But social practices and institutions that are partly constituted by certain ways of talking about them are not so identifiable. We have to understand the language, the underlying meanings that constitute them.

We can allow, once we accept a certain set of institutions or practices as our starting point and not as objects of further questioning, that we can easily take as brute data that certain acts are judged to take place or certain states judged to hold within the semantic field of these practices—for instance, that someone has voted Liberal or signed the petition. We can then go on to correlate certain subjective meanings—beliefs, attitudes, and so forth—with this behavior or its lack. But this means that we give up trying to define further just what these practices and institutions are, what the meanings are which they require and hence sustain. For these meanings do not fit into the grid; they are not subjective beliefs or values, but are constitutive of social reality. In order to get at them we have to drop the basic premise that social reality is made up of brute data alone. For any characterization of the meanings underlying these practices is open to question by someone offering an alternative interpretation. The negation of this is what was meant as brute data. We have to admit that intersubjective social reality has to be partly defined in terms of meanings; that meanings as subjective are not just in causal interaction with a social reality made up of brute data, but that as intersubjective they are constitutive of this reality.

We have been talking here of intersubjective meanings. And earlier I was contrasting the question of intersubjective meaning with that of consensus as convergence of opinions. But there is another kind of nonsubjective meaning which is also often inadequately discussed under the head of "consensus." In a society with a strong web of intersubjective meanings, there can be a more or less powerful set of common meanings. By these I mean notions of what is significant that are not just shared in the sense that everyone has them, but are also common in the sense of being in the common reference world. Thus, almost everyone in our society may share a susceptibility to a certain kind of feminine beauty, but this may not be a common meaning. It may be known to no one, except perhaps market researchers, who play on it in their advertisements. But the survival of a national identity as francophones is a common meaning of *Québeçois*; for it is not just shared, and not just known to be shared, but its being a common aspiration is one of the common reference points of all debate, communication, and all public life in the society.

We can speak of a shared belief, aspiration, and so forth when there is convergence between the subjective beliefs, aspirations, of many individuals. But it is part of the meaning of a common aspiration, belief, celebration, that it be not just shared but part of the common reference world. Or to put it another way, its being shared is a collective act; it is a consciousness which is communally sustained, whereas sharing is something we do each on his own, as it were, even if each of us is influenced by the others.

Common meanings are the basis of community. Intersubjective meaning gives a people a common language to talk about social reality and a common understanding of certain norms, but only with common meanings does this common reference world contain significant common actions, celebrations, and feelings. These are objects in the world that everybody shares. This is what makes community.

Once again, we cannot really understand this phenomenon through the usual definition of consensus as convergence of opinion and value. For what is meant here is something more than convergence. Convergence is what happens when our values are shared. But what is required for common meanings is that this shared value be part of the common world, that this sharing be shared. But we could also say that common meanings are quite other than consensus, for they can subsist with a high degree of cleavage; this is what happens when a common meaning comes to be lived and understood differently by different groups in a society. It remains a common meaning, because there is the reference point which is the common purpose, aspiration, celebration. Such is, for example, the American Way, or freedom as understood in the United States. But this common meaning is differently articulated by different groups. This is the basis of the bitterest fights in a society, and this we are also seeing in the United States today. Perhaps one might say that a common meaning is very often the cause of the most bitter lack of consensus. It thus must not be confused with convergence of opinion, value, attitude.

Of course, common meanings and intersubjective meanings are closely interwoven. There must be a powerful net of intersubjective meanings for there to be common meanings; and the result of powerful common meanings is the development of a greater web of intersubjective meanings as people live in community.

On the other hand, when common meanings wither, which they can do through the kind of deep dissensus we described earlier, the groups tend to grow apart and develop different languages of social reality, hence to share less intersubjective meanings.

To take our above example again, there has been a powerful common meaning in our civilization around a certain vision of the free society in which bargaining has a central place. This has helped to entrench the social practice of negotiation which makes us participate in this intersubjective meaning. But there is a severe challenge to this common meaning today, as we have seen. Should those who object to it really succeed in building up an alternative society, there would develop a gap between those who remain in the present type of society and those who had founded the new one.

Common meanings, as well as intersubjective ones, fall through the net of mainstream social science. They can find no place in its categories. For they are not simply a converging set of subjective reactions, but part of the common world. What the ontology of mainstream social science lacks is the notion of meaning as not simply for an individual subject; of a subject who can be a "we" as well as an "I." The exclusion of this possibility, of the communal, comes once again from the baleful influence of the epistemological tradition for which all knowledge has to be reconstructed from the impressions imprinted on the individual subject. But if we free ourselves from the hold of these prejudices, this seems a wildly implausible view about the development of human consciousness; we are aware of the world through a "we" before we are through an "I." Hence we need the distinction between what is just shared in the sense that each of us has it in our individual worlds, and that which is in the common world. But the very idea of something that exists in the common world in contradistinction to what exists in all the individual worlds is totally opaque to empiricist epistemology, and so finds no place in mainstream social science. What this results in must now be seen.

III

i

To sum up the last pages: a social science that wishes to fulfill the requirements of the empiricist tradition naturally tries to reconstruct social reality as consisting of brute data alone. These data are the acts of people (behavior) as identified supposedly beyond interpretation either by physical descriptions or by descriptions clearly defined by institutions and practices; and second, they include the subjective reality of individuals' beliefs, attitudes, values, as attested by their responses to certain forms of words, or in some cases their overt nonverbal behavior.

What this excludes is a consideration of social reality as characterized by intersubjective and common meanings. It excludes for instance, and attempt to understand our civilization, in which negotiation plays such a central part both in fact and in justificatory theory, by probing the self-definitions of agent, other, and social relatedness which it embodies. Such definitions which deal with the meaning for agents of their own and others' action, and of the social relations in which they stand, do not in any sense record brute data, in the sense that this term is being used in this argument; that is, they are in no sense beyond challenge by those who would quarrel with our interpretations of these meanings.

I tried to adumbrate above the vision implicit in the practice of negotiation by reference to certain notions of autonomy and rationality. But this reading will undoubtedly be challenged by those who have different fundamental conceptions of man, human motivation, the human condition; or even by those who judge other features of our present predicament to have greater importance. If we wish to avoid these disputes, and have a science grounded in verification as this is understood by the logical empiricists, then we have to avoid this level of study altogether and hope to make do with a correlation of behavior that is brute-data-identifiable.

A similar point goes for the distinction between common meanings and shared subjective meanings. We can hope to identify the subjective meanings of individuals if we take these in the sense in which there are adequate criteria for them in people's dissent or assent to verbal formulas or their brute-data-identifiable behavior. But once we allow the distinction between such subjective meanings which are widely shared and genuine common meanings, then we can no longer make do with brute-data identification. We are in a domain where our definitions can be challenged by those with another reading.

The profound bias of mainstream social scientists in favor of the empiricist conception of knowledge and science makes it inevitable that they should accept the verification model of political science and the categorical principles that this entails. This means in turn that a study of our civilization in terms of its intersubjective and common meanings is ruled out. Rather this whole level of study is made invisible.

On the mainstream view, therefore, the different practices and institutions of different societies are not seen as related to different clusters of intersubjective or common meanings; rather, we should be able to differentiate them by different clusters of "behavior" and/or subjective meaning. The comparison between societies requires on this view that we elaborate a universal vocabulary of behavior which will allow us to present the different forms and practices of different societies in the same conceptual web.

Now present-day political science is contemptuous of the older attempt at comparative politics through a comparison of institutions. An influential school of our day has therefore shifted comparison to certain practices, or very general classes of practices, and proposes to compare societies according to the different ways in which these practices are carried on. Such are the "functions" of the influential "developmental approach."[11] But it is epistemologically crucial that such functions be identified independently of those intersubjective meanings which are different in different societies; for otherwise, they will not be genuinely universal; or will be universal only in the loose and unilluminating sense that the function name can be given application in every society but with varying, and often widely varying meaning—the same term being "glossed" very differently by different sets of practices and intersubjective meanings. The danger that such universality might not hold is not even suspected by mainstream political scientists since they are unaware that there is such a level of description as that which defines intersubjective meaning and are convinced that functions and the various structures that perform them can be identified in terms of brute data behavior.

But the result of ignoring the difference in intersubjective meanings can be disastrous to a science of comparative politics, namely, that we interpret all other societies in the categories of our own. Ironically, this is what seems to have happened to American political science. Having strongly criticized the old institution-focused comparative politics for its ethnocentricity (or Western bias), it proposed to understand the politics of all society in terms of such functions, for instance, as "interest articulation" and "interest aggregation" whose definition is strongly influenced by the bargaining culture of our "civilization, but which is far from being guaranteed appropriateness elsewhere. The unsurprising result is a theory of political development which places the Atlantic-type polity at the summit of human political achievement.

Much can be said in this area of comparative politics (interestingly explored by Alasdair MacIntyre in a recently published paper).[12] But I should like to illustrate the significance of these two rival approaches in connection with another common problem area of politics. This is the question of what is called "legitimacy."[13]

ii

It is an obvious fact, with which politics has been concerned since at least Plato, that some societies enjoy an easier, more spontaneous cohesion which relies less on the use of force than others. It has been an important question of political theory to understand what underlies this difference. Among others, Aristotle, Machiavelli, Montesquieu, de Tocqueville have dealt with it.

Contemporary mainstream political scientists approach this question with the concept "legitimacy." The use of the word here can be easily understood. Those societies that are more spontaneously cohesive can be thought to enjoy a greater sense of legitimacy among their members. But the application of the term has been shifted. "Legitimacy" is a term in which we discuss the authority of the state or policy, its right to our allegiance. However we conceive of this legitimacy, it can only be attributed to a polity in the light of a number of surrounding conceptions—for example, that it provides men freedom, that it emanates from their will, that it secures them order, the rule of law, or that it is founded on tradition, or commands obedience by its superior qualities. These conceptions are all such that they rely on definitions of what is significant for men in general or in some particular society or circumstances, definitions of paradigmatic meaning which cannot be identifiable as brute data. Even where some of these terms might be given an "operational definition" in terms of brute data—a term like "freedom," for instance, can be defined in terms of the absence of legal restriction, à la Hobbes—this definition would not carry the full force of the term, and in particular that whereby it could be considered significant for men.

According to the empiricist paradigm, this latter aspect of the meaning of such a term is labeled "evaluative" and is thought to be utterly heterogeneous from the "descriptive" aspect. But this analysis is far from firmly established, no more so in fact than the empiricist paradigm of knowledge itself with which it is closely bound up. A challenge to this paradigm in the name of a hermeneutical science is also a challenge to the distinction between "descriptive" and "evaluative" and the entire conception of *Wertfreiheit* which goes with it.

In any case, whether because it is "evaluative" or can only be applied in connection with definitions of meaning, "legitimate" is not a word which can be used in the description of social reality according to the conceptions of mainstream social science. It can only be used as a description of subjective meaning. What enters into scientific consideration is thus not the legitimacy of a polity but the opinions or feelings of its member individuals concerning its legitimacy. The differences between different societies in their manner of spontaneous cohesion and sense of community are to be understood by correlations between the beliefs and feelings of their members toward them on one hand and the prevalence of certain brute data identifiable indices of stability in them on the other.

Robert Dahl in *Modern Political Analysis* speaks of the different ways in which leaders gain "compliance" for their policies.[14] The more citizens comply because of "internal rewards and deprivations," the less leaders need to use "external rewards and deprivations." But if citizens believe a government is legitimate, then their conscience will bind them to obey it; they will be internally punished if they disobey; hence government will have to use less external resources, including force.

Less crude is the discussion of Seymour Lipset in *Political Man*, but it is founded on the same basic ideas, namely, that legitimacy defined as subjective meaning is correlated with stability. "Legitimacy involves the capacity of the system to engender and maintain the belief that the existing political institutions are the most appropriate ones for the society."[15]

Lipset is engaged in a discussion of the determinants of stability in modern polities. He singles out two important ones, effectiveness and legitimacy. "Effectiveness means actual performance, the extent to which the system satisfies the basic functions of government as most of the population and such powerful groups within it as big business or the armed forces see them" (ibid.). Thus we have one factor which has to do with objective reality, what the government has actually done; and the other which has to do with subjective beliefs and "values." "While effectiveness is primarily instrumental, legitimacy is evaluative" (ibid.). Hence from the beginning the stage is set by a distinction between social reality and what men think and feel about it.

Lipset sees two types of crisis of legitimacy that modern societies have affronted more or less well. One concerns the status of major conservative institutions that may be under threat from the development of modern industrial democracies. The second concerns the degree to which all political groups have access to the political process. Under the first head, some traditional groups, such as landed aristocracy or clericals, have been roughly handled in a society like France, and have remained alienated from the democratic system for decades afterward; whereas in England the traditional classes were more gently handled, themselves were willing to compromise and have been slowly integrated and transformed into the new order. Under the second head, some societies managed to integrate the working class or bourgeoisie into the political process at an early stage, whereas in others they have been kept out until quite recently, and consequently, have developed a deep sense of alienation from the system, have tended to adopt extremist ideologies, and have generally contributed to instability. One of the determinants of a society's performance on these two heads is whether it is forced to affront the different conflicts of democratic development all at once or one at a time. Another important determinant of legitimacy is effectiveness.

This approach that sees stability as partly the result of legitimacy beliefs, and these in turn as resulting partly from the way the status, welfare, access to political life of different groups fare, seems at first blush eminently sensible and well designed to help us understand the history of the last century or two. But this approach has no place for a study of the intersubjective and common meanings which are constitutive of modern civilization. And we may doubt whether we can understand the cohesion of modern societies or their present crisis if we leave these out of account.

Let us take the winning of the allegiance of the working class to the new industrial regimes in the nineteenth and the early twentieth century. This is far from being a matter simply or even perhaps most significantly of the speed with which this class was integrated into the political process and the effectiveness of the regime. Rather the consideration of the granting of access to the political process as an independent variable may be misleading.

It is not just that we often find ourselves invited by historians to account for class cohesion in particular countries in terms of other factors, such as the impact of Methodism in early nineteenth-century England (Elie Halévy)[16] or the draw of Germany's newly successful nationalism. These factors could be assimilated to the social scientist's grid by being classed as "ideologies" or widely held "value systems" or some other such concatenations of subjective meaning.

But perhaps the most important such "ideology" in accounting for the cohesion of industrial democratic societies has been that of the society of work, the vision of society as a large-scale enterprise of production in which widely different functions are integrated into interdependence; a vision of society in which economic relations are considered as primary, as it is not only in Marxism (and in a sense not really with Marxism) but above all with the tradition of classical utilitarianism. In line with this vision there is a fundamental solidarity between all members of society that labor (to use Arendt's language),[17] for they are all engaged in producing what is indispensable to life and happiness in far-reaching interdependence.

This is the "ideology" that has frequently presided over the integration of the working class into industrial democracies, at first directed polemically against the "unproductive" classes, for example, in England with the Anti-Corn-Law League, and later with the campaigns of Joseph Chamberlain ("when Adam delved and Eve span/ who was then the gentlemen"), but later as a support for social cohesion and solidarity.

Of course the reason for putting "ideology" in quotes above is that this definition of things, which has been well integrated with the conception of social life as based on negotiation, cannot be understood in the terms of mainstream social science as beliefs and "values" held by a large number of individuals. For the great interdependent matrix of labor is not just a set of ideas in people's heads but is an important aspect of the reality that we live in modern society. And at the same time these ideas are embedded in this matrix in that they are constitutive of it; that is, we would not be able to live in this type of society unless we were imbued with these ideas or some others that could call forth the discipline and voluntary coordination needed to operate this kind of economy. All industrial civilizations have required a huge wrench from the traditional peasant populations on which they have been imposed; for they require an entirely unprecedented level of disciplined, sustained, monotonous effort, long hours unpunctuated by any meaningful rhythm, such as that of seasons or festivals. In the end this way of life can only be accepted when the idea of making a living is endowed with more significance than that of just avoiding starvation; and this it is in the civilization of labor.

Now this civilization of work is only one aspect of modern societies, along with the society based on negotiation and willed relations (in Anglo-Saxon countries), and other common and intersubjective meanings which have different importance in different countries. My point is that it is certainly not implausible to say that it has some importance in explaining the integration of the working class in modern industrial democratic society. But it can only be called a cluster of intersubjective meaning. As such it cannot come into the purview of mainstream political science; and an author like Lipset cannot take it into consideration when discussing this very problem.

But, of course, such a massive fact does not escape notice. What happens rather is that it is reinterpreted. And what has generally happened is that the interdependent productive and negotiating society has been recognized by political science, but not as one structure of intersubjective meaning among others, rather as the inescapable background of social action as such. In this guise it no longer need be an object of study. Rather it retreats to the middle distance, where its general outline takes the role of universal framework, within which (it is hoped) actions and structures will be brute-data-identifiable, and this for any society at any time. The view is then that the political actions of men in all societies can be understood as variants of the processing of "demands" which is an important part of our political life. The inability to recognize the specificity of our intersubjective meanings is thus inseparably linked with the belief in the universality of North Atlantic behavior types or "functions" which vitiates so much of contemporary comparative politics.

The notion is that what politics is about perennially is the adjustment of differences, or the production of symbolic and effective "outputs" on the basis of demand and support "inputs." The rise of the intersubjective meaning of the civilization of work is seen as the increase of correct perception of the political process at the expense of "ideology." Thus Almond and Powell introduce the concept of "political secularization" to describe "the emergence of a pragmatic, empirical orientation" to politics. A secular political culture is opposed not only to a traditional one, but also to an "ideological" culture, which is characterized by "an inflexible image of political life, closed to conflicting information" and "fails to develop the open, bargaining attitudes associated with full secularization."[18] The clear understanding here is that a secularized culture is one which essentially depends less on illusion, which sees things as they are, which is not infected with the "false consciousness" of traditional or ideological culture (to use a term which is not in the mainstream vocabulary).

iii

This way of looking at the civilization of work, as resulting from the retreat of illusion before the correct perception of what politics perennially and really is, is closely bound up with the epistemological premises of mainstream political science and its resultant inability to recognize the historical specificity of this civilization's intersubjective meanings. But the weakness of this approach, already visible in the attempts to explain the rise of this civilization and its relation to others, becomes even more painful when we try to account for its present malaise, even crisis.

The strains in contemporary society, the breakdown of civility, the rise of deep alienation, which is translated into even more destructive action, tend to shake the basic categories of our social science. It is not just that such a development was quite unpredicted by this science, which saw in the rise of affluence the cause rather of a further entrenching of the bargaining culture, a reduction of irrational cleavage, an increase of tolerance, in short, "the end of ideology." For prediction, as we shall see below, cannot be a goal of social science as it is of natural science. It is rather that this mainstream science does not have the categories to explain this breakdown. It is forced to look on extremism either as a bargaining gambit of the desperate, deliberately raising the ante in order to force a hearing. Alternatively, it can recognize the novelty of the rebellion by accepting the hypothesis that heightened demands are being made on the system owing to a revolution of "expectations," or else to the eruption of new desires or aspirations which hitherto had no place in the bargaining process. But these new desires or aspirations must be in the domain of individual psychology, that is, they must be such that their arousal and satisfaction is to be understood in terms of states of individuals rather than in terms of the intersubjective meanings in which they live. For these latter have no place in the categories of the mainstream, which cannot accommodate a genuine historical psychology.

But some of the more extreme protests and acts of rebellion in our society cannot be interpreted as bargaining gambits in the name of any demands, old or new. These can only be interpreted within the accepted framework of our social science as a return to ideology, and hence as irrational. Now in the case of some of the more bizarre and bloody forms of protest, there will be little disagreement; they will be judged irrational by all but their protagonists. But within the accepted categories this irrationality can only be understood in terms of individual psychology; it is the public eruption of private pathology; it cannot be understood as a malady of society itself, a malaise which afflicts its constitutive meanings.[19]

No one can claim to begin to have an adequate explanation for these major changes which our civilization is undergoing. But in contrast to the incapacity of a science which remains within the accepted categories, a hermeneutical science of man that has a place for a study of intersubjective meaning can at least begin to explore fruitful avenues. Plainly the discipline that was integral to the civilization of work and bargaining is beginning to fail. The structures of this civilization, interdependent work, bargaining, mutual adjustment of individual ends, are beginning to change their meaning for many, and are beginning to be felt not as normal and best suited to man, but as hateful or empty. And yet we are all caught in these intersubjective meanings insofar as we live in this society, and in a sense more and more all-pervasively as it progresses. Hence the virulence and tension of the critique of our society which is always in some real sense a self-rejection (in a way that the old socialist opposition never was).

Why has this set of meanings gone sour? Plainly, we have to accept that they are not to be understood at their face value. The free, productive, bargaining culture claimed to be sufficient for man. If it was not, then we have to assume that while it did hold our allegiance, it also had other meanings for us that commanded this allegiance and that have now gone.

This is the starting point of a set of hypotheses which attempt to redefine our past in order to make our present and future intelligible. We might think that in the past the productive, bargaining culture offered common meanings (even though there was no place for them in its philosophy), and hence a basis for community, that were essentially linked with its being in the process of building. It linked men who could see themselves as breaking with the past to build a new happiness in America, for instance. But in all essentials that future is built; the notion of a horizon to be attained by future greater production (as against social transformation) verges on the absurd in contemporary America. Suddenly the horizon that was essential to the sense of meaningful purpose has collapsed, which would show that like so many other Enlightenment-based dreams the free, productive, bargaining society can only sustain man as a goal, not as a reality.

Or we can look at this development in terms of identity. A sense of building their future through the civilization of work can sustain men as long as they see themselves as having broken with a millennial past of injustice and hardship in order to create qualitatively different conditions for their children. All the requirements of a humanly acceptable identity can be met by this predicament, a relation to the past (one soars above it but preserves it in folkloric memory), to the social world (the interdependent world of free, productive men), to the earth (the raw material which awaits shaping), to the future and one's own death (the everlasting monument in the lives of prosperous children), to the absolute (the absolute values of freedom, integrity, dignity).

But at some point the children will be unable to sustain this forward thrust into the future. This effort has placed them in a private haven of security, within which they are unable to reach and recover touch with the great realities: their parents have only a negated past, lives which have been oriented wholly to the future; the social world is distant and without shape; rather one can only insert oneself into it by taking one's place in the future-oriented productive juggernaut. But this now seems without any sense; the relation to the earth as raw material is therefore experienced as empty and alienating, but the recovery of a valid relation to the earth is the hardest thing once lost; and there is no relation to the absolute where we are caught in the web of meanings which have gone dead for us. Hence past, future, earth, world, and absolute are in some way or another occluded; and what must arise is an identity crisis of frightening proportions.

These two hypotheses are mainly focused on the crisis in American society, and they would perhaps help account for the fact that the United States is in some sense going first through this crisis of all Atlantic nations; not, that is, only because it is the most affluent, but more because it has been more fully based on the civilization of work than European countries that retained something of more traditional common meanings.

But they might also help us to understand why alienation is most severe among groups which have been but marginal in affluent bargaining societies. These have had the greatest strain in living in this civilization while their identity was in some ways antithetical to it. Such are blacks in the United States, and the community of French-speaking Canadians, each in different ways. For many immigrant groups the strain was also great, but they forced themselves to surmount the obstacles, and the new identity is sealed in the blood of the old, as it were.

But for those who would not or could not succeed in transforming themselves, but always lived a life of strain on the defensive, the breakdown of the central, powerful identity is the trigger to a deep turnover. It can be thought of as a liberation, but at the same time it is deeply unsettling, because the basic parameters of former life are being changed, and there are not yet the new images and definitions to live a new fully acceptable identity. In a sense we are in a condition where a new social compact (rather the first social compact) has to be made between these groups and those they live with, and no one knows where to start.

In the last pages, I have presented some hypotheses which may appear very speculative; and they may indeed turn out to be without foundation, even without much interest. But their aim was mainly illustrative. My principal claim is that we can only come to grips with this phenomenon of breakdown by trying to understand more clearly and profoundly the common and intersubjective meanings of the society in which we have been living. For it is these which no longer hold us, and to understand this change we have to have an adequate grasp of these meanings. But this we cannot do as long as we remain within the ambit of mainstream social science, for it will not recognize intersubjective meaning, and is forced to look at the central ones of our society as though they were the inescapable background of all political action. Breakdown is thus inexplicable in political terms; it is an outbreak of irrationality which must ultimately be explained by some form of psychological illness.

Mainstream science may thus venture into the area explored by the above hypotheses, but after its own fashion, by forcing the psychohistorical facts of identity into the grid of an individual psychology, in short, by reinterpreting all meanings as subjective. The result might be a psychological theory of emotional maladjustment, perhaps traced to certain features of family background, analogous to the theories of the authoritarian personality and the California F-scale. But this would no longer be a political or a social theory. We would be giving up the attempt to understand the change in social reality at the level of its constitutive intersubjective meanings.

IV

It can be argued, then, that mainstream social science is kept within certain limits by its categorical principles which are rooted in the traditional epistemology of empiricism; and second, that these restrictions are a severe handicap and prevent us from coming to grips with important problems of our day, which should be the object of political science. We need to go beyond the bounds of a science based on verification to one which would study the intersubjective and common meanings embedded in social reality.

But this science would be hermeneutical in the sense that has been developed in this chapter. It would not be founded on brute data; its most primitive data would be readings of meanings, and its object would have the three properties mentioned above: the meanings are for a subject in a field or fields; they are, moreover, meanings which are partially constituted by self-definitions, which are in this sense already interpretations, and which can thus be reexpressed or made explicit by a science of politics. In our case, the subject may be a society or community; but the intersubjective meanings, as we saw, embody a certain self-definition, a vision of the agent and his society, which is that of the society or community.

But then the difficulties which the proponents of the verification model foresee will arise. If we have a science that has no brute data, that relies on readings, then it cannot but move in a hermeneutical circle. A given reading of the intersubjective meanings of a society, or of given institutions or practices, may seem well founded, because it makes sense of these practices or the development of that society. But the conviction that it does make sense of this history itself is founded on further related readings. Thus, what I said above on the identity crisis which is generated by our society makes sense and holds together only if one accepts this reading of the intersubjective meanings of our society, and if one accepts this reading of the rebellion against our society by many young people (sc. the reading in terms of identity crisis). These two readings make sense together, so that in a sense the explanation as a whole reposes on the readings, and the readings in their turn are strengthened by the explanation as a whole.

But if these readings seem implausible, or even more, if they are not understood by our interlocutor, there is no verification procedure that we can fall back on. We can only continue to offer interpretations; we are in an interpretative circle.

But the ideal of a science of verification is to find an appeal beyond differences of interpretation. Insight will always be useful in discovery, but should not have to play any part in establishing the truth of its findings. This ideal can be said to have been met by our natural sciences. But a hermeneutic science cannot but rely on insight. It requires that one have the sensibility and understanding necessary to be able to make and comprehend the readings by which we can explain the reality concerned. In physics we might argue that if someone does not accept a true theory, then either he has not been shown enough (brute data) evidence (perhaps not enough is yet available), or he cannot understand and apply some formalized language. But in the sciences of man conceived as hermeneutical, the nonacceptance of a true or illuminating theory may come from neither of these, indeed is unlikely to be due to either of these, but rather from a failure to grasp the meaning field in question, an inability to make and understand readings of this field.

In other words, in a hermeneutical science, a certain measure of insight is indispensable, and this insight cannot be communicated by the gathering of brute data, or initiation in modes of formal reasoning or some combination of these. It is unformalizable. But this is a scandalous result according to the authoritative conception of science in our tradition, which is shared even by many of those who are highly critical of the approach of mainstream psychology, or sociology, or political science. For it means that this is not a study in which anyone can engage, regardless of their level of insight; that some claims of the form, "If you don't understand, then your intuitions are at fault, are blind or inadequate," some claims of this form will be justified; that some differences will be nonarbitrable by further evidence, but that each side can only make appeal to deeper insight on the part of the other. The superiority of one position over another will thus consist in this, that from the more adequate position one can understand one's own stand and that of one's opponent, but not the other way around. It goes without saying that this argument can only have weight for those in the superior position.

Thus, a hermeneutical science encounters a gap in intuitions, which is the other side, as it were, of the hermeneutical circle. But the situation is graver than this; for this gap is bound up with our divergent options in politics and life.

We speak of a gap when some cannot understand the kind of self-definition which others are proposing as underlying a certain society or set of institutions. Thus some positivistically minded thinkers will find the language of identity theory quite opaque: and some thinkers will not recognize any theory which does not fit with the categorical presuppositions of empiricism. But self-definitions are not only important to us as scientists who are trying to understand some, perhaps distant, social reality. As men we are self-defining beings, and we are partly what we are in virtue of the self-definitions which we have accepted, however we have come by them. What self-definitions we understand and what ones we do not understand are closely linked with the self-definitions that help to constitute what we are. If it is too simple to say that one only understands an "ideology" which one subscribes to, it is nevertheless hard to deny that we have great difficulty grasping definitions whose terms structure the world in ways that are utterly different from, incompatible with, our own.

Hence the gap in intuitions does not just divide different theoretical positions; it also tends to divide different fundamental options in life. The practical and the theoretical are inextricably joined here. It may not just be that to understand a certain explanation one has to sharpen one's intuitions; it may be that one has to change one's orientation —if not in adopting another orientation, at least in living one's own in a way which allows for greater comprehension of others. Thus, in the sciences of man insofar as they are hermeneutical there can be a valid response to, "I don't understand," which takes the form, not only, "develop your intuitions," but more radically, "change yourself." This puts an end to any aspiration to a value-free or "ideology-free" science of man. A study of the science of man is inseparable from an examination of the options between which men must choose.

This means that we can speak here not only of error, but of illusion. We speak of "illusion" when we are dealing with something of greater substance than error, error that in a sense builds a counterfeit reality of its own. But errors of interpretation of meaning, which are also self-definitions of those who interpret and hence inform their lives, are more than errors in this sense: they are sustained by certain practices of which they are constitutive. It is not implausible to single out as examples two rampant illusions in our present society. One is that of the proponents of the bargaining society who can recognize nothing but either bargaining gambits or madness in those who rebel against this society. Here the error is sustained by the practices of the bargaining culture, and given a semblance of reality by the refusal to treat any protests on other terms; it hence acquires the more substantive reality of illusion. The second example is provided by much "revolutionary" activity in our society which in desperate search for an alternative mode of life purports to see its situation in that of an Andean guerrilla or Chinese peasants. Lived out, this passes from the stage of laughable error to tragic illusion. One illusion cannot recognize the possibility of human variation; the other cannot see any limits to mankind's ability to transform itself. Both make a valid science of man impossible.

In face of all this, we might be so scandalized by the prospect of such a hermeneutical science that we will want to go back to the verification model. Why can we not take our understanding of meaning as part of the logic of discovery, as the logical empiricists suggest for our unformalizable insights, and still found our science on the exactness of our predictions? Our insightful understanding of the intersubjective meanings of our society will then serve to elaborate fruitful hypotheses, but the proof of these puddings will remain in the degree to which they enable us to predict.

The answer is that if the epistemological views underlying the science of interpretation are right, such exact prediction is radically impossible—this, for three reasons of ascending order of fundamentalness.

The first is the well-known "open system" predicament, one shared by human life and meteorology, that we cannot shield a certain domain of human events, the psychological economic, political, from external interference; it is impossible to delineate a closed system.

The second, more fundamental, is that if we are to understand men by a science of interpretation, we cannot achieve the degree of fine exactitude of a science based on brute data. The data of natural science admit of measurement to virtually any degree of exactitude. But different interpretations cannot be judged in this way. At the same time different nuances of interpretation may lead to different predictions in some circumstances, and these different outcomes may eventually create widely varying futures. Hence it is more than easy to be wide of the mark.

But the third and most fundamental reason for the impossibility of hard prediction is that man is a self-defining animal. With changes in his self-definition go changes in what man is, such that he has to be understood in different terms. But the conceptual mutations in human history can and frequently do produce conceptual webs which are incommensurable, that is, where the terms cannot be defined in relation to a common stratum of expressions. The entirely different notions of bargaining in our society and in some primitive ones provide an example. Each will be glossed in terms of practices, institutions, ideas in each society which have nothing corresponding to them in the other.

The success of prediction in the natural sciences is bound up with the fact that all states of the system, past and future, can be described in the same range of concepts, as values, say, of the same variables. Hence all future states of the solar system can be characterized, as past ones are, in the language of Newtonian mechanics. This is far from being a sufficient condition of exact prediction, but it is a necessary one in this sense, that only if past and future are brought under the same conceptual net can one understand the states of the latter as some function of the states of the former, and hence predict.

This conceptual unity is vitiated in the sciences of man by the fact of conceptual innovation, which in turn alters human reality. The very terms in which the future will have to be characterized if we are to understand it properly are not all available to us at present. Hence we have such radically unpredictable events as the culture of youth today, the Puritan rebellion of the sixteenth and seventeenth centuries, the development of Soviet society, and so forth.

And thus, it is much easier to understand after the fact than it is to predict. Human science is largely ex post understanding. Or often one has the sense of impending change, of some big reorganization, but is powerless to make clear what it will consist in: one lacks the vocabulary. But there is a clear asymmetry here, which there is not (or not supposed to be) in natural science, where events are said to be predicted from the theory with exactly the same ease with which one explains past events and by exactly the same process. In human science this will never be the case.

Of course, we strive ex post to understand the changes, and to do this we try to develop a language in which we can situate the incommensurable webs of concepts. We see the rise of Puritanism, for instance, as a shift in man's stance to the sacred; and thus, we have a language in which we can express both stances—the earlier medieval Catholic one and the Puritan rebellion—as "glosses" on this fundamental term. We thus have a language in which to talk of the transition. But think how we acquired it. This general category of the sacred is acquired not only from our experience of the shift that came in the Reformation, but from the study of human religion in general, including primitive religion, and with the detachment that came with secularization. It would be conceivable, but unthinkable, that a medieval Catholic could have this conception—or for that matter a Puritan. These two protagonists only had a language of condemnation for each other: "heretic," "idolater." The place for such a concept was preempted by a certain way of living the sacred. After a big change has happened, and the trauma has been resorbed, it is possible to try to understand it, because one now has available the new language, the transformed meaning world. But hard prediction before just makes one a laughingstock. Really to be able to predict the future would be to have explicated so clearly the human condition that one would already have preempted all cultural innovation and transformation. This is hardly in the bounds of the possible.

Sometimes men show amazing prescience: the myth of Faust, for instance, which is treated several times at the beginning of the modern era. There is a kind of prophesy here, a premonition. But what characterizes these bursts of foresight is that they see through a glass darkly, for the see in terms of the old language: Faust sells his soul to the devil. They are in no sense hard predictions. Human science looks backward. It is inescapably historical.

There are thus good grounds both in epistemological arguments and in their greater fruitfulness for opting for hermeneutical sciences of man. But we cannot hide from ourselves how greatly this option breaks with certain commonly held notions about our scientific tradition. We cannot measure such sciences against the requirements of a science of verification: we cannot judge them by their predictive capacity. We have to accept that they are founded on intuitions which all do not share, and what is worse, that these intuitions are closely bound up with our fundamental options. These sciences cannot be *wertfrei*; they are moral sciences in a more radical sense than the eighteenth century understood. Finally, their successful prosecution requires a high degree of self-knowledge, a freedom from illusion, in the sense of error which is rooted and expressed in one's way of life; for our incapacity to understand is rooted in our own self-definitions, hence in what we are. To say this is not to say anything new: Aristotle makes a similar point in Book I of the *Ethics*. But it is still radically shocking and unassimilable to the mainstream of modern science.

Notes

1. See, e.g., H. G. Gadamer, *Wahrheit und Methode* (Tübingen, 1960).
2. See Paul Ricoeur, *De l'interprétation* (Paris, 1965).
3. See, e.g., J. Hambermas, *Erkenntnis und Interesse* (Frankfurt, 1968).
4. The notion of brute data here has some relation to, but is not at all the same as, the brute facts discussed by Elizabeth Anscombe, "On Brute Facts," *Analysis* 18 (1957–1958): 69–72, and John Searle, *Speech Acts: An Essay in the Philosophy of Language* (Cambridge, 1969), 50–53. For Anscombe and Searle, brute facts are contrasted to what may be called "institutional facts," to use Searle's term, facts which presuppose the existence of certain institutions. Voting would be an example. But as we shall see below, some institutional facts, such as Xs have voted Liberal, can be verified as brute data in the sense used here, and thus find a place in the category of political behavior. What cannot as easily be described in terms of brute data are the institutions themselves.
5. See the discussion in M. Minsky, *Computation* (Englewood Cliffs, N.J., 1967), 104–107, where Minsky explicitly argues that an effective procedure that no longer requires intuition or interpretation is one which can be realized by a machine.
6. Cf. H. McClosky, "Consensus and Ideology in American Politics," *American Political Science Review* 58 (1964): 361–82.
7. Gabriel A. Almond and G. Bingham Powell, *Comparative Politics: A Developmental Approach* (Boston and Toronto, 1966), 23.
8. Cf. Thomas C. Smith, *The Agrarian Origins of Modern Japan* (Stanford, 1959), chap. 5. This type of consensus is also found in other traditional societies. Cf., for instance, the *desa* system of the Indonesian village.
9. Searle, *Speech Acts*, 33–42.
10. Cf. the discussion in Stanley Cavell, *Must We Mean What We Say?* (New York, 1969), 21–31.
11. Cf. Almond and Powell, *Comparative Politics*.

12. "How Is a Comparative Science of Politics Possible?" in Alasdair MacIntyre, *Against the Self-Images of the Age* (London, 1971).
13. MacIntyre's article also contains an interesting discussion of legitimacy from a different, although I think related, angle.
14. Robert Dahl, *Modern Political Analysis* (Englewood Cliffs, N.J., 1963), 31–32.
15. Seymour Lipset, *Political Man* (New York, 1963), 64.
16. *Histoire du peuple anglais au XIXe siècle* (Paris, 1913).
17. *The Human Condition* (New York, 1959).
18. *Comparative Politics*, 58, 61.
19. Thus Lewis Feuer, in *The Conflict of Generations* (New York, 1969), attempts to account for the "misperception of social reality" in the Berkeley student uprising in terms of a generational conflict (466–70), which in turn is rooted in the psychology of adolescence and attaining adulthood. Yet Feuer himself in his first chapter notes the comparative recency of self-defining political generations, a phenomenon which dates from the post-Napoleonic era (33). But an adequate attempt to explain his historical shift, which after all underlies the Berkeley rising and many others, would, I believe, have taken us beyond the ambit of individual psychology to psychohistory, to a study of the intrication of psychological conflict and intersubjective meanings. A variant of this form of study has been adumbrated in the work of Erik Erikson.

Chapter 14

Thick Description: Toward an Interpretive Theory of Culture

Clifford Geertz

I

In her book, *Philosophy in a New Key,* Susanne Langer remarks that certain ideas burst upon the intellectual landscape with a tremendous force. They resolve so many fundamental problems at once that they seem also to promise that they will resolve all fundamental problems, clarify all obscure issues. Everyone snaps them up as the open sesame of some new positive science, the conceptual center-point around which a comprehensive system of analysis can be built. The sudden vogue of such a *grande idée,* crowding out almost everything else for a while, is due, she says, "to the fact that all sensitive and active minds turn at once to exploiting it. We try it in every connection, for every purpose, experiment with possible stretches of its strict meaning, with generalizations and derivatives."

After we have become familiar with the new idea, however, after it has become part of our general stock of theoretical concepts, our expectations are brought more into balance with its actual uses, and its excessive popularity is ended. A few zealots persist in the old key-to-the-universe view of it; but less driven thinkers settle down after a while to the problems the idea has really generated. They try to apply it and extend it where it applies and where it is capable of extension; and they desist where it does not apply or cannot be extended. It becomes, if it was, in truth, a seminal idea in the first place, a permanent and enduring part of our intellectual armory. But it no longer has the grandiose, all-promising scope, the infinite versatility of apparent application, it once had. The second law of thermodynamics, or the principle of natural selection, or the notion of unconscious motivation, or the organization of the means of production does not explain everything, not even everything human, but it still explains something; and our attention shifts to isolating just what that something is, to disentangling ourselves from a lot of pseudoscience to which, in the first flush of its celebrity, it has also given rise.

Whether or not this is, in fact, the way all centrally important scientific concepts develop, I do not know. But certainly this pattern fits the concept of culture, around which the whole discipline of anthropology arose, and whose domination that discipline has been increasingly concerned to limit, specify, focus, and contain. It is to this cutting of the culture concept down to size, therefore actually ensuring its continued importance rather than undermining it, that [this essay is] ... dedicated. [It argues] ... for a narrowed, specialized, and, so I imagine, theoretically more powerful concept of culture to replace E. B. Tylor's famous "most complex whole," which, its originative power not denied, seems to me to have reached the point where it obscures a good deal more than it reveals.

The conceptual morass into which the Tylorean kind of pot-au-feu theorizing about culture can lead, is evident in what is still one of the better general introductions to anthropology, Clyde Kluckohn's *Mirror for Man*. In some twenty-seven pages of his chapter on the concept, Kluckhohn managed to define culture in turn as: (1) "the total way of life of a people"; (2) "the social legacy the individual acquires from his group"; (3) "a way of thinking, feeling, and believing"; (4) "an abstraction from behavior"; (5) a theory on the part of the anthropologist about the way in which a group of people in fact behave; (6) a "storehouse of pooled learning"; (7) "a set of standardized orientations to recurrent problem"; (8) "learned behavior"; (9) a mechanism for the normative regulation of behavior; (10) "a set of techniques for adjusting both to the external environment and to other men"; (11) "a precipitate of history"; and turning, perhaps in desperation, to similes, as a map, as a sieve, and as a matrix. In the fact of this sort of theoretical diffusion, even a somewhat constricted and not entirely standard concept of culture, which is at least internally coherent and, more important, which has a definable argument to make is (as, to be fair, Kluckhohn himself keenly realized) an improvement. Eclecticism is self-defeating not because there is only one direction in which it is useful to move, but because there are so many: it is necessary to choose.

The concept of culture I espouse ... is essentially a semiotic one. Believing, with Max Weber, that man is an animal suspended in webs of significance he himself has spun, I take culture to be those webs, and the analysis of it to be therefore not an experimental science in search of law but an interpretive one in search of meaning. It is explication I am after, construing social expressions on their surface enigmatical. But this pronouncement, a doctrine in a clause, demands itself some explication.

II

Operationalism as a methodological dogma never made much sense so far as the social sciences are concerned, and except for a few rather too well-swept corners—Skinnerian behaviorism, intelligence testing, and so on—it is largely dead now. But it had, for all that, an important point to make, which, however we may feel about trying to define charisma or alienation in terms of operations, retains a certain force: if you want to understand what a science is, you should look in the first instance not at its theories or its findings, and certainly not at what its apologists say about it; you should look at what the practitioners of it do.

In anthropology, or anyway social anthropology, what the practitioners do is ethnography. And it is in understanding what ethnography is, or more exactly *what doing ethnography is*, that a start can be made toward grasping what anthropological analysis amounts to as a form of knowledge. This, it must immediately be said, is not a matter of methods. From one point of view, that of the textbook, doing ethnography is establishing rapport, selecting informants, transcribing texts, taking genealogies, mapping fields, keeping a diary, and so on. But it is not these things, techniques and received procedures, that define the enterprise. What defines it is the kind of intellectual effort it is: an elaborate venture in, to borrow a notion from Gilbert Ryle, "thick description."

Ryle's discussion of "thick description" appears in two recent essays of his (now reprinted in the second volume of his *Collected Papers*) addressed to the general question of what, as he puts it, "*Le Penseur*" is doing: "Thinking and Reflecting" and "The Thinking of Thoughts." Consider, he says, two boys rapidly contracting the eyelids of their right eyes. In one, this is an involuntary twitch; in the other, a conspiratorial signal to a friend. The two movements are, as movements, identical; from an I-am-a-camera, "phenomenalistic" observation of them alone, one could not tell which was twitch and which was wink, or indeed whether both or either was twitch or wink. Yet the difference, however unphotographable, between a twitch and wink is vast, as anyone unfortunate enough to have had the first taken for the second knows. The winker is communicating, and indeed communicating in a quite precise and special way: (1) deliberately, (2) to someone in particular, (3) to impart a particular message, (4) according to a socially established code, and (5) without cognizance of the rest of the company, As Ryle points out, the winker has not done two things, contracted his eyelids and winked, while the twitcher has done only one, contracted his eyelids. Contracting your eyelids on purpose when there exists a public code in which so doing counts as a conspiratorial signal *is* winking. That's all there is to it: a speck of behavior, a fleck of culture, and—voilà!—a gesture.

That, however, is just the beginning. Suppose, he continues, there is a third boy, who, "to give malicious amusement to his cronies," parodies the first boy's wink, as amateurish, clumsy, obvious, and so on. He, of course, does this in the same way the second boy winked and the first twitched: by contracting his right eyelids. Only this boy is neither winking nor twitching; he is parodying someone else's, as he takes it, laughable, attempt at winking. Here, too, a socially established code exists (he will "wink" laboriously, overobviously, perhaps adding a grimace—the usual artifices of the clown), and so also does a message. Only now it is not conspiracy but ridicule that is in the air. If the others think he is actually winking, his whole project misfires as completely, though with somewhat different results, as if they think he is twitching. One can go further: uncertain of his mimicking abilities, the would-be satirist may practice at home before the mirror, in which case he is not twitching, winking, or parodying, but rehearsing; though so far as what a camera, a radical behaviorist, or a believer in protocol sentences would record he is just rapidly contracting his right eyelids like all the others. Complexities are possible, if not practically without end, at least logically so. The original winker might, for example, actually have been fake-winking, say, to mislead outsiders into imagining there was a conspiracy afoot when there in fact was not, in which case our descriptions of what the parodist is parodying and the rehearser rehearsing of course shift accordingly. But the point is that between what Ryle calls the "thin description" of what the rehearser (parodist, winker, twitcher ...) is doing ("rapidly contracting his right eyelids") and the "thick description" of what he is doing ("practicing a burlesque of a friend faking a wink to deceive an innocent into thinking a conspiracy is in motion") lies the object of ethnography: a stratified hierarchy of meaningful structures in terms of which twitches, winks, fake-winks, parodies, rehearsals of parodies are produced, perceived, and interpreted, and without which they would not (not even the zero-form twitches, which, *as a cultural category*, are as much nonwinks as winks are nontwitches) in fact exist, no matter what anyone did or didn't do with his eyelids.

Like so many of the little stories Oxford philosophers like to make up for themselves, all this winking, fake-winking, burlesque-fake-winking, rehearsed-burlesque-fake-winking, may seem a bit artificial. In way of adding a more empirical note, let me give, deliberately unpreceded by any prior explanatory comment at all, a not untypical excerpt from my own field journal to demonstrate that, however evened off for didactic purposes, Ryle's example presents an image only too exact of the sort of piled-up structures of inference and implication through which an ethnographer is continually trying to pick his way:

> The French [the informant said] had only just arrived. They set up twenty or so small forts between here, the town, and the Marmusha area up in the middle of the mountains, placing them on promontories so they could survey the countryside. But for all this they couldn't guarantee safety, especially at night, so although the *mezrag*, trade-pact, system was supposed to be legally abolished it in fact continued as before.
>
> One night, when Cohen (who speaks fluent Berber), was up there, at Marmusha, two other Jews who were traders to a neighboring tribe came by to purchase some goods from him. Some Berbers, from yet another neighboring tribe, tried to break into Cohen's place, but he fired his rifle in the air. (Traditionally, Jews were not allowed to carry weapons; but at this period things were so unsettled many did so anyway.) This attracted the attention of the French and the marauders fled.
>
> The next night, however, they came back, one of them disguised as a woman who knocked on the door with some sort of a story. Cohen was suspicious and didn't want to let "her" in, but the other Jews said, "oh, it's all right, it's only a woman." So they opened the door and the whole lot came pouring in. They killed the two visiting Jews, but Cohen managed to barricade himself in an adjoining room. He heard the robbers planning to burn him alive in the shop after they removed his goods, and so he opened the door and, laying about him wildly with a club, managed to escape through a window.
>
> He went up to the fort, then, to have his wounds dressed, and complained to the local commandant, one Captain Dumari, saying he wanted his *'ar*—i.e., four or five times the value of the merchandise stolen from him. The robbers were from a tribe which had not yet submitted to French authority and were in open rebellion against it, and he wanted authorization to go with his *mezrag*-holder, the Marmusha tribal *sheikh*, to collect the indemnity that, under traditional rules, he had coming to him. Captain Dumari couldn't officially give him permission to do this, because of the French prohibition of the *mezrag* relationship, but he gave him verbal authorization, saying, "If you get killed, it's your problem."
>
> So the *sheikh*, the Jew, and a small company of armed Marmushans went off ten or fifteen kilometers up into the rebellious area, where there were of course no French, and, sneaking up, captured the thief-tribe's shepherd and stole its herds. The other tribe soon came riding out on horses after them, armed with rifles and ready to attack. But when they saw who the "sheep thieves" were, they thought better of it and said, "all right, we'll talk." They couldn't really deny what had happened—that some of their men had robbed Cohen and killed the two visitors—and they weren't prepared to start the serious feud with the Marmusha a scuffle with the invading party would bring on. So the two groups talked, and talked, and talked, there on the plain amid the thousands of sheep, and decided finally on five-hundred-sheep damages. The two armed Berber groups then lined up on their horses at opposite ends of the plain, with the sheep herded between them, and Cohen, in his black gown, pillbox hat, and flapping slippers, went out alone among the sheep, picking out, one by one and at his own good speed, the best ones for his payment.

> So Cohen got his sheep and drove them back to Marmusha. The French, up in their fort, heard them coming from some distance ("Ba, ba, ba" said Cohen, happily, recalling the image) and said, "What the hell is that?" And Cohen said, "That is my *'ar*." The French couldn't believe he had actually done what he said he had done, and accused him of being a spy for the rebellious Berbers, put him in prison, and took his sheep. In the town, his family, not having heard from him in so long a time, thought he was dead. But after a while the French released him and he came back home, but without his sheep. He then went to the Colonel in the town, the Frenchman in charge of the whole region, to complain. But the Colonel said, "I can't do anything about the matter. It's not my problem."

Quoted raw, a note in a bottle, this passage conveys, as any similar one similarly presented would do, a fair sense of how much goes into ethnographic description of even the most elmental sort—how extraordinarily "thick" it is. In finished anthropological writings, including those collected here, this fact—that what we call our data are really our own constructions of other people's constructions of what they and their compatriots are up to—is obscured because most of what we need to comprehend a particular event, ritual, custom, idea, or whatever is insinuated as background information before the thing itself is directly examined. (Even to reveal that this little drama took place in the highlands of central Morocco in 1912—and was recounted there in 1968—is to determine much of our understanding of it.) There is nothing particularly wrong with this, and it is in any case inevitable. But it does lead to a view of anthropological research as rather more of an observational and rather less of an interpretive activity than it really is. Right down at the factual base, the hard rock, insofar as there is any, of the whole enterprise, we are already explicating: and worse, explicating explications. Winks upon winks upon winks.

Analysis, then, is sorting out the structures of signification—what Ryle called established codes, a somewhat misleading expression, for it makes the enterprise sound too much like that of the cipher clerk when it is much more like that of the literary critic—and determining their social ground and import. Here, in our text, such sorting would begin with distinguishing the three unlike frames of interpretation ingredient in the situation, Jewish, Berber, and French, and would then move on to show how (and why) at that time, in that place, their copresence produced a situation in which systematic misunderstanding reduced traditional form to social farce. What ripped Cohen up, and with him the whole, ancient pattern of social and economic relationships within which he functioned, was a confusion of tongues.

I shall come back to this too-compacted aphorism later, as well as to the details of the text itself. The point for now is only that ethnography is thick description. What the ethnographer is in fact faced with—except when (as, of course, he must do) he is pursuing the more automatized routines of data collection—is a multiplicity of complex conceptual structures, many of them superimposed upon or knotted into one another, which are at once strange, irregular, and inexplicit, and which he must contrive somehow first to grasp and then to render. And this is true at the most down-to-earth, jungle field work levels of his activity: interviewing informants, observing rituals, eliciting kin terms, tracing property lines, censusing households ... writing his journal. Doing ethnography is like trying to read (in the sense of "construct a reading of") a manuscript—foreign, faded, full of ellipses, incoherencies, suspicious emendations, and tendentious commentaries, but written not in conventionalized graphs of sound but in transient examples of shaped behavior.

III

Culture, this acted document, thus is public, like a burlesqued wink or a mock sheep raid. Though ideational, it does not exist in someone's head; though unphysical, it is not an occult entity. The interminable, because unterminable, debate within anthropology as to whether culture is "subjective" or "objective," together with the mutual exchange of intellectual insults ("idealist!"—"materialist!"; "mentalist!"—"behaviorist!"; "impressionist!"—"positivist!") which accompanies it, is wholly misconceived. Once human behavior is seen as (most of the time there *are* true twitches) symbolic action—action which, like phonation in speech, pigment in painting, line in writing, or sonance in music, signifies—the question as to whether culture is patterned conduct or a frame of mind, or even the two somehow mixed together, loses sense. The thing to ask about a burlesqued wink or a mock sheep raid is not what their ontological status is. It is the same as that of rocks on the one hand and dreams on the other—they are things of this world. The thing to ask is what their import is: what it is, ridicule or challenge, irony or anger, snobbery or pride, that, in their occurrence and through their agency, is getting said.

This may seem like an obvious truth, but there are a number of ways to obscure it. One is to imagine that culture is a self-contained "superorganic" reality with forces and purposes of its own; that is, to reify it. Another is to claim that it consists in the brute pattern of behavioral events we observe in fact to occur in some identifiable community or other; that is, to reduce it. But though both these confusions still exist, and doubtless will be always with us, the main source of theoretical muddlement in contemporary anthropology is a view which developed in reaction to them and is right now very widely held—namely, that, to quote Ward Goodenough, perhaps its leading proponent, "culture [is located] in the minds and hearts of men."

Variously called ethnoscience, componential analysis, or cognitive anthropology (a terminological wavering which reflects a deeper uncertainty), this school of thought holds that culture is composed of psychological structures by means of which individuals or groups of individuals guide their behavior. "A society's culture," to quote Goodenough again, this time in a passage which has become the locus classicus of the whole movement, "consists of whatever it is one has to know or believe in order to operate in a manner acceptable to its members." And from this view of what culture is follows a view, equally assured, of what describing it is—the writing out of systematic rules, an ethnographic algorithm, which, if followed, would make it possible so to operate, to pass (physical appearance aside) for a native. In such a way, extreme subjectivism is married to extreme formalism, with the expected result: an explosion of debate as to whether particular analyses (which come in the form of taxonomies, paradigms, tables, trees, and other ingenuities) reflect what the natives "really" think or are merely clever simulations, logically equivalent but substantively different, of what they think.

As, on first glance, this approach may look close enough to the one being developed here to be mistaken for it, it is useful to be explicit as to what divides them. If, leaving our winks and sheep behind for the moment, we take, say, a Beethoven quartet as an, admittedly rather special but, for these purposes, nicely illustrative, sample of culture, no one would, I think, identify it with its score, with the skills and knowledge needed to play it, with the understanding of it possessed by its performers or auditors, nor, to take care, *en passant*, of the reductionists and reifiers, with a particular performance of it or with some mysterious entity transcending material existence. The "no one" is perhaps too strong here, for there are always incorrigibles. But that a Beethoven quartet is a temporally developed tonal structure, a coherent sequence of modeled sound—in a word, music—and not anybody's knowledge of or belief about anything, including how to play it, is a proposition to which most people are, upon reflection, likely to assent.

To play the violin it is necessary to possess certain habits, skills, knowledge, and talents, to be in the mood to play, and (as the old joke goes) to have a violin. But violin playing is neither the habits, skills, knowledge, and so on, nor the mood, nor (the notion believers in "material culture" apparently embrace) the violin. To make a trade pact in Morocco, you have to do certain things in certain ways (among others, cut, while chanting Quranic Arabic, the throat of a lamb before the assembled, undeformed, adult male members of your tribe) and to be possessed of certain psychological characteristics (among others, a desire for distant things). But a trade pact is neither the throat cutting nor the desire, though it is real enough, as seven kinsmen of our Marmusha sheikh discovered when, on an earlier occasion, they were executed by him following the theft of one mangy, essentially valueless sheepskin from Cohen.

Culture is public because meaning is. You can't wink (or burlesque one) without knowing what counts as winking or how, physically, to contract your eyelids, and you can't conduct a sheep raid (or mimic one) without knowing what it is to steal a sheep and how practically to go about it. But to draw from such truths the conclusion that knowing how to wink is winking and knowing how to steal a sheep is sheep raiding is to betray as deep a confusion as, taking thin descriptions for thick, to identify winking with eyelid contractions or sheep raiding with chasing woolly animals out of pastures. The cognitivist fallacy—that culture consists (to quote another spokesman for the movement, Stephen Tyler) of "mental phenomena which can [he means "should"] be analyzed by formal methods similar to those of mathematics and logic"—is as destructive of an effective use of the concept as are the behaviorist and idealist fallacies to which it is a misdrawn correction. Perhaps, as its errors are more sophisticated and its distortions subtler, it is even more so.

The generalized attack on privacy theories of meaning is, since early Husserl and late Wittgenstein, so much a part of modern thought that it need not be developed once more here. What is necessary is to see to it that the news of it reaches anthropology; and in particular that it is made clear that to say that culture consists of socially established structures of meaning in terms of which people do such things as signal conspiracies and join them or perceive insults and answer them, is no more to say that it is a psychological phenomenon, a characteristic of someone's mind, personality, cognitive structure, or whatever, than to say that Tantrism, genetics, the progressive form of the verb, the classification of wines, the Common Law, or the notion of "a conditional curse" (as Westermarck defined the concept of *'ar* in terms of which Cohen pressed his claim to damages) is. What, in a place like Morocco, most prevents those of us who grew up winking other winks or attending other sheep from grasping what people are up to is not ignorance as to how cognition works (though, especially as, one assumes, it works the same among them as it does among us, it would greatly help to have less of that too) as a lack of familiarity with the imaginative universe within which their acts are signs. As Wittgenstein has been invoked, he may as well be quoted:

> We ... say of some people that they are transparent to us. It is, however, important as regards this observation that one human being can be a complete enigma to another. We learn this when we come into a strange country with entirely strange traditions; and, what is more, even given a mastery of the country's language. We do not *understand* the people. (And not because of not knowing what they are saying to themselves.) We cannot find our feet with them.

IV

Finding our feet, an unnerving business which never more than distantly succeeds, is what ethnographic research consists of as personal experience; trying to formulate the basis on which one imagines, always excessively, one has found them is what anthropological writing consists of as a scientific endeavor. We are not, or at least I am not, seeking either to become natives (a compromised word in any case) or to mimic them. Only romantics or spies would seem to find point in that. We are seeking, in the widened sense of the term in which it encompasses very much more than talk, to converse with them, a matter a great deal more difficult, and not only with strangers, than is commonly recognized. "If speaking *for* someone else seems to be a mysterious process," Stanley Cavell has remarked, "that may be because speaking *to* someone does not seem mysterious enough."

Looked at in this way, the aim of anthropology is the enlargement of the universe of human discourse. That is not, of course, its only aim—instruction, amusement, practical counsel, moral advance, and the discovery of natural order in human behavior are others; nor is anthropology the only discipline which pursues it. But it is an aim to which a semiotic concept of culture is peculiarly well adapted. As interworked systems of construable signs (what, ignoring provincial usages, I would call symbols), culture is not a power, something to which social events, behaviors, institutions, or processes can be causally attributed; it is a context, something within which they can be intelligibly—that is, thickly—described.

The famous anthropological absorption with the (to us) exotic—Berber horsemen, Jewish peddlers, French Legionnaires—is, thus, essentially a device for displacing the dulling sense of familiarity with which the mysteriousness of our own ability to relate perceptively to one another is concealed from us. Looking at the ordinary in places where it takes unaccustomed forms brings out not, as has so often been claimed, the arbitrariness of human behavior (there is nothing especially arbitrary about taking sheep theft for insolence in Morocco), but the degree to which its meaning varies according to the pattern of life by which it is informed. Understanding a people's culture exposes their normalness without reducing their particularity. (The more I manage to follow what the Moroccans are up to, the more logical, and the more singular, they seem.) It renders them accessible: setting them in the frame of their own banalities, it dissolves their opacity.

It is this maneuver, usually too casually referred to as "seeing things from the actor's point of view," too bookishly as "the *verstehen* approach," or too technically as "emic analysis," that so often leads to the notion that anthropology is a variety of either long-distance mind reading or cannibal-isle fantasizing, and which, for someone anxious to navigate past the wrecks of a dozen sunken philosophies, must therefore be executed with a great deal of care. Nothing is more necessary to comprehending what anthropological interpretation is, and the degree to which it *is* interpretation, than an exact understanding of what it means—and what it does not mean—to say that our formulations of other peoples' symbol systems must be actor oriented.[1]

What it means is that descriptions of Berber, Jewish, or French culture must be cast in terms of the constructions we imagine Berbers, Jews, or Frenchmen to place upon what they live through, the formulas they use to define what happens to them. What it does not mean is that such descriptions are themselves Berber, Jewish, or French—that is, part of the reality they are ostensibly describing; they are anthropological—that is, part of a developing system of scientific analysis. They must be cast in terms of the interpretations to which persons of a particular denomination subject their experience, because that is what they profess to be descriptions of; they are anthropological because it is, in fact, anthropologists who profess them. Normally, it is not necessary to point out quite so laboriously that the object of study is one thing and the study of it another. It is clear enough that the physical world is not physics and *A Skeleton Key to Finnegan's Wake* not *Finnegan's Wake*. But, as, in the study of culture, analysis penetrates into the very body of the object—that is, *we begin with our own interpretations of what our informants are up to, or think they are up to, and then systematize those*—the line between (Moroccan) culture as a natural fact and (Moroccan) culture as a theoretical entity tends to get blurred. All the more so, as the latter is presented in the form of an actor's-eye description of (Moroccan) conceptions of everything from violence, honor, divinity, and justice, to tribe, property, patronage, and chiefship.

In short, anthropological writings are themselves interpretations, and second- and third-order ones to boot. (By definition, only a "native" makes first-order ones: it's *his* culture.)[2] They are, thus, fictions; fictions, in the sense that they are "something made," "something fashioned"—the original meaning of *fictiō*—not that they are false, unfactual, or merely "as if" thought experiments. To construct actor-oriented descriptions of the involvements of a Berber chieftain, a Jewish merchant, and a French soldier with one another in 1912 Morocco is clearly an imaginative act, not all that different from constructing similar descriptions of, say, the involvements with one another of a provincial French doctor, his silly, adulterous wife, and her feckless lover in nineteenth-century France. In the latter case, the actors are represented as not having existed and the events as not having happened, while in the former they are represented as actual, or as having been so. This is a difference of no mean importance, indeed, precisely the one Madame Bovary had difficulty grasping. But the importance does not lie in the fact that her story was created while Cohen's was only noted. The conditions of their creation, and the point of it (to say nothing of the manner and the quality) differ. But the one is as much a *fictiō*—"a making"—as the other.

Anthropologists have not always been as aware as they might be of this fact: that although culture exists in the trading post, the hill fort, or the sheep run, anthropology exists in the book, the article, the lecture, the museum display, or, sometimes nowadays, the film. To become aware of it is to realize that the line between mode of representation and substantive content is as undrawable in cultural analysis as it is in painting; and that fact in turn seems to threaten the objective status of anthropological knowledge by suggesting that its source is not social reality but scholarly artifice.

It does threaten it, but the threat is hollow. The claim to attention of an ethnographic account does not rest on its author's abiity to capture primitive facts in faraway places and carry them home like a mask or a carving, but on the degree to which he is able to clarify what goes on in such places, to reduce the puzzlement—

what manner of men are these?—to which unfamiliar acts emerging out of unknown backgrounds naturally give rise. This raises some serious problems of verification, all right—or, if "verification" is too strong a word for so soft a science (I, myself, would prefer "appraisal"), of how you can tell a better account from a worse one. But that is precisely the virtue of it. If ethnography is thick description and ethnographers those who are doing the describing, then the determining question for any given example of it, whether a field journal squib or a Malinowski-sized monograph, is whether it sorts winks from twitches and real winks from mimicked ones. It is not against a body of uninterpreted data, radically thinned descriptions, that we must measure the cogency of our explications, but against the power of the scientific imagination to bring us into touch with the lives of strangers. It is not worth it, as Thoreau said, to go round the world to count the cats in Zanzibar.

V

Now, this proposition, that it is not in our interest to bleach human behavior of the very properties that interest us before we begin to examine it, has sometimes been escalated into a larger claim: namely, that as it is only those properties that interest us, we need not attend, save cursorily, to behavior at all. Culture is most effectively treated, the argument goes, purely as a symbolic system (the catch phrase is, "in its own terms"), by isolating its elements, specifying the internal relationships among those elements, and then characterizing the whole system in some general way—according to the core symbols around which it is organized, the underlying structures of which it is a surface expression, or the ideological principles upon which it is based. Though a distinct improvement over "learned behavior" and "mental phenomena" notions of what culture is, and the source of some of the most powerful theoretical ideas in contemporary anthropology, this hermetical approach to things seems to me to run the danger (and increasingly to have been overtaken by it) of locking cultural analysis away from its proper object, the informal logic of actual life. There is little profit in extricating a concept from the defects of psychologism only to plunge it immediately into those of schematicism.

Behavior must be attended to, and with some exactness, because it is through the flow of behavior—or, more precisely, social action—that cultural forms find articulation. They find it as well, of course, in various sorts of artifacts, and various states of consciousness; but these draw their meaning from the role they play (Wittgenstein would say their "use") in an ongoing pattern of life, not from any intrinsic relationships they bear to one another. It is what Cohen, the sheikh, and "Captain Dumari" were doing when they tripped over one another's purposes—pursuing trade, defending honor, establishing dominance—that created our pastoral drama, and that is what the drama is, therefore, "about." Whatever, or wherever, symbol systems "in their own terms" may be, we gain empirical access to them by inspecting events, not by arranging abstracted entities into unified patterns.

A further implication of this is that coherence cannot be the major test of validity for a cultural description. Cultural systems must have a minimal degree of coherence, else we would not call them systems; and, by observation, they normally have a great deal more. But there is nothing so coherent as a paranoid's delusion or a swindler's story. The force of our interpretations cannot rest, as they are now so often made to do, on the tightness with which they hold together, or the assurance with which they are argued. Nothing has done more, I think, to discredit cultural analysis than the construction of impeccable depictions of formal order in whose actual existence nobody can quite believe.

If anthropological interpretation is constructing a reading of what happens, then to divorce it from what happens—from what, in this time or that place, specific people say, what they do, what is done to them, from the whole vast business of the world—is to divorce it from its applications and render it vacant. A good interpretation of anything—a poem, a person, a history, a ritual, an institution, a society—takes us into the heart of that of which it is the interpretation. When it does not do that, but leads us instead somewhere else—into an admiration of its own elegance, of its author's cleverness, or of the beauties of Euclidean order—it may have its intrinsic charms; but it is something else than what the task at hand—figuring out what all that rigamarole with the sheep is about—calls for.

The rigamarole with the sheep—the sham theft of them, the reparative transfer of them, the political confiscation of them—is (or was) essentially a social discourse, even if, as I suggested earlier, one conducted in multiple tongues and as much in action as in words.

Claiming his *'ar*, Cohen invoked the trade pact; recognizing the claim, the sheikh challenged the offenders' tribe; accepting responsibility, the offenders' tribe paid the indemnity; anxious to make clear to sheikhs and peddlers alike who was now in charge here, the French showed the imperial hand. As in any discourse, code does not determine conduct, and what was actually said need not have been. Cohen might not have, given its illegitimacy in Protectorate eyes, chosen to press his claim. The sheikh might, for similar reasons, have rejected it. The offenders' tribe, still resisting French authority, might have decided to regard the raid as "real" and fight rather than negotiate. The French, were they more *habile* and less *dur* (as, under Mareschal Lyautey's seigniorial tutelage, they later in fact became), might have permitted Cohen to keep his sheep, winking—as we say—at the continuance of the trade pattern and its limitation to their authority. And there are other possibilities: the Marmushans might have regarded the French action as too great an insult to bear and gone into dissidence themselves; the French might have attempted not just to clamp down on Cohen but to bring the sheikh himself more closely to heel; and Cohen might have concluded that between renegade Berbers and Beau Geste soldiers, driving trade in the Atlas highlands was no longer worth the candle and retired to the better-governed confines of the town. This, indeed, is more or less what happened, somewhat further along, as the Protectorate moved toward genuine sovereignty. but the point here is not to describe what did or did not take place in Morocco. (From this simple incident one can widen out into enormous complexities of social experience.) It is to demonstrate what a piece of anthropological interpretation consists in: tracing the curve of a social discourse; fixing it into an inspectable form.

The ethnographer "inscribes" social discourse; *he writes it down*. In so doing, he turns it from a passing event, which exists only in its own moment of occurrence, into an account, which exists in its inscriptions and can be reconsulted. The sheikh is long dead, killed in the process of being, as the French called it, "pacified"; "Captain Dumari," his pacifier, lives, retired to his souvenirs, in the south of France; and Cohen went last year, part refugee, part pilgrim, part dying patriarch, "home" to Israel. But what they, in my extended sense, "said" to one another on an Atlas plateau sixty years ago is—very far from perfectly—preserved for study. "What," Paul Ricoeur, from whom this whole idea of the inscription of action is borrowed and somewhat twisted, asks, "what does writing fix?"

> Not the event of speaking, but the "said" of speaking, where we understand by the "said" of speaking that intentional exteriorization constitutive of the aim of discourse thanks to which the *sagen*—the saying—wants to become *Aus-sage*—the enunciation, the enunciated. In short, what we write is the *noema* ["thought," "content," "gist"] of the speaking. It is the meaning of the speech event, not the event as event.

This is not itself so very "said—if Oxford philosophers run to little stories, phenomenological ones run to large sentences; but it brings us anyway to a more precise answer to our generative question, "What does the ethnographer do?"—he writes.[3] This, too, may seem a less than startling discovery, and to someone familiar with the current "literature," an implausible one. But as the standard answer to our question has been, "He observes, he records, he analyzes"—a kind of *veni, vidi, vici* conception of the matter—it may have more deep-going consequences than are at first apparent, not the least of which is that distinguishing these three phases of knowledge seeking may not, as a matter of fact, normally be possible; and, indeed, as autonomous "operations" they may not in fact exist.

The situation is even more delicate, because, as already noted, what we inscribe (or try to) is not raw social discourse, to which, because, save very marginally or very specially, we are not actors, we do not have direct access, but only that small part of it which our informants can lead us into understanding.[4] This is not as fatal as it sounds, for, in fact, not all Cretans are liars, and it is not necessary to know everything in order to understand something. But it does make the view of anthropological analysis as the conceptual manipulation of discovered facts, a logical reconstruction of a mere reality, seem rather lame. To set forth symmetrical crystals of significance, purified of the material complexity in which they were located, and then attribute their existence to autogenous principles of order, universal properties of the human mind, or vast, a priori *weltanschauungen*, is to pretend a science that does not exist and imagine a reality that cannot be found. Cultural analysis is (or should be) guessing at meanings, assessing the guesses, and drawing explanatory conclusions from the better guesses, not discovering the Continent of Meaning and mapping out its bodiless landscape.

VI

So, there are three characteristics of ethnographic description: it is interpretive; what it is interpretive of is the flow of social discourse; and the interpreting involved consists in trying to rescue the "said" of such discourse from its perishing occasions and fix it in perusable terms. The *kula* is gone or altered; but, for better or worse, *The Argonauts of the Western Pacific* remains. But there is, in addition, a fourth characteristic of such description, at least as I practice it: it is microscopic.

This is not to say that there are no large-scale anthropological interpretations of whole societies, civilizations, world events, and so on. Indeed, it is such extension of our analyses to wider contexts that, along with their theoretical implications, recommends them to general attention and justifies our constructing them. No one really cares anymore, not even Cohen (well ... maybe, Cohen), about those sheep as such.

History may have its unobtrusive turning points, "great noises in a little room"; but this little go-round was surely not one of them.

It is merely to say that the anthropologist characteristically approaches such broader interpretations and more abstract analyses from the direction of exceedingly extended acquaintances with extremely small matters. He confronts the same grand realities others—historians, economists, political scientists, sociologists—confront in more fateful settings: Power, Change, Faith, Oppression, Work, Passion, Authority, Beauty, Violence, Love, Prestige; but he confronts them in contexts obscure enough—places like Marmusha and lives like Cohen's—to take the capital letters off them. These all-too-human constancies, "those big words that make us all afraid," take a homely form in such homely contexts. But that is exactly the advantage. There are enough profundities in the world already.

Yet, the problem of how to get from a collection of ethnographic miniatures on the order of our sheep story—an assortment of remarks and anecdotes—to wall-sized culturescapes of the nation, the epoch, the continent, or the civilization is not so easily passed over with vague allusions to the virtues of concreteness and the down-to-earth mind. For a science born in Indian tribes, Pacific islands, and African lineages and subsequently seized with grander ambitions, this has come to be a major methodological problem, and for the most part a badly handled one. The models that anthropologists have themselves worked out to justify their moving from local truths to general visions have been, in fact, as responsible for undermining the effort as anything their critics—sociologists obsessed with sample sizes, psychologists with measures, or economists with aggregates—have been able to devise against them.

Of these, the two main ones have been: the Jonesville-is-the-USA "microcosmic" model; and the Easter-Island-is-a-testing-case "natural experiment" model. Either heaven in a grain of sand, or the farther shores of possibility.

The Jonesville-is-America writ small (or America-is-Jonesville writ large) fallacy is so obviously one that the only thing that needs explanation is how people have managed to believe it and expected others to believe it. The notion that one can find the essence of national societies, civilizations, great religions, or whatever summed up and simplified in so-called typical small towns and villages is palpable nonsense. What one finds in small towns and villages is (alas) small-town or village life. If localized, microscopic studies were really dependent for their greater relevance upon such a premise—that they captured the great world in the little—they wouldn't have any relevance.

But, of course, they are not. The locus of study is not the object of study. Anthropologists don't study villages (tribes, towns, neighborhoods ...); they study *in* villages. You can study different things in different places, and some things—for example, what colonial domination does to established frames of moral expectation—you can best study in confined localities. But that doesn't make the place what it is you are studying. In the remoter provinces of Morocco and Indonesia I have wrestled with the same questions other social scientists have wrestled with in more central locations—for example, how comes it that men's most importunate claims to humanity are cast in the accents of group pride?—and with about the same conclusiveness. One can add a dimension—one much needed in the present climate of size-up-and-solve social science; but that is all. There is a certain value, if you are going to run on about the exploitation of the masses in having seen a Javanese sharecropper turning earth in a tropical downpour or a Moroccan tailor embroidering kaftans by the light of a twenty-watt bulb. But the notion that this gives you the thing entire (and elevates you to some moral vantage ground from which you can look down upon the ethically less privileged) is an idea which only someone too long in the bush could possibly entertain.

The "natural laboratory" notion has been equally pernicious, not only because the analogy is false—what kind of a laboratory is it where *none* of the parameters are manipulable?—but because it leads to a notion that the data derived from ethnographic studies are purer, or more fundamental, or more solid, or less conditioned (the most favored word is "elementary") than those derived from other sorts of social inquiry. The great natural variation of cultural forms is, of course, not only anthropology's great (and wasting) resource, but the ground of its deepest theoretical dilemma: how is such variation to be squared with the biological unity of the human species? But it is not, even metaphorically, experimental variation, because the context in which it occurs varies along with it, and it is not possible (though there are those who try) to isolate the *y*'s from *x*'s to write a proper function.

The famous studies purporting to show that the Oedipus complex was backward in the Trobriands, sex roles were upside down in Tchambuli, and the Pueblo Indians lacked aggression (it is characteristic that they were all negative—"but not in the South"), are, whatever their empirical validity may or may not be, not "scientifically tested and approved" hypotheses. They are interpretations, or misinterpretations, like any others, arrived at in the same way as any others, and as inherently inconclusive as any others, and the attempt to invest them with the authority of physical experimentation is but methodological sleight of hand. Ethnographic findings are not privileged, just particular: another country heard from. To regard them as anything more (*or anything less*) than that distorts both them and their implications, which are far profounder than mere primitivity, for social theory.

Another country heard from: the reason that protracted descriptions of distant sheep raids (and a really good ethnographer would have gone into what kind of sheep they were) have general relevance is that they present the sociological mind with bodied stuff on which to feed. The important thing about the anthropologist's findings is their complex specificness, their circumstantiality. It is with the kind of material produced by long-term, mainly (though not exclusively) qualitative, highly participative, and almost obsessively fine-comb field study in confined contexts that the mega-concepts with which contemporary social science is afflicted—legitimacy, modernization, integration, conflict, charisma, structure, ... meaning—can be given the sort of sensible actuality that makes it possible to think not only realistically and concretely *about* them, but, what is more important, creatively and imaginatively *with* them.

The methodological problem which the microscopic nature of ethnography presents is both real and critical. But it is not to be resolved by regarding a remote locality as the world in a teacup or as the sociological equivalent of a cloud chamber. It is to be resolved—or, anyway, decently kept at bay—by realizing that social actions are comments on more than themselves; that where an interpretation comes from does not determine where it can be impelled to go. Small facts speak to large issues, winks to epistemology, or sheep raids to revolution, because they are made to.

VII

Which brings us, finally, to theory. The besetting sin of interpretive approaches to anything—literature, dreams, symptoms, culture—is that they tend to resist, or to be permitted to resist, conceptual articulation and thus to escape systematic modes of assessment. You either grasp an interpretation or you do not, see the point of it or you do not, accept it or you do not. Imprisoned in the immediacy of its own detail, it is presented as self-validating, or, worse, as validated by the supposedly developed sensitivities of the person who presents it; any attempt to cast what it says in terms other than its own is regarded as a travesty—as, the anthropologist's severest term of moral abuse, ethnocentric.

For a field of study which, however timidly (though I, myself, am not timid about the matter at all), asserts itself to be a science, this just will not do. There is no reason why the conceptual structure of a cultural interpretation should be any less formulable, and thus less susceptible to explicit canons of appraisal, than that of, say, a biological observation or a physical experiment—no reason except that the terms in which such formulations can be cast are, if not wholly nonexistent, very nearly so. We are reduced to insinuating theories because we lack the power to state them.

At the same time, it must be admitted that there are a number of characteristics of cultural interpretation which make the theoretical development of it more than usually difficult. The first is the need for theory to stay rather closer to the ground than tends to be the case in sciences more able to give themselves over to imaginative abstraction. Only short flights of ratiocination tend to be effective in anthropology; longer ones tend to drift off into logical dreams, academic bemusements with formal symmetry. The whole point of a semiotic approach to culture is, as I have said, to aid us in gaining access to the conceptual world in which our subjects live so that we can, in some extended sense of the term, converse with them. The tension between the pull of this need to penetrate an unfamiliar universe of symbolic action and the requirements of technical advance in the theory of culture, between the need to grasp and the need to analyze, is, as a result, both necessarily great and essentially irremovable. Indeed, the further theoretical development goes, the deeper the tension gets. This is the first condition for cultural theory: it is not its own master. As it is unseverable from the immediacies thick description presents, its freedom to shape itself in terms of its internal logic is rather limited. What generality it contrives to achieve grows out of the delicacy of its distinctions, not the sweep of its abstractions.

And from this follows a peculiarity in the way, as a simple matter of empirical fact, our knowledge of culture ... cultures ... a culture ... grows: in spurts. Rather than following a rising curve of cumulative findings, cultural analysis breaks up into a disconnected yet coherent sequence of bolder and bolder sorties. Studies do build on other studies, not in the sense that they take up where the others leave off, but in the sense that, better informed and better conceptualized, they plunge more deeply into the same things. Every serious cultural analysis starts from a sheer beginning and ends where it manages to get before exhausting its intellectual impulse. Previously discovered facts are mobilized, previously developed concepts used, previously formulated hypotheses tried out; but the movement is not from already proven theorems to newly proven ones, it is from an awkward fumbling for the most elementary understanding to a supported claim that one has achieved that and surpassed it. A study is an advance if it is more incisive—whatever that may mean—than those that preceded it; but it less stands on their shoulders than, challenged and challenging, runs by their side.

It is for this reason, among others, that the essay, whether of thirty pages or three hundred, has seemed the natural genre in which to present cultural interpretations and the theories sustaining them, and why, if one looks for systematic treatises in the field, one is so soon disappointed, the more so if one finds any. Even inventory articles are rare here, and anyway of hardly more than bibliographical interest. The major theoretical contributions not only lie in specific studies—that is true in almost any field—but they are very difficult to abstract from such studies and integrate into anything one might call "culture theory" as such. Theoretical formulations hover so low over the interpretations they govern that they don't make much sense or hold much interest apart from them. This is so, not because they are not general (if they are not general, they are not theoretical), but because, stated independently of their applications, they seem either commonplace or vacant. One can, and this in fact is how the field progresses conceptually, take a line of theoretical attack developed in connection with one exercise in ethnographic interpretation and employ it in another, pushing it forward to greater precision and broader relevance; but one cannot write a "General Theory of Cultural Interpretation." Or, rather, one can, but there appears to be little profit in it, because the essential task of theory building here is not to codify abstract regularities but to make thick description possible, not to generalize across cases but to generalize within them.

To generalize within cases is usually called, at least in medicine and depth psychology, clinical inference. Rather than beginning with a set of observations and attempting to subsume them under a governing law, such inference begins with a set of (presumptive) signifiers and attempts to place them within an intelligible frame. Measures are matched to theoretical predictions, but symptoms (even when they are measured) are scanned for theoretical peculiarities—that is, they are diagnosed. In the study of culture the signifiers are not symptoms or clusters of symptoms, but symbolic acts or clusters of symbolic acts, and the aim is not therapy but the analysis of social discourse. But the way in which theory is used—to ferret out the unapparent import of things—is the same.

Thus we are lead to the second condition of cultural theory: it is not, at least in the strict meaning of the term, predictive. The diagnostician does not predict measles; he decides that someone has them, or at the very most *anticipates* that someone is rather likely shortly to get them. But this limitation, which is real enough, has commonly been both misunderstood and exaggerated, because it has been taken to mean that cultural interpretation is merely post facto: that, like the peasant in the old story, we first shoot the holes in the fence and then paint the bull's-eyes around them. It is hardly to be denied that there is a good deal of that sort of thing around, some of it in prominent places. It is to be denied, however, that it is the inevitable outcome of a clinical approach to the use of theory.

It is true that in the clinical style of theoretical formulation, conceptualization is directed toward the task of generating interpretations of matters already in hand, not toward projecting outcomes of experimental manipulations or deducing future states of a determined system. But that does not mean that theory has only to fit (or, more carefully, to generate cogent interpretations of) realities past; it has also to survive —intellectually survive—realities to come. Although we formulate our interpretation of an outburst of winking or an instance of sheep-raiding after its occurrence, sometimes long after, the theoretical framework in terms of which such an interpretation is made must be capable of continuing to yield defensible interpretations as new social phenomena swim into view. Although one starts any effort at thick description, beyond the obvious and superficial, from a state of general bewilderment as to what the devil is going on—trying to find one's feet—one does not start (or ought not) intellectually empty-handed. Theoretical ideas are not created wholly anew in each study; as I have said, they are adopted from other, related studies, and, refined in the process, applied to new interpretive problems. If they cease being useful with respect to such problems, they tend to stop being used and are more or less abandoned. If they continue being useful, throwing up new understandings, they are further elaborated and go on being used.[5]

Such a view of how theory functions in an interpretive science suggests that the distinction, relative in any case, that appears in the experimental or observational sciences between "description" and "explanation" appears here as one, even more relative, between "inscription" ("thick description") and "specification" ("diagnosis")—between setting down the meaning particular social actions have for the actors whose actions they are, and stating, as explicitly as we can manage, what the knowledge thus attained demonstrates about the society in which it is found and, beyond that, about social life as such. Our double task is to uncover the conceptual structures that inform our subjects' acts, the "said" of social discourse, and to construct a system of analysis in whose terms what is generic to those structures, what belongs to them because they are what they are, will stand out against the other determinants of human behavior. In ethnography, the office of theory is to provide a vocabulary in which what symbolic action has to say about itself—that is, about the role of culture in human life—can be expressed.

... A repertoire of very general, made-in-the-academy concepts and systems of concepts—"integration," "rationalization," "symbol," "ideology," "ethos," "revolution," "identity," "metaphor," "structure," "ritual," "worldview," "actor," "function," "sacred," and, of course, "culture" itself—is woven into the body of thick-description ethnography in the hope of rendering mere occurrences scientifically eloquent. The aim is to draw large conclusions from small, but very densely textured facts; to support broad assertions about the role of culture in the construction of collective life by engaging them exactly with complex specifics.

Thus it is not only interpretation that goes all the way down to the most immediate observational level: the theory upon which such interpretation conceptually depends does so also. My interest in Cohen's story, like Ryle's in winks, grew out of some very general notions indeed. The "confusion of tongues" model—the view that social conflict is not something that happens when, out of weakness, indefiniteness, obsolescence, or neglect, cultural forms cease to operate, but rather something which happens when, like burlesqued winks, such forms are pressed by unusual situations or unusual intentions to operate in unusual ways—is not an idea I got from Cohen's story. It is one, instructed by colleagues, students, and predecessors, I brought to it.

Our innocent-looking "note in a bottle" is more than a portrayal of the frames of meaning of Jewish peddlers, Berber warriors, and French proconsuls, or even of their mutual interference. It is an argument that to rework the pattern of social relationships is to rearrange the coordinates of the experienced world. Society's forms are culture's substance.

VIII

There is an Indian story—at least I heard it as an Indian story—about an Englishman who, having been told that the world rested on a platform which rested on the back of an elephant which rested in turn on the back of a turtle, asked (perhaps he was an ethnographer; it is the way they behave), What did the turtle rest on? Another turtle. And that turtle? "Ah, Sahib, after that it is turtles all the way down."

Such, indeed, is the condition of things. I do not know how long it would be profitable to meditate on the encounter of Cohen, the sheikh, and "Dumari" (the period has perhaps already been exceeded); but I do know that however long I did so I would not get anywhere near to the bottom of it. Nor have I ever gotten anywhere near to the bottom of anything I have ever written about.... Cultural analysis is intrinsically incomplete. And, worse than that, the more deeply it goes the less complete it is. It is a strange science whose most telling assertions are its most tremulously based, in which to get somewhere with the matter at hand is to intensify the suspicion, both your own and that of others, that you are not quite getting it right. But that, along with plaguing subtle people with obtuse questions, is what being an ethnographer is like.

There are a number of ways to escape this—turning culture into folklore and collecting it, turning it into traits and counting it, turning it into institutions and classifying it, turning it into structures and toying with it. But they *are* escapes. The fact is that to commit oneself to a semiotic concept of culture and an interpretive approach to the study of it is to commit oneself to a view of ethnographic assertion as, to borrow W. B. Gallie's by now famous phrase, "essentially contestable." Anthropology, or at least interpretive anthropology, is a science whose progress is marked less by a perfection of consensus than by a refinement of debate. What gets better is the precision with which we vex each other.

This is very difficult to see when one's attention is being monopolized by a single party to the argument. Monologues are of little value here, because there are no conclusions to be reported; there is merely a discussion to be sustained.... [There has been] an enormous increase in interest, not only in anthropology, but in social studies generally, in the role of symbolic forms in human life. Meaning, that elusive and ill-defined pseudoentity we were once more than content to leave philosophers and literary critics to fumble with, has now come back into the heart of our discipline. Even Marxists are quoting Cassirer; even positivists, Kenneth Burke.

My own position in the midst of all this has been to try to resist subjectivism on the one hand and cabalism on the other, to try to keep the analysis of symbolic forms as closely tied as I could to concrete social events and occasions, the public world of common life, and to organize it in such a way that the connections between theoretical formulations and descriptive interpretations were unobscured by appeals to dark sciences. I have never been impressed by the argument that, as complete objectivity is impossible in these matters (as, of course, it is), one might as well let one's sentiments run loose. As Robert Solow has remarked, that is like saying that as a perfectly aseptic environment is impossible, one might as well conduct surgery in a sewer. Nor, on the other hand, have I been impressed with claims that structural linguistics, computer engineering, or some other advanced form of thought is going to enable us to understand men without knowing them. Nothing will discredit a semiotic approach to culture more quickly than allowing it to drift into a combination of intuitionism and alchemy, no matter how elegantly the intuitions are expressed or how modern the alchemy is made to look.

The danger that cultural analysis, in search of all-too-deep-lying turtles, will lose touch with the hard surfaces of life—with the political, economic, stratificatory realities within which men are everywhere contained—and with the biological and physical necessities on which those surfaces rest, is an ever-present one. The only defense against it, and against, thus, turning cultural analysis into a kind of sociological aestheticism, is to train such analysis on such realities and such necessities in the first place. It is thus that I have written about nationalism, about violence, about identity, about human nature, about legitimacy, about revolution, about ethnicity, about urbanization, about status, about death, about time, and most of all about particular attempts by particular peoples to place these things in some sort of comprehensible, meaningful frame.

To look at the symbolic dimensions of social action—art, religion, ideology, science, law, morality, common sense—is not to turn away from the existential dilemmas of life for some empyrean realm of deemotionalized forms; it is to plunge into the midst of them. The essential vocation of interpretive anthropology is not to answer our deepest questions, but to make available to us answers that others, guarding other sheep in other valleys, have given, and thus to include them in the consultable record of what man has said.

Notes

1. Not only other peoples': anthropology *can* be trained on the culture of which it is itself a part, and it increasingly is; a fact of profound importance, but which, as it raises a few tricky and rather special second order problems, I shall put to the side for the moment.
2. The order problem is, again, complex. Anthropological works based on other anthropological works (Lévi-Strauss', for example) may, of course, be fourth order or higher and informants frequently, even habitually, make second-order interpretations—what have come to be known as "native models." In literate cultures, where "native" interpretation can proceed to higher levels—in connection with the Maghreb, one has only to think of Ibn Khaldun; with the United States, Margaret Mead—these matters become intricate indeed.
3. Or, again, more exactly, "inscribes." Most ethnography is in fact to be found in books and articles, rather than in films, records, museum displays, or whatever; but even in them there are, of course, photographs, drawings, diagrams, tables, and so on. Self-consciousness about modes of representation (not to speak of experiments with them) has been very lacking in anthropology.
4. So far as it has reinforced the anthropologist's impulse to engage himself with his informants as persons rather than as objects, the notion of "participant observation" has been a valuable one. But, to the degree it has led the anthropologist to block from his view the very special, culturally bracketed nature of his own role and to imagine himself something more than an interested (in both senses of that word) sojourner, it has been our most powerful source of bad faith.
5. Admittedly, this is something of an idealization. Because theories are seldom if ever decisively disproved in clinical use but merely grow increasingly awkward, unproductive, strained, or vacuous, they often persist long after all but a handful of people (though *they* are often most passionate) have lost much interest in them. Indeed, so far as anthropology is concerned, it is almost more of a problem to get exhausted ideas out of the literature than it is to get productive ones in, and so a great deal more of theoretical discussion than one would prefer is critical rather than constructive, and whole careers have been devoted to hastening the demise of moribund notions. As the field advances one would hope that this sort of intellectual weed control would become a less prominent part of our activities. But, for the moment, it remains true that old theories tend less to die than to go into second editions.

Chapter 15

Hermeneutics and the Hypothetico-Deductive Method

Dagfinn Føllesdal

1 *Hermeneutics*

Hermeneutics was originally the method used for interpreting theological and legal texts. These two kinds of texts had two features whose combination made the problems of interpretation acute. The texts had arisen in times and circumstances that were often very different from those in which they were to be applied. And second: it was considered important to give a correct interpretation of these texts.

Through the work of Ast, Schleiermacher, Dilthey and many others, the methods of hermeneutics were improved and its scope was extended to cover the interpretation of all kinds of texts and ultimately "all manifestations of the human spirit" to use Dilthey's phrase, including not only texts, but also paintings, sculptures, social institutions, human actions and so on.

In this chapter, I will construe hermeneutics in this broad sense as the general method of interpretation of human actions and all products of such actions.

During the development of hermeneutics, and especially through Dilthey, the view gradually took form that hermeneutics is a method specific to the humanities and completely different from the hypothetico-deductive method used in the natural sciences. Some philosophers have made a major point out of this alleged opposition between the method of understanding of hermeneutics and the method of explanation in the natural sciences. Thus, for example, Jürgen Habermas argues in *Erkenntnis und Interesse*[1] that each of the three main areas of scientific research, natural science, the humanities and the social sciences, has its specific method; the natural sciences use the hypothetico-deductive method, the humanities the hermeneutic method and the social sciences the so-called critical method. I shall not discuss the last method here, but will concentrate on the first two methods. My thesis will be the following: *the hermeneutic method is the hypothetico-deductive method applied to meaningful material* (*texts, works of art, actions, etc.*). By meaningful material, I mean anything that expresses an agent's beliefs and/or values. What this in turn means will have to be clarified for each type of material. An account of the interrelation between actions, beliefs and values will be given in section 6.

2 *The Hypothetico-Deductive Method*

Habermas and others who restrict the hypothetico-deductive method to natural science usually do not explain what they understand by hypothetico-deductive method. When I have tried to ferret out their arguments, I have found it very difficult to recognize anything there that bears a similarity to the hypothetico-deductive method.

For this reason, I will begin by giving a short exposition of the hypothetico-deductive method. I will thereafter endeavor to show, by help of an example, that the hermeneutic method is the hypothetico-deductive method applied to meaningful material.

In the later parts of the chapter, I will then utilize this subsumption of the hermeneutic method under the hypothetico-deductive method in order to bring out some features of the hermeneutic method that are not always well heeded.

First then: What is the hypothetico-deductive method?

As the name indicates, it is an application of two operations: the formation of *hypotheses* and the *deduction* of consequences from them in order to arrive at beliefs which—although they are hypothetical—are well supported, through the way their deductive consequences fit in with our experiences and with our other well-supported beliefs.

The hypothetico-deductive method aims at establishing a set of hypotheses concerning the subject matter that we are studying. Together with our beliefs these hypotheses form a comprehensive hypothetico-deductive system which is logically consistent and fits in with all our experience.

The beliefs that make up such a hypothetico-deductive system are not justified "from above," as they are in an axiomatic system, where the axioms are supposed to be justified by some special kind of insight or necessity. Instead, they are justified from below, through their consequences. In a hypothetico-deductive system, the hypotheses are never known with certainty. From a system of hypotheses an infinite number of consequences follow and there is always a risk that some of these consequences may turn out not to fit in with our experience. Some of the consequences relate, for example, to our future experiences—these are the predictions of the theory—and only time may show whether they are true. Further, even if all the consequences should fit in with our experience, the same consequences may be derived also from other hypotheses, as was observed by several philosophers already in antiquity and the Middle Ages, as, e.g., Simplicius and Thomas Aquinas. The question then arises as to which of these hypotheses we should believe. As you know, the simplicity of our total set of hypotheses, i.e., our theory, is generally considered decisive. The notion of simplicity includes several different factors that we regard as important for the evaluation of a theory, as, for example, the variety of different data that are accounted for by the same set of hypotheses.

The important questions in the philosophy of science are therefore: (1) What do we mean when we say that one theory is *simpler* than another? and (2) Why is a simpler theory more trustworthy than one which is more complex?

These two important questions will not be discussed in this chapter. The notion of simplicity will, however, come up later in connection with interpretation of texts, for as we shall see, simplicity seems to be crucial also for the choice between competing interpretations, as we should expect if hermeneutics proceeds hypothetico-deductively.

3 *An Example from the Interpretation of Literature*

Having completed this outline of the hypothetico-deductive method, I will now by help of an example from the interpretation of literature show how one uses the hypothetico-deductive method in this discipline. In a textbook on the theory of argumentation that was published in Norwegian some years ago and appeared in German in 1980 (in de Gruyter's series "Grundlagen der Kommunikation"), my co-author, Lars Walløe, and I give a number of other such examples from the study of history, from social science, grammar and ethics. For reasons of space I will here include only this one example from the interpretation of literature.

The example that I have chosen is the interpretation of the Stranger in Henrik Ibsen's *Peer Gynt*.

This peculiar person appears twice in *Peer Gynt*, both times in the fifth act. First he comes up to Peer's side as Peer is standing nervously aft on the ship during the storm. Then, the second time, he swims beside the overturned lifeboat on which Peer is riding.

Five different interpretations have been proposed for the stranger:

1. The first, and oldest, is that the stranger represents anxiety. In support of this interpretation there are particularly two observations that have been made: one is that the stranger invariably appears in situations where Peer is anxious (that he shall die). The other observation in favor of this interpretation is that Ibsen was very interested in and influenced by Kierkegaard, the philosopher of anxiety.

This interpretation was originally set forth as a criticism; it was held that it was a weakness in a play to let persons represent abstract ideas. Ibsen defended himself by saying that this interpretation had never even occurred to his mind. This disclaimer is relevant only if one wants to understand the author's intentions, not if one wants to arrive at the most reasonable interpretation of the work per se. We shall not enter into this issue here; a decisive objection against the interpretation is that it accounts for very little of what is said about the stranger in the play, and that there are other interpretations that do better than this one.

2. A somewhat better interpretation is that the stranger represents *death*. This fits in with the fact that the anxiety that Peer has when the stranger appears is anxiety that he shall die. However, this interpretation, too, accounts for too little of what is said about the strange passenger in the play.

3. A third interpretation, proposed by Martin Svendsen in 1922, was that the strange passenger is *Ibsen* himself. This interpretation has gained wide support. Martin Svendsen listed eight considerations that count in its favor:

> 1. The passenger, who is "white as a sheet," stays indoor during the day. Ibsen, who according to Bjørnson's poem "Old Heltberg" was "tense and meager, with the color of gypsum," was mostly indoors working during the time when *Brand* and *Peer Gynt* were written. 2. The passenger enjoyed tempest and shipwreck. Ibsen has a similar liking for that which overthrows the game. 3. The passenger has a scientific interest in anatomy and wants Peer's permission to perform an autopsy on "corpse." Ibsen uses in several places words like "anatomize" and "anatomy" about his work. 4. The passenger wants to be a moral guide for Peer; Ibsen intends to waken the Norwegian people. 5. The passenger states that "time so very often will alter things"; Ibsen writes in several letters of the development that his views have undergone. 6. According to Peer, the stranger is something of a freethinker; Ibsen often heard similar things about himself. 7. Where the passenger comes from, they think a smile worth quite as much as any pathos; the same could be said for the poet's workshop where Peer Gynt was created; in this drama Ibsen could use both tragic and comic elements. 8. The passenger says to comfort Peer that one does not die in the middle of the fifth act; thereby he uses a terminology which by the nature of things is otherwise only that of the dramatist—there is nobody but *him* who may have at his disposal the persons

> of the poem and their future. In addition, it did not suit Ibsen to let Peer meet his end in this way; the drawn should end with reconciliation. Beside these points of similarity Svendsen claims that a poet is in fact always a fellow passenger of his persons.[2]

More arguments have been added to his list by other Ibsen scholars. It has, for example, been pointed out that other authors have performed similar guest roles in their works.

The pattern of this interpretation is clearly hypothetico-deductive. One sets forth a hypothesis, that the strange passenger is Ibsen, and then deduces a number of consequences from it that are shown to fit in with the text. In addition to the hypothesis, one also makes use of, as premises in the deduction, several other pieces of information and theories, as, for example, information about Ibsen's looks, his interests and experiences, a theory to the effect that when a person in a literary work makes statements about what is going to happen in the work, this person must be the author himself; he is the only one who may have at his disposal the persons of the work and their future (point 8 in Svendsen's argumentation).

As in natural science, the following factors are decisive for our evaluation of the hypothesis:

> a. How well does the hypothesis fit in with the data that are mentioned? Are, for example, the assumptions and theories that one makes use of when one deduces consequences from the hypothesis reasonable, or do they seem ad hoc, that is, do they seem like "epicycles" that are introduced only in order to "save" the hypothesis, or can they be supported by arguments and evidence?
> b. How well does the hypothesis fit in with data that are not mentioned, e.g. other passages in the text?
> c. Even when the hypothesis fits in with all the data, we should also ask: Are there other hypotheses and theories which fit in at least equally well with all the data and which are simpler?

As for (a), the theory that Svendsen sets forth in point 8 above seems doubtful. There exist very many literary works where a person makes remarks *about* the work of the kind that is mentioned above, but where it is quite clear that the person does not represent the author of the work. As we shall see in a moment, this kind of "romantic irony" has been quite common, e.g. in Byron.

As regards (b), there are a number of very specific pieces of information in the text that Svendsen does not mention and that it is difficult to make compatible with his hypothesis—for example, the stranger says: "I swim quite well with my left leg."

It is a considerable weakness of Svendsen's interpretation that many of those passages in the text that he lists in support of his interpretation are considerably less specific than those which he does not include. Passage number 5 is, for example, so little specific as to be compatible with almost any interpretation.

These weaknesses suffice to make us strongly dissatisfied with this interpretation, at least until it has been supplemented with reasonable additional hypotheses which make it fit in with these other passages as well.

4. Let us therefore, in accordance with (c), consider some of the other interpretations that have been proposed. An interpretation that was put forward by Albert Morey Sturtevant in 1914 is that the stranger is the Devil. This interpretation has gained many adherents and much can be said in its favor.

1. First and foremost the stranger's statement that "I swim quite well with my left leg" fits well with the fact that the devil traditionally has a horse's hoof instead of a right foot.
2. The Sailor's answer to Peer's question concerning who went into the cabin—"The ship's dog, sir!"—fits in with the fact that the devil according to popular belief and in many literary works often appears in the shape of a dog.
3. Peer's exclamations to the stranger—"Get out of here!" and "Get out of here, scarecrow!"—remind of the words Jesus used against the devil when he was tempted. (But of course, this indicates only that Peer took the stranger to be the devil, and it is possible that he was wrong.)
4. The stranger's statement—"I'll float if only I insert my finger-tip into this crack"—may be an allusion to the adage: "If you give the devil a finger he will soon take the whole hand." It may also be an allusion to the fact that the devil is often equipped with claws, not fingers. In that case a link is created to the scene with the thin person further out in the fifth act, to whom Peer says: "Your nails seem most remarkably developed" and who, by the way, also has a hoof. A connection between these scenes increases the possibilities of finding a "unitary interpretation" of the work. A "unitary interpretation" seems to correspond approximately to what we earlier have called a "simple" theory, which in a unitary way brings all the data into connection with one another. We have some intuitions concerning what this is, but it is not easy to make these clear. It is not unreasonable to expect that the literary analyst's attempt to clarify what is meant by a "unitary interpretation" and the philosopher of science's attempt to clarify what is meant by "simplicity" may support one another.

There are, however, some difficulties with this interpretation, for example the following exchange:

The Stranger: What do you think?
Don't you know anyone that's like me?
Peer Gynt: I know the Devil—
The Stranger: (lowering his voice): Is he wont
To light us on the darkest paths
Of life when we're beset by fear?

It has been proposed that the stranger's answer is intended to mislead Peer, and reasons have been given to show that this is a reasonable proposal and not only ad hoc, designed to "save" the interpretation.

5. A fifth interpretation has been proposed by Daniel Haakonsen, Professor of Norwegian literature in the University of Oslo, in his book on Peer Gynt[3] in order, as Haakonsen expresses it ironically, "to find out how much or little we can get to fit in with the text if we choose a quite arbitrary starting point." Haakonsen proposes that the strange passenger is the ghost of Lord Byron. Haakonsen shows that this interpretation, which initially may seem quite unreasonable, fits remarkably well with numerous details in the text.

Byron was, for example, a good swimmer although his right foot was clubfoot; he was ardently devoted to romantic irony, etc.. Besides, Ibsen was strongly engaged with Byron and his work. Haakonsen also points out that there exists a parallel to the procedure Ibsen uses when—or if—he inserts a camouflaged but nevertheless recognizable portrait of Byron into Peer Gynt: Goethe confirmed in his conversations with Eckermann that Euphorion in *Faust* was created in Byron's image.

One might think that even if the Byron hypothesis fits in with what is said about the strange passenger, it does not help us to see the function of the stranger in the play. The hypothesis should then be in a weak position with respect to the desire for a unitary interpretation. Haakonsen argues, however, that the Byron hypothesis also fits in well with other parts of the play. Among other things, he looks upon the whole of the fourth act in relation to Byron as the model Peer is matched with and looses against. Thereby Ibsen's idea of Byron's ghost becomes better understandable, and the connection between the fourth and fifth act becomes better, which helps to strengthen the disputed fourth act.

Haakonsen points out, however, that the Byronism is quite transient and soon forgotten and that it therefore becomes a foreign element in *Peer Gynt*. He concludes from this that "Ibsen has probably made a miscalculation that weakens this part of the play."

Note that this kind of a criticism of a work of literature counts against an interpretation if one accepts as a working hypothesis in interpreting a work of art that one should prefer the interpretation that makes the work more interesting and artistically more satisfactory. If one has a choice between the Byron interpretation and another interpretation which is equally satisfactory, for example, as concerns the fourth act, but which is not as transient as the Byron interpretation, one should give up the Byron interpretation. In the absence such an interpretation, the Byron interpretation may be of a certain psychological interest. It explains how Ibsen, through his interest in Byron, has been misled into writing a play with a fourth act which we in our time find weak.

We shall not weigh against one another these and other hypotheses concerning the strange passenger. The point of the example has been to show how, when one interprets a text like this one, one proceeds hypothetico-deductively. We set forth a hypothesis concerning the text or possibly the work as a whole and test this hypothesis by checking how its consequences fit in with the various details in the text. These details may be rhyme, rhythm and other literary devices. If, for example, a work is kept strictly in a rhythm that suddenly is broken, a satisfactory interpretation of the text, together with a stylistic theory of rhythm, should enable us to derive a conclusion which fits in with the break in the rhythm.

If I am right that the method of interpretation is the hypothetico-deductive method, we now know what we should do in order to evaluate interpretations of works of literature: we ask the three questions (a), (b) and (c) that I formulated earlier. In order to preclude misunderstandings, it should be noted that numbering of the various features of the text that fit in with the various hypotheses does by no means mean that it is the *number* of features that count. The *nature* of the features is much more important. We have, for example, noted that the more specific the features are that the hypothesis accounts for, the better. Above all, however, a premium is put on a unitary interpretation that makes a large number of very different and specific features in all parts of the work all "fit into place."

Note also that theories of literature and of style are needed when we are going to derive consequences from the hypotheses that shall be tested against the text. These theories must be tested like other theories, through their consequences. The theories are not always explicitly formulated, but come in as tacit premises in the derivation of the various consequences. A first step toward testing the theories is to formulate them explicitly.

Finally some reservations: In interpreting literary texts one can of course not take it for granted that there is some interpretation that fits in with the data in a reasonable way. It is not at all obvious that the literary value of a text depends on there being a consistent interpretation of the whole text. Also, in the case of some texts, one or more interpretations may seem to be equally plausible. In some such situations one feels convinced that also the author has intended the text to be interpretable in several ways and hence simultaneously give associations in several directions. It has, for example, been proposed (by Daniel Haakonsen) that the strange passenger in *Peer Gynt* should be interpreted both as the Devil and as Lord Byron. The two interpretations are in this case interconnected; Byron was the black sheep of romanticism and was feared like the Evil One in certain circles.

4 *Are All Methods Variants of the Hypothetico-Deductive Method?*

The example from *Peer Gynt* that I have discussed fits into a hypothetico-deductive pattern. In this example the problem was to identify one of the persons of a play. There are numerous other aims that one may have in interpreting a work of literature. However, my guess, based on a number of different samples that I will not get into here, is that the hypothetico-deductive method is used wherever interpretation takes place. Numerous examples of interpretation by means of the hypothetico-deductive method can also be found in the other humanities and in the social sciences. This shows that the hypothetico-deductive method is used in all these sciences. It does, however, not show that the hypothetico-deductive method is the only method that is being used in these sciences. From examples alone one may not derive such a conclusion. Nor do I know any argument to the effect that all human knowledge is hypothetico-deductive. We can, of course, not have any argument to the effect that knowledge *must* be hypothetico-deductive; this would be incompatible with the main point of the method: that all our insight is hypothetical and tentative.

One might, of course, try to argue in the opposite direction and try to give an argument that the hypothetico-deductive method cannot be universal. We shall now consider quickly some such arguments that are sometimes put forth:

i. The hypothetico-deductive method is specific for the natural sciences.

This statement, which is merely a claim and not an argument, is refuted by the example I gave concerning *Peer Gynt*.

ii. The hypothetico-deductive method can only be used in experimental sciences.

This is often put forth as a justification of (i). This view has probably arisen because the hypothetico-deductive method is used in the experimental natural sciences. However, the method is also used in nonexperimental sciences, like astronomy or, as we have seen, in interpretation of literature.

iii. Hypothetico-deductive systems consist of "if-then" sentences.

This claim is sometimes used in support of (ii) above. The idea then is that in order to find out whether an if-then sentence is true, one must in some way or other arrange to make the antecedent true and then notice what happens to the consequent. For example:" If one mixes two parts of hydrogen and one part of oxygen and ignites, an explosion takes place." This may be the idea behind Habermas' claim that the hypothetico-deductive method is guided by a manipulative interest. Habermas' arguments for this claim are difficult to make coherent unless one imputes to him a view like this. However, what exactly Habermas means by the hypothetico-deductive method is hard to tell since his arguments are not very perspicuous. In any case, Habermas' colleague in the Frankfurt school, Max Horkheimer, seems to hold something like (iii) fairly explicitly. Thus, for example, in "Traditionelle und kritische Theorie" Horkheimer says that the sciences are characterized by their hypothetical character and he adds:"Between the forms of judgment and the epochs of society there are connections that I will permit myself to make a short indication. The categorical judgment is typical for the pre-bourgeois society; so it is, man can do nothing about it. The hypothetical and also the disjunctive form of judgment belongs particularly to the bourgeois world; under certain circumstances this effect may take place, either it is so or otherwise. Critical theory declares: it must not be so, man can change what is, the circumstances for this are present now."[4] Horkheimer may slip so easily back and forth between speaking of a sentence that has the epistemological status of a hypothesis and a sentence that has if-then form, because in German the same phrase "hypothetischer Satz" is used for them both. However, as with Habermas, it is hard to tell what his position is. It is largely because of the influence of the Frankfurt school that many students in German-speaking countries and parts of Scandinavia have developed an antipathy to the hypothetico-deductive method that makes it necessary to explain not only what this method is but also what it is not, and has led me in this chapter to elaborate points that to philosophers in the "analytic" tradition and others who are acquainted with the method may seem obvious.

As should be clear from section 2, there are no restrictions on the form of the sentences in a hypothetico-deductive system. They may be existential—"there is a planet outside Neptun"—or universal—"all mammals are warm-blooded"—and they may of course also be if-then sentences. When one applies the hypothetico-deductive method in experimental sciences, one often tries to derive within the system some sentences which have if-then form, and one then tries to set up an experiment which makes the antecedent true. However, one would do the same with any sentence that one wants to test experimentally, whether it belongs to a hypothetico-deductive system or not.

iv. The hypothetico-deductive method presupposes that the object of study is considered as a thing and the method can therefore not be applied in the study of man.

As a justification for (iv) one often adduces

v. The hypothetico-deductive method presupposes that the researcher who applies the method does not affect that which is being investigated, as happens, for example, in the social sciences, and/or

vi. The hypothetico-deductive method should not be applied in the social sciences since it does not have room for the possibility that the researcher himself is a part of the society that is being studied.

However, the hypotheses we use when we apply the hypothetico-deductive method may very well be to the effect that the object that is being studied is a person, that the researcher affects the object of study, that the person or society that is being studied in its turn influences the researcher, that the researcher is a part of the society that is being studied, and so on. These hypotheses will be connected with other hypotheses, perhaps concerning how this reciprocal influence takes place, i.e., how the activity and theorizing of the researcher affects the object of study and society, etc. All these hypotheses are tested and confirmed or disconfirmed by observing how well they fit in with one another, with our other views and with our experience.

> vii. The hypothetico-deductive method does not give room for and is not compatible with self-reflection.

Again there seems to be nothing to prevent one from including among the sentences in a hypothetico-deductive system sentences about oneself and one's own activity. If one is a holist, it seems inevitable to include such sentences, since all one's opinions (also one's opinions concerning oneself) are part of the whole and must fit in with one's other opinions. One may, of course, also reflect on one's own reflection, etc. A hypothetico-deductive system is always open, in the sense that it may be supplemented by ever new hypotheses that one gets through sense experience, through reflection on the hypothetico-deductive system and on how it may be made simpler and more coherent, through self-reflection, on self-reflection, and so on.

Finally:

> viii. While the hypothetico-deductive or "explaining" method is used in the natural sciences, one uses in the humanities, and partly also in the social sciences, the "hermeneutic" or "understanding" method.

As we saw in the example from *Peer Gynt*, one proceeds, however, hypothetically-deductively when one uses the hermeneutic method, and instead of contrasting the two methods, we have found it natural to say that the hermeneutic method is the hypothetico-deductive method applied to meaningful material.

We have, of course, not thereby shown that all acceptable methods within all areas are variants of the hypothetico-deductive method. Within most sciences there are no doubt often set forth theories and arguments that are difficult to fit into the hypothetico-deductive pattern. As I mentioned earlier, it is important that one approaches these with an open mind. To the extent that they seem to give us reasons to believe in the views that are set forth, we should ask ourselves: What are these reasons? Exactly what do the arguments amount to? Can we see in these arguments a general type of argument which we should think through more carefully and which possibly lead to methods that are different from the hypothetico-deductive method but which are nevertheless fully acceptable?

5 *Differences between the Natural Sciences and the Humanities and Social Sciences*

The fact that the hypothetico-deductive method is used in the humanities and the social sciences as well as in the natural sciences does not mean that the distinction between man and nature gets lost or becomes softened. There are many and important differences, but these have to do with the object of study, man and his actions and creations. When one sees clearly the similarities between the methods of natural science and humanities and social science, then one sees also better what the differences are between man and nature and between the study of man and society on the one hand and nature on the other.

We have already noted some such differences, for example, the influence of the observer on the subject that is being studied and also the influence of the subject on the researcher. Further also, man's capacity for self-reflection.

There are several other important differences, of which I will mention only two:

1. Already from the natural sciences we know how our theories influence our observations that give us our data, such that changes in the theories often bring with them small changes in the data. This holds not only for our description of the data, but also for what we describe—that is, what we see or observe. This is also the case in the humanities and the social sciences and to a much larger extent. Thus, for example, in grammar and in ethics our grammatical and ethical intuitions, which make up our data, are strongly influenced by our theories. When our theories change, then also many of our intuitions change, so that we often must go repeatedly back and forth between our intuitions and our theories before we arrive at a reflective equilibrium, as Rawls has called it,[5] where our theories and our intuitions fit in with one another.

This is also the case when we interpret literary texts. Our interpretation of the words and sentences is influenced by our interpretation of the whole work, but the interpretation of the whole work depends, of course, on the interpretation of the individual parts, so that we often have to go for a long time back and forth between whole and part before we arrive at a "reflective equilibrium," that is, a satisfactory interpretation, where the interpretation of the whole and the interpretation of the parts fit in with one another. This movement back and forth, which is so conspicuous in the humanities and the social sciences, is what is usually called the "hermeneutic circle." It is particularly striking in the humanities, but as we noted, we find it in the natural sciences as well.

2. Another and more important difference between natural science and the part of the humanities where one is concerned with the interpretation of literary texts is that while nature is inexhaustible, a literary text is a finite collection of data. When one has constructed a scientific theory, one may test it again and again; one may always make new observations. When one has to choose between two competing interpretations of a work of literature, one can usually not find new data that falsify one of the interpretations. All the data are already there. One must make the choice on the basis of the simplicity of the interpretations, how much of the work they comprise, how specific they are, etc..

Especially when one has very few data—for example, only some fragments—it is very simple to find several interpretations which fit in with all the data, and one is then skeptical toward claims that one particular of these interpretations is *the* right one. This makes it understandable why Michael Ventris' now so famous decipherment of Linear B was not immediately accepted as the final solution. Only when more inscriptions were found that fit in with his decipherment was it generally acknowledged as *the* correct solution. In many areas of the humanities one is not in a situation where one may find new data after the presentation of the interpretation. When one interprets a novel or a poem, the full text is usually there, and every interpretation that is fairly carefully worked out will fit in with the data, since this is a requirement for any acceptable interpretation. All one can do in order to sort the good interpretations from the poor ones is, then, as we have noted, to use criteria of simplicity, comprehensiveness, etc.

6 *Hermeneutics and Rationality*

Finally, I will mention some differences between the natural sciences and the humanities and social sciences that have to do with the rationality of human beings. I will discuss very briefly which role assumptions concerning rationality play when we apply the hypothetico-deductive method in the humanities and social sciences.

Let us begin with the explanation of actions. As I have argued in a paper on the reasons and causes of actions, which was published in German in 1979,[6] it seems natural to explain a person's action by help of decision theory. That is, we use a hypothesis that says that when on the basis of our psychological, physiological, and other theories of man we have no reason to think that a person's behavior springs from causes that are not reasons or from unconscious motives, we shall assume that he acts as a rational agent in the sense of decision theory. This means that he goes through a two-step procedure. First he considers the alternatives which he believes to be possible in the situation at hand. Thereafter he chooses from among these alternatives one which in view of his values and beliefs concerning probabilities maximizes his expected utility. There are a number of studies by psychologists, Amos Tversky and others,[7] which show that such a decision theoretic model is suitable for explaining situations that occur in practical life.

However, the decision theoretic model is not all that we need in order to explain and understand action. We must also make other assumptions concerning the rationality of the agent. I shall now mention four such assumptions:[8]

1. *Consistency at a time.* Although the decision-theories pattern of explanation is often called the rational choice theory of human behavior, almost any agent could be made to fit into this pattern without seeming very rational.

Trying to determine a person's preferences, we may, for example, find that he prefers *a* to *b*, and *b* to *c*, but *c* to *a*. This seems irrational—and also difficult to understand, as we should expect if assumptions about rationality seem to be a prerequisite for understanding. It is common, therefore, to require for rationality also the following: a rational individual who is placed between a number of alternatives to be realized at time *t makes* his choice according to transitive preferences at *t*. If a person satisfies this condition and also the various others that we have been and will be discussing, then we feel that we understand him; if not, we do not yet understand him and we must continue our search for an explanation of his behavior, perhaps by no longer regarding the behavior as action but as something that in fact springs from psychological or neurological causes and calls for physical explanation.

This seems to me to be the most important difference between understanding and causal explanation: *Understanding presupposes rationality hypotheses.* However, both, understanding and causal explanation, proceed hypothetico-deductively.

2. *Consistency over time.* In order to understand, we also have to make assumptions concerning consistency over time. The rationality condition I just mentioned concerned only a person's preference at a given instant of time *t*. He could satisfy such a condition at every instant of time and yet appear quite irrational and hence difficult to understand. Thus, for example, given a choice among three options, *a*, *b*, and *c*, he might first prefer *a*, in the next moment *b*, and then immediately thereafter *c*. This kind of inconstancy would tend to disturb us. We expect a person's preferences to be consistent not only at a given time, but also over time. This does not mean that we expect a person never to change his preferences, but we would like to understand why he changes them. We would like to have an explanation, by reasons or by causes, for

such changes, while we at other times may want explanations for lack of changes. This, by the way, is an example which illustrates how an appeal to reasoning and information acquisition, e.g., through perception, comes into explanation and understanding of action. There are numerous such ties between a person's actions and the other features of a person we want to understand, like his thoughts, beliefs, feelings, fears, and other mental states.

3. *Concern for the future.* Further, we expect a rational person to let his preferences be guided not only by his present desires but also his future ones. A certain concern for our own future seems to be part of rationality. This means that it would be irrational to discount completely our future. It would, however, also be irrational not to discount it at all, given human mortality and the uncertainties of life. Hence both very rapid discounting of the future and very slow discounting are phenomena that may defy understanding and call for special explanation.

4. *Interaction between agents.* In order to understand human action, we must take into account that many of the beliefs that guide our actions are beliefs about the beliefs and values of other persons and about how they will act as a result of our acts. The standard framework for the study of such interaction between agents is game theory. Since so many of the human actions that we seek to understand are cases of such interaction, it seems to me that no satisfactory study of man can take place without game theory. Game theory is one of the important tools of hermeneutics. Many cases of group behavior that initially may seem puzzling and irrational, e.g. people throwing litter and Chinese peasants cutting down the forests and thereby bringing about erosion, get simple, rational explanations in terms of game theory, and we thereby come to *understand* them.

Game theory also seems to me to be an excellent framework for analyzing speech acts and for determining the various beliefs and values that may go into such acts. However, this is a topic for itself, which I shall not enter here.

7 *Understanding and Prediction*

We should note that the decision theoretic model makes it clear what we may predict and what we may not predict in the area of the humanities.

There are, I think, difficulties in principle in predicting in general which alternatives a person is going to consider when he is deliberating what to do. Thus, for example, when Ibsen wrote *Peer Gynt*, it was difficult even for him to predict that he would get the idea of the strange passenger, that he should write such and such a verse, and so on. However, if one knows which ideas, i.e., alternatives, a person has considered, it is then easier to judge which of these ideas are good and which not so good. The hypothetico-deductive method itself may serve to illustrate this. It has often been observed that the invention of new hypotheses and theories requires ideas and creative power and cannot normally be predicted, while once a hypothesis has been put forth, the testing of it is a considerably simpler phenomenon.

In physical nature there does not seem to be anything like putting forth and considering possibilities that are not actualized. Explanation in natural science does therefore not make use of the decision theoretic model. The possibility of prediction in natural science therefore comes much closer to the second step in the decision theoretic model; given certain propensities, what will now happen? For this reason one can normally not predict the exact outcome in the humanities and social sciences.

8 *Conclusion*

After I have given this all too brief sketch of the method of understanding and how it is an application of the hypothetico-deductive method, I will now at the end mention that some of the main ingredients that go into the method of understanding have already been fairly thoroughly worked out in the literature. Economists and also some philosophers have, for example, done quite a lot to develop decision theory and game theory. The different rationality conditions that I mentioned have been discussed in several articles and books by economists.

Philosophers like Quine[9] and Davidson[10] have pursued the issue of how understanding is based on rationality assumptions. Quine's "principle of charity" and Davidson's "maxim of maximizing agreement" express this idea. Philosophers who call themselves hermeneuticists have not discussed as clearly and fully as these two philosophers what understanding is and how it is connected with rationality. However, one finds similar ideas in many of the hermeneuticists. Thus, for example, Gadamer in *Wahrheit und Methode*[11] says that agreement is the basis of understanding. Wittgenstein, too, emphasized this point. It seems that philosophers belonging to very different traditions are here working on the same problems and have much to learn from one another.

Notes

1. Jürgen Habermas, *Erkenntnis und Interesse*. Frankfurt a. M.: Suhrkamp, 1968.
2. Quoted from Asbjørn Aarseth's summary in *Dyret i mennesket: et bidrag til tolkning av Henrik Ibsens "Peer Gynt."* Oslo: Oslo University Press, 1975.
3. Daniel Haakonsen, *Henrik Ibsens "Peer Gynt."* Oslo: Gyldendal, 1967.
4. Max Horkheimer, "Traditionelle und kritische Theorie." Here quoted from Max Horkheimer, *Die Gesellschaftliche Funktion der Philosophie*. Frankfurt a. M.: Suhrkamp, 1974, p. 184 n.
5. John Rawls, A *Theory of Justice*. Cambridge, Mass.: Harward University Press, 1971, esp. pp. 17–22, 46–53, 577–587.
6. "Handlungen, ihre Gründe und Ursachen." In Hans Lenk (ed.), *Handlungstheorien-interdisziplinär*. München: Fink, 1979, vol. 2, pp. 431–444.
7. Amos Tversky, "A critique of expected utility theory: descriptive and normative considerations." *Erkenntnis* 9 (1975), 163–174, and several other articles.
8. In the following I am indebted to several articles and books by Jon Elster, especially his *Ulysses and the Sirens: Studies in Rationality and Irrationality*. Cambridge: Cambridge University Press, 1979.
9. W. V. Quine, *Word and Object*. Cambridge, Mass.: MIT Press, 1960, and several later works.
10. Donald Davidson, "Truth and meaning." *Synthese* 17 (1967), 304–323, and a large number of later articles.
11. Hans-Geory Gadamer, *Wahrheit und Methode*. Tübingen: Mohr, 1975.

Chapter 16

Another Look at the Doctrine of *Verstehen*

Jane Roland Martin

The long-standing doctrine that a special kind of understanding—empathic understanding—is required in the field of human behavior has been given new life in recent years. Traditionally, this doctrine has been part of the general thesis that the natural and social sciences differ fundamentally because of their different subject matter. Roughly, the claim is that to understand human beings and their actions we must put ourselves in their position, e.g., think their thoughts, feel their feelings. We are able to do this because we too are human; that is to say, we understand others on an analogy with ourselves. This doctrine, usually called the theory of *Verstehen,* has found wide support among philosophers, social scientists, and historians. In the past it has been expressed in vague, metaphorical language and supported by metaphysical arguments. Contemporary philosophers have tried to strip it of these trappings while preserving the core of the doctrine.

The standard criticism of the *Verstehen* doctrine has been that empathic understanding is nothing more than a useful heuristic device for thinking up explanations; it is in no sense required for understanding human behavior. Empathic understanding, it is objected, does not guarantee that the explanation of behavior so obtained is sound; nor is it necessary for obtaining sound explanations of behavior (Hempel, 1965, pp. 239–40). In effect the critics of the doctrine have maintained that empathic understanding belongs to the context of discovery rather than to the context of validation and that proponents of the *Verstehen* doctrine have failed to distinguish these two contexts adequately.[1] Some have gone so far as to assert that the method of empathy is available to anyone; that it is not the affair of the social scientist and historian exclusively since physicists can use it in studying atoms (Popper, 1957, p. 138). But even when it is granted that empathy is exclusively the affair of social scientists, the criticism that its status is that of a heuristic device hits home, at least with modern *Verstehen* theorists. In reformulating the doctrine their concern is not simply to state it in the relatively precise language of contemporary philosophy but to give the doctrine "logical force."

In this chapter I shall be concerned with one such reformulation of the *Verstehen* doctrine, namely, William Dray's (1957, 1963, 1964, ch. 2). I select Dray's version of the doctrine for examination here for several reasons: it is relatively clear; it is more modest in its claims than some contemporary reformulations without being so modest as to become uninteresting; it has been the subject of much philosophical discussion but has not been assessed as a theory of understanding human actions.[2] My procedure will be to set forth and clarify Dray's position and then to criticize it, first as a general theory of understanding human actions and next as a theory of historical understanding of human actions. I shall, in other words, be considering Dray's version of the *Verstehen* doctrine in its own right. In effect, Dray takes the standard criticism of the doctrine as given; rather than try to defend traditional formulations against the criticism, he seeks a formulation which will get around it. For the present purpose I too shall take it as given, but I want to make it clear that my discussion of Dray leaves open the question of whether the standard criticism does in fact do justice to traditional formulations of the *Verstehen* doctrine.

Rational Explanation and the Actor's Point of View

Traditionally, *Verstehen* theorists have held that laws and causal explanations are inappropriate to the understanding of human actions. Dray adopts a modified form of this thesis: he does not deny that human actions fall under law, but he argues that even if they do, "discovery of the law would still not enable us to understand them in the sense proper to this special subject-matter" (1957, p. 118). Causal explanation looks at phenomena from the outside, but human actions, according to Dray, can only be understood from "the actor's point of view."

"When we ask for the explanation of an action," says Dray, "what we very often want is a reconstruction of the agent's *calculation* of means to be adopted toward his chosen end in the light of the circumstances in which he found himself" (1957, p. 122). This does not mean that every action is performed with a plan consciously preformulated; Dray points out that the historian may have to construct, rather than reconstruct, a calculation:

> But such an admission need not affect the main point; for insofar as we say an action is purposive at all, no matter at what level of conscious deliberation, there is a calculation which could be constructed for it: the one the agent would have gone through if he had had time, if he had not seen what to do in a flash, if he had been called upon to account for what he did after the event, etc. And it is by eliciting some such calculation that we explain the action. (1957, p. 123)

Explanations of this sort are called "rational" since they display the point or rationale of what was done. There is an element of appraisal in them, moreover, for they show in what way an action was appropriate; that is to say, they show an action as the thing to have done under the circumstances. The notions of explanation and justification therefore overlap in rational explanations. Reasons for acting have generality, of course, but this generality is the generality of principles of action rather than empirical laws. These principles state the thing to do in certain circumstances, not the thing that everyone does. Dray believes that there is a standing presumption in favor of giving a rational explanation of an action: "Rational and nonrational explanations are alternatives—and alternatives sought in a certain order. We give reasons if we can, and turn to empirical laws if we must" (1957, p. 138; 1963, pp. 115–16).

"Only by putting yourself in the agent's position can you *understand* why he did what he did." Thus formulated by Dray (1957, p. 128), the doctrine of *Verstehen* does not merely give methodological advice; such advice would be worded, "Only by putting yourself in the agent's position can you *find out* why he did what he did." Rather, it sets forth certain conditions which must be satisfied before a historian is prepared to say: "Now I have the explanation." By the phrase, "putting yourself in the agent's position," Dray means "giving an explanation from the agent's point of view," i.e., giving a rational explanation. In other words, the doctrine of *Verstehen* as formulated by Dray consists in the claim that rational explanations alone allow us to understand human actions. This is not to say that every rational explanation is correct or self-evidently true merely by virtue of its form; rational explanations have "an inductive, empirical side" and are built up from the evidence. But rational structure is necessary for understanding human actions; hence the doctrine is interpreted as a logical one.

The doctrine of *Verstehen* is often interpreted as a theory of explanation:

1. There can be no causal explanations of human actions.

Dray, however, formulates it as a theory of understanding. Now as we have seen, Dray says, "Only by putting yourself in the agent's position can you understand why he did what he did." In this chapter, however, I shall take him to be making the following claim:

2. Only by putting yourself in the agent's position can you understand his action.

A thesis about understanding why an agent did what he did would be, of course, a more modest one that (2), for it would leave open the possibility of understanding, for example, what he did or how he could possibly have done what he did, without taking the agent's point of view. Yet it seems to me that (2) preserves the core of the traditional *Verstehen* doctrine while capturing Dray's intent; for it is far from obvious that *Verstehen* theorists in general, and Dray in particular, want to grant that there can be *any* sort of understanding of human actions if the actor's position is not taken. I shall proceed therefore to take Dray to be presenting a theory of understanding actions. If it is thought that I have taken too many liberties in this respect, let me hasten to add that much, although not all, of the discussion of (2) which follows is applicable to the more modest thesis—a thesis about understanding why he did what he did—as well as to (2).

To put oneself in the agent's position, according to Dray, is to give a "rational" explanation of his action; it is to show the point of the action, to construct the agent's calculation. It seems quite clear, however, that at least in the ordinary sense of the phrase, one can put oneself in an agent's position without giving or being able to give a rational explanation of his action. Moreover, one can give a rational explanation, i.e., construct an agent's calculation, without putting oneself in his position. It seems advisable for purposes of analysis, therefore, to distinguish between the notions of rational explanation and putting yourself in the agent's position. If we do this, Dray's formulation of the *Verstehen* doctrine can be given two separate interpretations:

2.1. Only by putting yourself in the agent's position can you understand his action.
2.2. Only by giving (or being given) a rational explanation can you understand his action.

Dray presumably holds both (2.1) and (2.2), but it is possible to subscribe to one without subscribing to the other.

Internal and External Understanding

Before we evaluate theses (2.1) and (2.2) it is necessary to say a few words about the notion of understanding something.[3] An adequate account of this notion is not at present available. For our purposes, however, one is not needed. What is needed is recognition of the fact that a given thing—the thing to be understood—can be treated in two quite different ways and that, depending on how it is treated, quite different sorts of understanding can be had of it. On the one hand a given thing, X, may be treated as a whole, a unity, and understanding it may involve connecting or relating it to something else, something apart from it. On the other hand, X may be taken in isolation—that is to say without relating it to other things, treated as a composite, and understanding it may involve singling out and relating parts or aspects of it. A work of art is often treated in this latter way where understanding is concerned: it is taken by itself, parts or elements are singled out for attention, and relationships are sought. All sorts of events are often treated in the former way: an event is taken as a unity, is related to another event, e.g., one thought to be the cause, and a relationship between the two distinct entities is pointed to, e.g., a causal one.

Let us call the sort of understanding that takes something X as a unity and relates it to something else *external understanding*, and let us call the sort of understanding that takes X as a composite, singles out parts or aspects of X, and discovers relationships between them *internal understanding*. It should be readily apparent that whether a given thing is treated as a unity, that is, as a simple whole, or as a composite, that is, as a complex whole, depends on one's interests and purposes and point of view. Thus, it does not follow from the fact that one has external understanding of an engine seizure and internal understanding of a novel, that one does not or could not also have internal understanding of that engine seizure and external understanding of that novel. One could, for example, connect the novel up with other novels or with the novelist's life or with the social conditions of the times, etc., and, other things being equal, have understanding of it. One could view the engine seizure in isolation and, other things being equal, understand it by singling out parts and, for example, seeing how they interact.

It should also be readily apparent that although two people may both have internal understanding of a given thing, X, their understanding may be different. We never understand a thing *per se*; rather, we understand it under some description. Thus, for example, we may describe the same "chunk of reality" as a war, a civil war, a revolution. Given different descriptions of some "chunk of reality," then, there is room for variation in people's understanding. For what is taken as a part or element of X may vary according to the way X is described, so that under different initial descriptions X may be taken to have different parts. For example, if X is described as a war, battles may be singled out as its parts; if it is described as a revolution, social relationships and class behavior may be singled out as its parts. It should not be thought, however, that for any given description, D, of a given thing, X, there is one and only one set of parts or elements of X. There is no reason why X under D cannot be broken down into quite different sets of parts. Thus, if something is described as a sonata, its parts may be taken to be its several movements, but other parts could be singled out, e.g. its rhythm, melody, harmony, texture.

If many factors combine to afford the possibility of wide variation in people's internal understanding of a given thing, X, they combine also to afford variation in their external understanding of X. Quite naturally, given different initial descriptions of X, people's external understanding of X will be different. But even given the same initial description of X, it is possible to connect X to different things and to see different connections between X and these things. Thus, it is possible for two people to differ in their external understanding of X although they agree in their initial description of X.[4]

Consider, for example, an event like the American Revolution. It can be understood externally in terms of its causes and also in terms of its consequences. Moreover, it can be understood in terms of things which stand to it as neither cause nor effect. Thus to understand the American Revolution as, e.g., a civil war, would be to relate it to things apart from it which were neither cause nor effect. A given thing bears all sorts of relationships to things outside it and although it is certainly not the case that every sort of relationship, if perceived, yields understanding in every context, I am not at all sure that any sort can in general be ruled out as yielding understanding in some context. External understanding of a given thing, X, then, can vary greatly in its details just as internal understanding can.

A Theory of Understanding Actions

There is no doubt that one good way to understand an action is in terms of a rational explanation of it. This sort of understanding would be a form of internal understanding. The action itself would be taken in isolation, so to speak, and looked into. Such elements or parts of the action as the agent's goal, his beliefs about means, his beliefs about his circumstances would be singled out for attention. These elements or parts of the action would be related to one another in the way that the parts of a plan or calculation are related. So far so good. The only question to raise about (2.2) is whether this is the *only* way to understand an action.

Let us grant all the things Dray wants us to grant about human behavior: that it is rule governed, that it is meaningful, that it has a point. It follows neither that we can only have internal understanding of actions nor that the only internal understanding we can have involves the singling out of the aforementioned elements and relationships. Actions, like everything else in the world, bear connections or relationships with things outside them, and I see no reason at all for supposing that at least some of these relationships could not in certain contexts be illuminating. As a matter of fact, we do at times consider it to be illuminating to be shown the connection between an action and some antecedent event, e.g., a childhood experience, or some standing condition, e.g., a state of society or of an individual, and we also consider it illuminating on occasion to be shown that an action fits into a larger context, e.g., a historical trend, or falls under some interesting and surprising category.[5]

The *Verstehen* doctrine as formulated in (2.2) in effect claims that external understanding is inapplicable to human actions.[6] Dray's arguments for this position, however, are few and far between. Thesis (I), that there can be no causal explanations of human actions, may be thought to supply support for (2.2), but then Dray does not advocate (I). Moreover, we may grant the truth of (I) without granting that external understanding of human actions is impossible. For even if there can be no causal explanations of actions, external understanding is not thereby ruled out. On the one hand, actions can still be classified as something else, can still be related to their effects, can still be related to noncausal antecedent conditions, and so on. Moreover, thesis (I) does not maintain that actions have no causes at all, but only that no causal explanations can be given for actions.

It is important to remember that the denial of (2.2) does not commit one to the thesis that external understanding of actions is *complete* understanding. This thesis is surely false, taken in the natural sense of denying the possibility of further understanding that is not external. To say that there can be external understanding of actions is not to say that there can only be such understanding or that such understanding is sufficient or that it is the best sort. To deny (2.2) is to make the very modest claim that external understanding of actions is possible. It is for this reason that one may grant the truth of (1) if one wishes—one may even go further and grant that actions have no causes—while denying that (2.2) is true. For neither causal explanation nor even the existence of causes is necessary for external understanding. External understanding can take many different forms.

An argument Dray might adduce in support of (2.2) is that to have external understanding of an action is to take the spectator's point of view and that one must take the agent's point of view if one is to understand an action. Let us for the moment grant Dray's distinction between the agent's and spectator's point of view. We may then ask why we must take the one point of view toward actions rather than the other. We are sometimes told that we are in a privileged position relative to human actions (as opposed to stones and stars) because we ourselves are agents. But supposing this is true, it does not follow from the fact that we *can* take the agent's point of view in connection with actions but not other things, that we *must* take it. What follows is that actions, unlike other things, are open to a double viewpoint. If so, and if to have external understanding is to take the spectator's point of view, then external understanding of an action would not by itself be complete understanding. But, of course, neither would the understanding specified by (2.2) be complete understanding.

It is normally taken for granted that our understanding of a thing will increase if we look at it in different ways. There is no reason why actions should be an exception to this dictum. Indeed, one would suppose that just because we ourselves are agents, it would be illuminating to get some other perspective on human actions. The *Verstehen* doctrine is disturbing in its implication that human actions are forever exempt from scrutiny except from one particular viewpoint. To say that *we* are in a privileged position with regard to actions as opposed to stones is one thing: to hold, as the doctrine does, that *actions* are in a privileged position with regard to us is something else. If we could not take a spectator's point of view toward actions they would be doubly different from stones and stars. Now Dray and other *Verstehen* theorists do not deny that we can take the spectator's point of view toward actions, just as we can toward stones and stars; they simply deny that such a point of view can yield understanding, and this, it seems to me, is simply arbitrary.

Now it might be argued that the *Verstehen* doctrine on some formulations, if not on Dray's, does not deny absolutely that the spectator's point of view can yield understanding, but holds merely that understanding acquired from this viewpoint "presupposes" or is "parasitical" upon understanding acquired form the actor's viewpoint. Perhaps in attributing to Dray the position that the possibility of such understanding is ruled out completely, I may be doing him an injustice. But even so, this apparently more moderate position seems to be no more defensible. To be sure, we are now allowed to look at actions form a double viewpoint, but we are still denied all hope of understanding if we do not take the agent's viewpoint. The agent's viewpoint, if not now the only legitimate one, is still the privileged one. But is it really the case that one who sees an action as a resultant of various social forces must also give a rational explanation or put himself in the agent's position if he is to understand it? That one who looks to child rearing patterns in an effort to understand an action is doomed to failure unless he also puts himself in the agent's position or gives a rational explanation? So far as I can see, this position, albeit moderate, is still an arbitrary one.

If (2.2) is to be condemned for being unduly restrictive, (2.1) must be condemned on the same grounds. For it, too, rules out external understanding. Indeed, it also rules out various forms of internal understanding, although it is perhaps not as restrictive in this respect as is (2.2). It will be recalled that a thing is never understood *per se*, but rather under some description and that alternative descriptions can be given for any particular thing. Now the *Verstehen* doctrine, as represented by theses (2.1) and (2.2), requires not only that actions be understood internally but that they be understood under a particular initial description, namely the one the agent himself would give. Thus for example, if an agent himself describes what he is doing as restoring the glory of Spain, we must understand his action under this description rather than, say, the description "massacring innocent Basques." Otherwise we could scarcely give his rationale for his action as (2.2) requires or put ourselves in his position as (2.1) requires.

Theses (2.1) and (2.2) restrict the sort of internal understanding we can have of actions, then, in that they limit the sort of initial description under which we can understand them. Now as we have seen, the initial description, *D*, given of a thing, *X*, does not uniquely determine the parts of *X* to be singled out for attention. However, (2.2), in making a rational explanation necessary, in effect specifies these parts: the agent's goal, his beliefs about means, his beliefs about his circumstances, and so on. It therefore rules out the possibility of internal understanding of an action under the description the agent would give but involving the singling out of nonrational elements of the action, e.g., the agent's feelings, his emotions, his crazy impulses. In this respect (2.1) is more liberal: to single out for attention the agent's feelings and the like is compatible with taking his position. But (2.1) is by no means liberal enough, for it presumably would rule out a priori someone's selecting as the parts of the action features of the situation the agent himself fails to see, e.g., the agent's unconscious motives.

Actually, the distinction which Dray and many others draw between the point of view of the agent and the point of view of the spectator is in many ways misleading. The impression is usually given that the spectator's point of view is unitary when actually the spectators of an action can be a varied lot, each one having quite different understanding of the action. Some spectators might single out external relationships for attention; others might single out internal relationships. Thus, it is not even possible to construe the spectator's point of view as that which yields external understanding. The impression given by Dray, moreover, is that to single out the agent's plan is to take his point of view. By Dray's own admission, however, the agent need not himself have consciously formulated a plan; a rational explanation can be a construction, not a reconstruction. But then, his notion of the agent's point of view does not involve taking the real agent's point of view at all. Indeed, in some cases at least there would be no difference between the point of view of Dray's agent and the point of view of a particular sort of spectator.

One suspects that Dray's agent, at least where rational explanation is concerned, is very close to being a nontechnical spectator and that it is only the technical or theoretically minded spectator that Dray wants to bar from all possibility of understanding human actions. Hence, his drawing of the line at the agent's point of view. It would seem, however, that distinctions need to be made within the class of spectators' viewpoints, and that Dray himself has allowed at least one sort of spectator to understand performances.

To put oneself in the agent's position, the requirement of (2.1), is, presumably, to take the agent's point of view. Dray could be interpreted, then, as holding that understanding an action requires taking the point of view of one particular sort of spectator, namely the sort who tries to construct the plan. Yet this leaves out of the picture a great variety of spectator viewpoints, each able to illuminate actions in its own way.

A Theory of Historical Understanding

So far I have interpreted Dray's reformulation of the doctrine of *Verstehen* as applying quite generally to the understanding of actions. It may be the case, however, that it is intended to apply only to the understanding of actions in a specific context. Dray may be willing to grant that actions cannot only be explained without empathic understanding, but can also be understood without empathic understanding. His claim may be the relatively weak one that it is necessary only for one sort of understanding.

Many supporters of the *Verstehen* doctrine would object to this weakened version of it on the grounds that the doctrine has "logical force" in relation to the understanding of actions in general. Dray's primary concern, however, is with historical explanation and understanding, and he does at times seem to be upholding a more limited version of the doctrine. Dray's version of the doctrine, when limited to the historical context can be formulated as follows:

> 3. Only by putting yourself in the agent's position can you have historical understanding of his action.

For reasons set forth earlier (3) can be broken down into two distinct theses:

> 3.1. Only by putting yourself in the agent's position can you have historical understanding of his action.
> 3.2. Only by giving (being given) a rational explanation can you have historical understanding of his action.

Let us consider theses (3.1) and (3.2). There is no doubt that in certain contexts certain ways of understanding things have preferred status. Thus, the anthropologist seeks to understand things in terms of their structure and function while the lawyer is forever searching for precedents. In the one case preferred status seems to be conferred by scientific, or quasi-scientific, theory; in the other case preferred status seems to be conferred simply by tradition. The historical context would seem to be more like the legal one than the anthropological one: there is no scientific theory conferring preferred status on either rational explanation or on the agent's point of view, but perhaps this status is conferred by long-established practice. To determine whether it is, hence whether (3.1) and (3.2) are true, requires, then, an extensive examination of historical practice. For the claims of (3.1) and (3.2) are ultimately factual ones: to determine their truth we must study the conditions under which historians and readers of history say "I understand."

Now it may be argued that the theory of *Verstehen*, although not scientific, is epistemological and that as such it does confer preferred status on putting oneself in the agent's position and on the act's rationale. It is true that, insofar as the theory is an attempt to clarify the notion of understanding actions, it is, or at least is intended to be, epistemological. But a proponent of (3.1) and (3.2) cannot appeal to this epistemological theory, in order to justify giving preferred status to empathy as a way of understanding actions, without begging the question. It is this theory whose validity is at issue. An appeal to it, therefore, does nothing to increase the warrant for the view that in the historical context actions must be understood in the way specified by (3.1) and (3.2). Similarly, if the validity of Freudian theory is questioned, it will not suffice to justify understanding dreams in relation to their latent meaning by appealing to that theory.

How is one to study the conditions under which historians and readers of history do in fact say, "I understand?" One way is to observe historians carefully, note when they say,"Now I understand," or something equivalent, and see what conditions obtain. Another way is to examine historical writing and see if the conditions specified by (3.1) and (3.2) are fulfilled there. For we may assume that, if putting yourself in the agent's position and giving or being given a rational explanation are necessary conditions of historical understanding, the historian's writing will reflect this. After all, by means of his writing he hopes to enable his readers to understand. It is reasonable to assume that he wants *them* to gain *historical* understanding, and that he will do what is necessary to achieve this.

We cannot undertake an investigation of this sort here, but the reader is urged to launch his own tests of (3.1) and (3.2). On the face of it, it seems easier to test (3.2) than (3.1). In the case of (3.2) all that is needed is to make an "objective" observation, i.e., see if a historian does in fact give rational explanations of actions. But in the case of (3.1) how is one to know if he has in fact put himself in the agent's position? And supposing this to be no problem, we must note that since readers vary in their ability to put themselves in another's position, a given work may fulfill (3.1) relative to some readers but not relative to others.

Thesis (3.2) is not quite so easily tested as one might suppose, however. How much of the agent's calculation must be presented if an explanation for the reader is to qualify as a rational explanation? Is it sufficient for the historian to refer to one aim of the agent, or give one reason, or is more than this required? Yet perhaps this is not so great a problem as it seems. One must expect to find borderline cases as well as easily decidable cases. A more serious problem of testing arises because (3.2) and, for that matter, (3.1) are not theses about historical understanding in general but rather about historical understanding of actions.[7] Where actions are not at issue in a historical work, then, we seem not to have a testing ground for (3.1) and (3.2).[8]

More needs to be said, however, about testing the *Verstehen* doctrine of historical understanding in the face of historical works which seem not to deal with actions. When we find the historian discussing the revival of commerce, the ascendancy of towns, the rise of a merchant class, we are tempted to say unequivocally that he is not dealing with actions at all and that such works constitute no rebuttal to theses (3.1) and (3.2). But let us recall that things are always understood under some description and that for any object of understanding alternative descriptions are available. We have no guarantee that what the historian has described as "the revival of commerce" or as "the rise of the merchant class" could not have been described by him in such a way that we would take him to be talking about actions; that what one historian has described as "the landlords abolished or modified serfdom," could not have been described by another as a nonaction, as "serfdom changed or disappeared." The point is that whether something qualifies as an action depends on the description given of it and that in some cases, at least, both action and nonaction descriptions can legitimately be given. Consciously or not, the historian chooses the description under which we are to understand a given thing, X. If he chooses a nonaction description, what are we to say about theses (3.1) and (3.2)?

If no action description were available for a given thing, X, then I should think it would be correct to conclude that failure to put oneself in the agent's position and to give a rational explanation did not count against (3.1) and (3.2). Were such a description available, however, I am not so sure that such failures could justifiably be ignored. In the last section we interpreted (2.1) and (2.2) as requiring that the initial description under which X was understood should be the description the agent himself would give of X. Now in those cases in which the historian gives a nonaction description of X, although an action description is available, he is most likely, although perhaps not necessarily, failing to describe X as the agent himself would describe it. But then accounts of the disappearance of serfdom, the rise of commerce, and the like would most likely constitute counterexamples to (3.1) and (3.2).

Of course, we can reject this interpretation of (2.1) and (2.2), hence of (3.1) and (3.2). We can refuse to look behind the historian's initial description and say that if he gives a nonaction description of X that is all that matters: for all intents and purposes X is not an action and (3.1) and (3.2) are unaffected by the historian's failure to put one in the actor's position and to give a rational explanation of X. This interpretation will serve to ward off numerous potential counterexamples to (3.1) and (3.2), but it will do so at a price. The *Verstehen* doctrine will on this construal be hypothetical. In effect it will say: if the historian views something as an action, he can only have historical understanding if he does such and such. Since it will not require him to view things as actions even if the ostensible agents of them did, it will be compatible with his taking standpoints the agents never dreamed of.

The question is not whether the doctrine so interpreted is correct or praiseworthy but whether it captures the intent of *Verstehen* theorists. I have my doubts on this latter score. For one thing, *Verstehen* theorists normally talk as if historians deal primarily, if not exclusively, with actions. Yet on this interpretation I think we would find that although some historical works are filled with action descriptions, in others they appear infrequently if at all. Moreover, on this interpretation the *Verstehen* doctrine allows the historian in describing human behavior to use technical and theoretical concepts. But this surely is at odds with the program of which the doctrine is a part, namely, that of drawing a sharp line between history and the sciences.[9] I do not think Dray is advocating this hypothetical version of (3.1) and (3.2), but if he is he would, I think, be well on the way to abandoning the "core" he wants to preserve of traditional formulations of the doctrine.

Let us assume now that the problems we have been discussing in connection with testing (3.1) and (3.2) against historical writing are resolved. One caution is still in order. (3.1) and (3.2) specify necessary, not sufficient, conditions for historical understanding. Thus, on the one hand one cannot assume that because they are met by a work, that work actually provides historical understanding; and on the other hand one cannot assume that because a work does not provide historical understanding they have not been met. We originally put the test of (3.1) and (3.2) in terms of discovering the conditions under which a historian or reader of history says, "Now I understand." There is nothing wrong with this as long as we keep in mind that (3.1) and (3.2) do not purport to give us all the conditions under which he says this and that his failure to say this cannot itself be held against (3.1) and (3.2).

Dray should be pleased to find us asking for studies of historical practice, for along with many other contemporary philosophers of history he prides himself on sticking fairly close to what the historian actually does, and criticizes his opponents for imposing their own preconceived ideas on historical practice instead of taking it at face value. I myself am far from convinced, however, that a study of historical practice will yield the results Dray apparently expects. As everyone knows, but as those philosophers and historians who write about history tend to forget, history is immensely various. I would expect to find theses (3.1) and (3.2) gaining confirmation from some but not all historical practice. That is to say, I would expect these results provided that the investigation ranged over the whole of what is naturally called historical practice.

It is all too easy to stack the evidence in favor of these theses by ruling out—as bad, perhaps, or as parasitic or as nonprimary or as nonstandard—all sorts of practice which, if included in the sample, would tend to refute theses (3.1) and (3.2). Unless there is some independent criterion for ruling out a particular sort of practice as not being historical, Dray and other advocates of the *Verstehen* doctrine, as represented by these theses, must rest content with an investigation which studies a wide-ranging sample of historical practice. Such a sample, may, of course, make it difficult for theses (3.1) and (3.2) to survive their testing by the real world, but such is the fate of theses of this sort. Had theses (3.1) and (3.2) been claiming merely that putting oneself in the agent's position, etc., was *one* way to understand actions historically, they would, needless to say, have an easier time of it when confronted by the facts of practice. But then, of course, the doctrine of *Verstehen* would lose much of its interest; for surely it is the very lack of modesty of its various claims that on the one hand endangers its truth but on the other hand makes it worthy of discussion and extended investigation.

Notes

1. For a discussion of the two contexts see Richard Rudner, 1966, pp. 5–7. See pp. 71–3 for Rudner's discussion of *Verstehen.*
2. Dray's account of human actions has been discussed and criticized primarily as a theory of explanation. See, for example, Hempel, 1963, and 1965, Essay 12, Section 10; and the other essays in Hook, 1963, including Dray's in which Dray's views are discussed.
3. It should be noted that I am talking here about *understanding* something and not about *being understanding toward* someone. This latter notion is, I think, relevant to a discussion of the *Verstehen* doctrine but limitations of space prevent our pursuing its relevance here.
4. And, of course, it is possible for two people to agree in their initial description of a thing, X, but for one to have internal understanding of X and the other to have external understanding of X.
5. This last would, I suppose, constitute understanding *what*; hence it would be compatible with an interpretation of the *Verstehen* doctrine as a theory of understanding why he did what he did. As the other examples illustrate, however, some cases of external understanding would not be compatible with this narrower theory.
6. As we will see, it also claims in effect that many forms of internal understanding are inapplicable.
7. That is to say, they are not theses about historical understanding in general unless one were to hold that actions alone are the historian's proper subject matter. R. G. Collingwood, 1956, part 5, for one, appears to take this position. Dray, however, rejects Collingwood's stand on this matter; see Dray, 1964, p. 13, and 1963, pp. 105–6.
8. Except insofar as the nonfulfilment of (3.1) or (3.2) in the case of a nonaction could be said to constitute confirmation of (3.1) or (3.2). In this connection see Israel Scheffler's discussion of the paradoxes of confirmation (1963, part 3).

9. See Morton White, 1965, p. 184, for a brief account of some methodological implications of the position he calls moralism, a position which bears a close resemblance to the *Verstehen* doctrine as formulated here in (3.2). It should be noted in this connection that White is perhaps more generous toward moralism than I am being toward the *Verstehen* doctrine. He points out that historians explain things other than actions (he is interested in it as a doctrine of explanation, not understanding), but does not hold nonaction descriptions of the historian's subject matter against moralism as I am suggesting they should be held against the *Verstehen* doctrine.

References

Collingwood, R. G. (1965) *The Idea of History*. New York: Oxford.
Dray, W. (1957) *Laws and Explanation in History*. London: Oxford.
Dray, W. (1963) The historical explanation of actions reconsidered. *Philosophy and History*. Ed. S. Hook. New York University Press.
Dray, W. (1964). *Philosophy of History*. Englewood Cliffs, N.J.: Prentice-Hall.
Hempel, C. G. (1963) Reasons and covering laws in historical explanation. In *Philosophy and History*. Ed. S. Hook. New York University Press.
Hempel, C. G. (1965) *Aspects of Scientific Explanation*. New York: The Free Press.
Hook, S., Ed. (1963) *Philosophy and History*. New York University Press.
Popper, K. R. (1957) *The Poverty of Historicism*. Boston: Beacon Press.
Rudner, R. (1966) *Philosophy of Social Science*. Englewood Cliffs, N.J.: Prentice-Hall.
Scheffler, I. (1963) *The Anatomy of Inquiry*. New York: Alfred A. Knopf.
White, M. (1965) *Foundations of Historical Knowledge*. New York: Harper & Row.

Chapter 17

Taylor on Interpretation and the Sciences of Man

Michael Martin

One of the most frequently cited[1] and often reprinted[2] papers in English advocating an interpretive approach to the social sciences is Charles Taylor's "Interpretation and the Sciences of Man."[3] There are several reasons for its popularity. First, Taylor does not explicitly rely on the European traditions of *Verstehen* and hermeneutics that English-speaking students often have difficulty understanding. Although he mentions Dilthey, Gadamer, Ricoeur, and Habermas in the first paragraph, he does not cite them again. Second, although Taylor's arguments are abstract and general, he cites social science literature, primarily from the field of political science, to illustrate his main points. Third, his position is an extreme one that can be sharply contrasted with the naturalistic, positivistic position that is usually associated with mainstream social science. Thus, he makes no attempt to argue, as David Braybrooke does, that the naturalistic and the interpretive approaches have complementary insights.[4]

Taylor presents a view that has appeal for many students of the social sciences. Drawing on the analogy of the interpretation of a literary text, he argues that the aim of the social sciences is to provide an interpretation of the social meanings connected with the social practices and institutions of particular societies rather than to furnish causal explanations and predictions. The interpretation of social meanings involves, on his view, clarifying the field of concepts—the interconnections of a system of notions—that constitutes a social practice (intersubjective meaning) and provides for shared values and a sense of community (common meanings). The meanings to be clarified are those of the social actors who engage in the practice or are members of the institution.

According to Taylor there is no objective way of validating such interpretations. The validation of an interpretation is a circular procedure (the hermeneutic circle), and disagreements about interpretations ultimately rest on conflicting intuitions. In particular, objective validation by appeal to brute data—that is, data that one cannot question by further interpretations—is not available in the social sciences, although it is in the natural sciences. Nevertheless, he holds that the social meanings revealed by interpretive procedures are nonsubjective and that one can contrast these with the subjective meanings appealed to, for example, by mainstream political scientists. Although one can infer these subjective meanings from brute data, they fail to capture the social meanings expressed in interpretations. Social meanings, however, do not seem to be based on causal considerations. Taylor maintains that if his views about the interpretive nature of the social sciences are correct, then exact prediction in the social sciences, as opposed to the natural, is impossible.

Despite the popularity of Taylor's paper, I know of no extended critique of it, an unfortunate situation, for the paper has many problems. I will argue, first, that Taylor's view of the scope of social science is too limited; second, that his thesis that the interpretation of social phenomena does not involve causal considerations is mistaken; third, that interpretations need not be circular; fourth, that his reliance on nonsubjective meaning sits uneasily with his subjective methodology of interpretation; fifth, that nonsubjective meaning need not be for the actor; and, finally, that he gives no good reasons why exact predictions cannot be made in the social sciences.

The Scope of Interpretive Social Science

Taylor maintains that the aim of the interpretation of a text or a social practice is "to bring to light an underlying coherence or sense" of the text or practice.[5] This aim, he says, presumes three things. First, there must be "an object or a field of objects, about which we can speak in terms of coherence or its absence, of making sense or non-sense."[6] What would this object or field of objects be in the social sciences? I take it that it would be a society, institutions within a society, or perhaps the social practice of actors within the society or institutions, Second, there must be a distinction between the meaning embodied in a text or a social practice and the expression of it that one gives in an interpretation. Third, the meaning of a text or social practice is for or by a subject. Taylor believes that this last requirement, when applied to a social practice, distinguishes the human from the natural sciences. One can speak of the meaning or coherence of a certain object studied in the natural sciences, such as a rock formation. But a rock formation neither has meaning for the rocks nor is an expression of meaning given by the rocks. In the social sciences, however, the meaning of a social practice has meaning for social actors, and social actors in turn might give expression to this meaning.

Is it misuse of language to speak of the meaning of a social practice? Do only linguistic expressions have meaning? Not according to Taylor. He admits that the meaning of a social practice is different from the meaning of a linguistic expression. However, he maintains that it would be "hard to argue that it is an illegitimate use of the term."[7] We speak of the meaning of certain social actions or situations, and we do so correctly, as, for example, when we say, "What is the meaning of this ritual?"

According to Taylor, in order to understand the meaning of a social practice, one must (1) distinguish between the meaning of the practice and the practice itself, (2) understand the meaning a practice has for a subject or a group of subjects, and (3) realize the meaning of a practice is only in a field; that is, it is connected with the meaning of other things. To use one of his examples, we might interpret some social practice in terms of the concept of shame. However, situations are shameful only for subjects. Further, there is a distinction between shame and the situation or action a subject feels shame toward. Finally, the meaning of the term "shame" can only be "explained by reference to other concepts which in turn cannot be understood without reference to shame."[8]

No one would deny that interpreting the meaning of social practices is something that social science should be doing. The crucial critical question for Taylor's theory is, How important is it? Several answers, among others, suggest themselves:

1. The only task of social science is to interpret the meaning of social institutions, practices, and so on.
2. The main task of social science is to interpret the meaning of social institutions, practices, and so on.

3. An important task of social science is to interpret the meaning of social institutions, practices, and so on.
4. One task of social science is to interpret the meaning of social institutions, practices, and so on.

Taylor's position, however, is unclear. Perhaps the most plausible reading is that Taylor holds either (1) or (2). There are two reasons for this. First, one would suppose that if he held a generous view of the scope of social sciences, he would have said so. To be sure, he never explicitly denies that the social sciences have other jobs besides the interpretation of meaning, but none except the interpretation of the meaning of social institutions and practices is mentioned. Second, he argues against the standard empiricist approach to the social sciences, which certainly attributed to the social sciences a larger scope than does Taylor. Taylor does not seem to maintain that the standard view is wrong only up to point, for example, that causal explanations are appropriate in certain areas of the social sciences and that a hermeneutic approach should replace an empiricist approach only in certain areas of inquiry. Indeed, he seems to be saying that the standard approach is wrong in all areas of the social sciences. It is plausible, therefore, to think that Taylor does not advocate (3) and (4) but maintains either (1) or (2).

Has Taylor exaggerated the importance of interpretation in the social sciences? There are good reasons to suppose so. To be sure, many social scientists are interested in interpreting the meaning of social institutions and practice, but they do not limit themselves to this task. Indeed, there are many important questions that social scientists need to ask once an interpretation of a social practice in Taylor's sense has been given.

Consider some of the questions social scientists might ask about—for example, menstruation taboos that have been interpreted in terms of a field of concepts. In anthropological literature menstrual blood and menstruating women are often interpreted as being considered polluting and dangerous, and men must avoid them in order to prevent contamination.[9] However, Marla N. Powers has recently argued that the standard negative picture of menstruation practices is the result of Western anthropologists' reading their biases onto non-Western tribal cultures and has given a different interpretation of menstruation in the American Indian society of the Oglala.[10] One can understand Powers's work as giving an alternative and presumably more adequate interpretation of the meaning of the menstrual taboos than the standard account, but it does more than this. Taking the concept of life stages from the work of Arnold Van Gennep[11] and deriving the concept of *communitas* or antistructure from that of Victor Turner,[12] Powers goes on to maintain that the function of the menstrual taboo "is to give structure to what otherwise is a period of antistructure."[13]

This type of functional investigation has played a large role in both sociology and anthropology. However, it is difficult to see that it would have any role at all to play if social science investigation were limited to the interpretation of meaning. It is one thing to consider what the meaning of the menstrual taboo in the Oglala society is and quite another to determine the function of this taboo. Powers does both. Yet Taylor's concept of the social sciences seems to include only the former. One major difficulty standing in the way of Taylor's incorporating a functional analysis into his scheme is this. Functional concepts need not be ones that members of the culture understand; hence, they might have no meaning in Taylor's sense for the social actors. Thus, there is no reason to suppose that the function of menstruation taboos that Powers attributes to American Indian tribal societies had meaning for the members of these societies.

Given the limited scope of her paper, Powers does not ask other important questions about the menstrual taboos, but despite the fact that these questions seem to be excluded by Taylor's scheme, there is no a priori reason why she could not. Powers seems to assume that the defilement interpretation of menstruation is correct for Western societies. If she is right, then one naturally wonders what the basis of the differences between Western and non-Western societies is. Why does menstruation have a negative meaning in the one sort and not in the other? Furthermore, if menstrual taboos have a function in non-Western societies, do they have a dysfunction in Western societies? If so, what is it? Another obvious line of inquiry has to do with the psychological effects of menstrual taboos on women. For example, do different cultural attitudes concerning menstruation bring about different effects in terms of the premenstrual syndrome? Social scientists have investigated similar questions in other contexts,[14] and there seems to be no a priori reason to suppose that they could not investigate them in the present case. However, these questions seem to be excluded a priori from Taylor's scheme. A third line of inquiry would be to pursue questions about the origins of the menstrual taboos.[15] Why have they taken the form they do? Why have they developed so differently in Western and non-Western societies? Social science should be able to pursue these questions, but it is not clear that Taylor's interpretive social science could.

The menstruation taboo example should make it clear that the strongest thesis, (1) above, excludes a variety of questions that social scientists could consider. Unless Taylor gives good reasons why they should not be included in the purview of social science, his position does not seem reasonable. Does he provide such arguments? At one point in his paper, he raises criticisms against what he calls "the influential 'developmental approach,'" which relies on the concept of function.[16] However, he does not show that functional questions do not belong to the social science.[17] His opposition to a functional approach seems to be that the developmental approach is ethnocentric and uses functions such as "interest articulation" and "interest aggregation" whose "definition is strongly influenced by the bargaining culture of our civilization, but which is far from being guaranteed appropriateness elsewhere."[18] However, there is no a priori reason why the functional approach must be ethnocentric in this sense. Social scientists obviously should attribute functions to social institutions that are appropriate to the cultural contexts they study. Powers's use of function concepts did precisely this.

One might well wonder if either (2) or (3) or (4) is a more reasonable position for Taylor to take. Certainly (2) also seems too strong, for it suggests that once one has interpreted the meaning of a social practice, little work of any importance remains for the social sciences. But as we saw in the menstruation taboo example, a number of important questions remain. One might object that although (2) is too strong, (3) and (4) are too weak in that (4) makes it seem as if the interpretation of meaning is just one among many tasks of social science, while (3) makes it seem as if it is simply one among many other important tasks. One might argue that both are misleading suggestions, that the interpretation of meaning is not merely a task among others—even a task among other important tasks—but rather a necessary step for the social sciences to take before they can begin their other tasks. In other words, one might argue that Taylor is saying this:

5. Interpreting the meaning of a social practice is a necessary condition that must be achieved before doing any other social scientific investigation.

It is surely mistaken, however, to maintain that social scientists must interpret a social practice before any other investigation can begin, for this can go hand in hand with other investigations such as the study of a practice's origin, its development, and its effects on personality. Thus, the detailed study of the meaning of menstruation taboos can proceed simultaneously with investigations into the function of these taboos, historical inquiries into their origins, and psychological studies of their effects.

Of course, Taylor's thesis might come down to:

6. The existence of meaningful social practice is a necessary condition of doing social science.

But (6) is hardly news. Without the meaning of social practices, there would be no society—hence, no social scientific investigation would be possible.

Clearly Taylor would not want his theory reduced to a truism. Since, however, the stronger interpretations run the risk of excluding legitimate questions and conflicting with acceptable social scientific practice, perhaps a way of understanding Taylor's position that is more modest than he intends but does not make it trivial can be found. One possibility is that he is maintaining that interpreting the meaning of social institutions, practices, and so on is important to social inquiry and that many positivistic social scientists have neglected this. On this reading of Taylor, although he may have exaggerated the importance of interpretation in social scientific inquiry, his views nevertheless bring to the light aspects of social inquiry that have been neglected. Taylor's approach would then be seen as redressing the balance and providing a complementary approach to the standard view.

Causality and Interpreting Meaning

The relevance of causality to Taylor's approach to the social sciences is difficult to determine since he hardly uses the term in his paper, but the most plausible reading is that he thinks that causality is unimportant. I say this for the following reasons. First, Taylor seems to believe not only that interpretation is the main task or even the only task of the social sciences but that interpretation has nothing to do with causality. Thus, he never mentions "cause" or "causality" in connection with interpreting social phenomena or even with explanations. He seems to think that interpretation is all there is to explanation in the social sciences Second, he associates causal analysis with empirical social science, precisely the position that he opposes.

Is Taylor correct to suppose that causality is unimportant? If one conceives of the task of social science to include more than merely interpreting the meaning of social practices, as one should, then causality inevitably plays an important role. For example, understanding which social or cultural conditions bring about a menstruation taboo or what effects a taboo has on women's personalities involves a consideration of causes. But what about the narrower social science task of interpreting social practice? Does this not involve considerations of causality?

There are three reasons to suppose that it does. First, if one looks at the natural sciences, one sees that the questions of the meaning of many concepts are closely connected with considerations of causality. Second, although Taylor provides no explicit reasons to suppose that the social sciences would be different from the natural sciences in this respect, the most plausible argument that one can construct on his behalf is unsound. Third, some of Taylor's own descriptions of the meaning of concepts in the social sciences implicitly assume causality.

We should not be surprised that the meaning of a field of concepts in the social science presupposes causal considerations. The same is true in the natural sciences. For example, in medicine an understanding of the field of concepts of cardiology presupposes understanding complex causal connections having to do with heart disease—its etiology, its symptoms, and its treatment. Thus, the meaning of coronary heart disease is closely connected with the formation of atheroma in the arteries that results in a lack of oxygen to the heart and angina. It is difficult to see how one could understand the conceptual connection of the disease, the formation of atheroma, and angina unless one understand the causal connections involved. By analogy, one might suppose that one could not understand the conceptual connections of a humiliating situation, the feeling of shame, and a behavioral disposition unless one understand the causal connections involved. Taylor just seems to assume knowledge of these causal connections as part of common sense and in his interpretation relies on them.

One might object that the analogy does not hold. Yet Taylor's presuppositions regarding determining the meaning of social practice have obvious analogues in natural science. Thus, for example, in cardiology there is a field of objects, such as heart patients, and there is a distinction between the meaning of, say, angina and its expression in a medical interpretation of a heart patient. To be sure, Taylor believes that the meaning of a social practice is for or by a subject and that this distinguishes the human from the natural sciences.

However, it is hard to see what relevance this has to the issue of causality. The argument from the premise that X has meaning for social actor A to the conclusion that X is not a cause of A's action or that X is not causally related to factors Y and Z that are closely connected with the meaning of X is invalid. Let us suppose that the meaning of angina is not for or by a heart patient, whereas the meaning of shame is for or by people who engage in certain social practices involving shame. It does not follow that the feeling of shame is not causally related to other items that are part of the field of meaning of shame or that in order to understand shame one would not have to understand these causal relations.

When one sees how Taylor in fact talks about the field of concepts of shame, it is clear that he implicitly assumes causal notions:

> An emotion term like "shame," for instance, essentially refers us to a certain kind of situation, the "shameful," or "humiliating," and a certain mode of response, that of hiding oneself, of covering up, or else "wiping out" the blot. That is, it is essential to this feeling's identification as shame that it be related to this situation and give rise to this type of disposition. But this situation in its turn can only be identified in relation to the feelings it provokes; and the disposition is to a goal that can similarly not be understood without reference to the feeling experience.[19]

This passage fairly bristles with causal concepts. Taylor assumes that the feeling of shame is brought about because of some humiliating situation, that is, it is caused by this situation; and this feeling brings about, that is, causes, a certain disposition to hide oneself; and so on. Thus, the field of concepts of shame involves causal considerations. This is hardly surprising since the concept of shame is found in theories of commonsense psychology that are implicitly causal. When psychologists such as Freud expand and explain the insights of commonsense psychology, the connection of causality and shame becomes explicit. For Freud, shame is causally connected to sexual inhibitions and to certain forms of sexual perversion.[20]

We can conclude that although Taylor might be justified in his attempt to exclude causality from social science, he has provided no good reasons to do so.

Objective Validation and the Hermeneutic Circle

According to Taylor, just as one should base one's understanding of texts on interpretations, so one should base one's understanding of societies and cultures on interpretations. But how can one validate these interpretations? Taylor says that the process of validation "cannot but move in a hermeneutical circle. A given reading of the intersubjective meaning of a society, or of a given institution or practice, may seem well founded, because it makes sense of these practices or the development of that society. But the conviction that it does make sense of this history itself is founded on further related readings."[21] This circularity of interpreting a text can be construed either in terms of passages or whole-part relations. In interpreting a text, we try to establish a certain reading of one passage by appealing to our reading of other passages. But to accept this reading of the other passages, we have to accept our reading of the first. Again, we try to establish an interpretation of a whole text by appealing to some part, but our interpretation of this part is dependent on our interpretation of the whole. Our understanding of a society is circular in an analogous way. One can construe the circularity either in respect to aspects of society or whole-part relations. An interpretation of one aspect A_1 of, for example, the political process of a society at a particular time is dependent on our understanding of another aspect A_2 at that time. But our understanding of aspect A_2 is dependent on our understanding of A_1. Similarly, to understand some part of a political culture, we must have some understanding of the whole; however, in order to understand the whole, we must understand the part.

Given this construal of interpretive process, it is small wonder that Taylor rejects appeals to rational argument to decide conflicts in interpretation and instead relies on intuition. He argues: "For it means that this is not a study in which anyone can engage, regardless of their level of insight; that some claims of the form 'If you don't understand, then your intuitions are at fault, are blind or inadequate,' some claims of this form will be justified; that some differences will be nonarbitrable by further evidence, but that each side can only make appeal to deeper insights on the part of the other."[22] But since different people's intuitions are based on different ways of life and value options, Taylor argues, "In the sciences of man insofar as they are hermeneutical there can be a valid response to 'I don't understand', which takes the form of, not only 'develop your intuitions,' but more radically 'change yourself.' This puts an end to any aspiration to a value-free or 'ideologyfree' science of man."[23] The clear implication of this approach—one that Taylor does not seem to shrink from—is that there is no objective way of validating interpretations of social phenomenon, and, hence, interpretations are ultimately based on rationally unsupported intuitions and value decisions.

Is Taylor correct? In what follows I will first show that there is a problem analogous to the hermeneutic circle in the natural sciences but that has not prevented natural scientists from objectively testing their theories. Next, I will maintain that the same constraints and checks that natural scientists use to test their theories objectively can be applied to the interpretation of texts, social institutions, and practices.[24] Finally, I will maintain that some methodologists of hermeneutics have argued for an objective approach to interpretation and that Taylor makes no attempt to refute their position.

Taylor does not suppose that circularity exists in the verification process of the natural sciences. Indeed, he makes a distinction between the social sciences and the natural sciences precisely on this ground: natural science can appeal to some neutral observational basis—what he calls "brute data"—but interpretive social science cannot. Taylor has a naive view of natural science practice. A circularity similar to that he pointed out for the social sciences has been alleged to exist in the natural sciences. Writing well before Taylor's paper appeared, Thomas Kuhn and other philosophers of science argued that observation in the natural sciences is theory laden and can provide only specious support for the theory it supposedly tests. This problem has led some philosophers of science to claim that theory acceptance in the natural sciences is based on irrational factors.

Once one takes the theory-laden nature of observation in the natural sciences into account, the difference cannot be made that Taylor alleges between the validation of interpretations in hermeneutics and the testing of hypotheses in the natural science. Moreover, since the theory-laden nature of observation in the natural sciences does not make objective testing impossible,[25] there is reason to suppose that objective validation might not be impossible in textual interpretations either.

In the case of natural sciences, one must distinguish between the categories in which observations are couched and the observational reports themselves. For example, the observation report, "An electron passed through the cloud chamber," is couched in the theoretic categories of "electron" and "cloud chamber." Let us grant, then, that since observational reports must be made in the categories of some theory or other, they are theory laden in this sense. But the existence of what one might call theoretical category influence does not mean that a theory predetermines the particular observational report that is made. For example, the theories of physics may predetermine that observational reports made in the context of physical experiments will be couched in the terms of physical theories. But this does not mean that someone who views an experiment in terms of the categories of a physical theory will observe that an electron is passing through a cloud chamber at some particular time, even if this is what the theories predict. An observational report, although stated in terms of the categories of a particular theory, can conflict with the premises of that theory. But then the existence of theoretical category influence does not necessarily undermine scientific objectivity. It does not mean that the support for a theory which an observational report provides is circular.

The theoretical dependency of observational reports would seem to be a more serious matter for scientific objectivity than the categories they are couched in, for if what is accepted as a true observational report is dependent on the observer's theoretical commitment to the premises of the theory, how can observation be an independent standard that can be appealed to in testing a theory? In this case, theoretical commitment would determine not only the categories of an observation report but the report itself. It is necessary, however, to distinguish several different theses about what I will call theoretical premise influence:

1. Theoretical premise influence is strong in some cases of observation.
2. Theoretical premise influence is strong in all cases of observations.
3. Strong theoretical premise influence, when it is present, can be detected and overcome.
4. Strong theoretical premise influence, when it is present, cannot be detected and overcome.

The history of science and psychological experiments show that thesis (1) is true. The history of science shows that negative evidence in relation to some theory is often not recognized as such by scientists because of their prior theoretical commitments. An analogue to this has been demonstrated in the well-known experiment of Bruner and Postman in which subjects were asked to identify a series of playing cards in short exposure.[26] Subjects find it easy to identify the normal cards but difficult to identify abnormal cards as abnormal—for example, a black four of hearts. The tendency was to identify these as normal—for example, the black four of hearts as a black four of spades. Only after a long exposure were the abnormal cards correctly identified as abnormal by the majority of subjects, but some of the subjects were unable to make correct identifications even after repeated exposures.

There is no reason at all, however, to suppose that (2) is true, for if it were, advocates of a theory would never make observational reports that were in conflict with it; a scientist's observational report would never be a shock to his or her expectations. But they sometimes are. Furthermore, even if (2) were true, this by itself would not undermine the objectivity of scientific theory testing by observation. Supposing that all observational reports were strongly influenced by the theoretical commitments of the observer, such theoretical influence might be detected and overcome. There is good reason to suppose that (4) is false, hence that (3) is true, for there do seem to be means of detecting and correcting premise influence. Certainly, for most subjects in the Bruner and Postman experiment, repeated exposure to an incongruity was enough to bring it to the light.

In addition to there being evidence that natural scientists are sometimes capable of recognizing observations that conflict with their theoretical beliefs, there are specific ways in which empirical observation can function as a constraint and a check on theory.[27] Suppose a doctor claims to observe that a patient has disease D—for example, measles—and that the observation is based on an alleged connection C between property D and symptom S. The extent to which this observation can function as a constraint on theory will depend on how justified we are in relying on C. This, in turn, might depend on whether our belief in C can be regarded as free from doubt given the context of inquiry in which C was tested.[28] Or it might be a function of the directness of the connection between D and S. Other things being equal, a connection that one establishes by long and complex causal chains gives less security than one that is short and simple.[29] Or it might turn on the type of connection. The ideal case would be one in which our background knowledge warrants our believing that C has a biconditional form linking the bases for the observation to what one supposedly observes; in the present case, $(x)\ (Dx \leftrightarrow Sx)$. Then, given that a patient has S, one could deduce that it has D. Weaker logical relations would, of course, give less security.

At least two other factors affect the ways in which theory-laden observation can constrain and check theory.[30] One is the independence of observation from the theory under test. Suppose that a geologist claims to observe that a rock was formed by a glacier and uses this observation to test some theory. If the geologist bases this observation entirely on theories that are not under test in the present context, the observation, although based on a theory, is independent of the theory under test and thus functions as a constraint on it.

Another factor is the convergence of independent and diverse sources of evidence that establish different but connected properties of an entity.[31] Suppose a doctor claims to observe that patient X has disease D that consists of interconnected properties $P_1, P_2, \ldots, P_n$ and that her observations are based on heterogeneous and independent sources of theory-laden evidence $E_1, E_2, \ldots, E_n$. One supposes that it would be improbable in the extreme—indeed, it would be a virtual miracle—that these independent sources would converge in this way if this patient did not have D. In such cases, too, the observation would be a constraint and check on theory.

Might the approaches just outlined be relevant to the validation of interpretations?[32] Certainly there is a distinction to be drawn between the categories in which the interpretation of a text is couched and the statements of the interpretation itself. For example, on a Freudian interpretation, one would describe Hamlet's actions in terms of, say, an Oedipus complex.[33] But unless psychoanalysis is an irrefutable theory, this does not mean that someone who accepts Freudian theory will necessarily interpret a given passage as being in accord with the premises of the theory. For example, a person might say, "In this scene with his mother, Hamlet does not manifest any Oedipal reaction," although this sentence might be in conflict with the premises of Freudian theory. A Freudian interpretation of Hamlet would entail that in this sort of situation, Hamlet would manifest an Oedipal reaction.

The theoretical dependency of interpretive statements would seem to raise more serious matters for the objectivity of interpretations than the theoretical dependency of categories they are couched in. Suppose one accepts as a particular interpretive statement, for example, "Hamlet is manifesting an Oedipal reaction in this scene," and that this is dependent on the interpreter's theoretical commitment to the premises of the theory. How can such a particular interpretation serve as independent grounds for accepting the Freudian interpretation of the whole play? Theoretical commitment in this case would determine not only the categories used in an interpreter's particular judgment but the judgment itself.

Just as experience shows that in the context of science premise influence sometimes is very strong, it shows that premise influence sometimes is very strong in hermeneutic contexts. People committed to a particular theory sometimes do not recognize the negative textual evidence relative to an interpretation. But equally, experience shows that it is not strong in all cases, for if it were, a person committed to an interpretation of a whole text would never make particular judgments that were not in accord with it; the interpretation of a particular passage would never be a shock to the interpreter's expectations.

Further, even if premise influence were strong in all cases, this would not undermine the objectivity of interpretations. Sometimes one can detect premise influence and overcome it. Literary critics who are strongly committed to some particular interpretation sometimes change their minds when their interpretation is exposed to criticism and/or they flounder in their attempts to justify it; they acknowledge that their interpretation was biased or one-sided or that a key passage can be looked at in a different and more adequate way.

In addition, the same sorts of consideration that enable theory-laden observation to be a check and restraint on theory can operate to allow interpretations of particular passages to be checks and constraints on interpretations of whole texts. The interpretation of a passage in which Hamlet allegedly shows Oedipal behavior O might be based on an alleged connection C between O and behavior B. Whether C is acceptable would depend on factors similar to those considered in natural science contexts: freedom from doubt of the background knowledge on which C is based, the length and complexity of C, and the specific nature of C—for example, whether it is unique and deterministic. The acceptability of C could also be based on two independent factors analogous to those that are relevant in natural science contexts. A reading of a particular passage, although based on some interpretation, might not be based on the particular interpretation that is being validated by this reading. If so, then this reading could serve as an independent check on the interpretation that is being validated. Or the connected properties that constitute a particular interpretation of the whole text might be based on variety of independent and heterogeneous sources, such as letters by the author, literary works by other authors, or literary conventions of the period whose chance convergence would be highly unlikely.

Everything said so far concerning literary interpretation applies to interpretation in the social sciences. Theoretical category influence will determine how one categorizes social scientific data, but it would not entail that someone who accepts an interpretation will always construe the data in terms of the premises of the theory. On the other hand, there is no a priori reason to suppose that theoretical premise influence is always strong or that, even when it is, one cannot overcome it. Furthermore, there is a posteriori reason to suppose that it is not always strong and that when it is, one can sometimes overcome it. Social scientists no less than literary critics sometimes find evidence against their interpretations and as a result have to reject or modify them.

Although Taylor is correct that there are no brute data in social science and that observations are based on interpretations, observation in the social sciences, as in the natural sciences, can sometimes function as a constraint and a check on theory. Suppose an archaeologist claims to observe that an artifact A belongs to culture C, which existed ten thousand years ago. Let us admit that this observation is an interpretation and is based on an alleged connection N between physical property of the artifact and process used in making artifacts in C. The extent that this interpretation can function to validate objectively another interpretation will depend on the same sorts of factors as those considered above.

The two independence factors are also relevant. Suppose that the archaeologist claims to observe that an artifact A belongs to culture C is used to validate theory T_1. If the archaeologist's observation is based entirely on theory T_2 that is not under test in the present context, then her observation, although based on a theory, is independent of T_1, the theory under test. Suppose, however, that the archaeologist claims to observe that a past culture C has properties $P_1, P_2, \ldots, P_n$ that make up a coherent and interconnected whole, and her observation or interpretation is based on heterogeneous and independent sources of theory-laden evidence $E_1, E_2, \ldots, E_n$. It would be improbable in the extreme that these independent sources would converge in this way if this coherent whole did not actually exist, that is, if her interpretation of C was not true.

This analogy between the objective testing of theories in natural science, despite the theory-laden nature of observation, and the objective validation of interpretations in hermeneutic disciplines, despite the hermeneutic circle, was drawn explicitly by two important methodologists of hermeneutics before the publication of Taylor's article. Taylor does not cite them, let alone attempt to rebut their arguments.

In 1962 Emilio Betti argued that hermeneutic knowing involves public objects that are constructions of the human mind and that the job of the interpreter is to grasp or reconstruct the ideas and intentions expressed in these objects.[34] Consequently, he opposed what he called the subjective approach of Gadamer in which there is a "fusion of horizons" between the text and the interpreter.[35]

Also rejecting the subjective approach of Gadamer,[36] in 1967 E. D. Hirsch[37] maintained that the validation of an interpretation is simply an application of the hypothetico-deductive method.[38] Although one cannot validate any interpretation with certainty, one can often eliminate alternatives in the light of the evidence. Starting with an initial hypothesis with low probability about the meaning of a text, the interpreter proposes and tests new hypotheses as more evidence is obtained. One determines probability judgments primarily used in this testing on the basis of three criteria: the narrowness of the class, the number of members of the class, and the frequency of the trait among those members. For example, suppose that the meaning of particular phrase in a nineteenth-century English novel is unclear to an interpreter. He or she might argue that the dozen or so nineteenth-century English novelists so far examined used this phrase in a certain sense about 70 percent of the time. Consequently, it is probable that this novelist did as well. This hypothesis could then be strengthened by narrowing the reference class. For instance, it might be possible to establish that this particular novelist always used the phrase in this certain sense in her other novels.

Hirsch argued that one can use genre to evaluate the probability of certain interpretations. For example, the conventions found in a particular literary genre might allow one to argue that it is likely that a symbol means one thing rather than something else. This judgment might, of course, be undermined by further evidence, for example, that this author might have indicated in her correspondence with her sister that she was trying to break out of this genre and was attempting to use the symbol in a new way. But if one could not narrow the reference class in this or similar ways, an interpreter would be justified in supposing that this symbol was used in the standard way found in the genre. This, in turn, would help the interpreter to eliminate alternative interpretations related to the symbol.

Unfortunately, Hirsch linked the possibility of the objectivity of interpretations with the possibility of being able to discern the author's meaning.[39] But there is no necessity in this linkage. The possibility of the objectivity of interpretations should be understood in terms of the falsifiability of interpretations and the ability to support interpretations by probability judgments. Nothing Taylor says undermines the claims of objectivity of literary interpretation in this sense. But if one can use objective considerations in the interpretation of literary texts, as Betti and Hirsch have argued, and the interpretation of society and social practice is analogous to the interpretation of literary texts, as Taylor has suggested, then this seems to weaken still further Taylor's thesis of the fundamental difference between social and natural science. Both the social and the natural sciences can use objective criteria of interpretation.

Nonsubjective Meanings

The alleged impossibility of objective interpretations affects Taylor's views on the ontology of social meaning. Taylor contrasts his interpretive approach to social science, and in particular his approach to political science, with the approach of "the empiricists." The empiricists' approach is based on verification in terms of "brute data," data "whose validity cannot be questioned by another interpretation or reading, data whose credibility cannot be founded or undermined by further reasoning."[40] Unlike natural science, which can be understood in term of brute data, social science cannot.

Consider political science. According to Taylor, mainstream political science identifies political behavior with action that is "brute data identified," that is, identified in terms of physical end states or of institutional rules that are closely tied to some physical end state, such as raising one's hand at an appropriate time when a vote is being taken. One determines the meaning of voting, in contrast to voting behavior, by ascertaining certain facts about the actor's subjective state through such techniques as opinion polls and content analysis. Thus, once political scientists know how people vote (which can be established by observing their behavior in certain institutional contexts), they can determine the subjective bases of their vote—their beliefs, attitudes, and values—by asking them questions and making inferences on the basis of their answers.

Taylor maintains that this construal wrongly makes the meaning of political behavior subjective in that it construes meaning in terms of the beliefs, attitudes, and values of the social actors. This is to overlook intersubjective and common meanings. Consider the concept of bargaining. It is part of a field of concepts that includes entering into negotiations, bargaining in good (bad) faith, and breaking off negotiations, which one uses to interpret social practices in this society but not in some other societies, such as Japanese. It would be a complete misconstrual, according to Taylor, to suppose that the reality that this field of concepts refers to is something that is purely subjective. In fact, our language of bargaining is constitutive of our social reality—our social world. Just as one would not have chess without certain rules that govern the movement of the queen, so one would not have our social reality without certain bargaining rules.

For Taylor, then, there are meanings that are intersubjective, and the language used to express them is constitutive of social reality. But he holds that intersubjective meaning is not the only type of nonsubjective meaning that empiricists have overlooked with their reliance on brute facts. There are also common meanings that provide shared values that are part of our common world and supply the locus for our common aspirations and sense of community. For example, he says that "there is a common meaning in our civilization connected around a vision of the free society in which bargain has a central place."[41] Although this vision of a free society has been challenged from several quarters, it nevertheless furnishes a locus for debate.

What can one say about Taylor's notions of intersubjective and common meaning? Let us suppose that his ontological thesis is correct: one cannot explain these meanings as mere psychological projections of the social actors. It does not follow that other aspects of his claim are correct. As we have seen, Taylor maintains that the nonsubjective meaning that is made clear in an interpretation is always meaning for a subject or subjects. But what exactly does this involve? With respect to intersubjective meaning IM of practice P, does this mean that social actors in P would tend under questioning to agree that IM is constitutive of P? With respect to common meaning CM of P, does this mean that they would tend under questioning to agree that CM provides a sense of shared values and community? Or is there a much more indirect relation between the intersubjective meaning and common meaning of a social practice and the reactions of the actors in the practice?

Whatever the relation between the reaction of social actors and the social meanings that one supposes is for social actors, one must question whether social meaning must always be for social actors. It is possible that the social actors of P might have no understanding or even awareness of the meaning of P that is postulated by a social scientist. Certainly, theories of ideology and false consciousness seem to assume this possibility. In addition, the Weltanschauung of a society as formulated through historical research and understood by scholars might not be comprehensible to a typical member of that society.[42] Since the commonsense categories of the actor may be imprecise and muddled, a reformulation, an explication, or even a replacement of them would certainly seem to be in order, as it is in other scientific fields.

To cite several examples, Tom Burns has shown that in several sociological studies, the explanation given of people's behavior is in sharp contrast to those given by the actors themselves.[43] For example, "Vilhelm Aubert's study of the judiciary in Norway, when it was first published, evoked violent reaction among the legal profession precisely because it pointed to the fact that, in giving sentences, judges appeared to be following a tacit code which contravened the explicit code of equality before the law."[44] Indeed, he argues that sociology has a crucial function: exposing the falsehoods and deceptions of the commonsense ideas of the actors about their own actions. Victor Turner, in his study of Ndembu ritual,[45] introduced theories and concepts based upon psychoanalytic and sociological theory that clearly transcend the primitive thinking of the natives concerning the use of symbols. Lévi-Strauss introduced explanatory notions that go well beyond those of the natives whose behavior he was attempting to understand. This, of course, is not to say that all of these social scientists' explanations are successful. Far from it. Lévi-Strauss's theory, for example, has been severely criticized.[46] But the problems with his work do not stem from the fact that they depart from the commonsense thinking of the natives and do not provide an interpretation of meaning that is for the social actors. On the contrary, the assumption has been that social scientists are perfectly justified in introducing explanatory categories and theories that are quite foreign to the people whose behavior they are trying to explain.[47]

Another obvious difficulty with Taylor's theory is that of reconciling his insistence on an objective ontology of meaning and a subjective epistemology for determining such meaning. Taylor calls both common and intersubjective meanings "nonsubjective meaning" because on this view they exist as part of the objective fabric of the social world and not merely in the minds of the social actors. Nevertheless, as we have seen, he denies that one can establish any judgment about social meaning by objective rational procedures and maintains in the end that disagreements about meaning must come down to conflicting intuitions. Thus, he wants social meanings to have an objective ontological basis but at the same time construes their epistemological basis as subjective. There is nothing inconsistent about this dualism, but a recent historical analogue does suggest some difficulties with it.

Although there are obvious differences,[48] early twentieth-century ethical intuitionism provides an analogy to Taylor's dualism of ontological objectivism and epistemological subjectivism. Intuitionists such as G. E. Moore claimed that ethical properties were part of the objective furniture of the world, although they could be discerned only by intuition. However, since the intuitions of different people conflicted, there seemed to be no rational way of reconciling them. Given this problem, critics argued that there was no good reason to suppose that ethical properties existed in the objective world. It was much more plausible to suppose that the phenomenological objectivity of ethical experience, for example, that the moral obligation of an action seems to impose on one from the outside, could be better explained as a psychological illusion than as an objective ontological fact. If human beings projected their values and subjective ethical views onto the world and wrongly read them as an objective part of it, this could explain why different people have conflicting intuitions about a supposedly objective ethical reality: different people are making different projections.

Can one raise the same sort of objection against Taylor's view? Can the phenomenological objectivity of the meaning of social institutions and practices be better explained as a psychological projection than as an objective ontological fact? It would seem that one could give the same argument against it as against ethical intuitionism. If human beings project their subjective beliefs and goals onto the world and wrongly read them as an objective part of it, this could explain why different people have conflicting intuitions about a supposedly objective realm of meaning: different people are making different projections. In order to avoid the problem Taylor would have to maintain (as we have seen he should have maintained) that the interpretation of meaning is objectively discernible.

The Impossibility of Exact Predictions

Taylor argues that "if the epistemological views underlying the science of interpretation are right," then "exact prediction is radically impossible."[49] He bases his claim on three different reasons.

The first, and least important, reason Taylor gives is the well-known "'open systems' predicament, one shared by human life and meteorology, that we cannot shield a certain domain of human events ... from external interference, it is impossible to delineate a closed system."[50] The second and more fundamental reason is that the human sciences do not admit of the degree of exactitude that is possible in the nature sciences where one can measure the data "to virtually any degree of exactitude." The third and most fundamental reason is that human beings are "self-defining animals." "With changes in self-definitions go changes in what man is, such that he has to be understood in different terms."[51] These conceptual mutations often produce changes that are "incommensurable," that "cannot be defined by a common stratum of expressions." As an example, Taylor cites the allegedly incommensurable notions of bargaining in our society and in some primitive ones: "Each can be glossed in terms of practices, institutions, ideas in each society which have nothing corresponding to them in the other."[52]

Before we evaluate these arguments, one should note that it is unclear what Taylor means by "exact prediction." He admits that different interpretations in the social sciences "may lead to different predictions in some circumstances," so he apparently thinks that some prediction is possible even in an interpretive social science.[53] What he denies is that exact social science prediction is possible. But what precisely is this?

Let us characterize one dimension of predictive exactitude that he might have in mind in terms of the degree of numerical definiteness of the prediction. Thus, one may predict that a certain magnitude (1) will change, (2) will increase, (3) will increase by at least a certain amount, (4) will increase within definite limits, or (5) will increase by a definite amount. Another independent dimension of prediction exactitude is definiteness of the time within which the prediction is supposed to come true. For example, a statement that predicts that a certain magnitude will change next week is less exact than one that specifies that this magnitude will change on Tuesday of next week.[54]

Given these dimensions of exactness, what does Taylor's thesis amount to? On a strong reading, his claim would be that any natural science can make more exact predictions than any social sciences on any of these dimensions of exactness. On a weaker reading, it would be that there are some natural sciences that make more exact predictions than any social sciences on any of these dimensions of prediction. On the weakest reading, he would be saying that there are some natural sciences that can make more exact predictions than some social sciences on any of these three dimensions.

Taylor surely is correct on the weakest reading. For example, physicists can surely make more exact predictions on all dimensions of exactness than clinical psychologists. But this is of little interest if one wants to draw a fundamental distinction between the exactness of prediction in the natural and social sciences. Taylor might also be correct on the less weak reading. Predictions in experimental physics, for example, are probably more exact than in any social science on all dimensions of exactness. But this is compatible with some social science being able to make more exact predictions than some natural sciences on all dimensions of exactness and all social science being able to make rather exact predictions on these dimensions. Thus, it hardly supports Taylor's view that "exact prediction [in the human sciences] is radically impossible."

The only thesis that seems to make any sense in the light of Taylor's own words is the strongest reading that any natural science can make more exact predictions than any social sciences on any of these dimensions of exactness. However, when it is read in this way, Taylor's claim is implausible. In the first place, Taylor admits that meteorology has the same predictive problems as the social sciences. But then he can hardly maintain no social science can make predictions with the same exactness as any natural sciences unless he does not consider meteorology a natural science. In addition, it seems likely that some social sciences can make as exact predictions as some natural sciences. Although political pollsters are often wrong, it is not implausible to suppose that they can make as accurate predictions about who will win an election and by how much as seismologists can predict when and where an earthquake will occur and how severe it will be.

Taylor's second argument does not show that in any natural science more exact prediction can be made than in any social science. It is not obviously true that one can measure the data of the natural sciences to virtually any degree of exactitude. The precision with which a magnitude can be measured is dependent on the technology and the theories available. For example, air pressure was not measurable to any precise degree before both the development of the theory that the earth is submerged in a sea of air and the invention of the mercury barometer.[55] Even today in some of the less well-developed natural sciences, certain magnitudes remain incapable of precise specification. In medical science, for example, the degree of severity of a disease is not capable of being specified with any precision. Doctors speak of a mild or severe case of measles. It would make little sense to say that X's case of measles is 2.3 times as severe as Y's. Furthermore, just because a natural science can measure a magnitude precisely, it does not mean that it can predict changes in magnitude precisely. Meteorology can measure precisely the amount of rain that fell on Boston on any given day last year, but it cannot predict precisely how much rain there will be on any given day. Moreover, some magnitudes in the social sciences seem to be capable of more precise measurement than some in the natural sciences. The degree of subjective probability that Mr. Jones holds a certain belief seems capable of being measured to a higher degree of exactitude in terms of his betting behavior than the severity of his measles.

It is unclear why Taylor believes that his third reason (humans are self-defining animals) provides a justification for believing that exact prediction in the social sciences is impossible. Perhaps his argument is this: since human beings have constructed incommensurable concepts of social class, political party, bargaining, and so on in different times and societies, it is impossible to construct generalizations in terms of these that transcend these times and societies and to predict on the basis of these generalizations.

Some philosophers of science have argued that concepts have changed meaning in the natural sciences, thereby making theories that are apparently about the same thing incommensurable. For example, one might argue that the concept of the sun was different for Tycho Brahe and Kepler and, consequently, that the concept of the sun was different for Tycho Brahe and Kepler and, consequently, that they were talking about different things.[56] So if there is a problem in the social sciences because of incommensurability, it is not unique to this field.

Does the possibility of future conceptual change—for example, concerning the meaning of bargaining—show that prediction about bargaining is impossible? No; it does not. Predictions in both the natural and the social sciences are almost always conditional. At most what Taylor's argument shows, then, is that another condition must be added in the case of the social sciences: the concepts used to predict the future will not change in meaning. Even if he is correct that conceptual change might occur, it does not follow that we cannot predict on the basis of evidence of present bargaining what will happen in the future given the stability of the concept of bargaining that one uses in the prediction.

But even if the concepts used to predict the future do change their meaning, that is, their connotation, this by itself does not mean that prediction is impossible since change in connotation is compatible with constancy of extension. Sameness of extension is sufficient for accurate prediction. For example, suppose that the concept of bargaining changes over the next hundred years. In the twentieth century bargaining has been defined by the properties A, B, C, while in the twenty-first century it will be defined by the properties D, E, F. Let us now suppose that extension of the concept of bargaining has not changed from the twentieth to the twenty-first century. Then it would still be possible to make predictions about twenty-first century bargaining using the twentieth century concept: the concept would still be about the same thing, that is, the same set of entities. In short, change in connotation is compatible with constancy of denotation and this constancy is sufficient for prediction.[57]

Moreover, even if the extension of the concept of bargaining changes from the twentieth to the twenty-first century, it may be possible to predict what will happen in the twenty-first century using laws constructed in twentieth century concepts so long as the changed extensions stand in certain relations.[58] For example, suppose that we have a law L couched in the twentieth-century concept of bargaining B_1 to the effect:

(L) For every x, if x is B_1, then x is P.

Suppose that the extension of bargaining changes in the twenty-first century from B_1 to B_2 such that:

(A) for every x, if x is B_2, then x is B_1.

Then given L and A one could predict that given an instance of bargaining in the twenty-first-century meaning of the term, it would have property P. That is, L and A and

(1) a is B_2

entails:

(2) a is P.

What we have said about prediction we can also say about retrodiction. Change in the connotation or even the extension of certain terms is compatible with retrodiction under certain circumstances. The mere fact that human beings are self-defining animals and change their concepts in itself need be no obstacle to prediction. So long as the change is of a certain kind, prediction is possible.

One might object, however, that the above argument assumes that in order to make predictions about societies at different times, one must have knowledge that the extension of concepts used in the prediction will not change or at least will not change in a certain way. But, one might argue, this is precisely what one cannot know. Human beings might modify concepts in radical ways that make prediction impossible; that is, human beings might change the extension of a concept in such a way that no relation such as that specified by A above exists.

There is no reason to suppose that science would not be able to predict that the extension of certain concepts will change in these ways. One would base this prediction on empirical investigation of conceptual change. Let us call the type of change of extension that is compatible with accurate prediction Change $\sum$. Such an investigation might come up with the inductive generalization of the following kind:

(G_1) Most changes of concepts of type T_1 are in the range of $\sum$.

Given the information that the concept of bargaining is of type T_1, one could infer that probably any change of extension of bargaining would be within $\sum$. This evidence would provide some grounds for thinking that predictions concerning bargaining in the twenty-first century would not simply be conditional on the extensions remaining within $\sum$ and would be accurate if predictions about bargaining in the twentieth century were accurate. Moreover, even if a concept such as bargaining changes beyond $\sum$, this would not mean that social science could not predict political behavior in the twenty-first century based on laws utilizing concepts developed in the twentieth century. To be sure, what social scientists would predict in the twenty-first century would no longer be called "bargaining." However, this might be of no concern to twentieth-century social scientists. Whether it was of concern would depend on the theoretical and practical aims of the predictors.

To illustrate the general point, consider an analogy from meteorology. Suppose for some reason that the extension of the concept of fair weather changes in the twenty-first century. What meteorologists mean by fair weather now is not what they will mean in the year 2091. Let us suppose that these changes are such that generalizations having to do with fair weather in the twentieth century would not apply to fair weather in the twenty-first century. One might suppose that this would not greatly affect the practical importance of meteorology's ability to predict fair weather in twentieth-century terms for the twenty-first century. Most of the practical decisions connected with weather in the next century that might have to be made—for example, those connected with long-term farming programs, city planning, and scientific research—would probably be unaffected: twentieth-century concepts would suffice. It is not out of the question that something similar might be true in the social sciences. A radical change in the extension of the concept of bargaining might have very little practical effect on what twentieth-century political scientists are trying to accomplish by their predictions of twenty-first-century political life. To suppose otherwise might be to pay too much importance to social actors' own concepts and not enough importance to the theoretical autonomy of scientists.

Conclusion

Despite the importance of Taylor's article, the interpretive approach that he presents has serious problems. First, the scope of the social sciences as he conceives of them is too narrow and excludes important questions that social science has been and should be concerned with. In particular, it excludes causal questions that any social science should deal with. Moreover, even if one limits one's concerns to the interpretation of meaning, causality is relevant since interpreting the meaning of a social practice often involves understanding the connections between causal notions. Second, despite what Taylor says, objective interpretations seem possible. Even if there is a hermeneutic circle, this does not rule out the use of evidence from either falsifying or supporting interpretations by probabilistic reasoning nor does it exclude the placing of objective constraints on interpretations. Third, Taylor's theory is restrictive in still another way: it bars social scientists from using theoretical concepts that would go beyond the understanding of social actors. Finally, Taylor fails to show that one cannot make exact predictions in the social sciences. In particular, the self-defining nature of human beings is compatible with exact predictions.

Notes

1. See, for example, David Braybrooke, *Philosophy of Social Science* (Englewood Cliffs, N.J.: Prentice-Hall, 1987), p. 3; David Thomas, *Naturalism and Social Science* (London and New York: Cambridge University Press, 1979), p. 23; Daniel Little, *Varieties of Social Explanation* (Boulder, Colo.: Westview Press, 1991), p. 68.
2. For example, Taylor's paper is reprinted in Paul Rabinow and William M. Sullivan, (eds.), *Interpretative Social Science: A Second Look* (Berkeley and Los Angeles: University of California Press, 1987), Fred R. Dallmayr and Thomas A. McCarthy (eds.), *Understanding and Social Inquiry* (Notre Dame: Notre Dame University Press, 1977), and Eric Bredo and Walter Feinberg (eds.), *Knowledge and Values in Social and Educational Research* (Philadelphia: Temple University Press, 1982). It appears as chapter 13 in this book.
3. This paper was originally published in *Review of Metaphysics* 25 (1971): 3–51.
4. Braybooke, *Philosophy of Social Science.*
5. Taylor, "Interpretation and the Sciences of Man," in *Interpretative Social Science: A Second Look*, p. 33.
6. Ibid., p. 34.
7. Ibid., p. 41.
8. Ibid., p. 43.
9. See, for example, Janice Delaney, *The Curse* (New York: E. P. Dutton, 1976).

10. Marla N. Powers, "Menstruation and Reproduction: An Oglala Case," *Signs* 6 (1980): 54–65.
11. See Arnold Van Gennep, *The Rites of Passage* (Chicago: University of Chicago Press, 1960).
12. See Victor Turner, *The Ritual Process* (Chicago: Aldine Publishing Co., 1969).
13. Powers, "Menstruation and Reproduction," p. 65.
14. See, for example, H. G. Gough, "Personality Factors Related to Reported Severity of Menstrual Distress," *Journal of Abnormal Psychology* 84 (1975) 59–65; P. Slade and F. A. Jenner, "Attitudes to Female Roles, Aspects of Menstruation, and Complaining about Menstrual Symptoms," *British Journal of Social and Clinical Psychology* 19 (1980): 109–113.
15. See Paula Weideger, *Menstruation and Menopause* (New York: Alfred A. Knopf, 1976), chap. 4.
16. Taylor, "Interpretation and the Sciences of Man," p. 63.
17. In criticizing Taylor's apparent rejection of the function approach, I do not wish to deny that some functional explanations of social phenomena are problematic. However, which functional accounts are problematic cannot be determined a priori. Little, *Varieties of Social Explanations*, pp. 91–102.
18. Taylor, "Interpretation and the Sciences of Man," p. 64.
19. Ibid., pp. 42–43.
20. See Sigmund Freud, "The Sexual Aberrations" in *The Basic Writings of Sigmund Freud*, ed. A. A. Brill (New York: Modern Library, 1938).
21. Taylor, "Interpretation and the Sciences of Man," p. 75.
22. Ibid., p. 76.
23. Ibid., p. 77.
24. For a defense of the objectivity of interpretation along somewhat different lines, see James Bohman, *New Philosophy of Social Science* (Cambridge: MIT Press, 1991), chap. 3.
25. See Michael Martin, *Concepts of Science Education* (Chicago: Scott-Foresman, 1972), pp. 116–121. My argument here is based on the analysis of Israel Scheffler, *Science and Subjectivity* (Indianapolis: Bobbs-Merrill, 1967).
26. Jerome Bruner and Leo Postman, "On the Perception of Incongruity: A Paradigm," *Journal of Personality* 18 (1949): 206–223.
27. I am indebted here to Alison Wylie's discussion in "Evidential Constraints: Pragmatic Objectivism in Archaeology," chapter 48 in this book.
28. On a similar point connected with theory-laden observation in the natural science, see Dudley Shapere, "Observation and the Scientific Enterprise," in *Observation, Experiment and Hypothesis in Modern Physical Science*, ed. P. Achinstein and O. Hannaway (Cambridge: MIT Press, 1985), p. 29, cited by Wylie, "Evidential Constraints."
29. On a similar point connected with the natural sciences, see Peter Kosso, "Dimensions of Observability," *British Journal for the Philosophy of Science* 39 (1988): p. 445, cited by Wylie, "Evidential Constraints."
30. On an analogous consideration in the natural sciences, see Ian Hacking, *Representing and Intervening* (Cambridge: Cambridge University Press, 1983), pp. 183–185, and Kosso, "Dimensions of Observability," p. 456, cited by Wylie, "Evidential Constraints."
31. See Peter Kosso, "Science and Objectivity," *Journal of Philosophy* 86 (1989): 247 On a similar point in the natural sciences, see Wylie, "Evidential Constraints."
32. For a defense of the thesis that hermeneutics is simply the application of the method of hypothesis applied to meaningful material, see Dagfinn Føllesdal, "Hermeneutics and the Hypothetico-Deductive Method," reprinted as chapter 15 in this book.
33. See Ernest Jones, *Hamlet and Oedipus* (Garden City, N.Y.: Doubleday, 1955).
34. Emilio Betti, "Hermeneutics as the General Methodology of the Gestewissentschaften" (1962), in Josef Bleicher, trans., *Contemporary Hermeneutics* (London: Routledge and Kegan Paul, 1980), pp. 51–94.
35. Ibid., p. 79.
36. See E. D. Hirsch, Jr., "Gadamer's Theory of Interpretation," *Review of Metaphysics* (March 1965). Reprinted in Hirsch, *Validation in Interpretation* (New Haven: Yale University Press, 1967), appendix II.
37. Hirsch, *Validation in Interpretation.*
38. Ibid., p. 264.
39. See the critique of Hirsch's appeal to the author's intention by David Couzens Hoy in *The Critical Circle* (Los Angeles: University of California Press, 1982), pp. 11–40.
40. Taylor, "Interpretation and the Sciences of Man," p. 38.
41. Ibid., p. 61.

42. See Thomas, *Naturalism and Social Science*, pp. 96–97.
43. Tom Burns, "Sociological Explanations," in Dorothy Emmet and Alasdair MacIntyre (eds.), *Sociological Theory and Philosophical Analysis* (London: MacMillan, 1970), pp. 55–76.
44. Ibid., p. 64.
45. Turner, "Symbols in Ndembu Ritual," in Emmet and MacIntyre, *Sociological Theory*, pp. 150–182.
46. For criticisms of Lévi-Strauss, see Edmund Leach, "Telstar and the Aborigines or Ia Pensée Sauvage," in Emmet and MacIntyre, *Sociological Theory*, pp. 183–203, and Peter Worsley, "Groote Eylandt Totemism and Le Totémisme Aujour d'hui," ibid., pp. 204–23.
47. For a similar point, see Jane Roland Martin, "Another Look at the Doctrine of *Verstehen*," *British Journal for the Philosophy of Science* 20 (1969): 53–67, reprinted as chapter 16 in this book. See also Jane Roland Martin, *Explaining, Understanding and Teaching* (New York: McGraw-Hill, 1970).
48. There are, of course, two basic differences between Taylor's position and the intuitionists'. First, ethical intuitionists claimed that ethical properties were not culturally relative. Taylor claims that meanings are relative and vary from culture to culture. Second, Taylor advocates intuition when interpretation in its usual sense fails to convince someone who does not accept your interpretation. Ethical intuitionists relied on intuition right from the start.
49. Taylor, "Interpretation and the Sciences of Man," p. 78.
50. Ibid.
51. Ibid., p. 79.
52. Ibid.
53. Ibid., p. 78.
54. See Fritz Machlup, "Are the Social Sciences Really Inferior?" in Maurice Natanson (ed.), *Philosophy of Social Science* (New York: Random House, 1961), p. 172. This paper is reprinted as chapter 1 in this book.
55. See C. G. Hempel, *The Philosophy of the Natural Science* (Englewood Cliffs, N.J.: Prentice-Hall, 1966), pp. 28–29.
56. See N. R. Hanson, *Patterns of Discovery* (New York: Cambridge University Press, 1958), p. 5.
57. This argument was first explicitly formulated by Scheffler, *Science and Sujectivity*, chap. 3.
58. See Michael Martin, "Referential Variance and Scientific Objectivity," *British Journal for the Philosophy of Science* 22 (1971): 17–26.

PART IV
Rationality

Introduction to Part IV

Social scientists make inferences in two directions: from social actors' beliefs and desires to their actions and from their actions to their beliefs and desires. An assumption of rationality guides the inferences in both cases. Thus, an economist will infer that social actors who want to maximize their profits and who believe that taking action A will achieve this will take action A. An anthropologist who observes a medicine man giving a sick child an herbal remedy will infer that he desires to cure the child and that he believes that this remedy brings about a cure. This rationality assumption raises a number of questions, however. What does "rationality" mean? What is the status of the presumption of rationality? How does rationality enter into social science explanations? What are the limitations of explanations that use rationality? These issues are addressed by Steven Lukes, Dagfinn Føllesdal, Jon Elster, and David Henderson in the selections that follow.

In "Some Problems about Rationality" (chapter 18), Steven Lukes surveys five different positions on the issue of what the attitude of an anthropologist should be toward a set of native beliefs that appears prima facie irrational: the theory of Firth and Beattie that seeming irrationality should be interpreted symbolically; the view of Best that there are universal criteria of rationality that, when applied to native beliefs, show them to be incomprehensible; the position of Horton that native magico-religious beliefs are primitive and unsuccessful scientific attempts to explain natural phenomena; the view of Lévy-Bruhl that native beliefs, at least in terms of our standards, are mystical and prelogical; the theory of Winch that rationality is context dependent and that judged in their own terms native beliefs are as rational as ours. After distinguishing different senses of "irrational," Lukes maintains that some criteria of rationality are universal—for example, those for distinguishing truth and falsehood—while others are context dependent—for example, those for deciding which beliefs are appropriate in a particular context. Using these ideas, he analyzes and evaluates the five positions he surveyed.

After distinguishing different senses of "rationality"—as logical consistency, as well-founded belief, as well-founded values, as rational action—Dagfinn Føllesdal, in "The Status of Rationality Assumptions in Interpretation and in the Explanation of Action" (chapter 19), argues that the assumption of rationality plays an important role in the interpretation and explanation of action. He holds that since one person never has direct access to another's beliefs and values, one attributes beliefs and values to someone else on the assumption of his or her rationality. But what is the status of this assumption? It has been regarded as necessary by Davidson and Dray, as empirical by Hempel, as superfluous by Popper, and as false by Donagan. Although Føllesdal argues that rationality is constitutive of belief, desire, and action and thus, that it is necessary, he maintains that one need not assume that human beings are perfectly rational. Saying that even where purely causal factors would be sufficient for an explanation one should include the reasons for an action in any explanation of it, he insists that this does not mean that one should try to ascribe beliefs and values that make a person's action as rational as possible. Humans are not always rational, according to Føllesdal, yet they have a "second-order disposition" toward rationality; that is, they are inclined to "mend" their ways when their own lack of rationality is pointed out to them in terms they can understand.

Addressing the question of how rational-choice explanations explain, Jon Elster, in "The Nature and Scope of Rational-Choice Explanation" (chapter 20), initially considers the basic elements of intentional explanations. He argues that in such explanations, one must make three assumptions: first, that given an actor's beliefs, some behavior is the best way to achieve his or her desires; second, that the actor's desires and beliefs cause the behavior; and third, that the causal desires and beliefs are the reasons for the action. Since rational-choice explanations put restrictions on beliefs that play a role in such explanations—for example, the beliefs must be consistent and be caused by the available evidence—they go beyond intentional explanations. But are rational-choice explanations always possible? No, according to Elster. He maintains that there is often no uniquely rational way to accomplish the agent's goal. In some cases, there are multiple optimal behaviors and in others no optimal behavior at all. Given the indeterminacy of rational-choice explanations, he argues that they must be supplemented with causal accounts. Moreover, rational-choice explanations may fail because people engage in wishful thinking, succumb to weakness of the will, and have inconsistent beliefs and desires; in other words, because they are irrational.

In "The Principle of Charity and the Problem of Irrationality (Translation and the Problem of Rationality)" (chapter 21), David Henderson argues that common formulations of the principle of charity—the maxim that any translation is mistaken if it entails that the speaker is uttering a contradiction or committing a logical error—prevents one from making attributions of irrationality to the speaker. He resolves this problem by distinguishing two complementary views of the principle of charity. He proposes, on the one hand, that it be considered as a preparatory stance in which a first-approximation translation manual is constructed. In later stages of manual construction, when the translation is refined and developed, the principle of charity is not constraining, and the translator can, given relevant evidence, attribute irrational beliefs to the speaker. On the other hand, Henderson considers the "weighted" principle of charity, a maxim that would lead a translator in constructing a first-approximation translation manual to construe the speaker "as commonly correct in cases where correct judgment and reasoning is likeliest on empirical grounds." He argues that this principle is reducible to a special case of "the principle of explicability," a maxim directing translators to attribute explicable beliefs and practices to speakers, not necessarily rational beliefs and practices.

Chapter 18

Some Problems about Rationality

Steven Lukes

In what follows I shall discuss a philosophical problem arising out of the practice of anthropologists and sociologists which may be stated, in a general and unanalyzed form, as follows: when I come across a set of beliefs which appear prima facie irrational, what should be my attitude toward them? Should I adopt a critical attitude, taking it as a fact about the beliefs that they *are* irrational, and seek to explain how they came to be held, how they manage to survive unprofaned by rational criticism, what their consequences are, etc? Or should I treat such beliefs charitably: should I begin from the assumption that what appears to me to be irrational may be interpreted as rational when fully understood in its context? More briefly, the problem comes down to whether there are alternative standards of rationality.

There are, of course, a number of different issues latent in the problem as I have stated it. In particular, it will be necessary to distinguish between the different ways in which beliefs may be said to be irrational. There are, for example, important differences and asymmetries between falsehood, inconsistency, and nonsense. Also there are different sorts of belief; indeed there are difficult problems about what is to count as a belief. Let us, however, leave the analysis of the problem until a later stage in the argument.

First, I shall set out a number of different answers to it that have been offered by anthropologists and philosophers with respect to primitive magical and religious beliefs. In doing so I make no claim to comprehensiveness. These and related issues have been widely debated throughout the history of anthropology; all I aim to do here is to compare a number of characteristic positions. It is, however, worth stressing at this point that I do not pose the problem as a problem *in* anthropology but rather as a philosophical problem[1] raised in a particularly acute form by the practice of anthropology. It is raised, though in a less clearcut form, by all sociological and historical inquiry that is concerned with beliefs.

Second, I shall try to separate out a number of distinct criteria of rationality which almost all discussions of these issues have confused. Finally, I shall make some attempt at showing which of these criteria are context dependent and which are universal, and why.

I

Let us compare for plausibility five different answers to the problem.

I. First, there is the view that the seeming irrationality of the beliefs involved in primitive religion and magic constitutes no problem, for those beliefs are to be interpreted as *symbolic*. Take, for instance, the following passages from Dr. Leach:

> ... A very large part of the anthropological literature on religion concerns itself almost wholly with a discussion of the content of belief and of the rationality or otherwise of that content. Most such arguments seem to me to be scholastic nonsense. As I see it, myth regarded as a statement in words "says" the same thing as ritual regarded as a statement in action. To ask questions about the content of belief which are not contained in the content of ritual is nonsense.
> ... In parts of this book I shall make frequent reference to Kachin mythology but I shall make no attempts to find any logical coherence in the myths to which I refer. Myths for me are simply one way of describing certain types of human behavior.[2]

And again,

> ... The various nats of Kachin religious ideology are, in the last analysis, nothing more than ways of describing the formal relationships that exist between real persons and real groups in ordinary human Kachin society.
> The gods denote the good relationships which carry honor and respect, the spooks and the witches denote the bad relationships of jealousy, malice and suspicion. Witchcraft becomes manifest when the moral constraints of the ideally correct social order lose their force.[3]

Professor Firth argues, in a similar fashion, that judgment about the rationality of beliefs is irrelevant to the purposes of the anthropologist. It is, he writes, "not important for an anthropological study whether witches exist or not ... we are dealing here only with human relations...."[4] Religious experience

> is essentially a product of human problems, dispositions and relationships.... In its own rather different way it is to some extent an alternative to art, symbolizing and attributing value to human existence and human endeavor.... At the level of human dilemma, creative activity and symbolic imagery, indeed, religious concepts and values can be taken as real; they are true in their context. With the claim that their basic postulates have an autonomous, absolute validity I do not agree. But to us anthropologists the important thing is their *affirmation* of their autonomy, their validity, their truth—not the metaphysical question whether they are correct in saying so. Basically, in an anthropological study of religion, as in studies of art, we are concerned with the relevance of such affirmations rather than with their ultimate validity.[5]

The most systematic recent statement of this position is by Dr. Beattie.[6] According to Beattie, beliefs associated with ritual are essentially expressive and symbolic. Thus, "[f]or the magician, as for the artist, the basic question is not whether his ritual is true in the sense of corresponding exactly with some empirically ascertainable reality, but rather whether it says, in apt symbolic language, what it is sought, and held important, to say."[7] More generally,

> Although not all of what we used to call "primitive" thought is mystical and symbolic, some is, just as some—though less—of "western" thought is. If it is "explanatory," it is so in a very different way from science. Thus it requires its own distinct kind of analysis. No sensible person subjects a sonnet or a sonata to the same kind of examination and testing as he does a scientific hypothesis, even though each contains its own kind of "truth." Likewise, the sensible student of

> myth, magic and religion will, I think, be well advised to recognize that their tenets are not scientific propositions, based on experience and on a belief in the uniformity of nature, and that they cannot be adequately understood as if they were. Rather, as symbolic statements, they are to be understood by a delicate investigation of the levels and varieties of meaning which they have for their practitioners, by eliciting, through comparative and contextual study, the principles of association in terms of which they are articulated, and by investigating the kinds of symbolic classifications which they imply.[8]

Thus the first answer to our problem amounts to the refusal to answer it, on the grounds that it is nonsensical (Leach), or irrelevant (Firth), or misdirected (Beattie).[9]

2. The second answer to the problem comes down to the claim that there are certain criteria which we can apply both to modern and to primitive beliefs which show the latter to be quite incomprehensible. (I leave until later the question of whether this claim is itself intelligible.)

As an example, take the following passage from Elsdon Best:

> The mentality of the Maori is of an intensely mystical nature.... We hear of many singular theories about Maori beliefs and Maori thought, but the truth is that we do not understand either, and, what is more, we never shall. We shall never know the inwardness of the native mind. For that would mean tracing our steps, for many centuries, back into the dim past, far back to the time when we also possessed the mind of primitive man. And the gates have long closed on that hidden road.[10]

A similar view was expressed by the Seligmans about the tribes of the Pagan Sudan:

> On this subject [of magic] the black man and the white regard each other with amazement: each considers the behavior of the other incomprehensible, totally unrelated to everyday experience, and entirely disregarding the known laws of cause and effect.[11]

3. The third answer amounts to the hypothesis that primitive magical and religious beliefs are attempted explanations of phenomena. This involves the claim that they satisfy certain given criteria of rationality by virtue of certain rational precedures of thought and observation being followed; on the other hand they are (more or less) mistaken and to be judged as (more or less) unsuccessful explanations against the canons of science (and modern common sense).

The classical exponents of this position were Tylor and Frazer, especially in their celebrated "intellectualist" theory of magic. Professor Evans-Pritchard has succinctly summarized their standpoint as follows:

> They considered that primitive man had reached his conclusions about the efficacy of magic from rational observation and deduction in much the same way as men of science reach their conclusions about natural laws. Underlying all magical ritual is a rational process of thought. The ritual of magic follows from its ideology. It is true that the deductions of a magician are false—had they been true they would have been scientific and not magical—but they are nevertheless based on genuine observation. For classification of phenomena by the similarities which exist between them is the procedure of science as well as of magic and is the first essential process of human knowledge. Where the magician goes wrong

> is in inferring that because things are alike in one or more respects they have a mystical link between them whereas in fact the link is not a real link but an ideal connection in the mind of the magician.... A causal relationship exists in his mind but not in nature. It is a subjective and not an objective connection. Hence the savage mistakes an ideal analogy for a real connection.[12]

Their theory of religion was likewise both rationalistic and derogatory: Frazer in particular held religion to be less rational (though more complex) than the occult science of magic because it postulated a world of capricious personal beings rather than a uniform law-governed nature.[13]

There has recently been elaborated a highly sophisticated version of this position on the part of a number of writers, who have stressed the explanatory purport of primitive magical and religious beliefs. In a brilliant paper,[14] Dr. Robin Horton treats traditional African religious systems as theoretical models akin to those of the sciences, arguing that many of the supposed differences between these two modes of thought result, more than anything else, from differences of idiom used in their respective theoretical models. His aim is to break down the contrast between traditional religious thought as "nonempirical" and scientific thought as "empirical."

Horton's case is not that traditional magico-religious thought is a variety of scientific thought but that both aim at and partially succeed in grasping causal connections. He also, of course, maintains that "scientific method is undoubtedly the surest and most efficient tool for arriving at beliefs that are successful in this respect"[15] and examines the different ways in which traditional and scientific thought relate to experience: his case is that these can ultimately be traced to the differences between "closed" traditional cultures "characterized by lack of awareness of alternatives, sacredness of beliefs, and anxiety about threats to them" and "open" scientifically orientated cultures "characterized by awareness of alternatives, diminished sacredness of beliefs, and diminished anxiety about threats to them."[16]

Thus the third answer to our problem involves the application of given rational criteria to prima facie irrational beliefs which shows them to be largely rational in method, purpose, and form, though unscientific and more or less (for Tylor and Frazer, entirely; for Horton, less than we thought) irrational in content. Durkheim put this case, with customary clarity, as follows:

> [I]t is through [primitive religion] that a first explanation of the world has been made possible.... When I learn that *A* regularly precedes *B*, my knowledge is enriched by a new item, but my understanding is not at all satisfied with a statement which does not appear rationally justified. I commence to *understand* only when it is possible for me to conceive *B* in a perspective that makes it appear to me as something that is not foreign to *A*, as united to *A* by some intelligible relationship. The great service that the religions have rendered to thought is that they have constructed a first representation of what these intelligible relationships between things might be. In the circumstances under which it was attempted, the enterprise could obviously attain only precarious results. But then, does it ever attain any that are definitive, and is it not necessary ceaselessly to reconsider them? And also, it is less important to succeed than to try.... The explanations of contemporary science are surer of being objective because they are more methodical and because they rest on more rigorously controlled observations, but they do not differ in nature from those which satisfy primitive thought.[17]

4. The fourth position we are to consider is that of Lucien Lévy-Bruhl (until the time of writing *Les Carnets*). This is, as well shall see, crucially ambiguous on the point of concern to us.[18]

Lévy-Bruhl's central theme was to emphasize the differences between the content of two types of beliefs (seen as Durkheimian *représentations collectives*):[19] those characteristic of primitive societies and those characteristic of "scientific" thinking. He tried to bring out those aspects in which these two types of belief differed: as he wrote, "I intended to bring fully to light the mystical *aspect* of primitive mentality in contrast with the rational *aspect* of the mentality of our societies."[20] Thus primitive beliefs were characteristically mystical, in the sense of being committed to "forces, influences, powers imperceptible to the senses, and never the less real."[21] Indeed,

> the reality in which primitives move is itself mystical. There is not a being, not an object, not a natural phenomenon that appears in their collective representations in the way that it appears to us. Almost all that we see therein escapes them, or is a matter of indifference to them. On the other hand, they see many things of which we are unaware.[22]

Furthermore, their thought is (in his confusing but revealing term) "prelogical":[23] that is,

> [it] is not constrained above all else, as ours is, to avoid contradictions. The same logical exigencies are not in its case always present. What to our eyes is impossible or absurd, it sometimes will admit without seeing any difficulty.[24]

Lévy-Bruhl endorsed Evans-Pritchard's account of his viewpoint as seeking "to understand the characteristics of mystical thought and to define these qualities and to compare them with the qualities of scientific thought":[25] "thus it is not in accord with reality and may also be mystical where it assumes the existence of suprasensible forces"[26] and is not "logical" in the sense in which a modern logician would use the term,[27] so that "primitive beliefs when tested by the rules of thought laid down by logicians are found to contravene those rules."[28] "Objects, beings, phenomena" could be "in a manner incomprehensible to us, at once both themselves and something other than themselves."[29] Thus according to given criteria derived from "scientific" thought, "mystical" and "prelogical" thought was to be judged unsuccessful. Yet Lévy-Bruhl also wants to say that there are criteria which it satisfies. Hence, he wants to say that there is a sense in which the suprasensible forces are "real." Thus, as we have seen, he writes of mystical forces as being "never the less real."[30] (On the other hand, he came to see that the primitive is not uniquely preoccupied with the mystical powers of beings and objects[31] and has a basic, practical notion of reality too). Again, he explicitly endorses Evans-Pritchard's interpretation that "primitive thought is eminently coherent, perhaps over-coherent.... Beliefs are co-ordinated with other beliefs and behavior into an organized system."[32] Yet he is crucially ambiguous about the nature of this coherence. On the one hand he writes that it is "logical": "[t]he fact that the *'patterns of thought'* are different does not, once the premises have been given, prevent the 'primitive' from reasoning like us and, in this sense, his thought is neither more nor less 'logical' than ours."[33] Yet on the other hand, he appears to accept the propositions that mystical thought is "intellectually consistent even if it is not logically consistent"[34] and that it is "organized into a coherent system with a logic of its own."[35]

Thus Lévy-Bruhl's position is an uneasy compromise, maintaining that primitive "mystical" and "prelogical" beliefs are on our standards irrational, but that on other (unspecified) standards they are about "real" phenomena and "logical."[36]

5. The fifth answer to our problem asserts that there s a strong case for assuming that, in principle, seemingly irrational belief systems in primitive societies are to be interpreted as rational. It has been most clearly stated by Professor Peter Winch,[37] and it has been claimed that Evans-Pritchard's book *Nuer Religion* supports it.[38] According to Winch's view, when an observer is faced with seemingly irrational beliefs in a primitive society, he should seek contextually given criteria according to which they may appear rational.

Winch objects to Evans-Pritchard's approach in *Witchcraft, Oracles and Magic among the Azande* on the grounds that the criteria of rationality which he applies there are alien to the context. According to Evans-Pritchard,

> It is an inevitable conclusion from Zande descriptions of witchcraft that it is not an objective reality. The physiological condition which is said to be the seat of witchcraft, and which I believe to be nothing more than food passing through the small intestine, is an objective condition, but the qualities they attribute to it and the rest of their beliefs about it are mystical. Witches, as Azande conceive them, cannot exist.[39]

Winch objects to this position on the ground that it relies upon a notion of "objective reality" provided by science: for Evans-Pritchard "the scientific conception agrees with what reality actually is like, whereas the magical conception does not,"[40] but, Winch maintains, it is a mistake to appeal to any such independent or objective reality. What counts as real depends on the context and the language used (thus "it is *within* the religious use of language that the conception of God's reality has its place");[41] moreover, "[w]hat is real and what is unreal shows itself *in* the sense that language has ... we could not in fact distinguish the real from the unreal without understanding the way this distinction operates in the language."[42] Thus European skepticism is misplaced and (we must suppose) Zande witchcraft is real.

Again, Winch objects to Evans-Pritchard's account of contradictions in the Zande belief system. The Zande believe that a suspect may be proved a witch by postmortem examination of his intestines for witchcraft substance; they also believe that this is inherited through the male line. Evans-Pritchard writes:

> To our minds it appears evident that if a man is proven a witch the whole of his clan are ipso facto witches, since the Zande clan is a group of persons related biologically to one another through the male line. Azande see the sense of this argument but they do not accept its conclusions, and it would involve the whole notion of witchcraft in contradiction were they to do so.... Azande do not perceive the contradiction as we perceive it because they have no theoretical interest in the subject, and those situations in which they express their belief in witchcraft do not force the problem upon them.[43]

Winch's comment on this passage is that

> the context from which the suggestion about the contradiction is made, the context of our scientific culture, is not on the same level as the context in which the beliefs about witchcraft operate. Zande notions of witchcraft do not

> constitute a theoretical system in terms of which Azande try to gain a quasi-scientific understanding of the world. This in its turn suggests that it is the European, obsessed with pressing Zande thought where it would not naturally go—to a contradiction—who is guilty of misunderstanding, not the Zande. The European is in fact committing a category-mistake.[44]

Thus Winch's complaint against Evans-Pritchard's treatment of the Azande is "that he did not take seriously enough the idea that the concepts used by primitive peoples can only be interpreted in the context of the way of life of these people":[45] thus we cannot legislate about what is real for them or what counts as a contradiction in their beliefs.[46] Moreover, Winch goes on to argue, rationality itself is context or culture dependent. "We start," he writes, "from the position that standards of rationality in different societies do not always coincide; from the possibility, therefore, that the standards of rationality current in S are different from our own.... What we are concerned with are differences in *criteria of rationality*."[47] He objects to the view, expressed by Professor MacIntyre, that "the beginning of an explanation of why certain criteria are taken to be rational in some societies is that they *are* rational. And since this last has to enter into our explanation we cannot explain social behavior independently of our own norms of rationality."[48] Winch's case against this is that rationality in the end comes down to "conformity to norms"; how this notion is to be applied to a given society "will depend on our reading of their conformity to norms—what counts for them as conformity and what does not."[49]

Let us see how Evans-Pritchard's *Nuer Religion* could be seen as an examplification of Winch's approach. In the chapter entitled "The Problem of Symbols" Evans-Pritchard attempts to show that the Nuer, although they *appear* to say contradictory and inconsistent things, do not really do so. Thus,

> It seems odd, if not absurd, to a European when he is told that a twin is a bird as though it were an obvious fact, for Nuer are not saying that a twin is like a bird, but that he is a bird. There seems to be a complete contradiction in the statement; and it was precisely on statements of this kind recorded by observers of primitive peoples that Lévy-Bruhl based his theory of the prelogical mentality of these peoples, its chief characteristic being, in his view, that it permits such evident contradictions—that a thing can be what it is and at the same time something altogether different.[50]

However, "no contradiction is involved in the statement which, on the contrary, appears quite sensible and even true, to one who presents the idea to himself in the Nuer language and within their system of religious thought.[51]

According to Evans-Pritchard,

> the Nuer do not make, or take, the statement that twins are birds in any ordinary sense.... In addition to being men and women they are of a twin-birth, and a twin-birth is a special revelation of Spirit; and Nuer express this special character of twins in the "twins are birds" formula because twins and birds, though for difference reasons, are both associated with Spirit and this makes twins, like birds, "people of the above" and "children of God," and hence a bird is a suitable symbol in which to express the special relationship in which a twin stands to God.[52]

Thus, it seems, Evans-Pritchard is claiming that according to Nuer criteria this statement is rational and consistent, indeed "quite sensible and even true." As he writes, toward the end of the book,

> It is in the nature of the subject that there should be ambiguity and paradox. I am aware that in consequence I have not been able to avoid *what must appear to the reader to be obscurities, and even contradictions, in my account.*[53]

We shall return below to this example and to the question of whether in fact it is a practical application of Winch's views. Here let us merely restate the fifth answer to our problem: that it is likely in principle that beliefs that appear to be irrational can be reinterpreted as rational, in the light of criteria of rationality to be discovered in the culture in which they occur. (Of course, individual beliefs may fail according to these criteria, but Winch seems to hold that no reasonably large set of beliefs could do so.)

II

The use of the word "rational" and its cognates has caused untold confusion and obscurity, especially in the writings of sociological theorists.[54] This, however, is not the best reason for seeking to break our problem down into different elements. There are strong reasons for suspecting that the first mistake is to suppose that there is a single answer to it; and this suspicion is only reinforced by the very plausibility of most of the statements cited in the foregoing section.

What is it for a belief or set of beliefs to be irrational? A belief may be characterized as a proposition accepted as true.[55] Beliefs, or sets of beliefs, are said to be irrational if they are inadequate in certain ways: (1) if they are illogical, e.g., inconsistent or (self-) contradictory, consisting of or relying on invalid inferences, etc.; (2) if they are, partially or wholly, false; (3) if they are nonsensical (though it may be questioned whether they would then qualify as propositions and thus as beliefs); (4) if they are situationally specific or ad hoc, i.e.: not universalized because bound to particular occasions;[56] (5) if the ways in which they come to be held or the manner in which they are held are seen as deficient in some respect. For example: (a) the beliefs may be based, partially or wholly, on irrelevant considerations; (b) they may be based on insufficient evidence; (c) they may be held uncritically, i.e.: not held open to refutation or modification by experience, regarded as "sacred" and protected by "secondary elaboration" against disconfirming evidence;[57] (d) the beliefs may be held unreflectively, without conscious consideration of their assumptions and implications, relations to other beliefs, etc. (though here the irrationality may be predicated of the believer rather than the belief).

In addition, there are other well-used senses of "rational" as applied to actions, such as (6) the widest sense of simply goal-directed action;[58] (7) the sense in which an action is said to be (maximally) rational if what is in fact the most efficient means is adopted to achieve a given end;[59] (8) the sense in which the means that is believed by the agent to be the most efficient is adopted to achieve the agent's end (whatever it may be); (9) the sense in which an action is in fact conducive to the agent's (expressed or unexpressed) "long-term" ends; (10) the sense in which the agent's ends are the ends he ought to have.[60]

III

In this section I shall suggest that some criteria of rationality[61] are universal, i.e., relevantly applicable to all beliefs, in any context, while others are context-dependent, i.e., are to be discovered by investigating the context and are only relevantly applicable to beliefs in that context. I shall argue (as against Winch) that beliefs are not only to be evaluated by the criteria that are to be discovered in the context in which they are held; they must also be evaluated by criteria of rationality that simply *are* criteria of rationality, as opposed to criteria of rationality in context (c). In what follows universal criteria will be called "rational (1) criteria" and context-dependent criteria "rational (2) criteria."

Let us assume we are discussing the beliefs of a society *S*. One can then draw a distinction between two sets of questions. One can ask, in the first place: (i) what for society *S* are the criteria of rationality *in general*? And, second, one can ask: (ii) what are the appropriate criteria to apply to a given class of beliefs within that society?

(i) Insofar as Winch seems to be saying that the answer to the first question is culture dependent, he must be wrong, or at least we could never know if he were right; indeed we cannot even conceive what it could *be* for him to be right. In the first place, the existence of a common *reality* is a necessary precondition of our understanding *S*'s language. This does not mean that we and the members of *S* must agree about all "the facts" (which are the joint products of language and reality); any given true statement in *S*'s language may be untranslatable into ours and vice versa. As Whorf wrote, "language dissects nature in many different ways." What must be the case is that *S* must have our distinction between truth and falsity if we are to understand its language, for, if *per impossibile* it did not, we would be unable even to agree about what counts as the successful indentification of public (spatiotemporally located) objects.[62] Moreover, any culture, scientific or not, which engages in successful prediction (and it is difficult to see how any society could survive which did not) must presuppose a given reality. Winch may write that "[o]ur idea of what belongs to the realm of reality is given for us in the language that we use"[63] and he may castigate Evans-Pritchard as "wrong, and crucially wrong, in his attempt to characterize the scientific in terms of that which is 'in accord with objective reality.'"[64] But, it is, so to speak, no accident that the predictions of both primitive and modern common sense and of science come off. Prediction would be absurd unless there were events to predict.[65] Both primitive and modern men predict in roughly the same ways; also they can learn each other's languages. Thus they each assume an independent reality, which they share.

In the second place, *S*'s language must have operable logical rules and not all of these can be pure matters of convention. Winch states that "logical relations between propositions ... depend on social relations between men."[66] Does this imply that the concept of negation and the laws of identity and noncontradiction need not operate in *S*'s language? If so, then it must be mistaken, for if the members of *S* do not possess even these, how could we ever understand their thought, their inferences and arguments? Could they even be credited with the possibility of inferring, arguing, or even thinking? If, for example, they were unable to see that the truth of *p* excludes the truth of its denial, how could they ever communicate truths to one another and reason from them to other truths? Winch half sees this point when he writes that "the possibilities of our grasping forms of rationality different from ours in an alien culture ... are limited by certain formal requirements centering round the demand for consistency. But these formal requirements tell us nothing about what in particular is to *count* as consistency, just as the rules of the propositional calculus limit, but do not themselves determine, what are to be proper values of *p*, *q*, etc."[67] But this is merely a (misleading) way of saying that it is the content of propositions, not the logical relations, between them, that is, "dependent on social relations between men."

It follows that if *S* has a language, it must, minimally, possess criteria of truth (as correspondence to reality) and logic, which we share with it and which simply *are* criteria of rationality. The only alternative conclusion is Elsdon Best's, indicated in position (2) of section I above, which seeks to state the (self-contradictory) proposition that *S*'s thought (and language) operate according to quite different criteria and that it is literally incomprehensible to us. But if the members of *S* really did not have our criteria of truth and logic, we would have no grounds for attributing to them language, thought, or beliefs and would a fortiori be unable to make any statements about these.

Thus the first two ways that beliefs may be irrational that are specified in section II are fundamental and result from the application of rational (1) criteria. Moreover, it can be shown that the other types of irrationality of belief indicated there are dependent on the use of such criteria. Thus nonsense (3) and the failure to universalize (4) may be seen as bad logic, (e.g. self-contradiction and bad reasoning). Whether this is the most *useful* way to characterize a particular belief in a given case is another question. Again, the types of irrationality relating to the ways of arriving at and of holding beliefs are dependent on rational (1) criteria. Thus (5) (a)–(d) are simply methodological inadequacies: they result from not following certain procedures that can be trusted to lead us to truths.[68] Again, in the senses of "rational" relating to actions, senses (7) and (9) require the application of rational (1) criteria.

Thus the general standpoint of position (3) in section I is vindicated. Insofar as primitive magico-religious beliefs are logical and follow methodologically sound procedures, they are, so far, rational (1); insofar as they are, partially or wholly, false, they are not. Also part of Lévy-Bruhl's position is vindicated. Insofar as "mystical" and "prelogical" can be interpreted as false and invalid, primitive (and analogous modern) beliefs are irrational (1).

(ii) What, now, about the question of whether there are any criteria which it is appropriate to apply to a given class of beliefs within *S*? In the first place, the context may provide criteria specifying which beliefs may acceptably go together. Such criteria may or may not violate the laws of logic. Where they do, the beliefs are characteristically labeled "mysterious." Then there are contextually provided criteria of *truth*:[69] thus a study of Nuer religion provides the means for deciding whether "twins are birds" is, for the Nuer, to be counted as "true." Such criteria may apply to beliefs (i.e., propositions accepted as true) which do not satisfy rational (1) criteria insofar as they do not and could not correspond with "reality": that is, insofar as they are *in principle* neither directly verifiable nor directly falsifiable by empirical means. (They may, of course, be said to relate to "reality" in another sense;[70] alternatively, they may be analyzed in terms of the coherence or pragmatist theories of truth.) This is to disagree with Leach and Beattie who seek to discount the fact that beliefs are accepted as true and argue that they must be interpreted metaphorically. But it is also to disagree with the Frazer-Tylor approach, which would simply count them false because they are "nonobjective."

There are (obviously) contextually provided criteria of *meaning*. Again, there are contextually provided criteria which make particular beliefs *appropriate* in particular circumstances. There are also contextually provided criteria which specify the best way to arrive at and hold beliefs. In general, there are contextually provided criteria for deciding what counts as a "good reason" for holding a belief.

Thus, reverting to our schema of the way that beliefs can be irrational in section II, it will be seen that, for any or all of a particular class of beliefs in a society, there may be contextually provided criteria according to which they are "consistent" or "inconsistent," "true" or "false," meaningful or nonsensical, appropriate or inappropriate in the circumstances, soundly or unsoundly reached, properly or improperly held, and in general based on good or bad reasons. Likewise, with respect to the rationality of actions, the context may provide criteria against which the agent's reason for acting and even the ends of his action may be judged adequate or inadequate.

Thus the first position in section I is largely vindicated, insofar as it is really pointing to the need to allow for contextual (e.g., symbolic) interpretation, but mistaken insofar as it ignores the fact that beliefs purport to be *true*[71] and relies exclusively upon the nonexplanatory notion of "metaphor."[72] The third position is mistaken (or inadequate) only insofar as it denies (or ignores) the relevance of rational (2) criteria. The fourth position foreshadows that advanced here, but it is misleading (as Lévy-Bruhl himself came to see) insofar as it suggests that rational (1) criteria are not universal and fundamental. The fifth position is ambiguous. Insofar as Winch is claiming that there are no rational (1) criteria, he appears mistaken. Insofar as he is claiming that there are rational (2) criteria, he appears correct. I take the quotations from *Nuer Religion* to support the latter claim.

One may conclude that all beliefs are to be evaluated by both rational (1) and rational (2) criteria. Sometimes, as in the case of religious beliefs, rational (1) truth criteria will not take the analysis very far. Often rational (1) criteria of logic do not reveal anything positive about relations between beliefs that are to be explicated in terms of "provides a reason for." Sometimes rational (1) criteria appear less important than "what the situation demands." In all these cases, rational (2) criteria are illuminating. But they do not make rational (1) criteria dispensable. They could not, for the latter, specify the ultimate constraints to which thought is subject: that is, they are fundamental and universal in the sense that any society which possesses what we may justifiably call a language must apply them *in general,* though particular beliefs, or sets of beliefs, may violate them.

If both sorts of criteria are required for the understanding of beliefs (for they enable us to grasp their truth conditions and their interrelations), they are equally necessary to the explanation of why they are held, how they operate and what their social consequences are. Thus only by the application of rational (1) criteria is it possible to see how beliefs which fail to satisfy them can come to be rationally criticized, or fail to be.[73] On the other hand, it is usually only by the application of rational (2) criteria that the point and significance that beliefs have for those that hold them can be grasped. Rational (1) and rational (2) criteria are necessary both to understand and to explain.

Notes

I am most grateful to Martin Hollis, John Beattie, Rodney Needham, Jean Floud, John Torrance, and Vernon Bodganor, among others, for their very kind and helpful criticisms of an earlier draft of this chapter.

1. Some have argued that its solution bears directly on anthropological practice (see, e.g., P. Winch, Understanding a Primitive Society, *American Philosophical Quarterly*, where Evans-Pritchard's account of witchcraft among the Azande is held to be partly vitiated by his supposedly mistaken answer to it). I agree with this position, but in this paper I do not seek to substantiate it.
2. E. Leach, *Political Systems of Highland Burma*, London, 1954, pp. 13–14.
3. Ibid., p. 182.
4. R. Firth, *Essays on Social Organization and Values*, London, 1964, p. 237.
5. Ibid., pp. 238–39.
6. See J. Beattie, *Other Cultures*, London, 1964, Chapters V and XII, and idem, "Ritual and Social Change," *Man: The Journal of the Royal Anthropological Institute*, I, 1966, 60–74.
7. J. Beattie, loc. cit. (1966), p. 68. Thus, magic is "the acting out of a situation, the expression of a desire in symblic terms; it is not the application of empirically acquired knowledge about the properties of natural substances" (Beattie, op. cit. (1964), p. 206). Cf. T. Parsons, *The Structure of Social Action*, New York and London, 1937, p. 431 (2nd edition 1949): "Ritual actions are not ... either simply irrational, or pseudo rational, based on prescientific erroneous knowledge, but are of a different character altogether and as such not to be measured by the standards of intrinsic rationality at all" (cited in Beattie, loc. cit. 1966). Parsons wrongly attributes this position to Durkheim: as I shall show, Durkheim did not see religion *as merely* symbolic.
8. Beattie, op. cit. (1966), p. 72. For Beattie magic and religion "both imply ritual, symbolic ideas and activities rather than practical, 'scientific' ones" idem. (1964, p. 212). For an example of the procedures Beattie advocates, see V. Turner, "Symbols in Ndembu Ritual" in M. Gluckman (ed.), *Closed Systems and Open Minds*, Edinburgh, 1964, pp. 20–51.
9. Beattie appeals to the authority of Suzanne Langer (Beattie, "Ritual and Social Change," loc. cit. p. 66), but I am unsure how far his allegiance to her views goes. I do not know whether he would wish to argue, as she does, that rationality and even logic can be ascribed to expressive symbolism and whether he would subscribe to her general view that "[rationality] is the essence of mind and symbolic transformation its elementary process. It is a fundamental error, therefore, to recognize it only in the phenomenon of systematic, explicit reasoning. That is a mature and precarious product. Rationality, however, is embodied in every mental act." (idem, *Philosophy in a New Key*, Harvard, 1942, p. 99; 3rd edition 1963. Miss Langer's is in any case a special sense of "rationality." As I hope to show, the fundamental meaning of rationality is essentially linked to the phenomenon of systematic, explicit reasoning.
10. "Maori Medical Lore," *Journal of Polynesian Society*, XIII, 1904, p. 219, cited in L. Lévy-Bruhl, *Les fonctions mentales dans les sociétés inférieures*, Paris, 1910, p. 69 (2d edition 1912).
11. C. G. and B. Z. Seligman, *Pagan Tribes of the Nilotic Sudan*, London, 1932, p. 25, cited in E. E. Evans-Pritchard, Lévy-Bruhl's Theory of Primitive Mentality, *Bulletin of the Faculty of Arts*, II, 1934, 1–36.
12. E. E. Evans-Pritchard, "The Intellectualist (English) Interpretation of Magic," *Bulletin of the Faculty of Arts*, I, 1933, 282–311. Cf. also idem, *Theories of Primitive Religion*, Oxford, 1965, Chapter II.
13. Cf. E. Leach, "Frazer and Malinowski," *Encounter*, XXV, 1965, 24–36: "For Frazer, all ritual is based on fallacy, either an erroneous belief in the magical powers of men or an equally erroneous belief in the imaginary powers of imaginary deities" (p. 29).
14. R. Horton, "African Traditional Thought and Western Science," above, pp. 131–71; Cf. also idem, "Destiny ad the Unconscious in West Africa," *Africa* XXXI (1961), 110–16; "The Kalabari World View: An Outline and Interpretation," ibid., XXXII, 1962, 197–220; "Ritual Man in Africa," ibid. XXXIV, 1964, 85–104. (For a symbolist critique of Horton, see Beattie, "Ritual and Social Change," loc. cit.). For other "neo-Frazerian" writings, see J. Goody, "Religion and Ritual: The Definitional Problem," *British Journal of Sociology*, XII, 1961, 142–64; I. C. Jarvie, *The Revolution in Anthropology*, London, 1964; I. C. Jarvie and J. Agassi, "The Rationality of Magic," above, pp. 172–93.
15. See Horton, above, p. 140.
16. Ibid. pp. 155–6.
17. E. Durkheim, *Les formes élémentaires de la vie religieuse*, Paris, 1912, pp. 339–41.
18. See *Les Carnets de Lucien Lévy-Bruhl*, Paris, 1949, passim, where it is made explicit and partially resolved.
19. It is worth noting that Durkheim differed crucially from Lévy-Bruhl, emphasizing the continuities rather than the differences between primitive and modern scientific thought: see E. Durkheim, *Les formes élémentaires de la vie religieuse*, op. cit. pp. 336–42, and Review of L. Lévy-Bruhl, *Les fonctions*

mentales dans les sociétés inférieures, and E. Durkheim, *Les formes élémentaires de la vie religieuse, in Année sociologique*, XII, 1913, 33–7.

20. L. Lévy-Bruhl, "A Letter to E. E. Evans-Pritchard," *British Journal of Sociology*, III, 1952, 117–23.
21. L. Lévy-Bruhl, *Les fonctions mentales dans les sociétés inférieures*, Paris, 1910, p. 30.
22. Ibid., pp. 30–31.
23. He eventually abandoned it: see *Les Carnets de L. Lévy-Bruhl*, op. cit., pp. 47–51, 60–62, 69–70, 129–35, etc.
24. L. Lévy-Bruhl, *La mentalité primitive* (Herbert Spencer Lecture), Oxford, 1931, p.21.
25. E. E. Evans-Pritchard, "Lévy-Bruhl's Theory of Primitive Mentality," loc. cit. Lévy-Bruhl's general endorsement of this article is to be found in Lévy-Bruhl, "A Letter to E. E. Evans-Pritchard," loc. cit.
26. E. E. Evans-Pritchard, "Lévy-Bruhl's Theory," loc. cit.
27. Ibid.
28. Ibid.
29. Lévy-Bruhl, *Les fonctions mentales*, op. cit., p. 77.
30. This position he did not abandon: see *Les Carnets de L. Lévy-Bruhl*, op. cit. (e.g. pp. 163–98), where it is strongly refirmed.
31. E. E. Evans-Pritchard, "Lévy-Bruhl's Theory of Primitive Mentality," loc. cit.
32. Ibid.
33. L. Lévy-Bruhl, "A Letter to E. E. Evans-Pritchard," loc. cit, p. 121.
34. E. E. Evans-Pritchard, "Lévy-Bruhl's Theory," loc. cit.
35. Ibid. Cf. *Les Carnets*, op. cit., p. 61, where he recalls that he had begun from the hypothesis that societies with different structures had different logics. The theory of the "prelogical" was a modified version of this hypothesis, which he only finally abandoned much later, when he came to hold that "the logical structure of the mind is the same in all known human societies" (ibid., p. 62).
36. Lévy-Bruhl's final position was as follows: "there is no primitive mentality which is distinguished from the other by *two* characteristic features (being mystical and prelogical). There is one mystical mentality that is more marked and more easily observable among 'primitives' than in our societies, but present in every human mind." (*Les Carnets*, p. 131.)
37. P. Winch, "Understanding a Primitive Society," above, pp. 78ff.
38. E. Gellner, "Concepts and Society," above, pp. 18ff.; and A. MacIntyre, "Is Understanding Religion Compatible with Believing?" above, pp. 62ff.
39. E. E. Evans-Pritchard, *Witchcraft, Oracles and Magic among the Azande*, Oxford, 1937, p. 63.
40. P. Winch, "Understanding a Primitive Society," loc. cit., p. 81 above.
41. Ibid., p. 82 above.
42. Ibid., p. 82.
43. *Witchcraft*, op. cit., pp. 24–5.
44. "Understanding a Primitive Society," above, p. 93.
45. Ibid.
46. The philosophical basis for this position is to be found in P. Winch, *The Idea of a Social Science and Its Relation to Philosophy*, London, 1958. Cf. in particular the following passage: "criteria of logic are not a direct gift of God, but arise out of, and are only intelligible in the context of, ways of living and modes of social life. It follows that one cannot apply criteria of logic to modes of social life as such. For instance, science is one such mode and religion is another; and each has criteria of intelligibility peculiar to itself. So within science or religion, actions can be logical or illogical: in science, for example, it would be illogical to refuse to be bound by the results of a properly carried out experiment; in religion it would be illogical to suppose that one could pit one's own strength against God's, and so on." (pp. 100–1).
47. P. Winch, "Understanding" loc. cit, p. 97.
48. A. MacIntyre, "A Mistake about Causality in Social Science," in P. Laslett and W. G. Runciman (eds.), *Philosophy, Politics and Society*, Second Series, Oxford, 1962, p. 61. This formulation suffers from its emphasis on the location of these norms rather than on their nature.
49. P. Winch, "Understanding," loc. cit., p. 100.
50. E. E. Evans-Pritchard, *Nuer Religion*, Oxford, 1956, p. 131.
51. Ibid.
52. Ibid., pp. 131–2.
53. Ibid., p. 318. Emphasis mine. Professor Gellner's comment on this approach is that it "absolves too many people of the charge of systematically illogical or false or self-deceptive thought." Moreover (E. Gellner, loc. cit., p. 36 above): "The trouble with such all-embracing logical charity is, for one thing,

that it is unwittingly quite *a priori*: it may delude anthropologists into thinking that they had *found* that no society upholds absurd or self-contradictory beliefs, whilst in fact the principle employed has ensured in advance of any inquiry that nothing may count as prelogical, inconsistent or categorically absurd though it may be. And this, apart from anything else, would blind one to at least one socially significant phenomenon: the social role of absurdity."

54. I think Max Weber is largely responsible for this. His uses of these terms is irredeemably opaque and shifting.
55. Philosophers have disputed over the question of whether "belief" involves reference to a state of mind. I agree with those who argue that it does not; thus I would offer a dispositional account of "acceptance." As will be evident, I take it that belief is by definition propositional. As to the philosophical status of propositions, this does not affect the argument.
56. This is the sense of rationality stressed by Professor Hare. See R. Hare, *Freedom and Reason*, Oxford, 1963.
57. Cf. R. Horton, "African Traditional Thought and Western Science," *Africa*, XXXVII (1967), 50–71, and 155–87, especially pp. 167–69 (above, pp. 162–64). For numerous examples of this, see E. E. Evans-Pritchard, *Witchcraft*, op. cit.
58. See e.g., I. C. Jarvie and J. Agassi, loc. cit.
59. Cf. e.g., Parsons, op. cit., pp. 19 and 698–99.
60. Cf., e.g., G. C. Homans, *Social Behaviour: Its Elementary Forms*, London, 1961, p. 60 for senses (9) and (10). It is perhaps worth adding here that I do not find Mr. Jonathan Bennett's stipulative definition of rationality germane to the present discussion ("whatever it is that humans possess which marks them off, in respect of intellectual capacity, sharply and importantly from all other known species," in J. Bennett, *Rationality*, London, 1964, p. 5.)
61. I take "criterion of rationality" to mean a rule specifying what would count as a reason for believing something (or acting). I assume that it is only by determining the relevant criteria of rationality, that the question, "Why did X believe *p*?" can be answered (though, of course, one may need to look for other explanatory factors. I merely claim that one must first look here).
62. Cf. P. Strawson, *Individuals*, London, 1959, and S. Hampshire, *Thought and Action*, London, 1959, Chapter I.
63. P. Winch, *The Idea of a Social Science*, op. cit., p. 15.
64. P. Winch, "Understanding," loc. cit., p. 80.
65. I owe this argument to Martin Hollis. I have profited greatly from his two papers, "Winchcraft and Witchcraft" and "Reason and Ritual" (below, pp. 221–39).
66. P. Winch, *the Idea*, op. cit., p. 126.
67. P. Winch, "Understanding," loc. cit., p. 100.
68. Though, as Horton shows, they may be unnecessary ("African Traditional Thought," loc. cit., 140).
69. Cf. *Les Carnets de L. Lévy-Bruhl*, op. cit., pp. 80–82 and 193–95.
70. Ibid., p. 194.
71. Beattie and Firth see the sense of this argument but do not accept its conclusions (see quotations in text above and J. Beattie, *Other Cultures*, London, 1964, pp. 206–7).
72. Cf. J. Goody, "Religion and Ritual: the Definitional Problem," *British Journal of Sociology*, XII, 1961, 142–64, especially pp. 156–57 and 161. As Evans-Pritchard (somewhat unfairly) says: "It was Durkheim and not the savage who made society into a god" (*Nuer Religion*, op. cit., p. 313).
73. Cf. E. E. Evans-Pritchard, *Witchcraft, Oracles and Magic among the Azande*, Oxford, 1937, pp. 475–78, where twenty-two reasons are given why the Azande "do not perceive the futility of their magic."

Chapter 19

The Status of Rationality Assumptions in Interpretation and in the Explanation of Action

Dagfinn Føllesdal

In discussions of the philosophy and methodology of interpretation and action explanation it is often argued that one has to assume that man is rational. This, supposedly, is just what distinguishes the study of man and the method of understanding from the study of nature and the method of causal explanation.

In this chapter, I will attempt to find a reasonable rendering of what is meant by this assumption and discuss its status. The paper has four sections. After a brief discussion of the question whether man is rational (sec. 1) I turn to the questions of what rationality is (sec. 2) and what role rationality assumptions play in interpretation and explanation of action (sec. 3). Finally I discuss the status of rationality assumptions (sec. 4).

1 *Is Man Rational?*

Aristotle maintained that to be rational is definitory of man; man is a rational animal. In our time, Donald Davidson is one of those who most vigorously has argued that in order to understand man and attribute beliefs, desires, and actions to him, we have to assume that he is rational:

> The satisfaction of conditions of consistency and rational coherence may be viewed as constitutive of the range of application of such concepts as those of belief, desire, intention and action.[1]

> If we are intelligibly to attribute attitudes and beliefs, or usefully to describe motions as behavior, then we are committed to finding, in the pattern of behavior, belief and desire, a large degree of rationality and consistency.[2]

Similarly, William H. Dray argues:

> Understanding is achieved when the historian can see the reasonableness of a man's doing what this agent did, given the beliefs and purposes he referred to (what the agent believed to be the facts of his situation, including the likely results of taking various courses of action considered open to him and what he wanted to accomplish: his purposes, goals, or motives).[3]

While Davidson regards rationality as necessary for the very applicability of concepts like belief, desire, intention, and action, Dray considers rationality as necessary at least for our knowing what the other's beliefs, purposes, goals, or motives are.

Carl G. Hempel, on the other hand, disagrees with Dray and regards the assumption that man is rational as merely an empirical hypothesis, which presumably may be false:

> Now, information to the effect that the agent A was in a situation of kind C, and that in such a situation the rational thing to do is x, affords grounds for believing that it would have been rational for A to do x; but not for believing that A did in fact do x. To justify this latter belief, we clearly need a further explanatory assumption, namely that—at least at the time in question—A was a rational agent and thus was disposed to do whatever was rational under the circumstances.
>
> But when this assumption is added, the answer to the question "Why did A do x?" takes on the following form:
>
> A was in a situation of type C
> A was a rational agent
> In a situation of type C any rational agent will do x
> ———————————————————————
> Therefore A did x.[4]

Hempel hence seems to regard the assumption that the agent is rational as an empirical premise which enters into explanation of action and which must be true if the explanation is to be correct, but which may conceivably be false, in which case our explanation of the person's action is faulty. A similar view is argued for, with more precision and detail, by Wolfgang Stegmüller, in *Probleme und Resultate der Wissenschaftstheorie und Analytischen Philosophie.*[5]

Karl Popper maintains that no rationality assumption is needed: "the method of applying a situational logic ... is not based on any psychological assumption concerning the rationality (or otherwise) of 'human nature.' "[6]

Alan Donagan, finally, regards the assumption of rationality as plainly false (and presumably as not required for an explanation of action):

> There is no reason to believe that all historical agents are rational, in any of the several senses of "rational" which Hempel has explored.[7]
>
> Considering what human history has been, a historian would be in a pretty pass if he were obliged to assume that the only actions he may succeed in understanding were rational. They must, indeed, be intelligible, but that is another thing.[8]

Opinions hence seem to differ widely, unless these authors mean very different things with rationality. Before trying to settle the dispute, let us now consider some main notions of rationality and the role they play in interpretation and explanation of action.

2 *What Is Rationality?*

Rationality is a multifarious notion. The literature abounds with different and often seemingly unrelated notions of rationality, from various kinds of "minimal rationality" to stronger notions. To exemplify one of the stronger notions it is enough to remind you of Rawls: "The rational individual does not suffer from envy."[9]

In a recent survey article on rationality[10] Jon Elster distinguishes more than twenty senses of rationality. These are not all in conflict—they often relate to different areas: rational behavior (rationality as efficiency), rational beliefs (rationality as consistency), etc.

In the following, I will concentrate on rational behavior, but I will also make some remarks on rational beliefs and rational values, and I will emphasize a distinction that Elster does not make, between normative and descriptive theory of rationality.

Let me first, however, mention briefly four main kinds of rationality and indicate how they are connected with interpretation and explanation of action. They are (1) rationality as logical consistency, (2) rationality as well-foundedness of beliefs, (3) rationality as well-foundedness of values, and (4) rationality of action.

1. Rationality as logical consistency is the requirement that a person's beliefs shall be logically consistent with one another.

This may mean several different things, depending on two factors that may vary independently: First, how much are we to include in a person's beliefs? Three options that immediately come to mind are (i) the beliefs we actively entertain in a given moment, i.e., those that we are presently "thinking about," (ii) those beliefs that play a role in determining the actions, if any, that we are presently carrying out (we are going to return to this in connection with rationality of action later), (iii) those beliefs that can be elicited by questioning a person, as, for example, in the questioning of the slave boy in Plato's *Meno*.

Note that while groups (ii) and (iii) presumably include group (i), groups (ii) and (iii) may be incomparable. Group (iii) obviously would include lots of beliefs that are not included in group (ii); many mathematical truths that are not yet even conjectured would, for example, belong in group (iii). Group (ii), however, in its turn, appears to include many beliefs that are not included in groups (iii). If Freud was right, each of us has many such beliefs which influence our actions, but which cannot be elicited by normal questioning.

The second factor that is of importance for what is meant by our beliefs being logically consistent is whether we just mean that none of our beliefs is the negation of another, or whether we mean that the set of our beliefs does not logically imply a contradiction. The latter condition is obviously stronger, and perhaps none of us satisfies this rationality condition.

2. The second kind of rationality I mentioned was rationality as well-foundedness of beliefs. This is a much stronger notion than the previous one. That our beliefs are not contradicting one another is necessary, but by no means sufficient for their being well-founded. Well-foundedness requires something more, namely, that our beliefs be well supported by the available evidence. Our beliefs may go well beyond the available evidence, as they do in the case of the more theoretical parts of scientific theories. But there should be no other competing theories that would be better supported by the available evidence. The specification of the criteria of well-foundedness would recapitulate epistemology and scientific methodology; we should only note that such a specification must also make precise the phrase "available evidence." Rationality as well-foundedness concerns not only what beliefs we should hold *given* a certain amount of evidence, but also to what extent it is rational to actively search for additional evidence before we allow our beliefs to become fixed.

Rationality as well-foundedness of beliefs is clearly a normative notion, not a descriptive one; most of us are not very rational in this sense most of the time. I shall come back to this normative-descriptive issue in the last section of this chapter.

3. Rationality as well-foundedness of values. It is often claimed that while one may choose means toward an end in a more or less rational way, there is no notion of rationality that applies to the evaluation or ends, or values. This view does not seem to me to be correct. I think that our normative considerations can also be more or less rational and that normative judgments, like factual judgments, can be founded with more or less rationality. I do not here have in mind Weber's notion of *Wertrationalität*,

which signifies little but a retreat to ultimate commitment, but a rational justification of norms and values.

The most promising approach to this issue is in my view the method of a reflective equilibrium that has been worked out by Goodman, Rawls, and others.[11] I shall not go into this view here, since I assume that [my readers] all know it. What is important from the perspective of our discussion of rationality is that this view on justification of normative statements gives us a way to judge the rationality of normative statements. Rationality in ethics becomes similar to rationality in science: the rationality, or well-foundedness, of our judgments depends up on the degree to which we have achieved a reflective equilibrium in which our general principles fit in with one another and with our adjusted judgments.

As in the case of rationality as well-foundedness of beliefs, rationality as well-foundedness of values is a normative notion, not a descriptive one. I shall come back to this issue later.

4. The fourth and last of the kinds of rationality that I wanted to discuss was rationality of action. Since I have discussed rationality of action more thoroughly in a couple of earlier papers, I shall be very brief here. I will only say that I find decision theory, as it has been developed by economists, mathematicians, and philosophers, the best framework currently available for discussing the rationality of actions.

According to this theory, an action is the end product of a two-step process. First, we consider which courses of action are open to us in our given situation. This is largely a matter of our beliefs concerning what is possible for us and what not. In praxis, however, we do not survey all possible alternatives; our search is limited partly by our imagination and the time at our disposal, and partly also by our conception of how much of a bearing the various alternatives have on what we desire or fear. We tend to focus on alternatives that it seems to be particularly important for us to realize or avoid, i.e., alternatives with particularly great positive or negative expected utility, as the economists say.

The second step in the process now consists in weighing the envisaged alternatives against one another, taking into account both our beliefs concerning the probabilities or their various consequences and the values that we assign to each of these consequences. Multiplying the probabilities with the values and adding up, we then arrive at the expected utility of each alternative, whereupon we choose the alternative with the highest expected utility. This multiplication and addition is subject to severe difficulties which have been extensively discussed in the literature of decision theory. The way in which one pools preferences and values into a resultant expected utility may turn out to have little to do with the arithmetical operations of multiplication and addition.

One comment may be in place here: Some versions of decision theory treat the set of alternatives and consequences in a platonistic way, as the set of all physically possible alternatives open to the agent in the given situation and all physically possible consequences of them. Clearly, this is not the appropriate conception if we want to clarify rationality. Rationality always has to do with what the agent ought to choose, given his or her limited, perspectival view on the situation, with a limited amount of information, limited imagination, and time for considering different alternatives and thinking through their various consequences. Rationality of action is normally a question of how to make the best use of one's resources, one's information-seeking capabilities, and one's ability to create good alternatives, and not a question of choosing from within a vast set of alternatives that lie there ready for one's inspection.

Even when understood in this weak sense, rationality of action is a normative rather than a descriptive notion. We shall return to this question later.

Explanation of action, according to this view, is characterized by the use of a premise to the effect that we are dealing with a rational agent, whose actions arise from his beliefs and values in the way described above. It is this premise of rationality that distinguishes reason explanation from causal explanation.

A final remark: it is noteworthy that when we say that a person is rational we tend to focus almost exclusively on the rationality of his or her beliefs and do not take his values into account. That is, we emphasize (1), (2) and (4) above, but do not pay much attention to (3). A person is regarded as rational if his beliefs seem rational in sense (2) and his actions seem to spring rationally, in sense (4), from his beliefs and values, regardless of how wild his values may be. This disregard of a person's values when we judge his rationality probably reflects the widespread tendency that I mentioned above to regard questions of ultimate values as beyond the realm of rational justification.

3 *The Role of Rationality in Interpretation and Explanation of Action*

The main aim of the study of man is to understand a person's beliefs and values. Two related aims are to understand his actions, which as I indicated above, depend upon these beliefs and values, and to understand the results of these actions, to the extent that these results manifest the agent's beliefs and values. Written texts, speech, and works of art are typical manifestations of this sort but so are various social institutions which have been brought into existence through an interplay between several agents.

Now, the simplest way to go about it would be to start with the person's beliefs and values and proceed to the actions and results of actions from there. This would be the natural approach if we had direct access to a person's beliefs and values. However, in my opinion we do not have such access to other people's beliefs and values. And even if we should have such direct access to some of our own beliefs and values, there would still remain the problem of communicating our self-insight to others. Communication is not based on direct reading off of the other's beliefs and values, i.e., telepathy, but on starting out from evidence that reaches us through our senses and constructing hypotheses about the other's beliefs and values.

This is where the rationality assumptions come in. In trying to understand the other person's actions I attribute beliefs and values to him on the assumption that he is rational. In trying to understand what the other person says or writes, I cannot take it for granted that he means by the words what I mean. Therefore, I cannot attribute to him the beliefs I would have if I assented to the same sentences. In the learning of the semantic aspects of language, belief and meaning are intertwined in such a way that there are not two elements there to be separated. This is an ontological point, not just an epistemological one. That is, the point is not that there are two elements there, meaning and belief, which we are unable to separate in a unique way, given the limited evidence we have available. The point is rather that meaning and belief are complementary. If we attribute one set of beliefs to a person, we come to interpret his sentences one way; if we attribute another set of beliefs to him, we come to interpret his sentences another way, and to the extent that the choice is undetermined by the evidence that forms the basis of language learning and communication, there is nothing to be right or wrong about. This is one way of putting Quine's doctrine of indeterminacy of translation, as has also been pointed out by Donald Davidson.

A basic methodological point is now the following, which is a variant of what Neil Wilson, W. V. Quine, Donald Davidson and others have called "the principle of charity" and Richard Grandy "the principle of humanity."[12] One should attribute to a person the beliefs that one considers it most likely that that person will have, given one's epistemology and given one's knowledge of the person's experiences, education, mental powers, and actions.

Similar considerations come in when we construct hypotheses concerning what normative views and values a person has. We use our theory of rationality in ethics, e.g., a theory of reflective equilibrium, and again take into account what we know about the person's experiences, education, upbringing, reflective inclinations, and actions.

Note how actions come in in both cases, both in connection with beliefs and in connection with values. Observation of action is a major source of evidence for our hypotheses concerning a person's beliefs and values, since both beliefs and values play a role in explaining a person's actions. Again we assume that the person is rational and then try to find out what beliefs and values we may attribute to him that are compatible with his being rational.

Before I go on to discuss this rationality assumption further, let me make a short remark about the observability of actions. Some behaviorists hold that I do not see actions, I see only bodily movements, and then *infer* that they are actions. Following Husserl, I regard this as wrong. I see actions, but I should be aware that in seeing an action, as in seeing anything else, there is a hypothetical element. If my understanding of the person I have in front of me were to change, I might come to see a quite different action where I now see this one, or I might even come to see only a bodily movement. When an economist reports that a person acts in such and such a way, which supports one of the economist's theories, he might later come to see that what the person did was something quite different, that from a mere physical point of view looked very much the same.

It has long been popular among economists studying so-called revealed preference to hold that the only way of determining a person's beliefs and values is to examine his actual choices; there is no nonchoice source of information concerning a person's beliefs and values. This gives us a very small circle. We explain a person's choices by appeal to his beliefs and values, and we attribute beliefs and values to him on the basis of his choices.

We clearly have other sources of information. We can ask him about his beliefs and values. However, there are difficulties here. First, we cannot always trust what a person says. He may give us a false story of what his reasons are for acting in the way he does. He may also lie to himself; he may rationalize. His behavior may be due to reasons that he does not know. It may be due to factors that might call for a Freudian or a physiological explanation. Thus, to use an example given by Patrick Suppes, a young boy who has just entered puberty and has an attractive female teacher may very frequently come up to the teacher after class to ask questions concerning his school work. When asked why he does this, he may answer that he has these questions concerning the school work and that he wants to learn. This may be his sincere answer, but we may want to give a different explanation of his actions.

Note, by the way, that just as our theory of explanation of action must have room for such deviant phenomena, so on the other hand the classification of something as deviant, as rationalization, as repression, sublimation, or something else is possible only on the basis of such a theory of how actions should be explained.

It is highly important to remember that the information we get from a person by asking him questions, listening to what he says, etc., is also ultimately based upon his behavior, notably his linguistic behavior. Asking him questions concerning his reasons for acting as he did broadens our base from comprising just choice behavior to comprising behaviour in general. We are therefore still caught in a circle; we explain a person's behavior in terms of his values and beliefs, which in turn are attributed to him only on the basis of his behavior.

As in the case of all circular explanation, we have a lot of leeway here. It is, for example, well known that when we explain a person's actions, we attribute different values to him depending on what beliefs we take him to have, and conversely different assumptions concerning his values will yield different conclusions concerning his beliefs. Davidson has pointed out that this interdependence between values and beliefs is similar to the interdependence that we noted earlier, in connection with Quine's thesis of indeterminacy of translation, between what we take a person's utterances to mean and what beliefs we attribute to him.[13]

Fortunately, there are various restrictions here, that help to cut down on all this under-determinateness. One such restriction, which is quite obvious, is due to the fact that the two interdependent pairs that we just considered, beliefs and values in the case of action and beliefs, and meaning in the case of utterances, have one member in common, belief. Clearly, the beliefs that we attribute to a person when we seek to explain his actions have to fit into the same belief system as the beliefs that we attribute to him when we try to interpret what he says.

Other restrictions, that I have discussed elsewhere[14] have to do with preception and with ostension. Further restrictions emerge when we give a detailed analysis of speech acts as a species of actions and incorporate into our interpretation of the other person the assumptions we make about his beliefs and values when we seek to explain his speech acts.

The decision theoretic pattern of explanation of action that I have just sketched is often regarded as the paradigm of rationality, the model of rational decision making. One always chooses what is best for one, or more accurately, what one believes to be best for one.

When I am proposing this pattern as a basis for explaining actions and thereby for understanding them, it is because I know no better theory. Some theory of action based on rationality is, I think, necessary in order to get the enterprise started of explaining and understanding actions. I shall return to this in the next section. However, although in this sense such a theory is a presupposition of understanding, I do not regard this particular theory as in any way necessary. It is a working hypothesis, which I am willing to replace by some other hypothesis as soon as something better comes along.

In any case, the pattern has to be refined and extended in several ways, which I have discussed in my contribution to the 1978 International Colloquium in Biel on Knowledge and Understanding.[15]

4 *What Is the Status of Rationality Assumptions?*

Let us now turn to the central question of this chapter: What is the status of the assumption that man is rational?

I shall state my position in the form of four theses.

First, I agree with Davidson that rationality is constitutive of belief, desire, action etc., and not just needed in order to find out what beliefs and desires a person has and what actions he performs. My basic reason for accepting Davidson's view is the following, which is similar to Davidson's own reasons:

As was noted at the end of section 2, explanation of action and interpretation invariably require the use of a premise to the effect that the agent is rational. This is just the crucial difference between explanation of action and interpretation on the one hand and causal explanation on the other. What qualifies a movement as an action is that it is explained by a reason explanation rather than by a purely causal explanation. Similarly, to be a desire or a belief is just to be a factor that can figure appropriately in reason explanation or in logical arguments.

The assumption that man is rational is therefore inseparable from other hypotheses that we make about man: that he has beliefs and values, that he acts, etc. We may in a given case be forced to give it up, but then we have to give up these other hypotheses, too.

However, we do obviously not have to assume that man is perfectly rational in order to ascribe beliefs and values to him. Rationality comes in degrees, and the crucial question is: how much rationality do we have to require in order to talk meaningfully about desires and other "intentional" notions?

My answer should be obvious from the reason I gave for accepting rationality. I shall formulate it as the following thesis:

> 1. In order for the intentional notions to make sense we must require enough rationality to let our pattern of explanation be reason explanation rather than merely causal explanation.

We may permit all kinds of interferences of a merely causal kind, but in order to say that we deal with beliefs, desires, actions, etc., rather than with mere physical phenomena, the underlying pattern of explanation must be reason explanation. That is, we must invoke rationality of one of the four kinds that we outlined in section 2.

Perfect rationality we have probably only in a very limited sense of the word, when rationality of kind 1, logical consistency of beliefs, is combined with belief of the first kind mentioned in section 2, belief that is actively entertained in a given moment. Presumably, we do not at a given time actively entertain both a belief and its negation. However, if we strengthen the notion of logical consistency to mean "implying no contradiction," then rationality even of kind 1 is probably very rare.

Rationality of kind 4, rationality of action, might seem to obtain quite often. We have a lot of freedom in attributing beliefs and values to a person in such a way as to make what he does and says come out rational. As Tversky and Davidson have pointed out,[16] normative decision theory may, for example, be made compatible with many kinds of behavior, since to quote Tversky, "the axioms of utility theory can be regarded as maxims of rational choice only in conjunction with an intended interpretation and the criteria for the selection of an interpretation are not part of utility theory."[17]

However, we are not completely free to make whatever assumptions we want concerning what alternatives and consequences an agent has taken into consideration. Our assumptions have to be reasonable, i.e., they must fit in with what we should expect on the basis of our knowledge about the agent, his beliefs, his imaginativeness, his past experience and performance, the possible influence of panic, pressure, etc. on his decision process, and so on. The same holds for our assumptions concerning his beliefs and values. They must accord with our theory of how beliefs and attitudes are formed and changed under the influence of experience, reflections, etc., together with the information we have concerning his past experience. All of these considerations, too, involve assumptions about rationality, about what beliefs and values a rational person can be expected to have, given his experiences.

In trying to understand a person and his actions, we must weigh against one another all these considerations that we make on the basis of observation of his actions and what goes on at his sensory surfaces, listening to what he says, etc. While a normative theory of rationality provides a framework for all of these considerations, causal, irrational factors come into the picture at a number of places, where our general theory of man makes us expect them to come in, as when a person gets his beliefs and values formed through propaganda, advertising or group pressure, or when he acts under the influence of hypnosis, drugs, drives with which he is not yet familiar, etc.

My view is that whenever a person experiences himself as carrying out an action, what he does should be explained in conformity with the pattern of reason explanation. Causal factors that are not reasons should in part be regarded as influencing the agent's consideration of alternatives, in part as affecting the alternatives themselves, and in part as dispositions that are to be weighed together with the person's beliefs and desires in the weighing process that I described above. Physiological urges, hypnotic instructions, and subconscious beliefs all seem to be amenable to this kind of treatment.

All these restrictions and further considerations make it extremely difficult to interpret all that a person does so that he comes out perfectly rational. If by rationality is meant perfect rationality, I therefore agree with Donagan that man is not rational in this sense. However, as we have seen, this would be acceptable to Davidson and presumably to Dray too. Although the passages I quoted from Davidson, Dray and Donagan might seem contradictory, there hence seems to be a way of reconciling them. Also Hempel's view, that rationality assumptions are empirical hypotheses, can be interpreted so as to accord with this view. One only has to accept that if A is not a rational agent, for at least much of the time, then it makes no sense to attribute beliefs and desires to A. It is not clear whether Hempel would accept this. The same holds for Stegmüller. Popper's position, however, is another matter. I will not venture to interpret him in such a way as to be able to agree with his views on the explanation of action.

Although I have now taken a stand on all the seemingly so conflicting passages that I quoted a the beginning of this chapter, more remains to be said about the status of rationality assumptions.

I will not turn to my second point, which has to do with rationality as a norm for the study of man. Given that we cannot expect perfect rationality we might take another approach: We might try to regard the rationality assumption as a normative methodological hypothesis in the following sense: In attempting to understand man we should always try to make him come out as rational as possible. We should always emphasize the reasons for an action, and only as a last resort, when we cannot make the agent come out completely rational, should we appeal to causes in our explanation of actions.

Suppes's example concerning the boy in early puberty might naturally lead us on to this. Are we not doing something wrong when we put aside the reasons given by the boy himself and instead offer a psychological explanation? The words "wrong" here and "should" above might be taken in both a methodological and an ethical sense. Methodologically: is not the agent himself the authority concerning his own desires and beliefs, at least when we have reason to regard him as sincere? And ethically: do we not violate a person's integrity when we regard him as acting from mere physical causes where he experiences himself as acting from reasons?

I think there is an important insight here, but it has to be disentangled from what I take to be a fallacy. First the insight: We should not easily set aside an agent's reason explanation in favor of a causal explanation. To give reasons for belief and reasons for action is to be concerned with justification and truth, autonomy, and responsibility. To set such reasons aside may be oppressive. It may also make us miss the point of what is going on. The example concerning the boy in early puberty may be of help here too. We should take heed to distinguish four possible cases in this example:

i. The reasons given by the boy were by themselves sufficient to overcome the countervailing factors (his shyness, his desire to go out and play, etc.) without the assistance of the urges adduced by the psychologist. However, these urges were not by themselves sufficient to overcome the countervailing factors.
ii. The reasons given by the boy were not by themselves sufficient to explain his behavior, nor did the urges by themselves suffice; it was the combination of both that did it.
iii. The reasons given by the boy were neither sufficient nor necessary to explain his behavior; the urges by themselves were sufficient.
iv. The reasons given by the boy were sufficient; so were his urges by themselves. That is, his behavior was overdetermined.

Granted that the reasons given by the boy were present at the time of his action and no wayward chain of reasons or causes occurred, the reasons would in all four cases be part of the picture and they should therefore in no case be brushed aside. In fact, it is in virtue of these factors that we would in all four cases speak of the boy's *action*, even in case (iii) where the behavior would have happened even if the reasons had not been present, but where it would not have happened for the reasons alone.

In cases (ii) and (iii) the psychological factors would have to be brought in in order to get a satisfactory explanation of what happened, and in cases (i) and (iv) the psychological factors would be needed in order to get a full picture of the situation.

We may sum this up in the following thesis:

2. When explaining actions, we should always include the reasons for the action, even in cases where purely causal factors would suffice to explain it.

But, as we have seen, we should not rest satisfied with this. We should always try to find out whether a person's reasons for an action are strong enough to outweigh the countervailing reasons and thereby explain the action. And even where the reasons suffice, as in cases (i) and (iv), we may want to complete the picture by also charting the other factors that are present.

So much for the positive side of the injunction for favoring reasons and rationality. The injunction may, however, be taken to mean something stronger, which I regard as fallacious, namely, the following: When ascribing beliefs and other propositional attitudes to a person on the basis of observation of what he does and says, always try to ascribe these beliefs and attitudes in such a way as to make him come out as rational as possible.

This I regard as too strong an injunction. The correct view seems to be the following:

> 3. When ascribing beliefs, desires and other propositional attitudes to a person on the basis of observation of what he does and says, do not try to maximize his rationality or his agreement with yourself, but use all your knowledge about how beliefs and attitudes are formed under the influence of causal factors, reflection, and so forth, and in particular your knowledge about his past experience and his various personality traits, such as credulity, alertness, and reflectiveness. Ascribe to him the beliefs and attitudes you should expect him to have on the basis of this whole theory of man in general and of him in particular.

This view is what I spelled out a little more in detail just after I stated thesis (1), in my discussion of deviations from rationality.

Finally, I come to the last observation I wanted to make, which has to do with rationality being a norm not for the study of man, but for man himself.

Man is not always rational, nor should we always and at all costs try to regard man as rational. But we should regard man as always being inclined to mend his ways toward more rationality when his lack of rationality is pointed out to him in terms that he can understand. We may, for example, like Aristotle in the Nichomachean ethics, be misled into inferring from "everything is striving toward an end" to "there is an end that everything is striving toward." (The fact that one may easily come to accept such an inference is, by the way, enough to refute both the view that man is invariably rational and the view that man ought to be regarded as invariably rational.) However, when it is pointed out to us that this inference is parallel to the inference from "everybody has a mother" to "there is a mother that everybody has," then we are likely not to trust the first inference any longer. There are differences between people as regards how much detail they need before they give up their irrationalities. However, this is my final thesis:

> 4. Man has rationality as a norm, as a second-order disposition of the following kind: once one becomes aware that one has fallen into irrationality, one will tend to adjust one's belief, attitudes, and actions such as to make them more rational.

This second-order disposition is the basis for the rationality assumptions we make in the study of man. Studying man, we may assume that he is rational unless he cannot help being irrational, e.g., because his irrationality is hidden from his view, because the situation is too complex, or because of other obstructions.

I began by alluding to Aristotle's definition of man as a rational animal. We have had to admit that we are far from rational. We can, however, give a new meaning to Aristotle's definition by observing that we are striving to be rational: Man is a rational animal in the sense that *man has rationality as a norm.*

Notes

This chapter is an offshoot of a major project on Husserl's phenomenology which has been supported by the Guggenheim Foundation, the Center for Advanced Study in the Behavioral Sciences, Deutscher Akademischer Austauschdienst, and the Norwegian Research Council for Science and the Humanities. I gratefully acknowledge all this support. I also thank Dr. Andrew Jones for several helpful comments.

1. Donald Davidson, "Psychology as Philosophy," p. 237 of the reprint in Donald Davidson, *Essays on Actions and Events*. Oxford: Clarendon Press, 1980.
2. Donald Davidson, *loc. cit.*
3. William H. Dray, "The Historical Explanation of Actions Reconsidered," p. 106 of the reprint in William H. Dray (ed.), *Philosophical Analysis and History*. New York: Harper & Row, 1966.
4. Carl G. Hempel, "Rational Action," *Proceedings and Addresses of the American Philosophical Association 1961–1962*, Vol. XXXV, October, 1962, p. 12. Yellow Springs, Ohio: Antioch Press, 1962.
5. Wolfgang Stegmüller, *Probleme und Resultate der Wissenschaftstheorie und Analytischen Philosophie*. Band IV, Erster Halbband: Personelle Wahrscheinlichkeit und Rationale Entscheidung. Berlin: Springer, 1973.
6. Karl Popper, *The Open Society and Its Enemies* (3 edition), p. 97. London: Routledge and Kegan Paul, 1957, vol. II. Cited in Alan Donagan, "The Popper-Hempel Theory Reconsidered," *History and Theory* 4 (1964), 3–26, pp. 147–148 of the reprint of this article in Dray (ed.), *Philosophical Analysis and History*.
7. Alan Donagan, *op. cit.*, p. 155 of the reprint in Dray.
8. Alan Donagan, *loc. cit.*
9. John Rawls, *A Theory of Justice*, p. 143. Cambridge, Mass.: Harvard University Press, and Oxford: Clarendon Press, 1971.
10. Jon Elster, "Rationality." In Guttorm Fløistad (ed.), *Contemporary Philosophy. A New Survey*. Vol. 2, pp. 11–131. The Hague: Nijhoff, 1982.
11. See John Rawls, *A Theory of Justice*, pp. 19–21, 48–51, 577–587.
12. Neil L. Wilson, "Substances without substrata," *Review of Metaphysics* 12 (1959), 521–539. W. V. Quine, *Word and Object*, pp. 59 and 69. Cambridge, Mass.: MIT Press, 1960. Donald Davidson, "On the very idea of a conceptual scheme," *Proceedings of the American Philosophical Association* (1974), 5–20, esp. p. 19, and "Thought and talk," in Samuel Guttenplan (ed.), *Mind and Language: Wolfson College Lectures 1974*, pp. 7–23, esp. pp. 20–22. Oxford: Clarendon Press, 1975. Richard Grandy, "Reference, meaning and belief," *Journal of Philosophy* 70 (1973), 439–452, esp. p. 443.
13. Donald Davidson, "Radical interpretation," *Dialectica* 27 (1973), 313–327.
14. In Guttenplan, *op. cit.*, pp. 25–44.
15. "Hermeneutics and the hypothetico-deductive Method," *Dialectica* 33 (1979), 319–336, esp. pp. 333–335.
16. Amos Tversky, "A critique of expected utility theory: Descriptive and normative considerations," *Erkenntnis* 9 (1975), 163–173. Donald Davidson, "Hempel on explaining action," *Erkenntnis* 10 (1976), 239–253.
17. Tversky, *op. cit.*, p. 172.

Chapter 20

The Nature and Scope of Rational-Choice Explanation

Jon Elster

How do rational-choice explanations explain? What are their limits and limitations? I want to discuss these questions in three steps. In section 1 the topic is the more general category of intentional explanation of behavior. Section 2 adds the specifications needed to generate rational-choice explanation. Section 3 considers more closely the power of rational-choice theory to yield unique deductions. In particular, this concerns the possible nonunicity and even nonexistence of optimal choice.

1 *Intentionality*

To explain a piece of behavior intentionally is to show that it derives from an intention of the individual exhibiting it. A successful intentional explanation establishes the behavior as *action* and the performer as an *agent*. An explanation of this form amounts to demonstrating a three-place relation between the behavior (B), a set of cognitions (C) entertained by the individual, and a set of desires (D) that can also be imputed to him. The relation is defined by three conditions, that form the topic of this section.

First, we must require that the desires and beliefs are *reasons* for the behaviour. By this I mean:

(1) Given C, B is the best means to realize D.

The presence of such reasons is not sufficient for the occurrence of the behavior for which they are reasons. An actor might be asked to shudder as part of a scene. Even with the requisite beliefs and desires, he might find himself unable to shudder at will. More important, even if the behavior does occur, the reasons do not suffice to explain it. The sight of a snake on the set might cause the actor to shudder involuntarily. This also holds if we assume that the actor is in fact able to shudder at will, if his intention to shudder is preempted by the sight of the snake. We must add, then, a clause ensuring that his behavior was actually caused by his intention to behave in that way:

(2) C and D caused B.

The reasons, that is, must also be causes of the action which they rationalize.[1] To see why this is also insufficient, we must look into the ways in which beliefs and desires can act as causes. Consider a rifleman aiming at target. He believes that only by hitting the target can he achieve some further goal that he values extremely highly. The belief and the desire provide reasons for a certain behavior: pulling the trigger when the rifle is pointed toward the target. They may, however, cause him to behave quite differently. If he is unnerved by the high stakes, his hand might shake so badly that he pulls the trigger at the wrong moment. If he cared less about hitting the target, he might have succeeded more easily. Here the strong desire to hit the target acts as a cause, but not qua reason. To act qua reason, it would at the very least have to be a cause of the behavior for which it is a reason.

Now consider Davidson's well-known example:

> A climber might want to rid himself of the weight and danger of holding another man on a rope, and he might know that by loosening his hold on the rope, he could rid himself of the weight and the danger. This belief and want might so unnerve him as to cause him to loosen his hold, and yet it might be the case that he never *chose* to loosen his hold, nor did he do so intentionally.[2]

Here conditions (1) and (2) are fulfilled, yet the beliefs and desires do not cause the behavior qua reasons. The example differs from that of the rifleman in that the beliefs and desires of the climber cause the very same behavior for which they are reasons, but is similar in that they do not cause it qua reasons. It is a mere accident that in the case of the climber they happen to cause the very same behavior for which they are reasons. Hence we must add:

(3) C and D caused B qua reasons

As in other cases, we may ask by virtue of which features the cause produced its effect. When the falling of the stone leads to the breaking of the ice, we point to the weight of the stone, not to its color, to explain what happened. When the desire of the rifleman causes him to miss the target, we point to something like psychic turbulence or emotional excitement, not the strength of the desire. The latter reflects the agent's evaluation of the importance of the goal, compared to other goals that he might entertain. Hence the strength of the desire is primarily relevant for its efficacy qua reason, and only to the extent that the desire causes behavior qua reason for the behavior is its strength also relevant for its causal efficacy. The emotional halo surrounding the desire is irrelevant for its efficacy qua reason, but may influence its efficacy qua nonrational cause. To be sure, these are loose and metaphorical manners of speaking. We do not yet have a good language for getting emotions and their relevance for action into focus. Yet I take it that no one would deny the phenomenological reality of the facts I am describing, or the need for something like clause (3) in order to exclude a certain kind of accidental coincidences, just as clause (2) was needed to exclude another kind of conincidence.

Although these clauses would have to be satisfied in a fully satisfactory intentional explanation, we usually make less stringent requirements. An analogy would be the detective story that proceeds by inquiring into motive and opportunity. When a person engages in a certain kind of behavior, we already know that he had the opportunity. If he did it, he could do it (in one sense of "could"). If in addition we find that he had a motive and also knowledge of the opportunity, we usually conclude that we have found an intentional explanation of the behavior, even if the kind of coincidences excluded by clauses (2) and (3) might conceivably have been operating. In some special cases we might want to reduce the likelihood of the first kind of coincidence, by also establishing that the agent had the ability to perform the behavior in question, e.g., the ability to shudder at will or the ability to hit a target. While this does not fully eliminate the possibility of coincidence, it does so for most practical purposes. The point is that satisfaction of clauses (2) and (3) requires us to scrutinize the actual mental machinery at work, which is something we are only exceptionally able to do.

By contrast, establishing motive, opportunity, knowledge, and ability is a much easier task (which is not to say that it is at all an easy one).

The nonsufficiency of clause (1) in establishing an intentional explanation is related to the difference between explaining and predicting action. If (1) were sufficient for explanation, we could also use it for prediction. There is, however, no regular lawlike connection between having certain desires and beliefs on the one hand, and performing a certain action on the other.[3] However, just as for practical purposes clause (1) goes a long way toward explaining behavior, one may with some practical confidence predict that motive, opportunity, etc. will result in action. This chapter, nevertheless, is mainly concerned with first-best explanation.

2 *Rationality*

Rational-choice explanation goes beyond intentionality in several respects. For one thing, we must insist that behavior, to be rational, must stem from desires and beliefs that are themselves in some sense rational. For another, we must require a somewhat more stringent relation between the beliefs and desires on the one hand and the action on the other.

Minimally, we require that:

(4) The set of beliefs C is internally consistent.

(5) The set of desires D is internally consistent.

One might think that these are required not just for rational-choice explanation, but for intentional explanation more generally. If for instance, there is *no* way of realizing a given desire, because it is internally inconsistent, how could anyone choose the *best* way to realize it? The answer, of course, is that the agent must believe that the desire is feasible. This belief, in turn, is internally inconsistent. For the belief that a certain goal is feasible to be consistent, there must be some possible world in which it is feasible. And that implies that there must be some further world in which it is realized, contrary to the assumption. Yet purposive action may spring from such inconsistent mental states. Someone may believe that the best way of trisecting the angle by means of ruler and compass is by first drawing a certain auxiliary construction. That drawing can then be explained in terms of the logically inconsistent goal of trisecting the angle in this way, and the belief that the goal is feasible and best attained by first taking that step. If this is not an intentional explanation, nothing is, but we might not want to call it a rational-choice explanation.

True, this example is controversial, because the implicit notion of rationality might seem to be too stringent. In fact, it seems to confuse irrationality with lack of mental competence. To this one may answer that while there need not be anything irrational in wanting to bring about a goal that happens to be logically inconsistent, rationality requires that we should be aware of the possibility that it might not be feasible. To believe, unconditionally, in the feasibility of a certain mathematical construction can be irrational, regardless of its actual feasibility. This however, pertains to the well-groundedness of the belief, not to its internal consistency; I return to this issue below. There are, however, other and more clear-cut examples of actions deriving from internally inconsistent desires or beliefs. The belief "It will rain if and only if I do not believe it will rain" is logically inconsistent,[4] yet people might decide, on the basis of this belief, to bring their umbrella along for a trip across the Sahara. Also one may cite the less exotic phenomena of intransitive preferences, inconsistent time preferences, subjective probabilities over exhaustive and exclusive events that do not add up to 1, etc.[5]

One might want to demand more rationality of the beliefs and desires than mere consistency. In particular, one might require that the beliefs be in some sense substantively well grounded, i.e., inductively justified by the available evidence. This, to be sure, is a highly problematic notion; yet here I assume throughout that it is a meaningful one. The analysis of rational belief then closely parallels that of intentional action. Again there are three conditions to be satisfied:

(1b) The belief must be the best belief, given the available evidence.

(2b) The belief must be caused by the available evidence.

(3b) The evidence must cause the belief "in the right way."

Of these, the first condition presupposes some rather strong rule of inductive inference. The second is needed to rule out the possibility that one has hit on the best belief merely by accident. It may be possible, for example, to arrive by wishful thinking at the belief which also happens to be the best.[6] The third condition is needed to exclude the possibility that by considering the evidence one might arrive at the belief which is in fact warranted by it—but by an incorrect process of reasoning. There could, for instance, be several compensating errors in the method of inference.[7] Once again, we may make the distinction between this first-best analysis of rational belief-formation, and the less demanding condition that only (1b) be satisfied.

Given the satisfaction of (1b), (2b), and (3b), the belief is explained by its well-groundedness with respect to the available evidence E. One might want to make this part of the definition of rational-choice explanations:

(W) The relation between C and E must satisfy (1b), (2b), and (3b).

For reasons set out in section 3, this proposal is incomplete. It needs to be supplemented by a condition about how much evidence it is rational to collect.

Could one, similarly, demand substantive rationality of the desires? If so, what requirements would one want to impose on the rational formation of desires and preferences? Although I believe it possible to suggest the beginning of an answer to these questions, the results are not sufficiently robust to be reported here.[8] We do need, however, an additional condition on the relation between desires and behavior. This is designed to exclude akratic behavior, or weakness of the will.

Consider the man who wants to stop smoking, and yet yields to temptation when offered a cigarette. In accepting it, he behaves in conformity with conditions (1) through (5). He desires to smoke: a perfectly consistent goal. He believes that he is offered a cigarette, not just a plastic imitation. Hence the best way to realize his desire is to accept it, which he does. This, however, gives only part of the picture. The account mentions that there are reasons for smoking, but omits the reasons against smoking. When discussing intentional explanation, I implicitly used an existential quantifier: there exist a set of beliefs and a set of desires that constitute reasons for the action and that actually, *qua* reasons, cause it. But these need not be all the reasons there are. The agent may have a desire to stay in good health that would provide a reason for not accepting the offer. Moreover, he might think that this desire outweighs the immediate wish to smoke: all things considered, he had better reject the offer. And yet he might take it. To exclude such akratic behavior from being considered rational, we must add the following condition;

(6) Given C, B is the best action with respect to the full set of weighed desires.

There are various accounts of how akratic behavior comes about. To my mind, the most plausible is offered by Donald Davidson, who argues that it occurs because of faulty causal wiring between the desires and the action.[9] The weaker reason may win out because it blocks the stronger ones from operating; or the stronger reasons might lose because they cause another behavior than that for which they are reasons. In either case, condition (1) fails to hold for the full set of desires. The action is intentional, but irrational.

Is there a cognitive analogy to condition (6)? This would have to be part of condition (1b). By considering only part of the evidence, one might form a belief that is the best relative to that part, but not the best relative to the whole evidence. A related, although different process is at work when one decides to stop collecting evidence at the point where it favors the belief that, on other grounds, one wants to be true. I return to this shortly.

3 *Optimality*

The explanatory force in condition (1) derives from the requirement that the explanandum "the best" means to accomplish the agent's goal. The enormous success of rational-choice models in economics and other sciences is due to their apparent ability to yield unique, determinate predictions in terms of maximizing behavior. Although, generally speaking, explanation may take the form of elimination as well as determination,[10] the explanatory ideal in science is always to form hypotheses from which a unique observational consequence can be deduced. In this section I want to consider some difficulties with this view when applied to the social sciences. For one thing, there may be several options that are equally and maximally good; for another, there may be no "best" option at all. One might retort that these are nonstandard cases that, like the problems underlying conditions (2) and (3), only arise in rather perverse situations. This replique is not valid. There exists a strong general argument to the effect that uniquely maximizing behavior is in general not possible.

Consider first the nonunicity of optimal choice, arising because the agent is indifferent between several options than which none better. There is then no room left for rational choice; yet typically the agent will be able at least to "pick" one of the options.[11] A fully satisfactory theory would then offer a causal supplement to the rational-choice explanation, by indicating how perceptual salience or some other value-neutral feature of the situation led to one option rather than another being "picked." Or, alternatively, one might redefine the choice situation by bunching the top-ranked alternatives into a single option. If I am indifferent between a red umbrella and a blue umbrella, but prefer both to a raincoat, the choice becomes determinate once we have bunched the first two options as "an umbrella." This way out, however, may be unavailable if the top-ranked alternatives differ along more than one dimension, since then the indifference between the options could be due to offsetting virtues rather than to value-neutrality.

The presence of multiple optima can create a good deal of embarrassment. General-equilibrium theory, for instance, is not really able to cope with this problem. In the simplest version of this theory, all optima in production and consumption are assumed to be unique. Given some additional assumptions, one can then show that there is a set of prices that will allow all markets to clear when agents optimize. In the more complex version, multiple optima are allowed. The equilibrium concept is correspondingly modified, to mean the existence of a set of prices and a set of optimizing acts that allow market clearing.[12] The difficulty is not that the choice of these acts rather than other optimizing acts would be a pure accident. Rather it is that the indeterminacy is essential for the existence proof to go through. In the actual world, there is no indeterminacy. One optimum will always be chosen. Clearly, if one had a theory that explained which of the maximally good options is chosen (or picked), it would be an improvement over a theory which leaves this indeterminate. Yet it would destroy the existence proof, by introducing a discontinuity in the reaction functions.

In game theory, multiple optima abound. In the wide class of noncooperative games that have an equilibrium point, many have equilibria that consist of mixed strategies. In any equilibrium point of mixed strategies any actor has many optimal strategies, given that all the others choose their equilibrium strategies. In fact, any pure strategy or linear combination of pure strategies is as good as any other. Why, then, should an actor choose the equilibrium strategy? John Harsanyi argues that the lack of any good answer to this question is a basic flaw in game theory as traditionally conceived. He proposes a substitute solution concept, according to which only "centroid" or equiprobabilistic mixed strategies are allowed. This corresponds to the idea that when there are several optima, one is chosen at random by "what amounts to an unconscious chance mechanism inside [the player's] nervous system."[13] This, of course, is essentially a causal concept.

Consider now the nonexistence of optimal behavior, which can arise in strategic as well as nonstrategic situations. A simple case obtains when an agent has incomplete preferences, so that for at least one pair of alternatives x and y it is neither true that he weakly prefers x to y nor that he weakly prefers y to x. If a pair of such noncomparable options is on the top of the agent's preference ranking, in the sense that for each of them it is true that there is none better, it will not be true that there is at least one alternative that is at least as good as all others. In actual cases it may seem hard to distinguish between incomparability and indifference, but the following test should help us. If there is an alternative (perhaps outside the feasible set) that is preferred to x, then it should also be preferred to y if the relation is one of indifference,[14] but this implication does not hold in cases of noncomparability.

As suggested by Sen and Williams, noncomparability may be especially important when our rankings are sensitive to the welfare of other people.[15] Assume that I have the choice between giving ten dollars to one of my children and giving them to another. I may well find myself unable to decide, and, moreover, find that I am equally unable to choose between giving eleven dollars to the first and ten dollars to the second, although I would rather give eleven than ten to the first. This would indicate that I simply am unable to assess the welfare they would derive from the money in a sufficiently precise way to allow me to make up my mind. Yet decisions will usually be made (although in this case paralysis of action is perhaps more plausible than in some other cases)[16] so for their explanation we must look beyond rational-choice theory.

Preference can be defined over outcomes or over actions. I shall assume that the latter are derived from the former, so that one prefers an action over another because one prefers the outcome it brings about.[17] I have just discussed the case in which the preferences among actions are incomplete, because the corresponding outcome-preferences are. Action-preferences may, however, be incomplete even when the outcome-preferences are complete, viz. if one is in the presence of uncertainty. Observe first that in condition (1) the notion of "best" is to be taken in a subjective sense—"best" relative to the beliefs of the agent. This includes the case of probabilistic beliefs, in which to act rationally means to maximize expected utility. Sometimes, however, it is not possible to establish subjective probabilities on which one can rationally rely on making up one's mind. In decisions concerning nuclear energy, for instance, it seems pointless to ask for the subjective probability attached to the event that a given democratic country some time in the next millennium turns into a military dictatorship that could use the reactor plutonium to make bombs.[18] And I believe the same problem arises in many cases of short-term planning as well. In decision making under uncertainty it is only under very special conditions that we can pick out the top-ranked action. Specifically, this requires that there is one option such that its worst consequence is better than the best consequence of any other option.[19] Failing this, rationality is no guide to action, and a fortiori not a guide to explanation of action.

Nonexistence of optimal choice may also stem from the strategic nature of the situation. There are two cases: either there is no equilibrium point, or there are several equilibria, none of which can be singled out as the solution.[20] The first can arise when the set of alternatives is unbounded or open. In the game "Pick a number—and the player who has picked the largest number wins," there is no equilibrium set of strategies because the strategy set is unbounded. Hyperinflation sometimes looks a bit like this game. In the game "Pick a number strictly smaller than 1—and the player who has picked the largest number wins," there is no equilibrium point because the set is open. One may illustrate this with a variant of the game of "Chicken," in which the point is to drive at top speed toward a wall and then stop as close to it as possible.

More central, probably, are games that do have equilibria but no unique solution. The standard version of "Chicken" illustrates this concept. Here two players are driving straight toward each other, and the point is not to be the first to swerve. There are two equilibria, in each of which one driver swerves and the other does not, but there is no way in which rationality alone will help the players converge toward the one or the other. An example of this interaction structure could be some forms of technical innovation, characterized by "Winner takes all."[21] The individual firm will have little incentive to invest in R&D if other firms invest heavily, and a strong incentive to do so if others do not. I want to insist that such cases illustrate the nonexistence of optimal choice, rather than its nonunicity. When there are multiple equilibria, individual agents cannot toss a coin between the various equilibrium strategies attached to them. True, by coordinating their actions they might toss a coin between the full equilibrium strategy sets, but in that case we have left the domain of individual rationality with which we are concerned here.

Let me pause at this point by drawing a little diagram, to summarize what has been said so far about nonunicity and nonexistence of optimal choice.

I have been arguing for the following phenomena:

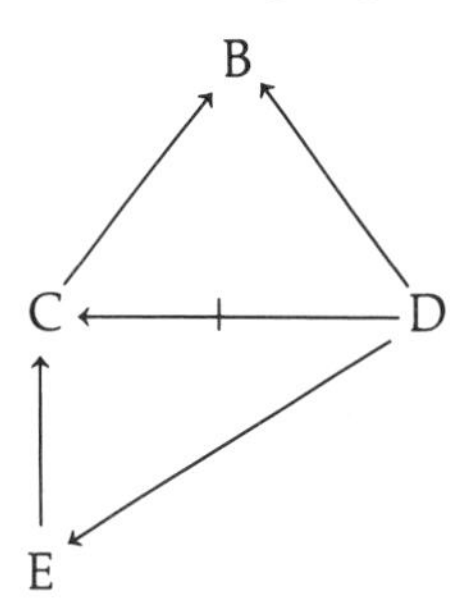

a. Nonunicity of optimal behavior, given D and C.
b. Nonexistence of optimal behavior, given D and C.
c. Nonexistence of optimal beliefs, given E.

Here, D and E have been considered as given. I said above that I did not want to enter into the speculative question whether D could also be subject to rationality criteria, but we surely have to ask this question concerning E. How much evidence is it rational to collect before forming the belief on the basis of which one decides to act? Every decision to act can be seen as accompanied by a *shadow decision*—the decision about when to stop collecting information. The former can be no more rational than the latter, on which it is based, although it may well be less rational if some other things go wrong in the sequence.

In most cases it will be equally irrational to spend no time on collecting evidence and to spend most of one's time doing so. In between there is some optimal amount of time that should be spent on information gathering. This, however, is true only in the objective sense that an observer who knew everything about the situation could assess the value of gathering information and find the point at which the marginal value of information equals marginal costs. But of course the agent who is groping toward a decision does not have the information needed to make an optimal decision with respect to information collecting.[22] He knows, from first principles, that information is costly and that there is a trade-off between collecting information and using it, but he does not know what that trade-off is.

> It is like going into a big forest to pick mushrooms. One may explore the possibilities in one limited region, but at some point one must stop the exploration and start picking because further exploration as to the possibilities of finding more and better mushrooms would defeat the purpose of the outing. One must decide on an intuitive basis, i.e. without actually investigating whether further exploration would have yielded better results.[23]

To repeat, this argument does not imply that any decision about when to stop information gathering is arbitrary. There will usually be many specific pieces of information that one knows it is worthwhile acquiring. One knows that in order to build a bridge there are some things one must know. These form a lower bound on information collection. An obvious upper bound is that one must not spend so much time gathering the information that it becomes pointless. If one wants to predict next day's weather, one cannot spend more than twenty-four hours gathering evidence. Sometimes the gap between the upper and the lower bound can be narrowed down considerably, notably in highly stereotyped situations like medical diagnostics. One then has a basis for estimating, with good approximation, the expected value of more information. In many everyday decisions, however, not to speak of military or business decisions, a combination of factors conspires to pull the lower and upper bounds apart from one another. The situation is novel, so that past experience is of limited help. It is changing rapidly, so that information runs the risk of becoming obsolete. If the decision is urgent and important, one may expect both the benefits and the opportunity costs of information collecting to be high, but this is not to say that one can estimate the relevant marginal equalities.

The upper and lower bounds on information collection are determined in part by the nature of the problem, in part by one's preferences. When building a bridge with profit as objective and safety as constraint, one will have different bounds than when using safety as objective and profit as constraint. There is nothing wrong, therefore, in the presence of a causal link between D and E, as drawn in the diagram. Note, however, that desires can determine the collection of information in another way, more related to wishful thinking. (Wishful thinking in the diagram is indicated by the line from D to C—blocked in order to indicate that this is not a proper causal influence.) One may stop collecting evidence at the point where the sum total of the evidence collected until then favors the belief that one would want to be true. Sometimes this is clearly irrational, viz if one is led to stop collecting evidence before the lower bound has been reached. But what if the wish for a certain belief to hold true leads one to collect an amount of evidence well between the lower and upper bounds? Imagine a general who is gathering information about the position of enemy troops. The information is potentially invaluable, but waiting to gather it exposes him to grave risks. He decides to attack when *and because* the net balance of information so far leads him (rationally) to believe that the enemy is highly vulnerable. I am not sure about this case, but I submit that his procedure is not irrational. The wish in this case functions merely as a heuristic device that allows him to make a decision. There is no reason to think that the causal influence of the wish tends to make the decision worse than it would have been had a different device been used.

In short, the only condition one can impose on E is rather vague:

(N) One should collect an amount of evidence that lies between the upper and lower bounds that are defined by the problem situation, including D.

Correspondingly, we may impose the following condition on the relation between evidence, belief and desires:

(7) The relation between C, D and E must satisfy (1b), (2b), (3b) and (N).

This concludes my account of rational-choice explanation.

4 *Summary*

Ideally, a fully satisfactory rational-choice explanation of an action would have the following structure. It would show that the action is the (unique) best way of satisfying the full set of the agent's desires, given the (uniquely) best beliefs the agent could form, relatively to the (uniquely determined) optimal amount of evidence. We may refer to this as the *optimality part* of the explanation. In addition the explanation would show that the action was caused (in the right way) by the desires and beliefs, and the beliefs caused (in the right way) by consideration of the evidence. We may refer to this as the *causal part* of the explanation. These two parts together yield a first-best rational-choice explanation of the action. The optimality part by itself yields a second-best explanation, which, however, for practical purposes may have to suffice, given the difficulty of access to the psychic causality of the agent.

If follows from section 3 that even the second-best explanation runs into serious difficulties. It rests on three uniqueness postulates: unique determination of the optimal evidence, of the optimal beliefs given the evidence and of the optimal action given the beliefs and the desires. Each of the links in the chain has been challenged, in the sense that both the unicity and the very existence of optimality have been shown to be problematic in certain cases. The most serious challenge arises at the level of information gathering, since it will only exceptionally be possible for the agent to determine the marginal cost and benefit of more information. The challenge at the next level arises in cases of uncertainty, i.e., when the evidence does not allow any belief, even a probabilistic one, to be formed. Finally, the link from mental states to action was shown to be problematic, both with respect to unicity and with respect to existence.

Given that one or more of these links fails to yield a unique optimum, the explanation cannot take the form of determination; rather it must consist in eliminating some of the abstractly possible actions. At each level, it is possible to eliminate some of the options in the feasible set. The nature of the problem sets upper and lower bounds on the amount of information one should collect. In cases of uncertainty one should at least not choose an action that has worse best consequences than the worse worst consequences of some other action. In cases of indifference or noncomparability one should not choose an option to which some other alternative is strictly preferred. In games without solutions it is less clear what options are eliminated.

Under the same assumption, the rational-choice explanation must be supplemented by a causal account. At the level of information gathering one may refer to the fact that people have different aspiration levels. Some people spend ten minutes, others two hours, looking for the best place for mushrooms. In decision making under uncertainty one may invoke such psychological features as optimism or pessimism to explain why people choose maximax or maximin strategies. When the indeterminacy occurs at the level of action, the explanation may involve perceptual salience (in the case of indifference or noncomparability) or a desire for security (if the maximin behavior is chosen in games without solution).

Hence rational-choice explanation may fail because the situation does not allow a unique behavioral prediction from the hypothesis that agents behave rationally. But we should not forget that it sometimes fails simply because people act irrationally. They yield to wishful thinking, in the sense of letting their desires determine their beliefs or the amount of evidence they collect before forming their beliefs (assuming that the result is below the lower bound). Or they succumb to weakness of will, in the sense of acting for the sake of a desire which they themselves value less highly than the remaining set of desires. Finally, their intentions and beliefs may be subject to various inconsistencies that are also incompatible with rational choice.

Let me point to a final consequence of this analysis. It has shown that there are many dimensions of latitude in the notion of rationality. Correspondingly, we get more degrees of freedom in our interpretation of other people. In trying to understand each other, we are guided and constrained by the idea that by and large others are as rational as ourselves. The slack in the concept of rationality implies that we are able to understand more, although it also implies that our understanding will be more diffuse.[24]

Notes

I thank Marcelo Dascal, Dagfinn Føllesdal and Michael Root for their comments on an earlier version of this chapter.

1. Here, as elsewhere, my debt to Donald Davidson's work will be obvious.
2. D. Davidson, *Essays on Actions and Events* (Clarendon Press, Oxford, 1980), p. 79.
3. Ibid., ch. 11.
4. It is inconsistent because there is no possible world in which the belief is both true *and believed* (J. Hintikka, *Knowledge and Belief*, Cornell University Press (Ithaca, N.Y., 1961)).
5. Cp. J. Elster, *Sour Grapes*, Cambridge University Press (Cambridge, 1983), ch. I, for more details.
6. This is contested by D. Pears, *Motivated Irrationality* (Oxford University Press, Oxford, 1984), ch. 5. He argues that motivated, irrational belief formation always takes the form of a failure to correct an irrational belief, not the positive form of directly producing it; hence there is never any superfluous irrationality. I disagree, but the point is not essential to my argument, since there are other ways in which a belief might be caused by something else than the available evidence. A person might be hypnotized into forming a belief for which he also has good evidence, without having formed the belief prior to the hypnosis, since we do not usually put together the pieces of information in our mind unless there is a need to do so.
7. R. Nisbett and L. Ross, *Human Inference: strategies and shortcomings of social judgment* (Prentice-Hall, Englewood Cliffs, N.J., 1980), pp. 267–8.
8. See Elster, *Sour Grapes*, ch. 1.3.
9. Davidson, *Essay on Action and Events*, ch. 2. An alternative account is that of Pears, *Motivated Irrationality*.
10. R. Ashby, *Introduction to Cybernetics* (Chapman & Hall, London, 1971), p. 130.
11. E. Ullmann-Margalit and S. Morgenbesser, "Picking and choosing," *Social Research*, 44 (1977), pp. 757–85.
12. See G. Debreu, *Theory of Value* (Wiley, New York, 1959), and many later expositions.
13. J. Harsany, *Rational Behavior and Bargaining Equilibrium in Games and Social Situations* (Cambridge University Press, Cambridge, 1977), p. 114.
14. This follows if we make the assumption of consistent preferences (K. Suzumura, *Rational Choice, Collective Decisions and Social Welfare*, p. 8 (Cambridge University Press, Cambridge, 1984), a somewhat weaker requirement than transitivity.
15. A. Sen and B. Williams, Introduction to *Utilitarianism and Beyond*, ed. A. Sen and B. Williams (Cambridge University Press, Cambridge, 1982), p. 17.
16. The alternatives are x: give to one child, y: give to the other child and z: give to neither. It may happen that *because* I weakly prefer neither x to y nor y to x, I strongly prefer z to both, perhaps because it would create family trouble if I selected one child without being able to justify my choice in terms of welfare. Yet in the absence of x (or y), I would strongly prefer y (or x) to z.
17. I do not, of course, deny that actions may be valued for themselves. The assumption is made only for the sake of simplifying the discussion.
18. J. Elster, *Explaining Technical Change* (Cambridge University Press, Cambridge, 1983), appendix 1, has a further discussion.
19. For the proof that in decision making under uncertainty one can rationally only take account of the best and the worst consequences of each action, see K. Arrow and L. Hurwicz, "An optimality criterion for decision-making under uncertainty," in *Uncertainty and Expectation in Economics*, ed. C. F. Carter and J. L. Ford (Kelley, Clifton, N.J., 1972). The proof turns upon the idea that rational choice should remain invariant under an arbitrary reclassification of "states of nature."
20. Recent work has raised a third possibility: even if there is only one equilibrium point in the game, there may be several strategy sets that are "rationalizable" (B. D. Bernheim, "Rationalizable strategic behavior," *Econometrica*, 52 (1984), pp. 1007–28; D. G. Pearce, "Rationalizable strategic behavior and the problem of perfection," *Econometrica*, 52 (1984), pp. 1029–50).
21. See Elster, *Explaining Technical Change*, pp. 109ff., drawing on P. Dasgupta and J. Stiglitz, "Uncertainty, industrial structure and the speed of R&D," *Bell Journal of Economics*, 11 (1980), pp. 1–28.

22. Since the point is crucial, let me clarify it by means of an analogous example. In the theory of induced factor-bias in technical change, the argument was put forward (C. Kennedy, "Induced bias in innovation and theory of distribution," *Economic Journal*, 74 (1964), pp. 541–7) that firms optimize with respect to an "innovation possibility frontier." Although one may agree, at least for the sake of argument, that an omniscient observer would know which innovations are possible at a given time, it is impossible to see how this would help to explain the behavior of the firms, since there is no way in which they could acquire the same knowledge. Rational-choice explanations turn upon what the agents *believe* to be the best action, not on an objective conception of the best. Any theory that neglects this constraint lacks microfoundations (W. Nordhaus, "Some sceptical thoughts on the theory of induced innovations," *Quarterly Journal of Economics*, 87 (1973), pp. 208–19).
23. L. Johansen, *Lectures on Macro-Economic Planning* (North-Holland, Amsterdam, 1977), p. 144. Ultimately the argument derives from Herbert Simon. For a strikingly provocative discussion, see also S. Winter, "Economic 'natural selection' and the theory of the firm," *Yale Economic Essays*, 4 (1964), pp. 225–72.
24. Davidson, *Essays on Actions and Events*, ch. 11.

Chapter 21

The Principle of Charity and the Problem of Irrationality (Translation and the Problem of Irrationality)

David K. Henderson

1 *Introduction*

The principle of charity in translation is best treated as a proposed codification of the methodology for constructing translation manuals. There are various formulations of the principle. Their common characteristic is that they urge the translator to end by counting the source-language speakers as correct, either quite generally or in some specific set of utterances. The topic of this chapter is the principle of charity and the epistemic status of the claim, often derived from that principle, that any translation manual which leads us to attribute irrational beliefs (or contradictory beliefs in particular) to the subjects of our investigations is inadequate (or is less likely to be adequate than otherwise). If correct formulations of the principle of charity support this derivative claim, then attributions of irrationality in psychology and the social sciences are improper (or are less likely to be proper than otherwise), being based upon evidence derived from a suspect translation manual. It is requisite, then, that such attributions be withdrawn (or held in abeyance) and that the attempt be made to revise the translation manual toward the end of attributing rationality instead. In this way, the principle of charity is taken to legislate against the possibility of irrational belief or behavior (or to make all attributions of such suspect), and this in spite of the substantial evidence for their fairly frequent occurrence. More disagreebly still, irrationality is ruled out (or militated against) not so much on empirical as on purely methodological grounds. That cogent considerations should seem to lead to such results, I call the "problem of attributions of irrationality" or simply, the "*problem of irrationality.*"

As is reflected above, the problem of irrationality varies in seriousness with the strength of the formulation of the principle of charity settled upon. Minimally, it involves taking attributions of obvious inconsistency to be evidence against the sponsoring translation manual. Thus, a formulation of the principle of charity gives rise to the problem of irrationality if it supports the following claims: Attributions of irrationality, at least of obvious inconsistency, detract from our confidence in the translation manual upon which they are based. Thus, such attributions count as evidence against psychological and social scientific accounts.

Among the writers who have discussed some principle of charity (under various rubrics) and who have concluded that attributions of irrationality, and of obvious inconsistency in particular, count against translation manuals on which they are based, are Donald Davidson, Martin Hollis, Stephen Turner, Peter Winch, and, on some readings, W. V. Quine.[1] This conclusion seems incompatible with what we know about ourselves and our own society. For not only do we attribute relatively obvious irrationality to individuals encountered on the street and in the psychological laboratory—a result which most of the above writers would be able to tolerate[2]—but we also occasionally use attributions of obvious inconsistency in explanatory accounts of large-scale social and political phenomena, such as religious behavior, voting behavior, and political toleration. Any account of the methodology of translation and interpretation that rules out, or undermines, apparently successful explanatory accounts deserves careful, critical scrutiny before being accepted.

Most writers recognize it would not be surprising or disturbing if there were, implicit within a complex belief system, inconsistencies for which no "proof" had been found. This leads them to count only attributions of "obvious inconsistency" as evidence against the sponsoring translation manual.[3] My concern is that even this does not sufficiently restrict the principle of charity. I am particularly interested in the status of attributions of obvious violations of basic norms of rationality, such as are codified in the first-order predicate calculus and in decision theory.

I argue in the next section that the problem of irrationality arises from a failure to appreciate the limitations upon the principle of charity following from the justifications for that principle's use in constructing translation manuals. In the third section I argue that the principle of charity, where it applies, is properly subsumable under a more fundamental principle that applies throughout the construction of translation manuals: so translate as to maximize the explicability of beliefs and actions attributed to subjects. I show that this "principle of explicability" allows us to attribute irrationality to our subjects in certain situations. I do not conceive of this chapter as repudiating the principle of charity so much as refining present formulations of that principle.

2 *Provisional Translation Manuals and the Principle of Charity as Preparatory*

I begin with the following rough distinction between the earlier and later stages in the construction of translation manuals. In the early stages, a rudimentary translation manual is constructed that is adequate to translating a wide range of sentences in "everyday usage."[4] In these early stages of inquiry the principle of charity is, in practice and in principle, necessary. I call a rudimentary translation manual that is adequate to most everyday contexts a "first-approximation translation manual"; I call the results of the application of such a translation manual "first-approximation translations."

As the first-approximation translation manual is used in the investigation of the society studied, we may be led to attribute apparently awkward sets of beliefs or even apparent inconsistencies to the source-language speakers. We might, for example, find ourselves attributing to them beliefs expressed using indicative sentences such as "Twins are birds,"[5] or "We are red macaws."[6] Or we might find ourselves attributing an inconsistency at the level of first-order logic in certain of their beliefs, as E. E. Evans-Pritchard seems to do when discussing the Azande.[7] When the first-approximation translation manual leads us to such attributions, alternative suggestions for modifying, or refining, the manual are often advanced. I will refer to the results of such fiddling to "fine tune" a first-approximation translation manual as a "refined translation manual," or simply as an "R-translation manual," and to the results of its application as "R-translations."[8]

It is important to note that, according to the received view, the principle of charity continues to constrain the translator in *all* phases of his or her work: both in the construction of first-approximation translation manuals and R-translation manuals. Thus, for example, Hollis discusses the anthropological study of ritual and myth as being, to the end, so constrained.[9] Quine seems to share this general approach.[10] Davidson believes that the principle of charity is constraining in interpreting the utterances of those for whom we already have a basic translation manual. It even constrains us when we have a durable homophonic first-approximation translation manual, for example, in our interpretation of subjects of psychological experiments and intimate friends.[11] This commitment to the application of the principle of charity in all stages of constructing translation manuals is the target of my central argument.

The line of argument I now present resolves the problem of irrationality at the level of R-translation by attending to differences between the earlier and later stages in constructing translation manuals. I show that while the principle of charity is constraining in the early stages of constructing translation manuals, it is not constraining in the later stages, where we become involved with R-translation. In effect, it does not apply there. Thus, in the construction of R-translation manuals, the principle of charity does not dictate avoidance of attributions of obvious inconsistency to the speakers of the source-language. Instead, in R-translation, the concern for attributing explicable beliefs and actions, which is repeatedly exhibited in actual social scientific studies, controls and may lead us to explanatory accounts that include attributions of obvious inconsistency.

One view of the principle of charity is that it is, in an important sense, *preparatory*. Davidson formulates this position when he remarks that the charitable method he advocates "is not designed to eliminate agreement, nor can it; its purpose is to make meaningful disagreement possible, and this depends entirely on a foundation—*some* foundation—in agreement."[12] This is, I think, correct: the charitable development of a first-approximation translation manual is undertaken in order to make meaningful disagreement possible. We want to be able to identify where we agree and disagree; we want to be able to investigate variations in beliefs—how beliefs, patterns of beliefs, and the courses of their development vary with the natural and social environment. To do this we must first construct a first-approximation translation manual with which we can identify apparent disagreement worthy of further investigation. Thus, the principle of charity in application to the earlier stages of translation is preparatory to subsequent investigation and explanation.

Quine's and Hollis's discussions of the principle of charity bring out the manner in which the principle is compelling in the early stages of constructing a translation manual. Quine argues with particular forcefulness that there are, among others, two sorts of behavioral evidence for the construction of translation manuals: observation sentences (paired according to stimulus meaning) and truth-functional operators (identified by patterns of assent and dissent). In breaking into the language—in beginning to construct a translation manual for the language—we must make use of such evidence in the only plausible way, a way that is on the whole charitable.[13] Hollis follows suit, citing Quine and writing of the need to find a way into the maze that is the source-language and of the need to establish a bridgehead of agreement for further translation.

Of course, any adequate (or roughly adequate) translation manual will provide for the translation of more than observation sentences and truth-functions. Quine, Davidson, and Hollis discuss the generally charitable treatment of other utterances in constructing first-approximation translation manuals.[14]

Constructing R-translation manuals proceeds against a background of first-approximation translation manuals that are *generally successful* in satisfying charitable constraints. I emphasize that later stages are concerned with refining a "generally successful" first-approximation manual because, although constructing such a first-approximation is constrained by a weighted global principle of charity,[15] first-approximation translation manuals are not so "successful" as to dispense with all attributions of error, nor should they be.

The reason that successful first-approximation translation manuals do not dispense with all attributions of error is that first-approximation translation manuals, as their label indicates, are provisional. They serve as a basis for further work, although not necessarily further charitable development. As Davidson notes, when using the principle of charity "[n]ot all the evidence can be expected to point in the same way."[16] Accordingly, unless we go on modifying our translation manual in extraordinary ways, we will find we are led by our manual to attribute errors to the speakers of the source-language. But such extraordinary elaboration is not appropriate in constructing first-approximation translation manuals. These are manuals that have been developed, according to a weighted principle of charity, to a point where attributions of egregious error can be taken seriously. At some point, our success in the tasks noted above will result in a translation manual that provides for the (more or less) charitable translation of a wide range of source-language constructions (in addition to observation sentences and truth-functions). It is only against the background of such translation manuals that attributions of egregious error or obvious inconsistency can be troubling. There are two reasons for this: (1) the identification of putative, flagrant error or conspicuous inconsistency can be made sufficiently precise and well grounded to warrant investigation only using such a manual, and (2) well-grounded information needed for the investigation of such attributions of error (for explicability) is accessible only through such a manual.

Attributions of egregious error and obvious inconsistency do raise questions about the adequacy of the translation manual on which they are based, but ones that I believe can be answered without ultimately undermining our confidence in the translation manual. Certain sorts of beliefs are initially surprising in that they are not generally expected, although *upon further examination* it may turn out that they are to be expected under the circumstances. Such (initially) *anomalous* beliefs call for further investigation in order to determine whether they are ultimately explicable or not. Attributions of egregious error generally comprise one type of anomaly. Attributions of surprising correct beliefs—say, for example, knowledge of parts of atomic physics possessed by a "primitive" tribe—comprise another type. It is with respect to a particular attribution of anomalous beliefs that translation manuals become first-approximation manuals. To designate a translation manual a "first-approximation" does not indicate that it will subsequently be changed in the face of attributions of serious error (or other anomalous belief). It indicates, rather, that the translation manual provides the context within which such putative error (or anomalous belief) can be examined.

There does not seem to be any precise way to specify how much or just where agreement must be provided under translation before a translation manual can serve as a first-approximation translation manual. This varies relative to the attributed beliefs to be explained and the sorts of information necessary to provide an explanatory account of them. Typically, the translation manual must be generally successful in charitably translating observation sentences, truth-functions, low-level empirical generalizations concerning technological matters, and occasion sentences and low-level generalizations concerning kinship and status relations.

More generally, it is clear that first-approximation translation manuals must be sufficiently developed to provide for the charitable translation of a wide range of native utterances. This is because most attributions of error that prompt significant further investigation involve translation of source-language utterances beyond observation sentences and truth-functions. Typically the translation of several standing sentences and various logical constructions beyond truth-functions is involved. Furthermore, constructing a translation manual for such utterances is a holistic task, as has been persuasively argued by Davidson and Quine.[17] Thus, confidence that a translation manual provides for the translation of sentences "not conveniently linked to observable circumstances" turns upon general success in construing the source-language speakers as expressing plausible messages. Without such confidence in the translation manual, when the putative error involves sentences beyond observation sentences and truth-function, the investigator will justifiably decide that there is no "probable cause" to believe that egregious error or inconsistency has been committed, and will not embark on further investigation.

The importance of having a well-founded first-approximation translation manual is not simply that it provides us a reason for taking seriously, and investigating, putative egregious error or obvious inconsistency. Only by using the first-approximation translation manual can we acquire much of the evidence in terms of which the ensuing explanatory endeavor proceeds. This is seen in Melford Spiro's and Edmund Leach's controversy concerning the construal of utterances by the Tully River blacks regarding human procreation.[18] By relying on many of the same sources, Spiro and Leach in effect rely on much the same first-approximation translation manual as the basis for their separate attempts to construct explanatory accounts of the beliefs of the Tully River blacks regarding procreation in humans. Leach's challenge to Spiro is framed in terms of information accessible only through using first-approximation translation manuals for various languages. Only using such a manual is it possible to have reasonably settled, as data for any account, such facts as that the Tully River natives say that hunting bull-frogs is one possible antecedent to human pregnancy, or that the natives also attribute pregnancy in animals to copulation, or that the natives cite these differences as indications of the special status of human beings, or that there are groups with whom they interact whose members cite copulation as the cause of human pregnancy. That Spiro accepts essentially the same manual is testified by his acceptance of these data as common ground in the controversy. Thus, not only the puzzling beliefs, but also much of the data from which an explanatory account for the puzzle is to be constructed, come by way of a commonly accepted first-approximation translation manual. In short, R-translation depends upon such a basis.

Another example of this dual importance of first-approximation translation manuals in R-translation is provided by Evans-Pritchard's discussion of what he calls "the futility" of Zande magic. Of course, this futility, which includes both contradictions within Zande beliefs about witchcraft and the empirical falsity of their generalizations, is identified using a first-approximation translation manual. Many of Evans-Pritchard's twenty-two reasons for the Azande failing to perceive the futility (and thus continuing to use magic) also depend upon his first-approximation translations. His reason (7), for example, is, in effect, a generalization regarding the flow of information: the flow of information is limited with the result that only an observer pooling data from various sources will note the inordinate variation, bias, and disagreement in Zande beliefs concerning instances of witchcraft and magic.[19] This generalization (together with the illustrations that Evans-Pritchard provides) depends on numerous translations using the same translation manual that was used to identify the apparent futility of Zande beliefs.

Given a first-approximation translation manual that leads to attributions of apparently irrational belief and provides information for explaining them, the investigator no longer requires the principle of charity. In these later stages of constructing the translation manual—in "constructing an R-translation manual"—construing the source-language users as correct or consistent is beside the point. With access to generally correct information regarding the beliefs of source-language speakers,[20] the time has come to get on with the social scientific business at hand: explanation. Accordingly, the constraints of the principle of charity, understood as an unreduced requirement that we find the source-language speakers correct and consistent, dissipate in later stages of constructing translation manuals. In these later stages, the principle of explicability guides the investigator.

Social scientific practice evidences this primary concern with explanation in R-translation. Consider, again, the Spiro-Leach controversy. Agreeing on a first-approximation translation manual, Leach and Spiro can argue over refinements. Leach argues for modifications in the manual in order to construe the Tully River blacks as symbolically expressing aspects of their social structure. He believes that taken literally the native's beliefs are inexplicable.[21] Spiro argues otherwise, preferring to take the native's beliefs literally (given the first-approximation translation manual) and to attempt explanations of those beliefs.[22]

Similarly, Leach is willing to attribute inconsistency to the Kachin when providing an explanatory account of their beliefs in terms of the political and social uses of religious utterances and rites.[23] Robin Horton attributes errors including inconsistency to adherents of traditional religions in the course of providing explanatory accounts of such phenomena primarily at the level of the sociology of knowledge.[24] This primary concern for explanation is justifiable, while a continued concern for charity in translation, as a constraint that would undermine otherwise explanatory accounts, is unjustifiable. To opt for the latter would be to forget the preparatory role of the principle of charity in developing first-approximation translation manuals as provisional bases from which to construct R-translation manuals. When engaged in this later task, the concern is with explanation.

This is not to say that we can come to understand the source-language users as generally indifferent to logical truth and inconsistency, as possessing a "prelogical" mentality. I do believe that Quine's and Hollis's arguments concerning logic under translation are sufficient to rule this out. But the issue here is the proper treatment of limited domains of discourse where the use of a translation manual based on the general pattern of utterances within the language leads to attributions of inconsistency. My point is that when such an attribution of inconsistency can be incorporated within an explanatory account, the inconsistency of the attributed beliefs does not count as evidence against the translation manual by which they are identified.

The view I am advancing situates the construction of translation manuals within the context of the ongoing social and psychological sciences. We are to construct translation manuals so as to construe the speakers of the source-language as, so far as, possible given our present theories, holding explicable beliefs and performing explicable actions. On this view, R-translation manuals are integral parts of low-level explanatory theories. Here R-translation manuals play an important role in the ongoing development of social-scientific and psychological theory. Translation manuals are needed for the cross-cultural application of more general theory. In application, a global theory comprised of a translation manual and some social-scientific and psychological theory is put to test. If the account is not adequate or explanatory, then modifications are needed either in the translation manual, the other theoretical components, or both. If the account is successful, then the components are together supported.

Adequate translation need not lead us to construe our subjects as rational, for psychological theory and social-scientific theory are composed of more than decision theory and serve more purposes than of rationalizing explanation. This, I believe, is where Turner and Putnam go wrong.[25] Thus, theories concerning the behavior of the human organism under the stress of various social and economic conditions can have an important role in translation-cum-explanation.[26] Theories having to do with inductive errors are also clearly relevant. Thus the work of cognitive psychologists such as that discussed by Nisbett and Ross[27] could usefully be employed in accounts both attributing to and explaining egregious inductive error by the speakers of some source-language. Jon Elster has made use of such cognitive psychological results in an attempt to provide the microfoundations for a theory of ideology.[28] This could prove useful in the explanation of egregious errors and incompatibility within belief systems of various peoples. The flexibility of oral traditions can be relevant to the persistence of traditional belief systems; thus Horton appeals to recent anthropological results concerning such flexibility in the course of his account of traditional religions.[29] Versions of a theory of psychological needs have commonly been used by a diverse group of anthropologists in theories of magic.[30] The list can obviously be extended, but the point has already been made: once R-translation is understood as an integral part of the social-scientific and psychological explanatory endeavor, and we recognize the range of theory relevant to R-translation, we see that the principle of explicability that guides R-translation can lead us to attribute flagrant error and conspicuous inconsistency to the speakers of the source-language.

The principle of explicability is not a recommendation that we hide our faces in the sand upon encountering phenomena we cannot explain. This is to say, it does not constitute a recommendation that we force the phenomena we observe into conceptual molds they do not fit, awkwardly making them out to be something we can explain and ignoring anomalies. Instead, it says that where our best scientific theories are up to explaining the phenomena, we should construct translation manuals so as to make use of those theories. But where our present theories are not up to the task of explaining the phenomena in conjunction with some translation manual or other, we should seek to formulate new, testable, theories. As different theories are formulated and receive substantial confirmation, we will want to revise further our R-translation manuals to take advantage of what then become our best theories. If the revisions are successful, and the relevant phenomena are explained, then the new theories receive further confirmation.

Where an R-translation manual is grounded in an account that makes use of theories with little confirmation, or where the resulting account seems, on the whole, unexplanatory, even if it is the best we have, little confidence is to be placed in the translation manual. Thus, far from advocating the evasion of recalcitrant phenomena, my view holds that when we cannot translate so as to construe our subjects as explicable, we should both seek to construct new theories and place commensurately less confidence in our R-translation manuals in the meantime. This view is forward-looking: it sees translation as a part of social and psychological science and science as the fallible and self-correcting search for explanation.

Finally, let me mention one further aspect of the principle of charity relevant to its preparatory role, namely, its being weighted according to empirical considerations. Not only is the principle of charity a preparatory concession, necessary in the early stages of constructing a translation manual, but the weighted principle of charity as formulated by Hollis, Quine, and Davidson is a matter of picking or tailoring occasions for that concession under the guidance of empirical considerations.

While there are variations in their weightings of the principle of charity, all three of the above-mentioned thinkers weight the principle similarly and provide similar reasons for their weightings. Their reasoning, which seems to me correct, is most explicit in Quine and Davidson's writings. Both argue that certain sorts of error are, as an empirical matter, quite unlikely, and that the principle of charity should be weighted accordingly.[31] Thus, in cases where the likelihood of error is low, we can afford to indulge the methodological need to be charitable without significantly biasing the results.

Thus, the translator following the weighted principle of charity in constructing a first-approximation translation manual is led to construe the speakers of the source-language as commonly correct in cases where correct judgment and reasoning is likeliest on empirical grounds. The result is also that, on the whole, attributions of error are confined to cases where correct judgment is unlikely, where we have little reason to expect correct judgment or reasoning, and may have good reason to expect error. A translation manual constructed according to such a weighted principle of charity performs *both* of the preparatory tasks of first-approximation translation manuals admirably. Insofar as such a manual leads us to attribute agreement where agreement is to be expected, our confidence in the manual is grounded (using a minimum of charitable constraint). At the same time, such a translation manual leads us to attributions of error that typically do not fly in the face of our best theories.

3 *The Reduction of the Principle of Charity to the Principle of Explicability*

In the preceding section, I elaborated one view of the principle of charity according to which that principle is preparatory, having application only in the development of first-approximation translation manuals. However, this view—in which charity in translation is understood as an unreduced constraint on the construction of translation manuals—is not the most fundamental view of the principle of charity possible. In this section I show that the weighted principle of charity can be understood as a special case of the principle of explicability; it is reducible to the application of the principle of explicability in the construction of first-approximation translation manuals.

Of course, translating charitably, where this is understood simply as maximizing agreement, is not the same as translating so as to construe what the speakers of the source-language say as explicable. I do not attempt to reduce such simple (and simple-minded) charity in translation to the application of the principle of explicability. Descriptively and normatively, a weighted principle of charity—directing the translator to *optimize*, rather than maximize, agreement—is used in constructing such translation manuals. I reduce only the weighted principle of charity to the principle of explicability.

The view that the weighted principle of charity is a fundamental constraint on the construction of translation manuals (even first-approximation translation manuals) had already begun to show signs of yielding to a reduction in the last part of the previous section. There I noted that an appropriately weighted principle of charity was informed by empirical knowledge of where error is to be expected on the one hand and where correct judgment on the other. The principle of charity should be weighted most heavily where we know that the speakers of the source-language are most often correct. Attributions of error will count most heavily against a translation manual when the error is of a sort that is unlikely. On the other hand, the principle of charity should be least heavily weighted in cases where we can expect error. Thus, attributions of error will count least against a translation manual when the error is of a sort that is likely. The following points are central to the general reduction of the weighted principle of charity to the principle of explicability: if the empirical knowledge on which the weighting of attributions of error is based is sufficient to support explanations, then the errors attributed to the speakers of the source-language will often be explicable in terms of that knowledge. For then the translation manual that is constructed will lead us generally to attribute error where it is expected according to an explanatory theory. The actual explanation of attributed error may need to wait on the acquisition of further information, which can serve as initial conditions. But the errors attributed to subjects will primarily be of sorts that we can explain in terms of the theories used in weighting attributions of error.

If the empirical basis for weighting attributions of error is of a less developed sort that supports little in the way of explanations, but that is nevertheless generally correct, then the errors attributed to the speakers of the source-language will often be explicable in terms of some successor theory. In such cases, by applying an appropriately weighted principle of charity, if we do not so translate that we attribute explicable beliefs to speakers of the source-language, we at least so translate that the attributed beliefs will relatively often prove explicable on further investigation.

It may be that, in addition to being too ill developed to be explanatory, our best present theories are not more likely correct than mistaken, but only more likely correct than any alternative theory we can presently formulate. Still, in following such theories in weighting the principle of charity, we are maximizing our chances (given our present alternatives) of attributing ultimately explicable beliefs to the speakers of the source-language.

Generally, then, the appropriately weighted principle of charity leads us to construct manuals of translation that construe source-language users in ways that are most likely to prove explicably right and explicably wrong.

Notice that the beliefs attributed to the speakers of the source-language when using a first-approximation translation manual complying with the weighted principle of charity may be only *prospectively* explicable. This may happen for either of two different reasons, both of which are significant in the earlier stages of constructing a translation manual. One has to do with limitations in our best present theories, the other with limitations in the information available to the investigator at a given point in his inquiry.

First, certain attributed beliefs may be only prospectively explicable because our best theory of such beliefs is too ill developed to be explanatory, although we believe it to be correct as far as it goes, or believe it more likely correct than any other presently constructible alternative theory. To seek to attribute explicable errors, and to settle for attributing errors that are only prospectively explicable due to the limits of our present theories, is to work within our best present theories. At the very least, in so doing we are maximizing our chances of using a theory that is "correct as far as it goes," and, indirectly, of attributing ultimately explicable beliefs to the speakers of the source-language.

Consider the following example of a prospectively explicable error. We may have noted that, instead of multiplying probabilities, people tend to average probabilities in casual contexts, unless trained to do otherwise. This observation itself constitutes an ill-developed theory. But, by itself, it could hardly be considered an explanatory theory. Still, if it is the best information that we have, we do well to use it in weighting the principle of charity. A translation manual might then be constructed that would readily attribute such inductive errors to speakers of the source-language. Suppose then that we are led by this manual to attribute the following error to our informants: an individual judged very likely to be a Republican but rather unlikely to be a lawyer is judged moderately likely to be a Republican lawyer. Were the rough generalization all we had, then the attributed inferential error would be only prospectively explicable. However, recent advances in cognitive psychology seem to yield explanations for such errors. Nisbett and Ross argue that such errors (and many other inferential errors) can be explained in terms of the overuse of certain useful "heuristics."[32] If their new theory holds up under testing, what was once only a prospectively explicable error will have become an explicable error.

Second, certain beliefs may be only prospectively explicable because they are not explicable without the help of information unavailable until there is further development and application of the translation manual. This is, of course, particularly characteristic of the early stages of constructing a translation manual. Here, regardless of the explanatory power of our present theories, we often lack the information needed to apply them.

For example, we have various ill-developed theories—generalizations—in which we account for some inferential errors in terms of strong motivational structures such as vested interests and emotional needs or desires. However, information regarding interests and desires often requires substantial translation or interpretations. Accordingly, in the early stages of constructing a translation manual, we will frequently be unable to bring such theories into play. If these ill-developed theories give rise to explanatory successor theories, these will probably also prove, on the whole, inapplicable in the early stages.

The weighted principle of charity, then, leads us to construe the speakers of the source-language as holding beliefs that are either explicable or relatively likely to prove explicable. This is the basis for the general reduction of charity in translation to the principle of explicability. Constructing a translation manual charitably is just what the principle of explicability requires in the early stages of constructing translation manuals. In such contexts, the investigator often lacks sufficient information to provide much in the way of explanation for a judgment, correct or erroneous. Accordingly, the principle of explicability directs the investigator to translate in such contexts in a way that holds the greatest prospect of leading to explanations. To do this, the investigator begins with broad sorts of sentences, such as observation sentences, where there are good reasons to believe that the speakers of the source-language are generally correct. In such basic cases, where error is likely to prove inexplicable and correct judgment explicable, the investigator strives to attribute correct beliefs. The result is also to concentrate attributions of error in sorts of utterances and cases where we have found error most readily explicable, or where we expect most error. Thus, the weighted principle of charity is a version of the principle of explicability.

In the remainder of this section, this reductive view of the principle of charity is examined in greater detail by considering the treatment of two sorts of utterances that are central in the early stages of constructing a translation manual: observation sentences and truth-functions.

An Illustration: The Treatment of Observation Sentences

Consider charity in providing for the translation of observation sentences. One charitable procedure is that one Quine discusses: matching observation sentences by stimulus meaning. In the earliest stages of constructing a translation manual, this is what is done and what should be done (subject to qualifications I discuss below). It is charitable insofar as we are commonly correct in assenting to and dissenting from observation sentences within the source-language.

When we consider how observation sentences are paired according to stimulus meaning, we see that this is not the fully mechanical process it might first appear to be. It involves a good amount of extrapolation from limited samples on the basis of psychological considerations. Several important points may be noted separately, then put together in an account of pairing observation sentences. First, in pairing observation sentences according to the ranges of stimuli that dispose the speakers of the languages to assent to and to dissent from them, the translator relies on a very limited sample of the speech behavior of source-language speakers, and of course, such samples constitute very limited indications of speech dispositions. Second, within the sample there may be mistakes by speakers of the source-language. Third, the stimulus meanings of both target-language and source-language observation sentences include stimuli that would induce mistaken assent and mistaken dissent. Fourth, we often are able, sometimes in commonsense terms, sometimes in more sophisticated psychological terms, to account for such mistakes as perceptual illusions, the results of inattention, and so on.

In order to pair observation sentences by matching similar stimulus meanings, the translator must work from reconstructions, or educated guesses, as to the stimulus meanings of the relevant sentences. The needed reconstructions are provided only with the help of more or less explicit knowledge of likely perceptual error and success. Using a theory of perception and beginning with the sample of verbal behavior, the investigator must arrive at both a pairing of observation sentences and an associated reconstruction of stimulus meanings. This will involve tentatively pairing sentences and testing these pairings in terms of expected (correct and incorrect) utterances, given our theories of perception.

Because the stimulus meaning of observation sentences includes (deceptive) stimuli that provoke mistaken assent and dissent, the investigator who pairs observation sentences according to stimulus meaning will rightly provide for the translation of the speakers of the source-language as mistaken in roughly those cases where the speakers of the target-language would be mistaken. But it is important to notice that the error so provided for will generally be (presently or prospectively) explicable. Basically, this is so because stimulus meanings for paired sentences are reconstructed from behavioral samples using (naive or sophisticated) psychological theory. More fully, deceptive stimuli are included in the stimulus meaning of source-language utterances in part on the basis of the translator's sample of speech behavior and in part on the basis of knowledge of perceptual error. If this knowledge of perceptual error is adequate for accurately reconstructing stimulus meanings from samples of behavior, then the source-language speakers will be disposed to just those utterances we expect, and their errors will be just those we expect, given our knowledge of perceptual error. Thus, the perceptual errors that do come to be attributed to source-language users on the basis of the constructed translation manual will be explicable in terms of the same knowledge of perceptual error that went into the construction of the translation manual.

Adapting Quine's example, suppose that the investigator pairs a source-language sentence, "Gavagai," with the one-word English sentence "Rabbit." If properly done on the basis of accurately reconstructed stimulus meanings, the source-language users will assent to "Gavagai?" in the presence of a sufficiently realistic rabbit facsimile placed in the bush, as would English speakers. Of course, the investigator probably does not test this implication of his translation manual. But this and other deceptive stimuli are included in the stimulus meaning of "Gavagai" as reconstructed. Equating "Gavagai" with "Rabbit" will provide for occasional attributions of error in less contrived situations as well, some of which are likely to obtain. And given the supposed accuracy of the reconstructed stimulus meanings, the speech behavior of speakers of the source-language will be construed as erroneous in such situations. Whatever knowledge led the investigator in his reconstruction of the stimulus meaning can be employed to give an explanation of such errors, provided that that knowledge is sufficiently developed to ground explanatory accounts.

The above account must be qualified in two ways. First, a role for ill-developed theories and prospective explicability should be recognized, for translation and science are programmatic and self-correcting. Second, room must be made in the basic account for the knowledge of differences in susceptibility to perceptual errors based in differences in acquired acuteness. Hunters may be less susceptible to the trickery of fake rabbits than city folk, and jewelers less readily taken in by simulated diamonds than others. This shows that the general rule of pairing observation sentences according to simulus meaning is not itself fundamental, as Quine also recognizes.[33] The proper treatment of observation sentences in the construction of a translation manual is *not* completely charitable, *nor* is it simply a matter of matching sentences according to stimulus agreement: it is rather a matter of matching sentences according to stimulus meanings whose differences are limited to explicible ones. Accordingly, we can attribute errors, or the tendency to error, differently to speakers of the source-language and speakers of the target-language, if there are situations where, according to our theories of perception, this is to be expected given different backgrounds. Here we see that the most fundamental consideration in providing for the translation of observation sentences is the explicability of what is attributed to the speakers of the source-language.

To summarize, in dealing with observation sentences the translator seeks to match stimulus meanings, allowing for explicable differences between communities. Doing so involves the use of a more or less explicit theory of perception both in reconstructing the stimulus meanings of sentences from samples of behavior and in allowing for differences in susceptibility to perceptual errors across communities. The resulting translation manual will lead us to attribute errors to source-language speakers, if their speech dispositions are as we suppose in constructing the translation manual. Most important for purposes of this chapter, those errors will also generally be explicable in terms of our theory of perception. If our theory is too underdeveloped to support explanations, then we will seek some better, successor theory (supposedly supporting many of the same attributions) to explain the errors attributed to the speakers of the source-language. Accordingly, the weighted charitable procedure used in providing for the translation of observation sentences leads the translator to attribute (presently or prospectively) explicable beliefs and utterances to the speakers of source-language observation sentences.

The treatment of observation sentences in the construction of translation manuals is a particularly striking instance of the weighted principle of charity's leading to attribution of explicable beliefs. Here theories of perception obviously play an important role in the recognizably charitable procedure for dealing with observation sentences, a role that quite directly ensures the (present or prospective) explicability of the errors and successes that come to be attributed under translation. In providing for the translation of many other broad sorts of utterances, our knowledge of explicable error does not play a direct role in the procedure for treating utterances of the broad sort. In such cases, knowledge of error and success applies more crudely. In these cases, our knowledge of the likelihood of error and success leads us to "weight disagreements" according to the broad sort of utterance involved.

As I have explained, this weighted charitable treatment of broad sorts of utterances in constructing a first-approximation translation manual is central to my general reduction of charity in translation to the principle of explicability. The treatment of truth-functions exemplifies this basic and central case.

An Illustration: The Treatment of Truth-Functions

The treatment of truth-functions in constructing a translation manual is more simply charitable than is the treatment of observation sentences. In providing for the translation of truth-functions, prevailing patterns of assent and dissent are matched with the truth-conditions for truth-functions (allowing, of course, for the sorts of ambiguity familiar to English speakers). To take a simple case, a source-language particle will be construed as the sign for conjunction if it is found that the source-language users almost universally assent to a compound sentence comprised of that particle and two sentences just in case they assent individually to both the sentences used in the compound sentence.[34] This charitable search for a good fit, maximizing agreement on truth-functions, is tempered in one major way: such fit is at a premium within a range of less complex source-language utterances (or sets of utterances); it is less important in dealing with complex utterances. The procedure described here is a matter of assigning a high negative weight to attributions of error involving the general sort of utterance, truth-functions, subject to one major qualification concerning complexity and simplicity. I believe that we can discern the principle of explicability in this weighted charitable treatment of truth-functions.

To begin with, truth-functions are behaviorally quite simple, and dispositions to the prevailing patterns of assent and dissent are effectively instilled. In this situation, we can either fit such patterns with the truth-conditions for truth-functions so as to construe the source-language users as almost always correct, or fit patterns and truth-conditions so as to construe them as very widely mistaken. The fact that the source-language speakers have evidently learned their lessons well does not alone decide for us which option to take. However, the charitable option is the obvious choice for the following reasons.

So to translate as to construe the source-language speakers as widely mistaken, although they have learned their society's truth-functional patterns well, would be to construe them either as mistaken concerning the meaning of their own truth-functional constructions or as suffering from a systematic logical incompetence (or both). As Quine observes of a similar case of perverse translation, to so construct our translation manual "reduces to nonsense when we reflect that there is nothing in meaning that is not in behavior."[35] When patterns of assent and dissent associated with certain source-language particles fit the truth-conditions for truth-functions, there is no behavioral basis for an uncharitable construal of these source-language constructions. To construct a translation manual uncharitably here would be unmotivated, and pointless.

This is a fine response, as far as it goes. But it only shows that there is no reason not to be charitable and that, consequently, to translate truth-functions uncharitably would be entirely gratuitous. More can be said. There is also a reason *not* to translate uncharitably. Put simply, to be uncharitable would be much more trouble than it is worth.

What leads us to choose the charitable option in translating truth-functions is the realization that a translation-cum-explanation for source-language speakers' behavior will be much more readily obtained when following the charitable option. This is because the representation of behavioral associations is the job of translation manuals *and* sociological and psychological theories. Choosing the charitable option, the theories we have developed for use on ourselves become available to account for minor deviations. Choosing the uncharitable option, we find what seem to be their extraordinary mistakes inexplicable, unexpected. We have no theories that will begin to take up the burden such a translation manual would place upon the theoretical components of translation-cum-explanation. Our theories would not combine with such an uncharitable translation manual to represent the associations we must represent.

We could, of course, attempt to develop theories that combined with uncharitable translation manuals (for truth-functions) to provide representations of the behaviors found in the full range language communities. Rather than embarking on such an unpromising and unmotivated attempt to construct wholly new psychological theories to explain putatively pervasive truth-functional logical errors, we choose to make use of present theories and their descendants. Thus, we choose the basically charitable option for dealing with truth-functions.

Notice that the above argument for constructing a translation manual so as to translate truth-functions charitably is grounded in the exigencies of constructing explanatory accounts (translation-cum-explanations) given our present and foreseeable stock of theories. This argument is of a piece with the view that charity in developing a first-approximation translation manual is preparatory, and with the general argument for the reduction of the principle of charity to the principle of explicability provided at the beginning of the present section.

Attention to the limits of our ability to account for truth-functional errors also uncovers the reasons for the one distinction in truth-functional utterances that is made in providing for their translation. We have relatively few ways of explaining truth-functional errors. One is to appeal to the complexity or length of the sentences or arguments in question, where they *are* complex or long. The length and complexity of truth-functional utterances can be gauged even in the early going. Thus, the relative explicability of error in such cases can be built into the charitable procedure used. By seeking to maximize agreement in simple cases, truth-functional errors attributed under translation come generally to be localized in cases involving complex utterances or sets of utterances. Again, this is as it should be according to the principle of explicability.

Other ways of explaining truth-functional errors include noting stressful circumstances in which the sentence or argument is encountered or contrived, and noting loyalties or biases. Significantly, such accounts of truth-functional error must often be supported by information about the reasoner that is not accessible in the early stages of constructing a translation manual. Accordingly, such explanations of error are not readily built into a weighted principle of charity for treating truth-functions. However, such accounts typically support explanations only of limited, or isolated, deviations from the general pattern being treated as indicative of the correct usage. Such effects can be ignored in charitably settling upon the basic translations for source-language truth-functions by attention to the prevailing patterns of assent and dissent. Thus, aside from the weighting by complexity, the principle of charity for dealing with truth-functions is a matter of maximizing agreement.

Once an approximate fit between truth-conditions for truth-functions and patterns of assent to and dissent from sentences involving source-language particles is established, and the source-language truth-functions are identified accordingly, errors will come to be attributed to certain individuals and, perhaps, in certain contexts, to many or most individuals. Most, but not all, of such apparent truth-functional errors will occur in cases where the utterances (or sets of utterances) are long or complex. As the translation manual is developed further, we will be able to bring to bear information having to do with stresses and biases in attempts to explain such errors (especially those where complexity or length is not a prominant factor). As explained in section 2, this is to become involved in R-translation.

I have repeatedly shown, both by general argument and by examining the treatment of two important broad sorts of utterances, that the weighted principle of charity, as applied in the construction of a first-approximation translation manual, leads to just the results indicated by my principle of explicability. However, the reduction of the weighted principle of charity to the principle of explicability does not depend simply on the fact that these principles counsel the same things in the construction of first-approximation translation manuals. It is also supported by the realization that the weighted principle of charity follows where the principle of explicability and our empirical theories lead. This is illustrated by the treatment of truth-functions. The generally charitable treatment of truth-functions can be understood as dictated by the search for combinations of translation manuals with our scientific theories that adequately represent the behavioral associations in source-language communities. The weighting of the principle of charity in dealing with truth-functions of differing complexity is the result of allowing for attributions of errors where they can most readily be explained.

As we move beyond truth-functions to the weighted charitable treatment of various other sorts of utterances, modifications in the basic charitable procedure begin to accumulate. As the modifications or weightings accumulate, it becomes at best inelegant to view the principle of charity as a fundamental statement of the constraints on the construction of translation manuals. The weighted principle of charity becomes a sort of methodological equivalent of Ptolemaic astronomy. A more perspicacious view is achieved by stating the fundamental constraints on translation in terms of the desideratum that induces modifications in other (less fundamental) constraints. Thus, the principle of charity comes to be understood as a special case of the principle of explicability.

Summarizing, I have presented a general argument for the reduction of the principle of charity to the principle of explicability. This is followed and illustrated by a detailed discussion of the charitable treatment of two important broad sorts of utterances—observation sentences and truth-functions—in constructing first-approximation translation manuals.

The general reduction of the principle of charity to the principle of explicability turns on the way the appropriate weighting of the principle of charity reflects our knowledge of the likelihood of successes and errors. This weighting not only reflects empirical knowledge, but also results in the construction of translation manuals that tend to lead us to attribute explicable error and success to the speakers of the source-languages. This result obtains because the appropriate weighting of attributions of error leads us to concentrate these attributions in the sorts of cases where errors are most readily explained. The result is just that called for by the principle of explicability. (Of course, given the present status of psychology and the social sciences, room must be made for the provisional use of theories that are too ill developed to provide explanations. However, this requires no more than a recognition of the provisional character of translation manuals as well as of the psychological and social sciences themselves).

The treatment of truth-functions in the construction of a first-approximation translation manual illustrates how attributions of errors involving a broad sort of utterance come to be weighted according to our knowledge of the likelihood of errors. In some cases knowledge of the antecedents of, and the relative likelihood of, errors and successes does more than determine a weighting of disagreements according to the broad sorts of utterances involved. Such knowledge can play a role in the process of constructing the translation manual. This is seen in the treatment of observation sentences, where theories of perception have an important role in reconstructing and matching relevantly similar stimulus meanings. In such cases, the theories that lead us to construe the source-language users as mistaken also serve to explain the errors that come to be attributed under translation. Or, they do so insofar as they are sufficiently developed to support explanations. Thus, not only is the weighted principle of charity preparatory—needed in the construction of first-approximation translation manuals that provide the basis for explanatory accounts at the level of R-translation—but it is itself a special case of the principle of explicability.

To apply the weighted principle of charity in the construction of first-approximation translation manuals is to construct, as best as one can in the early going, a translation manual suitable for translating the source-language speakers as holding explicable beliefs and performing explicable actions. Accordingly, it would be nothing less than absurd to see the principle of charity as undermining otherwise explanatory accounts in which obvious inconsistency or other egregious error is attributed to the speakers of the source-language.

Notes

I am grateful to Robert Barrett, Roger Gibson, and Paul Roth for their generous assistance.

1. Hollis, M.: 1982, "Reason and Ritual, The Limits of Irrationality," in *Rationality*, pp. 214–30; "The Social Destruction of Reality," in M. Hollis and S. Lukes (eds.), *Rationality and Relativism*, MIT Press, Cambridge, Massachusetts, pp. 67–86; D. Davidson: 1984, "Truth and Meaning," in *Inquiries into*

Truth and Interpretation, pp. 17–36; 1984b, "Radical Interpretation," in *Inquiries into Truth and Meaning*, pp. 125–39; 1980, "Towards a Unified Theory of Meaning and Action," *Grazer Philosophische Studien* 11, 1–12; 1976, "Paradoxes of Irrationality," *Review of Metaphysics* 29, 289–305. Davidson believes that there are better forms for a theory of meaning to take than the form of a translation manual; thus he discusses theories of meaning and of interpretation rather than translation manuals. The distinctions he draws between his Tarski-style theories and translation manuals are not of importance for my concerns here (see Davidson, "Radical Interpretation," p. 129). I believe that the arguments that I develop apply, *mutatis mutandis*, to charity in the construction of Davidsonian theories of meaning. Accordingly, I will generally use the familiar rubric of translation: Quine. W. V. O.: 1960, *Word and Object*, MIT Press, Cambridge, Massachusetts, pp. 57–60, 69; 1970, *Philosophy of Logic*, Prentice-Hall, Englewood cliffs, NJ, pp. 80–83; 1970, "Philosophical Progress in Language Theory," *Metaphilosophy* 1, 2–19. Turner, S.: 1980, *Sociological Explanation as Translation*, Cambridge University Press, Cambridge, Winch, P.: 1964, "Understanding a Primitive Society," *American Philosophical Quarterly* 1, 307–24.

2. Davidson seems to be an exception (Davidson, "Paradoxes of Irrationality") For my purposes here, "irrationality" is the violation of basic norms of rationality, such as are codified in the first-order predicate calculus and in decision theory. Whether there is some more refined and normative sense of "rationality" is not my concern here. My concern here is with the tendency to rule out attributions of (obvious) violations of basic norms of rationality.
3. Hollis, "Reason and Ritual," p. 234; Quine, *Philosophy of Logic*, pp. 82–83.
4. The phrase is Hollis's ("Reason and Ritual," pp. 222–23). The class includes logical truths, most observation sentences, and a good deal more.
5. Evans-Pritchard, E. E.: 1956, *Neur Religion*, Clarendon Press, Oxford, p. 131.
6. Crocker, C.: 1977, "My Brother the Parrot," in J. D. Sapir and J. C. Crocker (eds.), *The Social Uses of Metaphor: Essays in the Anthropology of Rhetoric*, University of Pennsylvania Press, Philadelphia, p. 192.
7. According to Evans-Pritchard, the Azande believe that witchcraft is a biologically inherited trait, passed along from father to son and from mother to daughter. However, this theoretical belief is contradicted by the limited willingness to attribute witchcraft to entire clans. The apparent contradiction cannot be resolved by Zande appeals to bastardy, which are evidently not sufficiently extensive (nor persuasive). It may be that the contradiction was not originally obvious to the Azande. However, the contradictions were made obvious, at least to a limited number of Azande, by Evans-Pritchard's intervention. He discussed the problem with his informants and he reports that "Azande see the sense of this argument but they do not accept its conclusions, and it would involve the whole notion of witchcraft in contradiction were they to do so." (Evans-Pritchard, *Witchcraft Among the Azande*, p. 24.) But the recalcitrant fact is that the Azande see the sense of the argument, and see that their beliefs lead to contradictions, yet apparently do not modify their beliefs about the inheritance of witchcraft. Short of getting someone to assent to a sentence of the form "*p* and not *p*," it is hard to imagine a clearer case of someone persisting in logical inconsistency once it has become obvious.
8. This broad usage of "translation" is becoming quite common. My thinking here is informed by Stephen Turner's writings and by D. Føllesdal: 1981 "Understanding and Rationality," in H. Parret and J. Bouveresse (eds.), *Meaning and Understanding*, de Gruyter, New York.
9. Hollis, "Reason and Ritual," p. 234.
10. Quine, *Word and Object*, p. 69.
11. Davidson, "Towards a Unified Theory of Meaning and Action"; "On the Very Idea of a Conceptual Scheme," in *Inquiries into Truth and Interpretation*, pp. 183–98, 195–96.
12. Davidson, "On the Very Idea of a Conceptual Scheme," pp. 196–97. See also, Quine, "Philosophical Progress in Language Theory," p. 17.
13. Quine, *Word and Object*, pp. 68–72.
14. Quine, "Philosophical Progress in Language Theory," p. 15; Hollis, "Reason and Ritual," pp. 232–34; Davidson, "Thought and Talk," p. 169; "Radical Interpretation," pp. 134–35.
15. Versions of the principle of charity typically include the provision for its being weighted so that some attributions of error count against the sponsoring translation manual more heavily than others. The weighting of the principle of charity according to the others is discussed in the last part of the present section and in Section 3.
16. Davidson, "Belief and the Basis of Meaning," p. 152.
17. Quine, "Philosophical Progress in Language Theory," p. 15; Davidson, "Radical Interpretation," pp. 134–35.

18. The Tully River blacks are reported by W. E. Roth to list four antecedents to (or "causes" for) pregnancy in humans. (Roth, W. E.: 1903, "Superstition, Magic and Medicine," *North Queensland Ethnographical Bulletin* 5, 22; quoted in Spiro, M.: "Virgin Birth: Parthenogenesis and Physiological Paternity," p. 242.) None of the four involves sexual relations. However, the natives are able to provide a rudimentary account of the biological basis of pregnancy in animals. Leach adopts the case as an example of symbolist reinterpretation. According to Leach: "The modern interpretation of the ritual described would be that in this society the relationship between the woman's child and the clansmen of the woman's husband stems from the public recognition of the bonds of marriage, rather than from the facts of cohabitation, which is a very normal state of affairs." Leach, E.: 1961, "Golden Bough or Gilded Twig?" *Daedalus*, p. 376.)

    Spiro holds rather that "the aborigines are indeed ignorant of physiological paternity, and that the four statements quoted by Roth are in fact proffered explanations for conception." (Spiro, M.: 1966, "Religion: Problems of Definition and Explanation," in M. Banton (ed.), *Anthropological Approaches to the Study of Religion*, Travistock, London, pp. 85–126, 112.)

    Leach objects strongly to Spiro's alternative interpretation, suggesting that it is not only wrong but, as an attribution of ignorance, also highly offensive. (Leach, E. (ed.): 1969, "Virgin Birth," in *Genesis as Myth and Other Essays*, Jonathan Cape, London, p. 92.) However, it turns out that Leach would discriminate among attributions of ignorance, ruling some out, but not others. Leach would not insist that attributions of ignorance regarding molecular biology or atomic physics are attributions of irrationality. Leach's central objection to Spiro is that he finds it "highly implausible on commonsense grounds that genuine 'ignorance' of the basic facts of physiological paternity should anywhere be a cultural fact." (Leach, "Virgin Birth," p. 93.) Thus, for Leach, those attributions of ignorance that constitute objectionable attributions of irrationality are those that seem improbable, given our *commonsense psychology*. Leach objects not to attributions of ignorance but to attributions of *inexplicable ignhorance*.
19. Evans-Pritchard, *Witchcraft Among the Azande*, p. 476.
20. In this chapter, I generally speak of translation manuals being "adequate" and "well-grounded," and of "confidence" in translation manuals. However, I do, at a few places, use the natural (but seemingly more realistic) formulations, talking of translation manuals "being basically correct," and of confidence in a translation manual as "on the whole correct." The reader may wonder what confidence in adequacy or in correctness amounts to in this context. The answer, I believe, is that the investigator who has a first-approximation translation manual of the sort discussed here can be confident that he or she is on as promising a road to the construction of explanatory accounts as possible, given present theories and prospective successor-theories. (See Section 3.)
21. Leach, "Virgin Birth," pp. 94–95.
22. Spiro, "Virgin Birth, Parthenogenesis and Physiological Paternity," pp. 255–57.
23. Leach, E.: 1954, *Political Systems of Highland Burma*, Athlone Press, London.
24. Horton, R.: 1970, "African Traditional Thought and Western Science," in B. Wilson (ed.), Rationality, Basil Blackwell, Oxford, pp. 131–71; 1982, "Tradition and Modernity Revisited," in M. Hollis and S. Lukes (eds.), *Rationality and Relativism*, MIT Press, Cambridge, Massachusetts, pp. 201–60.
25. Turner, Sociological Explanation as Translation; Putnam, H.: 1978, *Meaning and the Moral Sciences*, Routledge and Kegan Paul, London, p. 41.
26. See, for example, Jahoda's discussion of Colin Turnbull's account of the Ik. (Jahoda, G.: 1982, *Psychology and Anthropology: A Psychological Perspective*, Academic Press, New York, pp. 102–105.)
27. Nisbett, R. and Ross, L.: 1980, *Human Inference: Strategies and Shortcomings of Social Judgment*, Prentice-Hall, Englewood Cliffs. Stephen Stich and Christopher Cherniak have appealed to this literature in their arguments against an ideal rationality requirement in psychology. Stich, S.: 1985, "Could Man be an Irrational Animal," *Synthese* 64, 115–35. Cherniak, C.: 1981, "Minimal Rationality," *Mind* xc, 161–83.
28. Elster, J.: 1982, "Belief, Bias and Ideology," in M. Hollis and S. Lukes (eds.), *Rationality and Relativism*, MIT Press, Cambridge, Massachusetts, pp. 123–48.
29. Horton, "African Traditional Thought and Wester Science" and "Tradition and Modernity Revisited."
30. Malinowski, B.: 1931, "Culture," in *Encyclopaedia of the Social Sciences*, Routledge, London, pp. 627–28, 634–36. Since Malinowski, the appeal to a psychological notion of needs has often taken on a more cognitive and less emotional caste; cf. Spiro, "Religion: Problems of Definition and Explanation," p. 107; Beattie, *Other Cultures*, pp. 205–206, and Geertz, C. (ed.): 1973, "Religion as a Cultural System," in *The Interpretation of Cultures*, Basic Books, New York, pp. 87–125, 100–103.

31. Quine, *Word and Object*, p. 59; Davidson "Radical Interpretation," p. 136.
32. Nisbett and Ross, *Human Inference: Strategies and Shortcomings of Social Judgment*, pp. 3–61, 141–50. The example is theirs as well (p. 146).
33. Quine, *Word and Object*, p. 37.
34. This case is somewhat more complicated when we consider abstentions as well as assent and dissent. See Quine: 1973, *Roots of Reference*, Open Court, Lasalle, IL, 75–78.
35. Quine "Philosophical Progress in Language Theory," 16–17.

PART V

Functional Explanation

Introduction to Part V

Although functional explanations—for example, accounting for the rain dance of a preliterate society by reference to the beneficial effect of reinforcing group identity—are used in the social sciences, a number of closely related questions arise about whether they are truly explanatory. The initial puzzle is this: functional explanations purport to explain something in terms of its beneficial consequences, yet how can the consequences of something explain that thing?

A similar question can be raised about functional explanations in biology. A biologist attempts to explain the persistence of dark-colored moths in a sooty environment by citing a beneficial consequence: the protection the color provides from predators. But how can the beneficial consequence of this color in this environment explain its persistence? In biology, an acceptable answer can be given in terms of the theory of evolution: a well-understood causal mechanism links advantageous traits with differential survival. Organisms with advantageous characteristics lacked by other organisms in the environment will tend to survive and therefore have a greater chance of passing their traits on to their progeny. Thus, dark-colored moths tend to be selected over light-colored moths in a sooty environment. Hence, the beneficial consequence of the dark color of the moths explains their persistence in this environment.

An important question to raise about a proffered functional explanation in the social sciences is whether there is any reason to suppose that a causal mechanism exists linking the beneficial consequence and the practice or institution. For example, does a causal mechanism link the beneficial consequences of the rain dance and its persistence? A closely related issue is whether one can know that such a causal mechanism exists without knowing at least in rough outline how it works. Another related question is whether the existence of a causal mechanism presupposes the existence of causal laws. If so, can functional explanation in the social sciences be construed as a type of covering law explanation? If it can be, does it meet the requirements of the covering law model?

The importance of functional explanations in the social sciences goes beyond their occasional use. Such theorists as Bronislaw Malinowski and Robert Merton have advocated functionalism, the view that a very wide range of social phenomena can be functionally explained. Indeed, some functionalists have maintained that any social phenomenon has beneficial consequences that explain it. Others have made the somewhat weaker claim that if a social phenomenon has beneficial consequences, then it can be given a functional explanation. Crucial questions can be raised about both the strong and weak forms of functionalism. Even if functional explanations apply in the social sciences, it does not follow that all phenomena or even those with beneficial consequences can be functionally explained. Is there any good independent reason to suppose that functional explanations have wide application? Many social scientists have become skeptical of the sweeping claims often made about the application of functional explanations. If this skepticism is justified and functional explanations have a much more limited application in the social sciences than functionalists have supposed, might not functionalism, understood heuristically, still be an interesting position? Instead of understanding functionalism to be making expansive claims about the application of functional explanations, could it be construed as advocating a program of research directing social scientists to look for functional explanations? So interpreted, might it not be judged most appropriately by its fruitfulness in guiding research?

In the selections by Carl G. Hempel, R. P. Dore, G. A. Cohen, Jon Elster, and Harold Kincaid that follow, these and other issues concerning functional explanations and functionalism are addressed.

Hempel, one of the leading advocates of the covering law model of explanation, argues in "The Logic of Functional Analysis" (chapter 22) that once functional explanations are put into the form of covering law explanations, their weaknesses are apparent. Consider a functional explanation of a rain dance. Put it into covering law form, Hempel would say, and one sees that one cannot explain the existence of the rain dance unless one assumes the empirically unsupported premise that the rain dance is the only way to reinforce group identity. At most, one can explain that the rain dance or some functional equivalent exists. Hempel also points out other problems that functional explanations have. Given these difficulties, he concludes with the suggestion that functionalism should be regarded as a program for research: a directive to seek how "various traits of a system might contribute to its particular mode of self-regulation."

Are functional explanations in the social sciences a type of causal explanation? This question, which is independent of whether they are reducible to covering law form, is discussed in chapter 23, "Function and Cause" by Dore, an English sociologist. Consider the rain dance again. Obviously, there is some not very clearly defined causal relation between a rain dance and group identity. The crucial question for functional analysis is whether there is a causal relationship between group identity and the rain dance's tendency to persist in a society. Dore discusses three possible causal mechanisms that account for a functional relation—one that is analogous to evolutionary selection—and argues that all three ways have serious problems. In particular, he maintains that with respect to evolutionary analogies, they tend to be far-fetched and in most cases have no clear application.

Maintaining that one should interpret many of the explanations given by Karl Marx as functional, Cohen in a selection from his book, *Karl Marx's Theory of History: A Defense,* that appears as chapter 24, defends functional explanations in general, and Marx's use of them in particular, against several popular criticisms. According to Cohen, functional explanations are a type of consequence explanations; that is, they are explanations that use consequence laws, laws that relate X's tendency to do Y to X's existence. However, knowledge of an underlying causal mechanism is not essential to functional explanation. Cohen maintains that although there might be several ways to explain why a relation holds between the beneficial consequences of a practice and the practice, one can know that such a relation exists without knowing *how* it works.

Elster, in a selection from his book *Explaining Technical Change,* that appears as chapter 25, argues for the importance of distinguishing explanatory and nonexplanatory consequence laws. According to him a basic problem with many functional accounts is that they fail to meet one key requirement: they do not show how a particular causal mechanism connects the beneficial consequence of a social practice to the social practice to be explained. Thus, in the rain dance, there is no reason to suppose that the tendency of the rain dance to reinforce group identity explains the rain dance. The correlation between the existence of the dance and its tendency to reinforce group identity, he would argue, might not indicate any causal relation. Thus, he rejects Cohen's thesis that functional explanations are a type of consequence law explanation.

Kincaid, in turn, in "Assessing Functional Explanation in the Social Sciences" (chapter 26) admits that in practice functional explanations are often problematic. However, he maintains that one can in principle answer criticisms leveled against them. This is not to say that functional explanations can answer all questions that social scientists ask about social activities and institutions. Functional explanations do not answer questions about the origins of an activity or institution; they do not answer the question of why one particular activity or institution rather than some functional equivalent is found in a society; and, despite what Elster claims, they do not necessarily reveal the mechanism involved in the causal connection between the effect of the institution or practice and its persistence. On the other hand, he rejects Cohen's view that functional explanations are a type of consequence explanation. Rather, he argues that well-confirmed consequence laws might be neither necessary nor sufficient for functional explanations.

Chapter 22

The Logic of Functional Analysis

Carl G. Hempel

Empirical science, in all its major branches, seeks not only to *describe* the phenomena in the world of our experience, but also to *explain* or *understand* their occurrence: it is concerned not just with the "what?" "when?" and "where?" but definitely, and often predominantly, with the "why?" of the phenomena it investigates.

That explanation and understanding constitute a common objective of the various scientific disciplines is widely recognized today. However, it is often held that there exist fundamental differences between the explanatory *methods* appropriate to the different fields of empirical science, and especially between those of the "exact" natural sciences and those required for an adequate understanding of the behavior of humans or other organisms, taken individually or in groups. In the exact natural sciences, according to this view, all explanation is achieved ultimately by reference to causal or correlational antecedents; whereas in psychology and the social and historical disciplines—and, according to some, even in biology—the establishment of causal or correlational connections, while desirable and important, is not sufficient. Proper understanding of the phenomena studied in these fields is held to require other types of explanation.

Perhaps the most important of the alternative methods that have been developed for this purpose is the method of functional analysis, which has found extensive use in biology, psychology, sociology, and anthropology. This procedure raises problems of considerable interest for the comparative methodology of empirical science. This essay is an attempt to clarify some of these problems; its object is to examine the logical structure of functional analysis and its explanatory and predictive significance by means of an explicit confrontation with the principal characteristics of the explanatory procedures used in the physical sciences. We begin by a brief examination of the latter.

1 *Nomological Explanation: Deductive and Inductive*

In a beaker filled to the brim with water at room temperature, there floats a chunk of ice which partly extends above the surface. As the ice gradually melts, one might expect the water in the beaker to overflow. Actually, however, the water level remains unchanged. How is this to be explained? The key to an answer is provided by Archimedes' principle, according to which a solid body floating in a liquid displaces a volume of liquid which has the same weight as the body itself. Hence the chunk of ice has the same weight as the volume of water its submerged portion displaces. Now, since melting does not affect the weights involved, the water into which the ice turns has the same weight as the ice itself, and hence, the same weight as the water initially displaced by the submerged portion of the ice. Having the same weight, it also has the same volume as the displaced water; hence the melting ice yields a volume of water that suffices exactly to fill the space initially occupied by the submerged part of the ice. Therefore, the water level remains unchanged.

This account (which deliberately disregards certain effects of small magnitude) is an example of an argument intended to explain a certain event. Like any explanatory argument, it falls into two parts, which will be called the *explanans* and the *explanandum*.[1] The latter is the statement, or set of statements, describing the phenomenon to be explained; the former is the statement, or set of statements, adduced to provide an explanation. In our illustration, the explanandum states that at the end of the process, the beaker contains only water, with its surface at the same level as at the beginning. To explain this, the explanans adduces, first of all, certain laws of physics; among them, Archimedes' principle; laws to the effect that at temperatures above 0°C and atmospheric pressure, a body of ice turns into a body of water having the same weight; and the law that, at any fixed temperature and pressure, amounts of water that are equal in weight are also equal in volume.

In addition to these laws, the explanans contains a second group of statements; these describe certain particular circumstances which, in the experiment, precede the outcome to be explained; such as the facts that at the beginning, there is a chunk of ice floating in a beaker filled with water; that the water is at room temperature; and that the beaker is surrounded by air at the same temperature and remains undisturbed until the end of the experiment.

The explanatory import of the whole argument lies in showing that the outcome described in the explanandum was to be expected in view of the antecedent circumstances and the general laws listed in the explanans. More precisely, the explanation may be construed as an argument in which the explanandum is deduced from the explanans. Our example then illustrates what we will call explanation by deductive subsumption under general laws, or briefly, *deductive nomological explanation*. The general form of such an explanation is given by the following schema:

(1.1) $L_1, L_2, \ldots, L_m$
$\underline{C_1, C_2, \ldots, C_n}$ Explanans
E Explanandum

Here, $L_1, L_2, \ldots, L_m$ are general laws and $C_1, C_2, \ldots, C_n$ are statements of particular fact; the horizontal line separating the conclusion *E* from the premises indicates that the former follows logically from the latter.

In our example, the phenomenon to be explained is a particular event that takes place at a certain place and time. But the method of deductive subsumption under general laws lends itself also to the explanation of what might be called "general facts" or uniformities, such as those expressed in laws of nature, For example, the question why Galileo's law holds for physical bodies falling freely near the earth's surface can be answered by showing that that law refers to a special case of accelerated motion under gravitational attraction, and that it can be deduced from the general laws for such motion (namely, Newton's laws of motion and of gravitation) by applying these to the special case where two bodies are involved, one of them the earth and the other the falling object, and where the distance between their centers of gravity equals the length of the earth's radius. Thus, an explanation of the regularities expressed by Galileo's law can be achieved by deducing the latter from the Newtonian laws and from statements specifying the mass and the radius of the earth; the latter two yield the value of the constant acceleration of free fall near the earth.

It might be helpful to mention one further illustration of the role of deductive nomological explanation in accounting for particular facts as well as for general uniformities or laws. The occurrence of a rainbow on a given occasion can be deductively explained by reference to (1) certain particular determining conditions, such as the presence of raindrops in the air, sunlight falling on these drops, the observer facing away from the sun, etc., and (2) certain general laws, especially those of optical reflection, refraction, and dispersion. The fact that these laws hold can be explained in turn by deduction from the more comprehensive principles of, say, the electromagnetic theory of light.

Thus, the method of deductive nomological explanation accounts for a particular even by subsuming it under general laws in the manner represented by the schema (1.1); and it can similarly serve to explain the fact that a given law holds by showing that the latter is subsumable, in the same fashion, under more comprehensive laws or theoretical principles. In fact, one of the main objectives of a theory (such as, say, the electromagnetic theory of light) is precisely to provide a set of principles—often expressed in terms of "hypothetical," not directly observable, entities (such as electric and magnetic field vectors)—which will deductively account for a group of antecedently established "empirical generalizations" (such as the laws of rectilinear propagation, reflection, and refraction of light). Frequently, a theoretical explanation will show that the empirical generalizations hold only approximately. For example, the application of Newtonian theory to free fall near the earth yields a law that is like Galileo's except that the acceleration of the fall is seen not to be strictly constant, but to vary slightly with geographical location, altitude above sea level, and certain other factors.

The general laws or theoretical principles that serve to account for empirical generalizations may in turn be deductively subsumable under even more comprehensive principles; for example, Newton's theory of gravitation can be subsumed, as an approximation, under that of the general theory of relativity. Obviously, this explanatory hierarchy has to end at some point. Thus, at any time in the development of empirical science, there will be certain facts which, at that time, are not explainable; these include the most comprehensive general laws and theoretical principles then known and, of course, many empirical generalizations and particular facts for which no explanatory principles are available at the time. But this does not imply that certain facts are intrinsically unexplainable and thus must remain unexplained forever: any particular fact as yet unexplainable, and any general principle, however comprehensive, may subsequently be found to be explainable by subsumption under even more inclusive principles.

Causal explanation is a special type of deductive nomological explanation; for a certain event or set of events can be said to have caused a specified "effect" only if there are general laws connecting the former with the latter in such a way that, given a description of the antecedent events, the occurrence of the effect can be deduced with the help of the laws. For example, the explanation of the lengthening of a given iron bar as having been caused by an increase in its temperature amounts to an argument of the form (1.1) whose explanans includes (a) statements specifying the initial length of the bar and indicating that the bar is made of iron and that its temperature was raised, (b) a law to the effect that the length of any iron bar increases with the temperature.[2]

Not every deductive nomological explanation is a causal explanation, however. We cannot properly say, for example, that the regularities expressed by Newton's laws of motion and of gravitation *cause* the free fall of bodies near the earth's surface to satisfy Galileo's laws.

Now we must give at least brief consideration to another type of explanation, which again accounts for a given phenomenon by reference to general laws, but in a manner which does not fit the deductive pattern (1.1). For example, when little Henry catches the mumps, this might be explained by pointing out that he contracted the disease from a friend with whom he played for several hours just a day before the latter was confined with a severe case of mumps. The particular antecedent factors involved in this argument are Henry's exposure and, let us assume, the fact that Henry had not had the mumps before. But to connect these with the event to be explained, we cannot invoke a general law to the effect that under the conditions just mentioned, the exposed person invariably contracts the mumps: what can be asserted is only a high statistical probability that the disease will be transmitted. Again, when a neurotic trait in an adult is psychoanalytically explained by reference to critical childhood experiences, the argument explicitly or implicitly claims that the case at hand is but an exemplification of certain general laws governing the development of neuroses. But, surely, whatever specific laws of this kind might be adduced at present can purport, at the very best, to express probabilistic trends rather than deterministic uniformities: they may be construed as *laws of statistical form,* or briefly as *statistical laws,* to the effect that, given the childhood experiences in question—plus, presumably, certain particular environmental conditions in later life—there is such and such a statistical probability that a specified kind of neurosis will develop. Such statistical laws differ in form from strictly universal laws of the kind adduced in our earlier examples of explanatory arguments. In the simplest case, a *law of strictly universal form,* or briefly, a *universal law,* is a statement to the effect that in *all* cases satisfying certain antecedent conditions A (e.g., heating of a gas under constant pressure), an event of a specified kind B (e.g., an increase in the volume of the gas) will occur; whereas a law of statistical form asserts that the probability for conditions A to be accompanied by an event of kind B has some specific value p.

Explanatory arguments which, in the manner just illustrated, account for a phenomenon by reference to statistical laws are not of the strictly deductive type (1.1). For example, the explanans consisting of information about Henry's exposure to the mumps and of a statistical law about the transmission of this disease does not logically imply the conclusion that Henry catches the mumps; it does not make that conclusion necessary, but, as we might say, more or less probable, depending upon the probability specified by the statistical laws. An argument of this kind, then, accounts for a phenomenon by showing that its occurrence is highly probable in view of certain particular facts and statistical laws specified in the explanans. An account of this type will be called an *explanation by inductive subsumption under statistical laws,* or briefly, an *inductive explanation.* For the purposes of the present essay, this sketchy characterization of the explanatory use of statistical laws will suffice; a precise analysis of the method, which requires an inquiry into rather complex technical issues in inductive logic and the theory of statistical inference, reveals certain fundamental differences between deductive and inductive explanation.[3]

The two types of explanation we have distinguished will both be said to be forms of *nomological explanation;* for either of them accounts for a given phenomenon by "subsuming it under laws," that is, by showing that its occurrence could have been inferred —either deductively or with a high probability—by applying certain laws of universal or of statistical from to specified antecedent circumstances. Thus, a nomological explanation shows that we might in fact have *predicted* the phenomenon at hand, either deductively or with a high probability, if, at an earlier time, we had taken cognizance of the facts stated in the explanans.

But the predictive power of a nomological explanation goes much further than this: precisely because its explanans contains general laws, it permits predictions concerning occurrences other than that referred to in the explanandum. These predictions provide a means of testing the empirical soundness of the explanans. For example, the laws invoked in a deductive explanation of the form (1.1) imply that the kind of event described in E will recur whenever and wherever circumstances of the kind described by $C_1, C_2, \ldots, C_n$ are realized; e.g., when the experiment with ice floating in water is repeated, the outcome will be the same. In addition, the laws will yield predictions as to what is going to happen under certain specifiable conditions which differ from those mentioned in $C_1, C_2, \ldots, C_n$. For example, the laws invoked in our illustration also yield the prediction that if a chunk of ice were floating in a beaker filled to the brim with concentrated brine, which has a greater specific gravity than water, some of the liquid would overflow as the ice was melting. Again, the Newtonian laws of motion and of gravitation, which may be used to explain various aspects of planetary motion, have predictive consequences for a variety of totally different phenomena, such as free fall near the earth, the motion of a pendulum, the tides, and many others.

This kind of account of further phenomena which is made possible by a nomological explanation is not limited to future events, but may refer to the past as well. For example, given certain information about the present locations and veolcities and of the celestial bodies involved, the principles of Newtonian mechanics and of optics yield not only predictions about future solar and lunar eclipses, but also "postdictions," or "retrodictions," about past ones. Analogously, the statistical laws of radioactive decay, which can function in various kinds of predictions, also lend themselves to retrodictive use; for example, in the dating, by means of the radiocarbon method, of a bow or an ax handle found in an archaeological site.

A proposed explanation is scientifically acceptable only if its explanans is capable of empirical test, i.e., roughly speaking, if it is possible to infer from it certain statements whose truth can be checked by means of suitable observational or experimental procedures. The predictive and postdictive implications of the laws invoked in a nomological explanation clearly afford an opportunity for empirical tests; the more extensive and varied the set of implications that have been borne out by empirical investigation, the better established will be the explanatory principles in question.

2 *The Basic Pattern of Functional Analysis*

Historically speaking, functional analysis is a modification of teleological explanation, i.e., of explanation not by reference to causes which "bring about" the event in question, but by reference to ends which determine its course. Intuitively, it seems quite plausible that a teleological approach might be required for an adequate understanding of purposive and other goal-directed behavior; and teleological explanation has always had its advocates in this context. The trouble with the idea is that in its more traditional forms, it fails to meet the minimum scientific requirement of empirical testability.

The neovitalistic idea of entelechy or of vital force is a case in point. It is meant to provide an explanation for various characteristically biological phenomena, such as regeneration and regulation, which according to neovitalism cannot be explained by physical and chemical laws alone. Entelechies are conceived as goal-directed non-physical agents which affect the course of physiological events in such a way as to restore an organism to a more or less normal state after a disturbance has occurred. However, this conception is stated in essentially metaphorical terms: no testable set of statements is provided (i) to specify the kinds of circumstances in which an entelechy will supervene as an agent directing the course of events otherwise governed by physical and chemical laws, and (ii) to indicate precisely what observable effects the action of an entelechy will have in such a case. And since neovitalism thus fails to state general laws as a to when and how entelechies act, it cannot explain any biological phenomena; it can give us no grounds to expect a given phenomenon, no reasons to say: "Now we see that the phenomenon had to occur." It yields neither predictions nor retrodictions: the attribution of a biological phenomenon to the supervenience of an entelechy has no testable implications at all. This theoretical defect can be thrown into relief by contrasting the idea of entelechy with that of a magnetic field generated by an electric current, which may be invoked to explain the deflection of a magnetic needle. A magnetic field is not directly observable any more than an entelechy; but the concept is governed by strictly specifiable laws concerning the strength and direction, at any point, of the magnetic field produced by a current flowing through a given wire, and by other laws determining the effect of such a field upon a magnetic needle in the magnetic field on the earth. And it is these laws which, by their predictive and retrodictive import, confer explanatory power upon the concept of magnetic field. Teleological accounts referring to entelechies are thus seen to be pseudoexplanations. Functional analysis, as will be seen, though often worded in teleological phraseology, need not appeal to such problematic entities and has a definitely empirical core.

The kind of phenomenon that a functional analysis[4] is invoked to explain is typically some recurrent activity or some behavior pattern in an individual or a group; it may be a physiological mechanism, a neurotic trait, a culture pattern, or a social institution, for example. And the principal objective of the analysis is to exhibit the contribution which the behavior pattern makes to the preservation or the development of the individual or the group in which it occurs. Thus, functional analysis seeks to understand a behavior pattern or a sociocultural institution in terms of the role it plays in keeping the given system in proper working order and thus maintaining it as a going concern.

By way of a simple and schematized illustration, consider first the statement:

> (2.1) The heartbeat in vertebrates has the function of circulating blood through the organism.

Before asking whether and how this statement might be used for explanatory purposes, we have to consider the preliminary question: what does the statement *mean?* What is being asserted by this attribution of function? It might be held that all the information conveyed by a sentence such as (2.1) can be expressed just as well by substituting the word "effect" for the word "function." But this construal would oblige us to assent also to the statement:

> (2.2) The heartbeat has the function of producing heart sounds; for the heartbeat has that effect.

Yet a proponent of functional analysis would refuse to assert (2.2), on the ground that heart sounds are an effect of the heartbeat which is of no importance to the functioning of the organism; whereas the circulation of the blood effects the transportation of nutriment to, and the removal of waste from, various parts of the organism—a process that is indispensable if the organism is to remain in proper working order, and indeed if it is to stay alive. Thus understood, the import of the functional statement (2.1) might be summarized as follows:

> (2.3) The heartbeat has the effect of circulating the blood, and this ensures the satisfaction of certain conditions (supply of nutriment and removal of waste) which are necessary for the proper working of the organism.

We should notice next that the heart will perform the task here attributed to it only if certain conditions are met by the organism and by its environment. For example, circulation will fail if there is a rupture of the aorta; the blood can carry oxygen only if the environment affords an adequate supply of available oxygen and the lungs are in proper condition; it will remove certain kinds of waste only if the kidneys are reasonably healthy; and so forth. Most of the conditions that would have to be specified here are usually left unmentioned, partly no doubt because they are assumed to be satisfied as a matter of course in situations in which the organism normally finds itself. But, in part, the omission reflects lack of relevant knowledge, for an explicit specification of the conditions in question would require a theory in which (a) the possible states of organisms and of their environments could be characterized by the values of certain physicochemical or perhaps biological "variables of state," and in which (b) the fundamental theoretical principles would permit the determination of that range of internal and external conditions within which the pulsations of the heart would perform the function referred to above.[5] At present, a general theory of this kind, or even one that could deal in this fashion with some particular kind of organism, is unavailable, of course.

Also, a full restatement of (2.1) in the manner of (2.3) calls for criteria of what constitutes "proper working," "normal functioning," and the like, of the organism at hand; for the function of a given trait is here construed in terms of its causal relevance to the satisfaction of certain necessary conditions of proper working or survival of the organism. Here again, the requisite criteria are often left unspecified—an aspect of functional analysis whose serious implications will be considered later (in section 5).

The considerations here outlined suggest the following schematic characterization of a functional analysis:

> (2.4) *Basic pattern of a functional analysis:* The object of the analysis is some "item" i, which is a relatively persistent trait or disposition (e.g., the beating of the heart) occurring in a system s (e.g., the body of a living vertebrate); and the analysis aims to show that s is in a state, or internal condition, c_i and in an environment presenting certain external conditions c_e such that under conditions c_i and c_e (jointly to be referred to as c) the trait i has effects which satisfy some "need" or "functional requirement" of s, i.e., a condition n which is necessary for the system's remaining in adequate, or effective, or proper, working order.

Let us briefly consider some examples of this type of analysis in psychology and in sociological and anthropological studies. In psychology, it is especially psychoanalysis which shows a strong functional orientation. One clear instance is Freud's functional characterization of the role of symptom formation. In *The Problem of Anxiety,* Freud expresses himself as favoring a conception according to which "all symptom formation would be brought about solely in order to avoid anxiety; the symptoms bind the psychic energy which otherwise would be discharged as anxiety."[6] In support of this view, Freud points out that if an agoraphobic who has usually been accompanied when going out is left alone in the street, he will suffer an attack of anxiety, as will the compulsion neurotic who, having touched something, is prevented from washing his hands. "It is clear, therefore, that the stipulation of being accompanied and the compulsion to wash has as their purpose, and also their result, the averting of an outbreak of anxiety."[7] In this account, which is put in strongly teleological terms, the system *s* is the individual under consideration; *i* his agoraphobic or compulsive behavior pattern; *n* the binding of anxiety, which is necessary to avert a serious psychological crisis that would make it impossible for the individual to function adequately.

In anthropology and sociology the object of functional analysis is, in Merton's words, "a standardized (i.e., patterned and repetitive) item, such as social roles, institutional patterns, social processes, cultural pattern, culturally patterned emotions, social norms, group organization, social structure, devices for social control, etc."[8] Here, as in psychology and biology, the function, i.e., the stabilizing or adjusting effect, of the item under study, may be one not consciously sought (and indeed, it might not even be consciously recognized) by the agents; in this case, Merton speaks of *latent* functions—in contradistinction to *manifest* functions, i.e., those stabilizing objective effects which are intended by participants in the system.[9] Thus, e.g., the rain-making ceremonials of the Hopi fail to achieve their manifest meteorological objective, but they "may fulfill the latent function of reinforcing the group identity by providing a periodic occasion on which the scattered members of a group assemble to engage in a common activity."[10]

Radcliffe-Brown's functional analysis of the totemic rites of certain Australian tribes illustrates the same point:

> To discover the social function of the totemic rites we have to consider the whole body of cosmological ideas of which each rite is a partial expression. I believe that it is possible to show that the social structure of an Australian tribe is connected in a very special way with these cosmological ideas and that the maintenance of its continuity depends on keeping them alive, by their regular expression in myth and rite.
>
> Thus, any satisfactory study of the totemic rites of Australia must be based not simply on the consideration of their ostensible purpose..., but on the discovery of their meaning and of their social function.[11]

Malinowski attributes important latent functions to religion and to magic: he argues that religious faith establishes and enhances mental attitudes such as reverence for tradition, harmony with environment, and confidence and courage in critical situations and at the prospect of death—attitudes which, embodied and maintained by cult and ceremonial, have "an immense biological value." He points out that magic, by providing man with certain ready-made rituals, techniques, and beliefs, enables him "to maintain his poise and his mental integrity in fits of anger, in the throes of hate, of unrequited love, of despair and anxiety. The function of magic is to ritualize man's optimism, to enhance his faith in the victory of hope over fear."[12]

There will soon be occasion to add to the preceding examples from psychoanalysis and anthropology some instances of functional analysis in sociology. To illustrate the general character of the procedure, however, the cases mentioned so far will suffice: they all exhibit the basic pattern outlined in (2.4). We now turn from our examination of the form of functional analysis to a scrutiny of its significance as a mode of explanation.

3 *The Explanatory Import of Functional Analysis*

Functional analysis is widely considered as achieving an *explanation* of the "items" whose functions it studies. Malinowski, for example, says of the functional analysis of culture that it "aims at the explanation of anthropological facts at all levels of development by their function"[13] and he adds, in the same context: "To explain any item of culture, material or moral, means to indicate its functional place within an institution,"[14] At another place, Malinowski speaks of the "functional explanation of art, recreation, and public ceremonials."[15]

Radcliffe-Brown, too, considers functional analysis as an explanatory method, though not as the only one suited for the social sciences: "Similarly one 'explanation' of a social system will be its history, where we know it—the detailed account of how it came to be what it is and where it is. Another 'explanation' of the same system is obtained by showing (as the functionalists attempt to do) that it is a special exemplification of laws of social physiology or social functioning. The two kinds of explanation do not conflict, but supplement one another."[16]

Apart from illustrating the attribution of explanatory import to functional analysis, this passage raises two points which bear on the general question as to the nature of explanation in empirical science. We will therefore digress briefly to comment on these points.

First, as Radcliffe-Brown stresses, a functional analysis has to refer to general laws. This is shown also in our schematic characterization: the statements that i, in the specified setting c, has effects that satisfy n, and that n is a necessary condition for the proper functioning of the system, both involve general laws. For a statement of causal connection this is well known; and the assertion that a condition n constitutes a functional prerequisite for a state of some specified kind (such as proper functioning) is tantamount to the statement of a law to the effect that whenever condition n fails to be satisfied, the state in question fails to occur. Thus, explanation by functional analysis requires reference to laws.[17]

The second point relates to a concept invoked by Radcliffe-Brown, of a historic-genetic explanation, which accounts for an item such as a social system or institution by tracing its origins. Clearly, the mere listing of a series of events preceding the given item cannot qualify as an explanation; temporal precedence does not in itself make an event relevant to the genesis of the item under consideration. Thus, a criterion of relevance is needed for the characterization of a sound historic-genetic explanation. As brief reflection shows, relevance here consists in causal or probabilistic determination. A historic-genetic explanation will normally proceed in stages, beginning with some initial set of circumstances which are said to have "brought about," or "led to," certain events at a later time; of these it is next argued that by virtue of, or in conjunction with, certain further conditions prevailing at that later time, they led to a specified further set of events in the historical development; these are in turn combined with additional factors then prevailing and lead to a still later stage, and so forth, until the final explanandum is reached. In a genetic account of this kind, the assertion that a given set of circumstances brought about certain specified subsequent conditions clearly has to be construed as claiming a nomological connection of causal, or more likely, of probabilistic, character. Thus, there is tacit reference to general laws of strictly universal or of statistical form; and a historic-genetic explanation can be construed schematically as a sequence of steps each of which has the character of a nomological explanation. However, while in each step but the first, some of the particular facts mentioned in the explanans will have been accounted for by preceding explanatory steps, the other particular facts invoked will be brought in simply by way of supplementary information. Thus, even in a highly schematic construal, a historic-genetic explanation cannot be viewed as proceeding from information about circumstances at some initial time, via certain statistical or causal laws alone, to the final explanandum: it is essential that, as the argument goes on, additional information is fed into it, concerning certain events which supervene "from the outside," as it were, at various stages of the process under study. Let us note that exactly the same procedure would be required in the case of the melting ice if, during the period of time under consideration, the system were subject to certain outside influences, such as someone's pushing the beaker and spilling some of the water, or salt being added to the water. Basically, then, historic-genetic explanation is nomological explanation.

Returning now to the main issue of the present section, we have to ask what explanatory import may properly be attributed to functional analysis. Suppose, then, that we are interested in explaining the occurrence of a trait *i* in a system *s* (at a certain time *t*), and that the following functional analysis is offered:

(3.1) (*a*) At *t*, *s* functions adequately in a setting of kind *c* (characterized by specific internal and external conditions).
(*b*) *s* functions adequately in a setting of kind *c* only if a certain necessary condition, *n*, is satisfied.
(*c*) If trait *i* were present in *s* then, as an effect, condition *n* would be satisfied.

(*d*) (Hence,) at *t*, trait *i* is present in *s*.

For the moment, we will leave aside the question as to what precisely is meant by statements of the types (*a*) and (*b*), and especially by the phrase "*s* functions adequately"; these matters will be examined in section 5. Right now, we will concern ourselves only with the *logic* of the argument; we will ask whether (*d*) formally follows from (*a*), (*b*), (*c*), just as in a deductive nomological explanation the explanadum follows from the explanans. The answer is obviously in the negative, for, to put it pedantically, the argument (3.1) involves the fallacy of affirming the consequent in regard to premise (*c*). More explicitly, the statement (*d*) could be validly inferred if (*c*) asserted that *only* the presence of trait *i* could effect satisfaction of condition *n*. As it is, we can infer merely that condition *n* must be satisfied in some way or other at time *t*; for otherwise, by reason of (*b*), the system *s* could not be functioning adequately in its setting, in contradiction to what (*a*) asserts. But it might well be that the occurrence of any one of a number of alternative items would suffice no less than the occurrence of *i* to satisfy requirement *n*, in which case the account provided by the premises of (3.1) simply fails to explain why the trait *i* rather than one of its alternatives is present in *s* at *t*.

As has just been noted, this objection would not apply if premise (*c*) could be replaced by the statement that requirement *n* can be met *only* by the presence of trait *i*. And indeed, some instances of functional analysis seem to involve the claim that the specific item under analysis is, in this sense, functionally indispensable for the satisfaction of *n*. For example, Malinowski makes this claim for magic when he asserts that "magic fulfills an indispensable function within culture. It satisfies of definite need which cannot be satisfied by any other factors of primitive civilization,"[18] and again when he says about magic that "without its power and guidance early man could not have mastered his practical difficulties as he has done, nor could man have advanced to the higher stages of culture. Hence the universal occurrence of magic in primitive societies and its enormous sway. Hence we do find magic an invariable adjunct of all important activities."[19]

However, the assumption of functional indispensability for a given item is highly questionable on empirical grounds: in all concrete cases of application, there do seem to exist alternatives. For example, the binding of anxiety in a given subject might be effected by an alternative symptom, as the experience of psychiatrists seems to confirm. Similarly, the function of the rain dance might be subserved by some other group ceremonial. And interestingly, Malinowski himself, in another context, invokes "the principle of limited possibilities, first laid down by Goldenweiser. Given a definite cultural need, the means of its satisfaction are small in number, and therefore the cultural arrangement which comes into being in response to the need is determined within narrow limits."[20] This principle obviously involves at least a moderate liberalization of the conception that every cultural item is functionally indispensable. But even so, it may still be too restrictive. At any rate, sociologists such as Parsons and Merton have assumed the existence of "functional equivalents" for certain cultural items; and Merton, in his general analysis of functionalism, has insisted that the conception of the functional indispensability of cultural items be replaced quite explicitly by the assumption of "functional alternatives, or functional equivalents, or functional substitutes."[21] This idea, incidentally, has an interesting parallel in the "principle of multiple solutions" for adaptational problems in evolution. This principle, which has been emphasized by functionally oriented biologists, states that for a given functional problem (such as that of perception of light) there are usually a variety of possible solutions, and many of these are actually used by different—and often closely related—groups of organisms.[22]

It should be noted here that, in any case of functional analysis, the question whether there are functional equivalents to a given item *i* has a definite meaning only if the internal and external conditions *c* in (3.1) are clearly specified. Otherwise, any proposed alternative to *i*, say *i'*, could be denied the status of a functional equivalent on the ground that, being different from *i*, the item *i'* would have certain effects on the internal state and the environment of *s* which would not be brought about by *i*; and that therefore, if *i'* rather than *i* were realized, *s* would not be functioning in the same internal and external situation.

Suppose, for example, that the system of magic of a given primitive group were replaced by an extension of its rational technology plus some modification of its religion, and that the group were to continue as a going concern. Would this establish the existence of a functional equivalent to the original system of magic? A negative answer might be defended on the grounds that as a result of adopting the modified pattern the group had changed so strongly in regard to some of its basic characteristics (i.e., its internal state, as characterized by c_i, had been so strongly modified) that it was not the original kind of primitive group any more; and that there simply was no functional equivalent to magic which would leave all the "essential" features of the group unimpaired. Consistent use of this type of argument would safeguard the postulate of the functional indispensability of every cultural item against any conceivable empirical disconfirmation, by turning it into a covert tautology.

Let I be the class of those items, *i*, *i'*, *i"*, . . . , any one of which, if present in *s* under conditions c, would effect satisfaction of condition *n*. Then those items are functional equivalents in Merton's sense, and what the premises of (3.1) entitle us to infer is only:

> (3.2) Some one of the items in class 1 is present in *s at t*. But the premises give us no grounds to expect *i* rather than one of its functional alternatives.

So far, we have viewed functional analysis only as a presumptive deductive explanation. Might it not be construed instead as an inductive argument which shows that the occurrence of *i* is highly probable in the circumstances described by the premises? Might it not be possible, for example, to add to the premises of (3.1) a further statement to the effect that the functional prerequisite *n* can be met only by *i* and by a few specifiable functional alternatives? And might not these premises make the presence of *i* highly probable? This course is hardly promising, for in most, if not all, concrete cases it would be impossible to specify with any precision the range of alternative behavior patterns, institutions, customs, or the like that would suffice to meet a given functional prerequisite or need. And even if that range could be characterized, there is no satisfactory method in sight for dividing it into some finite number of cases and assigning a probability to each of these.

Assume, for example, that Malinowski's general view of the function of magic is correct: how are we to determine, when trying to explain the system of magic of a given group, all the different systems of magic and alternative cultural patterns any one of which would satisfy the same functional requirements for the group as does the actually existing system of magic? And how are we to ascribe probabilities of occurrence to each of these potential functional equivalents? Clearly, there is no satisfactory way of answering these questions, and practitioners of functional analysis do not claim to achieve their explanation in this extremely problematic fashion.

Nor is it any help to construe the general laws implicit in the statements (*b*) and (*c*) in (3.1) as statistical rather than strictly universal in form, i.e., as expressing connections that are very probable, but do not hold universally; for the premises thus obtained would still allow for functional alternatives of *i* (each of which would make satisfaction of *n* highly probable), and thus the basic difficulty would remain: the premises taken jointly could still not be said to make the presence just of *i* highly probable.

In sum then, the information typically provided by a functional analysis of an item *i* affords neither deductively nor inductively adequate grounds for expecting *i* rather than one of its alternatives. The impression that a functional analysis does provide such grounds, and thus explains the occurrence of *i*, is no doubt at least partly due to the benefit of hindsight: when we seek to explain an item *i*, we presumably know already that *i* has occurred.

But, as was briefly noted earlier, a functional analysis provides, in principle, the basis for an explanation with a weaker explanandum; for the premises (*a*) and (*b*) of (3.1) imply the consequence that the necessary condition *n* must be fulfilled in some way or other. This much more modest kind of functional explanation may be schematized as follows:

(3.3) (*a*) At time *t* system *s* functions adequately in a setting of kind *c*.
(*b*) *s* functions adequately in a setting of kind *c* only if condition *n* is satisfied.

(*e*) Some one of the items in class I is present in *s* at *t*.

This kind of inference, while sound, is rather trivial, however, except in cases where we have additional knowledge about the items contained in class I. Suppose, for example, that at time *t*, a certain dog (system *s*) is in good health in a "normal" kind of setting *c* which precludes the use of such devices as artificial hearts, lungs, and kidneys. Suppose further that in a setting of kind *c*, the dog can be in good health only if his blood circulates properly (condition *n*). Then schema (3.3) leads only to the conclusion that in some way or other, the blood must be kept circulating properly in the dog at *t*—hardly a very illuminating result. If, however, we have additional knowledge of the ways in which the blood may be kept circulating under the circumstances and if we know, for example, that the only feature that would ensure proper circulation (the only item in class I) is a properly working heart, then we may draw the much more specific conclusion that at *t* the dog has a properly working heart. But if we make explicit the additional knowledge here used by expressing it as a third premise, then our argument assumes a form considered earlier, namely, that of a functional analysis which is of the type (3.1), except that premise (*c*) has been replaced by the statement that *i* is the *only* trait by which *n* can be satisfied in setting *c*; and, as was pointed out above, the conclusion (*d*) of (3.1) does follow in this case; in our case, (*d*) is the sentence stating that the dog has a properly working heart at *t*.

In general, however, additional knowledge of the kind here referred to is not available, and the explanatory import of functional analysis is then limited to the precarious role schematized in (3.3).

4 *The Predictive Import of Functional Analysis*

We noted earlier the predictive significance of nomological explanation; now we will ask whether functional analysis can be put to predictive use.

First of all, the preceding discussion shows that the information which is typically provided by a functional analysis yields at best premises of the forms (*a*), (*b*), (*c*) in (3.1); and these afford no adequate basis for the deductive or inductive prediction of a sentence of the form (*d*) in (3.1). Thus, functional analysis no more enables us to predict than it enables us to explain the occurrence of a particular one of the items by which a given functional requirement can be met.

Second, even the much less ambitious explanatory schema (3.3) cannot readily be put to predictive use; for the derivation of the weak conclusion (*e*) relies on the premise (*a*); and if we wish to infer (*e*) with respect to some future time *t*, that premise is not available, for we do not know whether *s* will or will not be functioning adequately at that future time. For example, consider a person developing increasingly severe anxieties, and suppose that a necessary condition for his adequate functioning is that his anxiety be bound by neurotic symptoms, or be overcome by other means. Can we predict that some one of the modes of "adjustment" in the class I thus roughly characterized will actually come to pass? Clearly not, for we do not know whether the person in question will in fact continue to function adequately or will suffer some more or less serious breakdown, perhaps to the point of self-destruction.

It is of interest to note here that a somewhat similar limitation exists also for the predictive use of nomological explanations, even in the most advanced branches of science. For example, if we are to predict, by means of the laws of classical mechanics, the state in which a given mechanical system will be at a specified future time t, it does not suffice to know the state of the system at some earlier time t_0, say the present; we also need information about the boundary conditions during the time interval from t_0 to t, i.e., about the external influences affecting the system during that time. Similarly, the "prediction," in our first example, that the water level in the beaker will remain unchanged as the ice melts assumes that the temperature of the surrounding air will remain constant, let us say, and that there will be no disturbing influences such as an earthquake or a person upsetting the beaker. Again when we predict for an object dropped from the top of the Empire State Building that it will strike the ground about eight seconds later, we assume that during the period of its fall, the object is acted upon by no forces other than the gravitational attraction of the earth. In a full and explicit formulation then, nomological predictions such as these would have to include among their premises statements specifying the boundary conditions obtaining from t_0 up to the time t to which the prediction refers. This shows that even the laws and theories of the physical sciences do not actually enable us to predict certain aspects of the future exclusively on the basis of certain aspects of the present: the prediction also requires certain assumptions about the future. But, in many cases of nomological prediction, there are good inductive grounds, available at t_0, for the assumption that during the time interval in question, the system under study will be practically "closed," i.e., not subject to significant outside interference (this case is illustrated, for example, by the prediction of eclipses) or that the boundary conditions will be of a certain specified kind—a situation illustrated by predictions of events occurring under experimentally controlled conditions.

Now, the predictive use of (3.3) likewise requires a premise concerning the future, namely, (*a*); but there is often considerable uncertainty as to whether (*a*) will in fact hold true at the future time t. Furthermore, if in a particular instance there should be good inductive grounds for considering (*a*) as true, the forecast yielded by (3.3) is still rather weak; for the argument then leads from the inductively warranted assumption that the system will be properly functioning at t to the "prediction" that a certain condition n, which is necessary for such functioning, will be satisfied at t in some way or other.

The need to include assumptions about the future among the premises of predictive arguments can be avoided, in nomological predictions as well as in those based on functional analysis, if we are satisfied with prediction conclusions which are not categorical, but only conditional, or hypothetical, in character. For example, (3.3) may be replaced by the following argument, in which premise (*a*) is avoided at the price of conditionalizing the conclusion:

(4.1) (*b*) System s functions adequately in a setting of kind c only if condition n is satisfied.

(*f*) If s functions adequately in a setting of kind c at time t, then some one of the items in class I is present in s at t.

This possibility deserves mention because it seems that at least some of the claims made by advocates of functional analysis may be construed as asserting no more than that functional analysis permits conditional predictions of the kind schematically represented by (4.1). This might be the intent, for example, of Malinowski's claim: "If such [a functional] analysis discloses to us that, taking an individual culture as a coherent whole, we can state a number of general determinants to which it has to conform, we shall be able to produce a number of predictive statements as guides for field-research, as yardsticks for comparative treatment, and as common measures in the process of cultural adaptation and change."[23] The statements specifying the determinants in question would presumably take the form of premises of type (*b*); and the "predictive statements" would then be of a hypothetical character.

Many of the predictions and generalizations made in the context of functional analysis, however, eschew the cautious conditional form just considered. They proceed from a statement of a functional prerequisite or need to the categorical assertion of the occurrence of some trait, institution, or other item suited to meet the requirement in question. Consider, for example, Sait's functional explanation of the emergence of the political boss: "Leadership is necessary; and *since* it does not develop readily within the constitutional framework, the boss provides it in a crude and irresponsible form from the outside";[24] or take Merton's characterization of one function of the political machine: Referring to various specific ways in which the political machine can serve the interests of business, he concludes, "These 'needs' of business, as presently constituted, are not adequately provided for by conventional and culturally approved social structures; *consequently*, the extra-legal but more-or-less efficient organization of the political machine comes to provide these services."[25] Each of these arguments, which are rather typical of the functionalist approach, is an inference from the existence of a certain functional prerequisite to the categorical assertion that the prerequisite will be satisfied in some way. What is the basis of these inferences, which are marked by the words, "since" and "consequently" in the passages just quoted? When we say that *since* the ice cube was put in warm water it melted; or that the current was turned on, and *consequently*, the ammeter in the circuit responded, these inferences can be explicated and justified by reference to certain general laws of which the particular cases at hand are simply special instances; and the logic of the inferences can be exhibited by putting them into the form of the schema (1.1). Similarly, each of the two functionalist arguments under consideration clearly seems to presuppose a general law to the effect that, within certain limits of tolerance or adaptability, a system of the kind under analysis will—either invariably or with high probability—satisfy, by developing appropriate traits, the various functional requirements (necessary conditions for its continued adequate operation) that may arise from changes in its internal state or in its environment. Any assertion of this kind, no matter whether of strictly universal or of statistical form, will be called a (*general*) *hypothesis of self-regulation*.

Unless functional analyses of the kind just illustrated are construed as implicitly proposing or invoking suitable hypotheses of self-regulation, it remains quite unclear what connections the expressions "since," "consequently," and others of the same character are meant to indicate, and how the existence of those connections in a given case is to be objectively established.

Conversely, if a precise hypothesis of self-regulation for systems of a specified kind is set forth, then it becomes possible to explain, and to predict categorically, the satisfaction of certain functional requirements simply on the basis of information concerning antecedent needs; and the hypothesis can then be objectively tested by an empirical check of its predictions. Take, for example, the statement that if a hydra is cut into several pieces, most of these will grow into complete hydras again. This statement may be considered as a hypothesis concerning a specific kind of self-regulation in a particular kind of biological system. It can clearly be used for explanatory and predictive purposes, and indeed the success of the predictions it yields confirms it to a high degree.

We see, then, that wherever functional analysis is to serve as a basis for categorical prediction or for generalizations of the type illustrated by the passages from Sait and from Merton, it is of crucial importance to establish appropriate hypotheses of self-regulation in an objectively testable form.

The functionalist literature does contain some explicitly formulated generalizations of the kind here referred to. Merton, for example, after citing the passage from Sait quoted above, comments thus: "Put in more generalized terms, *the functional deficiencies of the official structure generate an alternative (unofficial) structure to fulfill existing needs somewhat more effectively*."[26] This statement seems clearly intended to make explicit a hypothesis of self-regulation that might be said to underlie Sait's specific analysis and to provide the rationale for his "since." Another hypothesis of this kind is suggested by Radcliffe-Brown: "it may be that we should say that ... a society that is thrown into a condition of functional disunity or inconsistency ... will not die, except in such comparatively rare instances as an Australian tribe overwhelmed by the white man's destructive force, but will continue to struggle toward ... some kind of social health."[27]

But, as was briefly suggested above, a formulation proposed as a hypothesis of self-regulation can serve as a basis for explanation or prediction only if it is a reasonably definite statement that permits of objective empirical test. And indeed many of the leading representatives of functional analysis have expressed very clearly their concern to develop hypotheses and theories which meet this requirement. Malinowski, for example, in his essay significantly entitled "A scientific theory of culture," insists that "each scientific theory must start from and lead to observation. It must be inductive and it must be verifiable by experience. In other words, it must refer to human experiences which can be defined, which are public, that is, accessible to any and every observer, and which are recurrent, hence fraught with inductive generalizations, that is, predictive."[28] Similarly, Murray and Kluckhohn have this to say about the basic objective of their functionally oriented theory, and indeed about any scientific "formulation," of personality: "the general purposes of formulation are three: (1) to *explain* past and present events; (2) to predict future events (the conditions being specified); and (3) to serve, if required, as a basis for the selection of effective measures of *control*."[29]

Unfortunately, however, the formulations offered in the context of concrete functional analyses quite often fall short of these general standards. Among the various ways in which those conditions may be violated, two call for special consideration because of their pervasiveness and central importance in functional analysis. They will be referred to as (i) *inadequate specification of scope*, and (ii) *nonempirical use of functionalist key terms* (such as "need," "functional requirement," "adaptation," and others). We will consider these two defects in turn: the former in the balance of the present section, the latter in the next.

Inadequate specification of scope consists in failure to indicate clearly the kind of system to which the hypothesis refers, or the range of situations (the limits of tolerance) within which those systems are claimed to develop traits that will satisfy their functional requirements. Merton's formulation, for example, does not specify the class of social systems and of situations to which the proposed generalization is meant to apply; as it stands, therefore, it cannot be put to an empirical test or to any predictive use.

The generalization tentatively set forth by Radcliffe-Brown has a similar shortcoming: Ostensibly, it refers to any society whatever, but the conditions under which social survival is claimed to occur are qualified by a highly indefinite "except" clause, which precludes the possibility of any reasonably clear-cut test. The clause might even be used to protect the proposed generalization against any conceivable disconfirmation: If a particular social group should "die," this very fact might be held to show that the disruptive forces were as overwhelming as in the case of the Australian tribe mentioned by Radcliffe-Brown. Systematic use of this methodological strategy would, of course, turn the hypothesis into a covert tautology. This would ensure its truth, but at the price of depriving it of empirical content: thus construed, the hypothesis can yield no explanation or prediction whatever.

A similar comment is applicable to the following pronouncement by Malinowski, in which we italicize the dubious qualifying clause: "When we consider any culture *which is not on the point of breaking down or completely disrupted, but which is a normal going concern* we find that need and response are directly related and tuned up to each other."[30]

To be sure, Radcliffe-Brown's and Malinowski's formulations do not *have to* be construed as covert tautologies, and their authors no doubt intended them as empirical assertions; but, in this case, the vagueness of the qualifying clauses still deprives them of the status of definite empirical hypotheses that might be used for explanation or prediction.

5 *The Empirical Import of Functionalist Terms and Hypotheses*

In the preceding section, we mentioned a second flaw that may vitiate the scientific role of a proposed hypothesis of self-regulation. It consists in using key terms of functional analysis, such as "need" and "adequate (proper) functioning"[31] in a non-empirical manner, i.e., without giving them a clear "operational definition," or more generally, without specifying objective criteria of application for them.[32] If functionalist terms are used in this manner, then the sentences containing them have no clear empirical meaning; they lead to no specific predictions and thus cannot be put to an objective test; nor, of course, can they be used for explanatory purposes.

A consideration of this point is all the more important here because the functionalist key terms occur not only in hypotheses of self-regulation, but also in functionalist sentences of various other kinds, such as those of the types (*a*), (*b*), and (*f*) in our schematizations (3.1), (3.3), and (4.1) of functionalist explanation and prediction. Non-empirical use of functionalist key terms may, therefore, bar sentences of these various kinds from the status of scientific hypotheses. We turn now to some examples.

Consider first the terms 'functional prerequisite' and 'need,' which are used as more or less synonymous in the functionalist literature, and which serve to define the term 'function' itself. "Embedded in every functional analysis is some conception, tacit or expressed, of the functional requirements of the system under observation";[33] and indeed, "a definition [of function] is provided by showing that human institutions, as well as partial activities within these, are related to primary, that is, biological, or derived, that is, cultural needs. Function means, therefore, always the satisfaction of a need."[34]

How is this concept of need defined? Malinowski gives a very explicit answer: "By need, then, I understand the system of conditions in the human organism, in the cultural setting, and in the relation of both to the natural environment, which are sufficient and necessary for the survival of group ad organism."[35] This definition sounds clear and straightforward; yet it is not even quite in accord with Malinowski's own use of the concept of need. For he distinguishes, very plausibly, a considerable number of different needs, which fall into two major groups: primary biological needs and derivative cultural ones; the latter include "technological, economic, legal, and even magical, religious, or ethical"[36] needs. But if every single one of these needs did actually represent not only a necessary condition of survival but also a sufficient one, then clearly the satisfaction of just one need would suffice to ensure survival, and the other needs could not constitute necessary conditions of survival at all. It seems reasonable to assume, therefore, that what Malinowski intended was to construe the needs of a group as a set of conditions which are individually necessary and jointly sufficient for its survival.[37]

However, this correction of a minor logical flaw does not remedy a more serious defect of Malinowski's definition, which lies in the deceptive appearance of clarity of the phrase "survival of group and organism." In reference to a biological organism, the term 'survival' has a fairly clear meaning, though even here, there is need for further clarification. For when we speak of biological needs or requirements—e.g., the minimum daily requirements, for human adults, of various vitamins and minerals—we construe these, not as conditions of just the barest survival but as conditions of persistence in, or return to, a "normal," or "healthy" state, or to a state in which the system is a "properly functioning whole." For the sake of objective testability of functionalist hypotheses, it is essential, therefore, that definitions of needs or functional prerequisites be supplemented by reasonably clear and objectively applicable criteria of what is to be considered a healthy state or a normal working order of the systems under consideration; and that the vague and sweeping notion of survival then be construed in the relativized sense of survival in a healthy state as specified. Otherwise, there is definite danger that different investigators will use the concept of functional prerequisite—and hence also that of function—in different ways, and with valuational overtones corresponding to their diverse conceptions of what are the most "essential" characteristics of "genuine" survival for a system of the kind under consideration.

Functional analyses in psychology, sociology, and anthropology are even more urgently in need of objective empirical criteria of the kind here referred to; for the characterization of needs as necessary conditions of psychological or emotional survival for an individual, or of survival of a group is so vague as to permit, and indeed invite, quite diverse subjective interpretations.

Some authors characterize the concept of functional prerequisite or the concept of function without making use of the term 'survival' with its misleading appearance of clarity. Merton, for example, states: "*Functions* are those observed consequences which make for the adaptation or adjustment of a given system; and *dysfunctions*, those observed consequences which lessen the adaptation or adjustment of the system."[38] And Radcliffe-Brown characterizes the function of an item as its contribution to the maintenance of a certain kind of unity of a social system, "which we may speak of as a functional unity. We may define it as a condition in which all parts of the social system work together with a sufficient degree of harmony or internal consistency, i.e., without producing persistent conflicts which can neither be resolved nor regulated."[39]

But like the definitions in terms of survival, these alternative characterizations, though suggestive, are far from giving clear empirical meanings to the key terms of functional analysis. The concepts of adjustment and adaptation, for example, require specification of some standard; otherwise, they have no definite meaning and are in danger of being used tautologically or else subjectively, with valuational overtones.

Tautological use could be based on construing *any* response of a given system as an adjustment, in which case it becomes a trivial truth that any system will adjust itself to any set of circumstances. Some instances of functional analysis seem to come dangerously close to this procedure, as is illustrated by the following assertion: "Thus we are provided with an explanation of suicide and of numerous other apparently antibiological effects as so many forms of relief from intolerable suffering. Suicide does not have *adaptive* (survival) value but it does have *adjustive* value for the organism. Suicide is *functional* because it abolishes painful tension."[40]

Or consider Merton's formulation of one of the assumptions of functional analysis: "when *the net balance of the aggregated of consequences* of an existing social structure is clearly dysfunctional, there develops a strong and insistent pressure for change."[41] In the absence of clear empirical criteria of adaptation and thus of dysfunction, it is possible to treat this formulation as a covert tautology and thus to render it immune to empirical disconfirmation. Merton is quite aware of such danger: in another context he remarks that the notion of functional requirements of a given system "remains one of the cloudiest and empirically most debatable concepts in functional theory. As utilized by sociologists, the concept of functional requirement tends to be tautological or *ex post facto*."[42] Similar warnings against tautological use and against ad hoc generalizations about functional prerequisites have been voiced by other writers, such as Malinowski[43] and Parsons.[44]

On the other hand, in the absence of empirical criteria of adjustment or adaptation, there is also the danger of each investigator's projecting into those concepts (and thus also into the concept of function) his own ethical standards of what would constitute a "proper" or "good" adjustment of a given system—a danger which has been pointed out very clearly by Levy.[45] This procedure would obviously deprive functionalist hypotheses of the status of precise objectively testable scientific assertions. And, as Merton notes, "If theory is to be productive, it must be sufficiently *precise* to be *determinate*. Precision is an integral element of the criterion of *testability*."[46]

It is essential, then, for functional analysis as a scientific procedure that its key concepts be explicitly construed as relative to some standard of survival or adjustment. This standard has to be specified for each functional analysis, and it will usually vary from case to case. In the functional study of a given system *s*, the standard would be indicated by specifying a certain class or range *R* of possible states of *s*, with the understanding that *s* was to be considered as "surviving in proper working order," or as "adjusting properly under changing conditions" just in case *s* remained in, or upon disturbance returned to, some state within the range *R*. A need, or functional requirement, of system *s* relative to *R* is then a necessary condition for the system's remaining in, or returning to, a state in *R*; and the function, relative to *R*; and the function, relative to *R*, of an item *i* in *s* consists in *i*'s effecting the satisfaction of some such functional requirement.

In the field of biology, Sommerhoff's analysis of adaptation, appropriateness, and related concepts is an excellent illustration of a formal study in which the relativization of the central functionalist concepts is entirely explicit.[47] The need of such relativization is made clear also by Nagel, who points out that "the claim that a given change is functional or dysfunctional must be understood as being relative to a specified *G* (or sets of *G*'s)"[48] where the *G*'s are traits whose preservation serves as the defining standard of adjustment or survival for the system under study. In sociology, Levy's analysis of the structure of society[49] clearly construes the functionalist key concepts as relative in the sense just outlined.

Only if the key concepts of functional analysis are thus relativized can hypotheses involving them have the status of determinate and objectively testable assumptions or assertions; only then can those hypotheses enter significantly into arguments such as those schematized in (3.1), (3.3), and (4.1).

But, although such relativization may give definite empirical content to the functionalist hypotheses that serve as premises or conclusions in those arguments, it leaves the explanatory and predictive import of the latter as limited as we found it in sections 4 and 5; for our verdict on the logical force of those arguments depended solely on their formal structure and not on the meanings of their premises and conclusions.

It remains true, therefore, even for a properly relativized version of functional analysis, that its explanatory force is rather limited; in particular, it does not provide an explanation of why a particular item *i* rather than some functional equivalent of it occurs in system *s*. And the predictive significance of functional analysis is practically nil—except in those cases where suitable hypotheses of self-regulation can be established. Such a hypothesis would be to the effect that within a specified range *C* of circumstances, a given system *s* (or: any system of a certain kind *S*, of which *s* is an instance) is self-regulating relative to a specified range *R* of states; i.e., after a disturbance which moves *s* into a state outside *R*, but which does not shift the internal and external circumstances of *s* out of the specified range *C*, the system *s* will return to a state in *R*. A system satisfying a hypothesis of this kind might be called *self-regulating with respect to R*.

Biological systems offer many illustrations of such self-regulation. For example, we mentioned earlier the regenerative ability of a hydra. Consider the case, then, where a more or less large segment of the animal is removed and the rest grows into a complete hydra again. The class *R* here consists of those states in which the hydra is complete; the characterization of range *C* would have to include (i) a specification of the temperature and the chemical composition of the water in which a hydra will perform its regenerative feat (clearly, this will not be just one unique composition, but a class of different ones: the concentrations of various salts, for example, will each be allowed to take some value within a specified, and perhaps narrow, range; the same will hold of the temperature of the water); and (ii) a statement as to the kind and size of segment that may be removed without preventing regeneration.

It will no doubt be one of the most important tasks of functional analysis in psychology and the social sciences to ascertain to what extent such phenomena of self-regulation can be found, and clearly represented by laws of self-regulation, in these fields.

6 *Functional Analysis and Teleology*

Whatever specific laws might be discovered by research along these lines, the kind of explanation and prediction made possible by them does not differ in its logical character from that of the physical sciences.

It is quite true that hypotheses of self-regulation, which would be characteristic results of successful functionalist research, seem to have a teleological character, asserting, as they do, that within specified conditions systems of some particular kind will tend toward a state within the class *R*, which thus assumes the appearance of a final cause determining the behavior of the system.

But, first of all, it would be simply untenable to say of a system *s* which is self-regulating with respect to *R* that the future event of its return to (a state in) *R* is a "final cause" which determines its present behavior. For even if *s* is self-regulating with respect to *R* and if it has been shifted into a state outside *R*, the future event of its return to *R* may never come about: in the process of its return toward *R*, *s* may be exposed to further disturbances, which may fall outside the permissible range *C* and lead to the destruction of *s*. For example, in a hydra that has just had a tentacle removed, certain regenerative processes will promptly set in; but these cannot be explained teleologically by reference to a final cause consisting in the future event of the hydra's being complete again. For that event may never actually come about since in the process of regeneration, and before its completing, the hydra may suffer new and irreparably severe, damage, and may die. Thus, what accounts for the present changes of a self-regulating system *s* is not the "future event" of *s* being in *R*, but rather the *present disposition* of *s* to return to *R*; and it is this disposition that is expressed by the hypothesis of self-regulation governing the system *s*.

Whatever teleological character may be attributed to a functionalist explanation or prediction invoking (properly relativized) hypotheses of self-regulation lies merely in the circumstance that such hypotheses assert a tendency of certain systems to maintain, or return to, a certain kind of state. But such laws attributing, as it were, a characteristic goal-directed behavior to systems of specified kinds are by no means alien to physics and chemistry. On the contrary, it is these latter fields which provide the most adequately understood instances of self-regulating systems and corresponding laws. For example, a liquid in a vessel will return to a state of equilibrium, with its surface horizontal, after a mechanical disturbance; an elastic band, after being stretched (within certain limits), will return to its original shape when it is released. Various systems controlled by negative feedback devices, such as a steam engine whose speed is regulated by a governor, or a homing torpedo, or a plane guided by an automatic pilot, show, within specifiable limits, self-regulation with respect to some particular class of states.

In all of these cases, the laws of self-regulation exhibited by the systems in question are capable of explanation by subsumption under general laws of a more obviously causal form. But this is not even essential, for the laws of self-regulation themselves are causal in the broad sense of asserting, essentially, that for systems of a specified kind, any one of a class of different "initial states" (any one of the permissible states of disturbance) will lead to the same kind of final state. Indeed, as our earlier formulations show, functionalist hypotheses, including those of self-regulation, can be expressed without the use of any teleological phraseology at all.[50]

There are, then, no systematic grounds for attributing to functional analysis a character sui generis not found in the hypotheses and theories of the natural sciences and in the explanations and predictions based on them. Yet, psychologically, the idea of function often remains closely associated with that of purpose, and some functionalist writing has no doubt encouraged this association, by using a phraseology which attributes to the self-regulatory behavior of a given system practically the character of a purposeful action. For example, Freud, in stating his theory of the relation of neurotic symptoms to anxiety, uses strongly teleological language, as when he says that "the symptoms are created in order to remove or rescue the ego from the situation of danger";[51] the quotations given in section 2 provide further illustrations. Some instructive examples of sociological and anthropological writings which confound the concepts of function and purpose are listed by Merton, who is very explicit and emphatic in rejecting this practice.[52]

It seems likely that precisely this psychological association of the concept of function with that of purpose, though systematically unwarranted, accounts to a large extent for the appeal and the apparent plausibility of functional analysis as a mode of explanation; for it seems to enable us to "understand" self-regulatory phenomena of all kinds in terms of purposes or motives, in much the same way in which we "understand" our own purposive behavior and that of others. Now, explanation by reference to motives, objectives, or the like may be perfectly legitimate in the case of purposive behavior and its effects. An explanation of this kind would be causal in character, listing among the causal antecedents of the given action, or of its outcome, certain purposes or motives on the part of the agent, as well as his beliefs as to the best means available to him for attaining his objectives. This kind of information about purposes and beliefs might even serve as a starting point in explaining a self-regulatory feature in a human artifact. For example, in an attempt to account for the presence of the governor in a steam engine, it may be quite reasonable to refer to the purpose its inventor intended it to serve, to his beliefs concerning matters of physics, and to the technological facilities available to him. Such an account, it should be noted, might conceivably give a probabilistic explanation for the presence of the governor, but it would not explain why it functioned as a speed-regulating safety device: to explain this latter fact, we would have to refer to the construction of the machine and to the laws of physics, not to the intentions and beliefs of the designer. (An explanation by reference to motives and beliefs can be given as well for certain items which do not, in fact, function as intended; e.g., some superstitious practices, unsuccessful flying machines, or ineffective economic policies, etc.) Furthermore—and this is the crucial point in our context—for most of the self-regulatory phenomena that come within the purview of functional analysis, the attribution of purposes is an illegitimate transfer of the concept of purpose from its domain of significant applicability to a much wider domain, where it is devoid of objective empirical import. In the context of purposive behavior of individuals or groups, there are various methods of testing whether the assumed motives or purposes are indeed present in a given situation; interviewing the agent in question might be one rather direct way, and there are various alternative "operational" procedures of a more indirect character. Hence, explanatory hypotheses in terms of purposes are here capable of reasonably objective test. But such empirical criteria for purposes and motives are lacking in other cases of self-regulating systems, and the attribution of purposes to them has therefore no scientific meaning. Yet, it tends to encourage the illusion that a profound type of understanding is achieved, that we gain an insight into the nature of these processes by likening them to a type of behavior with which we are thoroughly familiar from daily experience. Consider, for example, the law of "adaptation to an obvious end" set forth by the sociologist L. Gumplowicz with the claim that it holds both in the natural and the social domains. For the latter, it asserts that "every social growth, every social entity, serves a definite end, however much its worth and morality may be questioned. For the universal law of adaptation signifies simply that no expenditure of effort, no change of condition, is purposeless on any domain of phenomena. Hence, the inherent reasonableness of all social facts and conditions must be conceded."[53] The suggestion is rather strong here that the alleged law enables us to understand social dynamics in close analogy to purposive behavior aimed at the achievement of some end. Yet the purported law is completely devoid of empirical meaning since no empirical interpretation has been given to such key terms as 'end,' 'purposeless,' and 'inherent reasonableness' for the contexts to which it is applied. The "law" asserts nothing whatever, therefore, and cannot possibly explain any social—or other—phenomena.

Gumplowicz's book antedates the writings of Malinowski and other leading functionalists by several decades, and certainly these more recent writers have been more cautious and sophisticated in stating their ideas. Yet, there are certain quite central assertions in the newer functionalist literature which are definitely reminiscent of Gumplowicz's formulation in that they suggest an understanding of functional phenomena in the image of deliberate purposive behavior or of systems working in accordance with a preconceived design. The following statements might illustrate this point: "[Culture] is a system of objects, activities, and attitudes in which every part exists as a means to an end,"[54] and "The functional view of culture insists therefore upon the principle that in every type of civilization, every custom, material object, idea and belief fulfills some vital function, has some task to accomplish, represents an indispensable part within a working whole."[55] These statements express what Merton, in a critical discussion, calls the postulate of universal functionalism.[56] Merton qualifies this postulate as premature;[57] the discussion presented in the previous section shows that, in the absence of a clear empirical interpretation of the functionalist key terms, it is even less than that, namely, empirically vacuous. Yet, formulations of this kind may evoke a sense of insight and understanding by likening sociocultural developments to purposive behavior and in this sense reducing them to phenomena with which we feel thoroughly familiar. But scientific explanation and understanding are not simply a reduction to the familiar: otherwise, science would not seek to explain familiar phenomena at all; besides, the most significant advances in our scientific understanding of the world are often achieved by means of new theories which, like quantum theory, assume some quite unfamiliar kinds of objects or processes which cannot be directly observed, and which sometimes are endowed with strange and even seemingly paradoxical characteristics. A class of phenomena has been scientifically understood to the extent that it can be fitted into a testable, and adequately confirmed, theory or a system of laws; and the merits of functional analysis will eventually have to be judged by its ability to lead to this kind of understanding.

7 *The Heuristic Role of Functional Analysis*

The preceding considerations suggest that what is often called "functionalism" is best viewed, not as a body of doctrine or theory advancing tremendously general principles such as the principle of universal functionalism, but rather as a program for research guided by certain heuristic maxims or "working hypotheses." The idea of universal functionalism, for example, which becomes untenable when formulated as a sweeping empirical law or theoretical principle, might more profitably be construed as expressing a directive for research, namely, to search for specific self-regulatory aspects of social and other systems and to examine the ways in which various traits of a system might contribute to its particular mode of self-regulation. (A similar construal as heuristic maxims for empirical research might be put upon all the "general axioms of functionalism" suggested by Malinowski, and considered by him as demonstrated by all the pertinent empirical evidence.[58])

In biology, for example, the contribution of the functionalist approach does not consist in the sweeping assertion that all traits of any organism satisfy some need and thus serve some function; in this generality, the claim is apt to be either meaningless or covertly tautologous or empirically false (depending on whether the concept of need is given no clear empirical interpretation at all, or is handled in a tautologizing fashion, or is given one definitive empirical interpretation). Instead, functional studies in biology have been aimed at showing, for example, how in different species, specific homeostatic and regenerative processes contribute to the maintenance and development of the living organism; and they have gone on (i) to examine more and more precisely the nature and limits of those processes (this amounts basically to establishing various specific empirical hypotheses or laws of self-regulation), and (ii) to explore the underlying physiological or physicochemical mechanisms, and the laws governing them, in an effort to achieve a more thorough theoretical understanding of the phenomena at hand.[59] Similar trends exist in the study of functional aspects of psychological processes, including, for example, symptom formation in neurosis.[60]

Functional analysis in psychology and in the social sciences no less than in biology may thus be conceived, at least ideally, as a program of inquiry aimed at determining the respects and the degrees in which various systems are self-regulating in the sense here indicated. This conception clearly underlies, for example, Nagel's essay, "A Formalization of Functionalism,"[61] a study which develops an analytic scheme inspired by, and similar to, Sommerhoff's formal analysis of self-regulation in biology[62] and uses it to exhibit and clarify the structure of functional analysis, especially in sociology and anthropology.

The functionalist mode of approach has proved highly illuminating, suggestive, and fruitful in many contexts. If the advantages it has to offer are to be reaped in full, it seems desirable and indeed necessary to pursue the investigation of specific functional relationships to the point where they can be expressed in terms of reasonably precise and objectively testable hypotheses. At least initially, these hypotheses will likely be of quite limited scope. But this would simply parallel the present situation in biology, where the kinds of self-regulation, and the uniformities they exhibit, vary from species to species. Eventually, such "empirical generalizations" of limited scope might provide a basis for a more general theory of self-regulating systems. To what extent these objectives can be reached cannot be decided in a priori fashion by logical analysis or philosophical reflection: the answer has to be found by intensive and rigorous scientific research.

Notes

I am greatly indebted to the Council of the Humanities at Princeton University for the award of a Senior Fellowship for the academic year 1956–57, which offered me an opportunity to do research while on a substantially reduced teaching schedule. This essay is part of the work done under the auspices of the Council.

1. These terms are given preference over the more familiar words "explicans" and "explicandum," in order to reserve the latter for use in the context of philosophical explication in the technical sense proposed by R. Carnap; see, for example, his *Logical Foundations of Probability* (Chicago: University of Chicago Press, 1950), secs. 1–3. The terms "explanans" and "explanandum" were introduced, for this

reason, in an earlier article: Carl G. Hempel and P. Oppenheim, "Studies in the Logic of Explanation," *Philosophy of Science*, 15 (1948), pp. 135–75. Reprinted in part in H. Feigl and M. Brodbeck, eds., *Readings in the Philosophy of Science* (New York: Appleton-Century-Crofts, 1953). While that article does not deal explicitly with inductive explanation, its first four sections contain various further considerations on deductive explanation that are relevant to the present study. For a careful critical examination of some points of detail discussed in the earlier article, such as especially the relation between explanation and prediction, see the essay by I. Scheffler, "Explanation, Prediction, and Abstraction," *British Journal for the Philosophy of Science*, 7 (1957), pp. 293–309, which also contains some interesting comments bearing on functional analysis.

2. An explanation by means of laws which are causal in the technical sense of theoretical physics also has the form (1.1) of a deductive nomological explanation. In this case, the laws invoked must meet certain conditions as to mathematical form, and $C_1, C_2, \ldots, C_n$ express so-called boundary conditions. For a fuller account of the concepts of causal law and of causality as understood in theoretical physics, see, for example, H. Margenau, *The Nature of Physical Reality* (New York: McGraw-Hill Book Company, 1950), ch. 19; or Ph. Frank, *Philosophy of Science* (Englewood Cliffs, N.J.: Prentice-Hall, 1957), chs. 11, 12.
3. Some brief but lucid and stimulating comments on explanation by means of statistical laws will be found in S. E. Gluck, "Do Statistical Laws Have Explanatory Efficacy?" *Philosophy of Science*, 22 (1955), pp. 34–38. For a much fuller analysis of the logic of statistical inference, see R. B. Braithwaite, *Scientific Explanation* (Cambridge: Cambridge University Press, 1953), chs. V, VI, VII. For a study of the logic of inductive inference in general, Carnap's *Logical Foundations of Probability*, op. cit., is of great importance.
4. In developing the characterization of functional analysis presented in this section, I have obtained much stimulation and information from the illuminating and richly documented essay "Manifest and Latent Functions" in R. K. Merton's book, *Social Theory and Social Structure* (Glencoe, Ill.: Free Press; revised and enlarged edition, 1957), pp. 19–84. Each of the passages from this work which is referred to in the present essay may also be found in the first edition (1949), on a page with approximately the same number.
5. For a fuller statement and further development of this point, see part I of the essay "A Formalization of Functionalism" in E. Nagel, *Logic Without Metaphysics* (Glencoe, Ill.: Free Press, 1957), pp. 247–83. Part I of this essay is a detailed analytical study of Merton's essay mentioned in note 4, and thus is of special significance for the methodology of the social sciences.
6. S. Freud, *The Problem of Anxiety* (Transl. by H. A. Bunker. New York: Psychoanalytic Quarterly Press, and W. W. Norton & Company, 1936), p. 111.
7. Ibid., p. 112.
8. Merton, op. cit., p. 50.
9. Ibid., p. 51. Merton defines manifest functions as those which are both intended and recognized, and latent functions as those which are neither intended nor recognized. But this characterization allows for functions which are neither manifest nor latent; e.g., those which are recognized though not intended. It would seem to be more in keeping with Merton's intentions, therefore, to base the distinction simply on whether the stabilizing effect of the given item was deliberately sought.
10. Ibid., pp. 64–65.
11. A. R. Radcliffe-Brown, *Structure and Function in Primitive Society* (London: Cohen and West Ltd., 1952), p. 145.
12. B. Malinowski, *Magic, Science and Religion, and Other Essays* (Garden City, N.Y.: Doubleday Anchor Books, 1954), p. 90. For an illuminating comparison of Malinowski's views on the functions of magic and religion with those advanced by Radcliffe-Brown, see G. C. Homans, *The Human Group* (New York: Harcourt, Brace & Company, 1950), pp. 321 ff. (Note also Homan's general comments on "the functional theory," ibid., pp. 268–72.) This issue and other aspects of functional analysis in anthropology are critically examined in the following article, which confronts some specific applications of the method with programmatic declarations by its proponents: Leon J. Goldstein, "The Logic of Explanation in Malinowskian Anthropology," *Philosophy of Science*, 24 (1957), pp. 156–66.
13. B. Malinowski, "Anthropology," *Encyclopaedia Britannica*, First Supplementary volume (London and New York: Encyclopaedia Britannica, 1926), p. 132.
14. Ibid., p. 139.
15. B. Malinowski, *A Scientific Theory of Culture, and Other Essays* (Chapel Hill: University of North Carolina Press, 1944), p. 174.

16. Radcliffe-Brown, op. cit., p. 186.
17. Malinowski, at one place in his writings, endorses a pronouncement which might appear to be at variance with this conclusion: "Description cannot be separated from explanation, since in the words of a great physicist, 'explanation is nothing but condensed description.'" (Malinowski, "Anthropology," op. cit., p. 132.) He seems to be referring here to the views of Ernst Mach or of Pierre Duhem, who took a similar position on this point. Mach conceived the basic objective of science as the brief and economic description of recurrent phenomena and considered laws as a highly efficient way of compressing, as it were, the description of an infinitude of potential particular occurrences into a simple and compact formula. But, thus understood, the statement approvingly quoted by Malinowski is, of course, entirely compatible with our point about the relevance of laws for functional explanation.

    Besides, a law can be called a description only in a Pickwickian sense. For even so simple a generalization as "All vertebrates have hearts" does not describe any particular individual, such as Rin-Tin-Tin—and of any other object, whether vertebrate or not—that *if* it is a vertebrate *then* it has a heart. Thus, the generalization has the import of an indefinite set of conditional statements about particular objects. In addition, a law might be said to imply statements about "potential events" which never actually take place. The gas law, for example, implies that if a given body of gas were to be heated under constant pressure at time *t*, its volume would increase. But if in fact the gas is not heated at *t* this statement can hardly be said to be a description of any particular event.
18. Malinowski, "Anthropology," op. cit., p. 136.
19. Malinowski, *Magic, Science and Religion, and Other Essays*, op. cit., p. 90. (Note the explanatory claim implicit in the use of the word "hence.")
20. B. Malinowski, "Culture," *Encyclopedia of the Social Sciences*, IV (New York: Macmillan Company, 1931), p. 626.
21. Merton, op. cit., p. 34. Cf. also T. Parsons, *Essays in Sociological Theory, Pure and Applied* (Glencoe, Ill.: Free Press, 1949), p. 58. For an interesting recent attempt to establish the existence of functional alternatives in a specific case, see R. D. Schwartz, "Functional Alternatives to Inequality," *American Sociological Review*, 20 (1955), pp. 424–30.
22. See G. G. Simpson, *The Meaning of Evolution* (New Haven: Yale University Press, 1949), pp. 164 ff., 190, 342–43; and G. G. Simpson, C. S. Pittendrigh, L. H. Tiffany, *Life* (New York: Harcourt, Brace & Company, 1957), p. 437.
23. Malinowski, *A Scientific Theory of Culture, and Other Essays*, op. cit., p. 38.
24. E. M. Sait, "Machine, Political," *Encyclopedia of the Social Sciences*, IX (New York: Macmillan Company, 1933), p. 659. (Italics supplied.)
25. Merton, op. cit., p. 76. (Italics supplied.)
26. Merton, op. cit., p. 73. (Italics the author's.)
27. Radcliffe-Brown, op. cit., p. 183.
28. Malinowski, *A Scientific Theory of Culture, and Other Essays*, op. cit., p. 67.
29. Henry A. Murray and Clyde Kluckhohn, "Outline of a Conception of Personality," in Clyde Kluckhohn and Henry A. Murray, eds., *Personality in Nature, Society, and Culture* (New York: Knopf, 1950), pp. 3–32; quotation from p. 7; italics the authors'.
30. Malinowski, *A Scientific Theory of Culture, and Other Essays*, op. cit., p. 94.
31. In accordance with a practice followed widely in contemporary logic, we will understand by terms certain kinds of words or other linguistic expressions, and we will say that a term expresses or signifies a concept. For example, we will say that the term 'need' signifies the concept of need. As this illustration shows, we refer to, or mention, a linguistic expression by using a name for it which is formed by simply enclosing the expression in single quotes.
32. A general discussion of the nature and significance of "operational" criteria of application for the terms used in empirical science, and references to further literature on the subject, may be found in C. G. Hempel, *Fundamentals of Concept Formation in Empirical Science* (Chicago: University of Chicago Press, 1952), secs. 5–8; and in the symposium papers on the present state of operationalism by G. Bergmann, P. W. Bridgman, A. Grunbaum, C. G. Hempel, R. B. Lindsay, H. Margenau, and R. J. Seeger, which form ch. II of Philipp G. Frank, ed., *The Validation of Scientific Theories* (Boston: Beacon Press, 1956).
33. Merton, op. cit., p. 52.
34. Malinowski, *A Scientific Theory of Culture, and Other Essays*, op. cit., p. 159.
35. Malinowski, ibid., p. 90.
36. Malinowski, ibid., p. 172; see also ibid., pp. 91ff.

37. In some of his statements Malinowski discards, by implication, even the notion of function as satisfaction of a condition that is at least *necessary* for the survival of group or organism. For example, in the same essay containing the two passages just quoted in the text, Malinowski comments as follows on the function of some complex cultural achievements: "Take the airplane, the submarine, or the steam engine. Obviously, man does not need to fly, nor yet to keep company with fishes, and move about within a medium for which he is neither anatomically adjusted nor physiologically prepared. In defining, therefore, the function of any of those contrivances, we cannot predicate the true course of their appearance in any terms of metaphysical necessity." (Ibid., pp. 118–19.)
38. Merton, op. cit., p. 51. (Italics the author's.)
39. Radcliffe-Brown, op. cit., p. 181.
40. Murray and Kluckhohn, op. cit., p. 15. (Italics the authors'.)
41. Merton, op. cit., p. 40.
42. Merton, op. cit., p. 52.
43. See, for example, Malinowski, *A Scientific Theory of Culture, and Other Essays,* op. cit., pp. 169–70; but also compare this with pp. 118–19 of the same work.
44. See, for example, T. Parsons, *The Social System* (Glencoe, Ill.: Free Press, 1951), p. 29, fn. 4.
45. Marion J. Levy, Jr., *The Structure of Society* (Princeton: Princeton University Press, 1952), pp. 76ff.
46. R. K. Merton, "The Bearing of Sociological Theory on Empirical Research" in Merton, *Social Theory and Social Structure,* op. cit., pp. 85–101; quotation from p. 98. (Italics the author's.)
47. See G. Sommerhoff, *Analytical Biology* (New York: Oxford University Press, 1950).
48. Nagel, "A Formalization of Functionalism," op. cit., p. 269. See also the concluding paragraph of the same essay (pp. 282–83).
49. Levy speaks of eufunction and dysfunction of a unit (i.e., a system) and characterizes these concepts as relative to "the unit as defined." He points out that this relativization is necessary "because it is to the definition of the unit that one must turn to determine whether or not 'adaptation or adjustment' making for the persistence or lack of persistence of the unit is taking place." (Levy, ibid., pp. 77–78.)
50. For illuminating discussions of further issues concerning "teleological explanation," especially with respect to self-regulating systems, see R. B. Braithwate, *Scientific Explanation* (Cambridge: Cambridge University Press, 1953), ch. X; and E. Nagel, "Teleological Explanation and Teleological Systems" in S. Ratner, ed., *Vision and Action: Essays in Honor of Horace Kallen on His Seventieth Birthday* (New Brunswick, N.J.: Rutgers University Press, 1953); reprinted in H. Feigl and M. Brodbeck, eds., *Readings in the Philosophy of Science* (New York: Appleton-Century-Crofts, 1953).
51. Freud, op. cit., p. 112.
52. Merton, "Manifest and Latent Functions," op. cit., pp. 23–25, 60ff.
53. L. Gumplowicz, *The Outlines of Sociology,* translated by F. W. Moore (Philadelphia: American Academy of Political and Social Science, 1899), pp. 79–80.
54. Malinowski, *A Scientific Theory of Culture, and Other Essays,* op. cit., p. 150.
55. Malinowski, "Anthropology," op. cit., p. 133.
56. Merton, "Manifest and Latent Functions," op. cit., pp. 30ff.
57. Ibid., p. 31.
58. Malinowski, *A Scientific Theory of Culture, and Other Essays,* op. cit., p. 150.
59. An illuminating general account of this kind of approach to homeostatic processes in the human body will be found in Walter B. Cannon, *The Wisdom of the Body* (New York: W. W. Norton & Company, revised edition 1939).
60. See, for example, J. Dollard and N. E. Miller, *Personality and Psychotherapy* (New York: McGraw-Hill Book Company, 1950), ch. XI, "How Symptoms Are Learned," and note particularly pp. 165–66.
61. Nagel, "A Formalization of Functionalism," op. cit. See also the more general discussion of functional analysis included in Nagel's paper, "Concept and Theory Formation in the Social Sciences," in *Science, Language, and Human Rights*: American Philosophical Association, Eastern Division, Volume 1 (Philadelphia: University of Pennsylvania Press, 1952), pp. 43–64. Reprinted in J. L. Jarrett and S. M. McMurrin, eds., *Contemporary Philosophy* (New York: Henry Holt & Co., 1954).
62. Sommerhoff, op. cit.

Chapter 23
Function and Cause
R. P. Dore

Kingsley Davis has argued that we should abandon the notion that functionalism is a special form of sociological analysis.[1] It *is* sociological analysis, albeit occasionally clouded by misleading terminology. In at least one reader the effect of his thoughtful and wise-ranging paper was to stimulate reflection on our notions of function and cause and on the relations between them. The starting point of these reflections was the question: does not Professor Davis's argument rest on a special and hardly universal view of what sociological analysis is or should be?

At one point he commends functionalism as having "helped to make a place in sociology and anthropology for those wishing to explain social phenomena in terms of social systems, as against those who wished to make no explanation at all, to explain things in terms of some other system or to plead a cause." Sociological analysis, in other words, is the explanation of social phenomena in terms of social systems. But surely cause pleading, explanation in terms of other systems, and so on are not the only alternatives. There is another position, equally sociological, equally analytical, which holds that sociologists should search for regularities in the concomitant occurrences of social phenomena, seek to induce causal laws from such regularities, and seek eventually to order such laws into comprehensive theory. According to this view, systematic theory (a logically consistent body of causal laws) is the end product of a long search for causal relations, not a heuristically useful starting point.

This, perhaps, betrays a preference for "neat single propositions whose validity is proved but whose significance is not," a preference which Professor Davis condemns as "scientific ritualism." It is comforting to reflect that in the natural sciences at least we would not have got far without our ritualists. Newton in developing his systematic theory of mechanics owed a good deal to Galileo's neat single proposition about the rate of acceleration of falling bodies.

The difference between these two views which we might characterize as the system approach and the piecemeal approach is not identical with the often imputed distinction between functional and nonfunctional analysis, but it does seem to be true that only the system approach encourages the use of the concept of function. The piecemeal approach is quite clearly bent on looking for causal relations. The system approach finds the concept of cause and causal law difficult to apply, and often finds functions easier to handle.

Perhaps the best way to justify this assertion would be to analyze closely the relations between the concepts of function and of cause. Let us take as starting point the question: in what ways can a statement about the function of an institution, a pattern of behavior, a role, or a norm be translated into a statement about causal relations?

Function—Effect

In the first place it is fairly obvious that "the function of X is to maintain Y" implies that X has some kind of causal influence on Y, and it is presumably this kind of "translation" Professor Davis had in mind when he denied that functional relations are noncausal. But an analysis of "causal influence" leads to difficulties. Can we say, "The assertion that, say, the system of stratification has the function of making the division of labor possible implies that among the causes of the division of labor is the system of stratification"? Obviously not if "the causes of the *origin* of the division of labor" is intended. We have to say something like "the causes of the persistence of the division of labor." This suggests that while one can legitimately ask the function of an institution, one cannot ask for the *cause* of an institution; one has to specify cause of origin or cause of persistence. It will be argued later that what this really amounts to is that one can legitimately ask only for the causes of *events*. Let us assume this argument for the moment and formulate this particular relation between function and cause as follows: "institution X has the function of maintaining institution Y" implies that the recurring events referred to as institution X are among the causes of other events integral to the institution Y (or, can be related by causal laws to other events integral to institution Y).

Function—Cause

But this is not the kind of causal relationship implied when it is said that an institution is "explained" in terms of its function. Here (less often explicitly than implicitly as a result of the ambiguities of the word "explain") the transition is suggested not from the function of X to the cause of something other than X, but from the function of X to the causes of X itself. When and how may this kind of transition be made?

A small boy's examination of the interior of a watch may lead him to conclude that the function of the balance spring is to control the movement of the balance wheel. He would have little difficulty in using his functional insight to arrive at a causal explanation of the spring's presence—it is there because the man who made the watch realized a need for something to control the movement of the wheel, and the process of ratiocination which ensued led him to put in the spring.

Sociologists are not always precluded from making the same kind of transition from function of X to cause of X. Human institutions are now purposefully designed on a scale rarely attempted before. An analysis of the functions of the Chinese communes leads easily to an explanation of the causes of their existence, for they were created by historically identifiable persons to perform these functions and there may well be minutes of committees which record the process of invention with constant reference to their intended consequences, both those which were to be manifest to the communed Chinese and those which were to be latent to them and manifest only to their leaders.

Perhaps more common is the case where human purpose, based on an awareness of function, is a causal factor not so much in the initiation of an institution as in its growth and development. The Roman circus started well before emperors realized the salutary political functions it shared with bread. It was not until the third century B.C. that, as Radcliffe-Brown has pointed out,[2] the Chinese sociologist, Hsun-tse, realized the latent psychological and social functions of ancestral rites, but his discovery certainly prompted later Confucian scholars to encourage the deluded masses in a continued belief in the reality of the manifest functions of those rites. Nowadays, with sociologists busily ferreting out latent functions in every nook and cranny of society and their writings gaining general currency, latent functions are not likely to stay latent for long. Here indeed is the complement of the self-fulfilling prophecy—the self-falsifying assertion. The sociologist who contends that X has such and such a latent function in his own society in fact makes that function manifest. The intervention of human purpose to preserve institutions so that they may continue to fulfill their *once* latent functions is likely to occur more frequently as a result.[3]

However, modern sociologists still probably have less direct influence in molding the institutions of their society than Hsun-tse had in his, and in any case most sociologists are not imputing such a causal chain when they imply a connection between latent function of X and cause of X. Merton, for instance, clearly is thinking of something else when he speaks, apropos of the Hopi rain dances, of the analysis of their latent function as an *alternative* to describing their persistence "only as an instance of 'inertia', 'survival' or 'manipulation by powerful subgroups.'"[4]

How then, without reference to human awareness of functions, can a statement about the function of an institution be translated into a causal statement about either the origin or the persistence of that institution?

Societal Integration

One way is to postulate an immanent tendency, universal in human societies, for the parts of the society all to be functionally integrated in the whole. Given such a tendency the function of an institution is its raison d'être and hence its cause. The logical grounds for such a postulate seem to be two. First there is the complementarity of roles and institutions; the role of wife implies the role of husband; the specialization of the executive to executive functions implies separate institutions for legislation and litigation, and so on. Such complementarities, however, are of limited range. Let an integrationist try his hand, for instance, at specifying the chains of complementarity which might link the institution of presidential elections with that of the burlesque show. The second basis for belief in the integration of societies rests on the supposed integration of the human personality. Since the same individual occupies numerous roles in a variety of institutional contexts and since all individuals are subject to a craving for consistency, it follows that all the institutions of a society must be permeated by the same value-preferences, the same modes of orientations to action, the same patterns of authority, the same worldview, the same sense of time, and so on. But how valid is the assumption of the consistent personality? Which of us, sophisticates that we are, could confidently claim that he has never been guilty of preferring value *A* to value *B* in one situation and reversing his preference in another? And even if this were not so, this argument would create an a priori expectation of social integration only in the case of very simple societies. In such simple societies the number of roles is limited. Every individual in the society may occupy at some stage of his life a high proportion of the total number of roles. In such a society integrated personalities might make for integrated institutions. But this is not the case in large complex societies, segmented into regional and class subcultures with specialized personality types and offering a vast multiplication of roles only a tiny fraction of which any one individual will ever find himself performing.

Obviously there are no grounds for expecting such societies to be perfectly integrated. To quote Professor Davis again, "it would be silly to regard such a proposition as literally true." And one might add that modification of the proposition from "always perfectly" to "usually somewhat" integrated (a) destroys the possibility of its empirical falsification and (b) destroys its value as an automatic means of transition from function to cause.[5]

Evolutionary Selection

There remains, however, at least one way in which the sociologist may move on from function to cause—by means of the notions of adaptation and selection developed in the theories of biological evolution. To take the example of stratification and the division of labor, the hypothesis would have to go something like this: for various reasons some societies which began the division of labor also had, or developed, a system of unequal privileges for different groups; others did not. Those which did functioned more efficiently as societies; perhaps they bred more rapidly than, acquired resources at the expense of, and eventually eliminated, the others. Perhaps (and this is an extension of the concept of selection not available to the biologist) their obvious superiority in wealth, power, the arts, standard of living, etc., induced the others to imitate their institutions wholesale, including the principle of stratification; or just conceivably (though here we slip "human awareness" back into the causal chain) the others bred sociologists who noted the importance of stratification to the superior societies and urged its adoption specifically. At any rate, by one, or a combination, of these processes it now happens that all societies with a division of labor have a system of stratification.

It is an unlikely story, but it seems to be the only kind of story which will make a statement about the latent function of X relevant to a causal explanation of X. And even this, of course, is not a complete causal explanation. The "various reasons" why some societies had stratification in the first place still need to be explained. For the biologist the place of these "various reasons" is taken by "random mutation" and some sociologists, too, are prepared to probe no further.[6]

But often the sociologist can think of specific "various reasons" which eliminate randomness. Dennis H. Wrong, in his assessment of Davis and Moore on stratification,[7] suggests, for instance, that when the division of labor takes place certain groups acquire greater power in the society by virtue of that division and consequently arrogate to themselves a larger share of material and other rewards. And in this case, if this hypothesis concerning one of the "various reasons" for the development of stratification in *a* society is historically validated, or accepted on the basis of what we know in general of human nature, then it could equally explain the development of stratification in any and all societies. The adaptive superiority of stratification due to its function in making the division of labor workable *may* still be relevant, too, but it is only one of a number of possible causal chains, the relative importance of which can only be assessed in the light of the historical evidence.

In any case, if one, is looking for the causes of (either the origination or the continuance of) X, it is better to look for causes as such; looking for the functions of X is never a necessary, and not always even a useful, first step.[8]

In point of fact we know from historical evidence that this evolutionary argument relating function to cause is irrelevant to certain social institutions which sociologists describe. The American boss-directed political machine, for instance, is said by Merton to have the functions of providing a centralization of power, of providing necessary services to those who need help rather than justice, of organizing essential, but morally disapproved, sectors of the economy, etc., and as such contributes to the maintenance of the social system as a whole. However, we know that the boss-system developed long after the United States was in direct and aggressive competition with other social groups for resources; we know from historical evidence that there has been no process of selective weeding out of societies involved.

In such cases one can still appeal to a weakened form of the evolutionary argument to relate function to cause by defining causally important conditions not for the original development of the institution but for its later transmission. It would have to go something like this: if the boss-system had not had these effects and so contributed to the smooth working of society, nor had these effects been neutral with respect to the smooth working of society, but had, on the contrary been positively detrimental to society's smooth working, people would have stopped doing it. In other words the fact that this feature was *not dysfunctional* to the workings of the society is a necessary condition of its present existence. It is also a necessary condition that all members of the society were not eliminated by an epidemic of bubonic plague. One could think of many more such negatively defined necessary conditions, all of which play a part, but only a small part, in a full causal explanation.

Summary

This seems to exhaust the possible methods by which assertions about the functions of X can be involved in assertions about the causes of the (origin or continuance of) X. The sociologist may not be the least bit interested in any of them. Having discovered that, say, social stratification has the function of making the division of labor workable, he may be content with saying just that—and with perfect justification provided he concedes that he has said nothing about *why* societies are stratified. He may go on to point a corollary of his assertion—that if stratification were abolished the division of labor would become unworkable. This is, indeed, an eminently useful social activity and the kind of analysis which can properly precede attempts at social reform. It is also, incidentally, the kind of activity in which a good many social anthropologists in particular have been professionally engaged in colonial administrations. The practical need to assess the probable effects of changes in institutions wrought by colonial policy has provided an important application of functional analysis which perhaps explains (causally) why so many anthropologists have been content with functional analysis as a legitimate final goal of their activities.

We may sum up the argument so far as follows:

1. In a not very clearly defined way the suggestion that institution X has the function of maintaining Y implies some causal influence of X on Y.
2. Assertions about the functions of an institution X are relevant to assertions about the causes of the origin or the persistence of that same institution X if, and only if: (a) one assumes that the function is manifest to the present actors in, to the present upholders of, or to former upholers or inventors of the institution in question, and as such has played a part in their motives for performing, or inducing others to perform, the institutionalized behavior involved; (b) one postulated an immanent tendency for the functional integration of a society; (c) one postulates an adaptive superiority conferred by the institution which permitted it, having developed in one society, to spread to others.

These ideas are not particularly new.[9] The reason why they need reiterating is, it seems, largely because of the ill-defined relation between function and cause suggested by the first of our two propositions. It is the main business of this chapter to try to improve the definition of that relation and, in the course of doing so, to make a few pertinent remarks about the use of analogies from natural science.

System and Event

Let us first examine the concept of system, "How else can data be interpreted," said Professor Davis in his paper, "except in relation to the larger structures in which they are implicated? How can data on the earth's orbit, for example, be understood except in relation to a system in which they are involved—in this case the solar system or the earth's climatic system?" Is this, however, a good analogy? These are indeed systems in nature, such as the solar system, the parts of which are in continuous interaction with each other in such a way that causal laws, expressed in the form of differential equations, allow one to predict one state of the system from another prior or later state. In human societies, however, though the money market might be somewhat similar, such systems are rare. Social systems (in the Parsonian manner) are not analogous in that the parts are not *simultaneously* affecting each other in the way in which the sun and the moon simultaneously affect each other by their gravitational attraction. The mutual relation of, say, the system of socialization to the system of political control is mediated by the personality structure, and as such it is a relation which requires a long time interval to work through the whole causal sequence. Parents may well train their children today in ways which are "significantly congruent" with the ways in which they behave politically today, but, in the other direction, the way in which they now behave politically is affected by the way in which they were trained, not today but a generation ago.

The analysis of systems such as the solar system can dispense with the notion of cause in favor of function—but this, be it noted, is strictly the mathematician's function, not that of the sociologist or of the physiologist.[10] It is not, however, impossible to apply the concept of causal law and causal event to such systems, and to do so might help to elucidate the nature of the distinction we have earlier made between the causes of the origin of, and the causes of the persistence of, institutions. If we are to give a causal explanation of the movement of the moon between 10:00 P.M. and 10:05 P.M. tonight, we would need to refer to the simultaneous events of the movements of the earth and the sun, etc., relating them to it by Newton's law of gravitation. We should also have to mention a previous event—the moon's motion at the point immediately prior to 10:00 P.M.—and relate it to the event in question (its movement between 10:00 and 10:05 P.M.) by means of Newton's first law of motion concerning momentum. Having started on this track we can regress almost indefinitely from event to event back through time (chopping our time continuum arbitrarily into "events"), the moon's velocity at any particular moment being affected by its velocity the preceding moment, until we get to an earlier traumatic event, namely, the moon's supposed wrenching off from the earth. In the whole of this process it is only events which are related to each other by causal laws and only of events that we ask: what are their causes?[11] Similarly—and this is the point of the example—when we talk of "the cause of the origin of an institution" and "the cause of the persistence of an institution" we are in both cases asking for the causes of events—in the first case the causes of the particular once-and-for-all events associated with the origin of the institution, and in the second of the recurring events which *are* the institution.

In the light of this view of causal relations, let us now look at the analogy between physiology and sociology often invoked by those who favor sociological explanation in terms of systems. It is often asserted that because the physiologist leaves questions concerning the origin of the heart to the student of evolution and concentrates on tracing its functions as it at present exists, he is not concerned with causes. But this is surely not so. "The function of the heart in the human being is to pump blood" implies "the cause of the flow of this blood at this time is the pump of that heart then" and this is as much a causal assertion as "one of the reasons why animals have hearts is because when random mutation produced the first primitive heart its possessors gained the ability to outbreed the heartless."

Physiologists and students of evolution have achieved a division of labor which is not formalized among sociologists. Consequently, among sociologists the search for an "explanation" of an institution is often ambiguous. "Why is there a system of unequal rewards in this society?" may be answered by some, "because parents tell [this parent and this parent told] their children that some positions in society are more worthy of respect than others, and because employers pay [this employer and this employer paid] more for some kinds of work than others, etc." This is the "physiological" explanation of the recurring events of rewarding particular people with particular acts of deference and so on which is what we mean by stratification. Alternatively the answer might be, "Because with the division of labor some groups became more powerful and arrogated privileges to themselves, or because differentiated societies which had systems of stratification proved more successful than those which did not"—the "evolutionary" explanation of the particular events which led to the institutionalization of certain patterns of behavior.

Institutions

It will be noted again that whichever way the question is taken it can be handled as if it were a question about particular events. It is the chief assertion of this chapter that ultimately these are the only terms in which causal questions can be framed. But if this is the case, what then is the relation between the particular events observed by the sociologist and his concepts such as stratification, marriage, or socialization—concepts of "institutions," "norms," "behavior patterns"? Is it not exactly the same as the relation between a particular human heart and *the* human heart for the physiologist? The physiologist's statement that "the function of the heart is to pump blood" is a summary generalization of statements about the causal relations between the particular events of heart pump and blood flow in particular human bodies—events which nowadays recur more than two billion times a second. If the sociologist's statements are to have any empirical reference it is difficult to see how they can be different from this; how, that is to say, the relation between "John kisses Mary" and "courtship," or between "farmer George touched his cap to the lord of the manor" and "stratification" can be other than the relation between "the pump of this heart" and "the pump of the human heart."

Even sociologists who accept this are often tempted to forget it, partly because while any single heart pump is very much like another, kisses can vary greatly in intensity, passion, and significance. This is also the reason why it is more important that the sociologist should *not* forget it; it matters very little to the physiologist if he forgets that his abstract human organ is a generalization from particular organs in particular people *because* they are all very much alike.

If it be accepted that the sociologist's "institutions" are summary generic terms for classes of particular recurring events, then it follows that his statements about the functional interrelations of institutions are generalizations about the causal relations between these recurring events. In other words that "the system of stratification functions to make the division of labor workable" is a generalized summation of a number of lower-order generalizations to the effect that, for instance, "men submit to a lengthy medical training because they have the prospect of greater rewards" etc., which are themselves generalizations from statements of particular events ("Jack submitted ... because he had ...").

We might emphasize this assertion that statements about the functional relations between institutions are *only* generalizations about relations between particular events by means of a mathematical analogy (offered only, it might be added, as a didactic illustration and not as a proof). If it is granted that events like "John (unmarried) passionately kisses Mary (unmarried)" (a) are summarily referred to by such a term as "romantic courtship" (Σa); and events like "John (married) hits Mary (married)" (b) are summarily referred to by such a term as "pattern of marital maladjustment" (Σb), then the statement "patterns of courtship affect patterns of marital adjustment" is a summary of statements of the nature "the way John kissed Mary then affects the way he hits her now," and as such is a statement of the nature Σab, *not* of the nature $\Sigma a \times \Sigma b$.

Social Facts and Reductionism

Some sociologists would part company at this point. They might agree with the above view of the logical nature of constructs like "institution" "behavior pattern," etc., but still hold that there *is* a $\Sigma a \times \Sigma b$ kind of sense in which institutions can be related over and above the relations of the particular events they describe. It is difficult to see how this can be so. More consistent is the position of those who would hold that concepts like institutions are not, or are not only, generalizations abut recurring events. Such arguments might well appeal to the Durkheimian characterization of norms and institutions as "social facts." But the position outlined above is in no way incompatible with one interpretation of the Durkheimian view. It is undoubtedly true that the members of a society do have reified concepts of, say, "marriage," "romantic love," "filial conduct" which are both more than and less than generalizations concerning particular relations between particular people. But these reified concepts are part of the *data* of sociology. Having a concept of marriage is (though normally less easily observed) as much an event in society as having a quarrel with one's wife and susceptible of the same kinds of questions and explanations. There is no more reason for the sociologist to adopt for his thinking *about* society the terms used for thinking *in* society (to take, in other words, his analytical tools straight out of his data) than there is for a carpenter to use nothing but wooden saws.

The point might be made clearer if it is stated in the terms of Maurice Mandelbaum's discussion of "societal facts."[12] His argument that societal facts are not reducible to statements concerning the actions of individuals rests on an identification of what one might call "societal (or cultural) concepts" with "sociologists' concepts." One can agree with his formulation—that there is a language *S*, in which concepts like marriage, the banking system, the presidency, etc., appear; that there is another language

P in which we refer to the thoughts and actions of individuals; and that sentences in *S* cannot be translated wholly into *P* because some of the thoughts and actions of individuals consist of *using S*. But the contention here is that the sociologist should be speaking in a different language—meta-*SP* if one likes—which certainly resembles *S* and was developed from *S* but is an artificial creation for the purpose of analyzing causal relations in society and can only be effective for this purpose if it *is* reducible to *P* (including all the necessary concepts of *S*—the words spoken and the thoughts thought by individuals—which *P* must incorporate). Another way of putting it would be to say that Mandelbaum's arguments that societal facts are not reducible to facts about individuals are really arguments to show that *language* is a necessary part of the sociologist's data for which there can be no substitute. And no one would wish to quarrel with that.[13]

The position outlined above is part of the thesis of "methodological individualism,"[14] the brief debate about which seems to have died down without much interest being shown by professional sociologists. The methodological individualist doctrine which holds that all sociological laws are bound to be such as can ultimately be reduced to laws of individual behavior is a hard one to refute,[15] but one which few sociologists find attractive. The reason is perhaps this: the examples we have given of the particular causal relations actually implied by statements of the functions of institutions were of the type: "Jack became a doctor because of the prospect of . . . ," "the way John kissed Mary then affects the way he hits her now." All imputations of a causal relation imply a causal law. In these cases the relevant laws are laws of individual behavior—"an individual of such and such training in such and such circumstances will orient present actions to remotely deferred gratifications," "behavioral dispositions toward individual others built up under the stress of strong biological urges tend to be modified after the satiation of those urges" might be examples. These can be stated in purely behavior terms. Nevertheless, when they are so stated the possibility of further reduction to laws of psychological processes becomes apparent. Psychological reductionism has never appealed to sociologists; it has usually been conceived as a threat to the integrity and importance of sociology. It is difficult to see why. It would be as absurd to argue that because all the laws of social behavior might ultimately be reducible to psychological terms sociologists should give up sociology and take to psychology, as to hold that chemists should all abandon chemistry since their laws might ultimately be reduced to laws of physics. The antipathy toward the reductionist thesis exists, however, and sociologists have for a long time been intermittently fighting a losing battle to prove (to themselves, it seems, since no one else seems to have been particularly interested) that there *are* irreducible sociological laws sui generis. Is not the resort to "function" in part a continuation of this warfare by more diplomatic means?

Professor Davis noted that in their studies of social change functionalists behave no differently from other sociologists who claim to be opposed to functionalism. Now, studies of social change are explicitly looking for causes—for the causes of the particular events associated with the origination and changing of institutions. To keep one's nose equally on the scent for causes in the analysis of stable systems, however, involves constant reference to the recurring events which make up the institutional units under study and poses the problem of the kind of reductionism outlined above. The concept of function offers an escape; it blurs the precise causal relations imputed and yet descriptions in terms of functions seem somehow to be causal; it makes it easier for institutions to be treated as ultimate units without constant reference to the empirical content of such concepts;[16] in this way the sociological integrity of sociology is preserved and grand theory concerning social systems becomes possible.

Various Sociologies

What, then, of functional*ism*? It is, as Kingsley Davis points out, a name for a variety of methodological and philosophical (following Davis, following Radcliffe-Brown, though "moral" might be more apposite) positions. It might be useful to elucidate these positions with reference to the two main theses of this chapter. These theses are: (1) There is a difference between questions about the functions of an institution and questions about the cause(s) of (the particular one-and-for-all events leading to the origin of, or the recurring events which make up) that institution, and answers to the first kind of question are relevant to answers to the second kind of question only (legitimately) via human motives or evolutionary selection, or (illegitimately) by use of the postulate of necessary integration. (2) Questions about the functions of an institution logically imply questions about the effects of recurring particular events which make up that institution as causes of other recurring particular events.

Functionalists, then, could be any of four types of sociologists. Type (a) sociologists easily accept both of these propositions but find the concept of function useful because they are chiefly concerned with the way in which changes in one institution in a particular society would affect other institutions, for example, the social reformer or the colonial anthropologist. Type (b) sociologists accept both of these propositions but hold the philosophical view that sociologists should concern themselves only with the kind of causal relationships which have a direct bearing on the equilibrium of the social system (i.e., are [eu]functional or dysfunctional) and not with other causal chains which, being in this special sense "nonfunctional," are "pragmatically unimportant."[17]

It is this particular philosophical view with its implication that "stability is all," together with the fact that functionalists of type (a) have usually tended in practice to give reasons for pessimism about the possible scope of social reform, which provide the basis for the charge of functionalist conservatism. Type (c) sociologists are mainly concerned to construct models of social systems and either deny the second of these propositions or occasionally ignore its implications in order to reduce the difficulties of their task. Type (d) sociologists would deny the first of these propositions (usually specifically the charge that the postulate of necessary integration is illegitimate)[18] and, in giving a description of the functions of an institution, would imply that this is also, automatically, a causal explanation of that institution.

A number of alternative positions are possible if these two propositions are accepted. There is the piecemeal approach, outlined at the beginning of this chapter, which suggests that sociologists should concern themselves with searching for regularities in the concomitant occurrence of social events with a view to inducing causal laws which might ultimately be ordered in some systematic theory. There is the historical approach which is largely concerned with discovering the causes of the particular once-and-for-all events which explain the origins of institutions. These is the the static approach which concentrates on societies which have been stable over long periods of time and seeks for the causal relations between the recurring events which make up their institutions. There is still possible scope for the model-system approach insofar as it seeks to build up a pattern of causal relations such as might pertain to an ideal and entirely stable society, without having recourse to the short cut of functionalists variety (c). There is, finally, the "issue" approach, the virtues of which have recently been argued with much vehemence by C. Wright Mills. This involves starting from practical questions which actually worry people, such as, "Who is likely to plunge us into a world war?" and using for the purpose of elucidation questions about the causes of recurring institutionalized events—so that by knowing why people do things we shall be in a better position to know how to stop them; questions about the once-and-for-all causes—so that by knowing how things got the way they are we shall be in a better position to judge whether that is the way they ought to be; and questions about functions of institutions—so that we would have a better idea of what we would be up against if we tried to change them. All kinds of questions are asked not as ends in themselves but as means to eliciting guides for judgment and action. This is not, perhaps, a scientific pursuit in the way that the other approaches outlined above are scientific, though it it one that has intermittently occupied a great many sociologists of repute.

The differences between these various positions are in part methodological—differences concerning the truth of the two propositions enunciated earlier. In part they are moral differences, about the proper scale of priorities which should guide the sociologist's use of this time. About the methodological issues there is legitimate ground for dispute. But about the "oughts" implied in these various positions, we can only preach at each other. In would be sad if we stopped preaching, but let us try to keep our sermons and our methodological discussions separate.

Notes

1. Kingsley Davis, "The Myth of Functional Analysis as a Special Method in Sociology and Anthropology," *American Sociological Review*, xxiv (Dec. 1959), 752–72.
2. A. R. Radcliffe-Brown, *Structure and Function in Primitive Society*, London: Cohen and West, 1952, pp. 157–9.
3. This raises moral as well as analytical problems when the institutions concerned involve factual beliefs. At the turn of the century John Morley's liberal conscience was somewhat exercised by "the question of a dual doctrine ... the question whether it is expedient that the more enlightened classes in a community should ... not only possess their light in silence, but whether they should openly encourage a doctrine for the less enlightened classes which they do not believe to be true for themselves while they regard it as indispensably useful in the case of less fortunate people" (*On Compromise*, London: Macmillan, 1908, p. 44).
4. Robert K. Merton, *Social Theory and Social Structure*, rev. ed., Glencoe, Ill.: Free Press, 1957, p. 65.
5. A less ambitious and more precise integrationist thesis such as, for instance, "the kinship structure and the occupational structure will always be integrated to the degree that the kinship structure does not impose obstacles to such free movement of individuals as the occupational structure requires" still does not allow for automatic transition from function to cause. (Would it be: the family is the way it is because of the occupational structure, or vice versa?) Such a thesis can, however, by specifying areas where causal relations are likely to be found, direct one's thinking toward such empirically testable hypotheses as those implicit in Parsons's discussion of the family. (See e.g. *The Social System*, Glencoe, Ill.: Free Press, 1951, p. 178.) Such for instance as "when industrialization proceeds the importance of the conjugal relation in the kinship structure increases."
6. Ruth Benedict, for instance, remarks that "the course of life and the pressure of the environment, not to speak of the fertility of the human imagination, provide an incredible number of possible leads, all of which, it appears, may serve a society to live by" and, the implication is, it is more or less beyond precise determination why they should utilize one lead rather than another (Ruth Benedict, *Patterns of Culture*, London: Routledge and Kegan Paul, 1949, p. 16).
7. Dennis H. Wrong, "The Functional Theory of Stratification," *American Sociological Review*, xxiv (Dec. 1959), 774.

8. Cf., for instance, George C. Homans and David M. Schneider, *Marriage, Authority and Final Causes*, Glencoe, Ill.: Free Press, 1955. In suggesting "one efficient cause" for the development of patrilateral cross-cousin marriage in societies of certain types, namely, that such a form of marriage best conforms to the personal interests of the members of such societies, they conclude, apropos of Lévy-Strauss's functional, or "final cause," explanation (that such an institution makes for a "better," because more organically integrated, society) "not ... that [it] is right or that it is wrong, but only that it is now unnecessary" (p. 59). Lévy-Strauss's functional explanation did not lead them to their own causal explanation in any sense except that it prompted them to challenge his assumption that it was all there was to be said on the subject.
9. The clear differentiation of causes and consequences, for instance, and the assertion that one may argue from consequence to cause only via (a) motive or (b) evolutionary theory is to be found in Harry C. Bredemeier, "The Methodology of Functionalism," *American Sociological Review*, xx (Apr, 1955), 173. He somewhat obscures his first point, however, with the discussion, in the latter part of his paper, of the precise ways in which motives *also* have to be considered even for a discussion of consequences.
10. See Bertrand Russell, *Mysticism and Logic*, London: Penguin Books, 1953, p. 184. It would be interesting to know whether Russell's well-known assertion in this paper that the concept of cause was useful "only in the infancy of a science" and that as a science developed it was replaced by function (mathematical) had any influence in causing sociologists and anthropologists to drop the unfashionable word "cause" and take up "function" instead. If so they were the rather naive victims of the ambiguity of the word "function," rather like the lady who heard that bearskins were replacing mink this year and though somewhat puzzled decided that fashion was fashion and went to the party naked. One might add that Russell's assertion that the concept of cause is useful only in the infancy of a science is not incompatible with the claim that "cause" is still useful for sociology.
11. On close analysis it becomes extremely difficult, as Bertrand Russell shows, to define the concept of "event" (*Mysticism and Logic*, London: Penguin Books, 1953, pp 176–8; *The Analysis of Mind*, London: Allen and Unwin, 1921, pp. 94–5). The difficulties are, however, not such as to prevent Russell himself from ignoring them in his later work (see, e.g., *Human Knowledge*, London: Allen and Unwin, 1948, p. 344), and the commonsense notion of "event" or "happening," widened slightly perhaps to include not only "the eclipse of the moon" but also "the movement of the moon between 10:00 and 10:01 P.M." (i.e., not only commonsensically discrete events, but also arbitrarily chopped-up units of continuous processes—a legitimate extension since all "events," even eclipses, have arbitrarily defined boundaries) is adequate for the purposes of this discussion and, for the moment at least, for the purposes of sociological inquiry.
12. Maurice Mandelbaum, "Societal Facts," *British Journal of Sociology*, vi (Dec. 1955), 305–16.
13. In Parsonian terms the contention here could be put in the form that the analytical categories of the social system are not identical with those of the cultural system. This is indeed what Parsons says (see *The Social System*, Glencoe, Ill.: Free Press, 1951, p. 15) but in actual practice—in, for instance, his analysis of medical practice in the same book—his method seems to be to take the definition of the role from the cultural system and "fill it out" with examples of concrete action.
14. The thesis is outlined in two articles by J. W. N. Watkins, "Ideal Types and Historical Explanations," *British Journal for the Philosophy of Science*, iii (May 1952), 22–43, and "The Principle of Methodological Individualism," ibid., iii (Aug, 1952), 186–9.
15. The two main attacks on Watkins's articles have been those of Leon J. Goldstein ("The Inadequacy of the Principle of Methodological Individualism," *Journal of Philosophy*, liii (Dec. 1956), 801–13) and E. A. Gellner ("Explanations in History" in *Dreams and Self-Knowledge*, Aristotelian Society Supplementary Volume xxx (1956), 157–76). Goldstein's objection is chiefly that the individualist position "would leave us with theories the entire content of which were the facts that suggested them in the first place, having no further power of prediction or generalization" and rests on such dubious arguments as that "to know that in such and such a society descent is reckoned in the female line or that residence is avunculocal provides no information about the aspirations and activities of particular persons." Gellner objects on several grounds; he uses the Durkheimian "social fact" argument, but eventually admits its irrelevance on approximately the same grounds as are indicated above; he argues also that in practice there may well be a principle of indeterminacy that makes it impossible to observe the precise individual causal sequences which account for events which can be generalized about in macroscopic statistical terms (trends in road accidents, etc.), but his main point is that statement in individualist terms *adds* nothing to a statement in holistic terms. There is, he suggests, only neatness and intelligibility to be lost and nothing to be gained in translating "the committee

made this decision" into statements about the processes that went on in the minds of the individual committee members. But the individualist thesis is not one about descriptive statements, but about laws. It holds that if we knew enough such a "law" as "committees composed of equal proportions of members of low and high status in societies where a stress is placed on harmonious unanimity will tend to reach unanimous decisions reflecting the wishes of those of higher status" is reducible to a number of "laws" about individuals, the way in which they are disposed to react when faced with individuals of higher and of similar status, when faced with the demand for an expression of opinion, etc. The *advantage* of these atomistic laws over the holistic one is that they have greater explanatory power; each is applicable to a wider range of situations than just committees, much as the theory of ionization has greater explanatory power than a "law" dealing with the electrolysis of water, which states that electrodes placed in water give off hydrogen and oxygen and applies only to the specific case of water.

16. E.g., Talcott Parsons, op. cit., p. 456. The universalism of the doctor's role is spoken of as having the function of protecting the doctor from involvement in particularistic personal relations with his patients. This sounds like a causal relation but would seem on closer inspection to be a matter of logic: "a doctor treats his patients all alike" logically implies that he does not treat them as individuals. If the universalism of the doctor's role were something *more* than his treating his patients all alike there would be more than this to Parsons's analysis; but it does not seem that this is the case.
17. R. K. Merton, op. cit., p. 51.
18. This is the position from which Hempel has recently attempted to analyze the logic of functional analysis. Carl G. Hempel, "The Logic of Functional Analysis" in Llewellyn Gross (ed.), *Symposium on Sociological Theory*, Evanston, Ill.: Row, Peterson, 1959, pp. 271–303.

Chapter 24

Functional Explanation: In Marxism

G. A. Cohen

Introduction

... The Preface to [Marx's] *The Critique of Political Economy* uses a number of explanatory expressions: relations of production *correspond* to productive forces; the legal and political superstructure *rises on* the real foundation; the social, political, and intellectual life process *is conditioned by* the mode of production of material life; consciousness is *determined by* social being. In each case Marx distinguishes two items, the second of which he asserts to be in some way explanatory of the first. He fails to say, here and everywhere else, what kind of explanation he is hypothesizing, and semantic analysis of the italicized phrases would not be a good way of discovering what he meant.... Central Marxian explanations are functional, which means, *very roughly*, that the character of what is explained is determined by its effect on what explains it. One reason for so interpreting Marx: if the *direction* of the explanatory tie is as he laid down, then the best account of the *nature* of the tie is that it is a functional one. For production relations profoundly affect productive forces, and superstructures strongly condition foundations. What Marx claims to explain has momentous impact on what he says explains it. Construing his explanations as functional makes for compatibility between the causal power of the explained phenomena and their secondary status in the order of explanation.

Thus to say that an economic structure *corresponds* to the achieved level of the productive forces means: the structure provides maximum scope for the fruitful use and development of the forces, and obtains *because* it provides such scope. To say that being *determines* consciousness means, at least in large part: the character of the leading ideas of a society is explained by their propensity, in virtue of that character, to sustain the structure of economic roles called for by the productive forces.[1]

Putting the two theses together, we get such hypotheses as that Protestantism arose when it did because it was a religion suited to stimulating capitalist enterprise and enforcing labor discipline at a time when the capital/labor relation was preeminently apt to develop new productive potentials of society. When Marx says that "Protestantism, by changing almost all the traditional holidays into workdays, plays an important role in the genesis of capital,"[2] he is not just assigning a certain effect to the new religion, but proposing a (partial) explanation of its rise in terms of that effect.

While Marx was inexplicit about the structure of the central explanations he hypothesized, there are some hints: "The first 'Statute of Labourers' (23 Edward III, 1349) found its immediate pretext (not its cause, for legislation of this kind lasts centuries after the pretext for it has disappeared) in the great plague that decimated the people."[3] The statute cannot be explained by the circumstances of its origin, but only by reference to the persisting effect of legislation of that kind on the developing social structure. We must avoid the error of the "English, who have a tendency to look upon the earliest form of appearance of a thing as the cause of its existence."[4]

There is no well-stated alternative to the view that major Marxian explanatory claims are functional in character. The functional construal is nevertheless not popular, for a number of bad reasons, which will shortly be exposed. In practice Marxists advance functional explanations, but they do not theorize their practice accurately. They recoil from the functional construal when it is made explicit, for reasons to be reviewed. They then have recourse to opaque ideas of "structural causality,"[5] to invocation of Engels's unexplained "determination in the last resort,"[6] to the facile suggestion that the priority of the base lies in the fact that it limits the superstructure, as though the converse were not also true; or they effectively abandon the master theses of explanatory priority by interpreting them as merely heuristic.

Marxists regard functional explanation as suspect for a variety of reasons, the most important of which are dealt with in the next two sections.

Conceptual Criticisms of Functional Explanation

... Let us begin with a simple functional explanation. In some industries there is, over a period of time, a marked increase in the median size of the producing units: small workshops grow into, or are replaced by, large factories. The increased scale reduces the costs of producing a given volume of output. It generates economies of scale. If we find that scale grows just when growth in scale would have that effect, and not otherwise, then it is a plausible explanatory hypothesis that scale grows *because* the growth brings economies. Note that we may be justified in proposing this explanation before we know *how* the fact that enlarged scale induces economies explains large scale. We can know that something operated in favor of large scale, because of its cost-effectiveness, without knowing what so operated. We may not know whether the increase was deliberately sought by wise managers, or came about through an economic analogue of chance variation and natural selection. We might be able to claim *that* the change is explained by its consequences without being able to say *how* it is so explained.

Let us now delineate that form of the explanation more carefully. We have a cause, increase of scale, and an effect, economies of scale. It is not proposed that the cause occurred because the effect occurred. Nor even—though this formulation is closer to the truth—that the cause occurred because it caused that effect. Instead, the cause occurred because of its propensity to have that effect: *the increase in scale occurred because the industry was of a sort in which increases in scale yield economies.*

This being the form of functional explanation, a common objection to it is misplaced. We take the objection as it is stated by Percy Cohen. His examples of functional explanation are

> that religion exists in order to sustain the moral foundations of society ... [and] ... that the State exists in order to coordinate the various activities which occur in complex societies. In both these cases a consequence is used to explain a cause; the end conditions of moral order and coordination are used to explain the existence of religion and the State.... Critics rightly argue that this type of explanation defies the laws of logic, for one thing cannot be the cause of another if it succeeds it in time.[7]

Now it is true (if not, perhaps, a law of logic) that what comes later does not explain what comes earlier. But it is false that the theses mentioned by Cohen violate that truth. It is a plausible generalization that a society develops and/or sustains a religion when a religion is necessary for (or would contribute to) its stability. The religion of a society might, then, be explained in terms of this feature of the society: it requires a religion to be viable. That feature is not a consequence of having a religion, and there would be no contortion of time order in the explanation.

Now suppose a society requires a religion for stability, and has a religion fulfilling that need. It does not follow that its need for a religion *explains* its having one. The society may indeed require a religion, but it is a further question whether it has one *because* it requires one. It may have one not at all because it needs one, but for other reasons. Imagine ten godless communities, each, because it lacks a religion, teetering on the brink of disintegration. A prophet visits all ten, but only one of them accepts his teaching. The other nine subsequently perish, and the single believing society survives. But they took up religion because they liked the prophet's looks, and not because they needed a religion (though they did need a religion). So the fact that there is a religion, and it is needed, does not show that there is a religion because it is needed. That demands further argument. Perhaps some sociologists mistake the need for further argument as a defect in functional explanation itself.

To say that

1. *f* occurred

is not to advance an explanation of why *e* occurred.[8] Yet it might be true that

2. *e* occurred because *f* occurred.

(2) may or may not be true, and if it is true, it is not true simply because (1) is true.

Analogous remarks apply to functional explanation. One does not propose an explanation of the existence of religion by saying that

3. Religion is required to sustain social order.[9]

Yet it might be true that

4. Religion exists because it is required to sustain social order.

(4) may or may not be true, and if it is true, it is not true simply because (3) is true.

The mere fact that *f* preceded *e* does not guarantee that *f* caused *e*, though it may be true that *f* caused *e*. Similarly, the mere fact that *g*'s propensities are beneficial does not guarantee that *g* is explained by those propensities, but it may be true that it is so explained. The existence of the fallacy *post hoc ergo propter hoc* does not disqualify all causal explanations. Neither does the comparable fallacy of supposing that, if something is functional, it is *explained* by its function(s), rule out all functional explanations.

So Percy Cohen is misled when he rejects a theory of religion solely because it explains religion functionally, and when he impugns Durkheim's account of the division of labor on the ground that it is functional in form.[10] There is nothing in principle wrong with functional explanations, though to identify a function something serves is not necessarily to provide one. Failure to recognize both truths generates confused debate in sociology, for many catch one truth only.

Thus while Cohen mistakenly argues that assigning a function to a phenomenon cannot be explanatory, others suppose that to show that a usage or institution is required or eufunctional is *ipso facto* to explain its existence. Merton's classic paper tends toward the supposition that to establish that an item has functions is automatically to contribute to explaining it. He never satisfactorily distinguishes between explaining something by reference to its function(s) (functional explanation proper), and explaining the function(s) of something. He identifies a function the Hawthorne experiment had but fails to note it is not a function which explains why the experiment took place.[11]

Sociologists often identify interesting functions, but it is always a further question, whose answer needs further evidence and argument, whether what they identify explains why something is so. Sometimes good evidence and argument is forthcoming, but not always.

Functionalism, Functional Explanation, and Marxism

Other objections to the functional explanation of social phenomena spring from the historical association between *functional explanation* and the theory of *functionalism*. Defects in the latter have affected the reputation of the former. This is regrettable since, as we shall see, there is not a necessary connection between them.

By functionalism we understand the trend in anthropology whose chief proponents were Malinowski and Radcliffe-Brown. It affirmed three theses,[12] here listed in ascending order of strength ((3) entails (2) and (2) entails (1)):

1. All elements of social life are interconnected. They strongly influence one another and in aggregate "form one inseparable whole" (*Interconnection Thesis*).[13]
2. All elements of social life support or reinforce one another, and hence too the whole society which in aggregate they constitute (*Functional Interconnection Thesis*).
3. Each element is as it is *because* of its contribution to the whole, as described in (2) *(Explanatory Functional Interconnection Thesis*).

Thesis (3) embodies a commitment to functional explanation, and it has therefore been criticized on grounds like those discussed and rejected in the last section. But there has also been separate criticism of thesis (2), which proposes no functional *explanations*, but asserts the universal eufunctionality of social elements. It is objected that (2) is falsified by the conflict, strain, and crisis so common in so many societies. How could Malinowski think that "in every type of civilisation, every custom, material object, idea and belief fulfils some vital function, has some task to accomplish, represents an indispensable part within a working whole"?[14]

(2) is widely thought to be not only false but also viciously conservative in its implications. Marxists have, accordingly, been strong opponents of functionalism, a fact which helps to explain their failure to acknowledge the functional nature of their own explanatory theses.

Whether functionalism is in truth inescapably conservative is a question we need not discuss, though we may note how natural it is to conclude that if everything serves a useful purpose or is, indeed, indispensable, then there is no scope for desirable social change. Radcliffe-Brown's principle of the "functional consistency of social systems"[15] seems hard to reconcile with the reality of class struggle, and whatever serves to deny the latter is a comfort to conservative convictions.

It should be obvious that a Marxist can assert functional explanations without endorsing any of theses (1) to (3). Functional explanation is compatible with rejection of the doctrine of functionalism, and functional explanation is not necessarily conservative. Functional explanation in historical materialism is, moreover, revolutionary, in two respects: it predicts large-scale social transformations, and it claims that their course is violent.

To say that forms of society rise and fall according as they advance and retard the development of the productive forces is to predict massive transformations of social structure as the productive forces progress. The master thesis of historical materialism puts the growth of human powers at the center of the historical process, and it is to this extrasocial development that society itself is constrained to adjust. The conservative tendency of functionalism lies in its functionally explaining institutions as sustaining (existing) society. There is no conservatism when institutions, and society itself, are explained as serving a development of power which prevails against forms of society resisting it.

The theory is also revolutionary in that the means whereby society is transformed is class conflict. Transitions do not occur quietly and easily. Society adjusts itself to nature through access to power of a new class. Class struggle is a large part of the answer to the question: *how* does the fact that a new economic structure would benefit the productive forces explain its actualization? We must now consider such "how-questions" more generally.

Elaborations

... Sound functional explanations apply to the development of biological species. The theory of chance variation and natural selection does not displace functional explanation in that domain. Instead, it shows, inter alia, why functional explanation is appropriate there. The theory entails that plants and animals have the useful equipment they do because of its usefulness, and specifies in what manner the utility of a feature accounts for its existence.

Now in the absence of such a theory we shall still observe provocative correlations between the requirements of living existence and the actual endowments of living things, correlations fine enough to suggest the thesis that they have those endowments because they minister to those requirements. We can rationally hypothesize functional explanations even when we lack an account which, like Darwin's, shows how the explanations work, or ... even when we lack *elaborations* of the explanations. A satisfying elaboration provides a fuller explanation and locates the functional fact within a longer story which specifies its explanatory role more precisely.

Now the fact that functional explanations may reasonably be proposed, in the light of suitable evidence, but in advance of an elaborating theory, is very important for social science and history. For functional explanations in those spheres often carry conviction in the absence of elaborative context. And it would be a mistake to refrain from taking those explanatory steps which are open to us, just because we should prefer to go farther than our current knowledge permits.[16] *If*, for example, the pattern of educational provision in a society evolves in a manner suitable to its changing economy, then it is reasonable to assert that education changes as it does because the changes sustain economic evolution, even when little is known about *how* the fact that an educational change would be economically propitious figures in explaining its occurrence. To be sure, there are grounds for caution pending acquisition of a plausible fuller story, but that is not especially true of functional explanations.

For it is not only explanations of functional cast which, though accepted as explanations, are yet felt to require further elaboration. We are frequently *certain* that p explains q yet unclear *how* it explains it. Someone ignorant of the contribution of oxygen to combustion may yet have overwhelming evidence that when a match, having been struck, bursts into flame, it bursts into flame *because* it has been struck, for all that his ignorance prevents him from saying how it is that the friction leads to ignition. So similarly, to return to functional explanation, one ignorant of genetics and evolutionary theory, will, when he finds species of insects regularly developing means of resisting pesticides introduced into their environments, naturally conclude that they develop those means because they are protective, although he can say nothing more. Perhaps historians and social scientists never record cases of adaptation as unarguable as the biological ones. But the rest of their explanatory hypothesizing is also based on less impressive evidence than what natural scientists are in a position to demand.

Functional explanations, then, have intellectual validity and value, even if it is said that "they raise more questions than they answer." For they answer some questions, and the further ones to which they give rise point research in the right direction.

But now let us examine some ways in which functional explanations may be elaborated.

Consider once again an industry in which average scale of production expands because of the economies large scale brings. We [imagine] this explanatory judgment being passed without detailed knowledge of a connection between the fact that scale yields economies and the (consequent) fact that scale expanded. Two elaborations readily suggest themselves.

First, we can suppose that the industry's decision makers knew that increased scale would yield economies, and that they enlarged their producing units out of awareness of that functional fact. The functional fact would then play its explanatory role by accounting for formation of the (correct) belief that an increase in scale would be beneficial, that belief, together with a desire for the relevant benefits, being a more proximate cause of the expansion in size. For obvious reasons, we call this a *purposive* elaboration of a functional explanation.

In the above elaboration we neither assert nor deny that the industrial units operate in a competitive environment. The decision makers might be Gosplanners, setting the course of an industry wholly subject to their will. But purposive elaboration can also apply in a competitive setting, in which case among the known benefits to be had from expanding scale might be the very survival of each of the firms in question.

In a competitive economy a purposive elaboration is, as noted, possible, but so is a second important form of elaboration. Imagine a competitive economy in which a certain industry would function more efficiently under increased scale, but suppose the managers of the industry's firms are ignorant of the fact. Then if mean scale expands, it is not because anyone seeks the economies increased scale promises. Still, some firms increase the scale of their producing units, perhaps because prestige is attached to size, or because the move is seen as a way of reducing tension between managers; or suppose that there is no intention to increase scale, but, in certain firms, an ungoverned drift in that direction. Then we could not say of any particular firm that its scale grew because of the associated economies. But the functional fact might still explain a change over time in the industry's scale profile, if only those firms which expanded (for whatever reason) would have succeeded, in virtue of having expanded, against the competition. Competition is bound to select in favor of firms whose practice is efficient, regardless of the inspiration of that practice. In the case described, we have what may be called a *Darwinian* elaboration of a functional explanation, for these are its salient elements: chance[17] variation (in scales of production), scarcity (in virtue of finite effective demand), and selection (on the market of those variants which, by chance, had a superior structure).

A third kind of elaboration may be called *Lamarckian*. In Lamarckian biological theory, by contrast with that of Darwin, the species evolves in virtue of evolution within the life history of its specimens, which acquire more adaptive characteristics, and transmit them to offspring.[18] An organ not fully suited to the creature's environment becomes more suited as a result of the struggle to use it in that environment. (An example would be teeth becoming sharper as a result of regular chewing on food best chewed by sharp teeth.) The suggested elaboration is not purposive, because it is not the intention of the organism so to alter its equipment: it is altered as a result of a use which is not intended to alter it, but which reflects the environment's demands. Nor is the elaboration Darwinian. The initial variations, which are then preserved, do not occur by chance relative to the environmental requirements, and there *need* not be any competitive pressure on the organism, expressing itself in differential survival rates as between well- and ill-equipped specimens.

A fourth form of elaboration—really a special case of the first—is appropriate in cases of *self-deception*. By contrast with the second and third forms, the functional fact operates through the minds of agents, but unlike paradigm purposive examples, it does so without the agents' full acknowledgment. An elaboration of this form for the economies of scale case would be quite fanciful, but it is relevant to Marxian theory, as will be seen.

The above classification is not exhaustive, and the types of elaboration reviewed admit of combination with one another: there are often several interlaced routes from the functional fact to the fact it explains. C. Wright Mills contrasted "drift" and "thrust" in social development,[19] and it is easy to envisage agglomerations of the two. Thus, returning once again to economies of scale, there could initially be an unplanned drift to greater average size, controlled by competition, and later a perception of the functional relationship, with increasing thrust as a result.

Marxian Illustrations

Our discussion will be confined to two central topics: the generation and propagation of ideology, and the adaptation of the economic structure to the productive forces.

When Marxists venture functional explanations of ideological and superstructural phenomena, they are often accused of espousing a "conspiracy theory of history." A Marxist says "it is no accident that" left-wing commentators receive little space in major American newspapers, or that British trade union leaders end their careers in the House of Lords. He is then criticized for imagining that an omnicompetent elite exercises fine control over these matters. He sometimes tries to forestall the response by disclaiming an assertion of conspiracy, but too commonly he fails to say in what other fashion phenomena like those mentioned are explained by the functions they serve.

Our discussion of nonpurposive elaborations of functional claims suggests ways of filling that lacuna, but it is also necessary to point out that Marxists can be too sensitive to the charge that they perceive conspiracies. There is more collective design in history than an inflexible rejection of "conspiracy theories" would allow, and richer scope for purposive elaboration of Marxian functional theses than that posture recognizes. Thus, while ideologies are not normally invented to fit the purposes they serve, a fairly deliberate and quite concerted effort to maintain and protect an *existing* ideology is not unusual. According to Christopher Hill, nobility and gentry in seventeenth-century England doubted they "would still be able to control the state without the help of the church," and, therefore, "rallied to the defence of episcopacy in 1641...for explicitly social reasons."[20] Ruling class persons of no special devotion to an Anglican God frankly professed that the established church was required to ensure political obedience, and acted on that inspiration. Or, to take another example, when a high state functionary, reflecting on the unequal distribution of information in society, concludes that "this inequality of knowledge has become necessary for the maintenance of all the social inequalities *which gave rise to it*,"[21] he may be expected to see to the persistence of an educational structure which reproduces ignorance in the right places.

Conspiracy is natural effect when men of like insight into the requirements of continued class domination get together, and such men do get together. But sentences beginning, "The ruling class have decided ... " do not entail the convocation of an assembly. Ruling class persons meet and instruct one another in overlapping milieux of government, recreation, and practical affairs, and a collective policy emerges even when they were never all in one place at one time.

There are, of course, many shades between the cynical[22] handling of ideology just emphasized and an unhypocritical commitment to it, and a division of labor between lucid and *engagé* defenders of dominant ideas can be quite functional. If awareness of the true name of the game penetrates too far down the elite, it could leak into the strata beneath them. There is always a mix of manipulation, self-deception, and blind conviction in adherence to an ideology, the optimal proportions varying with circumstance.

All classes are receptive to whatever ideas are likely to benefit them, and ruling classes are well placed to propagate ideologies particularly congenial to themselves. But before an ideology is received or broadcast it has to be formed. And on that point there are traces in Marx of a Darwinian mechanism, a notion that thought-systems are produced in comparative independence from social constraint, but persist and gain social life following a filtration process which selects those well adapted for ideological service. Thus it is true but in one respect unimportant that the idea of communism has been projected time and again in history,[23] for only when the idea can assist a viable social purpose, as it can now, by figuring in the liberation of the proletariat, will it achieve social significance. There is a kind of "ideological pool" which yields elements in different configurations as social requirements change.

Yet it is unlikely that ideas fashioned in disconnection from their possible social use will endorse and reject *exactly* what suits classes receptive to them. Here a Lamarckian element may enter, to make the picture more plausible.[24] In Lamarck's theory the equipment of the individual organism is somewhat plastic, for it changes under environmental challenge when it is put to a novel use. Because of the delicacy of intellectual constructions, sets of ideas enjoy a partly similar plasticity: one change of emphasis, one slurred inference, etc., can alter the import of the whole. Such

"Lamarckian" possibilities are intimated in Marx's review of the numerous uses to which a self-same Christianity is liable,[25] and it is not because "liberalism" is an ambiguous term that its presumed teaching varies across space and time. And if it is true of revolutionaries that

> just when they seem engaged in revolutionising themselves and things, in creating something that has never yet existed, precisely in such periods of revolutionary crisis they anxiously conjure up the spirits of the past to their service and borrow from them names, battle cries and costumes in order to present the new scene of world history in this time-honoured disguise and this borrowed language,[26]

then it is perhaps not only for the reason Marx states that they so behave, but also because the only symbols and thought-forms available are those which come from the past, and which they must now adopt and adapt.

... Transformations of economic structure are responses to developments within the productive forces. Production relations reflect the character of the productive forces, a character which makes a certain type of structure propitious for their further development. We [deny] that this formulation removed class struggle from the center of history; ... instead ... it was a chief means whereby the forces assert themselves over the relations, and [we challenge] those who assign a more basic role to class struggle to explain what else determines the rise and fall of classes. Those remarks constitute a preliminary elaboration of the functional explanation of forms of economy, which must now be extended.

Classes are permanently poised against one another, and that class tends to prevail whose rule would best meet the demands of production. But how does the fact that production would prosper under a certain class ensure its dominion? Part of the answer is that there is a general stake in stable and thriving production, so that the class best placed to deliver it attracts allies from other strata in society. Prospective ruling classes are often able to raise support among the classes subjected to the ruling class they would displace. Contrariwise, classes unsuited to the task of governing society tend to lack the confidence political hegemony requires, and if they do seize power, they tend not to hold it for long.

Sometimes, too, as in the gradual formation of capitalism, the capacity of a new class to administer production expresses itself in nascent forms of the society it will build, which, being more effective than the old forms, tend to supplant them. Purposive and competitive elements mingle as early growths of capitalism encroach upon and defeat feudal institutions that would restrict them. There is also adaptive metamorphosis. For example: a precapitalist landed ruling class in an epoch of commercialization requires finance from a not yet industrial bourgeoisie. When landlords cannot meet the commitments engendered by their new connections, they lose their holdings, so others, in fear of a similar fate, place their operations on a capitalist basis. Some see what is required for survival, and undergo an alteration of class character; others fail to understand the times, or, too attached to an outmoded ideology and way of life, fight against the new order, and disappear.

The ideological and superstructural supports of the old order lose their authority. The sense of oppression and injustice always latent in the underclass becomes more manifest, encouraged by the class whose hour of glory is at hand, and the dominating illusions become pallid. Marx supposed that the ideological defences of existing conditions begin to collapse when those conditions no longer accord with productive growth. Thus

> when the illusion about competition as the so-called absolute form of free individuality vanishes, this is evidence that the conditions of competition, i.e. of production founded on capital, are already felt and thought of as *barriers*, and hence already *are such*, and more and more become such.[27]

In similar spirit, Engels opined[28] that ideas of equality and rectification of injustice are perennial, but that they achieve historical power only when and because there is contradiction between the productive forces and the production relations. The class able to take hold of the forces rides up on the resentment of the exploited producers.

[There is a] distinction between a change in economic structure which institutes a new dominant relation of production, thus altering the economic structure's type, and a lesser change, which leaves the latter intact. We have been looking at the more dramatic case, the replacement of one type of economic structure by another. But adaptations of the economic structure falling short of total transformation also occur. One such change was the legislated reduction of the working day in Britain, which modified the dominant relation of production, by altering the space in which the wage bargain between bourgeois and proletarians was to be struck.

Marx states two reasons for the contraction of the working day effected by the Factory Acts, and he establishes no bond uniting them. Having presented his reasons, we shall sketch a possible connection between them, which will suggest a generalization about the way functional requirements sometimes assert themselves.

The reasons: "apart from the working-class movement that daily grew more threatening, the limiting of factory labour was dictated by" the need to "curb the passion of capital for a limitless draining of labour power."[29] These are cited as separate forces, whose confluence produced the Factory Acts. (Marx does not say whether either would have been enough without the other.)

Marx conceived that the health of the system required a brake on capitalist exploitation, which was reaching a pitch inimical to the reproduction of the labor force:

> ... capital is reckless of the health or length of life of the labourer, unless under compulsion from society. To the outcry as to the physical and mental degradation, the premature death, the torture of overwork, it answers: Ought these to trouble us since they increase our profits?[30]

Here "capital" is capital embodied in the individual capitalist, and the behavior ascribed to him is imposed by the regime of competition, a set of "external coercive laws having power over every individual capitalist."[31] The coercion of competition can be countered only by the coercion of society, in the shape of its political guardian, the relatively responsive capitalist state. The state must intervene because, despite the behavior forced upon the capitalist, "the interest of capital itself points in the direction of a normal working day."[32] "Capital" in this last excerpt refers to the system as opposed to its members, or to the capitalist as stakeholder in, not puppet of, the system. Large capitalists, whose positions are relatively secure, and whose enterprises would survive—or improve—under additional state constraint, often rise to the stakeholding attitude, and press upon the state the need for reform.

The capitalist state, legislator of the Factory Acts, is, then, the eye of the otherwise blind capitalist, the stabilizer of a system capitalist activity itself endangers. The needs of the system cannot be attended to by dispersed entrepreneurs severally driven to maximize individual profit. Collected in the state. they may see, and see to, those needs, and may respond to working-class demands which suit those needs but which they necessarily repel in civil society.

Let us now take stock. The workers demand remission of exploitation because they want to live; the state grants it because capital needs living labor. The suggested generalization (not a universal law) is that substantial changes in economic structure which favor the immediate welfare of the subordinate class occur when the class fights for them *and* they increase—or at least preserve—the stability of the system (for reasons independent of allaying a felt grievance of the exploited).[33] The elements are connected because ruling class perception of the need for change is quickened by the pressure of underclass demand, and the latter gets bigger in consequence.

Class insurgency is more likely to achieve its object when the object has functional value, a fact which bears on the undialectical question whether it was systemic need, or, on the supposed contrary, militant struggle, which accounted for the coming of welfare capitalism. A reform essential to capital's survival can also qualify as a "victory of the political economy of labor over the political economy of property."[34] There is victory when capitalism is able to sustain itself only under the modification the reform imposes on it.[35]

Notes

1. For extended commentary on "Social being determines consciousness," see my "Being, Consciousness, and Roles." In C. Abramsky (ed.), *Essays in Honour of E. H. Carr* (London, 1974).
2. *Capital,* i. 276. Protestant reformers themselves stressed that the abolition of Saints' days would have a salutary effect on industry. See Hill, *Puritanism and Revolution,* (London, 1968) p. 51. and *Change and Continuity in Seventeenth-Century England* (London, 1974). p. 81.
3. *Capital,* i. 272.
4. Ibid., p. 403. See also the passages quoted on pp. 232–3 above.
5. Althusser, whom we may associate with this phrase, himself employs functional explanations when dealing with actual social phenomena. See, for example, "Ideology and the Ideological State Apparatuses," in *Lenin and Philosophy.* (London, 1971).
6. Engels to Bloch, 21–22 Sept. 1890, *Selected Correspondence,* (Moscow, 1975), p. 394.
7. *Modern Social Theory,* (London, 1968), pp. 47–8. Cohen's other criticisms are implicitly answered in the last chapter.
8. Unless (1) is offered in response to "Why did *e* occur?" in which case uttering (1) is tantamount to uttering (2).
9. Again, unless (3) is a response to, "Why is there religion?"
10. *Modern Social Theory,* pp. 35–6.
11. It might be claimed in Merton's defense that he was concerned only with *identifying* functions of social patterns and institutions, not with functionally *explaining* them. This is a highly implausible reading, and if it is correct, then we may object that in an article recommending the study of functions Merton neglected their explanatory significance.
12. Not often clearly distinguished from one another.
13. Malinowski, *Argonauts of the West Pacific,* (London, 1922). p. 515.
14. Malinowski, *A Scientific Theory of Culture,* (New York, 1960).
15. See *A Natural Science of Society,* (Glencoe, Illinois, 1957) pp. 124–8, and *Structure and Function in Primitive Society,* (London, 1952) p. 43.
16. Cf. G. V. Plekhanov, *The Development of the Monist View of History* (Moscow, 1956). p. 330.
17. This designation does not imply that the variation is uncaused or inexplicable. What is meant by "chance" is that the explanation of the variation is unconnected with the functional value of greater scale. Darwin calls genetic variation *chance* only because it is not controlled by the requirements of the environment.

18. Following Ritterbush (*Overtures to Biology*, New Haven, 1964, p. 175), we may distinguish between the *acquisition* of inheritable characteristics and the *inheritance* of acquired characteristics, and it is the former which interests us here: we are not concerned with the transmission of features from one social entity to another. Lamarck is relevant for his concept of an adaptation to the environment which is not mediated by a prior chance variation. The movement toward the adaptation is from the beginning controlled by the environment's demands.

    Lamarck's specification of the mechanism of adaptation, in terms of "the influx of subtle fluids," is also irrelevant here. What has social application is the concept of plasticity, of organs being able to develop new uses under new constraints.
19. See *The Causes of World War Three* (New York, 1958).
20. *Reformation to Industrial Revolution*, (London, 1968), pp. 153, 92. Cf. *Change and Continuity*, p. 191.
21. Jacques Necker, as quoted at Karl Marx, *Theories of Surplus Value* (1862–3), (Moscow, 1969) i. 307.
22. What was cynical was not the belief that the existing order ought to be defended, but the use of religion in its defense.
23. *German Ideology*, p. 51.
24. Plekhanov invoked Lamarck in the service of historical materialism: "In the same way must also be understood the influence of economic requirements, and of others following from them, on the psychology of a people. Here there takes place a slow adaptation by exercise or non-exercise." *The Monist View*, Marx and Engels, *The German Ideology* (1846), (Moscow, 1956) pp. 217–18.
25. See "The Communism of the Paper *Rheinischer Beobachter*" (1847), in Marx and Engels, *On Religion* (Moscow, 1957), p. 82.
26. Marx "The Eighteenth Brumaire of Louis Bonaparte" (1852), in *Marx-Engels Selected Works*, vol. 1, (Moscow, 1958) p. 247.
27. Marx, *Grondrisse* (1857–8), (Harmondsworth, 1973) p. 652.
28. *Anti-Dühring* (1878), (Moscow, 1954) p. 369.
29. Karl Marx, *Capital* (1867), (Moscow, 1961) i. 239.
30. Ibid., i. 270. In comparable fashion, overintensive capitalist farming threatens the productivity of the soil: see ibid., i. 239, 265, 507, iii. 603, 792.
31. Ibid., i. 270.
32. *Capital*, i. 266.
33. That is, the change is functional for the system other than because it reduces the anger of the proletariat....
34. "Inaugural Address of the W. M. I. A." (1864), in *Marx-Engels Selected Works*, (Moscow, 1958) p. 383. That is how Marx described the Ten Hours Bill.
35. Merton ("Manifest and Latent Functions," p. 104) requires of items to which functions may be assigned that they be "standardized, i.e. patterned and repetitive," for example: "social roles, institutional patterns, social processes, cultural patterns, culturally patterned emotions, social norms, group organization, social structure, devices for social control, etc." There is no good reason for this restriction, as the case of the Factory Acts shows. It is possible to offer functional explanation of a particular event, such as the passage of a bill, or, for that matter, an enactment of a social role on a particular occasion, or a change in a cultural pattern, which is not itself a cultural pattern, and which may occur because of its salutary consequences for the culture.

    In our view, no special type of phenomenon or fact is by nature an object of functional explanation.

Chapter 25

Functional Explanation: In Social Science

Jon Elster

My discussion of functional explanation in the social sciences will proceed in two steps. First I shall set out an argument against such explanation that I have developed elsewhere, and that I still believe to be basically valid.[1] Second, however, I shall explain some of the reasons that have made me see that the issue is more complex than I used to think. Before I enter into the detail of these arguments, I want to say a few words about the immense attraction that functional explanation seems to have for many social scientists, quite independently of the serious arguments that can be marshaled in its defense. The attraction stems, I believe, from the implicit assumption that all social and psychological phenomena must have a *meaning*, i.e. that there must be *some* sense, *some* perspective in which they are beneficial for someone or something; and that furthermore these beneficial effects are what explain the phenomena in question. This mode of thought is wholly foreign to the idea that there may be elements of sound and fury in social life, unintended and accidental consequences that have no meaning whatsoever. Even when the tale appears to be told by an idiot, it is assumed that there exists a code that, when found, will enable us to decipher the real meaning.

This attitude has two main roots in the history of ideas. The first is the theological tradition culminating in Leibniz's *Theodicy*, with the argument that all the apparent evils in the world have beneficial consequences for the larger pattern that justify and explain them. True, this is not the only possible form of the theodicy, for there is also the alternative tradition that explains evil as the inevitable by-product of the good rather than a necessary means to it.[2] The breaking of the eggs does not contribute anything to the taste of the omelette; it just cannot be helped. Moreover, the theodicy cannot serve as a deductive basis for the sociodicy, to use a term coined by Raymond Aron, only as an analogy. There is no reason why the best of all possible worlds should also include the best of all possible societies. Indeed, the whole point of the theodicy is that suboptimality in the part may be a condition for the optimality of the whole, and this may hold also if the part in question is the corner of the universe in which human history unfolds itself. These logical niceties notwithstanding, the legacy of the theological tradition to the social sciences was a strong presumption that private vices will turn out to be public benefits.

Second, the search for meaning derives from modern biology. Pre-Darwinian biology also found a pervasive meaning in organic phenomena, but this was a meaning bestowed by the divine creator and not one that could serve as an independent inspiration for sociology. Darwin, however, gave biological adaptation a solid foundation in causal analysis and thereby provided a substitute for the theological tradition to the demolition of which he also contributed. Formerly, both sociodicy and biodicy derived directly from the theodicy, but now sociodicy could invoke an independent biodicy. Once again, the biodicy served not as a deductive basis (except for some forms of social Darwinism and more recently for sociobiology), but as an analogy. In forms sometimes crude and sometimes subtle, social scientists studied society as if the presumptions of adaptation and stability had the same validity as in the animal realm. In the cabinet of horrors of scientific thought the biological excesses of many social scientists around the turn of the century have a prominent place.[3] The situation is less disastrous today, but the biological paradigm retains an importance out of proportion with its merits.

We may distinguish between the strong and the weak program of functionalist sociology. The strong program can be summed up in

> *Malinowski's Principle:* All social phenomena have beneficial consequences (intended or unintended, recognized or unrecognized) that explain them.

This principle can be harnessed to conservative as well as to radical ideologies: the former will explain social facts in terms of their contribution to social cohesion, the latter according to their contribution to oppression and class rule.[4]

This theory was ably criticized by Merton,[5] who suggested instead the weaker program that may be expressed in

> *Merton's Principle:* Whenever social phenomena have consequences that are beneficial, unintended and unrecognized, they can also be explained by these consequences.

To locate the fallacy in this principle, let me set out what would be a valid, if rarely instantiated, form of functional explanation, in order to show how Merton's Principle deviates from it. From such standard sources as Merton's *Social Theory and Social Structure* and Stinchcombe's *Constructing Social Theories,* we may extract the following account of what a valid functional explanation in sociology would look like:[6]

> An institution or a behavioral pattern X is explained by its function Y for group Z if and only if:
>
> 1. Y is an effect of X;
> 2. Y is beneficial for Z;
> 3. Y is unintended by the actors producing X;
> 4. Y—or at least the causal relation between X and Y—is unrecognized by the actors in Z;
> 5. Y maintains X by a causal feedback loop passing through Z.[7]

There are some cases in the social sciences that satisfy all these criteria. The best known is the attempt by the Chicago school of economists to explain profit-maximizing behavior as a result of the "natural selection" of firms by the market. The anomaly that motivated this attempt was the following. On the one hand, the observed external behavior of firms, such as the choice of factor combinations and of output level, seems to indicate that they adopt a profit-maximizing stance, by adjusting optimally to the market situation. On the other hand, studies of the internal decision-making process of the firm did not find that it was guided by this objective; rather some rough-and-ready rules of thumb were typical. To bridge this gap between the output of the black box and its internal workings, one postulated that some firms just happen to use profit-maximizing rules of thumb and others not; that the former survive whereas the latter go extinct; that the profit-maximizing routines tend to spread in the population of firms, either by imitation or by takeovers. If we set X equal to a certain rule of thumb, Y to profit maximizing, and Z to the set of firms, we have an example of successful functional analysis, in the sense that it has the right kind of explanatory structure. Note, however, that condition (4) is fulfilled only if the rules spread by takeovers, not if they spread by imitation.

Russell Hardin has persuaded me that this example is not as unique as I formerly thought.[8] He gives, among others, the following ingenious example: the growth of the American bureaucracy can be explained by its beneficial consequences for the incumbent congressmen, since more bureaucracy means more bureaucratic problems for the voter, and more complaints to their congressmen, who are then reelected because they are better able than new candidates to provide this service, but this also means that congressmen have less time for legislative work, which then devolves on the bureaucracy, which therefore grows. Similarly, Skinnerian reinforcement may provide a mechanism that can sustain functional explanations, although once again it is doubtful whether condition (4) is satisfied, for "if a causal link is so subtle that its perception is beyond the [beneficiary's] cognitive powers, it can play no role in reinforcement."[9] If we drop condition (4), we get the class of explanations that might be called *filter-explanations*, in which the beneficiary is able to perceive and reinforce (or adopt) the pattern benefiting him, although in the first place these benefits played no role in its emergence. These explanations while empirically important,[10] cannot serve as examples of successful functional explanation as the term is used here.

My main concern, however, is not with the rarity or frequency of successful instances of the paradigm set out above. Rather I want to argue that many purported cases of functional explanation fail because the feedback loop of criterion (5) is postulated rather than demonstrated. Or perhaps "postulated" is too strong, a better term being "tacitly presupposed." Functionalist sociologists argue *as if* (which is not to argue *that*) criterion (5) is automatically fulfilled whenever the other criteria are. Since the demonstration that a phenomenon has unintended, unperceived, and beneficial consequences seems to bestow some kind of meaning on it, and since to bestow meaning is to explain, the sociologist tends to assume that his job is over when the first four criteria are shown to be satisfied. This, at any rate, is the only way in which I can explain the actual practice of functionalist sociology, some samples of which will now be given.

Consider first an argument by Lewis Coser to the effect that "conflict within and between bureaucratic structures provides the means for avoiding the ossification and ritualism which threatens their form of organization."[11] The phrasing is characteristically ambiguous, but it is hard not to retain an impression that the prevention of ossification *explains* bureaucratic conflict. If no explanatory claims are made, why did not Coser write "has the effect of reducing" instead of "provides the means for avoiding"? The term "means" strongly suggests the complementary notion of an "end," with the implied idea that the means is there to serve the end. But of course no actor deploying the means or defining the end is postulated: we are dealing with an objective teleology, a process that has no subject, yet has a goal.

As my next example I shall take the following passage from the third volume of Marx's *Capital*:

> The circumstance that a man without fortune but possessing energy, solidity, stability and business acumen may become a capitalist in this manner...is greatly admired by apologists of the capitalist system. Although this circumstance continually brings an unwelcome number of new soldiers of fortune into the field and into competition with the already existing individual capitalists, it also reinforces the supremacy of capital itself, expands its base and enables it to recruit ever new forces for itself out of the substratum of society. In a similar way, the circumstance that the Catholic Church, in the Middle Ages, formed its hierarchy out of the best brains in the land, regardless of their estate, birth or fortune, was one of the principal means of consolidating ecclesiastical rule and suppressing the laity.[12]

Once again we note the tell-tale use of the word "means," as well as the suggestion that "capital"—not to be confused with the set of individual capitalists—has eyes that see and hands that move.[13] True, the text may be construed so as to understand Marx as merely describing a happy coincidence—but in the light of his Hegelian background and persistent inclination to functional explanation, I cannot take this suggestion very seriously.[14]

Later Marxists have continued in the tradition of objective teleology. It is standard procedure among Marxist social scientists to explain any given institution, policy, or behavior by, first, searching for the class whose objective interest they serve and, next, explaining them by these interests. Or, more often than not it is assumed that all social phenomena serve the interest of the capitalist class, and then the issue becomes one of finding a suitable sense in which this is true. This is made very easy by the multiple meanings of the notion of class interest, which is ambiguous with respect to the distinctions between the interest of the individual class member and the interest of the class as a whole; between short-term and long-term interest; between transitional and steady-state interest; between immediate and fundamental interest; between economic and political interest. But of course the mere fact that some class interest is in some sense served does not provide an explanation. It is true, for instance, that internal cleavages in the working class serve the interest of the capitalist class, but from this we should not conclude that they occur because they have this effect.[15] To do so is to neglect Simmel's important distinction between *tertius gaudens* and *divide et impera:* one may benefit from the conflict between one's enemies, and yet not be instrumental in bringing about that conflict.[16]

Marxist social scientists tend to compound the general functionalist fallacy with another one, the assumption that long-term consequences can explain their causes even when there is no intentional action (or selection).[17] There are, for instance, certain Marxist theories of the state that (i) reject the instrumental conception of the state as a tool in the hand of the capitalist class, (ii) accept that the state often acts in a way that is detrimental to the short-term interest of the capitalist class, (iii) argue that it is in the long-term interest of that class to have a state that does not always and everywhere act in its short-term interest, and (iv) assert that this long-term interest explains the actions of the state, including those which are against the short-term interest.[18] Now in the first place, the notion of long-term interest is so elastic and ambiguous that it can be used to prove almost anything; and in the second place one cannot invoke the pattern "one step backward, two steps forward" without also invoking the existence of an intentional agent. One cannot have it both ways: both invoke an objective teleology that does not require an intentional agent, and ascribe to this teleology a pattern that only makes sense for subjective intentionality.[19]

This concludes my argument—or should I say diatribe—against functional explanation of the more unreflective kind. It might seem unfair to include Merton among, and even as the main exemplar of, the adherents of this procedure. He does in fact leave it an open question whether his functional analyses are intended to *explain*, or perhaps are meant only as a paradigm for the study of unintended consequences in general. But since sympathetic readers have taken his aim to be one of explanation,[20] and since this is certainly the interpretation that has been most influential, I feel that my presentation is largely justified.

The arguments given above have rested on two tacit premises. First, a functional explanation can succeed only if there are reasons for believing in a feedback loop from the consequence to the phenomenon to be explained. Second, these reasons can only be the exhibition of a specific feedback mechanism in each particular case. The second premise is not needed in the case of functional explanation in biology, for here we have general knowledge—the theory of evolution through natural selection—that ensures the existence of some feedback mechanism, even though in a given case we may be unable to exhibit it. But there is no social-science analogue to the theory of evolution, and therefore social scientists are constrained to show, in each particular case, how the feedback operates. I now go on to discuss two defenses of functional explanation that rest on the denial of respectively the one and the other of these premises. I begin with the attempt by Arthur Stinchcombe to show that there *is* a social science analogue to the theory of natural selection, and then go on to discuss the more radical proposal of G. A. Cohen that for a successful functional explanation there is no need for *any* knowledge—general or specific—of a mechanism.

Stinchcombe suggests that we should look at social change as an absorbing Markov chain.[21] To explain this idea, I shall use an example from a passage quoted above, concerning the function of conflicts within and between bureaucracies. We assume that the bureaucratic system can be in one of two states: R (for rigid) and F (for flexible). A rigid bureaucracy has a hierarchical structure that does not permit the expression of conflicts. This leads to the accumulation of tension in the organization, which will have difficulties in adapting to changing conditions and problems. A flexible structure, on the other hand, permits the day-to-day-enactment of conflict and ensures adaptation. We then ask the Markov question: given that the organization at time t is in one of the states R and F, what are the probabilities that at time $t + 1$ it will be in one or the other of the states?

The central assumption embodied in table 25.1 is that the state F is an absorbing one. Once the organization has entered this state, it never leaves it, since there is no accumulation of tension making for change. If the organization starts in the state of F,

Table 25.1

		State at time t	
		R	F
State at	R	$0 < p < 1$	0
time $t + 1$	F	$1 - p$	1

it will remain there. If it starts in the state of R, we know from the theory of Markov chains—and it is in fact obvious—that it will sooner or later end up in state F with a 100 percent probability and remain there. This shows, the argument goes, that there is a *presumption for equilibrium* in social systems. Nonequilibrium states are not durable, and so are replaced by others, that may or may not be in equilibrium. So long as there is a nonzero probability that the nonequilibrium state will be replaced by an equilibrium, the latter will sooner or later be attained.

The Markov-chain theory of social evolution differs from the natural-selection theory of biological evolution in that there is no competition between coexisting solutions to the same functional problems. Rather there is sequence of successive solutions that comes to a halt once a satisfactory arrangement has emerged. In the language used above: once the machine has said Yes to one input, it stops scanning further inputs (until a change in the environment again makes it necessary). Or again, the theory assumes that social evolution is based on *satisficing* rather than maximizing....

What is the explanatory power of this ambitious and interesting attempt to provide a general basis for functional explanation in the social sciences? In my view it is feeble, since it is open to the two objections that I shall now state.

First, the model fails as a basis for functional explanation, since it does not explain social phenomena in terms of their beneficial or stabilizing consequences. Rather, the explanatory burden is shifted to the absence of destabilizing consequences. One implication is that the model violates our intuitive notions about functional explanation, which surely must be related somehow to *functions*, not just to the lack of dysfunctions. On the Markov-chain approach, for instance, we might say that the position of a sleeping person can be explained as an absorbing state: we toss around in sleep until we find a position in which there are no pressures toward further change.[22] This may well be a valid explanation of the position, but it is not a functional explanation. This, in itself, is not a serious objection. In the more general case, however, a state of functional neutrality or indifference will often induce drift,[23] which over time may accumulate to produce great changes. Tradition is a case in point. Unlike traditionalism, tradition has a short and inaccurate memory: it involves doing approximately what your parents did, not doing exactly what people in your society have been doing from time immemorial.[24] Traditional behavior, therefore, is often in a state of incessant if imperceptible change. To explain traditional behavior, such as the rain dance of the Trobrianders,[25] by the absence of destabilizing consequences is, therefore, to be a victim of myopia, or to accept unthinkingly the local description of the practice as ancient and unchanging. For tradition to be unchanging, there must be forces acting on it to keep the behavior constant.[26]

In many cases, however, the absence of a positive effect will count as the presence of a negative one. For instance, it is not always true that those who pay the bureaucrats will continue to support them even if they do not deliver at least some of the goods. Bureaucracies sometimes survive merely by doing no harm, but this will hardly do as a general statement. By concentrating on stable equilibria, I can go on to state my second objection to Stinchcombe's model.

The Markov chain model applies to specific institutions, not to whole societies. For any given institution, what counts as an equilibrium will depend on the current state of all other institutions, just as in the biological case. This means that as in the case of biological evolution, we are dealing with a moving target. In the social case, however, there are no reasons for believing that the speed of the process of adaptation much, if at all, exceeds that of the change of the criteria of adaptation. On the contrary, the very example considered above brings out very well that social changes do not have the slow and incremental character of biological evolution. Once the bureaucratic system of a society changes as radically as from R to F, or from F to R, all other subsystems may be thrown out of—or move away from—equilibrium, and there may not be the time to move to (or substantially toward) a new one.

True, this is an empirical matter that admits of degrees. In traditional peasant societies it may more nearly be the case that all subsystems are in a state of mutual adaptation to each other, though once again this could well be an illusion due to the observer's limited view (or the local ideology). Edmund Leach, in his study of the Kachin, argues for a very long cycle of social change in this traditional society, invoking an analogy with population cycles in ecology.[27] Be this as it may, it seems clear to me that in modern industrial societies too much is changing for equilibria to have the time to work themselves out in the way suggested by Stinchcombe. The association between social anthropology and functional explanation, therefore, may not be accidental.

Cohen's defense of functional explanation rests on epistemological considerations, not on a substantive sociological theory.[28] He argues that while knowledge of a mechanism is a sufficient condition for a successful functional explanation, and the existence of a mechanism a necessary condition, the knowledge is not a necessary condition. A functional explanation must be backed by something beyond the mere observation that the explanandum has beneficial consequences, but we need not invoke a specific mechanism for the backing. An alternative backing is provided by *consequence laws*, e.g. general lawlike statements to the effect that whenever some institution or behavior would have beneficial effects, it is in fact observed.

This idea may be spelled out by means of an example. In his analysis of the variety and number of organizations in the United States, Tocqueville offers an explanation that, in paraphrased quotation, goes as follows. In democracies the citizens do not differ much from each other, and are constantly thrown together in a vast mass. There arise, therefore, a number of artificial and arbitrary classifications by means of which each seeks to set himself apart, fearing that otherwise he would be absorbed by the multitude.[29] Similarly, without referring to Tocqueville, Paul Veyne offers the same analysis of the Roman colleges, whose "latent function" was to permit festivities in a group that was small enough for the intimacy required, while the manifest function was some arbitrarily chosen and irrelevant goal.[30] Now it is hard to see how these unintended, unrecognized, and beneficial effects of the groupings can also explain them, and in any case neither author offers a mechanism that could provide an explanation. Cohen, however, would argue that the analysis might yet be valid. Consider the following proposition that sums up Veyne's discussion:

> I. The Roman colleges can be explained by their beneficial effects on the participants.

It will not be contested by anyone, I believe, that we cannot uphold this proposition simply by pointing out the existence of the benefits. A possible backing could be the following:

> II. The Roman colleges can be explained by their beneficial effects on the participants, through the feedback mechanism X.

While accepting that this justification is as good as any, and indeed better than any, Cohen argues that we may also invoke a backing of the following form:

> III. By virtue of the consequence law Y, the Roman colleges can be explained by their beneficial effects on the participants.

This law, in turn, might have the following form:

> Y. In every case where the emergence of colleges (or similar associations) would have beneficial effects on the participants, such associations in fact emerge.

The structure of this law, as of all consequence laws, is "IF (if A, then B), then A."[31] Such a law being established, we can, if in a given case we observe A occurring and leading to B, invoke the law and say that A is *explained* by its consequence B. The explanation is not in any way invalidated by our lack of knowledge of the underlying mechanism, though we do, of course, assume that there is such a mechanism by virtue of which the explanation, if valid, is valid. Thus we may proceed as in the standard case: (i) propose a hypothesis; (ii) seek to verify it in as many specific cases as possible, (iii) actively seek to falsify it by counterinstances; (iv) after successful confirmation and unsuccessful disconfirmation accord it the (provisional) status of a law; and (v) use the law to explain further cases. The explanations suggested by Tocqueville and Veyne are, therefore, more convincing taken together than they are in isolation, since each case may enter into the law explaining the other.

Cohen argues for this thesis by citing the state of biology before Darwin. At that time biologists had sufficient knowledge to be justified in formulating a consequence law to the effect, roughly speaking, that the needs of the organism tend to be satisfied, even though they had not proposed a correct account of the underlying mechanism. Lamarck, for instance, was quite justified and indeed correct in explaining the structure of organisms through their useful consequences, even if wrong in his sketch of a mechanism. Similarly Cohen argues that Marxism today is in a pre-Darwinian situation. We do have the knowledge required to explain, say, the productive relations in a society in terms of their consequences for the productive forces, even though we are as yet unable to provide a detailed analysis of the feedback operating.

My first objection to this account is that it makes it impossible to distinguish between explanatory and nonexplanatory correlations. Whenever we have established a consequence law "If (if A, then B), then A," this may express some underlying relationship that does in fact provide an explanation of A in terms of its consequence B, or in terms of its propensity to bring about that consequence.[32] There is, however, always the possibility that there is a third factor C, which explains both the presence of A and its tendency to generate B. In fact, I believe this to be true for Lamarck's account of biological adaptation. He did not establish an explanatory correlation, for he wrongly thought that ecological rather than reproductive adaptation was the crucial fact about organisms. Darwin showed that reproductive adaptation is the "third factor," which explains both the features of the organism and their tendency to be ecologically adaptive.

An example from the social sciences is more to the point. In his analysis of authority relations in Classical Antiquity, Paul Veyne invokes Festinger's theory of cognitive dissonance in order to explain why the subjects so readily accepted their submission.[33] Not to accept the natural superiority of the rulers would have created an acutely unpleasant state—"cognitive dissonance"—in the subjects, who, therefore, resigned themselves to a state of submission from which in any case they could not escape. The resignation, to be sure, was useful to the rulers, though far from indispensable. If necessary, they could have upheld—and when necessary did uphold—their rule by force. The fact remains, however, that it was useful; moreover, it took place in precisely those circumstances where it would be useful, those characterized by severe social inequality. A moderate degree of inequality would not induce resignation—but neither would it lead to revolt. We may formulate, then, a consequence law to the effect that, "Whenever resignation would be useful for the rulers, resignation occurs," and yet we are not entitled to explain the resignation by these consequences, since there is a "third factor"—severe social inequality—which explains both the tendency to resignation and the benefits to the rulers. We must conclude that what explains the resignation of the subjects to the status quo is that it had benefits for them—the benefits to the rulers being merely incidental. Cohen's account of functional explanation in terms of consequence laws falls victim to the problem of epiphenomena. To clinch the argument, we may invoke Herbert Simon's argument, ... that it is impossible in principle to distinguish between real and spurious correlations without some a priori assumptions about the causal mechanism at work.

Similarly, the problem of preemptive explanation undermines Cohen's account. There could well be a nonspurious consequence law of the form given above, and yet the presence of A in some specific instance might be due to a quite different mechanism, which preempted the mechanism underlying the consequence law. One could modify the example used in the previous paragraph to illustrate this possibility. Imagine that in general the rulers have to indoctrinate the subjects in order to make the latter believe in the legitimacy of the rule, and that they in fact do this when they need to. In some cases, however, they might not need to, viz. if the subjects spontaneously invent this ideology for the sake of their peace of mind. The consequence law cited in the previous paragraph would then *necessitate* resignation, but would not *explain* it in the cases where the subjects preempt the indoctrination by creating their own ideology. Although Cohen is acutely aware of the difficulties that epiphenomena and preemption create for the Hempelian model of explanation,[34] he is strangely unaware of the extent to which they destroy his own theory of functional explanation.

Independently of these matters of principle, there are strong pragmatic reasons for being skeptical of the use of consequence laws to back functional explanation. These are related to some basic differences between functional explanation in biology and in sociology. First, biology relies on the idea of optimal consequences, whereas sociology rests on the much vaguer notion of beneficial consequences. Second, biology invokes the same consequence in all cases, viz. reproductive adaptation, whereas in the social sciences the explanatory benefits differ from case to case. Given this latitude in the notion of beneficial consequences, it appears likely that for any important social and historical phenomenon one may find a consequence to which it is linked in a spurious consequence law. There are so few historical instances of such phenomena as the transition from one mode of production to another, to take Cohen's major example, that little ingenuity would be required to find a spurious consequence law confirmed by them all. This objection, then, states that in the social sciences it may be difficult to distinquish between lawlike and accidental generalizations of the kind embodied in consequence laws. The objections argued in the previous paragraphs state that even a lawlike generalization may fail to explain, because of the possibility that we may be dealing with epiphenomena or preemption.

Notes

1. Elster (1979), Ch.1.5.
2. This, in fact, was the version of the theodicy proposed by Malebranche; see Elster (1975), Ch. 5.
3. Surveys are Stark (1962) and Schlanger (1971).
4. One might, perhaps, distinguish between the conservative, the Radical, and the Marxist versions of functionalism by the following schematic division. Let us look at the total net product of society as divided into a surplus that goes to the ruling classes and a subsistence component for the rest of society. The conservative functionalism then explains all social arrangements in terms of maximizing the total product (North and Thomas 1973, Posner 1977); Marxist functionalism assumes that the effect is to maximize the first component; and Radical functionalism that the effect is to minimize the second component. To be sure, Radical sociologists (Foucault 1975, Bourdieu 1979) are not mainly concerned with the division of the economic product, but in their work there is much more emphasis on the victims of oppression than on the beneficiaries, and at times one gets the impression that there are in fact no beneficiaries. Since social life is not a zero-sum game, minimizing the welfare of the worst-off may not amount to the same thing as maximizing that of the best-off.
5. Merton (1957), pp. 30ff.
6. Cohen (1982) argues, convincingly, that this paradigm is less closely related to evolutionary explanation in biology than I asserted in Elster (1979). He suggests, moreover, that "Cohen consequence explanations" may be better renderings of the intuitive notion of functional explanation than are "Elster loops." But my account is not designed to capture "intuition" generally, but to make explicit the paradigms used by the foremost defenders and practitioners of functionalist sociology.
7. Of these criteria, (1) through (4) are emphasized by Merton (1957), while criterion (5) is most explicitly stated by Stinchcombe (1968).
8. Hardin (1980).
9. van Parijs (1981), p. 101.
10. See Elster (1979), p. 30.
11. Coser (1971), p. 60.
12. Marx (1894), pp. 600–1.
13. This idea is at the centre of the "capital logic" school of Marxism, surveyed in Jessop (1977) and in the editorial introduction to Holloway and Picciotta (1978).
14. For a fuller argument, see Elster (1982).
15. This conclusion is drawn in Bowles and Gintis (1977) and Roemer (1979).
16. Simmel (1908), pp. 76ff.
17. Filter-explanations, based on the mechanism of deliberate selection among random options, form a subset of the general class of intentional explanations, and one might argue that they have a weaker intentional structure than other members of that class. In particular, intentional selection will not always lead to the global optimum if that option does not come up in the random mechanism, or if it comes up so late in the run that the selector chooses an inferior option that blocks the further development. Yet an intentional selector does have the capacity for waiting or for following indirect strategies, and so is not constrained to local maxima in all cases, contrary to what is suggested by van Parijs (1981), p. 56. Thus it could be legitimate to explain phenomena in terms of long-term positive consequences if there is an intentional actor screening the options that are produced by the random mechanism.
18. A classic instance of this argument is Marx's *Eighteenth Brumaire*; a modern version is found in O'Connor (1973), p. 70.
19. Thus Leibniz was perfectly entitled, given his theological setting, to explain world history as an instance of "reculer pour mieux sauter" (Leibniz 1875–90, vol. III, pp. 346, 578; vol. VII, p. 568). But without a divine creator and guide, one cannot consistently explain, as did Hegel and Marx, the historical process of alienation through its beneficial consequences for the attainment of a higher unity.
20. In particular, this is how it seems to be understood by Stinchcombe (1974). Cohen (1978, p. 258, note 2) thinks differently, but the context is that of a one-shot case (the Hawthorne experiment) in which it is indeed implausible to think that the explanans can succeed the explanandum. Other of Merton's examples, however, concern ongoing activities such as conspicuous consumption, and here it is much more plausible to see his use of the term "function" as being explanatory.
21. Stinchcombe (1974); see also Stinchcombe (1980).

22. This example, and the skeptical conclusion it suggests, were pointed out to me by Philippe van Parijs.
23. For the corresponding problem about genetic drift and "non-Darwinian" evolution in biology, see Kimura (1979).
24. For a brilliant study of traditionalism, see Levenson (1968), vol. I, pp. 26ff.
25. This example is discussed in Stinchcombe (1968), pp. 82–3 and in Stinchcombe (1980).
26. The mechanism could rely on intentional imitation of a constant model, or the superiority of the practice to all nearby alternatives.
27. Leach (1964).
28. Cohen (1978, 1982).
29. Tocqueville (1969), p. 605.
30. Veyne (1976), pp. 292–3. For a discussion of the occasional functionalism in Veyne's outstanding work, see Elster (1980).
31. See Cohen (1978), pp. 259ff. for a more precise statement.
32. Cohen (1982) argues that feedback loops are not crucial for the causal mechanism by virtue of which a functional explanation, if valid, is valid. Rather, he argues, a consequence law of the form specified above has explanatory power because (or if) A has a propensity or disposition to bring about B. It is, say, a dispositional fact about giraffes at an early stage in their evolution that, were they to develop longer necks, their fitness would be enhanced. To be precise, it is a dispositional feature of the environment of the giraffes that, were giraffes to develop longer necks in that environment, their fitness would increase. Moreover, the environment causes the necks to lengthen by virtue of this dispositional feature. Against this I tend to agree with Ruben (1980) that a constant environment cannot act as a cause, except in the sense of simultaneous causation. The issue, however, is in need of further clarification.
33. Festinger (1957, 1964), quoted in Veyne (1976), p. 311 and passim.
34. In unpublished work on Hempel's philosophy of scientific explanation.

References

Bourdieu, P. (1979) *La Distinction*: Paris: Editions de Minuit.

Bowles, S., and Gintis. H. (1977) "The Marxian theory of value and heterogeneous labour." *Cambridge Journal of Economics.*

Cohen, G. A. (1978) *Karl Marx's Theory of History: A Defence.* Oxford: Oxford University Press.

Cohen, G. A. (1982) "Functional explanation, consequence explanation, and Marxism." *Inquiry* 25, 27–56.

Coser, L. (1971) "Social conflict and the theory of social change." In C. G. Smith (ed.), *Conflict Resolution: Contributions of the Behavioral Sciences.* Notre Dame, Ind.: University of Notre Dame Press.

Elster, J. (1975) *Leibniz et la Formation de l'Esprit Capitaliste.* Paris: Aubier-Montaigne.

Elster, J. (1980) "Un historien devant l'irrationel: lecture de Paul Veyne." *Social Science Information* 19, 773–804.

Elster, J. (1982) "Marxism, functionalism and game theory." *Theory and Society* 11, 453–582.

Festinger, L. (1957) *A Theory of Cognitive Dissonance.* Stanford: Stanford University Press.

Festinger, L. (1964) *Conflict, Decision and Dissonance.* Stanford: Stanford University Press.

Foucault, M. (1975) *Surveiller et Punir.* Paris: Gallimard.

Hardin, R. (1980) "Rationality, irrationality and functional explanation." *Social Science Information* 19, 775–82.

Holloway, J. and Picciotta, S. (eds.) (1978) *State and Capital.* London: Edward Arnold.

Jessop, B. (1977) "Recent theories of the capitalist state." *Cambridge Journal of Economics* 353–74.

Kimura, M. (1979) "The molecular theory of neutral evolution." *Scientific American* 241 (5).

Leach, E. (1964) Political-Systems of Highland Burma, 2d ed. London: Bell.

Leibniz, G. W. (1875–90) *Die philosophische Schriften,* ed. Gerhardt. 7 vols. Berlin: Weidmannsche Buchhandlung.

Levenson, J. (1968) *Confucian China and its Modern Fate.* Berkeley and Los Angeles: University of California Press.

Marx, K. (1884) *Capital II.* New York: International Publishers.

Marx, K. (1894) *Capital III.* New York: International Publishers.

Merton, R. (1957) *Social Theory and Social Structure.* Glencoe, Ill.: Free Press.

North, D., and Thomas, R. (1973) *The Rise of the Western World.* Cambridge: Cambridge University Press.

O'Connor, J. (1973) *The Fiscal Crisis of the State,* New York: St. Martin's Press.

Parijs, P. van (1981) *Evolutionary Explanation in the Social Sciences.* Totowa, N.J.: Rowmand and Littlefield.
Posner R. (1977) *Economic Analysis of the Law,* 2nd ed., Boston: Little, Brown.
Roemer, J. (1979) "Divide and conquer: microfoundations of a Marxian theory of wage discrimination" *Bell Journal of Economics* 10, 695–705.
Ruben, D. H. (1980) Review of Cohen (1978) in *British Journal of Political Science* 11, 227–34.
Schlanger, J. (1971) *Les Métaphores de l'Organisme.* Paris: Vrin.
Simmel, G. (1908) *Soziologie.* Berlin: Duncker und Humblot.
Stark, W. (1962) *The Fundamental Forms of Social Thought.* London: Routledge and Kegan Paul.
Stinchcombe, A. (1968) *Constructing Social Theories.* New York: Harcourt, Brace and World.
Stinchcombe, A. (1974) "Merton's theory of social structure." In L. Coser (ed.), *The Idea of Social Structure: Papers in Honor of Robert Merton.* New York: Harcourt, Brace Jovanovich.
Stinchcombe, A. (1980) "Is the Prisoner's Dilemma all of sociology?" *Inquiry* 23, 187–92.
Sundt, E. (1862) "Nordlandsbaden", In *Verker i Utvalg,* vol. VII. Oslo: Gyldendal 1967.
Sundt, E. (1980) *On Marriage in Norway.* Cambridge: Cambridge University Press.
Tocqueville. A. (1969) *Democracy in America.* New York: Anchor Books.
Veyne, P. (1976) *Le Pain et le Cirque.* Paris: Seuil.

Chapter 26

Assessing Functional Explanations in the Social Sciences

Harold Kincaid

Despite decades of controversy, functionalism continues to be a lively research tradition in the social sciences. Furthermore, social scientists who reject Parsonian functionalism and its variants frequently make liberal use of functional explanations. Yet, neither social scientists nor philosophers have reached any consensus on when and where functional accounts are legitimate. Critics doubt that functionalism is really explanatory or testable; advocates are unfortunately vague on their claims and the evidence for them. In what follows I try to clarify what functionalism entails, discuss some ways in which functional hypotheses can be confirmed, and evaluate criticisms. While I think much functionalism is bad science, I hope to show that the problem is not one of principle but of practice.

1 *Functionalism and Its Critics*

Social scientists use the term "functionalism" in diverse ways. At times the term refers to a specific theoretical movement in sociology and anthropology—the movement identified with Malinowski and Parsons. At other times, functionalism refers to a general kind or schema of explanation, not to a specific theory. I will sort out these differences with care below. For the moment we can stick with a simple definition. In its broadest terms, functionalism explains social phenomena by means of their functions. Less trivially, functional accounts begin by identifying specific causal effects of a practice or institution and then proceed to argue that the practice exists in order to promote those effects. Initiation rites are a traditional example: anthropologists have argued that initiation rites exist because they promote social cohesion (Allen 1964; Yalmon 1963). Taken to the extreme—by Parsons and Malinowski, for example—functionalism becomes a total account of society. Essential social needs are identified, and then more or less every social institution is explained as existing in order to promote those needs.

Anyone familiar with sociology and anthropology knows just how important functionalism has been and continues to be. Mainstream sociological and anthropological theory for most of this century was dominated by functionalist theory. Not only Malinowski and Parsons but also Radcliffe-Brown, Durkheim, and Merton were functionalists. Despite his differences with this crowd, Marx also invoked functionalist claims: on his view, the relations of production exist in order to promote the forces of production and the state exists in order to protect the interests of the capitalist class. Current social theory is likewise strongly influenced by functionalism. "Neo-functionalism" is a trendy form of abstract sociological theory (Alexander 1983; Eisenstadt 1985). On a more empirical level, functionalism permeates much work in sociology and anthropology. Faia's *Dynamic Functionalism,* which primarily discusses methods for testing certain kinds of causal models, recently won a major prize in sociology. Current work (e.g., Rappaport 1974; Harris 1977) in anthropology on small-scale societies and their environments explains social practices in terms of their functions. Current Marxist accounts of the state (Milliband 1969) and welfare (Piven and Cloward 1971) explain these institutions by their functions.

Functionalism is nonetheless as controversial as it is prevalent. Critics come from both the social sciences and philosophy; their criticisms are far from minor and usually conclude that functionalism is bankrupt. The complaints fall into roughly three classes: those that deny that functional claims have adequate evidence, those that deny that functional claims are sufficiently explanatory even if confirmed, and those that claim there can be no real analogue of natural selection in the social realm. There are multiple worries about evidence. Elster (1983) and others (e.g., Vayda 1987) argue that most functionalist social science is unconfirmed because functionalists do not cite the mechanisms connecting beneficial practices with their persistence. Others doubt that functionalist analyses are really falsifiable: identifying the benefits of a social practice is too easy, as is finding "functional prerequisites" (Hallpike 1986; Elster 1983). Similarly, some argue that picking out useful practices is essentially value laden, for it presupposes some notion of proper functioning (Turner 1979). Doubts have also been raised about what little statistical evidence functionalists do offer. The needed correlations seem to be lacking. Finally, practicing anthropologists sometimes claim that functional accounts are unsupported because alternative, ordinary causal explanations are available.

Equally numerous are worries that functionalism really does not explain. As we saw above, functionalists frequently do not cite the mechanism underlying their explanation. Elster thinks this deficiency fatal, for there can be no explanation without one. Functional accounts also do not explain how institutions arose, only why they currently persist; they likewise do not tell us why one of many different and equally useful practices exist. According to the critics, both complaints show that functionalism really does not adequately explain (Hallpike 1986; Elster 1983).

A last batch of worries comes from functionalism's use of biological analogies. Functionalists frequently cast their models in evolutionary terms. However, it is not obvious that they are entitled to do so. Societies and social entities seemingly do not reproduce, so it is not clear how natural selection-like mechanisms could operate (Hallpike 1986; Young 1988). Furthermore, social entities do not have obvious boundaries, making talk of survival and death thus seem arbitrary. New cultural traits also do not originate randomly, as they allegedly must if social processes are to parallel evolutionary ones (Ellen 1982). Finally, cultural change does not show any evolutionary pattern (Hallpike 1986). So, functionalists apparently borrow illegitimately from the theory of evolution. If reasonable, these criticisms would be overwhelming reason to reject functionalism in the social sciences. Since functionalism is at the core of past and present social science, they would also raise serious doubts about the prospects for a science of society. In what follows, I begin the large task of evaluating these diverse objections.

2 *What Is Functionalism?*

To evaluate the above criticisms, we first need to get clear on just what functionalism is. Philosophers have spent much time trying to give a unitary account of functional language. Something more modest will do here. For the tasks of this chapter, I shall only propose an account that captures paradigm cases of functionalism, particularly in the social sciences. To develop that account, it will be helpful to look at previous attempts to analyze functional explanation in the social realm. While those attempts ultimately fail, they do point us to a more adequate picture.

Earlier I distinguished "functionalism" from "functional explanations" in the social sciences. Functionalism as a specific sociological theory can perhaps best be seen as a particular application of functional explanations. Functional explanations involve two broad claims: (1) that some social practice or institution has some characteristic effect and (2) that the practice or institution exists in order to promote that effect. Functionalism à la Parsons or Radcliffe-Brown asserts that (1) and (2) hold of most or all institutions and that the relevant effect is maintenance of "social equilibrium" or societal survival. Functionalism is thus just a special case of functional explanation. Since functional explanation is primary and often invoked independently of functionalism as a grand theory, my direct concern throughout this chapter is the former.

Before turning to positive accounts, let me mention one variant of functionalism that is not of interest here. Functionalism sometimes simply involves no more than identifying the causal roles or functions of social institutions. Early functionalist anthropology spent much time, for example, classifying and noting the effects of various native institutions. While such analysis may be both important and difficult, it is in principle at least a straightforward form of causal explanation. So, the real issue about functional explanation lies elsewhere—it concerns assertions that practices exist in order to promote their effects, not just that practices have effects.

Faia in *Dynamic Functionalism* argues that the essence of functional explanation in the social sciences lies in feedback relations. One factor A exists in order to promote another B when A's effects on B in turn result in a causal process influencing A. For example, social contacts outside one's group lower prejudice; reduced prejudice increases contacts. Faia concludes that such explanations are rife in the social sciences and thus that most social scientists are (frequently unwittingly) functionalists. However, it is not hard to see that Faia's account is inadequate. Many natural processes exhibit feedback relations and yet clearly do not require or allow functional explanations. Feedback, as Faia describes it, ultimately is just circular causation. Given that every physical entity in the universe is causally interconnected with every other, numerous physical processes will be amenable to functionalist explanation. Surely Faia has cast the net too broadly.

Faia's account faces another fundamental problem: it allows functional accounts to be "inverted." Functional explanations should not be invertible—if A exists in order to promote B, then B does not exist in order to promote A. This condition is motivated by some prime examples of functional explanation. In two of the most obvious cases where something exists because of its function—consciously designed artifacts and naturally selected traits—the explanations cannot be reversed. We have chairs in order to sit down but we do not sit in order to have chairs. Similarly, some finches have large beaks in order to crack seeds, but the opposite does not hold. And the same holds in the social realm: if initiation rites exist in order to promote social cohesion, the opposite certainly is not automatically true. Thus functional explanations with content are invertible. Feedback and circular causation are invertible, however. For these reasons, Faia's account fails. It may be that all function accounts involve circular causation and that the statistical tools Faia cites are relevant. Still, it is clear that not all circular causes make for functional explanations.

A related account grounds functional explanations in homeostatic systems (e.g, Stincombe 1968). Imagine that the function of the kidneys is to purify the blood. The homeostatic analysis says that kidneys increase their functioning when blood levels are too high, decrease their levels when blood levels are too low. Put more generally, an entity A has a function B if whenever some outside factor interferes to prevent B, then A would change in such a way that B is reestablished. To have a function is to play this role in a homeostatic system.

Nonetheless, something can be functionally explained without being part of a homeostatic system. The homeostatic account fails for several key cases. Some organ systems, for example, the skeleton, do not change in the face of interfering outside influences. More drastically, traits produced by natural selection cannot be handled either. Large finch beaks exist in order to ensure adequate nutrition. Nonetheless, if an outside force intervenes to drastically change the food sources, we (and the finches) have no guarantee that finch beaks will change accordingly. Since many functional explanations in the social sciences are based on natural selection analogies, this problem is particularly pressing. Homeostatic requirements may suffice to pick out some functional explanations; they are nonetheless not necessary.

A final and more helpful account is that of Cohen (1978). According to Cohen, functional explanations are a subspecies of consequence explanations—explanations that explain causes by their consequences. To establish that a social practice A exists in order to do B, we must establish a law—a law that relates A's disposition to do B with A's existence. In short, we must show that it is a law that when A would be useful (or serve its function), A comes to exist. A functional explanation then is an explanation that invokes a particular sort of consequence law. Cohen's account is on the right track. It focuses more directly than previous accounts on the idea that a practice exists because of its function. Nonetheless, I think Cohen's analysis is neither sufficient nor necessary as it stands.

It is possible to have a well-confirmed consequence law (correlating A's function and its existence) without having an adequate functional explanation. The problem is simply that such correlations do not establish causation. A biological trait could exist only when it is useful without thereby being functionally explained. Since phenotypic traits can be tightly linked for a given species, it is possible for a trait to exist when it is useful without it existing because it is useful. Similarly, we might identify social practices that exist only when they serve useful functions; yet that does not prove they exist because of those benefits. Until we rule out third factors, correlations between effects and existence are not sufficient.

Consequence laws may also not be necessary for functional explanation. Organisms that are mobile probably are so because mobility contributed to fitness. Yet, there is no reason to believe that when mobility would be useful, it comes to exist. Plants no doubt would be better off if they could move to the sun, but structural constraints make that impossible. So consequence laws are also too strong a requirement for functional explanation.

Cohen's account fails, but something like it is plausible. We are trying to clarify claims that some social practice A exists in order to do B or that the function of A is to do B. Two preconditions for such claims seem to be:

1. A causes B.
2. A persists because it causes B.[1]

The second condition, of course, does most of the work. However, the basic idea is relatively straightforward. A given social practice has a certain effect. When it has that effect, there is some causal mechanism that ensures A continues to exist. When the practice stops having that effect, that mechanism stops operating. The second condition is thus an ordinary causal claim. Focusing on these conditions avoids some problems seen earlier. Adaptive traits of organism, I argued, do not necessarily exhibit homeostasis—we have no guarantee that the trait in question will change correctly in the face of disturbing factors. Conditions 1 and 2 have just this result. If something interferes and prevents A from causing B, nothing guarantees that A changes appropriately. And unlike Cohen's dispositional account, the problem of spurious correlations is avoided, because condition 2 is a causal one. Furthermore, these conditions are not easily invertible—they do not entail that B exists in order to promote A. While these conditions may not be the whole story, they suffice for my purposes here.

Confirming Functional Accounts

Using the rough account just sketched we can now turn to look at just how functionalist claims can be tested. Debates over functionalism generally proceed at a high theoretical level. When evidence is cited, it is usually anecdotal. Yet, real progress will come only if social scientists are much more explicit about what functionalism actually claims and about how such claims might be tested. In what follows I try to make some contribution to that latter task.

On one level, testing functional explanations seems quite simple. Suppose we have a particular social practice A. To confirm that A exists in order to fulfill its function B, we need to show that (1) when A exists it brings about B and (2) that A persists because condition 1 holds. Since these claims are causal claims, then the usual probabilistic or statistical evidence is relevant. When A exists, it raises the probability that B exists. Furthermore, A persists only when it raises the probability of B. What does this mean in statistical terms? That question is complex and difficult. A simple first approximation at an answer is that functionalist claims must confirm something like the following causal model:

1. $B = x_1 A + e_1$
2. $P = x_2 E + e_2$

where A is the trait or practice to be functionally explained, B is A's function or effect, P is a variable measuring A's persistence, E measures whether equation 1 held at some time prior to P, x is the regression coefficient, and e is an error term. If we have good evidence for these two equations, we have good evidence that A causes B and persists because it does so. No doubt this model is statistically crude. But it does provide a rough generic model to use as a framework for thinking about tests of functionalist claims.

In principle at least this model could be tested directly. Imagine that the Human Relations Area File contained not just cross-sectional data on hundreds of societies but also time series data measuring the duration of various practices. Then we might be able to identify and estimate the above model directly. Unfortunately, the HRAF does not contain the appropriate time series information. Still, direct statistical tests of equations like these might well be possible in other contexts.

Another direct test of functionalist claims would come from showing differential sorting or survival by traits. If we can show that some trait A of a social entity S causally contributes to S's survival, then we have initial evidence that A exists because of its function. Using the generic model above, think of A as initiation rites and B as some measure of survival. Then showing a nonspurious correlation between A and B is evidence for equation 1 and thus for the hypothesis that initiation rites cause survival. Still, differential sorting does not entail persistence. Initiation rites could cause differential survival and still not persist—because the surviving social entities might rapidly change traits. So we still need evidence for equation 2. In natural selection accounts, that evidence comes from establishing inheritance. Analogues in the social sciences would thus need some such mechanism linking contributions to survival with persistence.

Differential sorting models are consequently really just a special case of the more general functionalist model described above. While functionalism in the social sciences thus need not establish differential sorting, doing so is good evidence for functionalist claims. Later I shall sketch some work in organizational ecology that proceeds in just this way.

Other, less direct kinds of evidence are also available. Biologists often test for natural selection by finding a correlation between a trait and environment (cf. Endler 1986). Strong correlations suggest that the trait exists because of the environment. This evidence is a first step toward establishing that the trait in question has been selected. Similar evidence may be possible in the social realm as well. If swidden agriculture occurs only in specific kinds of environments, then we have reason to think that there is something about swidden techniques that allow them to persist while other practices do not. In short, trait-environment correlation suggests equations 1 and 2 might hold when specified as a differential sorting model. Ecological anthropology makes most frequent use of such arguments.

"Environment," however, need not be understood only biologically. We might also identify "social environments"—overarching social structures that serve as an environment for lesser social entities. We could then look for correlations between these environments and characteristics of social organizations. For example, organizational strategies may be closely correlated to the kind of social environment they face. Depending on our background information, such correlations may be good evidence that strategies persist because they allow organizations to survive. So, correlational evidence can in principle help test functionalism in the social sciences.

A third way to confirm functional claims again parallels testing in evolutionary biology. Biologists find evidence that a trait is adaptive by doing design analyses—by showing that a given trait would best solve the environmental problems an organism faces. If the analysis and the actual trait correspond, then we have a prima facie case for thinking that the trait exists in order to solve the relevant problem.

Design analyses generally take the form of optimality arguments, arguments that go roughly like this:

1. Trait of type A would be the best available solution to the identified environmental problem.
2. Selection of traits would result in the optimal trait being established.
3. The observed trait is of type A.
C. Thus the observed trait exists in order to serve its function.

Are such arguments possible in the social sciences? Economics, for one, can sometimes provide design evidence. Assuming that a competitive environment would select profit maximizers, we can derive optimal firm behavior (under specified constraints). Firms, for example, would equate marginal cost and price. If we do in fact find that firms equate price and marginal cost (intentionally or unintentionally), we have some evidence that such behavior functions to maximize profits. Work by anthropologists on hunter-gather subsistence techniques also allow for design evidence. Details of a clan's environment are analyzed and the best means for solving subsistence problems are then determined. If observed and predicted practices correspond, we have evidence for selection and thus functionality. And Harris's (1979) charming explanations of seemingly irrational taboos and practices likewise work by showing that the practice in question is really optimal, given the local environment.

So there is good reason to believe that functionalism is testable: its main claims can be coherently formulated and diverse evidence at least in principle adduced. Whether the "in principle" can be made an "in fact" is an open question. All the testing procedures discussed above are fraught with complications. Those complications sometimes may entirely undercut the tests I have cited. At other times, however, such tests will appear to disconfirm what are really plausible hypotheses because important complexities are ignored. Let me sketch some of these complications.

Design analysis is controversial in biology, and we might expect it to be even more so in the social sciences. Three major problems confront design analyses: (a) making judgments about the possible alternatives and the optimal among them is difficult, (b) even when we know that selection is present, it may not pick optimal traits, and (c) optimal traits might come about by a nonselective process. Each problem is sufficient to raise doubts about an optimality argument.

Are such obstacles inevitable and fatal for this kind of evidence in the social sciences? A detailed answer is beyond the scope of this chapter. However, we can point to some areas where the answer may be affirmative, others negative. Traditionally, functionalism made much of finding functional prerequisites—the basic needs any society had to satisfy in order to survive. General social problems were identified, and then broad traits were deduced that would solve those problems. When these traits were found in some form, researchers concluded they existed in order to ensure social survival or equilibrium. Such arguments are a kind of design analysis. They work by making predictions about what an entity must be like to survive and then comparing that prediction with actual cases.

Unfortunately, prerequisite analysis exemplifies all the potential problems of design analysis. Little or no evidence is given that the prerequisite in question best solves the alleged problem. Little or no evidence is given that the "prerequisite" is really required rather than merely a possible solution. Finally, little or no evidence is given that the best trait will be selected. Such problems, for example, undermine Harris's functionalist explanation of seemingly irrational "foodways" (Vayda 1987).

Design analysis may fair better elsewhere. Some applied microeconomics has a fairly good handle on possible alternatives—for example, possible production techniques. At least for some markets, strong competition can be shown and thus we can be relatively sure that the optimal will survive over alternatives. Similarly, small scale societies facing extreme environmental pressure may also provide a better case for optimality arguments. So design evidence in the social sciences may not always be beyond reach.

Trait-environment correlations have an equivalent prognosis. By themselves, they do not rule out numerous nonfunctional explanations. Ultimately, they are conclusive evidence only if we are certain that some kind of selective process underlies the tie between trait and environment. Controlling for other, nonselective factors that might make the correlation spurious can help. Nonetheless, doing so is much easier to recommend than it is to realize in practice.

A final set of complications result from the overly simple models employed in functional explanations. Recall our elementary causal model for functional accounts. That model assumes the world is a very simple causal place. Each equation invokes only one independent variable. However, functional processes can be much more complicated. Multiple independent variables may be at work. Functional processes may interact with ordinary causal ones. Testing simple models when more complex ones are called for can produce misleading results—usually results falsely disconfirming the functionalist hypothesis.

Figure 26.1 depicts some complications frequently ignored. Situations (b) and (c) result because functional effects may happen on multiple levels. Some practice A might have useful effects at an organization, class, and social level for example—or its effects at one level might counteract those at another. (b) depicts just this situation. When the

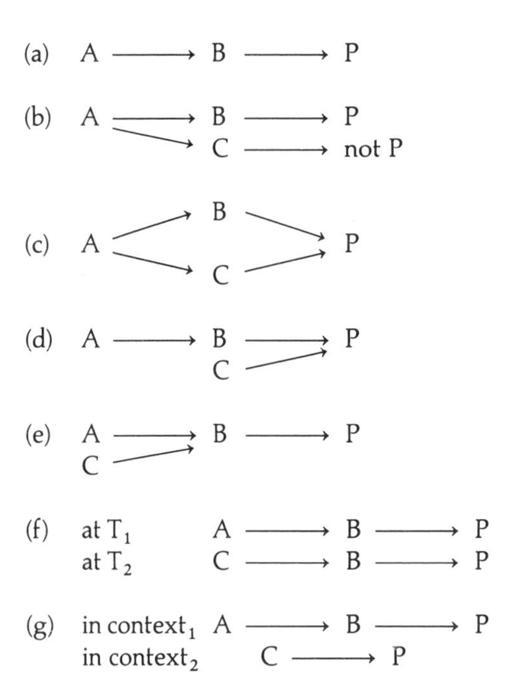

Figure 26.1
Various ways the simple model of functional explanation can be complicated. (a) is the simple model, where A is to be functionally explained, B is its effect and P is the persistence of A. (b) A exists in order to B but also has effect C which is dysfunctional for A. (c) multiple functions are present: A persists because of effect B and effect C. (d) A exists in order to B, but A also persists for reasons other than causing B. (e) A exists in order to B, but A is not the sole cause of effect B. (f) functional equivalents: at one time, A exists in order to B, at another C exists in order to B. (g) A is functionally explained in one causal context but exists for nonfunctional reasons in another.

true model is as in (b) but we test for the elementary model, we may find equation 2 statistically insignificant. C is screening off the effect on persistence. If the real world is as in (c), similar but less severe confusion results from assuming the simple model: equation 2 again may be statistically insignificant, because we are ignoring other causes of A's persistence. The second condition can also fail in situation (d). Here A persists because of its effects but does not persist solely for that reason. Once again, employing the elementary model may cause us to miss a real relation. Better tests, of course, require adding in the complicating factors to equation 2 and using multiple regression.

Situations (e), (f) and (g) cause more serious problems when we assume an elementary model. In (e), A is only partly responsible for its effect and using a simple model may lead us to reject equation 1 as statistically insignificant. It will also understate equation 2 for the same reasons as cases (b)–(d). Again, the other causal factors have to be added to the test. Situation (f) depicts the problem of functional equivalents: at one time the useful effect results from A, at another from C. Testing for only the elementary model will result in both weak evidence that A causes B and that A's effect and its persistence occur at rates greater than chance. The elementary model also assumes that A has a uniform effect across all contexts. Situation (g) illustrates circumstances where that assumption fails: in some contexts A persists because of its useful effects, in others for other reasons. Lumping these two contexts together means that both equations may be weak. It will look as if A is weakly correlated with B and A's effect is weakly connected with its persistence.

What do these complexities tell us? One moral is that simple models of functional explanation can lead to hasty rejection of functional claims. In particular, verbal arguments using anecdotal evidence are likely to be inconclusive. For example, Marxists claim that the state exists in order to promote class interests. Criticizing those accounts by pointing to other causes of the state make error (c) above and are thus inconclusive. A second moral, I hope, is that functional explanations are testable in principle. All the problems illustrated by (a)–(e) can be tamed by using more sophisticated models and more sophisticated statistical techniques. Functionalists typically do not do so, but that is not because there is not work to be done.

4 *Successful Functional Explanations*

So far I have tried to give a rough account of functional explanation as well as relevant evidence. I want now to argue that at least some functional accounts in the social sciences are relatively well confirmed. I do so by looking briefly at "ecological" accounts of organizational diversity.

Developed in the last decade [1980s] organization ecology tries to explain the change and diversity in organizations. As the name suggests, the theory borrows heavily from evolutionary biology and ecology. Organizational ecology holds that the kinds of organizations found at any given time are largely the product of differential birth and death of organizations. Some organizations survive longer than others; some types are founded at greater rates. Differences in both variables are determined by the social environment. In particular, organizations compete for resources: financing, members, legitimacy, customers, etc. Resources, however, are not unlimited. As a result, some organizations survive over others. Similarly, potential resources also influence foundings: individuals are most likely to found organizations when the needed resources are apparently available. Thus organizations, interacting with the environment, have different birth and survival prospects. In short, the fundamental concepts of evolutionary biology—differential mortality and fecundity determined by competition over resources—can be approximately applied to organizational change.

This work constitutes a prime case of functional explanation. Traits of organizations are identified that have positive effects on survival. Those traits are then shown to persist because of their positive effects; a selective mechanism is identified that underlies that connection. Thus when organizational ecology explains why organizations have the properties they do, they are providing functional explanations.

Of course, much previous functionalism in the social sciences appealed implicitly to the idea of selection. However, the appeals were only implicit. Details on possible mechanism were few, rigorous evidence scarce. Claims were not precisely formulated; the level and kinds of selective forces were left ambiguous.

Organizational ecology does a much better job than its functionalist predecessors of avoiding these problems. Mechanisms are explicitly cited and fairly convincing data provided. The general mechanism for selection is clear: differential birth and survival determined by environmental interactions. Which entities benefit (or are harmed) and what counts as the environment are clear specified. And, organizational ecology can sometimes be much more specific, citing the precise environmental and organizational factors responsible for differential survival. The evidence goes far beyond the anecdotal, case study approach found in Radciffe-Brown or Parsons and beyond the simple cross-sectional statistics of quantitative anthropology. Dynamic time-series data are used to test hypotheses about the causes of differential sorting among organizational types.

Some of the best work comes from Hannan and Freeman (1989). They have examined in detail changes in three kinds of organizations: unions, semiconductor firms, and restaurants. Data bases in each instance are extensive, covering the entire population of U.S. unions since 1830, all U.S. semiconductor firms, and all restaurants in a large urban area. Statistical tests look for differences in probability of survival and founding, controlling for numerous variables and looking for environmental causes. These include not just social and political variables but also the presence of other populations of organizations. The data employed thus largely provide direct test of the general differential sorting model discussed in section 3. Hannan and Freeman are able to show (a) that specific organizational characteristics cause differential survival in specific environments and (b) that surviving organizations rarely change their basic characteristics. Those traits thus persist because of their effects. In short, Hannan and Freeman's explanations meet all the requirements sketched earlier for functional explanation.

The organization ecology literature is, of course, not beyond criticism. Like any complex statistical inference, the studies cited here are not definitive proof. However, the difficulties are not, I would argue, any more severe than beset most work in evolutionary biology, for example.

5 *The Critics Answered*

It is time finally to answer the criticisms raised in section 1. Elster and others have claimed that no functional explanation is confirmed until we have cited the specific mechanism involved; however, much functional explanation in the social sciences provides no such mechanism. While we can share Elster's concern for rigor, his claim is nonetheless false. "The mechanism" is, of course, vague—vague in ways that immediately belie Elster's claim. Is knowing that there exists a selective mechanism sufficient? Or must we know the precise process? And, if so, in how much detail? Do we need to trace each causal step from effects to persistence? At what level of description? Once we begin asking such questions, it should become obvious that the demand for mechanisms has no natural stopping point. Certainly the more detail, the better—yet something short of the whole story has to be sufficient, or else we would never confirm any causal claim.

We can in fact have fairly good evidence for a functional claim without knowing the precise mechanism. To know that A persists because of its effects, I need to establish a nonspurious correlation between A's effect and its persistence. Elster is right that such a correlation might be spurious and that identifying underlying mechanisms will help avoid that problem. He is wrong, however, that identifying mechanisms is the only way to eliminate spurious causation. Spurious causation can be eliminated by controlling for all relevant possible causes that might make the correlation between effects and persistence spurious. We need not know the precise mechanism, whatever that means.

Elster's claim is perhaps more reasonable for indirect evidence, for it is weaker and relies more heavily on background assumptions. Optimality arguments and trait-environment correlations are compelling only if they rule out alternative, nonfunctional explanations. So this evidence requires us to know that some mechanism ties persistence to useful effects. We do need to know something about mechanisms in this case, unlike the general case of confirming functionalist hypotheses. Nonetheless, we still do not need to know the precise mechanism—only that some selective mechanism or other is at work.

A second complaint about functionalism is that functional prerequisites are either empty or nonexistent. No doubt much actual functional explanation employing this notion fares badly. However, the problem is not inherent. Design analysis is at least possible and perhaps actual in parts of economics, for example. Certainly, looking for completely universal factors that contribute to social survival is unhelpful, but there is no reason functional explanation must do so—any more than design analysis in biology needs to find universal elements of fitness. Furthermore, design analysis is just one of several possible sources of evidence.

Some social scientists have doubted whether the idea of beneficial effects can be objectively evaluated—doesn't it presuppose the idea of proper functioning? Maybe it does so, but not in any unacceptable way that I can see. Positive effects must be clearly defined; they will frequently concern survival (and/or reproduction). However, survival is not a value-laden term like "proper functioning." Organizational ecology studies differential survival of organizations. Newspapers, for example, either survive or do not—deciding that issue does not seem a value judgment about what constitutes a good newspaper.

Are the boundaries of social entities perhaps so vague that no objective decision can be made about when one survives? No doubt there are borderline cases, but that does not make it all subjective. Individuating biological species likewise involves borderline cases and can be done in multiple ways; that does not mean that any individuation is as good as another. Biologists must make clear how they individuate species. But, once that is done, they can go on to look for evidence for their claims. Organizational ecologists can do the same.

Some believe functional accounts too easy—almost any social practice is bound to have some positive effects. No doubt this claim is true and much functional explanation has been weak in this way. Nonetheless, there is no reason it must be. Organizational ecology is a case in point. Hannan and Freeman use standard statistical techniques to find traits that seem to cause differential survival, controlling for alternative explanations. So positive effects are measured and the tie to survival is quite directly demonstrated.

Similar complaints—that statistical evidence is weak or that alternative, nonfunctional causes can be cited—get similar answers. Statistical evidence is sometimes weak or even nonexistent. Yet that is not inevitable, as the work of Hannan and Freeman indicates. And, as I argued above, the critic's statistical techniques themselves may be primitive and entirely inconclusive. Likewise, nonfunctional causes can be a problem, but not because they show a functional explanation otiose. Functional accounts do not exclude multiple causes, even where the other causes have nothing to do with selection by consequences. Of course, if alternative explanations are not ruled out or incorporated, a functional account has weak evidence. Then, however, the problem is with the singer, not the song.

Is there something about functional accounts that does not satisfy strictures on explanations (beyond problems of evidence)? Functional explanations are, I have been arguing, a species of causal explanation. So in form at least they seem perfectly acceptable.

Do we perhaps need to cite mechanisms before a functional account explains? Since we can search for mechanisms on indefinitely many levels, this demand seems too strong—for most causal explanations do not completely cite the underlying mechanisms (whatever that would mean). If I have good evidence that some trait causes differential survival, then I can say why that trait exists—even if I do not know precisely how it causes differential survival. Mechanisms are not sine qua non for explanation.

Hallpike was worried that functional explanations are inadequate because they do not explain the origins of institutions. His complaint can be granted, properly understood. Darwin could not explain how new traits arose. To that extent his theory was incomplete. However, an incomplete theory is not necessarily an inadequate one. In other words, when a theory cannot answer some questions it does not follow that it cannot satisfactorily answer any. Social scientists might be able to show selection among practices or institutions—and yet have no account of how those differences first originated. Still, the former account would be perfectly adequate to ground a functional explanation. That explanation, however, would not be the whole story, just as Darwin's account was not.

Hallpike is also worried that functional accounts are not sufficiently specific. In many cases, the positive effects of a given trait could be fulfilled by other social practices. Thus it seems an appeal to functions does not say why this practice exists rather than some other equivalent institution. Hallpike's conclusion seems plausible, but I am not sure what it shows. We may not be able to explain why A rather than B originally arose. That does not mean, however, that we cannot explain why A persists rather than not. Two different questions are at issue here. Because functional explanations cannot answer the former, it does not follow they cannot answer the latter. Does functionalism trade on an illegitimate analogy with natural selection? Hallpike (1986) and Young (1988) argue that social entities do not reproduce; thus functionalism cannot be grounded on mechanisms analogous to natural selection. The criticism assumes not only that social entities cannot reproduce but also that they must for selection to operate. Neither assumption is reasonable. In some instances, social entities do in effect leave copies of themselves—Boyd and Richerson (1986), for example, describe a clear sense in which cultural practices "replicate." Furthermore, as Sober (1984) has pointed out, natural selection mechanisms do not require literal copying—only differential representation in future generations. Hannan and Freeman, for example, describe a process of differential longevity and foundings for organizations, even though organizations do not literally make copies of themselves. Such differential representation suffices for selection to operate. Finally, selection can occur even if births are not determined by selective causes—so long as mortality is. Similarly, worries (Ellen 1982) that new practices do not originate in a random fashion are not decisive. Simple models of natural selection require such randomness; sophisticated ones do not. Selection can remain an important force and thus an important explainer.

Finally, attacks on the idea of cultural evolution are not necessarily attacks on functional explanation. Many practices might be functionally explained without social change sowing any overall direction. Evolutionism is added baggage to the functional program. Just as Darwinian functional analyses do not entail that biological change has any innate direction, so too for functional accounts in the social sciences.

6 *Conclusion*

We have seen that functionalism is capable of coherent formulation, that at least some types of evidence for functional claims can be given and that the standard complaints about functionalism are largely misplaced. To that extent functionalism can be defended. Whether functional explanations ultimately gain any important and established place in social explanation remains, nonetheless, up for grabs. Much can be done to both clarify functional hypotheses and provide sophisticated evidence for their evaluation. If there is any compelling criticism of functionalism in the social sciences, it is one applicable to most social science: debates take place on a theoretical and ideological plane while potentially telling empirical work goes undone.

Note

1. I borrow here from Wright's (1976) consequence-etiological account, though I do not claim as does Wright that this account provides necessary and sufficient conditions. The "persists" formulation is also mine, not Wright's. I use "persists" rather than "exists" to avoid some confusions common in the literature. "Exists" implies that when the functional claim is made, A exists at that moment because it is currently causing A, which is typically false, or that it exists because of what it would do, an equally problematic formulation. Persists has no such implications.

References

Alexander, J. C. (1983), "The Modern Reconstruction of Classical Thought: Talcott Parsons," *Theoretical Logic in Sociology*. Volume IV. Berkeley: University of California Press.

Allen, M. R. (1967), *Male Cults and Secret Initiations in Melanesia*. Victoria: Melbourne University Press.

Boyd, R., and Richerson, P. (1985), *Culture and the Evolutionary Process*. Chicago: University of Chicago Press.

Cohen, G. A. (1978), *Karl Marx's Theory of History: A Defence*. Princeton: Princeton University Press.

Eisenstadt, S. N. (1985), "Systematic Qualities and Boundaries of Societies," in *Neofunctionalism*, J. C. Alexander (ed). Beverly Hills: Sage Publications.

Ellen, R. (1982), *Environment, Subsistence and System: the Ecology of Small-Scale Social Formations*. Cambridge: Cambridge University Press.
Elster, J. (1983), *Explaining Technical Change*. Cambridge: Cambridge University Press.
Endler, J. A. (1986), *Natural Selection in the Wild*. Princeton: Princeton University Press.
Faia, M. (1986), *Dynamic Functionalism: Strategy and Tactics*. Cambridge: Cambridge University Press.
Hallpike, C. (1986), *Principles of Social Evolution*. New York: Oxford University Press.
Hannan, M. T., and Freeman, J. (1989), *Organizational Ecology*. Cambridge: Harvard University Press.
Harris, M. (1979), *Cannibals and Kings*. New York: Random House.
Milliband, R. (1969), *The State in Capitalist Society*. New York: Basic Books.
Piven, F. F., and Cloward, R. A. (1971), *Regulating the Poor: The Functions of Public Welfare*. New York: Pantheon Books.
Rappaport, R. A. (1984), *Pigs for the Ancestors: Ritual in the Ecology of a New Guinea People*. New Haven: Yale University Press.
Sober, E. (1984), *The Nature of Selection*. Cambridge: MIT Press.
Stincombe, A. (1968), *Constructing Social Theories*. New York: Harcourt, Brace and World.
Turner, J. H., and Maryanski, A. (1979), *Functionalism*. Menlo Park: Benjamin/Cummings Publishing.
Vayda, A. P. (1987), "Explaining What People Eat: A Review Article," *Human Ecology* 15: 493–510.
Wright, L. (1976), *Teleological Explanations*. Berkeley: University California Press.
Yalmon, N. (1963), "On the Purity f Women in the Castes of Ceylon and Malabar," *Journal of the Royal Anthropological Institute* 93:25–58.
Young, R. (1988), "Is Population Ecology a Useful Paradigm for the Study of Organizations?" *American Journal of Sociology* 94:124.

PART VI

Reductionism, Individualism, and Holism

Introduction to Part VI

What level of inquiry is appropriate to the study of human systems? How should they be described? Should we aim at the explanation of human action in terms of individuals or that of groups? What role does reductionism play in the study of human behavior? These questions frame the issues taken up by the next seven selections.

Emile Durkheim's essay "Social Facts" (chapter 27) opens this part with the question, What sorts of things are "social facts"? The criteria by which we delimit social phenomena are explored as a prelude to the question of what level of decription—the individual or the collective—is necessary for the explanation of social events.

With J. W. N. Watkins's "Historical Explanation in the Social Sciences" (chapter 28) we arrive squarely at this question. Are there sociological "wholes"—systems that are sui generis and that, although they are dependent on individual action, cannot be explained by it? Or is it more appropriate to think that social phenomena must be explained in terms of the individual behavior underlying large-scale social events? Watkins outlines a powerful defense of "methodological individualism" and considers the circumstances in which it is appropriate.

Steven Lukes, in "Methodological Individualism Reconsidered" (chapter 29), sharpens this debate considerably by exploring several conceptual confusions and unclarities behind methodological individualism and distinguishes a more focused account of it to assess its plausibility. In chapter 30, Richard Miller's "Methodological Individualism and Social Explanation" carries this process further, exploring the constraints that methodological individualism allegedly puts on social explanations. But how far should it go? Ultimately, Miller leaves us with several empirical examples that have trouble fitting into the constraints of methodological individualism and leaves us wondering whether certain large-scale social hypotheses are nonetheless worthy of empirical consideration.

Just such a paradigm case is considered in more detail in the following chapter Daniel Little's "Microfoundations of Marxism." Little considers what Miller calls the bête noire of methodological individualism: Marxism. Little's careful dissection leads to the conclusion that Marxist social theories are not necessarily incompatible with the goals of methodological individualism, thus ratifying Miller's hypothesis that the issue must be decided on the basis of empirical content.

But what is the status of social scientific explanations that hope to be consistent with the strictures of methodological individualism? Must they be reductionist? Must we forego understanding of social uniformities until they are reduced to the rock-bottom facts about individuals? Or can we allow sociological explanations of macrophenomena, which we feel are supervenient upon individual action, without reduction to the microlevel? What is the status of the laws that may govern social events, and at what level of explanation should they be offered?

This set of issues is taken up by Harold Kincaid in chapter 32: "Reduction, Explanation, and Individualism." Kincaid explores the problems with reductionist interpretations of methodological individualism, even while admitting that reference to individuals is an important aspect of social explanations. That is, he finds incomplete those accounts of social behavior that refer only to social entities, recognizing the legitimacy of claims that call for us to root social facts in those about individuals, but he denies that in order to do this all social facts would somehow have to be reduced to claims about individuals. Thus, Kincaid believes that there is nothing mysterious about social facts that are over and above those about individuals—he believes that social facts supervene on those about individuals—but nonetheless upholds the idea that sociological explanations can be given at a level where we give social descriptions of human events, in lieu of the explanatory demands of reductionism.

Alan Nelson, in "Social Science and the Mental" (chapter 33), examines the debate about individualism more obliquely, through an examination of how the debate about individualism in the social sciences is related to contemporary debates about the status of the mental in the philosophy of psychology. He analyzes the "social" with respect to the "mental," hoping to show that while debates in the philosophy of psychology can inform our thinking about social explanation, progress on the issue of methodological individualism does not necessarily await resolutions within the philosophy of psychology.

Returning to Durkheim's conception of social facts, Nelson asks whether social phenomena can be explanatorily, though not ontologically, autonomous. After untangling all of the possible conceptual confusions about methodological individualism that have framed the previous six selections, this is still the most basic rendering of the issue.

Chapter 27

Social Facts

Emile Durkheim

Before inquiring into the method suited to the study of social facts, it is important to know which facts are commonly called "social." This information is all the more necessary since the designation "social" is used with little precision. It is currently employed for practically all phenomena generally diffused within society, however small their social interest. But on that basis, there are, as it were, no human events that may not be called social. Each individual drinks, sleeps, eats, reasons; and it is to society's interest that these functions be exercised in an orderly manner. If, then, all thses facts are counted as "social" facts, sociology would have no subject matter exclusively its own, and its domain would be confused with that of biology and psychology.

But in reality there is in every society a certain group of phenomena which may be differentiated from those studied by the other natural sciences. When I fulfil my obligations as brother, husband, or citizen, when I execute my contracts, I perform duties which are defined, externally to myself and my acts, in law and in custom. Even if they conform to my own sentiments and I feel their reality subjectively, such reality is still objective, for I did not create them; I merely inherited them through my education. How many times it happens, moreover, that we are ignorant of the details of the obligations incumbent upon us, and that in order to acquaint ourselves with them we must consult the law and its authorized interpreters? Similarly, the church-member finds the beliefs and practices of his religious life ready-made at birth; their existence prior to his own implies their existence outside of himself. The system of signs I use to express my thought, the system of currency I employ to pay my debts, the instruments of credit I utilize in my commerical relations, the practices followed in my profession, etc., function independently of my own use of them. And these statements can be repeated for each member of society. Here, then, are ways of acting, thinking, and feeling that present the noteworthy property of existing outside the individual consciousness.

These types of conduct or thought are not only external to the individual but are, moreover, endowed with coercive power, by virtue of which they impose themselves upon him, independent of his individual will. Of course, when I fully consent and conform to them, this constraint is felt only slightly, if at all, and is therefore unnecessary. But it is, nonetheless, an intrinsic characteristic of these facts, the proof thereof being that it asserts itself as soon as I attempt to resist it. If I attempt to violate the law, it reacts against me so as to prevent my act before its accomplishment, or to nullify my violation by restoring the damage, if it is acomplished and reparable, or to make me expiate it if it cannot be compensated for otherwise.

In the case of purely moral maxims, the public conscience exercises a check on every act which offends it by means of the surveillance it exercises over the conduct of citizens, and the appropriate penalities at its disposal. In many cases the constraint is less violent, but nevertheless it always exists. If I do not submit to the conventions of society, if in my dress I do not conform to the customs observed in my country and in my class, the ridicule I provoke, the social isolation in which I am kept, produce, although in an attenuated form, the same effects as a punishment in the strict sense of the word. The constraint is nonetheless efficacious for being indirect. I am not obliged to speak French with my fellow-countrymen nor to use the legal currency, but I cannot possibly do otherwise. If I tried to escape this necessity, my attempt would fail miserably. As an industrialist, I am free to apply the technical methods of former centuries; but by doing so, I should invite certain ruin. Even when I free myself from these rules and violate them successfully, I am always compelled to struggle with them. When finally overcome, they make their constraining power sufficiently felt by the resistance they offer. The enterprises of all innovators, including successful ones, come up against resistance of this kind.

Here, then, is a category of facts with very distinctive characteristics: it consists of ways of acting, thinking, and feeling, external to the individual, and endowed with a power of coercion, by reason of which they control him. These ways of thinking could not be confused with biological phenomena, since they consist of representations and of actions; nor with psychological phenomena, which exist only in the individual consciousness and through it. They constitute, thus, a new variety of phenomena; and it is to them exclusively that the term "social" ought to be applied. And this term fits them quite well, for it is clear that, since their source is not in the individual, their substratum can be no other than society, either the political society as a whole or some one of the partical groups it includes, such as religious denominations, political, literary, and occupational associations, etc. On the other hand, this term "social" applies to them exclusively, for it has a distinct meaning only if it designates exclusively the phenomena which are not included in any of the categories of facts that have already been established and classified. These ways of thinking and acting therefore constitute the proper domain of sociology. It is true that, when we define them with this word "constraint," we risk shocking the zealous partisans of absolute individualism. For those who profess the complete autonomy of the individual, man's dignity is diminished whenever he is made to feel that he is not completely self-determinant. It is generally accepted today, however, that most of our ideas and our tendencies are not developed by ourselves but come to us from without. How can they become a part of us except by imposing themselves upon us? This is the whole meaning of our definition. And it is generally accepted, moreover, that social constraint is not necessarily incompatible with the individual personality.[1]

Since the examples that we have just cited (legal and moral regulations, religious faith, financial systems, etc.) all consist of established beliefs and practices, one might be led to believe that social facts exist only where there is some social organization. But there are other facts without such crystallized form which have the same objectivity and the same ascendancy over the individual. These are called "social currents." Thus the great movements of enthusiasm, indignation, and pity in a crowd do not originate in any one of the particular individual consciousnesses. They come to each one of us from without and can carry us away in spite of ourselves. Of course, it may happen that, in abandoning myself to them unreservedly, I do not feel the pressure they exert upon me. But it is revealed as soon as I try to resist them. Let an individual attempt to oppose one of these collective manifestations, and the emotions that he denies will turn against him. Now, if this power of external coercion asserts itself so clearly in cases of resistance, it must exist also in the first-mentioned cases, although we are unconscious of it. We are then victims of the illusion of having ourselves created that which actually forced itself from without. If the complacency with which we permit ourselves to be carried along conceals the pressure undergone, nevertheless it does not abolish it. Thus, air is no less heavy because we do not detect its weight. So, even if we ourselves have spontaneously contributed to the production of the common emotion, the impression we have received differs markedly from that which we would have exprienced if we had been alone. Also, once the crowd has dispersed, that is, once these social influences have ceased to act upon us and we are alone again, the emotions which have passed through the mind appear strange to us, and we no longer recognize them as ours. We realize that these feelings have been impressed upon us to a much greater extent than they were created by us. It may even happen that they horrify us, so much were they contrary to our nature. Thus, a group of individuals, most of whom are perfectly inoffensive, may, when gathered in a crowd, be drawn into acts of atrocity. And what we say of these transitory outbursts applies similarly to those more permanent currents of opinion on religious, political, literary, or artistic matters which are constantly being formed around us, whether in society as a whole or in more limited circles.

To confirm this definition of the social fact by a characteristic illustration from common experience, one need only observe the manner in which children are brought up. Considering the facts as they are and as they have always been, it becomes immediately evident that all education is a continuous effort to impose on the child ways of seeing, feeling, and acting which he could not have arrived at spontaneously. From the very first hours of his life, we compel him to eat, drink, and sleep at regular hours; we constrain him to cleanliness, calmness, and obedience; later we exert pressure upon him in order that he may learn proper consideration for others, respect for customs and conventions, the need for work, etc. If, in time, this constraint ceases to be felt, it is because it gradually gives rise to habits and to internal tendencies that render constraint unnecessary; but nevertheless it is not abolished, for it is still the source from which these habits were derived. It is true that, according to Spencer, a rational education ought to reject such methods, allowing the child to act in complete liberty; but as this pedagogic theory has never been applied by any known people, it must be accepted only as an expression of personal opinion, not as a fact which can contradict the aforementioned observations. What makes these facts particularly instructive is that the aim of education is, precisely, the socialization of the human being; the process of education, therefore, gives us in a nutshell the historical fashion in which the social being is constituted. This unremitting pressure to which the child is subjected is the very pressure of the social milieu which tends to fashion him in its own image, and of which parents and teachers are merely the representatives and intermediaries.

It follows that sociological phenomena cannot be defined by their universality. A thought which we find in every individual consciousness, a movement repeated by all individuals, is not thereby a social fact. If sociologists have been satisfied with defining them by this characteristic, it is because they confused them with what one might call their reincarnation in the individual. It is, however, the collective aspects of the beliefs, tendencies, and practices of a group that characterize truly social phenomena. As for the forms that the collective states assume when refracted in the individual, these are things of another sort. This duality is clearly demonstrated by the fact that these two orders of phenomena are frequently found dissociated from one another. Indeed, certain of these social manners of acting and thinking acquire, by reason of their repetition, a certain rigidity which on its own account crystallizes them, so to speak, and isolates them from the particular events which reflect them. They thus acquire a body, a tangible form, and constitute a reality in their own right, quite distinct from the individual facts which produce it. Collective habits are inherent not only in the successive acts which they determine but, by a privilege of which we find no example in the biological realm, they are given permanent expression in a formula which is repeated from mouth to mouth, transmitted by education, and fixed even in writing. Such is the origin and nature of legal and moral rules, popular aphorisms and proverbs, articles of faith wherein religious or political groups condense their beliefs, standards of taste established by literary schools, etc. None of these can be found entirely reproduced in the applications made of them by individuals, since they can exist even without being actually applied.

No doubt, this dissociation does not always manifest itself with equal distinctness, but its obvious existence in the important and numerous cases just cited is sufficient to prove that the social fact is a thing distinct from its individual manifestations. Moreover, even when this dissociation is not immediately apparent, it may often be disclosed by certain devices of method. Such dissociation is indispensable if one wishes to separate social facts from their alloys in order to observe them in a state of purity. Currents of opinion, with an intensity varying according to the time and place, impel certain groups either to more marriages, for example, or to more suicides, or to a higher or lower birthrate. These currents are plainly social facts At first sight they seem inseparable from the forms they take in individual cases. But statistics furnish us with the means of isolating them. They are, in fact, represented with considerable exactness by the rates of births, marriages, and suicides, that is, by the number obtained by dividing the average annual total of marriages, births, suicides, by the number of persons whose ages lie within the range in which marriages, births, and suicides occur.[2] Since each of these figures contains all the individual cases indiscriminately, the individual circumstances which may have had a share in the production of the phenomenon are neutralized and, consequently, do not contribute to its determination. The average, then, expresses a certain state of the group mind (*l'âme collective*).

Such are social phenomena, when disentangled from all foreign matter. As for their individual manifestations, these are indeed, to a certain extent, social, since they partly reproduce a social model. Each of them also depends, and to a large extent, on the organopsychological constitution of the individual and on the particular circumstances in which he is placed. Thus they are not sociological phenomena in the strict sense of the word. They belong to two realms at once; one could call them sociopsychological. They interest the sociologist without constituting the immediate subject matter of sociology. There exist in the interior of organisms similar phenomena, compound in their nature, which form in their turn the subject matter of the "hybrid sciences," such as physiological chemistry, for example.

The objection may be raised that a phenomenon is collective only if it is common to all members of society, or at least to most of them—in other words, if it is truly general. This may be true; but it is general because it is collective (that is, more or less obligatory), and certainly not collective because general. It is a group condition repeated in the individual because imposed on him. It is to be found in each part because it exists in the whole, rather than in the whole because it exists in the parts. This becomes conspicuously evident in those beliefs and practices which are transmitted to us ready-made by previous generations; we receive and adopt them because, being both collective and ancient, they are invested with a particular authority that education has taught us to recognize and respect. It is, of course, true that a vast portion of our social culture is transmitted to us in this way; but even when the social fact is due in part to our direct collaboration, its nature is not different. A collective emotion which bursts forth suddenly and violently in a crowd does not express merely what all the individual sentiments had in common; it is somthing entirely different, as we have shown. It results from their being together, a product of the actions and reactions which take place between individual consciousnesses; and if each individual consciousness echoes the collective sentiment, it is by virtue of the special energy resident in its collective origin. If all hearts beat in unison, this is not the result of a spontaneous and preestablished harmony but rather because an identical force propels them in the same direction. Each is carried along by all.

We thus arrive at the point where we can formulate and delimit in a precise way the domain of sociology. It comprises only a limited group of phenomena. A social fact is to be recognized by the power of external coercion which it exercises or is capable of exercising over individuals, and the presence of this power may be recognized in its turn either by the existence of some specific sanction or by the resistance offered against every individual effort that tends to violate it. One can, however, define it also by its diffusion within the group, provided that, in conformity with our previous remarks, one takes care to add as a second and essential characteristic that its own existence is independent of the individual forms it assumes in its diffusion. This last criterion is perhaps, in certain cases, easier to apply than the preceding one. In fact, the constraint is easy to ascertain when it expresses itself externally by some direct reaction of society, as is the case in law, morals, beliefs, customs, and even fashions. But when it is only indirect, like the constraint which an economic organization exercises, it cannot always be so easily detected. Generality combined with extenality may, then, be easier to establish. Moreover, this second definition is but another form of the first; for if a mode of behavior whose existence is external to individual consciousnesses becomes general, this can only be brought about by its being imposed upon them.[3]

But these several phenomena present the same characteristic by which we defined the others. These "ways of existing" are imposed on the individual precisely in the same fashion as the "ways of acting" of which we have spoken. Indeed, when we wish to know how a society is divided politically, of what these divisions themselves are composed, and how complete is the fusion existing between them, we shall not achieve our purpose by physical inspection and by geographical observations; for these phenomena are social, even when they have some basis in physical nature. It is only by a study of public law that a comprehension of this organization is possible, for it is this law that determines the organization, as it equally determines our domestic and civil relations. This political organization is, then, no less obligatory than the social facts mentioned above. If the population crowds into our cities instead of scattering into the country, this is due to a trend of public opinion, a collective drive that imposes this concentration upon the individuals. We can no more choose the style of our houses than of our clothing—at least, both are equally obligatory. The channels of communication prescribe the direction of internal migrations and commerce, etc., and even their extent. Consequently, at the very most, it should be necessary to add to the list of phenomena which we have enumerated as presenting the distinctive criterion of a social fact only one additional category, "ways of existing"; and, as this enumeration was not meant to be rigorously exhaustive, the addition would not be absolutely necessary.

Such an addition is perhaps not necessary, for these "ways of existing" are only crystallized "ways of acting." The political structure of a society is merely the way in which its component segments have become accustomed to live with one another. If their relations are traditionally intimate, the segments tend to fuse with one another, or, in the contrary case, to retain their identity. The type of habitation imposed upon us is merely the way in which our contemporaries and our ancestors have been accustomed to construct their houses. The methods of communication are merely the channels which the regular currents of commerce and migrations have dug, by flowing in the same direction. To be sure, if the phenomena of structural character alone presented this permanence, one might believe that they constituted a distinct species. A legal regulation is an arrangement no less permanent than a type of architecture, and yet the regulation is a "physiological" fact. A simple moral maxim is assuredly somewhat more malleable, but it is much more rigid than a simple professional custom or a fashion. There is thus a whole series of degrees without a break in continuity between the facts of the most articulated structure and those free currents of social life which are not yet definitely molded. The differences between them are, therefore, only differences in the degree of consolidation they present. Both are simply life, more or less crystallized. No doubt, it may be of some advantage to reserve the term "morphological" for those social facts which concern the social substratum, but only on condition of not overlooking the fact that they are of the same nature as the others. Our definition will then include the whole relevant range of facts if we say: *A social fact is every way of acting, fixed or not, capable of exercising on the individual an external constraint;* or again, *every way of acting which is general throughout a given society, while at the same time existing in its own right independent of its individual manifestations.*[4]

... Social phenomena are things and ought to be treated as things. To demonstrate this proposition, it is unnecessary to philosophize on their nature and to discuss the analogies they present with the phenomena of lower realms of existence. It is sufficient to note that they are the unique data of the sociologist. All that is given, all that is subject to observation, has thereby the character of a thing. To treat phenomena as things is to treat them as data, and these constitute the point of departure of science. Now, social phenomena present this character incontestably. What is given is not the idea that men form of value, for that is inaccessible, but only the values established in the course of economic relations; not conceptions of the moral ideal, but the totality of rules which actually determine conduct; not the idea of utility or wealth, but all the details of economic organization. Even assuming the possibility that social life is merely the development of certain ideas, these ideas are nevertheless not immediately given. They cannot be perceived or known directly, but only through the phenomenal reality expressing them. We do not know a priori whether ideas form the basis of the diverse currents of social life, nor what they are. Only after having traced these currents back to their sources shall we know whence they issue.

We must, therefore, consider social phenomena in themselves as distinct from the consciously formed representations of them in the mind; we must study them objectively as external things, for it is this character that they present to us. If this exteriority should prove to be only apparent, the advance of science will bring the disillusionment and we shall see our conception of social phenomena change, as it were, from the objective to the subjective. But in any case, the solution cannot be anticipated; and even if we finally arrive at the result that social phenomena do not possess all the intrinsic characteristics of the thing, we ought at first to treat them as if they had. This rule is applicable, then, to all social reality without exception. Even phenomena which give the strongest impression of being arbitrary arrangements ought to be thus considered. *The voluntary character of a practice or an institution should never be assumed beforehand.* Moreover, if we may introduce our personal observation, it has always been our experience that, when this procedure is followed, facts most arbitrary in appearance will come to present, after more attentive observation, qualities of consistency and regularity that are symptomatic of their objectivity.

The foregoing statements concerning the distinctive characteristics of the social fact give us sufficient assurance about the nature of this objectivity to prove that it is not illusory. Indeed, the most important characteristic of a "thing" is the impossibility of its modification by a simple effort of the will. Not that the thing is refractory to all modification, but a mere act of the will is insufficient to produce a change in it; it requires a more or less strenuous effort due to the resistance which it offers, and, moreover, the effort is not always successful. We have already seen that social facts have this characteristic. Far from being a product of the will, they determine it from without; they are like molds in which our actions are inevitably shaped. This necessity is often inescapable. But even when we triumph over it, the opposition encountered signifies clearly to us the presence of something not depending upon ourselves. Thus, in considering social phenomena as things, we merely adjust our conceptions in conformity to their nature.

Clearly, the reform needed in sociology is at all points identical with that which has transformed psychology in the last thirty years. Just as Comte and Spencer declare that social facts are facts of nature, without, however, treating them as things, so the different empirical schools had long recognized the natural character of psychological phenomena, but continued to apply to them a purely ideological method. In fact, the empiricists, not less than their adversaries, proceeded exclusively by introspection. Now, the facts obtained thereby are too few in number, too fleeting and plastic, to be able to control and to correct the corresponding ideas fixed in us by habit. If they are not subjected to some other check, nothing counterbalances them; consequently, they take the place of facts and become the subject matter of science. Thus, neither Locke nor Condillac studied psychological phenomena objectively. They did not study sensation in itself but their particular idea of it. Therefore, although in certain respects they prepared the way for scientific psychology, its actual origin is to be dated much later, when it had finally been established that states of consciousness can and ought to be considered from without, and not from the point of view of the consciousness experiencing them. Such is the great revolution accomplished in this branch of studies. All the specific procedures and all the new methods by which this science has been enriched are only diverse means of realizing more completely this fundamental idea. It remains for sociology to make this same advance, to pass from the subjective stage, which it has still scarcely outgrown, to the objective.

Fortunately, this transformation is less difficult to effect here than in psychology. Indeed, psychological facts are naturally given as conscious states of the individual, from whom they do not seem to be even separable. Internal by definition, it seems that they can be treated as external only by doing violence to their nature. Not only is an effort of abstraction necessary, but in addition a whole series of procedures and artifices in order to hold them continuously within this point of view. Social facts, on the contrary, qualify far more naturally and immediately as things. Law is embodied in codes; the currents of daily life are recorded in statistical figures and historical monuments; fashions are preserved in costumes; and taste in works of art. By their very nature they tend toward an independent existence outside the individual consciousnesses, which they dominate. In order to disclose their character as things, it is unnecessary to manipulate them ingeniously. From this point of view, sociology has a significant advantage over psychology, an advantage not hitherto perceived, and one which should hasten ts development. Its facts are perhaps more difficult to interpret because more complex, but they are more easily arrived at. Psychology, on the contrary, has difficulties not only in the manipulation of its facts but also in rendering them explicit. Consequently, we believe that, once this principle of sociological method is generally recognized and practiced, sociology will progress with a rapidity difficult to forecast from its present tardiness of development and will even overtake psychology, whose present relative advantage is due solely to historical priority.[5]

Notes

1. We do not intend to imply, however, that all constraint is normal. We shall return to this point later.
2. Suicides do not occur at every age, and they take place with varying intensity at the different ages in which they occur.
3. It will be seen how this definition of the social fact diverges from that which forms the basis of the ingenious system of M. Tarde. First of all, we wish to state that our researches have nowhere led us to observe that preponderant influence in the genesis of collective facts which M. Tarde attributes to imitation. Moreover, from the preceding definition, which is not a theory but simply a résumé of the immediate data of observation, it seems indeed to follow, not only that imitation does not always express the essential and characteristic features of the social fact, but even that it never expresses them. No doubt, every social fact is imitated; it has, as we have just shown, a tendency to become general, but that is because it is social, i.e., obligatory. Its power of expansion is not the cause but the consequence of its sociological character. If, further, only social facts produced this consequence, imitation could perhaps serve, if not to explain them, at least to define them. But an individual condition which produces a whole series of effects remains individual nevertheless. Moreover, one may ask whether the word "imitation" is indeed fitted to designate an effect due to a coercive influence. Thus, by this single expression, very different phenomena, which ought to be distinguished, are confused.
4. This close connection between life and structure, organ and function, may be easily proved in sociology because between these two extreme terms there exists a whole series of immediately observable intermediate stages which show the bond between them. Biology is not in the same favorable position. But we may well believe that the inductions on this subject made by sociology are applicable to biology and that, in organisms as well as in societies, only differences in degree exist between these two orders of facts.
5. It is true that the greater complexity of social facts makes the science more difficult. But, in compensation, precisely because sociology is the latest comer, it is in a position to profit by the progress made in the sciences concerned with lower stages of existence and to learn from them. This utilization of previous experiments will certainly accelerate its development.

Chapter 28

Historical Explanation in the Social Sciences

J. W. N. Watkins

The hope which originally inspired methodology was the hope of finding a method of inquiry which would be both necessary and sufficient to guide the scientist unerringly to truth. This hope has died a natural death. Today, methodology has the more modest task of establishing certain rules and requirements which are necessary to prohibit some wrong-headed moves but insufficient to guarantee success. These rules and requirements, which circumscribe scientific enquiries without steering them in any specific direction, are of the two main kinds, formal and material. So far as I can see, the formal rules of scientific method (which comprise both logical rules and certain realistic and fruitful stipulations) are equally applicable to all the empirical sciences. You cannot, for example, deduce a universal law from a finite number of observations whether you are a physicist, a biologist, or an anthropologist. Again, a single comprehensive explanation of a whole range of phenomena is preferable to isolated explanations of each of those phenomena, whatever your field of inquiry. I shall therefore confine myself to the more disputable (I had nearly said "more disreputable") and metaphysically impregnated part of methodology which tries to establish the appropriate *material* requirements which the *contents* of the premises of an explanatory theory in a particular field ought to satisfy. These requirements may be called regulative principles. Fundamental differences in the subject-matters of different sciences—differences to which formal methodological rules are impervious—ought, presumably, to be reflected in the regulative principles appropriate to each science. It is here that the student of the methods of the social sciences may be expected to have something distinctive to say.

An example of a regulative principle is mechanism, a metaphysical theory which governed thinking in the physical sciences from the seventeenth century until it was largely superseded by a wave of field worldview. According to mechanism, the ultimate constituents of the physical world are impenetrable particles which obey simple mechanical laws. The existence of these particles cannot be explained—at any rate by science. On the other hand, every complex physical thing or event is the result of a particular configuration of particles and can be explained in terms of the laws governing their behavior in conjunction with a description of their relative positions, masses, momenta, etc. There may be what might be described as unfinished or halfway explanations of large-scale phenomena (say, the pressure inside a gas-container) in terms of other large-scale factors (the volume and temperature of the gas); but we shall not have arrived at rock-bottom explanations of such large-scale phenomena until we have deduced their behavior from statements about the properties and relations of particles.

This is a typically metaphysical idea (by which I intend nothing derogatory). True, it is confirmed, even massively confirmed, by the huge success of mechanical theories which conform to its requirements. On the other hand, it is untestable. No experiment could overthrow it. If certain phenomena—say, electromagnetic phenomena—seem refractory to this mechanistic sort of explanation, this refractoriness can always (and perhaps rightly) be attributed to our inability to find a successful mechanical model rather than to an error in our metaphysical intuition about the ultimate constitution of the physical world. But while mechanism is weak enough to be compatible with any *observation* whatever, while it is an untestable and unempirical principle, it is strong enough to be incompatible with various conceivable physcial *theories*. It is this which makes it a *regulative*, nonvacuous metaphysical principle. If it were compatible with everything it would regulate nothing. Some people complain that regulative principles discourage research in certain directions, but that is a part of their purpose. You cannot encourage research in one direction without discouraging reasearch in rival directions.

I am not an advocate of mechanism but I have mentioned it because I am an advocate of an analogous principle in social science, the principle of methodological individualism.[1] According to this principle, the ultimate constituents of the social world are individual people who act more or less appropriately in the light of their dispositions and understanding of their situation. Every complex social situation, institution, or event is the result of a particular configuration of individuals, their dispositions, situations, beliefs, and physical resources and environment. There may be unfinished or halfway explanations of large-scale social phenomena (say, inflation) in terms of other large-scale phenomena (say, full employment); but we shall not have arrived at rock-bottom explanations of such large-scale phenomena until we have deduced an account of them from statements about the dispositions, beliefs, resources, and interrelations of individuals. (The individuals may remain anonymous and only typical dispositions, etc., may be attributed to them.) And just as mechanism is contrasted with the organicist idea of physical fields, so methodological individualism is contrasted with sociological holism or organicism. On this latter view, social systems constitute "wholes" at least in the sense that some of their large-scale behavior is governed by macrolaws which are essentially *sociological* in the sense that they are sui generis and not to be explained as mere regularities or tendencies resulting from the behavior of interacting individuals. On the contrary, the behavior of individuals should (according to sociological holism) be explained at least partly in terms of such laws (perhaps in conjunction with an account, first of individuals' roles within institutions and second, of the functions of institutions within the whole social system). If methodological individualism means that human beings are supposed to be the only moving agents in history, and if sociological holism means that some superhuman agents or factors are supposed to be at work in history, then these two alternatives are exhaustive. An example of such a superhuman, sociological factor is the alleged long-term cyclical wave in economic life which is supposed to be self-propelling, uncontrollable, and inexplicable in terms of human activity, but in terms of the fluctuations of which such large-scale phenomena as wars, revolutions, and mass emigration, and such psychological factors as scientific and technological inventiveness can, it is claimed, be explained and predicted.

I say "and predicted" because the irreducible sociological laws postulated by holists are usually regarded by them as laws of social development, as laws governing the dynamics of a society. This makes holism well-nigh equivalent to historicism, to the idea that a society is impelled along a predetermined route by historical laws which cannot be resisted but which can be discerned by the sociologist. The holist-historicist position has, in my view, been irretrievably damaged by Popper's attacks on it. I shall criticize this position only insofar as this will help me to elucidate and defend the individualistic alternative to it. The central assumption of the individualistic position—an assumption which is admittedly counter-factual and metaphysical—is that no social tendency exists which could not be altered if the individuals concerned both wanted to alter it and possessed the appropriate information. (They might want to alter the tendency but, through ignorance of the facts and/or failure to work out some of the implications of their action, fail to alter it, or perhaps even intensify it.) This assumption could also be expressed by saying that no social tendency is somehow imposed on human beings "from above" (or "from below")—social tendencies are the product (usually undesigned) of human characteristics as well as of their knowledge and ambition. (An example of a social tendency is the tendency of industrial units to grow larger. I do not call "social" those tendencies which are determined by uncontrollable physical factors, such as the alleged tendency for more male babies to be born in times of disease or war.)[2]

My procedure will be: first, to delimit the sphere in which methodological individualism works in two directions; second, to clear methodological individualism of certain misunderstandings; third, to indicate how fruitful and surprising individualistic explanations can be and how individualistic social theories can lead to sociological discoveries; and fourth, to consider in somewhat more detail how, according to methodological individualism, we should frame explanations, first for social regularities or repeatable processes, and second for unique historical constellations of events.

Where Methodological Individualism Does Not Work

There are two areas in which methodological individualism does not work.

The first is a probability situation where accidental and unpredictable irregularities in human behavior have a fairly regular and predictable overall result.[3] Suppose I successively place 1,000 individuals facing north in the center of a symmetrical room with two exits, one east, the other west. If about 500 leave by one exit and about 500 by the other I would not try to explain this in terms of tiny undetectable west-inclining and east-inclining differences in the individuals, for the same reason that Popper would not try to explain the fact that about 500 balls will topple over to the west and about 500 to the east, if 1,000 balls are dropped from immediately above a north-south blade, in terms of tiny undetectable west-inclining and east-inclining differences in the balls. For in both cases such an "explanation" would merely raise the further problem: why should these west-inclining and east-inclining differences be distributed approximately *equally* among the individuals and among the balls?

Those statistical regularities in social life which are inexplicable in individualistic terms for the sort of reason I have quoted here are, in a sense, inhuman, the outcome of a large number of sheer *accidents*. The outcome of a large number of decisions is usually much less regular and predictable because variable human factors (changes of taste, new ideas, swings from optimism to pessimism) which have little or no influence on accident rates are influential here. Thus stock exchange prices fluctuate widely from year to year, whereas the number of road accidents does not fluctuate widely. But the existence of these actuarial regularities does not, as has often been alleged, support the historicist idea that defenseless individuals like you and me are at the chance mercy of the inhuman and uncontrollable tendencies of our society. It does not support a secularized version of the Calvinist idea of an Almighty Providence who picks people at random to fill His fixed damnation quota. For we can control these statistical regularities insofar as we can alter the conditions on which they depend. For example, we cound obviously abolish road accidents if we were prepared to prohibit motor traffic.

The second kind of social phenomenon to which methodological individualism is inapplicable is where some kind of physical connection between people's nervous systems short-circuits their intelligent control and causes automatic, and perhaps in some sense appropriate, bodily responses. I think that a man may more or less literally smell danger and instinctively back away from unseen ambushers; and individuality seems to be temporarily submerged beneath a collective physical rapport at jive sessions and revivalist meetings and among panicking crowds. But I do not think that these spasmodic mob organisms lend much support to holism or constitute a very serious exception to methodological individualism. They have a fleeting existence which ends when their members put on their mufflers and catch the bus or otherwise disperse, whereas holists have conceived of a social whole as something which endures through generations of men; and whatever holds together typical long-lived institutions, like a bank or a legal system or a church, it certainly is not the physical proximity of their members.

Misunderstandings of Methodological Individualism

I will now clear methodological individualism of two rather widespread misunderstandings.

It has been objected that in making individual dispositions and beliefs and situations the terminus of an explanation in social science, methodological individualism implies that a person's psychological make-up is, so to speak, God-given, whereas it is in fact conditioned by, and ought to be explained in terms of, his social inheritance and environment.[4] Now methodological individualism certainly does not prohibit attempts to explain the formation of psychological characteristics; it only requires that such explanations should in turn be *individualistic*, explaining the formation as the result of a series of conscious or unconscious responses by an individual to his changing situation. For example, I have heard Professor Paul Sweezey, the Harvard economist, explain that he became a Marxist because his father, a Wall Street broker, sent him in the 1930s to the London School of Economics to study under those staunch liberal economists, Professors Hayek and Robbins. This explanation is perfectly compatible with methodological individualism (though hardly compatible, I should have thought, with the Marxist idea that ideologies reflect class positions) because it interprets his ideological development as a human response to his situation. It is, I suppose, psychoanalysts who have most systematically worked the idea of a thorough individualist and historical explanation of the formation of dispositions, unconscious fears and beliefs, and subsequent defense mechanisms, in terms of responses to emotionally charged, and especially childhood, situations.

My point could be put by saying that methodological individualism encourages *innocent* explanations but forbids *sinister* explanations of the widespread existence of a disposition among the members of a social group. Let me illustrate this by quoting from a reply I made to Goldstein's criticisms:

> Suppose that it is established that Huguenot traders were relatively prosperous in seventeenth-century France and that this is explained in terms of a widespread disposition among them (a disposition for which there is independent evidence) to plough back into their businesses a larger proportion of their profits than

was customary among their Catholic competitors. Now this explanatory disposition might very well be explained in its turn—perhaps in terms of the general thriftiness which Calvinism is said to encourage, and/or in terms of the fewer alternative outlets for the cash resources of people whose religious disabilities prevented them from buying landed estates or political offices. (I cannot vouch for the historical accuracy of this example.)

I agree that methodological individualism allows the formation, or "cultural conditioning," of a widespread disposition to be explained only in terms of other human factors and not in terms of something *in*human, such as an alleged historicist law which impels people willynilly along some predetermined course. But this is just the antihistoricist point of methodological individualism.

Unfortunately, it is typically a part of the program of Marxist and other historicist sociologies to try to account for the formation of ideologies and other psychological characteristics in strictly sociological and nonpsychological terms. Marx, for instance, professed to believe that feudal ideas and bourgeois ideas are more or less literally generated by the water mill and the steam engine. But no description, however complete, of the productive apparatus of a society, or of any other nonpsychological factors, will enable you to deduce a single psychological conclusion from it, because psychological statements logically cannot be deduced from wholly nonpsychological statements. Thus whereas the mechanistic idea that explanations in physics cannot go behind the impenetrable particles is a prejudice (though a very understandable prejudice), the analogous idea that an explanation which begins by imputing some social phenomenon to human factors cannot go on to explain those factors in terms of some inhuman determinant of them is a necessary truth. That the human mind develops under various influences the methodological individualist does not, of course, deny. He only insists that such development must be explained "innocently" as a series of responses by the individual to situations and not "sinisterly" and illogically as a direct causal outcome of nonpsychological factors, whether these are neurological factors, or impersonal sociological factors alleged to be at work in history.

Another cause of complaint against methodological individualism is that it has been confused with a narrow species of itself (Popper calls it "psychologism") and even, on occasion, with a still narrower subspecies of this (Popper calls it the "Conspiracy Theory of Society")[5] Psychologism says that all large-scale social characteristics are not merely the intended or unintended result of, but a *reflection* of, individual characteristics.[6] Thus Plato said that the character and make-up of a polis is a reflection of the character and make-up of the kind of soul predominant in it. The conspiracy theory says that all large-scale social phenomena (do not merely reflect individual characteristics but) are deliberately brought about by individuals or groups of individuals.

Now there are social phenomena, like mass unemployment, which it would not have been in anyone's interest deliberately to bring about and which do not appear to be large-scale social reflections or magnified duplicates of some individual characteristic. The practical or technological or therapeutic importance of social science largely consists in explaining, and thereby perhaps rendering politically manageable, the unintended and unfortunate consequences of the behavior of interacting individuals. From this pragmatic point of view, psychologism and the conspiracy theory are unrewarding doctrines. Psychologism says that only a change of heart can put a stop to, for example, war (I think that this is Bertrand Russell's view). The conspiracy theory, faced with a big bad social event, leads to a hunt for scapegoats. But methodological individualism, by imputing unwanted social phenomena to individuals' responses to their situations, in the light of their dispositions and beliefs, suggests that we may be able to make the phenomena disappear, not by recruiting good men to fill the posts hitherto occupied by bad men, or by trying to destroy men's socially unfortunate dispositions while fostering their socially beneficial dispositions, but simply by altering the situations they confront. To give a current example, by confronting individuals with dearer money and reduced credit the government may (I do not say will) succeed in halting inflation without requiring a new self-denying attitude on the part of consumers and without sending anyone to prison.

Factual Discoveries in Social Science

To explain the unintended but *beneficial* consequences of individual activities—by "beneficial consequences" I mean social consequences which the individuals affected *would* endorse *if* they were called on to choose between their continuation or discontinuation—is usually a task of less practical urgency than the explanation of undesirable consequences. On the other hand, this task may be of greater theoretical interest. I say this because people who are painfully aware of the existence of unwanted social phenomena may be oblivious of the unintended but beneficial consequences of men's actions, rather as a man may be oblivious of the good health to which the smooth functioning of his digestion, nervous system, circulation, etc., give rise. Here, an explanatory social theory may surprise and enlighten us not only with regard to the connections between causes and effect but with regard to the existence of the effect itself. By showing that a certain economic system contains positive feedback leading to increasingly violent oscillations and crises an economist may explain a range of well-advertised phenomena which have long been the subject of strenuous political agitation. But the economists who first showed that a certain kind of economic system contains negative feedback which tends to iron out disturbances and restore equilibrium not only explained, but also revealed the existence of, phenomena which had hardly been remarked upon before.[7]

I will speak of organic-like social behavior where members of some social system (that is, a collection of people whose activities disturb and influence each other) mutually adjust themselves to the situations created by the others in a way which, without direction from above, conduces to the equilibrium or preservation or development of the system. (These are again evaluative notions, but they can also be given a "would-be-endorsed-if" definition.) Now such far-flung organic-like behavior, involving people widely separated in space and largely ignorant of each other, cannot be simply observed. It can only be theoretically reconstructed—by deducing the distant social consequences of the typical responses of a large number of interacting people to certain repetitive situations. This explains why individualistic-minded economists and anthropologists, who deny that societies really are organisms, have succeeded in piecing together a good deal of unsuspected organic-like social behavior, from an examination of individual dispositions and situations, whereas sociological holists, who insist that societies really are organisms, have been noticeably unsuccessful in convincingly displaying any organic-like social behavior—they cannot observe it and they do not try to reconstruct it individualistically.

There is a parallel holism and psychologism which explains their common failure to make surprising discoveries. A large-scale social characteristic should be explained, according to psychologism, as the manifestation of analogous small-scale psychological tendencies in individuals, and according to holism as the manifestation of a large-scale tendency in the social whole. In both cases, the explicans does little more than duplicate the explicandum. The methodological individualist, on the other hand, will try to explain the large-sale effect as the *indirect*, unexpected, complex product of individual factors none of which, singly, may bear any resemblance to it at all. To use hackneyed examples, he may show that a longing for peace led, in a certain international situation, to war, or that a government's desire to improve a bad economic situation by balancing its budget only worsened the situation. Since Mandeville's *Fable of the Bees* was published in 1714, individualistic social science, with its emphasis on unintended consequences, has largely been a sophisticated elaboration on the simple theme that, in certain situations, selfish private motives may have good social consequences and good political intentions bad social consequences.[8]

Holists draw comfort from the example of biology, but I think that the parallel is really between the biologist and the methodological individualist. The biologist does not, I take it, explain the large changes which occur during, say, pregnancy, in terms of corresponding large teleological tendencies in the organism, but physically, in terms of small chemical, cellular, neurological, etc., changes, none of which bears any resemblance to their joint and seemingly planful outcome.

How Social Explanations Should Be Framed

I will now consider how regularities in social life, such as the trade cycle, should be explained according to methodological individualism. The explanation should be in terms of individuals and their situations; and since the process to be explained is repeatable, liable to recur at various times and in various parts of the world, it follows that only very general assumptions about human dispositions can be employed in its explanation. It is no use looking to abnormal psychology for an explanation of the structure of interest rates—everyday experience must contain the raw material for the dispositional (as opposed to the situational) assumptions required by such an explanation. It may require a stroke of genius to detect, isolate, and formulate precisely the dispositional premises of an explanation of a social regularity. These premises may state what no one had noticed before, or give a sharp articulation to what had hitherto been loosely described. But once stated they will seem obvious enough. It took years of groping by brilliant minds before a precise formulation was found for the principle of diminishing marginal utility. But once stated, the principle—that the less, relatively, a man has of one divisible commodity the more compensation he will be disposed to require for foregoing a small fixed amount of it—is a principle to which pretty well everyone will give his consent. Yet this simple and almost platitudinous principle is the magic key to the economics of distribution and exchange.

The social scientist is, here, in a position analogous to that of the Cartesian mechanist.[9] The latter never set out to discover new and unheard-of physical principles because he believed that his own principle of action-by-contact was self-evidently ultimate. His problem was to discover the typical physical configurations, the mechanisms, which, operating according to this principle, produce the observed regularities of nature. His theories took the form of models which exhibited such regularities as the outcome of "self-evident" physical principles operating in some hypothetical physical situation. Similarly, the social scientist does not make daring innovations in psychology but relies on familiar, almost "self-evident" psychological material. His skill consists, first, in spotting the relevant dispositions, and, second, in inventing a simple but realistic model which shows how, in a precise type of situation, those dispositions generate some typical regularity or process. (His model, by the way, will also show that in this situation certain things cannot happen. His negative predictions of the form, "If you've got this you can't have that as well," may be of great practical importance.) The social scientist can now explain in principle historical examples of this regular process, provided his model does in fact fit the historical situation.

This view of the explanation of social regularties incidentally clears up the old question on which so much ink has been spilled about whether the so-called laws of economics apply universally or only to a particular "stage" of economic development. The simple answer is that the economic principles displayed by economists' models apply only to those situations which correspond with their models; but a single model may very well correspond with a very large number of historical situations widely separated in space and time.

In the explanation of regularities the same situational scheme or model is used to reconstruct a number of historical situations with a similar structure in a way which reveals how typical dispositions and beliefs of anonymous individuals generated, on each occasion, the same regularity.[10] In the explanation of a unique constellation of events the individualistic method is again to reconstruct the historical situation, or connected sequene of situations, in a way which reveals how (usually both named and anonymous) individuals, with their beliefs and dispositions (which may include peculiar personal dispositions as well as typical human dispositions), generated in this particular situation, the joint product to be explained. I emphasize *dispositions*, which are open and lawlike, as opposed to *decisions*, which are occurences, for this reason. A person's set of dispositions ought, under varying conditions, to give rise to appropriately varying decisions. The subsequent occurrence of an appropriate decision will both confirm, and be explained by, the existence of the dispositions. Suppose that a historical explanation (of, say, the growth of the early Catholic church) largely relies on a particular decision (say, the decision of Emperor Constantine to give Pope Silvester extensive temporal rights in Italy). The explanation is, so far, rather ad hoc: an apparently arbitrary fiat plays a key role in it. But if this decision can in turn be explained as the offspring of a marriage of a set of dispositions (for instance, the emperor's disposition to subordinate all rival power to himself) to set of circumstances (for instance, the emperor's recognition that Christianity could not be crushed but could be tamed if it became the official religion of the empire), and if the existence of these dispositions and circumstances is convincingly supported by independent evidence, then the area of the arbitrarily given, of sheer brute fact in history, although it can never be made to vanish, will have been significantly reduced.

Notes

1. Both of these analogous principles go back at least to Epicurus. In recent times methodological individualism has been powerfully defended by Professor F. A. Hayek in his *Individualism and Economic Order* and *The Counter-Revolution of Science*, and by Professor K. R. Popper in his *The Open Society and Its Enemies* and "The Poverty of Historicism," *Economica*, 1944–45, 11–12. Following in their footsteps I have also attempted to defend methodological individualism in "Ideal Types and

Historical Explanation," *British Journal for the Philosophy of Science*, 1952, 3, 22, reprinted in *Readings in the Philosophy of Science*, ed. Feigl and Brodbeck, New York, 1953. This article has come in for a good deal of criticism, the chief items of which I shall try to rebut in wha follows.

2. The issue of holism versus individualism in social science has recently been presented as though it were a question of the existence or nonexistence of irreducibly social facts rather than of irreducibly sociological *laws*. (See M. Mandelbaum, "Societal Facts," *British Jounal of Sociology*, 1955, 6, and E. A. Gellner, "Explanations in History," *Aristotelian Society*, Supplementary Volume 30, 1956.) This way of presenting the issue seems to me to empty it of most of its interest. If a new kind of beast is discovered, what we want to know is not so much whether it falls outside existing zoological categories, but how it behaves. People who insist on the existence of social facts but who do not say whether they are governed by sociological laws are like people who claim to have discovered an unclassified kind of animal but who do not tell us whether it is tame or dangerous, whether it can be domesticated or is unmanageable. If an answer to the question of social facts could throw light on the serious and interesting question of sociological laws, then the question of social facts would also be serious and interesting. But this is not so. On the one hand, a holist may readily admit (as I pointed out in my "Ideal Types" paper, which Gellner criticizes) that all observable social facts *are* reducible to individual facts and yet hold that the latter are invisibly governed by irreducibly sociological laws. On the other hand, an individualist may readily admit (as Gellner himself says) that some large social facts are simply too complex for a full reduction of them to be feasible, and yet hold that individualistic explanatons of them are in principle possible, just as a physicist may readily admit that some physical facts (for instance, the precise blast-effects of a bomb-explosion in a built-up area) are just too complex for accurate prediction or explanation of them to be feasible and yet hold that precise explanations and predictions of them in terms of existing scientific laws are in principle possible.

   This revised way of presenting the holism versus individualsim issue does not only divert attention from the important question. It also tends to turn the dispute into a purely verbal issue. Thus Mandelbaum is able to prove the existence of what he calls "societal facts" because he defines psychological facts very narrowly as "facts concerning the thoughts and actions of specific human beings" (op. cit., p. 307). Consequently, the *dispositions of anonymous* individuals which play such an important role in individualistic explanations in social science are "societal facts" merely by definition.
3. Failure to exclude probability-situations from the ambit of methodological individualism was an important defect of my "Ideal Types" paper. Here, Gellner's criticism (op. cit., p. 163) does hit the nail on the head.
4. Thus Gellner writes: "The real oddity of the reductionist [i.e., the methodological individualist's] case is that it seems to preclude a priori the possibility of human dispositions being the dependent variable in a historical explanation—when in fact they often or always are" (op. cit., p. 165). And Mr. Leon J. Goldstein says that in making human dispositions methodologically primary I ignore their cultural conditioning. (*Journal of Philosophy*, 1956, 53, 807.)
5. See K. R. Popper, *The Open Society and Its Enemies*, 2d ed., 1952, ch. 14.
6. I am at a loss to understand how Gellner came to make the following strange assertion: "Popper refers to both 'psychologism' which he condemns, and 'methodological individualism,' which he commends. When in the articles discussed [i.e., my 'Ideal Types' paper] 'methodological individualism' is worked out more fully than is the case in Popper's book, it seems to me to be indistinguishable from 'Psychologism'." Finding no difference between methodological individualism and a caricature of methodological individualism, Gellner has no difficulty in poking fun at the whole idea: "Certain tribes I know have what anthropologists call a segmentary patrilineal structure, which moreover maintains itself very well over time. I could 'explain' this by saying that the tribesmen have, all or most of them, dispositions whose effect is to maintain the system. But, of course, no only have they never given the matter much thought, but it also might very well be impossible to isolate anything in the characters and conduct of the individual tribesmen which *explains* how they come to maintain the system" (op. cit., p. 176). Yet this example actually suggests the lines along which an individualistic explanation might be found. The very fact that the tribesmen *have never given the matter much thought*, the fact that they accept their inherited system uncritically, may constitute an important part of an explanation of its stability. The explanation might go on to pinpont certain rules—that is, firm and widespread dispositons—about marriage, inheritance, etc., which help to regularize the tribesmen's behavior toward their kinsmen. How they come to share these common dispositions could also be explained individualistcally in the same sort of way that I can explain why my young children are already developing a typically English attitude toward policemen.

7. This sentence, as I have since learned from Dr. A. W. Phillips, is unduly complacent, for it is very doubtful whether an economist can ever *show* that an economic system containing negative feedback will be stable. For negative feedback may produce either a tendency toward equilibrium, or increasing oscillations, according to the numerical values of the parameters of the system. But numerical values are just what economic measurements, which are usually ordinal rather than cardinal, seldom yield. The belief that a system which contains negative feedback, but whose variables cannot be described quantitatively, is stable may be based on faith or experience, but it cannot be shown mathematically. See A. W. Phillips, "Stabilisation Policy and the Time-Forms of Lagged Responses," *Economic Journal*, 1957, 67.
8. A good deal of unmerited opposition to methodological individualism seems to spring from the recognition of the undoubted fact that individuals often run into social obstacles. Thus the conclusion at which Mandelbaum arrives is "that there are societal facts which exercise external constraints over individuals" (op. cit., p. 317). This conclusion is perfectly harmonious with the methodological individualist's insistence that plans often miscarry (and that even when they do succeed, they almost invariably have other important and unanticipated effects). The methodological individualist only insists that the social environment by which any particular individual is confronted and frustrated and sometimes manipulated and occasionally destroyed is, if we ignore its physical ingredients, made up of other *people*, their habits, inertia, loyalties, rivalries, and so on. What the methodological individualist denies is that an individual is ever frustrated, manipulated or destroyed, or borne along by irreducible sociological or historical *laws*.
9. I owe this analogy to Professor Popper.
10. This should rebut Gellner's conclusion that methodological individualism would transform social scientists into "biographers *en grande série*" (op. cit., p. 176).

Chapter 29

Methodological Individualism Reconsidered

Steven Lukes

1

In what follows I discuss and (hopefully) render harmless a doctrine which has a very long ancestry, has constantly reappeared in the history of sociology, and still appears to haunt the scene. It was, we might say, conceived by Hobbes, who held that "it is necessary that we know the things that are to be compounded before we can know the whole compound" for "everything is best understood by its constitutive causes," the causes of the social compound residing in "men as if but even now sprung out of the earth, and suddenly, like mushrooms, come to full maturity without all kinds of engagement to each other."[1] It was begotten by the thinkers of the Enlightenment, among whom, with a few important exceptions (such as Vico and Montesquieu) an individualist mode of explanation became preeminent, though with wide divergences as to what was included in the characterization of the explanatory elements. It was confronted by a wide range of thinkers in the early nineteenth century, who brought to the understanding of social life a new perspective, in which collective phenomena were accorded priority in explanation. As de Bonald wrote, it is "society that constitutes man, that is, it forms him by social education."[2] or, in Comte's words, a society was "no more decomposable into individuals than a geometric surface is into lines, or a line into points."[3] For others, however, such as Mill and the Utilitarians, "the Laws of the phenomena of society are, and can be, nothing but the actions and passions of human beings," namely "the laws of individual human nature."[4] This debate has recurred in many different guises—in the dispute between the "historical" school in economics and the "abstract" theory of classical economics, in endless debates among philosophers of history and between sociologists and psychologists,[5] and, above all, in the celebrated controversy between Durkheim and Gabriel Tarde.[6] Among others, Simmer[7] and Cooley[8] tried to resolve the issue, as did Gurvitch[9] and Ginsberg,[10] but in constantly reappears, for example, in reactions to the extravagantly macroscopic theorizing of Parsons and his followers[11] and in extraordinarily muddled debate provoked by the wide-ranging methodological polemics of Hayek and Popper.[12]

What I shall try to do here is, first, to distinguish what I take to be the central tenet of methodological individualism from a number of different theses from which it has not normally been distinguished; and second, to show why, even in the most vacuous sense, methodological individualism is implausible.

Let us begin with a set of truisms. Society consists of people. Groups consist of people. Institutions consist of people plus rules and roles. Rules are followed (or alternatively not followed) by people and roles are filled by people. Also there are traditions, customs, ideologies, kinship systems, languages: these are ways people act, think, and talk. At the risk of pomposity, these truisms may be said to constitute a theory (let us call it "Truistic Social Atomism") made up of banal propositions about the world that are analytically true, i.e., in virtue of the meaning of words.

Some thinkers have held it to be equally truistic (indeed, sometimes, to amount to the same thing) to say that facts about society and social phenomena are to be explained solely in terms of facts about individuals. This is the doctrine of methodological individualism. For example, Hayek writes:"There is no other way toward an understanding of social phenomena but through our understanding of individual actions directed toward other people and guided by their expected behavior."[13] Similarly, according to Popper,

> ... all social phenomena, and especially the functioning of all social institutions, should always be understood as resulting from the decisions, actions, attitudes, etc. of human individuals, and ... we should never be satisfied by an explanation in terms of so-called collectives.[14]

Finally we may quote Watkins's account of "the principle of methodological individualism":

> According to this principle, the ultimate constituents of the social world are individual people who act more or less appropriately in the light of their dispositions and understanding of their situation. Every complex social situation or event is the result of a particular configuration of individuals, their dispositions, situations, beliefs, and physical resources and environment.

It is worth noticing, incidentally, that the first sentence here is simply a (refined) statement of Truistic Social Atomism. Watkins continues:

> There may be unfinished or half-way explanation of large-scale social phenomena (say, inflation) in terms of other large-scale phenomena (say full employment); but we shall not have arrived at rock-bottom explanations of such large scale phenomena until we have deduced an account of them from statements about the dispositions, beliefs, resources and inter-relations of individuals. (The individuals may remain anonymous and only typical dispositions, etc., may be attributed to them.) And just as mechanism is contrasted with the organicist idea of physical fields, so methodological individualism is contrasted with sociological holism or organicism. On this latter view, social systems constitute "wholes" at least in the sense that some of their large-scale behaviour is governed by macro-laws which are essentially sociological in the sense that they are sui generis and not to be explained as mere regularities or tendencies resulting from the behaviour of interacting individuals. On the contrary, the behaviour of individuals should (according to sociological holism) be explained at least partly in terms of such laws (perhaps in conjunction with an account, first of individuals' roles within institutions, and secondly of the functions of institutions with the whole social system). If methodological individualism means that human beings are supposed to be the only moving agents in history, and if sociological holism means that some superhuman agents or factors are supposed to be at work in history, then these two alternatives are exhaustive.[15]

Methodological individualism, therefore, is a prescription for explanation, asserting that no purported explanations of social (or individual) phenomena are to count as explanations, or (in Watkins's version) as rock-bottom explanations, unless they are couched wholly in terms of facts about individuals.

It is now necessary to distinguish this from a number of others, from which it us usually not distinguished. It has been taken to be the same as any or all or the following:

1. Truistic Social Atomism. We have seen that Watkins, for example, seems to equate this with methodological individualism proper.

2. A theory of meaning to the effect that every statement about social phenomena is either a statement about individual human beings or else it is unintelligible and therefore not a statement at all. This theory entails that all predicates which range over social phenomena are definable in terms of predicates which range only over individual phenomena and that all statements about social phenomena are translatable without loss of meaning into statements that are wholly about individuals. As Jarvie has put it, "'Army' is merely a plural of soldier and *all* statements about the Army can be reduced to statements about the particular soldiers comprising the Army."[16]

It is worth noticing that this theory is only plausible on a crude verificationist theory of meaning (to the effect that the meaning of p is what confirms the truth of p). Otherwise, although statements about armies are true only in virtue of the fact that other statements about individuals are true, the former are not equivalent in meaning to the latter, nor a fortiori are they "about" the subject of the latter.

3. A theory of ontology to the effect that in the social world only individuals are real. This usually carries the correlative doctrine that social phenomena are constructions of the mind and "do not exist in reality." Thus Hayek writes,

> The social sciences ... do not deal with "given" wholes but their task is to constitute these wholes by constructing models from the familiar elements—models which reproduce the structure of relationships between some of the many phenomena which we always simultaneously observe in real life. This is no less true of the popular concepts of social wholes which are represented by the terms current in ordinary language; they too refer to mental models.[17]

Similarly, Popper holds that "social entities such as institutions or associations" are "abstract models constructed to interpret certain selected abstract relations between individuals."[18]

If this theory means that in the social world only individuals are observable, it is evidently false. Some social phenomena simply can be observed (as both trees and forests can): and indeed, many features of social phenomena are observable (e.g., the procedure of court) while many features of individuals are not (e.g., intentions). Both individual and social phenomena have observable and nonobservable features. If it means that individual phenomena are easy to understand, while social phenomena are not (which is Hayek's view), this is highly implausible: compare the procedure of the court with the motives of the criminal. If the theory means that individuals exist independently of, e.g., groups and institutions, this is also false, since just as facts about social phenomena are contingent upon facts about individuals, the reverse is also true. Thus, as we have seen, we can only speak of soldiers because we can speak of armies: only if certain statements are true of armies are others true of soldiers. If the theory means that all social phenomena are fictional and all individual phenomena are factual, that would entail that all assertions about social phenomena are false or else neither true nor false, which is absurd. Finally, the theory may mean that only facts about individuals are explanatory, which alone would make this theory equivalent to methodological individualism.

4. A negative theory to the effect that sociological laws are impossible, or that lawlike statements about social phenomena are always false. Hayek and Popper sometimes seem to believe this, but Watkins clearly repudiates it, asserting merely that such statements form part of "halfway" as opposed to "rock-bottom" explanations.

This theory, like all dogmas of the form "x is impossible" is open to refutation by a single counter-instance. Since such counter-instances are readily available[19] there is nothing left to say on this score.

A doctrine that may be called "social individualism" which (ambiguously) asserts that society has as its end the good of individuals. When unpacked, this may be taken to mean any or all of the following: (a) social institutions are to be understood as founded and maintained by individuals to fulfill their ends (as in, e.g., social contract theory); (b) social institutions in fact satisfy individual ends; (c) social institutions ought to satisfy individual ends. (a) is not widely held today, though it is not extinct; (b) is certainly held by Hayek with respect to the market, as though it followed from methodological individualism; and (c) which, interpreting "social institutions" and "individual ends" as a noninterventionist state and express preferences, becomes political liberalism, is clearly held by Popper to be uniquely consonant with methodological individualism.

However, neither (b) nor (c) is logically or conceptually related to methodological individualism, while (a) is a version of it.

2

What I hope so far to have shown is what the central tenet of methodological individualism is and what it is not. It remains to assess its plausibility.

It asserts (to repeat) that all attempts to explain social and individual phenomena are to be rejected (or, for Watkins, rejected as rock-bottom explanations) unless they refer exclusively to facts about individuals. There are thus two matters to investigate: (1) what is meant by "facts about individuals," and (2) what is meant by "explanation"?

1. What is a fact about an individual? Or, more clearly, what predicates may be applied to individuals? Consider the following examples:

i. genetic make-up; brain-states,
ii. aggression; gratification; stimulus-response,
iii. cooperation; power; esteem,
iv. cashing cheques; saluting; voting.

What this exceedingly rudimentary list shows is at least this: that there is a continuum of what I shall henceforth call individual predicates from what one might call the most nonsocial to the most social. Propositions incorporating only predicates of type (i) are about human beings qua material objects and make no reference to and presuppose nothing about consciousness or any feature of any social group or institution. Propositions incorporating only individual predicates of type (ii) presuppose consciousness but still make no reference to and presuppose nothing about any feature of any social group or institution. Propositions incorporating only predicates of type (iii) do have a minimal social reference; they presuppose a social context in which certain actions, social relations, and/or mental states are picked out and given a particular significance (which makes social relations of certain sorts count as "cooperative," which makes certain social positions count as positions of "power" and a certain set of attitudes count as "esteem"). They still do not presuppose or entail any particular propositions about any particular form of group of institution. Finally, propositions incorporating only individual predicates of type (iv) are maximally social, in that they presuppose and sometimes directly entail propositions about particular types of group and institution. ("Voting Labor" is at an even further point on the continuum.)

Methodological individualism has frequently been taken to confine its favored explanations to any or all of these sorts of individual predicates. We may distinguish the following four possibilities:

i. Attempts to explain in terms of type (i) predicates. A good example is H. J. Eysenck's *Psychology of Politics.*[20] According to Eysenck, "Political actions are actions of human beings; the study of the direct cause of these actions is the field of the study of psychology. All other social sciences deal with variables which affect political action indirectly."[21] (Compare this with Durkheim's famous statement that "every time that a social phenomenon is directly explained by a psychological phenomenon, we may be sure that the explanation is false.")[22] Eysenck sets out to classify attitudes along two dimensions—the Radical-Conservative and the Tough-minded-Tender-minded—on the basis of evidence elicited by carefully constructed questionnaires. Then, having classified the attitudes, his aim is to *explain* them by reference to antecedent conditions and his interest here is centered upon the modifications of the central nervous system.
ii. Attempts to explain in terms of type (ii) predicates. Examples are Hobbes's appeal to appetites and aversions, Pareto's residues and those Freudian theories in which sexual activity is seen as a type of undifferentiated activity that is (subsequently) channeled on particular social directions.
iii. Attempts to explain in terms of type (iii) predicates. Examples are those sociologists and social psychologists (from Tarde to Homans)[23] who favor explanations in terms of general and "elementary" forms of social behavior, which do invoke some minimal social reference but are unspecific as to any particular form of group or institution.
iv. Attempts to explain in terms of type (iv) predicates. Examples of these are extremely widespread, comprising all those who appeal to facts about concrete and specifically located individuals in order to explain. Here the relevant features of the social context are, so to speak, built into the individual. Open almost any empirical (though not theoretical) work of sociology, or history, and explanations of this sort leap to the eye.

Merely to state these four alternative possibilities is to suggest that their differences are more important than their similarities. What do they show about the plausibility of methodological individualism? To answer this it is necessary to turn to the meaning of "explanation."

2. To explain something is (at least) to overcome an obstacle—to make what was unintelligible intelligible. There is more than one way of doing this.

It is important to see, and it is often forgotten, that to *identify* a piece of behavior, a set of beliefs, etc., is sometimes to explain it. This may involve seeing it in a new way, picking out hidden structural features. Consider an anthropologist's interpretation of ritual or a sociological study of (say) bureaucracy. Often explanation resides precisely in a successful and sufficiently wide-ranging identification of behavior or types of behavior (often in terms of a set of beliefs). Again, to take an example from

Mandelbaum,[24] a Martian visiting earth sees one man mark a piece of paper that another has handed to him through some iron bars: on his being told that the bank teller is certifying the withdrawal slip he has had the action explained, through its being identified. If the methodological individualist is saying that no explanations are possible (or rock-bottom) except those framed exclusively in terms of individual predicates of types (i), (ii), and (iii), i.e., those not presupposing or entailing propositions about particular institutions and organizations, then he is arbitrarily ruling out (or denying finality to) most ordinarily acceptable explanations, as used in everyday life, but also by most sociologists and anthropologists for most of the time. If he is prepared to include individual predicates of type (iv), he seems to be proposing nothing more than a futile linguistic purism. Why should we be compelled to talk about the tribesman but not the tribe, the bank teller but not the bank? And let no one underestimate the difficulty or the importance of explanation by identification. Indeed, a whole methodological tradition (from Dilthey through Weber to Winch) holds this to be the characteristic mode of explanation in social science.

Another way of explaining is to deduce the specific and particular from the general and universal. If I have a body of coherent, economical, well-confirmed, and unfalsified general laws from which, given the specifications of boundary and initial conditions, I predict (or retrodict) x and x occurs, then, in one very respectable sense, I have certainly explained x.[25] This is the form of explanation which methodological individualists characteristically seem to advocate, though they vary as to whether the individual predicates which are uniquely to constitute the general laws and specifications of particular circumstances are to be of types (i), (ii), (iii), or (iv).

If they are to be of type (i), either of two equally unacceptable consequences follows. Eysenck writes, "It is fully realized that most of the problems discussed must ultimately be seen in their historical, economic, sociological, and perhaps even anthropological context, but little is to be gained at the present time by complicating the picture too much."[26] But the picture is already so complicated at the very beginning (and the attitudes Eysenck is studying are only identifiable in social terms); the problem is how to simplify it. This could logically be achieved either by developing a theory which will explain the "historical, economic, sociological ... anthropological context" exclusively in terms of (e.g.) the central nervous system or by demonstrating that this "context" is simply a backdrop against which quasi-mechanical psychological forces are the sole causal influences at work. Since, apart from quaint efforts that are of interest only to the intellectual historian, no one has given the slightest clue as to how either alternative might plausibly be achieved, there seems to be little point in taking it seriously, except as a problem in philosophy. Neurophysiology may be the queen of the social sciences, but her claim remains entirely speculative.

If the individual predicates are to be of type (ii), there is again no positive reason to find the methodological individualist's claim plausible. Parallel arguments to those for type (i) predicates apply: no one has yet provided any plausible reason for supposing that, e.g., (logically) presocial drives uniquely determine the social context or that this context is causally irrelevant to their operation. As Freud himself saw, and many neo-Freudians have insisted, the process of social channeling is a crucial part of the explanation of behavior, involving reference to features of both small groups and the wider social structure.

If the individual predicates are to be of type (iii), there is still no positive reason to find the methodological individualist's claim plausible. There may indeed be valid and useful explanations of this type, but the claim we are considering asserts that all proper, or rock-bottom, explanations must be. Why rule out as possible candidates for inclusion in an explicans (statement of general laws + statement of boundary and initial conditions) statements that are about, or that presuppose or entail other statements that are about, social phenomena? One reason for doing so might be a belief that, in Hume's words, "mankind are ... much the same in all times in all places."[27] As Homans puts it, the characteristics of "elementary social behaviour, far more than those of institutionalised behaviour, are shared by all mankind":

> Institutions, whether they are things like the physician's role or things like the bureaucracy, have a long history behind them of development within a particular society; and in institutions, societies differ greatly. But within institutions, in the face-to-face relations between individuals ... characteristics of behaviour appear in which mankind gives away its lost unity.[28]

This may be so, but then there are still the differences between institutions and societies to explain.

Finally, if the claim is that the individual predicates must be of type (iv), then it appears harmless, but also pointless. Explanations, both in the sense we are considering now and in the sense of identifications, may be wholly couched in such predicates but what uniquely special status do they possess? For, as we have already seen, propositions incorporating them presuppose and/or entail other propositions about social phenomena. Thus the latter have not really been eliminated; they have merely been swept under the carpet.

It is worth adding that since Popper and Watkins allow "situations" and "interrelations between individuals" to enter into explanations, it is difficult to see why they insist on calling their doctrine "methodological individualism." In fact the burden of their concerns and their arguments is to oppose certain sorts of explanations in terms of social phenomena. They are against "holism" and "historicism," but opposition to these doctrines does not entail acceptance of methodological individualism. For, in the first place, "situations" and "interrelations between individuals" can be described in terms which do not refer to individuals without holist or historicist implications. And second, it may be impossible to describe them in terms which do refer to individuals,[29] and yet they may be indispensable to an explanation, either as part of an identifying explanation in the statement of a general law, or of initial and boundary conditions.

Notes

1. *The English Works of Thomas Hobbes*, ed. Sir William Molesworth. London, 1839, i. 67; ii. xiv; ii. 109.
2. L. de Bonald, *Théorie du pouvoir*, Paris, 1854, i. 103.
3. A. Comte, *Système de politique positive*, Paris, 1851, ii. 181.
4. J. S. Mill, *A System of Logic*, 9th ed., London, 1875, ii. 469. "Men are not," Mill continues, "when brought together, converted into another kind of substance. with different properties."
5. See D. Essertier, *Psychologie et Sociologie*, Paris, 1927.
6. Cf. E. Durkheim, *Les Règles de la méthode sociologique*, Paris, 1895: 2nd ed., 1901, and G. Tarde, *Les Lois sociales*, Paris, 1898.
7. See *The Sociology of Georg Simmel*, trans and ed. with introd. by K. H. Wolff, Glencoe, Ill., 1950, esp. Ch. i, ii, and iv (e.g., Let us grant for the moment that only individuals 'really' exist. Even then, only a false conception of science could infer from this 'fact' that any knowledge which somehow aims at synthesizing these individuals deals with merely speculative abstractions and unrealities," pp. 4–5.

8. See C. H. Cooley, *Human Nature and the Social Order*, New York, 1902. For Cooley, society and the individual are merely "the collective and distributive aspects of the same thing" (pp. 1–2).
9. See G. Gurvitch, "Les Faux Problèmes de la sociologie au XIXe siècle," in *La Vocation actuelle de la sociologie*, Paris, 1950, esp. pp. 25–37.
10. See M, Ginsberg, "The Individual and Society," in *On the Diversity of Morals*, London, 1956.
11. See G. C. Homans, "Bringing Men Back In," *American Sociological Review*, xxix (1964) and D. H. Wrong,"The Oversocialised Conception of Man in Modern Sociology," *American Sociological Review*, xxvi (1961).
12. See the following discussions: F. A. Hayek, *The Counter-Revolution of Science*, Glencoe, Ill., 1952, Chs. 4, 6, and 8; K. R. Popper, *The Open Society and Its Enemies*, London, 1945, Ch. 14, and *The Poverty of Historicism*, London, 1957, Chs. 7, 23, 24, and 31; J. W. N. Watkins, "Ideal Types and Historical Explanation," *British Journal for the Philosophy of Science*, iii (1952) (reprinted in H. Feigl and M. Brodbeck, *Readings in the Philosophy of Science*, New York, 1953), "Methodological Individualism." (note), ibid., "Historical Explanation in the Social Sciences," ibid., viii (1957); M. Mandelbaum, "Societal Laws," ibid. (1957). L. J. Goldstein, "The Two Theses of Methodological Individualism" (note), ibid., ix (1958); Watkins, "The Two Theses of Methodological Individualism" (note), ibid., x (1959); Goldstein, "Mr Watkins on the Two Theses" (note), ibid., Watkins "Third Reply to Mr Goldstein" (note), ibid.; R. J. Scott, 'Methodological and Epistemological Individualism' (note), ibid. xi (1961); Mandelbaum, "Societal Facts," *British Journal of Sociology*, vi (1955); E. Gellner, Explanations in History," *Proceedings of the Aristotelian Society*, xxx (1956) (these last two articles together with Watkins's 1957 article above are reprinted in P. Gardiner (ed.), *Theories of History*, Glencoe, Ill., and London, 1959, together with a reply to Watkins by Gellner. Gellner's paper is here retitled "Holism and Individualism in History and Sociology"), M. Brodbeck, "Philosophy of Social Science," *Philosophy of Science*, xxi (1954); Watkins, "Methodological Individualism: A Reply" (note), ibid., xxii (1955); Brodbeck, "Methodological Individualisms: Definition and Reduction" ibid., xxv (1958); Goldstein, "The Inadequacy of the Principle of Methodological Individualism," *Journal of Philosophy*, liii (1956); Watkins, "The Alleged Inadequacy of Methodological Individualism" (note), ibid., lv (1958); C. Taylor, "The Poverty of Historicism," *Universities and Left Review* (Summer 1958), followed by replies from I. Jarvie and Watkins, ibid. (Spring 1959); J. Agassi, "Methodological Individualism," *British Journal of Sociology*, xi (1960); E. Nagel, *The Structure of Science*, London, 1961, pp. 535–46; A. C. Danto, *Analytical Philosophy of History*, Cambridge, 1965, Ch. xii; and W. H. Dray, "Holism and Individualism in History and Social Science," in P. Edwards (ed.), *The Encyclopedia of Philosophy*, New York 1967.
13. *Individualism and Economic Order*, London, 1949, p. 6.
14. *The Open Society*, 4th ed. ii. 98.
15. "Historical Explanation in the Social Sciences" in Gardiner (ed.), *Theories of History*, p. 505. Cf. "large-scale *social* phenomena must be accounted for by the situations, dispositions and beliefs of *individuals*. This I call methodological individualism." Watkins, "Methodological Individualism: A Reply," *Philosophy of Science*, xxii (1955), 58 (see n. 12 above).
16. *Universities and Left Review* (Spring 1959), 57.
17. *The Counter-Revolution of Science*, p. 56.
18. *The Poverty of Historicism*, p.140.
19. Popper himself provides some: see *The Poverty of Historicism*, pp. 62–3.
20. London, 1960.
21. *Psychology of Politics*. p. 10.
22. *Les Règles de la méthode sociologique*, p.103.
23. See *Social Behaviour*, London, 1961.
24. *British Journal of Sociology* (1955).
25. E.g., Hempel calls this "deductive-nomological explanation." For a recent defense of this type of explanation in social science, see R. Rudner, *Philosophy of Social Science*, Englewood Cliffs, N.J., 1965. I have not discussed "probabilistic explanation," in which the general laws are not universal and the explicans only makes the explicandum highly probable, in the text; such explanations pose no special problems for my argument.
26. *Psychology of Politics*, p. 5.
27. D. Hume, *Essays Moral and Political*, ed. T. H. Green and T. H. Grose, London, 1875, ii. 68.
28. *Social Behaviour*, p. 6.
29. E.g., in the cases of rules and terminologies of kinship or of language generally.

Chapter 30

Methodological Individualism and Social Explanation

Richard W. Miller

For over twenty years, Karl Popper, J. W. N. Watkins, and others have argued for methodological individualism, the doctrine that social phenomena must be explainable in terms of the psychologies and situations of the participants in those phenomena. This statement of methodological individualism is vague, because the claims put forward in the name of that doctrine have seemed to many readers to be extremely diverse. Is there, however, a version of methodological individualism, figuring prominently in writings of the individualists themselves, which is both plausible (in that a reasonable person might, on reflection, accept it as true) and nontrivial (in that there are sociological claims of significant popularity which would not be put forward if their proponents were fully conscious of the truth of methodological individualism)? The majority of writers on methodological individualism claim that no such version exists. According to these critics, methodological individualism either consists of doctrines which no reasonable person could accept once he fully understands their implications, or consists of doctrines which fail to exclude any current sociological theses, including the Marxist explanations which are the individualists' modern bête noire.[1] The continued attractiveness of methodological individualism is typically ascribed to a muddled and unconscious shifting between the implausible and the trivial versions of the doctrine.

These critics of methodological individualism are, I shall argue, mistaken. There is a version of methodological individualism that is both plausible and nontrivial. At the same time, this version of methodological individualism, plausible though it is, is not, in fact, a valid methodological principle. When I argue for the nontriviality and the nonvalidity of the relevant version of methodological individualism, Marxist sociology will be my main case of a source of nonindividualist explanations. I shall argue that the individualist principle in question ought not to be accepted in the relatively a priori spirit in which it is offered. If my criticisms are fair, any nontrivial version of methodological individualism must exclude appeals to nonrational processes which certainly do control behavior in small-group interactions and may well do so in historically significant large-scale social phenomena.

If my argument is right, the two decades of attack on methodological individualism have largely been a misfortune for the social sciences. The critics of methodological individualism have concentrated their fire on extremely implausible versions of methodological individualism, which in practice constrain no one working in the social sciences. Meanwhile, an individualist doctrine that exercises a real restraining influence has remained unscathed.

The main text I shall rely on as a source of individualist doctrines is Watkins's concise and relatively clear exposition of methodological individualism, "Historical Explanation in the Social Sciences" [13]. In this essay, Watkins says, "There may be unfinished or half-way explanations of large-scale social phenomena (say, inflation) in terms of other large-scale phenomena (say, full employment); but we shall not have arrived at rock-bottom explanations of such large-scale phenomena until we have deduced an account of them from statements about the dispositions, beliefs, resources, and interrelations of individuals. (The individuals may remain anonymous and only typical dispositions, etc., may be attributed to them)" ([13], p. 271). In context, this remark of Watkins seems to amount to the following claim:

> *Proposition I (The Individualist Constraint on Explanation):* There must be a rock-bottom explanation of every large-scale social phenomenon which explains the phenomenon as solely due to the beliefs and dispositions of actual or typical individuals and the situations to which they respond in accordance with their beliefs and dispositions.

With certain clarifications, Proposition I appears a much more plausible doctrine than methodological individualism is usually said to be. This proposition imposes a constraint on explanation in the social sciences, but not on the ultimate vocabulary of the social sciences. It does not require that the claims of social scientists be expressible in a language, no individual term of which refers to a phenomenon entailing the existence of a society. Very likely, no individualistic *definition* of "marriage," for example, can be given. But if a marriage custom can be explained as due to participants' beliefs about marriage, the individualistic constraint on explanation is still satisfied.

Proposition I is restricted in scope to "rock-bottom" explanations of "large-scale social phenomena." By the latter phrase, Watkins seems to mean the relatively complex social phenomena, involving long-lived institutions and affecting the lives of many people, which are objects of investigation for historians, historically minded sociologists, and cultural anthropologists. Thus, at one point, Watkins says that no statements compatible with methodological individualism can explain reflex-like group behavior in which "some kind of physical connection between people's nervous systems ... causes automatic, and perhaps in some sense appropriate, bodily responses." But he remarks that such actions are not a sufficient basis for "typical long-lived institutions, like a bank, or a legal system or a church" and do not "endure ... through generations of men" ([13], pp. 273f.)

In restricting the scope of his constraint to the "rock-bottom" level, Watkins is allowing that explanations which do not, superficially, meet his constraint may do so by a certain indirect route. To satisfy Proposition I, a sociological explanation either must show how the phenomenon in question is due to beliefs, dispositions, and situations of actual or typical individuals (I shall call such explanations "wholly individualistic"), or it must solely refer to processes, tendencies, or causal links which are explainable, in turn, as due to such beliefs, dispositions, and situations (the indirect route). Thus, every sociological explanation must either be wholly individualistic, or rest on the rock of a further, wholly individualistic explanation.

As Watkins makes clear at several points in his essay, "rock-bottom" does not mean "final" ([13], pp. 280, 275f.). To adapt an example of Watkins, an individualist must require that an explanation of a population pattern in terms of tribal marriage customs rest on an individualistic explanation of those customs, perhaps in terms of beliefs and dispositions concerning incest. But the existence of the latter beliefs and dispositions might well stand in need of further explanation. And the further explanation that initially comes to mind need not be wholly individualistic.[2]

In at least one other way, the demand for a rock-bottom individualistic explanation is weaker than it might seem. To make a justifiable claim that a wholly individualistic explanation of a phenomenon exists, one need not be prepared to construct one, or even to claim that it is humanly possible to do so. To employ another example of Watkins, one might make one's claim in the spirit of a physicist's assertion that there is an explanation of blast effects of a particular bomb explosion in terms of the trajectories and velocities of individual molecules. This claim could be valid, and even justifiable, although it is not humanly possible to construct such an explanation.

Our discussions so far have tended to show that Proposition I is less implausible than it might at first appear to be. But can this proposition avoid the opposite defect of triviality? To answer this question, we need a more precise understanding of "disposition," as the term is used in Proposition I. This interpretative problem is at once the most difficult and the most important one, for a proper understanding of Watkins's individualism.

Philosophers sometimes use "disposition" in a sense, which I shall call "the philosophy of science sense," in which "x has a disposition to F in circumstances C" is fully paraphrased by "should x be in circumstances C, it will F." When Watkins speaks of dispositions, in imposing his individualistic constraint on explanation, is he using the term in this broad philosophy of science sense? Clearly, he is not. For one thing, Watkins, as previously noted, admits that reflex-like, purely automatic behavior has no explanation satisfying the constraint he imposes. By way of citing examples of such nonindividualistic phenomena he says, "I think that a man may more or less literally smell danger and instinctively back away from unseen ambushers; and individuality seems to be temporarily submerged beneath a collective physical rapport ... among panicking crowds" ([13], pp. 273f.). Perhaps it is true, as Watkins seems to imagine, that panic behavior sometimes has no individual mental cause, not even the sudden onset of fear in the panicking individual. Still, if "disposition" were understood in the philosophy of science sense, then Watkins's examples of automatic responses, and any other examples, could be explained within the confines of Proposition I, the possibility Watkins denies. A crowd's flight from fire, no matter how automatic, could be explained as due to their disposition to run away from the vicinity of intense heat and smoke. There is also a further reason to suppose that the relevant sense of "disposition" is fairly narrow. Watkins subsequently makes it clear that the explanations characteristic of Marxist social theory are, in his view, incompatible with methodological individualism. Marx's idea that the view of social reality taken by a typical member of an economic class is determined by the economic interests of that class appears to be especially objectionable to Watkins ([13], p. 275). But Watkins could not use Proposition I to rule out Marxist explanations if "disposition" were used in the philosophy of science sense. Otherwise, one would, for example, be giving a rock-bottom individualist explanation in saying, "The belief of a typical capitalist that strikes are bad is due to the disposition of a typical member of an economically dominant class to have beliefs which justify the pursuit of the economic interests typical of his class." In general, the appropriate reading of "disposition" must be narrower than the philosophy of science sense, or the individualist constraint on explanation will lapse into triviality.

What narrower concept of "disposition" can we adopt in Proposition I, without making the latter proposition implausibly restrictive? The question is a difficult one. Watkins's own intentions are obscure. But in one of his articles he does offer a valuable clue to the plausible, nontrivial interpretation of individualist "dispositions," when he identifies the methodological chapters of Max Weber's *Economy and Society* as the classic presentation of methodological individualism by a major sociologist.[3]

Sociology, for Weber, is the science "which attempts the interpretative understanding of social action in order thereby to arrive at a causal interpretation of its course and effects. In 'action' is included all human behavior when and insofar as the acting individual attaches a subjective meaning to it.... Action is social insofar as, by virtue of the subjective meaning attached to it by the acting individual (or individuals), it takes account of the behavior of others and is thereby oriented in its course" ([17], p. 88). Some of Weber's subsequent examples of a subjective meaning (*subjecktiv gemeinter Sinn*) attached to an act are: someone's reasons for carrying out a multiplication or proving a theorem in a certain way, the goals of someone "trying to achieve certain ends by choosing appropriate means," "anxiety, anger, ambition, envy, jealousy, love, enthusiasm, pride, vengefulness, loyalty, devotion and appetites of all sorts" ([17], pp. 91f.).

In the previously cited article, Watkins makes it clear that individualistic explanations in terms of "dispositions" are Weberian explanations in terms of "subjective meanings" that agents attach to their actions. This is fully in the spirit of *Economy and Society*, where Weber himself employs the definition of sociology in terms of subjective meaning to argue for individualist constraints on social science. (For example, he says, "collectivities must be treated as solely the resultants and modes of organization of the particular acts of individual persons, since these alone can be treated as agents in a course of subjectively understandable action" ([17], p. 101).) But what is a subjective meaning of an act?

The definition of sociology, and numerous parallel discussions, together with the quoted examples, and many others that Weber gives, all point to the following notion of subjective meaning: a subjective meaning is a subject's reason for acting as he did. Y is a subjective meaning X attaches to his action, Z, at time t, just in case Y is a reason for action that X has at time t, and X did Z at that time because he had this reason.

For the purposes of this definition, X need not have consciously formulated the reason, Y. On the other hand, a desire, goal, or need is to be counted as someone's reason for action if, were he asked whether it is, he would respond that it was, if he were sincere, his memory sufficiently clear and sharp, his analytical skills adequate, and if no psychological mechanism of repression were operating. (I shall assume throughout that the attribution of repression is otiose if it is not reflected in felt tension, anxiety, or some other symptom.) Finally, "action," as used in the definition, is meant very broadly. In particular, X's "action" may be the formation or maintenance of a belief.

This emphasis on reasons is perfectly compatible with Weber's insistence that subjectively meaningful behavior may be irrational (see, e.g., [17], p. 92). When I irrationally flee from a harmless garden snake my reason for running away may be an irrational fear of snakes.

While the realm of subjective meaning is extremely broad, it has its limits. One obvious limit is suggested by Weber's distinction between "meaningful action and merely reactive behavior to which no subjective meaning is attached" ([17], p. 90). Actions performed purely out of habit are not subjectively meaningful. Once, long ago, I decided I would look better if I parted my hair on the left. But this morning I did so quite automatically and routinely. It was not true this morning that my reason for doing so was that I would look better that way. My typical hair partings are no longer subjectively meaningful.

Even when someone behaves in a certain way on account of desires, goals, or needs that he has, those desires, goals, or needs might not constitute subjective meanings that he attaches to his behavior. Ex hypothesi, they are *reasons why* he acted as he did, i.e., explanations of why he so acted. But they need not be *his reasons* for so acting.

There is nothing esoteric about this distinction between an agent's desires, needs, or goals which are the reason why he acted as he did and the agent's reason for so acting. Consider this banal episode. Bill is grocer. John, who lives above the shop, comes down and asks for credit. "I'm broke and out of work now," he says, but adds, with all signs of sincerity, "I'm sure to get a job within a month." Bill knows that John possesses overwhelming evidence that he will probably be out of work for many months. When he proposes to John, "You're just saying you'll get a job soon because you want credit," his neighbor responds with what looks like the most honest hurt at an insult, and leaves the shop.

How should Bill interpret his neighbor's saying that he will soon have a job? Bill might have independent grounds for believing that John is an extremely honest person, who, if he lied under extreme pressure, would show much anxiety and hesitation, no signs of which were, in fact, present in the scene at the store. Bill might also know that John is astute enough to comprehend the evidence available to him that he won't soon get a job. Given his knowledge of John's situation, character, and behavior, Bill might choose the following as the best explanation of why John described his job chances as he did: John really believed he would get a job soon, when he said he would. This belief was due to John's need to obtain credit together with his need to continue to see himself as an honest person. The satisfaction of these needs was not *his* reason for believing he was sure to get a job soon. He sincerely rejected this hypothesis. But satisfaction of these needs was *the* reason why he believed what he did. Or, as Bill might, more likely, put it, "He believed what he said. But when they need credit, people believe the strangest things about their finances."

I have described this case in detail to make it clear that no extraordinary or special insight is involved in the justified assertion that the reason for an agent's action consists of needs, desires, or goals which were not *his* reason for so acting. Such an assertion might be the best explanation in light of quite ordinary knowledge of his character, his needs, his situation, his behavior, and how others with similar character have behaved in similar cases. (Bill would have been justifiably reluctant to explain as he did were this the first time he had seen someone assert what that person ought to have known to be false when relevant dishonesty or stupidity were unlikely, on independent grounds.) As Marx and Engels say, in the *German Ideology*, the distinction between agents' reasons for actions and actual reasons for actions is one that every shopkeeper can make in ordinary life ([10], p. 67).

Of course, social scientists rarely have the detailed knowledge of someone's character, needs, situation, and behavior that Bill has of John's. But they usually pursue explanations of why a typical person of a certain kind behaves in a certain way. And here, there may well be knowledge concerning the typical situation, character, behavior and needs, goals or desires of a person of that kind sufficient to justify an explanation of the behavior in question as typically due to needs, goals, or desires which are not the agent's reasons for so behaving.

Again, the sort of inference made is not specialized or esoteric. Suppose I were to believe, as many people do, that most nuclear engineers possess evidence which shows that nuclear reactors are unsafe, given present safeguards, and also to believe that nuclear engineers are, typically, honest and intelligent. I might justify these beliefs in a variety of ways. There is no difficulty in principle, here. In light of these background beliefs, I seek to explain why most nuclear engineers say reactors are safe, with present safeguards. The best explanation is, "Nuclear engineers say reactors are safe because they want to regard their life-work as of great social utility." Here, I might, rely, in part, on a warranted belief about professionals in our society, that they want to regard their work as important to society. I need not rely on particular acquaintance with particular nuclear engineers. And my claim is quite compatible with the existence of atypical engineers who regard reactors as safe out of ignorance or stupidity, or who make safety claims which they do not believe out of corporate pressure.

As I shall soon emphasize, Marxist social theory depends, to a high degree, on the assertion that a typical occupant of a social role has certain beliefs because of desires, needs, or goals associated with that role which are the reasons for his belief, but not his reasons for his belief. The explanations of this sort which Marx gives are sometimes original and surprising, always controversial. But I hope to have shown that they are not original or controversial in form. Quite apart from controversial or specialized social theory, it is a common and legitimate practice to propose that an explanation of action in terms of needs, desires, or goals which are not the subject's reasons for action is the best explanation in light of a constellation of relevant facts.

If an explanation of a pattern of behavior explains it as due to desires, goals, or needs of the agent which are not his reasons for so behaving, I shall call it an explanation of behavior as due to *objective interests*. (Robert Stalnaker has pointed out to me that this expression is somewhat misleading, since the relevant desire or goal might be self-destructive. But some short phrase is needed. And the actual explanations to which I shall appeal involve no self-destructive tendencies.) Weberian sociology excludes patterns of behavior which are solely explainable as due to objective interests. I shall subsequently argue that this limitation ought not to be adopted as a methodological principle.

At one point in his later writings, Weber comes close to explicitly distinguishing subjective meanings from desires, goals, and needs which are the reasons, but not the agents' reasons, for actions. In an article which he subsequently cites, in *Economy and Society* ([17], p. 87), as a stricter and more detailed explanation of basic concepts of meaningful social action, he distinguishes subjective rationality in terms of goals (*subjektive Zweckrationalität*) as something utterly different from objective rationality in terms of correctness (*objektive Richtigkeitsrationalität*). An action carried out under the guidance of goals which are the agent's reasons for action has the former, subjective rationality. An action or other phenomenon which actually attains ends, which may not be agent's reasons for actions, but are a concern of the social scientist studying the action, has the latter, objective rationality. After a rough sketch of the distinction, Weber continues:

> Apart from certain elements of psychoanalysis which have this characteristic [i.e., reference to objective rationality] a construction such as Nietzsche's theory of envy involves an explanation which out of the practical features of a constellation of interests displays the objective rationality of external behavior, a rationality which is noticed barely, if at all, because it is "unintelligible" on a subjectively meaningful basis. It is precisely the same, methodologically speaking, with economic materialism. ([15], p. 434)

Thus, explanations asserting the objective rationality of actions, as fulfilling objective interests, need not explain them as due to subjective meanings, i.e., agents' reasons.

In the rest of this essay, I shall assume that dispositions in the sense of Proposition I are subjective meanings, agents' reasons for actions. I shall sometimes speak of them as "motives," sometimes as "psychological dispositions," but the Weberian sense will always be intended.[4]

Proposition, I, as I have interpreted it, makes a plausible claim. Weber, for example, in a previously cited passage, proposes, in effect, that sociology be confined to the study of social action explainable within the constraints of Proposition I. A commitment to the Weberian program surely is not, in itself, a mark of unreasonableness or lack of understanding.

Assuming that Proposition I expresses a plausible principle, does it, in fact, express a valid methodological principle? I shall argue that it does not. By a valid methodological principle, I mean a principle with at least two features: (a) commitment to it does not prevent one from attaining a true picture of reality in the sciences in question; (b) if one discovers that one cannot be in a position to put forward a given claim while still maintaining the methodological principle in question, that fact is always, in itself, a good reason for abandoning that claim. Thus, if a principle is methodologically valid, it is reasonable to employ it as a plausibility criterion. If acceptance of a theory would require violation of the principle, the theory is, by that token, too far-fetched to merit further appraisal. (Note that a plausible principle may be a bad criterion of plausibility. To take the most obvious kind of obstacle, there may be a rival, incompatible principle which is also plausible.) Watkins clearly believes that his constraint on explanation is a valid methodological principle in this sense. For when he recommends it, he characterizes it as a methodological principle, a regulative principle, and a principle which encourages research in some directions while discouraging research in others ([13], pp. 269f.).

I shall argue for the methodological invalidity of Proposition I by means of counterexamples. My main counterexample is taken from Marxist social theory. I shall argue that Marxists are not in a position to put forward certain characteristic claims about the bourgeoisie while maintaining Proposition I, but that this is not, in itself, a good reason for them to abandon any aspect of Marxist social theory. If so, Proposition I fails to possess feature (b).

Marxists and many other people believe that the following large-scale social phenomenon is characteristic of modern societies: a typical major, active capitalist regards the interests of big business as coinciding with the interests of the nation as a whole. In other words, he believes that actions and policies which maximize, on the whole, the wealth and power of the large firms which dominate his nation's economy also maximize, on the whole, the welfare of the people of the nation. For purposes of convenience, I shall sometimes refer to this belief as the identification of the bourgeois and the national interest.[5] Throughout his long career, which extended through the darkest days of the Great Depression, Alfred Sloan, chairman of the board of General Motors, used to express this identification with the pungency of proverb: "What is good for General Motors," he would declare, "is good for America."

The belief phenomenon I have sketched is large-scale and enduring. If it cannot be explained as due to the beliefs and psychological dispositions of actual or typical individuals, many other phenomena asserted to exist in Marxist discussions of ideology will similarly conflict with Proposition I. Can it be so explained? Not if Marxists are correct in their view of the relevant facts.

It might be felt that a typical capitalist identifies the bourgeois and the national interests as a result of his encounters with evidence indicating that this identity holds. The belief in question could then be explained individualistically, as due to the businessman's reasons for forming the belief, namely, his possession of other, evidential beliefs and his desire to form an accurate notion of the national interest on the basis of the evidence available to him. If Marxists are right, however, in their conception of the activities of the bourgeoisie, this explanation is not typically true. Especially in advanced capitalist societies such as our own, major active capitalists are seen as strikebreakers, war makers, and instigators of periodic political repression, who are in possession of overwhelming evidence to the effect that the bourgeois interest and the interest of most people in the nation are not identical. Thus, Marxists cannot explain the formation of the belief in question as due to standard learning processes.

Given the view of the relevant facts characteristic of Marxist and many non-Marxist theorists, the belief-phenomenon in question is best explained as due to objective interests, not to psychological dispositions. A typical major capitalist identifies the bourgeois and the national interest because such belief serves a variety of his desires and goals. For one thing, he has a goal of promoting this belief in others. And it is easier and less tense to encourage a belief in others if you share it. Also, he possesses overwhelming evidence that policies of layoff, speed-up, pollution, and war which he instigates or encourages hurt most people. If he were to accept this conclusion, he would feel much the worse for it. So, to achieve the peace of mind he desires, he must encounter the evidence strongly prejudiced toward the belief that the interests of big business actually coincide with the interests of most people, despite apparent evidence to the contrary. Of course, these desires and goals are not *his* reasons for making the crucial identification. He would emphatically, honestly, and serenely reject this explanation of his belief.

Marxism conflicts with Proposition I in that a Marxist is in no position to suppose that a certain belief-phenomenon crucial to Marxist social theory has an individualistic explanation. Watkins would gladly accept this conclusion and would claim that this conflict is a good reason for abandoning Marxism. If he did not view the conflict in this way, he could not propose Proposition I as a methodologically valid principle. Watkins, moreover, offers reasons for taking conflict with Proposition I as grounds for dismissing a theory, by presenting two arguments to the effect that Proposition I must be true. Both arguments, I shall maintain, are bad ones.

Shortly after presenting the constraint on explanation that has been under discussion, Watkins presents the following proposition, and calls it "the central assumption of the individualist position":

> [No] social tendency exists which could not be altered *if* the individuals concerned both wanted to alter it, and possessed the appropriate information. (They might want to alter the tendency but, through ignorance of the facts and/or failure to work out some of the implications of their action, fail to alter it, or perhaps even intensify it.) ... (I do not call "social" those tendencies which are determined by uncontrollable physical factors, such as the alleged tendency for more male babies to be born in times of disease or war). ([13], p. 271f.)

The claim Watkins is making might be state more briefly as follows:

Proposition II (The Alteration Principle): Any social tendency would be altered if the individuals concerned had the appropriate beliefs and desires as their reasons for action, and if there were no unremovable physical obstacles to change.

Watkins presents the Alteration Principle without argument, as if it were obviously true. Yet its truth is by no means obvious. There seem to be people who want to give up smoking, possess all appropriate information about giving up smoking, yet cannot give up smoking. Doctors label their condition "*psychological* addiction," suggesting that the obstacle to change is no more physical than any other force guiding human conduct. Very likely, this smokers' syndrome depends too little on enduring institutions to be counted as a large-scale social phenomenon by Watkins. But pessimistic social scientists have sometimes portrayed large-scale social phenomena as reflecting something like psychological addiction on a ghastly scale. For example, given the pervasive desire for peace throughout most societies and the failure of nearly all societies to avoid war, some social scientists explain the prevalence of war as a result of tendencies toward aggressive response, territoriality, and xenophobia too deep-seated to be overcome by appropriate knowledge and a desire for peace. Such pessimism may be wrong. Watkins would find Marx a perhaps unwelcome ally in rejecting it. But the pessimists' apparent denial of the Alteration Principle is not so obviously wrong as to be rejected a priori, and, indeed, without argument of any kind.

Still, the Alteration Principle might achieve methological validity if suitably qualified and refined. Perhaps "appropriate desires" should be expanded to "appropriate desires of strong enough intensity." Perhaps "unremovable physical obstacles" are usefully understood as including all consequences of human genetic structure. I shall not pursue these possibilities, though I suspect such modifications would leave the Alteration Principle either too dubious or too tautologous for Watkins's purpose. There is a simpler problem with Watkins's use of Proposition II. Even if it were methodologically valid, the truth (much less the methodological validity) of Proposition I would not follow. Even if different beliefs and desires functioning as agents' reasons for action could change any large-scale social tendency, there might be large-scale social phenomena which are not explainable as due to agents' reasons.

Suppose, as the Alteration Principle requires, that any social phenomenon *would* be otherwise, if people's reasons for action *were* to include certain beliefs and desires. It by no means follows that every social phenomenon is *in actuality* explainable in terms of such factors. It may be that a large-scale social phenomenon cannot be explained by appealing to agents' reasons, even though its existence depends on the *absence* of certain agents' reasons.

The following example shows that the kind of possibility I have sketched is more than a bare logical possibility. According to Watkins, a crowd in a theater which panics and runs in a purely automatic reflex reaction to smoke and fire is acting in a way which cannot be explained solely by reference to their beliefs and psychological dispositions in the situation at hand. Their action, as it were, short-circuits the belief-and-psychological-disposition mechanism. But it is certainly true, nonetheless, that such a group of people would act differently if they had different beliefs and psychological dispositions, e.g., if they believed that the smoke and fire were part of the show and had, as their reason for staying, the desire to watch the spectacle. By the same token, someone who denies that the identification of the bourgeois and the national interests is due to beliefs and dispositions can also accept that this identification would break down if capitalists' reasons for action were to include certain beliefs and dispositions now characteristic of militant trade unionists.

Watkins's second argument for his constraint on explanation occurs in the course of a criticism (and, as I argue in note 6, a misrepresentation) of the Marxist theory of ideology.

> Marx for instance professed to believe that feudal ideas and bourgeois ideas are more or less literally generated by the water-mill and the steam-engine. But no description, however complete, of the productive apparatus of a society, or of any other nonpsychological factors, will enable you to deduce a single psychological conclusion from it, because psychological statements logically cannot be deduced from wholly nonpsychological statements. Thus ... the idea that an explanation which begins by imputing some social phenomenon to human factors cannot go on to explain those factors in terms of some inhuman determinant of them is a necessary truth. That the human mind develops under various influences the methodological individualist does not, of course, deny. He only insists that such development must be explained "innocently" as a series of responses by the individual to situations and not "sinisterly" and illogically as a direct causal outcome of nonpsychological factors, whether these are neurological factors, or impersonal sociological factors alleged to be at work in history.[6] ([13], p. 275).

Here, Watkins appears to be arguing that the individualist theory of explanation is true, and, indeed, a necessary truth. If he is not, his words are certainly quite misleading. If this is his goal, his reasoning surely includes the following argument for Proposition I: all large-scale social phenomena are caused by the intentional actions of individuals, actions guided by beliefs and psychological dispositions. A phenomenon of this sort cannot be explained using wholly nonpsychological statements, which make no reference to beliefs and psychological dispositions. Therefore, the individualistic constraint on explanation must be valid. Every explanation of a large-scale social phenomenon must solely refer to beliefs and dispositions of individuals and situations to which they respond in accordance with their beliefs and dispositions.

This argument is simply fallacious. Even granting the initial steps (which might themselves be questioned), it follows that every explanation of a large-scale social phenomenon must, when fully spelled out, refer *in part* to individualistic psychological phenomena. It by no means follows that such an explanation must solely refer to such phenomena. Obviously a statement about social phenomena caused by intentional actions can be deduced from a set of statements that only partly refer to individualistic psychological phenomena. Lenin's explanation of the origins of World War I mostly consists of descriptions of objective class interests. But Lenin is certainly committed to the view that objective class interests, in a certain objective setting, make major capitalists *want* war and make such desires their reasons for action. Marx's theory of ideology is largely concerned with descriptions of objective interests and objective work situations. But he is certainly committed to propositions stating that class-interests mold *beliefs*.

Marxism is far from the only source of explanations which Proposition I might lead someone illegitimately to dismiss. Social anthropology is another common source of anti-individualistic explanations. Consider, for example, the following aspect of Evans-

Pritchard's account of beliefs concerning witches among the Azande, a people of the southern Sudan ([5], pp. 23ff.). Azande believe that being a witch is a heritable biological property, invariably inherited by the sons of a witch father and the daughters of a witch mother. Yet when, as frequently happens, the Azande are convinced that a man is a witch, they at most conclude that males in his immediate family are witches, never that all his patrilineal male kin are witches, as their theory requires. The insight and ingenuity that Azande show in political, agricultural, and other pursuits rules out stupidity as the explanation for this failure to accept the implications of Azande theories. Evans-Pritchard proposes that the Azandes' failure to regard all patrilineal male kin of a man they regard as a witch as witches is due to the fact that the Azande, if they accepted this inference (and a parallel one for women), would soon find that nearly everyone was a witch, so that accusations of witchcraft would lose their valuable social function of reducing economic inequalities which would otherwise tear apart Azande village communities. (Excessive striving promotes resentment in other Azande. Given the nature of the oracles that certify people as witches, an object of widespread resentment in a village is bound to be so certified.)

Most anthropologists would accept that the Azande fail to acknowledge entailments of their own beliefs about witches because this failure is strongly in their interests. But this interest clearly does not constitute a typical Azande's reason for not accepting the entailment. In a sense, it could not. If witchcraft-belief were regarded by most Azande as a piece of social engineering, such belief would lose its hold, and fail as social engineering.

My argument so far is, I have discovered, unsatisfying to many readers. The main source of this dissatisfaction is, I shall now argue, a confusion between explanation and the description of causes.

Many people insist on the validity of Proposition I, using arguments that might be spelled out as follows: every large-scale social phenomenon is created and maintained by the acts of individuals. These acts are, typically, the results of the beliefs and motives of the respective agents at the time of the action, beliefs and motives which are the respective agents' reasons for the individual acts. Acts not explainable in this manner, for example, slips of the tongue, play no crucial role in social causation. So every social phenomenon is caused by agents' reasons. This individualism about causes is common to every widely held social theory. A Marxist, for example, regards ideology as the product of a kind of informal training in habits of belief, a training consisting of acts guided by inculcators' beliefs and dispositions at the moments of action. Capitalists, he thinks, generally identify the bourgeois and the national interests because of a multitude of acts like the following: a future capitalist's being told by his father, "Smart, energetic people succeed. The poor are just lazy"; a fledgling capitalist's being congratulated by a superior for making an eloquent speech on the convergence of the interests of their corporation and the needs of the nation. The huge number of acts causing the tendency the Marxist asserts to exist are due to a huge number of agents' reasons, for example, the father's desire to discourage his son's sentimentalism and encourage him to succeed, the company president's admiration for the eloquence and good sense of the subordinate. An explanation—Marxist or otherwise—need not actually describe the individualist causes of a social phenomenon. But if one cannot hold a given theory and, at the same time, accept that a true description of this sort can in principle be constructed, that is a basis for rejecting the theory as implausible. Hence, methodological individualism is methodologically valid, though trivial in that it excludes no widely held current doctrine.

This objection contains many grains of truth. For one thing, the following principle does seem to be methodologically valid:

> *Proposition III (The Individualist Constraint on Causation).* Every social phenomenon is caused by the acts of individuals. Except for atypical and noncrucial cases, these acts are caused, in turn, by the beliefs and psychological dispositions of the agents.

I argued that Proposition I was not methodologically valid and was incompatible with Marxism. But Proposition III is methodologically valid and (a second grain of truth in the objection) quite compatible with Marxism. When Engels said, "Men make their own history ... in that each individual follows his own consciously desired end" ([4], p. 366), he meant to accept the individualist constraint on causation.

Given all these concessions, I can only reject the objection at hand if I deny that Proposition III entails Proposition I. And that is precisely my claim. The entailment fails because in social science explaining a large-scale social phenomenon often requires explaining why it would have happened even in the absence of the sequence of individual actions, beliefs, or dispositions which actually caused it.[7] More specifically, social scientists observe the following principle, in pursuing explanations:

> *Proposition IV (The Necessity Constraint):* If, given conditions obtaining at the time, X would have happened anyway, even in the absence of the sequence of individual actions, beliefs, or dispositions which actually did cause X, then an explanation of X must explain why X would have happened under the circumstances, even in the absence of that particular sequence.

Where the necessity constraint applies, the description of individualist causes the availability of which is guaranteed by Proposition III will not provide an explanation. For an explanation must show why the phenomenon in question would have happened anyway, even if that description were false. For example, if the necessity constraint applies to the capitalist belief-phenomenon that we have been examining, the individualist saga of indoctrination and encouragement previously sketched will not explain that phenomenon. The individualist story may still answer the question, "Why do the particular people who actually are major capitalists, namely, John D. Rockefeller III, David Lindsay, Walter Wriston, and others, typically identify the bourgeois and the national interests?" But it is not even intended to answer the question singled out by the necessity constraint, "Why, in a modern capitalist society, would major capitalists, whoever they might be, typically identify the bourgeois and the national interests, even if Rockefeller, Lindsay, and the other actual capitalists were to have different life histories?"

Three aspects of the necessity constraint require elaboration, or else that constraint will seem to refer only to rigidly deterministic processes. In the first place, the necessity sought in Proposition IV is relative to social conditions in the background of the large-scale social phenomenon to be explained. The object of explanation is why X would have happened anyway, in the absence of the individualist causal chain that actually produced it, but in the presence of conditions obtaining at the time in question. This relative necessity is, surely, what a social scientist demands when he asks, "Why would X have happened, even if Y had not produced it?" He is not presupposing that X was bound to occur from the beginning of time, or, less metaphysically, from the time the planet earth came into being. He is not supposing that X would have happened, no matter what the nature of the society in which it happened. Rather, he is assuming that under some set of conditions, current at the time in question, *X* would have arisen in some other way, if it were not produced by *Y*. For example, when a Marxist claims that a world war would have broken out around 1914, even in the absence of the chain of events involving the assassination of Archduke Rudolf, he obviously means that a world war would have occurred anyway, given the political and economic relations among major capitalist powers at the time. In Lenin's *Imperialism*, the classic Marxist argument that a world war would have broken out in the absence of the chain of individualist episodes that actually caused it, Lenin's whole case is built upon the description of new forms of capitalist activity, creating distinctive conditions for international relations, which arose at the turn of the nineteenth century. Moreover, the Marxist might regard the existence of the conditions in question as noninevitable, because, for example, the rise of capitalist societies on the planet earth was not inevitable, given the total state of the planet in certain precapitalist times.[8]

When social necessity is made relative to a particular social setting in this way, asserting that a large-social phenomenon would have occurred in the absence of the acts, beliefs, and psychological dispositions which actually produced it does not violate Proposition II, the alteration principle. For assuming that under the social conditions of the time a phenomenon would have occurred even in the absence of the beliefs and dispositions which actually did bring it about does not mean assuming that phenomenon to be compatible with just any beliefs and dispositions. The mechanisms guaranteeing that phenomenon may do so, in part, by preventing the occurrence of beliefs and dispositions incompatible with its existence. For example, the mechanisms responsible for the identification of the bourgeois and national interests by major capitalists must prevent the social role, major capitalist, from largely being filled by individuals whose beliefs and dispositions are, on the whole, those now characteristic of communists.

The necessity pursued according to Proposition IV is weak in one other respect. It might be true that the phenomenon in question would have occurred anyway, in the absence of the chain of individualistic episodes that actually caused it, even though there is a small but significant probability that this phenomenon might not have occurred under the conditions obtaining at the time. The crucial counterfactual condition is true, provided that only an extraordinary occurrence (such as did not take place in actuality) could have prevented the phenomenon at issue, in the circumstances obtaining at the time. But "extraordinary" does not mean "impossible." In using the crucial counterfactual in this way, I am conforming to the actual usage of social scientists, which conforms, in turn, to the ordinary usage of conditionals of the "*X* would have happened even if *Y* had not happened" variety.

The Nazi seizure of power was the result of many individualist episodes, including Hindenburg's inviting Hitler to become chancellor, an invitation which reflected Hindenburg's illusion that Hitler could readily be discarded later on, in favor of a nationalist politician of a more traditionalist sort. Many historians believe that the Nazis would have come to power anyway, in the absence of these particular individualist episodes. On their view, a Nazi government was worth the real attendent risks, from the standpoint of the German establishment, given the Nazis' unique ability, if installed in power, to repress domestic unrest while mobilizing for war. Had Hindenburg been less naive, he or his successor would have come to appreciate these facts, and would have installed a Nazi government anyway. Other historians regard the Nazi seizure of power as fortuitous in that they refuse to accept the claim that the seizure of power would have occurred in the absence of the causal chain including Hindenburg's naive assessment of Hitler.

Neither side in this dispute would regard the following consideration as showing that the "inevitabilist" side is wrong: "If Hindenburg had been more realistic, his anxieties would have prevented him from inviting Hitler to assume the Chancellorship as soon as he did, i.e., on January 30, 1933. As a result, Hitler, Goebbels, and Goering would have been crossing different street corners under different circumstances, breathing different air, and drinking different fluid in February 1933. There is a small probability they would have died, from accident, disease, or poison, as a result. This leadership vacuum could not have been filled. Hence, the Nazi seizure of power was not inevitable." This consideration is irrelevant, because the "inevitabilists" are only claiming that late-Weimar circumstances would have brought about the Nazi seizure of power in some way, if not through the individualist episodes which actually caused it, provided that nothing *extraordinary*, of a sort that never did occur, prevented the seizure of power. The coincidental death of the whole Nazi leadership in February 1933 would have been extraordinary, in the extreme.

In their dismissal of counterexamples depending on nonactual extraordinary possibilities, historians are faithful to ordinary usage of conditional of the form, "This would have happened anyway, if that had not brought it about." Suppose I say, in a postmortem of a bridge game, "Jones would have won, even if the finesse had not worked. He could have gotten to the board in the next round, and established his spades." I mean, then, that something extraordinary, for example, a lapse of memory remarkable in a player of Jones's experience and temperament, would have been required to prevent his carrying through the indicated winning strategy, which would have existed after the failed finesse. That is why I would not make this claim of a nervous beginner, in whom an incapacity to recover from the failed finesse would not be extraordinary.

I have used the vague phrase "extraordinary," the applicability of which heavily depends on the purposes and standards assumed by participants in the discussions in which the phrase is employed. This vagueness and context dependence fits the facts of social scientists' uses of "*X* would have happened anyway, even if *Y* had not brought it about." Suppose a historian believes that in January 1933, the German establishment was rapidly coming to the realization that a Nazi government was worth the real attendant risks. He also believes that the Nazi party would have rapidly disintegrated into warring factions without Hitler's leadership. Should he agree with the inevitabilists, that Hindenburg's naiveté only hastened the Nazi seizure of power? Or should he agree with the noninevitabilists, on the grounds that Hitler was more a prey to accidental death and assassination while out of office, and the Nazi seizure of power depended on his survival? While the coincidental death of the whole leadership of a major political party would be regarded as extraordinary in any dispute, it is not clear that the same can be said of the death of one leader. Without knowing the standards for the dismissal of hypotheses as too extraordinary which are shared by both sides, we do not know what side the historian should take. Perhaps there is no shared standard which dictates that he accept or reject the claim that the Nazis would have come to power, even in the absence of the chain of individualist episodes which brought them to power.

Fortunately, there are judgments as to what would be extraordinary on which all reasonable social scientists engaged in research and argument would agree. My subsequent uses of the necessity constraint will rely on such cases.

Counterfactuals of the form, "X would have happened even in the absence of the chain of individualistic episodes which brought it about," are too wordy to bear much further repetition. In what follows, I shall often abbreviate them by such phrases as "X was bound to happen" or "X was guaranteed by the circumstances in which it occurred." These paraphrases are not simply proposed, ad hoc, for purposes of convenience. As actually used by social scientists, the respective phrases are equivalent. Reflection on previous and subsequent examples (for example, Lenin on World War I, the "inevitabilists" on the Nazi seizure of power) should make this clear. Again no special social scientific usage is in question. My bridge postmortem might alternatively have been, "Even if the finesse had failed, Jones was bound to win."

I have emphasized respects in which the language of Proposition IV should be taken to conform to ordinary usage, though not to usages characteristic of many abstract discussions of necessity and contingency. In one crucial respect, though, the necessity constraint is more specific to social scientific usage. It is a constraint on what counts as an explanation *in the social sciences*, not in all (in particular, all extrascientific) contexts in which explanations are sought. It reflects the scientist's desire to reduce the extent to which reality seems the product of chance or, rather, to reduce this appearance to the greatest extent that the facts allow. We are not surprised by outbursts like the remark of the historian, E. H. Carr: "The shape of Cleopatra's nose, Bajazet's attack of gout, the monkey-bite that killed King Alexander, the death of Lenin—these were accidents which modified the course of history.... On the other hand, so far as they were accidental, they do not enter into any rational interpretation of history, or into the historian's hierarchy of significant causes" ([2], p. 135). Proposition IV is another expression—compared with Carr's a mild one—of this professional interest in reducing the realm of accident.

The scientist's interest in explanation is not, however, the only legitimate one. And when the pursuit of explanations is dominated by other interests, the necessity constraint may be inappropriate. Suppose that Simon tied Pauline to the railroad tracks, but, just before the 9:14 rolled by, Walter, equally villainous, saw Pauline and shot her in the head. At Walter's trial, the jury might properly accept the fact that Walter shot Pauline in the head as the explanation of why she died that day, even though she was bound to die that day, in any case. But juries are concerned to ascertain individual responsibility, that is, to locate people in individual causal chains of certain kinds leading up to individual events of certain kinds. They are not concerned to reveal the maximum extent to which necessity governs human affairs.

Here are two typical cases which would lead social scientists to acknowledge the general validity of the necessity constraint. In each case it is at least plausible that this constraint is applicable. By this, I mean that the claim of relative necessity ("Under the social conditions obtaining at the time, X would have happened, even in the absence of the individual causes that actually produced it") is at least worthy of empirical investigation, in each case. Such plausibility is sufficient for showing that Proposition III cannot be used to defend the methodological validity of Proposition I.

The first example is presented in an essay of Weber's which deserves to be much more widely read, "The Logic of the Cultural Sciences," a discussion of the methodological views of the historian, Eduard Meyer. Weber writes

> Eduard Meyer himself very nearly applies this procedure [viz., reflection on which might have happened] to the two shots which in the Berlin March days directly provoked the outbreak of the street fighting. The question as to who fired them is, he says, "historically irrelevant." Why is it more irrelevant than the discussion of the decisions of Hannibal, Frederick the Great, and Bismarck? "The situation was such that *any* accident whatever would have caused the conflict to break out"(!) Here we see Eduard Meyer himself answering the allegedly "idle" question as to what "would" have happened without these shots; thus their historical "significance" (in this case, irrelevance) is decided.... The judgment that, if a single historical fact is conceived of as absent from or modified in a complex of historical conditions, it *would* condition a course of *historically* important respects, seems to be of considerable value for the determination of the historical significance of those facts. [Weber's emphasis. It is clear in context that "seems to be of considerable value" is a piece of ironic understatement, on Weber's part.] ([16], pp. 165 f.)

The facts are these, as Meyer saw them: in March 1848, an uprising broke out in the workers' quarter of Berlin. The immediate cause was the firing of two shots at a crowd, which provoked a wave of rumors. The political and social tensions in Berlin, in 1848, were so extreme and so deep-seated that had those two shots not rung out, some other incident would have occurred and triggered a workers' uprising.

Surely, Meyer's view of the facts is not obviously wrong. Given this assessment of the facts, Weber and Meyer are both agreed that the firing of the two shots would be "historically irrelevant." Why? Because given the assumed conditions, a workers' uprising in Berlin in 1848 was bound to occur, even in the absence of the sequence of actions, beliefs, and psychological dispositions (the shots and the resultant rumors) that actually led up to it. Weber and Meyer both assume that the historian's question, "Why did X happen?" should be answered by a description of why X was bound to happen, if such a description is available. Implicitly, they both accept the necessity constraint.

Note that the acceptance of the necessity constraint in this case is obviously dependent on the interests characteristic of social science. In a trial for incitement to riot arising from the March 1848 uprising in Berlin, a jury might quite properly accept that rioting broke out because of speeches provoked by the two shots, refusing to acquit agitators on the grounds that an uprising would have occurred anyway. Note, too, that an extraordinary coincidence, for example, the absence of any provocative incident for week after week, despite the high level of discontent and the nervousness of the troops, might have forestalled an uprising, if the shots had not been fired. As we have seen, such hypothetical interventions of the extraordinary are irrelevant to historians' claims that something would have happened anyway, in the absence of the sequence of episodes that actually produced it.

A different example of social necessity will be helpful as bringing us closer to the case ultimately at issue, the tendency for businessmen to identify the interests of big business with the national interest. The change from carbon steel to stainless steel as the main material for knives would normally be explained as bound to happen, due to the greater capacity of stainless steel to keep its edge, together with the great reduction in the relative cost of stainless steel, as a result of technological advances in the 1920s. It is possible, in this case, to give a wholly individualistic description of the causes of the change, describing the episodes in which individual cutlery executives made actual decisions to switch production to stainless steel. But if this particular causal chain had not existed, stainless steel would still replace carbon steel as the main material for knives. Suppose that the executives in question had not formed beliefs to the effect that the material basis of knife production should be changed. Other people would have been smart enough to perceive the implications of the technological advances and to use this perception either to drive those executives out of office or to drive their companies out of the market. That Mr. Jones formed the belief, "We ought to switch to stainless," on May 18, 1927, that Mr. Smith formed that belief on April 2, etc., does not explain why stainless steel was bound to replace carbon steel as the main basis of knife production. For had they not formed those beliefs, the industrial capitalist system of production would, as it were, spontaneously correct for their stupidity.

Here again, the social scientist's interest in explaining what is bound to happen is so great, that the individualist saga fails to provide a social scientific explanation of why something happened. Economists would not regard the chronicle of executive decision as an explanation of why stainless steel has replaced carbon steel. And most would express their discontent by saying, "Even if these executives had been stupid, the structure of capitalist production, together with facts about cost and utility, would have guaranteed that stainless would win out, anyway."

I hope these example make it plausible that the necessity constraint governs social science. Surely, they show that answering the question, "Why would this have happened anyway, even in the absence of the individual causes leading up to it?" is a characteristic and important explanatory goal in social science. The response to the original objection based on the individualist sage of encouragement and indoctrination is now complete if the following hypothesis can be made plausible (that is, worthy of empirical investigation): in a capitalist society, typical major capitalists, whoever they might be, would identify the interests of big business with the national interest, even in the absence of the actual beliefs and dispositions which have produced this belief in actual individual capitalists.

We can, I think, regard a description of the individualist causes of the identification of the bourgeois and the national interests in the same light as the description of the individualist causes of the preponderance of stainless steel knives. The individualist causes of the fact that a capitalist typically identifies the bourgeois interest and the national interest may include this father's saying this to his son, who will become a capitalist, this company president's congratulating the second vice-president for this speech, and so on. But if the father had not said anything like that (if he had been raising his son to be a communist), if the president had not said anything like that (if he had been on the verge of renouncing the presidency and joining a hippie commune), the schools, media, and less formal training processes of our society would, in a Marxist's view, insure that others who became capitalists generally identify the bourgeois and the national interests. Perhaps the changes on the level of individual action would require no change in the nature of the overall training process in order to maintain the belief-tendency. But if there is a need for such a change, it will be forthcoming, on the Marxist's view. Thus, to take a small example, when the Indochina War reduced the number of students who identified the bourgeois and national interests and increased the number who contrasted them, businessmen launched a large-scale advertising campaign in student newspapers extolling the virtues of capitalism.

On a larger scale, Marxists such as Althusser have argued that advanced capitalist nations have dealt with the increased disenchantment with capitalism in the course of the nineteenth century by developing a greatly·elaborated ideological apparatus, centered on the public schools.

The explanation I have just sketched would still be compatible with the individualist constraint on explanation, if conditions which guarantee the belief-phenomenon in question themselves consisted of agent's reasons. But there is no basis, a priori, for the assumption that they take this individualist form. In particular, objective interests associated with social roles may provide the relevant "guarantee." Take the case of the ideological function of the mass media. Marxists (and others) claim that a managing editor of a major newspaper in a capitalist society will have an objective interest in encouraging beliefs which promote acquiescence to capitalism. Were he repeatedly to fail to do so, his desires and goals would be thwarted in that he would be fired, demoted, or otherwise removed from participation in decision making which is important to him. But for these desires and goals to lead an editor to promote pro-bourgeois beliefs, they need not provide his reasons for so acting. He might honestly deny that he ever promotes political-economic attitudes out of a desire to keep his job and personal influence. In general, what guarantees the perpetuation of certain ideologies may be the association of certain objective interests, with certain powers, an association which does not depend on the existence of any particular set of agents' reasons.

Many social scientists and philosophers regard Watkins-Weber individualism as too extreme, but accept some more moderate individualist constraint on explanation. They believe that every large-scale social phenomenon must have a rock-bottom explanation in terms of the psychological characteristics of individual participants. But they do not limit those psychological characteristics to participants' reasons for acting as they do. For example, the assumption is sometimes made that social phenomena are all explainable in terms of the beliefs, desires, and attitudes of actual or typical individuals regardless of whether those factors play the role of participants' reasons for action.[9]

Moderate individualism appears to be trivial, in the special sense introduced at the beginning of this chapter. The current theories violating Weber-Watkins individualism appeal, at rock-bottom, to the beliefs, desires, and attitudes of typical occupants of social roles. Short of extremely arbitrary restrictions on the scope of "psychological characteristic," it does not seem that moderate individualism excludes any current explanatory hypotheses.

It is all the more surprising that our discussion of causation and explanation casts doubt on the methodological validity of moderate individualism, i.e., the legitimacy of its use as a regulative principle excluding hypotheses as unworthy of investigation. Presumably, the rationale for so using moderate individualism is a variant of the argument from Proposition III to Proposition I. The first clause of Proposition III, according to which social phenomena are caused by acts of individuals, is assumed to be valid. So is some broadened version of the second clause, asserting that those acts are due, in turn, to psychological characteristics of the individuals, defined in some broader-than-Weberian way. On this basis, the availability of an explanation in terms of those psychological characteristics is asserted.

It is hard to see what other methodological rationale there could be for moderate individualism, aside from invalid variants of Watkins's invalid arguments for his position. But if I have correctly identified the basis of moderate individualism, it is doubly untenable. In the first place, there is no reason to be moderately individualist (i.e., to adopt the moderate position as a regulative principle), while stopping short of Watkins-Weber individualism. The argument from causation to explanation sketched in the last paragraph only works if the inference of Proposition I from Proposition III is sound. And the premise of the moderate argument is no more obviously true than Proposition III. Large-scale social phenomena surely are produced by acts which are due to beliefs and motives at the time of action which are the respective agents' reasons for action. Moderate and extreme individualism are equally justifiable as regulative devices.

In any case, our discussion suggests that moderate individualism, plausible as it sounds, lacks an a priori justification. Social phenomena are brought about by acts which are the result, in turn, of the psychological characteristics of the agents. But a description of those characteristics need not explain the social phenomenon in question.

Perhaps the grain of truth in moderate individualism is this: As new kinds of sociological explanations are accepted, the scope of "psychological characteristic" is broadened to include factors necessary to explain any new social tendencies asserted in terms of agents' psychological characteristics. Thus, Marxists often seem to treat being bourgeois as a psychological characteristic. Durkheimians might be held to treat anomie in the same fashion. But even if this speculation is right, individualism, moderate or extreme, appears to furnish no valid *constraint* on explanation. If the scope of "psychological characteristic" is defined in advance of empirical research, a requirement of rock-bottom explanations in terms of such characteristics ought not to regulate such research, ruling out hypotheses as unworthy of investigation.

Notes

I am extremely indebted to an anonymous *Philosophy of Science* referee for his numerous helpful criticisms. I also benefited from Robert Stalnaker's comments and advice and from the suggestions of Richard Boyd, Alan Garfinkel, Alan Gilbert and Carl Ginet.

1. Typical and influential criticisms of the first sort are offered by Gellner [6], Goldstein [7], Lukes [8], and Mandelbaum [9]. The compatibility of methodological individualism with Marxism is asserted by Cunningham [3].
2. Cf. the criticism of methodological individualism for ignoring cultural conditioning in [7].
3. [14]. See especially pp. 83f. As Watkins emphasizes in this article, methodological individualism is characteristic of Weber's final outlook, which received its most encyclopedic and influential expression in *Economy and Society* (written in 1916–1919, published posthumously in 1922). Such earlier writings as "Objectivity in the Social Sciences" (1903) present a less individualistic methodology. My own view is that Weber's approach to social science became increasingly individualistic and psychologistic as his opposition to Marxism, both in theorizing and in political practice, intensified.
4. A slight difference between Weber's usages and Watkins's creates a minor problem, here. Beliefs can be agents' reasons for action, just as much as desires. Such beliefs are "subjective meanings" for Weber. But Watkins seems to restrict "dispositions" to agent's reasons which are desires, wants, or goals. I shall use "disposition" in the broad sense of "agent's reason," to avoid the need sharply to distinguish the cognitive from the conative aspect of reasons. But I shall often speak, redundantly, of "beliefs and psychological dispositions" to preserve parallelism with the terminology of principles such as Proposition I.
5. This same phrase might be used to describe a different, though related, belief-phenomenon, the tendency, whenever an action or policy is in fact in the interests of the bourgeoisie, to regard it as in the national interest. Marx regards the major active capitalists of modern capitalist countries as sufficiently class conscious to have both tendencies. But tendencies of these respective sorts need not go together. Perhaps there have been peasantries from whom these tendencies split. When a policy or action is in the interests of the peasantry, they regard it as in the interests of society as a whole. But they are so unused or reluctant to think in anything like class terms that they do not have the belief that what is in the interests of the peasantry is in the interests of society as a whole.

6. In his initial statement about Marx, Watkins must have in mind the concluding sentence from the following paragraph in Marx's early polemic against Proudhon, *The Poverty of Philosophy*: "M. Proudhon the economist understands very well that men make cloth, linen or silk materials in definite relations of production. But what he has not understood is that these definite social relations are just as much produced by men as linen, flax, etc. Social relations are closely bound up with productive forces. In acquiring new productive forces men change their mode of production; and in changing their mode of production, in changing the way of earning their living, they change all their social relations. The handmill gives you society with the feudal lord, the steam-mill, society with the industrial capitalist" ([11], p. 109). The last sentence is the only passage in Marx which might, in isolation, suggest a belief in the more or less literal generation of feudal ideas and bourgeois ideas from kinds of machinery. I leave it to the reader to determine whether this is a reasonable interpretation when the sentence is read in context.
7. I was helped to see this point by discussions with Alan Garfinkel, I am not sure that Garfinkel would agree with this specific claim.
8. Marx's discussion of the stable, noncapitalist, "Asiatic" mode of production, in the *Grundrisse*, suggests that he did not regard it as inevitable that capitalism should have arisen on this planet. Indeed, the story of the rise of capitalism out of West European feudalism, in volume one of *Capital*, depends on several fortuitous coincidences, e.g., the occurrence of long-term inflation stimulated by gold from the New World soon after the Wars of the Roses had decimated the old English aristocracy and transferred dominance over the countryside to a new, more mercantile nobility.
9. Explicit expressions of moderate individualism are often found among the "culture and personality" school in anthropology (see, for example, [1], p. 24) and in the writings of Alfred Schutz (e.g., [12], p. 335). Where he does not embrace the Weber-Watkins extreme, Peter Winch, in [18], seems to assert moderate individualism (see p. 45).

References

[1] Benedict, R. "Configurations of Culture in North America." *American Anthropologist* 34 (1932): 1–27.
[2] Carr, E. H. *What is History?* New York: Random House, 1961.
[3] Cunningham, F. "Practice and Some Muddles about the Methodology of Historical Materialism." *Canadian Journal of Philosopohy* 3 (1973): 235–248.
[4] Engels, F. *Ludwig Feuerbach and the End of Classical German Philosophy*. In K. Marx and F. Engels, *Selected Works*, vol. 3. Moscow: Progress Publishers 1970. pp. 335–376.
[5] Evans-Pritchard, E. E. *Witchcraft, Oracles, and Magic Among the Azanda*. Oxford: Oxford University Press 1937.
[6] Gellner, E. "Explanations in History." *Proceedings of the Aristotelian Society* Suppl. vol. 30 (1956): 157–176.
[7] Goldstein, L. J. "The Inadequacy of the Principle of Methodological Individualism." *Journal of Philosophy* 53 (1956): 801–816.
[8] Lukes, S. "Methodological Individualism Reconsidered," *British Journal of Sociology* 19 (1968): 119–129.
[9] Mandelbaum, M. "Societal Facts." *British Journal of Sociology* 6 (1955): 305–317.
[10] Marx, K., and F. Engels. *The German Ideology*. New York: International Publishers 1970.
[11] Marx, K. *The Poverty of Philosophy*. New York: International Publishers 1963.
[12] Schutz, A. "Common Sense and Scientific Interpretation of Human Action." In M. Natanson, *Philosophy of the Social Sciences*. New York: Free Press 1963. pp. 330–341.
[13] Watkins, J. W. N. "Historical Explanations in the Social Sciences" *British Journal for the Philosophy of Science* 9 (1957): 104–117. Reprinted as "Methodological Individualism and Social Tendencies." In M. Brodbeck, *Readings in the Philosophy of the Social Sciences*. New York: Macmillan 1968. pp. 269–280. Page references are to the Brodbeck volume.
[14] Watkins, J. W. N. "Ideal Types and Historical Explanation." In A. Ryan, ed. *The Philosophy of Social Explanation*. Oxford University Press 1973. pp. 82–104.
[15] Weber, M. *Gesammelte Aufsätze zur Wissenschaftslehre*. Tübingen: J. C. B. Mohr 1968.
[16] Weber, M. *The Methodology of the Social Sciences*. New York: Free Press 1949.
[17] Weber, M. *Theory of Economic and Social Organization* [*Economy and Society*, vol. 1]. Translated by Henderson and Parsons. New York: Free Press 1947.
[18] Winch, P. *The Idea of a Social Science and Its Relation to Philosophy*. London: Routledge Kegan Paul 1958.

Chapter 31

Microfoundations of Marxism

Daniel Little

Marxist social science may be distinguished from other research programs in social science on a variety of grounds, but one especially important feature of it is its interest in providing "macroexplanations" of social phenomena and an associated mistrust of the doctrine of methodological individualism. Thus, Marxist economists are concerned to discover and explain the large-scale patterns of development of the capitalist economy; Marxist political scientists are interested in the ways in which state policy in a capitalist democracy serves the needs of the capitalist economic structure; Marxist sociologists are interested in analyzing patterns of class activity in terms of objective interests (e.g., labor organization, mass protest, or organized revolutionary activity); and so forth. In each case, the object of investigation is a large-scale characteristic of capitalist society or a supraindividual entity (e.g., a social class or a state), and the explanans is a description of some other set of macrophenomena.

Macroexplanations have long been the object of criticism within the philosophy of social science. Orthodox social scientists and philosophers of science have argued that social science must be subject to the principle of methodological individualism.[1] This doctrine may be formulated in several related but distinct forms: as a thesis about the ontological standing of social entities, as a thesis about the meaning of social concepts, and as a thesis about social explanation. The ontological thesis holds that all social entities must be reducible to logical compounds of individuals. The thesis about meaning asserts that all social concepts must be definable in terms of concepts that relate only to individuals. And the thesis about explanation holds that all social facts, patterns, and entities must ultimately be explicable in terms of facts about individuals—their motives, powers, beliefs, and capacities.[2]

Unhappily, the debate over methodological individualism has tended to proceed at a highly abstract level, based on general philosophical assumptions about metaphysics, the nature of explanation, or the nature of language and meaning. As a result, the debate has been inconclusive in its application to actual social scientific theory formation. More recently, however, a new set of criticisms of macroexplanations have been put forward within Marxist social science itself. Some Marxist thinkers have argued that macroexplanations stand in need of *microfoundations*, detailed accounts of the pathways by which macro-level social patterns come about.[3] These theorists have held that it is necessary to provide an account of the circumstances of individual choice and action that give rise to aggregate patterns if macroexplanations are to be adequate. Thus, in order to explain the policies of the capitalist state, it is insufficient to observe that this state tends to serve capitalist interests; we need to have an account of the processes through which state policies are shaped or controlled so as to produce this outcome.[4] Moreover, the arguments put forward in this recent literature turn on specific features of existing Marxist explanations—in particular, problems in the logic of explanation of collective phenomena and concerning functionalist explanation. These criticisms operate at a level of abstraction that is closer to the actual practice of social science, and consequently, they appear to have more immediate relevance to the methodology of social science.

This chapter has two aims. First, I will present somewhat briefly several grounds that have given new impetus to the call for microfoundations for Marxism, because I regard these as important contributions to the philosophy of social science. And second, I will offer a reconstruction of Marx's explanations in his political economy and his analysis of political behavior that will show that Marx's central explanations are in fact consistent with the call for microfoundations in social science: Marx's explanations depend on identifying the micropathways by which macrophenomena occur. Therefore I will hold that Marx's own conception of social explanation offers further support to recent doctrines of "micro-Marxism."

Grounds of "Micro-Marxism"

Let us begin with a brief discussion of several issues that have stimulated the "microfoundations" program. One important family of criticisms of macro-Marxism involves the status of functional explanations in social science. A functional explanation is one that explains a social institution in terms of its beneficial consequences for the social system as a whole or some important subsystem. Functional explanations have often played a prominent role in Marxist theory, and they have acquired fresh importance through the influential writings of Gerald Cohen, who argues that historical materialism depends crucially on functional explanations.[5] In considering the explanatory relationship between the forces and relations of production, Cohen writes: "To say that an economic structure *corresponds* to the achieved level of the productive forces means: the structure provides maximum scope for the fruitful use and development of the forces, and obtains *because* it provides such scope" (Cohen 1986, p. 221). According to this account, then, capitalist property relations existed in 1800 in England *because* they were suitable to the use and expansion of the productive forces available in 1800. As Cohen fully recognizes, this thesis invites some account of the causal mechanism establishing this functional fittedness, which he refers to as an "elaboration" of a functional explanation. But Cohen believes that functional relationships such as these may be identified and confirmed independently from an account of the underlying causal mechanism.[6]

Consider an example. Why did the American state enact the New Deal legislation of the 1930s, which gave rise to various forms of social welfare programs? This is a particular problem for Marxist political theory because social welfare programs appear to conflict with capitalist interests through higher taxation, a less pliable working class, and so forth. Various explanations are possible of this phenomenon, but a classical Marxist explanation goes along these lines: the New Deal was enacted because it contributed to the overall stability of capitalist property relations in a time of serious social and economic crisis.[7] This is a functional explanation. It explains an event (the enactment of the New Deal) in terms of its beneficial effects on the workings of a larger system (the capitalist social and economic system).

In a series of writings Jon Elster has argued that functional explanations are inherently incomplete in social science (though not in biology); they must be supplemented by detailed accounts of the social processes through which the needs of social and economic systems influence other social processes so as to elicit responses that satisfy those needs.[8] One problem is that it is almost always possible to come up with some beneficial consequences of a given institution; so in order to justify the judgment that the institution exists *because* of its beneficial consequences, we need to have an account of the mechanisms that created and reproduced the institution that shows how the needs of the system as a whole influenced the development of the institution. Second, there appears to be a problem of temporal order implicit in functional explanation; the benefit attributed to the effect (in virtue of which the event is to be explained) come *after* the event. We therefore need some account of a mechanism through which effects may influence their causes. In order for a functional relationship to count as explanatory, therefore, it is necessary to have some idea of the causal mechanisms that establish and preserve the functional relationship. In the case of functional explanations in biology, we have such an account: natural selection provides the causal mechanism through which an organism comes to be functionally adapted to its environment (through its evolutionary development in a relatively stable environment). But we do not have a comparable background mechanism to which we may refer in justifying functional explanations in social science.

In the New Deal case, therefore, we must identify a possible set of mechanisms through which the need for stability of the economic system might influence the process of legislation. Once we begin looking for such mechanisms, several different sorts of candidates appear possible. We might observe that legislators and lobbyists have economic interests; they can perceive that existing economic crisis endangers their economic interests; and they can undertake a process of legislation designed to dampen the social tensions that have been generating social and economic crisis. On this approach, the functional relation exists because it was designed to obtain by rational individual agents. Alternatively, we might imagine a social analogue to the principle of natural selection in biology: various legislative packages are put forward, and those that best serve the needs of the economic structure are selected over those that do not serve those needs well.[9] But without some account of the mechanisms that give rise to the functional relation, the functional explanation is unsatisfactory.

These remarks suggest a moral for macro-Marxism: in order to explain a phenomenon, it is not sufficient to demonstrate that the phenomenon has consequences beneficial for the economy or for the interests of a particular class. Rather, it is necessary to provide a detailed account of the micropathways by which the needs of the economy or the interests of a powerful class, are imposed upon other social phenomena so as to elicit beneficial consequences. Thus, the macroexplanation is insufficient unless it is accompanied by an analysis at the level of individual activity—a microanalysis—that reveals the mechanisms that give rise to the pattern to be explained.

Let us turn now to a second issue underlying the call for "microfoundations" for Marxian explanations: the problem of conflict of interest between a group as a whole and the individuals who compose the group. (John Roemer refers to this issue as the "aggregation gap." Roemer 1982, p. 516.) "Rational action" explanations depend upon identifying an individual's interests and then explaining this person's behavior as his or her rational attempt to best serve those interests. This model of explanation is often extended to attempt to explain collective behavior of groups as well: to explain the group's action, it is sufficient to identify the collective interests of the group and then account for the group's behavior as a rational attempt to further those interests. Thus, one might assert that members of the Organization of Petroleum Exporting Countries respect production quotas because it is in their collective interest to do so. However, Mancur Olson (1965) and others have made it plain that it is not sufficient to refer to collective interests in order to explain individual behavior. The members of a group may have a collective interest that all recognize, and yet it may not be the case that any member of the group has an individual interest in acting so as to further that collective interest (choosing instead to be a free rider). But some Marxist macro-explanations appear to depend precisely on such a conflation of collective interest and private interest. Thus, we cannot infer from the fact that a group has an interest in F (e.g., land reform) that the group will act in such a way as to bring about F.

This problem has been discussed in recent years in connection with the problem of revolutionary motivation. Why should Marxist theory expect that workers will contribute to a revolutionary overthrow of the capitalist system? The orthodox Marxist answer is that revolution is in the collective interest of the working class: "The proletarians have nothing to lose but their chains. They have a world to win" (Marx and Engels 1848, p. 98). Marx's view appears to be that it is in the rational self-interest of the working class as a whole to overthrow capitalism and establish socialism. There is an unseen difficulty here, however: that of getting from a circumstance's being in the collective interest of a group to its being in the individual interest of members of the group. In an important article, Allen Buchanan argues that "even if revolution is in the best interest of the proletariat, and even if every member of the proletariat realizes that this is so, so far as its members act rationally, this class will not achieve concerted revolutionary action" (Buchanan 1980, p. 268). The problem here is that it may be that the collective good to the working class as a whole is outweighted in each individual case by the cost of participation for the individual, with the result that the individual is rationally compelled to refrain from participation.

The point of Buchanan's arguments is not that collective action is impossible; on the contrary, examples of successful collective action are to be found throughout social life: industrial strikes, consumer boycotts, peasant rebellions, and others. These arguments are intended rather to show that collective action cannot be explained merely by alluding to collective interests shared by rational individuals. Thus, it is necessary to have a fine-grained account of the variety of motives that might lead individuals to act out of regard for collective interests rather than narrowly defined self-interest. Once we begin to look for appropriate microlevel mechanisms that might serve to bridge the aggregation gap, we will find a variety of possible solutions. First, we may discover that successful organizations have mechanisms for translating collective interests into individual interests—such as through in-process benefits to encourage cooperation or sanctions to discourage free riders.[10] Second, we may find that shared moral values within a group play an important causal role in collective action—such as, the values of solidarity, fairness, reciprocity, or unwillingness to be a free rider.[11] Third, we may turn to the findings of social psychologists, political scientists, and others who have studied problems having to do with motivation within small social groups. Here we will find that individuals are subject to a richer variety of motivations than a simple model of economic rationality would allow: ties of friendship and kinship, loyalty to the village or political party, religious motivations, or surges of "crowd psychology" (manifested in spontaneous peasant uprisings, for example).[12]

Consider an example. In his study of the Chinese Communist party's experience in northwest China, the Marxist sociologist Mark Selden (1971) attempted to explain the party's success in mobilizing peasant support. However, his account glosses over the distinction between collective and private peasant interests. Selden shows that peasant farmers in Shensi were in extremely harsh circumstances in the early twentieth century and that they had an objective shared interest in achieving land reform. Though subtle in his analysis of the military and political dimensions of this struggle, Selden generally treats the problem of peasant mobilization lightly, on the assumption that peasants recognize their collective interests and are prepared to act on them if a sufficiently plausible political option is presented to them. The Communist party presented such an option, and Selden assumes that peasant support followed more or less directly. This analysis, though, ignores the conflict between private and collective interests: peasants may find that individual strategies of survival (emigration, banditry, collaboration with the Japanese army) are more promising than strategies of collective self-defense.[13]

The lesson for Marxism to be gained from this application of collective action theory is simple but important: in reasoning about the behavior of a group—whether a social class, a labor union, or a group of oligopolists—it is not sufficient to consider only the collective interests of the group. Rather, it is necessary to provide a microanalysis of the motives, interests , and incentives that motivate the individuals of the group and an account of the various ways in which collective interests may shape individual interests or may be frustrated by individual interests. A macroexplanation concluding that a factory will be easily unionized because it is in the plain and evident interest of the workers as a whole that they be represented by a union fails on this criterion; it does not establish that individual workers will have sufficient motivation to contribute to this public good.

There is a final justification for the microfoundations thesis that derives from the weak nature of the regularities commonly available among social phenomena. The microfoundations thesis is not simply a general preference for intertheoretical reductionism—a preference sometimes found within the philosophy of biology and the philosophy of psychology as well. It depends rather on a specific feature of social causation and social regularities. In cognitive psychology there are strong regularities among cognitive phenomena, so it is credible to hold that various elements of the cognitive system are causally related to other elements without having specific knowledge about the neurophysiological mechanisms that underlie them.[14] In social science, however, we do not often find the strong types of regularities and laws that would make us confident in the causal connectedness of social phenomena; instead, we find laws of tendency and exception-laden regularities.[15] It is therefore mandatory to drop from the weak regularities at the social level to an analysis of underlying causal mechanisms at the individual level if we are to be able to discern causal relations among social phenomena at all. In brief, social causation always and unavoidably works through structured individual action, and causal relations among social phenomena can be established only through analysis of the latter because of the weakness of the causal regularities at the social level.

Phenomena of popular politics illustrate this point. Study of a variety of examples of exploitative class societies shows that we do not find the regularity that classical Marxism would predict at the macrolevel; exploited groups eventually come to support popular movements aimed at assaulting the class system. However, when we move to a lower-level description of the phenomena of popular politics—one that analyzes group behavior in terms of the specific class arrangements in which it finds itself—the variable character of individual political motivation, the forms of political culture available to the group, and the forms of organization and leadership available to the group—we find that there are regularities that emerge that divide the phenomena into a number of subclasses:

> Exploited groups with strong political cultures and ample organizational resources tend to be politically active, tenacious, and effective.
> Exploited groups with weak political cultures and ample organizational resources tend to be only moderately active, vacillatory, and ineffective.
> Exploited groups with strong political cultures and no organizational resources tend to be vigorously active, tenacious, but ineffective.
> Exploited groups with weak political cultures and no organizational resources tend to be inactive, vacillatory, and collectively ineffective.

It is clear that the search for microfoundations is even more critical in the case where no apparent social regularities exist at all than in the case where there are discernible tendencies or regularities at the social level. In the latter case, the demand for microfoundations stems from a need to bolster an explanation already available at the macro-level; but in the former, we need to identify the microfoundations of the social phenomena in question if we are even to be in a position to identify the macro-level regularities at all. Attention to some of the concrete mechanisms through which political behavior is shaped, then, should permit the construction of a more empirically adequate theory of popular politics from a Marxian perspective.

The Microfoundations Thesis

These arguments lead, then, to a central tenet of analytical Marxism: that macroexplanations in social science require microfoundations.[16] This doctrine maintains that macroexplanations of social phenomena must be supported by an account of the mechanisms at the individual level through which the postulated social processes work. More specifically, the thesis holds that an assertion of an explanatory relationship at the social level (causal, functional, structural) must be supplemented by knowledge about what it is about the local circumstances of the typical individual that leads him or her to act in such a way as to bring about this relationship and knowledge of the aggregative processes that lead from individual actions of that sort to an explanatory social relationship of this sort.[17] This doctrine may be put in a weak and a strong version: weakly, social explanations must be compatible with there being microfoundations of the postulated social regularities, which may, however, be entirely unknown; and more strongly, social explanations must be explicitly grounded on an account of the microfoundations that produce them. I will assume the stronger version: that we must have at least an approximate idea of the underlying mechanisms at the individual level if we are to have a credible hypothesis about explanatory social regularities at all. A putative explanation couched at the level of high-level social factors whose underlying individual-level mechanisms are entirely unknown is no explanation at all.

The objection being advanced to macro-Marxism here is grounded in a recognition of the point that there are no supraindividual actors in a society. Classes are not corporate entities that determine their interests collectively and act in concert; states are not rational, deliberative entities standing over and above the persons who occupy office; and technologies and economic systems do not literally seek out social institutions that best suit them. Rather, all social entities act through the individual men and women who occupy roles within them. If we are concerned to know why a given social entity undergoes a particular process of change, we need to have specific information about the motives and opportunities defining the positions of the persons who make up that entity. Explanations of social phenomena are unsatisfactory unless they identify the microprocesses through which explanandum and explanans are linked together.[18] In this way, the call for microfoundations represents a variant of methodological individualism. Unlike classical arguments for methodological individualism, however, this stricture on explanation does not rest upon a general ontological thesis about the relation between individuals and social entities. Further, it is not merely an application of a general program of reductionism in science, according to which higher-level laws must be grounded in lower-level laws. Rather, the microfoundational requirement depends on specific problems that arise in two common forms of macroexplanation: functional explanations and collective action explanations. One consequence of this point is that there may be legitimate forms of macroexplanations in social science that are not subject to these specific criticisms, and in that case microfoundational arguments would be silent.

Once we accept that macroexplanations require microfoundations, we must next ask what types of individual-level processes we should look for, and here there are two broad families of answers: rational-action models and social-psychology motivational models. The first approach attempts to explain a given social process as the aggregate result of large numbers of individuals pursuing individually rational strategies. The second approach attempts to explain the social phenomenon as the complex outcome of a variety of motives, rational and nonrational, that propel individual action. These observations suggest two separate issues to be considered in connection with Marxist social science. First, is Marx's system on the whole consistent with the requirement of microfoundations for macroexplanations? And second, are the sorts of microprocesses to which Marx generally alludes grounded in a rational-choice model, or do his accounts of individual behavior refer to nonrational types of motives? I hold that Marx's most characteristic explanations in *Capital* are rational-choice explanations based on individual self-interest. Thus, Marx's political economy is microfoundational and is grounded in an analysis of rational choice. In the final section, however, I will suggest that Marx's account of political behavior and class conflict is somewhat more complex than this, in that it takes account of the workings of ideology and class consciousness, factors that interfere with calculations based on narrow self-interest.

Marx's Explanatory Paradigm

Let us now turn now to Marx's own explanatory practice in *Capital* in the light of these points. Consider an example. In *Capital* Marx attempts to explain one of the most central characteristics of the capitalist mode of production: its revolutionary tendency toward technical innovation. Whereas classical slavery and medieval feudalism made use of technologies that scarcely changed over centuries, capitalism since its inception has witnessed a rapid stream of new technologies, each more productive and efficient than its predecessor. Why does capitalism display this dramatic tendency toward technical innovation?

There is a tempting macroexplanation of this tendency: increasing productivity in wage-goods industries leads to a fall in the wage, which in turn (other things being equal) leads to an increase in the average rate of profit within the capitalist economy as a whole. Consequently, one might propose that productivity-enhancing technical innovations occur within the capitalist economy *because* they increase the average rate of profit by cheapening the wage.[19] This is a functional explanation of the tendency; it explains the macropattern in terms of its beneficial consequences for another macropattern (increase in the average rate of profit). However, this explanation lurches into the aggregation gap immediately, for the fact that productivity-enhancing innovations would lead to a collective good for the capitalist class does not suffice to show that these innovations will come into being; rather, it is necessary to show that the particular circumstances, interests, and opportunities of representative capitalists would lead them to adopt such innovations.[20] In other words, it is necessary to provide a microfoundation for this macroexplanation.

Significantly, Marx's own explanation rests upon just such a microanalysis.[21] Marx's explanation of this tendency is not based on a consideration of the collective interests of the capitalist class but rather on the fact that capitalists are rational and goal directed, and their fundamental purpose is to organize their enterprises so as to create the greatest amount of profit. Marx considers the result for the individual capitalist of introducing a labor-saving technical innovation into the manufacture of shirts.

> The individual value of these articles is now below their social value.... The real value of a commodity, however, is not its individual, but its social value.... If, therefore, the capitalist who applies the new method sells his commodity at its social value of one shilling, he sells it for 3d. above its individual value, and thus he realizes an extra surplus value of 3d.... Hence, quite independently of [whether the commodity is a wage good], there is a motive for each individual capitalist to cheapen his commodities by increasing the productivity of labor.[22]

The profitability of an enterprise within a competitive market is determined, on the one hand, by the technical characteristics of the process of production employed by the enterprise itself—its intrinsic efficiency or productivity—and on the other hand, the average productivity of the industry as a whole which is involved in the production of the good. The price of the good is determined by the average productivity of firms in the given industry. Firms that produce the good at the average rate of productivity earn the average rate of profit; those that produce with less than average efficiency earn a lower rate of profit; and those that produce with higher than average productivity earn a higher rate of profit. Consequently, each capitalist has an immediate incentive to seek out technical innovations that will allow him or her to produce the good at higher than average productivity (or lower than average net cost of production). Thus, each individual capitalist is under a positive incentive toward discovering technical innovations that increase the productivity of the enterprise: that of achieving superprofits (until the new technique spreads throughout the industry as a whole). Given this incentive, each capitalist will energetically seek out ever-new techniques of production with higher productivity. The collective result of this incentive for individual capitalists is a powerful tendency within capitalism toward technical innovation.

This explanation, unlike those considered above, is not a functional explanation, and it does not fall into the aggregation gap. Marx does not explain the economy-wide tendency toward technical innovation in terms of the aggregate benefits that this tendency confers on the economy as a whole or on the capitalist class; instead, his explanation is founded on an analysis of the microcircumstances of the representative capitalist and an account of the aggregate consequences that the resulting individual choices have for the development of capitalism as a whole.

Let us turn now to a second example: Marx's explanation of the formation of a pool of unemployed workers in volume 1 of *Capital*. Marx takes it as an empirical fact that capitalism creates such a fund of workers (which he refers to as the "industrial reserve army"), even in times of prosperity, and the empirical evidence of the past century has tended to confirm this judgment. In this case, a functional explanation of this fact is even more attractive in the previous case, for the existence of a pool of unemployed workers serves an obvious beneficial function from the point of view of the capitalist class: it serves to depress the wage, since employed workers face competition for jobs from unemployed workers. And it serves to reduce the militancy of the working class, for much the same reason; it reduces the bargaining position of the proletariat. One might hold, therefore, that the industrial reserve army exists within capitalism because it confers these beneficial consequences on the economy.[23]

This is not Marx's strategy, however. Instead, Marx considers the incentives and opportunities that define the position of the representative capitalist in order to discover what leads to a relative surplus of workers.[24] Marx finds that there are two relevant incentives at the level of the individual capitalist. Each capitalist has an incentive to reduce his or her own labor costs and an incentive to reduce dependence on the available fund of workers in order to improve his or her own bargaining position with respect to them. Second, Marx observes that the capitalist has the opportunity to discover labor-saving innovations (through research and development) and the capacity to introduce these innovations (through his or her rights as property owner). The decision problem is now easily solved: the capitalist will pursue a strategy aimed at discovering and introducing labor-saving innovations into the production process. The aggregate consequence of this individual tendency, however, is that demand for labor tends to fall periodically; in other words, there is a tendency within capitalism toward the creation of a pool of unemployed and underemployed workers. Once again we find that Marx avoids making use of available functional explanations, preferring instead to explain the macrophenomenon in question in terms of the circumstances of choice within which the representative capitalist acts.

I believe that these examples are representative of Marx's explanatory practice throughout his political economy. If this is legitimate, then Marx's economic explanations typically involve two types of analysis, each compatible with the requirement that macroexplanations be provided with microfoundations: a rational-choice component and an aggregative-explanation component. First, Marx's explanations depend on analysis of the circumstances of choice of rational individuals. Marx identifies a set of motivational factors and constraints on action for a representative capitalist and then tries to work out an account of the most rational strategies available to the capitalist in these circumstances of choice. In the example of technical innovation, Marx isolates the dominant motive that guides the hypothetical capitalist—the desire to increase the amount of surplus value realized within his or her unit of production. The capitalist then canvasses available strategies for increasing the amount of surplus value, such as intensification of the labor process or productivity-enhancing technical innovation. And he or she holds that the capitalist, as a rational agent with the specified motives, will arrive at one such strategy and organize the unit of production accordingly by seeking out technical innovations that reduce the cost of production.

A second component of Marx's explanations is an attempt to work out the consequences for the system as a whole of the forms of activity attributed to the hypothetical capitalist at the preceding stage of analysis. That is, every capitalist is subject to essentially the same motives and is in roughly the same set of institutional constraints; on these assumptions, what sorts of collective patterns are likely to emerge as a result of a large number of actors making the choices attributed to the hypothetical actor? In some cases, this analysis is quite simple. If every capitalist is striving to discover productivity-enhancing technical innovations, then the collective result will be an increase in the productivity of labor in the economy as a whole. This increase will lead to a fall in the value of the wage, which will, in turn, lead to an increase in the rate of surplus value. Other aggregate consequences are less obvious. For example, each individual capitalist introduces technical innovations in order to increase the firm's rate of profit. The collective consequence of this incentive, however, is a fall in the average rate of profit (due to the attendant rise in the organic composition of capital). This form of analysis may be described as an aggregative explanation: one that attempts to discover the aggregate consequences of a particular structure of individual action.[25]

Here, then, is the model of explanation that appears to be at work in *Capital*. The capitalist economy is defined by a set of social relations of production (property relations). These relations define relatively clear circumstances of choice for the various representative actors (the capitalist, the worker, the official, the financier). These circumstances are both motivational and conditioning: they establish what each party's interests are, determine the means available to each, and lay down constraints on action that limit choice. The problem of explanation that confronts Marx is this: for a given characteristic of the capitalist mode of production (e.g., the falling rate of profit), to provide a demonstration that this characteristic follows from Marx's account of the defining institutions of capitalism through reasoning about rational behavior within the circumstances of choice which they define. Thus, Marx's basic pattern of explanation may be summarized in the following way: a given feature of capitalism occurs (e.g., the tendency toward a falling rate of profit), because capitalists are rational and are subject to a particular set of incentives, prohibitions, and opportunities. When capitalists collectively pursue the optimal individual strategies corresponding to these incentives, prohibitions, and opportunities, the explanans emerges as the expression or consequence of the resulting collective behavior.

Marxist Theories of Political Behavior

This account of Marx's explanatory paradigm in *Capital* gives central place to the assumption of individual rationality in explaining action. Up to this point, then, my analysis supports the rational-choice Marxism defended by Elster, Roemer, and others. Marx's shared legacy with classical political economy is visible here, for Marx derives the central laws of motion of capitalism from a model of rational capitalist decision making within the institutional arrangements of capitalism. However, Marxism is not solely concerned with economic behavior. It is also concerned with the politics of class: the success or failure of working-class organizational efforts, the occurrence of collective action in defense of class interests, the logic of working-class electoral politics, and the occurrence of revolution. In his own writings Marx attempted to analyze and explain a variety of political phenomena:

The struggle by the English bourgeoisie for control of the state in the eighteenth century.
The forms that working-class political action took in 1948 in France.
The reasons for Napoleon III's overwhelming electoral victory in 1849.
The goals and strategies of the leadership of the Paris Commune.
The efforts by organizations of the English working class to achieve the Ten Hours Bill.

What assumptions underlie Marx's analysis of the politics of class? To what extent does the rational-choice model carry over to Marx's extra-economic explanations—in particular, his explanations of political behavior?

It is not difficult to extend the rational-choice model to a theory of political behavior. This is a much-employed model in political science today, described under the umbrella of "public choice theory."[26] Such an approach postulates that individuals' political behavior is a calculated attempt to further a given set of individual interests—income, security, prestige, office, or something else. Dennis Mueller describes this approach in these terms: "Public choice can be defined as the economic study of nonmarket decision-making, or, simply the application of economics to political science. The basic behavioral postulate of public choice, as for economics, is that man is an egoistic, rational, utility maximizer."[27] One might suppose that such an approach is unavoidably bourgeois, depending upon a materialistic egoism characteristic of market society. However, Marx's own emphasis on class conflict based on material interests suggests a rational-choice model of political behavior: classes have objective material interests that are in opposition to one another, and members of classes are disposed to act so as to defend and extend those interests. And in fact, the rational-choice model has been effectively applied to political behavior from a Marxist point of view in the work of Adam Przeworski, who analyzes workers' political behavior in just these terms: "I assume, therefore, that workers under capitalism have an interest in improving their material conditions. The question is whether the pursuit of this interest, and only of this interest, would necessarily lead workers to opt for socialism as a superior system for satisfying material needs" (Przeworski 1984, p. 164).

Marx's theory of political behavior, like his theory of capitalist behavior, is ultimately grounded in a theory of individual rationality. But this theory is somewhat more comprehensive than public choice theory in that it describes the resources needed to permit groups to overcome atomization and privatization of action. Roughly, agents as members of classes behave in ways calculated to advance their perceived material interests; these interests are perceived as class interests (interests shared with other members of the class), and class organizations and features of class consciousness permit classes to overcome implicit conflicts of interest between private interest and class interest. This model may be extended to a variety of political phenomena: the concerted activity of members of the bourgeoisie to control the state, the effort of workers to win the right to form unions, or the eruption of rural social violence, for example. In each case, a political movement is analyzed in terms of the deliberate, calculated efforts of its supporters to forward a set of shared material interests.

So far, I have analyzed Marx's political theory in terms quite compatible with public choice theory, a somewhat surprising finding, given the proximity of that paradigm to neoclassical economic theory and a frequently conservative set of political views. However, Marx's theory of political behavior is substantially richer than the public choice paradigm because it contains motivational resources over and above narrow calculation of self-interest. In addition to material class interests, Marx refers to two other elements of political motivation: ideology and class consciousness. The former concept is used to explain apparent failures to achieve rational collective action, while the latter explains how conflicts between private and group interest are surmounted. Ideology sometimes leads individuals to act in ways that are contrary to their objective interests, while class consciousness leads individuals to act in ways that conform to their shared group interests at the possible expense of their individual interests.

Consider first the concept of ideology. How does the theory of ideology contribute to a Marxist theory of consciousness and political motivation? A clue may be found in the associated idea of "false consciousness": an ideology affects the worker's political behavior by instilling false *beliefs* and self-defeating *values* in the worker. First, an ideology may instill a set of values or preferences that propel individual behavior in ways that are contrary to the individual's objective material interests. Thus, German and French workers were induced by appeals to patriotism and national identity to support their national governments in World War I, contrary to their objective shared class interests.[28] This political behavior may be analyzed in this way: French workers place a high value on the national interests of France; they believe that Germany is waging aggressive war against France, and they act to support the French military effort against Germany. This behavior may be an example of rational action in pursuit of misguided ends: ends that are objectively contrary to the worker's class interests.

Second, ideologies modify rational individual action by instilling a set of false beliefs about the causal properties of the social world and about how existing arrangements affect one's objective interests. The belief that capitalism provides opportunities for advancement through education and individual initiative, for example, may lead workers to believe that a strategy of individual striving better serves their material interests than a strategy of collective action. The belief that inequalities in the United States now are less extensive than in 1940 may lead workers to believe that the direction of change within capitalism favors their interests. And the belief that the police powers of the capitalist state are overwhelmingly competent may lead them to believe that collective action is dangerous and ineffective. Each of these beliefs, when factored into a rational decision-making process in the circumstances in which the worker finds himself or herself, has the effect of inhibiting militant collective action.

The theory of ideology, then, can be fairly directly assimilated to a rational-choice model of political action. Ideologies modify the political behavior of individuals within a class society by instilling false beliefs about the environment of choice and by modifying the value system of the worker away from the worker's objective material interests. Rational individuals, operating under the grip of an ideology, will undertake actions that are contrary to their objective material interests but are fully rational given the false beliefs they hold about the social world they inhabit and their mistaken assumptions about their real interests and values. An ideology is an effective instrument, then, in shaping political behavior within a class system; it induces members of exploited classes to refrain from political action directed at overthrowing the class system. This is indeed Marx's use of the concept: an ideology functions as an instrument of class conflict, permitting a dominant class to manipulate the political behavior of subordinate classes.

The concept of class consciousness functions somewhat differently in Marx's writings. The term refers to a set of motivations, beliefs, values, and the like that are specific and distinctive for a given class (peasantry, proletariat, petty bourgeoisie). Marx holds that these motivational factors serve to bind together the members of a class and to facilitate their collective activities. Class consciousness takes the form of such motives as loyalty to other members of one's class, solidarity with partners in a political struggle, and commitment to a future social order in which the interests of one's class are better served. Marx describes such a complex of psychological properties, and their social foundation, in *The Eighteenth Brumaire*:

> A whole superstructure of different and specifically formed feelings, illusions, modes of thought and views of life arises on the basis of the different forms of property, of the social conditions of existence. The whole class creates and forms these out of its material foundations and the corresponding social relations. The single individual, who derives these feeling, etc. through tradition and upbringing, may well imagine that they form the real determinants and the starting-point of his activity. (Marx, 1974 [1852], pp. 173–74)

Thus, a class is supposed to develop its own conscious identity of itself as a class. Insofar as a group of people who constitute a structurally defined class fails to acquire such attitudes, Marx denies that the group is a class in the full sense at all (a class-for-itself as well as-in-itself). Thus, in his famous view of peasants as a "sack of potatoes," Marx writes:

> In so far as millions of families live under economic conditions of existence that separate their mode of life, their interests and their cultural formation from those of the other classes, they form a class. In so far as these small peasant proprietors are merely connected on a local basis, and the identity of their interests fails to produce a feeling of community, national links, or a political organization, they do not form a class. (Marx, 1974 [1852], pp. 239)

Here Marx's point is not that the peasantry fails to constitute a class in the objective sense—a group of persons sharing a distinct position within the property and production relations—but rather that the conditions of life that characterize peasant existence systematically undermine the emergence of collective action and political consciousness. In other words, peasants fail to arrive at a state of class consciousness.[29] (Michael Taylor [1986, p. 7] argues that this conception of peasant politics is badly off the mark. "In all [peasant societies] there was cooperation amongst the peasants in the agricultural work which dominated their lives and usually communal regulation of the use of communal land." Taylor reminds us that peasants have historically been more, not less, capable of collective action in rebellion than workers.

Marx does not provide an extensive analysis of the process through which class consciousness emerges, even within capitalism, but he suggests that it takes form through a historical process of class struggle. As workers or peasants come to identify their shared interests and gain experience working together to defend their shared interests, they develop concrete ties within their political groups that provide motivational resources for future collective action. Thus Marx writes in the *Communist Manifesto* in these terms: "This organization of the proletarians into a class, and consequently into a political party, is continually being upset again by the competition between the workers themselves. But it ever rises up again, stronger, firmer, mightier" (Marx and Engels 1974 [1848], p. 76). Finally, a class's accurate perception of its material interests depends upon a crucial historical development: the more complete development of the economic structure that defines these interests and the disappearance of the vestiges of old systems of production. In *The Class Struggles in France*, Marx diagnoses the Parisian proletariat in 1848 as immature and deceived about its true material interests (Marx and Engels 1974 [1850], pp. 45 ff.).

Marx's own writings provide only a schematic notion of this broader theory of political motivation. However, twentieth-century Marxist historians—such as E. P. Thompson, Eric Hobsbawm, and Eugene Genovese—have made fruitful efforts toward analyzing this side of the basis of underclass political behavior.[30] Thompson refers to what he calls a "moral economy" of class: a set of shared values and commitments that lead persons to behavior furthering their shared group interests even at the expense of their individual interests. And Thompson analyzes in great detail the processes through which a group of people can, over an extended period of time, develop the intragroup strands of commitment that give rise to loyalty and solidarity. Similarly, Hobsbawm provides a detailed account of some of the processes through which European workers and peasants established political communities that gave them the collective capacity to struggle against employers and landlords. And Genovese offers a fine-grained account of slave culture in the United States that is intended to show how slaves used local value systems to protect their interests as far as was possible within the context of the slave system.

The central role of class consciousness in Marx's political theory is to explain the moral capacity of members of exploited classes to join in prolonged, risky struggles in defense of their material interests. The concept of class consciousness thus functions as a bridge between individual interests and collective interests in classical Marxist analysis of political behavior. It gives workers effective motivation to undertake actions and strategies that favor their group interests, and it gives them motivational resources allowing them to persist in these strategies even in the face of risk and deprivation (that is, in circumstances where the collective strategy imposes costs on the individual's interests). This treatment of class consciousness shows a sensitivity to the point that political behavior is often driven by a set of motives that are richer than a narrow calculus of self-interest. However, this assumption does not run contrary to the main insights of contemporary rational-choice theory. A. K. Sen's critique of the narrow construct of economic rationality is instructive in this context. For Sen argues (in "Rational Fools" and elsewhere) that narrow economic rationality (egoism) is an impossible basis for any form of social cooperation. Instead, Sen holds that an adequate account of practical reason must take into account the workings of commitments, as well as self-interest: loyalties, moral values, adherence to a shared set of rules, and solidarity to one's fellows in a joint enterprise.[31] On Sen's account, practical rationality involves both calculation of individual interest and responsiveness to one's antecedent and ongoing commitments; the action selected is a complex function of both sorts of considerations. The idea of class consciousness is one way of referring to the motivations that work against narrow self-interest. Significantly, however, Sen's criticisms of the concept of narrow economic rationality do not invalidate the rational-choice paradigm; they rather militate for a somewhat richer conception of practical rationality, one that permits rational decision makers to take account of interests and commitments that are more comprehensive than narrow self-interest.[32]

Neither ideology nor class consciousness, then, requires the abandonment of a rational-choice analysis of Marx's central explanatory paradigm. This paradigm represents an idealization of individual behavior, of course, but subject to the terms of the idealization, individual capitalists behave rationally within the context of the capitalist economy. And members of classes behave rationally in defense of their material interests, with two provisos: dominant ideologies may give them false beliefs about both their objective interests and the causal properties of the social system they inhabit, and strands of class consciousness may give them the motivational resources needed to transcend implicit conflicts between individual and collective interests.

This analysis of Marx's political explanations permits us to provide further support for the microfoundations thesis, for we have found that the grasp of Marxist social science is substantively strengthened by attention to the microfoundations of popular politics. We have seen that the macrolevel predictions concerning popular politics of classical Marxism are not borne out; there is no strict or observable correlation between exploitative economic structures and vigorous popular protest and revolution. However, we have also seen that an account that provides analysis of the underlying mechanics of popular politics—including the micromechanisms of class as well as the features of political culture, organization, and leadership that facilitate collective action—constitutes an analysis of popular politics that assigns class and exploitation a central causal role. This level of analysis permits us to discern regularities among the chaos of the phenomena of popular politics that are not visible if we pay attention only to macrolevel factors. This constitutes, then, a powerful new reason for adopting a microfoundational approach to Marxist social science.

Conclusion

The theory of explanation I have attributed to Marx offers support to the view that Marxism depends on a microanalysis of social processes through consideration of the structured activities of rational individuals. Marx's explanatory purpose in his economics is to explain macrolevel social facts like the falling rate of profit, the tendency toward technical innovation, and the like. And his theory of politics is designed to provide a basis for analyzing such mass collective action as strikes or rebellions. But his treatment of these problems depends on a consideration of the factors that constrain and motivate individual activity and the organizational and normative resources that exist to facilitate collective action. Marx's explanatory paradigm thus indicates that Marx accepts the view that the ultimate source of social change is the active individual within specific social relations. And this conclusion suggests in turn that there is no fundamental incompatibility between Marxism and an enlightened individualism.[33]

Notes

1. Debate over methodological individualism has a long history, extending back at least to Mill's discussion of social science in his *System of Logic* and forwarded by Karl Popper and J. W. N. Watkins in the 1950s. Especially important is J. W. N. Watkins, "Methodological Individualism and Social Tendencies," in Brodbeck (1968).
2. "Methodological individualism is the view that all institutions, behavioral patterns, and social processes can in principle be explained in terms of individuals only: their actions, properties, and relations. It is a form of reductionism, which is to say that it enjoins us to explain complex phenomena in terms of their simpler components" (Elster 1986, p. 22). Note the ambiguity, however, between "Social processes can only be explained in terms of individuals" and "It is possible to explain social processes in terms that only refer to individuals." Elster plainly means the former.

3. Analytical Marxism represents a striking new development in Marxist thought: the marriage of some of the foundational ideas of classical Marxism with the methods and tools of rational-choice theory. Important contributions to this approach to Marxism include Roemer (1982b); Elster (1985); Brenner (1976); and Przeworski (1985).
4. Ralph Miliband provides this sort of detailed account of the behavior of the capitalist state and the pathways by which state policy comes to reflect capitalist interests throughout his work. See particularly *Capitalist Democracy in Britain* (1982) and *The State in Capitalist Society* (1969).
5. *Karl Marx's Theory of History: A Defence* (1978, ch. 9, 10). Several important criticisms of this view occur in a special number of *Theory and Society* (1982): John Elster, Marxism, Functionalism, and Game Theory," G. A. Cohen, "Reply to Elster," John Roemer, "Methodological Individualism and Deductive Marxism," and Philippe Van Parijs, "Functionalist Marxism Rehabilitated."
6. Cohen (1978, pp. 285ff., 1982, pp. 33–35, 50ff.).
7. Consider Szymanski's (1978, p. 271) brief account of the causes of New Deal legislation.
8. In this vein Elster (1982) writes that "by assimilating the principles of functionalist sociology ... Marxist social analysis has acquired an apparently powerful theory that in fact encourages lazy and frictionless thinking" (p. 453). See also Elster (1983, chap. 2).
9. Elster (1983, pp. 141ff.) provides an example of such an account based on an analogy with natural selection.
10. This is the approach Buchanan takes (1980, pp. 273ff.). This approach remains within the "rational action" model and attempts to show that in certain cases organizations can make it prudentially rational for their members to cooperate.
11. Nagel (1970) argues for a conception of practical reason according to which altruistic behavior is fully rational. Sen (1982) criticizes the maximizing conception of economic rationality and self-interest because it cannot take account of normative commitments that may rationally limit the range of accessible alternatives. And Elster (1982, pp. 468ff.) argues for a conception of "conditional altruism" as a possible solution of repeated problems of collective action.
12. Consider, for example, Scott's (1976) research on peasant behavior and the circumstances that led to collective action, resistance, and rebellion within Southeast Asian peasant communities. Scott argues that there is a distinctive "moral economy of subsistence" that underlies patterns of peasant resistance and acquiescence and that bonds of solidarity, kinship, and shared moral judgments play a central role in explaining peasant behavior in the face of changes in systems of land tenure, rent, or taxation. Cf. Popkin (1979).
13. For a work that devotes extensive attention to the problem, how did the CCP mobilize peasant support in the face of conflict between private, local, and class interests within the peasantry, see Chen (1986).
14. Fodor (1979, 1980) provides an important discussion of some of the relations between levels of psychological theory and brain science.
15. For developed arguments to this effect in connection with Marx's economic analysis, see Little (1986, pp. 24–29, 167–72).
16. In a representative vein, Roemer (1982b) writes that " class analysis must have individualist foundations.... Class analysis requires microfoundations at the level of the individual to explain why and when classes are the relevant unit of analysis" (p. 513).
17. We may refer to explanations of this type as "aggregative explanations." Schelling (1978) provides a developed treatment and numerous examples of this model of social explanation.
18. "On this account, what distinguishes Marxist from non-Marxist social science is not the postulate of methodological individualism, but the level of aggregation deduced as applicable in social analysis" (Roemer 1982, p. 518).
19. Sweezy (1968) appears to suggest this explanation. And O'Connor (1973, p. 103) offers an account of technical change that also depends upon the aggregate effects of increasing efficiency.
20. Brenner's (1982, pp. 17ff.) discussion of medieval agrarian relations illustrates just the opposite result in medieval society. He shows that technical innovations were available that could not be implemented because no social group had both the power and the incentive to introduce the innovation.
21. For Marx's explanation of technical change, see Marx (1977 [1867]), pp. 429ff.).
22. Ibid. (p. 435).
23. Sweezy (1968) appears to take this view: "In Marx's theory, however, the system of production includes both Industrial Employment and the Reserve Army.... The fact remains that [the capitalist system] contains within itself a mechanism for regulating the wage level and hence for maintaining profits" (p. 91).

24. For Marx's discussion of the industrial reserve army, see Marx (1977, pp. 781–94).
25. Schelling (1978) provides a wide range of examples of this sort of analysis.
26. See, for example, Frohlich, Oppenheimer and Young (1971). For an application of this model from a Marxist point of view, see Przeworski (1985). And for an application of this model to peasant behavior, see Popkin (1979).
27. Mueller (1976, p. 395).
28. See Ferro (1973) for an analysis of the role of patriotism in motivating European workers to support their national governments in their war policies.
29. Croix (1984) refers to this passage in similar terms; he takes Marx's position to mean not that peasant are not a class, but that they lack the features of political activity and consciousness that constitute a class-for-itself.
30. Thompson (1963); Hobsbawm (1959); Genovese (1974).
31. Sen (1982b). See also Sen (1973, 1982a).
32. Parfit's (1984) analysis of the prisoners' dilemma is an example of a current effort along these lines.
33. For a sharply opposed view of the relevance of methodological individualism to Marxism, see Miller (1978, pp. 387–414).

References

Block, Ned, ed. 1980. *Readings in Philosophy of Psychology*. Vol. 1. Cambridge: Harvard University Press.

Brenner, Robert. 1976. "Agrarian Class Structure and Economic Development in Pre-Industrial Europe." *Past and Present* 70: 30–75.

Brenner, Robert. 1982. "The Agrarian Roots of European Capitalism." *Past and Present* 97: 16–113.

Brodbeck, May, ed. 1968. *Readings in the Philosophy of the Social Sciences*. New York: Macmillan.

Buchanan, Allen. 1980. "Revolutionary Motivation and Rationality." In *Marx, Justice, and History*. In Cohen (1978).

Chen, Yung-fa. 1986. *Making Revolution: The Communist Movement in Eastern and Central China, 1937–1945*. Berkeley: University of California Press.

Cohen, G. A. 1978. *Karl Marx's Theory of History: A Defense*. Princeton: Princetion University Press.

Cohen, G. A. 1982. "Functional Explanation, Consequence Explanation, and Marxism." *Inquiry* 25: 27–56.

Cohen, G. A. 1986. "Marxism and functional explanation." In *Analytical Marxism*. See J. Roemer. Cambridge: Cambridge University Press.

Cohen, Marshall, ed. 1980. *Marx, Justice, and History*. Princeton: Princeton University Press.

Croix, Geoffrey De Ste. 1984. "Class in Marx's Conception of History." *New Left Review* 146.

Elster, Jon. 1982. "Marxism, Functionalism, and Game Theory." *Theory and Society* 11: 453–482.

Elster, Jon. 1983. *Explaining Technical Change*. Cambridge: Cambridge University Press.

Elster, Jon. 1985. *Making Sense of Marx*. Cambridge: Cambridge University Press.

Elster, Jon. 1986. "Three Challenges to Class." In *Analytical Marxism*. See Roemer (1982b).

Fernbach, David, ed. 1974. *The Revolutions of 1848: Political Writings*. Vol. 1. New York: Vintage.

Ferro, Marc. 1973. *The Great War, 1914–1918*. Boston: Routledge & Kegan Paul.

Fodor, Jerry. 1979. *The Language of Thought*. Cambridge: Harvard University Press.

Fodor, Jerry. 1980. "Special Sciences, or the Disunity of Science as a Working Hypothesis." In *Readings in Philosophy of Psychology*, Vol. 1. See N. Block.

Frohlich, Norman, Joe Oppenheimer, and Oran R. Young. 1971. *Political Leadership and Collective Goods*. Princeton: Princeton University Press.

Genovese, Eugene. 1974. *Roll, Jordan, Roll*. New York: Pantheon.

Hobsbawm, E. J. 1959. *Primitive Rebels: Studies in Archaic Forms of Social Movement in the 19th and 20th Centuries*. New York: Praeger.

Little, Daniel. 1986. *The Scientific Marx*. Minneapolis: University of Minnesota Press.

Marx, Karl. 1977 [1867]. *Capital*. Vol. 1. New York: Vintage.

Marx, Karl. and Frederick Engels. 1974 [1848]. The Communist Manifesto. In *The Revolutions of 1848: Political Writings*. Vol. 1. See Fernbach (1974).

Miliband, Ralph. 1969. *The State in Capitalist Society*. New York: Basic Books.

Miliband, Ralph. 1982. *Capitalist Democracy in Britain*. Oxford: Oxford University Press.

Mill, John Stuart. 1950. *Philosophy of Scientific Method*. New York: Hafner.

Miller, Richard. 1978. "Methodological Individualism and Social Explanation." *Philosophy of Science* 45: 387–414.

Mueller, Dennis C. 1976. "Public Choice: A Survey." *Journal of Economic Literature* 14(2): 395–433.
Nagel, Thomas. 1970. *The Possibility of Altruism*. Oxford: Oxford University Press.
O'Connor, James. 1973. *The Fiscal Crisis of the State*. New York: St. Martin's Press.
Olson, Mancur. 1965. *The Logic of Collective Action: Public Goods and the Theory of Groups*. Cambridge: Harvard University Press.
Parfit, Derek. 1984. *Reasons and Persons*. Oxford: Oxford University Press.
Popkin, Samuel L. 1979. *The Rational Peasant*. Berkeley: University of California Press.
Przeworski, Adam. 1985. *Capitalism and Social Democraacy*. Cambridge: Cambridge University Press.
Przeworski, Adam. 1986. "Material Interests, Class Compromise, and Socialism." In Roemer (1986).
Roemer, John. 1982a. "Methodological Individualism and Deductive Marxism." *Theory and Society* 11: 513–520.
Roemer, John. 1982b. *A General Theory of Exploitation and Class*. Cambridge: Harvard University Press.
Roemer, John, ed. 1986. *Analytical Marxism*. Cambridge: Cambridge University Press.
Schelling, Thomas C. 1978. *Micromotives and Macrobehavior*. New York: Norton.
Scott, James C. 1976. *The Moral Economy of the Peasant*. New Haven: Yale University Press.
Selden, Mark. 1971. *The Yenan Way in Revolutionary China*. Cambridge: Harvard University Press.
Sen, Amartya. 1973. "Behaviour and the Concept of Preference." In *Choice, Welfare and Measurement*. See Sen (1982a).
Sen, Amartya. 1982a. *Choice, Welfare and Measurement*. Cambridge: MIT Press.
Sen, A. K. 1982b. "Rational Fools." In *Choice, Welfare and Measurement*. See Sen (1982a).
Sweezy, Paul. 1968. *The Theory of Capitalist Development*. New York: Monthly Review.
Szymanski, Albert. 1978. *The Capitalist State and the Politics of Class*. Cambridge, Mass.: Winthrop.
Taylor, Michael. 1986. "Elster's Marx." *Inquiry* 29(1): 3–10.
Taylor, Michael. 1987. *The Possibility of Cooperation*. Cambridge: Cambridge University Press.
Thompson, E. P. 1963. *The Making of the English Working Class*. New York: Vintage.

Chapter 32

Reduction, Explanation, and Individualism

Harold Kincaid

Methodological individualism (MI) was hotly debated in the 1950s and 1960s, first being proclaimed an important and undeniable truth and then being criticized as trivially true or as an obviously implausible reductionism. When the debate subsided in the early 1970s, little agreement resulted about what had been achieved or clarified on the issues. Many philosophers and social scientists, particularly economists, left the debate believing that MI's opponents had focused on its extreme reductionist versions and that some important individualist claims remained untouched. In the last few years, that sentiment has again come to the fore as several authors have defended MI as both plausible and significant (Nozick 1977; Mellor 1982).

This chapter aims to contribute to the renewed debate, both by criticizing some individualist theses and defending others and by trying to avoid old confusions that surrounded MI. MI involves a host of claims, and I try below to sort them out carefully. Furthermore, debates in the philosophy of mind and biology have given us a better grip on the relation between reduction and explanation than was available to the discussants of MI in the 1950s and 1960s. Consequently, the arguments in this chapter borrow significantly from recent work on explanation and reduction outside the social sciences.

In what follows I explicate, evaluate, and consider the connections among the following theses, each of which has been advocated by individualists or might be charitably attributed to them:

(1) social theories are reducible to individualist theories;

(2) any explanation of social phenomena must refer solely to individuals, their relations, dispositions, etc.;

(3) any fully adequate explanation of social phenomena must refer solely to individuals, their relations, dispositions, etc.;

(4) individualist theory suffices to fully explain social phenomena;

(5) individualist theory suffices to partially explain social phenomena;

(6) some reference to individuals is a necessary condition for any explanation of social phenomena;

(7) some reference to individuals is a necessary condition for any full explanation of social phenomena.

Taken together, (1)–(7) capture most variants of MI.[1] I shall argue that (1), (2), (3), (4), and (6) above are highly implausible; that (5) is an open question; and that (7) is both plausible and much more interesting than is initially apparent.

I

Despite frequent denials by some adherents, MI involved from the start an assertion that sociological laws referring to social entities are reducible to theories referring only to individuals. For Watkins, MI is the thesis that no large-scale social events are explained "until we have deduced an account of them from statements about the dispositions, beliefs, resources, and interrelations of individuals" (Watkins 1973a). Dore (1973, p. 77) read MI as the thesis that "all sociological laws are bound to be such as can ultimately be reduced to laws of individual behavior," and Martin (1972, pp. 67–68) and Hempel (1966, p. 110) agreed with this reading. Finally, Mellor (1982) has recently defended versions of MI asserting that nonindividualist theories are reducible to individualist accounts.

Exactly what sort of reductionist claim is the individualist making? Initially, we can take the individualist to hold that:

1. social theory referring to social entities, events, etc., is reducible to theory or theories referring only to individuals;
2. such reduction is perhaps not possible now but is *in principle* possible; and
3. reduction requires lawlike co-extensionality between the primitive predicates of social theory and some predicate in the reducing theory.

Some clarification of these requirements is in order. (1) While I cannot exhaustively define "social entity," obvious examples include classes, castes, nations, churches, bureaucracies, peer groups, etc. Social events and processes are then events and processes involving social entities; social theories are ones whose elementary or primitive terms refer to social entities, events, or processes. (2) Reduction is qualified with "in principle" for two reasons. First, the individualist need claim neither that our current theories support reduction (but only that some future, well-developed theories will be reducible), nor that such reduction must be carried out in complete detail, for it might be extremely cumbersome, albeit possible. Thus, the individualist only alleges that reduction can in principle be done. "In principle" must not, however, be taken too broadly for fear of trivializing the individualist's claim. The individualist thus must be asserting more than simply that it is logically possible for social science to be done completely individualistically. For our purposes, one theory is in principle reducible to another if it is reasonable to believe that our current theories could be extended or replaced with well-developed ones that allowed for reduction. (3) Reduction does not require equivalence of meaning, but I shall assume, following standard accounts (for example, Quine 1964; Schaffner 1967) that it does require biconditional bridge laws connecting primitive terminology of social theories with terminology of individualist theory.

The notion of reducibility employed here is a traditional one that has recently been challenged (for example, Hull 1973, 1974). While I shall argue that those challenges are misguided, even if my arguments fail, MI as a traditional reductionist thesis still merits discussion because (1) past and present defenders of MI themselves assert that social theory is reducible in the traditional sense; (2) several important individualist claims about *explanation* (rather than reduction) ultimately turn out to presuppose a strict reductionism; and (3) some social scientists themselves explicitly advocate this version of individualism (for example, Homans 1974; Arrow 1968, p. 641); the attempt, for example, to provide microfoundations for macroeconomics is a reductionist program in the traditional sense (see Nelson 1984). Thus, thesis (1) as interpreted here retains an important place in any analysis of individualism.

My argument against the reductionist version of MI proceeds as follows. I begin by criticizing reductionist arguments claiming that certain obvious facts entail that reduction must be achievable. After undercutting such positive arguments, a variety of arguments will be given to show that reduction, as defined above, is highly *improbable*. While this conclusion falls sort of showing that no possible scientific development could lead to reduction—a thesis that seems to me just as implausible as its opposite —it does establish that social theory is not in principle reducible, given our previous definition of the term. Finally, this section ends by considering and rejecting various responses to my argument, responses that depend upon weakening the requirements for reduction.

Individualists have thought social theory to be obviously reducible in principle because reduction appears guaranteed by several undeniable truths about social events. Watkins (1973b, p. 179), for instance, cites two "metaphysical commonplaces" supporting reducibility: (1) "the ultimate constituents of the social world are individuals" and (2) "social events are brought about by people" or "it is people who determine history." These commonplaces, Watkins claims (1973b, p. 179), have the "methodological implication that large-scale phenomena ... should be explained in terms of the situations, dispositions, and beliefs of individuals." This methodological injunction presupposes that reduction is in principle possible, and we saw earlier that Watkins affirmed the latter claim. Thus, reducibility allegedly follows from the fact that societies neither exist nor act independently of individuals. This reasoning occurs again and again in the MI literature.

Watkins's principles are of course plausible, although identifying social institutions with sums of individuals may be somewhat troublesome. However, even if we grant the individualist these principles, they do not of themselves *entail* reducibility as Watkins implicitly assumes. Let me tighten up Watkins's principles and then explain why they imply nothing about reducibility.

Social wholes are both composed of individuals and determined by their actions. Borrowing from Hellman and Thompson's (1975) discussion of physicalism, we can describe two principles here: an *exhaustion* and a *determination* principle. Individuals exhaust the social world in that every entity in the social realm is either an individual or a sum of such individuals. Individuals determine the social world in the intuitive sense that once all the relevant facts (expressed in the preferred individualist vocabulary) about individuals are set, then so too are all the facts about social entities, events, etc. Or, to put this idea in terms of supervenient properties, the social supervenes on the individual in the sense that any two social domains exactly alike in terms of the individuals and individual relations composing them would share the same social properties.

Both principles can be made more precise, but for our purposes they will do as formulated. The question at issue is whether the social realm can be exhausted and determined by individuals yet social theory be irreducible to individualist theory. As we saw, Watkins believes that these principles imply that social phenomena should be explained individualistically. Assuming that should implies can, Watkins is in effect claiming that supervenience entails reducibility. Mellor (1982, p. 70) also argues that social theory is reducible in the traditional sense and he likewise seems to think reducibility follows from supervenience: group laws "relate attributes *supervenient* on its members' actions and attitudes. The [group] law ... must *therefore be derivable* from some true explanatory psychological theory" (my italics). Derivability is, of course, the goal of reduction as traditionally understood, and Mellor explicitly endorses this traditional account of reduction (1982, p. 51).

Both Watkins and Mellor are wrong, for determination and exhaustion do not *entail* reducibility of language, as has been aptly demonstrated in the philosophy of mind (for example, Hellman and Thompson 1975). The same mental state, the same program, the same camshaft, and the same social event can all be realized in very different configurations of composing elements. For example, it is logically possible that any number of different relations between individuals realize such social predicates as "peer group," "bureaucracy," etc., just as programs and camshafts can be embodied in indefinitely many physical materials and states. Thus, although each of the composing configurations determines and exhausts the events they realized, these multiple realizations could mean that there is no one configuration of composing elements co-extensive with predicates of the reduced theory. However, reduction of language requires just such co-extensionality. Exhaustion and determination of themselves, without further assumptions, do not show that reduction must be possible.

Much of MI's intuitive appeal comes from the fact that the exhaustion and determination principles are so plausible. We can now see that such appeal is in part ungrounded. I want next to argue for the stronger thesis that reduction is not just ungrounded but highly unlikely.

There are three good reasons to think that reduction will fail on any likely development of social sciences: (1) as already mentioned, multiple realizations of social events are likely; (2) individual actions have indefinitely many social descriptions depending on context; and (3) any workable individualist social theory will in all likelihood presuppose social facts. Each of these claims, if true, rules out reduction as defined here. The first two claims would prevent lawlike coextensionality of predicates, while the last prevents reduction even if coextensionality is established. Let me discuss each in turn.

We have already seen the problem that multiple realizations pose for reduction. Reduction requires equivalences between social and individual terms. However, if one social term refers to an event or entity that can be realized by many different configurations of individuals, then no single individualist term will be forthcoming for any given social term. In short, one side of the required biconditional will fail.

Should we expect multiple realizations to be more than a logical possibility? I think so. Consider the following list of social terms:

revolution	peer group
primary group	bureaucracy
power elite	

It seems obvious that any number of different relations between individuals, individual psychological states, beliefs, etc., could realize the referent of these terms. This point becomes even more compelling when we realize that even particular institutions (tokens) persist through significant structural changes in the configurations realizing them. If necessary and sufficient conditions can be given for these social predicates, it will generally be by means of their functional vis-à-vis other social institutions and events—much as psychological states might be defined in terms of their functional role in a cognitive system. For example, Cooley's (1956, pp. 23–29) classic discussion of "primary groups" defines them in just this way. Social predicates may be definable, but generally not individualistically.

Individualists might grant multiple realizations but deny that they necessarily cause problems. Defenders of type materialism have argued that multiple realizations of mental states do not rule out reduction if the various physical states can be grouped into an appropriate kind (for example, Hill 1984). A like-minded individualist could argue that although social predicates have multiple instantiations in configurations of individuals, those individuals may be similar enough to constitute a kind that would be coextensive with the social predicate in question. While this possibility is real, in all likelihood it will not help the individualist. Similarity is relative to some respect, and the respect in which the different instantiating configurations of individuals may be similar will probably be *social* in nature. Individuals become a social group if they, individually and/or collectively, play a certain *social* role vis-à-vis each other, other *social* groups, larger *social* institutions, etc. But appeal to these social facts generally reintroduces social predicates, thereby making reduction impossible. Thus multiple realizations in this case do undercut reduction.

Multiple realizations of social events make biconditionals between social and individual predicates unlikely, for each social event is not coextensive with any individual predicate. I want now to argue that a similar indeterminacy exists in the other direction: individual acts, relations, etc., described solely in individualist terms, do not uniquely determine their social description, for they are context sensitive. For example, a worker who shoots his foreman may be involved in an act of terrorism in one case, religious conflict in another, and class conflict in yet a third. Such examples can be multiplied at will. They should cause us to look askance at any claim that individual behavior, described solely in individualist terms, will be uniquely correlated with some social description.[2] In fact, this sort of problem plagues reduction elsewhere—witness Searle's claim (1981) that intentional and functional states cannot be identified because a particular functional state can be correlated with different intentional states.

It may be objected that this latter problem arises only because the individual behavior in question is too narrowly specified. If we more broadly conceive individuated kinds of behavior by including the larger context, then we could eliminate any indeterminacy between individual and social events.[3] That may well be so, but invoking the "context" raises new problems. How is the context to be spelled out? The most obvious method of identifying the context is *socially*. The violent worker expressing religious hate is a member of the Catholic church, the foreman, a Protestant church. However, these contexts appeal explicitly to social institutions, and thus the individualist must look for an individualist description of them as well. Such a quest seems unpromising, for (1) these social predicates, as we saw before, probably will not have a unique description in individualist terms because of multiple realization; and (2) the "surrounding" individual acts constituting the context may themselves not determine a unique social description. Appealing to the "context" thus threatens only to push the problems back a step instead of solving it.

The problem just raised leads us to yet another. The relevant context for describing an individual action often refers to a *social role*; two identical acts of physical violence may nonetheless be differentiated by the kinds of individuals involved: e.g., foreman, Catholic, etc. Social theory naturally invokes such roles, since it must refer to types of individuals, not specific individuals themselves. However, many social role predicates are apparently unavailable to the reductionist, for they have an essential social content.

Predicates such as teacher, employee, inmate, soldier, citizen, etc., do refer to individuals, but it is reasonable to believe they implicitly involve social terminology as well. To have true statements employing these role predicates, we must also have true statements about social entities, for there are presumably no inmates without prisons, a judicial system, laws and norms, and no teachers without schools, etc. Applying any of these role predicates to someone seems to presuppose or entail a host of further facts about the social institutions that give them meaning. Elimination of social predicates thus becomes quite unlikely. This point is the gist of Mandelbaum's (1973) "societal facts" and has also been made by Lukes (1973) and McCarthy (1975).

Whether multiple realizations, context sensitivity and social role predicates rule out reduction is, of course, ultimately an empirical question.[4] The considerations raised above do show, I think, the enormous difficulty facing any individualist reductionist program. For example, future theories might dispense altogether with social role predicates that involve institutional membership; but given the social science we know now and its likely permutations, that prospect seems remote. Similarly, we might find purely individualist predicates that group multiple realizations or specify social contexts, but current social science, so far as I can see, is in no position to do so. We may not know a priori that social theory is irreducible to the individual any more that we know the same about mental predicates vis-à-vis the physical, but in both cases the prospects look dim. The burden of proof lies with the individualist.

A more convincing case for the irreducibility of social theory would require looking in detail at proposed individualist construals of social phenomena. While that project cannot be undertaken here, I do want to look briefly at a concrete example that allegedly supports the individualist position. George Homans, an early advocate of MI, developed a partial theory of small groups, a theory he thought supported the individualist program. Homans attempted to expand behaviorist and rational preference theory to include group dynamics, and he claimed to be able to analyze group phenomena in terms of individual interaction motivated by "psychic profit" and the like. Among Homans's (1975, p. 166) results are statements such as:

(1) Groups control their members by creating rewards they can withdraw.

(2) Groups ostracize deviants.

Homans denies these statements make any essential commitment to social entities, for he can cite individual actions that are involved in "group rewarding."

As we shall see in section III, while there is a sense in which Homans has given an individualist explanation, nothing he has said shows that group laws are reducible. "Social groups control their members by withdrawing rewards" faces all three blocks to reduction discussed above.

Multiple realizations: There will be innumerable social arrangements for rewarding and punishing. Material incentives, social esteem, religious salvation, political power, etc., all can do the job, and do it in various ways. No one, set description of what individuals do to bring about "group rewarding" seems available. For example, we cannot group these various realizations simply by whether individuals are rewarded by their interactions, for not all rewarding interactions between individuals—for example—asocial behavior, realizes group rewarding.

Multiple descriptions of individual behavior: Not just any rewarding activity brings about "group rewarding"; whether a relation instantiates this predicate will depend on the social context. Some rewarding personal interaction will in some circumstances have nothing to do with group control or will even be counterproductive (for example, some sexual relations).

Implicit social content: explaining which personal interactions do count as "group rewarding" will naturally lead us to invoke the social roles which individuals play. Rewarding interactions between individuals will contribute to group control only if they bear some relation to group structure. But one natural way for that connection to be made is by way of social roles—interactions that reward as teacher, inmate, or employee, etc. However, these designations presuppose social truths—about schools, prisons, etc.—as argued before. In fact, other sociologists (Bates and Harvey 1975, p. 166) have raised criticisms of individualist small-group theory, based in large part on this point and the preceding one. Thus, while individual interactions, perhaps even ones based on psychic profit, may bring about all social phenomena as Homans claims, reduction remains improbable.

We have seen so far that reduction is likely to fail because (1) social wholes and events have multiple realizations, (2) individual acts have multiple social descriptions, and (3) individualist accounts describing social roles have implicit social content. However, the first two claims are telling only on the assumption that reduction requires lawlike coextensionality between predicates. I want now to briefly consider some challenges to that assumption.

Individualists might grant both that social events have multiple realizations and that individual acts have multiple descriptions, and yet be unimpressed. Why not, we might ask, simply define social predicates disjunctively by listing all the individual configurations that realize some social event? While the details might be complex, so the reasoning goes, we can always in principle handle multiple realizations and descriptions by equating the predicate in question with the totality of different individual configurations realizing it. Mellor (1982, p. 53), for instance, apparently thinks such disjunctive definitions can in principle handle multiple realizations. In this fashion the problems for reduction appear surmountable.

The appearance is deceptive. Disjunctive definitions as proposed here do not support reduction, for they will not preserve the truth value of counterfactuals contained in the social theory to be reduced. Given that social events have indefinitely many realizations, our disjunctive definition must proceed by listing all the actual individual configurations that have constituted the social entity or event in question. However, because such a definition refers only to actual realizations, it will be unable to handle counterfactuals, since they do not go beyond actual cases. Take, for instance, the claim from Marxist sociology that "If the French Revolution had failed, then there would still have been a bourgeois revolution in France." Defining "bourgeois revolution" disjunctively, we can list all the revolutions, described in individualist terms, which have been instances of the term "bourgeois revolution." The latter is thus replaced by "the French revolution or the English revolution or . . . ," with each of these particular revolutions described in individualist terms. But obviously the counterfactual claim under consideration cannot be true once reduced by means of this definition, since its truth would require that the English revolution or some other one occurred in France! (Note that it is the counterfactual nature of the higher-order statement—not its Marxist character—that causes the problem. Statements, such as, "If the stock market crash of

1929 had not taken place, there still would have been a depression in the near future," would also become false when translated.)

Such disjunctive definitions provide only accidental coextensionality; and as these examples illustrate, that does not provide an acceptable reduction.[5] The individualist cannot avoid the problem of multiple realizations by this route.

If disjunctive definitions fail for reduction, we might try to deny as some have done that reduction requires biconditionals between predicates. While this issue cannot be discussed exhaustively here, let me comment briefly on one such proposal.

Mellor (1982) has claimed that reduction demands only approximations relating reduced and reducing terminology and that approximations of this sort are readily available. He is wrong on both counts. A reduction, roughly put, takes place when one more fundamental or extensive theory can be shown to do in principle everything—explain, predict, etc.—that the reduced theory does. But, of course, to show that some social event is in *x* percent of the cases correlated with some type of individualist event is equally to admit that in 100-*x* percent of the cases individualist theory cannot do what social theory can. In short, *so long as* there are exceptions unexplained by individualist theory and explained by social theory, the former has failed to replace the latter. Approximations in such cases do not suffice for reduction.[6]

Furthermore, even if approximations do work for reduction, I see no reason to agree with Mellor that we already have many such approximations in the social sciences. Mellor (1982, p. 53) offers in support of his claim the following: "There are obvious general links between social and psychological phenomena: between language and perceptual ability, for example, and between economics and desires ... enough laws may link psychology and sociology to reduce the latter to the former." While there may be such laws as Mellor cites they have little to do with reduction. Laws may link phenomena without providing a reduction of one to the other. To use Mellor's example, language may influence perceptual ability in regular ways, but that hardly means we can define language in terms of those perceptual abilities. Similarly, the rate of inflation may be linked to people's desires for savings versus consumption, but it would be silly to think we can equate, for the purposes of reduction, inflation with those desires. Not just any link will do. (And even if we had the right approximations—between primitive predicates—we would still face the problem of implicit social content discussed above.)

Richardson (1979) has proposed the even stronger thesis that reduction requires no bridge laws whatsoever—neither biconditionals nor conditionals linking terminology are essential to reduction. Hull (1974) seems to support a similar position. Citing historical cases as evidence, these authors argue that reduction has been misunderstood by philosophers of science. Mendelian genetics, for example, has clearly been reduced to molecular generics: we have molecular explanations even if no bridge laws are forthcoming. Therefore, they conclude, standard accounts of reduction must be given up. Because this thesis is essentially about explanation, I shall postpone discussing it until the next section which considers explanatory versions of MI.

II

Rather than a thesis about reduction of theories, MI might be a thesis about *explanation*. Danto (1973, p. 328), for instance, reads MI as the claim that social events and entities "can only be explained by reference to [individuals]." Watkins (1973a) also insists that we have a "full" or "rock-bottom" explanation only when we have explained things solely in terms of individuals, their dispositions, etc. This section accordingly looks at MI as the claim that individualist theories can explain the social world without loss.

"Can explain," of course, hides several different theses. In this section, I want to consider theses (2)–(5) on the original list; namely, the theses that:

(2) any explanation of social phenomena must refer solely to individuals;

(3) any fully adequate explanation of social phenomena must refer solely to individuals;

(4) individualist theory suffices to fully explain social phenomena;

(5) individualist theory suffices to partially explain social phenomena.

Of these theses, (2) is clearly the strongest and (5) the weakest. Theses (2)–(4) are, I shall argue, false; (5) is true, but only on its weakest interpretation.

Anyone advocating (2) is a radical individualist: they hold that no theory making any reference to social entities ever explains, even if the variables of the theory range for the most part over individuals. Radical views are not necessarily suspect, but this one surely is. It entails that nearly all sociology, anthropology, and even microeconomics—which refers not only to individuals but also to corporations, a social entity—are pseudo-explanations. No matter how suspicious one is of these "soft" sciences, it is hard to deny that they do on occasion explain. Economic laws relating supply, demand, and the behavior of firms, for example, are probably as well confirmed as much in evolutionary biology, geology, etc. Some social theories apparently do explain.

Theses (3) and (4) can be considered as a unit. Since thesis (4) claims only that social phenomena *can* be fully explained individualistically while (3) makes individualism a *must* for full explanation, a refutation of (4) will inter alia be a refutation of (3). However, the idea that individualist theory suffices to explain fully seems to me clearly implausible.

Thesis (4) faces the following dilemma: either social phenomena are to be explained as types, in which case thesis (4) turns out to be nothing but an already rejected claim about theory reduction, *or* it claims that social *tokens* can be fully explained individualistically. Individualist explanations of the latter sort, however, are barely explanations at all and certainly not *fully adequate* as the individualist asserts.[7]

Let me unravel the above argument by considering the claim that every social type (for example, recessions) can be fully explained individualistically. Explanation is done by theories, and the individualist is now claiming that individualist theory can in principle do all the work of social theory. Thus, there is allegedly for each kind of social event or entity an explanation in solely individualist terms. However, the individualist is now committed to supplying an individualist equivalent for the types picked out by the social kind terms. Therefore, this explanatory version turns out to be, or at least require, reduction of theories—in short, the strict reductionist thesis (1). That thesis, however, has already been shown implausible.

A second alternative is to explain social phenomena in effect case by case. While we might not be able to give a single individualist account of social event kinds, we could still explain the social realm completely in that we can explain every particular occurrence of social events, or so the individualist might reason. In what sense can we provide a full explanation by explaining social tokens one by one? I can see two possibilities. On the one hand, the individualist may think that social phenomena are fully explainable in that we can simply describe—for every particular social event—the individuals involved, their interrelations, etc. Because social entities and events are realized in or supervenient upon individuals, there must be a description of that event referring only to individuals. We can presumably name and describe the particular individuals involved and their interrelations in solely individual terms—in physical terms if need be.

The second way in which particular events might be explainable in individualist terms is considerably more interesting. There might be individualist laws describing human behavior that apply to the particular behaviors realizing a specific social event, even though the predicates of those laws were not coextensive with ones in social theory. In other words, the configuration of individuals as a whole might fall under a social description, while the individualist laws, for example, the laws of psychology, might apply to individuals one by one in a way that allowed no equivalence of social and individual terms. Such a situation would be analogous, for example, to computer programs that have potentially infinitely many realizations: we may not be able to define programs in terms of machine states, but there are still laws to explain any particular realization in physical terms. Thus we seem to have here a real possibility for individualist explanation.

Neither of these two possibilities turns out to provide full explanations in individualist terms. In order to defend this claim, I need first to say something about what I take a full explanation to be. Following much recent work (van Fraassen 1980; Garfinkel 1981; Achinstein 1983), an explanation can be considered in part an answer to a why-question.[8] Questions are not answered simpliciter, but are rather given answers relative to a number of contextual parameters. The same question may be given different answers depending upon the contrast class involved ("Why did John die—in contrast to Bill, to recovering, etc.?"). Furthermore, once we set the contrast class of possible answers, some restriction on kinds of answers must be specified—for example, the immediate cause, the structural cause, the microdeterminants, etc. Leaving important details aside, I shall thus assume that an explanation is an answer to a why-question and that such questions can be given multiple correct answers depending on what counts as the contrast class and a relevant answer.

This brief sketch suggests that the explanatory power of a theory can be evaluated along at least two dimensions: (1) the extent to which it can adequately answer any given, fully specified question, and (2) the number of relevant questions it can answer.[9] Most of the standard literature on explanation analyzes the first dimension. The work of Hempel, Salmon, and many others can be seen as trying to specify when an answer to a question fully accounts for the facts cited by the question. The second dimension appeals to our sense that a theory that cannot answer important questions is incomplete. It also reflects in part the idea that good theories unify and provide general laws. Combining these two dimensions, we can thus say that a fully explanatory theory is at least one that can fully answer all relevant questions.

The above discussion is obviously very sketchy. However, for my purposes this sketch will suffice to evaluate the individualist's claim to explain fully by proceeding case by case.

Individualist theory cannot provide full explanations of social phenomena by explaining only social tokens largely because such token explanations leave important questions unanswered. More specifically, for any social phenomenon or event *S*, two fundamental kinds of questions are:

(1) Why did *S* occur, that is, what are the causal connections between the kind of social events (events involving social entities) preceding *S* and the onset of *S*?

(2) Why do these kind of events (the kind to which *S* belongs) occur, that is, what other kinds of antecedent social events might bring about this kind of social event?

Both questions are the general form of a great many specific questions asked in social explanation and explanation in general. Question (1) could include questions about what kind of preceding changes in political, religious, intellectual, economic, class, educational, etc., institutions causally influenced *S*. If functional explanations are not a species of causal explanations, then (1) could likewise include questions about the functional roles *S* plays vis-à-vis other social events, institutions, etc. Question (2) seeks to place *S* in a more unified framework by citing the kinds of preceding social events that could cause the event but were not in fact actually present.[10] Any theory of social phenomena that cannot answer these questions is prima facie not a complete explanation.

Individualist explanations of social tokens will not be able to answer questions falling under (1) and (2) for a simple reason: it has no way of referring to kinds of social events. Both (1) and (2) involve situating a particular social event in a web of causal laws connecting different social events. Causal laws, however, are about kinds of events, not event tokens. Thus the individualist who claims to give a full explanation—despite having no individualist terms coextensive with social kinds—must be in part wrong. Without tools to specify social event kinds, questions like (1) and (2) cannot be answered. Any theory that fails in this regard cannot claim to be full or complete.

Thus the supervenience of the social upon the individual does not entail that social events can be fully explained individualistically as has been claimed (Macdonald and Pettit 1981, p. 125). Supervenience, of course, insures that we can describe what individuals did, for example, in bringing about the French Revolution and we might go on to invoke laws of psychology or other laws about individuals to say why they behaved as they did. Such a story would be explanatory, but it surely would fall far short of being a full explanation. We would have no way to understand this event as a kind and, in particular, as a social kind. Consequently, we could not explain its causal connection to preceding changes in classes, religious institutions etc., of French society nor could we understand this revolution by relating it to other revolutions, other kinds of revolutions, political and economic transformations, etc., that have or might occur elsewhere. In short, many standard questions could not be answered.

The above argument against the explanatory thesis presupposes one crucial assumption: that social theory provides at least some laws relating kinds of events, properties of social entities, etc. If social theory never got beyond piecemeal and isolated descriptions or classifications of social phenomena, then individualist explanation of social tokens would look relatively good. Like individualist theory, a weak social theory would not be able to answer questions (1)–(3) above and consequently we would be missing nothing by giving individualist explanations that referred only to social tokens. The individualist could thus defend the explanatory thesis roughly as Churchland (1978) has defended eliminative materialism—by arguing that there is no real theory to reduce.

How plausible this defense is depends on one's estimation of current social science and its foreseeable progress. However, it seems to me that social science has produced sufficiently many rough generalizations, tendency laws, etc., about social phenomena to rule out an individualism motivated by an alleged failure of social theory. Parts of macroeconomics, conflict theory, work on bureaucracy, social stratification, deviance, etc., certainly seem to give at least occasional explanations of social kinds, explanations that explanatory individualism cannot handle. Unless the individualist can give an argument showing that all such purported explanations are fake, an "eliminative individualism" is implausible. I know no such arguments, at least none that do not already presuppose the truth of MI. The explanatory version of MI does not win by default.

We can now also see what is wrong with Richardson and Hull's redefinition of reduction mentioned in section I. If one event token is realized in another, then we can in a weak sense explain the former in terms of the latter. But if (1) the higher-level theory really does provide explanations of social phenomena and (2) the predicates of those explanations have no equivalent in individualist theory, then the latter theory cannot claim to replace or supersede social theory, for there will be much that only the higher-level account may do.[11] In other words, if reduction fails, social theory will provide explanatory kinds and relations that cannot be found at the individual level. In fact, this conclusion holds generally for higher- and lower-order theories lacking coextensional predicates. Kitcher (1984), for example, has argued that molecular genetics has not reduced (in the sense of completely replaced) Mendelian genetics for similar reasons.

So we can now conclude that MI thesis (4)—that individualist theory suffices to fully explain—fails. Either this thesis turns out simply to be the reduction thesis if we are talking about types, or it tries unsuccessfully to fully explain case by case. But if (4) fails, then so does the claim that any fully explanatory theory *must* refer only to individuals. To deny that individualist theory can fully explain is equally to deny that *only* individualist theories fully explain.

Having rejected theses (2)–(4), we need lastly to consider the weakest individualist claim for explanatory adequacy, namely, that individualist theory suffices to explain, even if not fully. It would seem that this anemic claim is guaranteed by supervenience. If the social supervenes on the individual, then at least every particular social event is determined by some relation between individuals. Any theory describing those relations will thus be a partial explanation of social phenomena. Although such an account will not answer questions concerning social events as kinds, it will nonetheless answer at least one sort of question—it will tell us why a particular event happened by citing the micro-events that caused it.

There is reason, however, to question whether a purely individualist account of individuals and their relations does suffice to describe the facts realizing the social. The individualist version of supervenience holds that once all the facts about individuals, described solely in individualist terms, are set, then so too are all the facts about social phenomena. However, it is unclear to me whether this minimal set of facts suffices to determine all the social facts. If, however, individual facts do not determine the social facts, then citing the former will not give us a micro-explanation of the latter.

There is at least one serious problem for the claim that all the individual facts determine the social facts. This individualist version of supervenience presupposes there can be laws of human behavior that are (1) sufficiently strong to fully explain the individual behaviors making up any particular social event and (2) also make no reference to or presuppose no facts about social entities. Are laws fulfilling both (1) and (2) likely to be developed? Obviously, we have nothing approaching them now, but this lack does not of itself rule out relevant future developments. However, an affirmative answer seems possible only on certain assumptions about human nature. If the *basic* patterns of human behavior depend significantly on social contexts—the institutions and culture one belongs to, the social roles one adopts—then any laws strong enough to explain will have to make reference to these social facts and thus will fail to be a purely individualistic explanation. of course, even if social contexts must be invoked, the possibility remains that those contexts could be cashed out individualistically. That move, however, threatens to reinstate the strict reductionist program. Whether human behavior can ultimately be explained without appeal to unreduced social facts is an empirical question. Affirming thesis (5) thus requires that question be answered in the affirmative.

III

So far we have given little ground to individualism. Nonetheless, the last section did make a concession: individualist explanations are possible, even if not fully adequate. I want to make further concessions in this section and say what seems true and interesting in MI.

Think, for the moment, about the most radical of the holist metaphysicians, namely, Hegel. When Hegel claims history is the necessary progression of human civilization toward greater human freedom and self-awareness, what exactly is wrong with his claim? Aside from questions of confirmation, a major problem comes from the fact that we do not see, as Dray has argues (1964), how or why these large-scale patterns should come about. In contrast, when Adam Smith says that markets develop and influence their participants like an invisible hand, our misgivings diminish, or at least they should. What is the difference? Smith gives us an account in terms of individuals, a mechanism based on individual self-interest that explains how large-scale social processes are realized. It is this legitimate demand for what Smith provides—an explanation of how social wholes and events are connected to individual behavior—that in part constitutes the intuitive appeal of individualism.

Let me formulate this demand more clearly. There are two things it does *not* involve. Above all, the need for an account in terms of individuals does not support any form of reduction considered above. Nothing in Smith's invisible hand, for example, implies that all social terms can be defined individualistically, for accounts of economic mechanisms presuppose truths about social-level patterns as Nozick (1974) has noted.

Second, the need to refer to individuals does not warrant the claim that some reference to individuals is a necessary condition for successful explanation (MI thesis (6)). Earlier I argued that it was unreasonable to hold that every theory making *some* reference to social entities, etc., failed to explain. It seems to me likewise true that *purely* social theories do explain or at least can in principle. Macro-economic theories, for instance, are often formulated so that they refer only to social entities—nations, corporations, industries, households, government institutions, etc. I see no reason in principle to deny that such theories adequately answer a range of questions —in other words, to deny they explain—simply because they do not refer directly to individuals.[12]

If reference to individuals is not necessary for social explanation, the individualist can much more plausibly claim that any social explanation which makes no reference to individuals, in particular to mechanisms involving individuals which bring about social events, has not given a *complete* or *full* explanation. We may give real explanations at the social level, but we have done so only partially.

This thesis—that some reference to individuals is a necessary condition for full explanation (MI thesis (7))—carries more punch than may be obvious at first glance. Presumably for every entity with composing elements, there is always something that can in principle be said about its micro-structure that adds somewhat to our comprehension or, in our earlier terminology, answers further why-questions. However, in a great many cases information about micro-structure seems far from essential for explanation and perhaps a positive hindrance. We can fully explain planetary motion, for example, despite the fact that we cannot cite the specific quantum-mechanical details that realize the planets. In this case information about micro-structure is not needed to explain adequately.

Thus, when the individualist asserts that full explanation in the social realm requires looking at the micro-structure—individuals and their relations—something nontrivial is being said. I have no theoretically grounded account of when micro-structure is essential and when it is not, but no doubt we find social explanation which makes no reference to individuals incomplete because of the "closeness" of the social and individual levels (unlike the planetary and quantum-physical levels), because much of our information about social entities comes from observing what individuals do, because many social predicates (like "needs," "interests," and "goals" of institutions) apply equally and paradigmatically to individuals, because of our natural practical interest in how things relate to ourselves not just to the "working class," etc. These factors and others lead us to hold that the important questions and the relevant kinds of answers for complete social explanation must make some reference to mechanisms involving individuals.

Individualists of course have not strictly separated this more plausible claim from the previous, less convincing version of MI. Still, when Homans (1975, p, 384) demands that social explanation cite the "underlying mechanisms of human behavior" and Popper (1950, p. 291) says "we should never be satisfied with an explanation in terms of so-called 'collectives,'" he is clearly endorsing something akin to the completeness thesis. MI, understood this way, gains some support from reductionism elsewhere. Mendelian genetics apparently cannot be reduced to molecular genetics or be completely replaced by it. Nonetheless, molecular genetics gives us what Kitcher (1984) calls an "explanatory extension" of classical genetics, An explanatory extension reveals the fine structure underlying the basic elements and processes of a theory (although not replacing or reducing that theory). Classical genetics remained in some sense incomplete until molecular biologists provided an explanatory extension. We can thus plausibly point to this example and others as lending credibility to the parallel individualist claim.

We have now discussed all the versions of MI listed in the introduction. As a claim about theory reduction, MI is highly implausible. It likewise fails when it restricts all explanation to the individualist level or even makes such reference only necessary for explanation. Much more plausible, however, is the individualist intuition that explanations that refer only to social entities remain incomplete.

Notes

Helpful comments were made on earlier drafts of this chapter by Geoffrey Hellman, George Graham, Terry Horgan, and Scott Arnold, and especially by two anonymous referees.

1. Individualists sometimes assert the epistemological claim that all knowledge of social phenomena is derived from knowledge about individuals and the ontological claim that groups are in some sense either fictions or derivative. The former thesis—which faces problems similar to those raised against phenomenalism—have been dealt with extensively elsewhere (McCarthy 1975; Mellor 1982). Many of the ontological issues may be spurious, since few holists believe wholes exist and act independently of their members and few individualists deny aggregates exist. While I do discuss in passing whether the social supervenes on individuals, other ontological theses are beyond the scope of this chapter.
2. Note that such indeterminacy does not contradict my earlier commitment to the supervenience of the social on the individual. Supervenience, as I formulated it, claims that once *all* the facts about individuals are set, then so too are social facts, But this does not entail that each individual fact or kind of fact uniquely determines some social fact or kind (except for the trivial case where one counts all the individual facts as one disjunctive fact or kind). The problem I am raising in effect argues that determination or supervenience relations between the social and individual must be relatively global, as opposed to local, in nature if they are to be plausible.
3. It might seem that specifying the beliefs involved would suffice. However, I doubt that this is a general solution for at least three reasons: (1) beliefs, desires, and mental states generally connect with the same act, physically described, in indefinitely many ways; (2) the social significance of individual behavior often goes beyond and does not depend on the intentions of the actor, thus giving acts different social specifications even when belief is specified and invariant; and (3) it may well be that individuating and/or identifying beliefs cannot be done without reference to the social context.
4. Sober (1980) nicely formulates some of the empirical issues here by contrasting the MI debate with the unit-of-selection controversy. Roughly put, group selection occurs if individual genotypes and phenotypes are irrelevant in determining fitness when compared to group membership. Similarly, the individualism debate turns on whether the individual behaviors bringing about social phenomena have properties resulting primarily from social membership or whether those properties can be accounted for by relations among individuals. The multiple realization and context-sensitivity problems suggest that in fact nonsocial properties and relations can be permutated in indefinitely many ways without altering the social phenomena to be explained and thus give reason for thinking that social membership is what counts. The importance of social role predicates and their apparent irreducibility would support a similar conclusion.
5. Kim (1984) has argued that supervenience entails that every higher-level property is *necessarily* coextensive with the disjunct of all its *possible* supervenience bases. However, as Kim acknowledges and as I argue in detail elsewhere (1986), coextensionality in this form does not suffice for reduction of theories for a variety of reasons.
6. Note that the kind of case Mellor cites must be distinguished from the situation where the reducing theory corrects the reduced theory. In the latter case the laws of the higher-level theory are known or shown to be partially inaccurate while the reducing theories avoid such inaccuracies. While we do have reduction here by means of only approximate connections between predicates, the approximation occurs because the social terms and laws are in fact misapplied and inaccurate in their own domain— so, reducing theory does everything the reduced theory does and does it better. However, when higher-level laws and predicates are known to accurately apply to their domain and are known to be only approximately correlated with lower-level predicates, we have an entirely different situation—the reducing theory cannot handle all events as the higher level.
7. In actual practice, one theory comes to do the explanatory work of another in much more complicated ways: some higher-level kinds are dropped altogether (because the laws invoking them are false), others are corrected and then equated with lower-level predicates, and others may be directly reduced without alteration. Although my argument simplifies greatly, the simplification makes no difference so long as there are *some* explanations involving social kinds that are successful.

8. I say "in part" because explanations may also be answers to other kinds of questions—what, how, etc. I stick here to why-questions to simplify the discussion.
9. Either (2) can be taken as the number of questions if we count every different specification of contrast classes and answer kinds as determining a new question, or (2) can be phrased in terms of the number of questions answered *and* the range of contrast classed and answer kinds handled for each question. Popper (1972) has also suggested the breadth of questions answered as a dimension for theory assessment.
10. Falling under (2) are also the kinds of counterfactuals that are typically involved in claims that such and such a structure is necessary. Garfinkel (1981) and Miller (1978) have emphasized this particular problem for MI.
11. Hull and Richardson try to support their weakened version of reduction by arguing that molecular genetics has clearly reduced classical genetics but without giving bridge-laws. etc., demanded by traditional accounts of reduction. However, as Patricia Kitcher has pointed out (1980, 1982), it is doubtful that molecular genetics really warrants a revision in the traditional account for reasons similar to those considered here: (a) classical genetics on the whole did not provide accurate laws, and thus molecular genetics could explain without providing a derivation of higher-level kinds and laws; and (b) where classical genetics does hold true, molecular biology does not generally replace such laws but rather provides an account of why the law holds true *in each particular case.* In short, molecular genetics appears to reduce only because there was either no explanatory theory to reduce or because reduction is confused with explanation case by case.
12. Aside from the bare fact that purely social theories do sometimes appear to explain, this thesis also faces more general problems. Unless we can say why explanation of social wholes differs essentially from macro-explanations in other domains, we seemed forced to the conclusion that no purely macro-explanation, no matter the domain, succeeds. That conclusions is, I assume wrong.

References

Achinstein, P. (1983), *The Nature of Explanation.* New York: Oxford University Press.

Arrow, K. (1968), "Mathematic Models in the Social Sciences," in Broadbeck (1968).

Bates, F., and Harvey, C. (1975), *The Structure of Social Systems.* New York: Gardner Press.

Broadbeck, M. (1968), *Readings in the Philosophy of the Social Sciences.* New York: Macmillan.

Churchland, P. (1978), "Eliminative Materialism and Propositional Attitudes," *Journal of Philosophy 78*: 67–91.

Cooley, C. (1956), *Social Organization.* Glencoe: Free Press.

Danto, A. (1973), "Methodological Individualism and Methodological Socialism," in O'Neill (1973).

Dore, R. (1973), "Function and Cause," in Ryan (1973), pp. 65–82.

Dray, W. (1964), *Philosophy of History.* Englewood Cliffs: Prentice-Hall.

Garfinkel, A. (1981). *Forms of Explanation.* New Haven: Yale University Press.

Haugeland, J. (1981), *Mind Design.* Montgomery: Bradford Books.

Hellman, G., and Thompson, F. (1975), "Physicalism: Ontology, Determination and Reduction," *Journal of Philosophy* 72: 551–64.

Hemple. C. (1966), *The Philosophy of the Natural Sciences.* New York: Prentice-Hall.

Hill. C. (1984), "In Defense of Type Materialism," *Synthese* 59: 295–321.

Homans, G. (1974), *Social Behavior: Its Elementary Forms.* New York: Harcourt Brace Jovanovich.

Hull, D. (1973), "Reduction in Genetics—Doing the Impossible," in Suppes (1973).

———. (1974), *Philosophy of Biology,* Englewood Cliffs: Prentice-Hall.

Kim, J. (1984), "Concepts of Supervenience," *Philosophy and Phenomenological Research* 45: 155–75.

Kincaid, H. (1986), "Rosenberg and the Reducibility of Biology," paper presented at the Central Division meetings of the A.P.A.

Kitcher, Patricia (1980), "How to Reduce a Functional Psychology?" *Philosophy of Science* 47: 134–60.

———. (1982), "Genetics, Reduction and Functional Psychology," *Philosophy of Science* 49: 633–36.

Kitcher, Philip (1984), "1953 and All That. The Tale of Two Sciences," *Philosophical Review 93*: 335–75.

Lukes, S. (1973), "Methodological Individualism Reconsidered," in Ryan (1973), pp. 119–29.

McDonald, G., and Pettit, P. (1981), *Semantics and Social Science.* London: Routledge and Kegan Paul.

McCarthy, M. (1975), "On Methodological Individualism," Ph.D. dissertation, Indiana University.

Mandelbaum, M. (1973), "Societal Facts," in O'Neill (1973), pp. 221–34.

Martin, M. (1972), "On Explanation in Social Science: Some Recent Work," *Philosophy of Social Science 2*: 67–68.
Mellor, D. (1982), "The Reduction of Society," *Philosophy 57*: 51–74.
Miller, R. (1978), "Methodological Individualism and Social Explanation," *Philosophy of Science 45*: 387–414.
Nagel, E. (1961), *The Structure of Science*. London: Routledge and Kegan Paul.
Nelson, A. (1984), "Some Issues Surrounding the Reduction of Macroeconomics to Microeconomics," *Philosophy of Science 51*: 573–94.
Nozick, R. (1974), *Anarchy State, and Utopia*. New York: Basic Books.
———. (1977), "On Austrian Methodology," *Synthese 36*: 354–92.
O'Neill, J. (ed.) (1973), *Modes of Individualism and Collectivism*. London: Heineman.
Popper, K. (1950), *The Open Society and Its Enemies*. Princeton: Princeton University Press.
———. (1972). *Objective Knowledge*. London: Oxford University Press.
Quine, W. V. (1964), "Ontological Reduction and the World of Numbers," *Journal of Philosophy 61*: 209–16.
Richardson, R. (1979), "Functionalism and Reduction," *Philosophy of Science 46*: 533–58.
Ryan, A. (ed.) (1973), *The Philosophy of Social Explanation*. Oxford: Oxford University Press.
Schaffner, K. (1967), "Approaches To Reduction," *Philosophy of Science 34*: 137–47.
Searle, J. (1981), "Minds, Brains and Programs," in Haugeland (1981).
Sober, E. (1980), "Holism, Individualism and the Units of Selection," in *PSA 1980*, Vol. 2. pp. 93–121.
Suppes P., et al. (eds.) (1973), *Logic, Methodology and Philosophy of Science IV*. Amsterdam: North-Holland.
van Fraassen, B. (1980), *The Scientific Image*. New York: Oxford University Press.
Watkins, J. (1973a), "Historical Explanation in the Social Sciences," in O'Neill (1973), pp. 166–78.
———. (1973b), "Methodological Individualism: A Reply," in O'Neill (1973).

Chapter 33

Social Science and the Mental

Alan J. Nelson

The solutions of many outstanding general problems in philosophy of social science seem to await progress in philosophy of mind. This has a close parallel in the practice of social science. Important social scientists have thought that their theories should be developed in concert with psychology, or at least with attention to what psychology has to say. This essay argues that it is easy to exaggerate the significance of the undeniable connections between the social and the mental. I am particularly concerned to advocate caution in using our understanding of the mental to place methodological constraints on social scientific theorizing, but I shall also advance a formulation of a weak individualistic constraint.

We need to begin with a simple and selective history of how the philosophy of social science has become entwined with the philosophy of mind and psychology.[1] Let us confine the discussion to social sciences that use intentional concepts such as preference, expectation, perception, belief, knowledge, and the like to formulate laws that explain social facts, institutions, and actions as being partially caused by things possessing those concepts. This includes many of the most important theories in social science when they are interpreted in natural ways. "Economic *agents attempt* to *satisfy* their *preferences* by efficiently allocating their resources in light of their *information* about *perceived* conditions." This encapsulation of a large part of economic theory is chock full of intentional concepts.[2] Some notes on usage: I shall henceforth use "social" and "social science" for the sort of intentional theory just described. I shall assume that social science is "naturalistic": it seeks laws[3] and does not require Cartesian substance dualism. "Mental" will refer to properties of what are commonly regarded as intentional states of individuals. Similarly, "psychology" will refer to a science of lawlike explanations making use of intentional concepts.

One uncontestable bond between social science and individual behavior (and, hence, the intentional explanation of that behavior) is established by the fact that many interesting social phenomena consist in individual behavior in social contexts or in response to social constraints. "Why do people use drugs recreationally?" asks a question about what individuals do, and those asking it are usually expecting an answer in terms of social conditons. Another example is intrafamily behavior. In these cases, familiar from countless texts in social psychology and cultural anthropology, we are explicitly dealing with the actions of individuals even though we may hope for interesting regularities or averages taken over individuals.

Another, more complex way the mental is clearly relevant to the social comes from some versions of the doctrine called methodological individualism (MI). MI has meant many different things, but one central meaning is that any acceptable scientific explanation of a social phenomenon must eventually be in terms of the individuals that make up the society.[4] MI is usually augmented with the requirement that individual actions, attitudes, dispositions, etc., themselves be explained in terms of individualistic concepts. There are some examples of social scientific explanations that appear to violate the explanation strictures of MI by making use of constructs describable only in macro-social terms.

But even for these cases there is a powerful argument concluding that the explanatory power of the macro-explanations rests upon mentalistic descriptions of the individuals. It might even be thought that such individualistic descriptions are a necessary *complement* of holistic explanations. I call this possible component of MI the Principle of the Individualistic Transmission of the Social (PITS).[5] Until near the end of this essay, PITS will remain implicitly defined by its application in the following example of an explanation that initially appears to violate MI.

A clear illustration is the casting into the PITS of some very influential metatheoretical theses of the famous holistic sociologist Emile Durkheim. Durkheim believed that there were social facts explanatorily autonomous (though not ontologically autonomous) with respect to the constitutive facts about individuals. Durkheim produced some philosophical considerations in favor of the concept of holistic social facts and made the disregard for facts about individuals into a methodological canon, but his best reasons for making use of social facts in sociological theory were empirical. A typical example is in his *Suicide* (1951) [1897] in which he argued that social facts concerning aggregate suicide rates could only be explained by the operation of social forces, and provided a theory of such forces. Individual psychology was inert in the explanation of suicide rates, he urged, because the data showed that suicide rates often varied dramatically among classes of people for whom the relevant individual motivations for suicide were similar.

The PITS are opened up before Durkheim by attending to the individual character of the act of suicide. Even granting that his account of social forces is correct, these forces must work on individual victims to cause them, at least partially, to commit suicide. The primary proximate cause of the particular suicide is the intention of the victim. Durkheim himself accepted that. So if he is right, it must be that these social forces are efficacious on individuals in virtue of effects on their states of motivation. But suicide rates are nothing more than aggregates of individual suicides, so it is natural to conclude that the complete explanation[6] of this social fact will involve mental states.[7] Generalizing from this example, we have PITS requiring that all causal connections at the social level operate through individualistic mechanisms.

PITS therefore also naturally creates a presumption that complete, lawlike explanations of the social are going to involve similarly *lawlike* explanations of whatever mental phenomena are involved. This is because interesting social regularities and generalizations will consist in aggregates of interesting regularities and generalizations concerning the actions of individuals and their chief causes, namely, the mental states serving as reasons for the actions. This presumption engages the notorious dispute in the philosophy of mind about the scientific status of the connections among reasons and actions. Since there are formidable philosophical arguments to the effect that explanations of individual acts in terms of their reasons cannot be scientific, and any unresolved difficulties for the scientific treatment of the mental will apparently infect social sciences as well, we are here obliged to make a foray into this part of the philosophy of mind and psychology.

The first group of arguments to consider are often called Wittgensteinian or, more circumspectly, neo-Wittgensteinian because they are either related to themes that were abroad because of Wittgenstein's later teachings or because they were propounded by those who had been associated with him.[8] The arguments all flow from considering the conceptual interplay among human actions, their causes, and their reasons. The conclusions of the arguments have the general form of denying that an explanation of an action in terms of the agent's reason for the action is a causal explanation. Since scientfic explanations are presumably in the business of citing causes, it is a corollary that explanations in terms of reasons cannot be scientific. These conclusions are arrived at via one or more of the following observations: reasons are not the right kind of entities to serve as causes (e.g., they are dispositions instead of events); free will would be impossible if reasons were causes; reasons cannot be characterized as such independently of the acts that they actually give rise to, hence they are logically connected to them, and so cannot be causes; actions cannot be characterized as such without appeal to the rules they accord with or to the agents' conceptions of these rules (that is, part of the agents' reasons), hence they are logically connected with them, and so cannot be caused by them; when we ask for the explanation of an act we are not typically asking for a cause, instead we are asking for a justification or interpretation of the act. Especially relevant to this essay are the explicit extensions of some of these points to social science. For instance, Dray (1957) and Winch (1958) use some of these arguments to buttress Weberian concerns by saying that individual action in a social setting cannot be subsumed under causal laws, the hallmark of explanation in the physical sciences, because of the essential role of reasons in dealing with action. Donald Davidson is generally acknowledged as having exploded these arguments, but not in a way that ameliorates the problems for the social sciences that the neo-Wittgensteinians introduced.[9] We must now consider his position.

Davidson's brilliant attack on skepticism about causes for actions can be crystallized for present purposes as utilizing two principles. Anomalous Monism (this name runs the two principles together) concedes the neo-Wittgensteinian point that there are no causal generalizations to be had between physcial and mental events or even among mental events for that matter (anomalism).[10] But, Davidson wrote, every physical event is lawfully connected with a physical cause and every particular mental event or "token" is identical to a particular physical event or "token" (monism). So the "logical connection" between reasons and actions is only an artifact of the logical connection between some of the linguistic descriptions of the events in question. The same events have other descriptions that make the contingency and causal nature of their connection clear. These principles enabled Davidson to say that although there cannot be psychological laws governing reasons (at least not "exact" ones like the ones found in natural sciences), there is no ground for rejecting what he took to be the intuitive view that reasons explain actions because they cause them. Since the causal relation requires lawlike connection, they cause them by courtesy of the physical laws subsuming the physical events they are token identical to.

It seems easy now to extend the conclusion of Davidson's analysis of individual action explanation into a convincing foreclosure of the possibility of scientific accounting for the causation of social phenomena. We need only consider the ways in which the social involve the mental. The case of social phenomena that consists in individual acts like the exchanging of gifts immediately requires laws governing individual actions—exactly what is ruled out by the Davidsonian system. With macro-social phenomena, on the other hand, it is first obvious that any strong reductionistic version of MI would have individual psychology being more basic than the social sciences so that laws in the latter would have to be derivable, in principle, from the nonexistent former. And even if these strong versions of MI are rejected, it is hard to avoid the much more reasonable PITS. The complete explanations of macro-social phenomena like unemployment rates or degrees of social cohesion will directly incorporate mechanisms that connect these to individual actions. In both cases, therefore, the required explanation is committed to including information about the mental that is not amenable to scientific treatment. The apparently inexorable drift of all this is that there cannot be social laws because there are not mental laws.

I shall summarize. We have seen that MI and PITS suggest that social sciences can exist only if there is a science of individual action that can, at least in principle, be called upon to complement their explanations. Some neo-Wittgensteinians thought that since psychology is impossible (because reasons are not causes), it followed that (naturalistic) social sciences are impossible and the attempt to practice them needed to be abandoned in favor of interpretive social studies. Davidson reinstated causality in intentional explanations, but also left in anomalism. The contemporary approach to philosophy of mind that accounts for the trouble intentional psychology is in (if it is indeed in trouble) seems to lead directly to corresponding trouble for social sciences that deal in the intentional—almost all of social science. This has certainly attracted attention from those who are inclined to think that social science *is* in trouble and have been looking for an explanation of the predicament. Of course, a satisfying explanation would have to show not only what is wrong with social science, but also why the pursuit of social laws has seemed so reasonable. The approach outlined above seems to do these things. Social science by its very nature requires a firm foundation on a science of individual action, but this foundation is lacking. The apparent reasonableness of attempting to produce social science stems from the fact that our ordinary intentional folk psychology gives the false initial appearance of being the core of an improvable scientific theory.

This state of affairs seems to leave the fate of the social sciences in the balance while the real battle is waged in the philosophy of mind to determine whether successful scientific psychology is forthcoming. I propose to kick over the teetering balance by disengaging the social sciences from the mental cleanly enough to ensure that the long-standing methodological and philosophical battle over the status of these disciplines can be settled on their own ground.

The greatest obstacle I face is that PITS all by itself seems to make the connection between the mental and the social quite tight. I find the prospect of denying the Principle repugnant, so I won't. I shall instead attempt to show how a satisfying version of PITS can be made compatible with anomalism. Despite PITS and the anomalism of the mental and the psychophysical, it is possible, I shall argue, that there be social sciences that give causes as well as reasons for aggregative social phenomena. Although I shall eventually grant it as an assumption that there can be no psychology of individual action, both the position and its application to social science have ramifications that should first be considered.

We should note to begin that even if Davidson is right about the multiple realizability and anomalousness of the mental, it seems that multiple realization by itself is compatible with nomologicity in other "special sciences" (J. Fodor's term for all sciences except "basic physics" [1975, 9–26]). The premise of the "holistic" and "hermaphroditic" character of intentional mental states is crucial for concluding that the mental is anomalous; a consideration of other examples shows this. For instance, even though "river" and "money" are multiply realized in physics, there may well be *nomological* relationships among such things in geology and economics. Or, in more Fodorian language, there may be natural kinds that are best described in the vocabulary of geology despite multiple realization, even if it turns out that there are no natural kinds best described in the vocabulary of intentional psychology—geology does not depend on psychology. Whether the special sciences of geology and economics will "finally" be vindicated is an empirical question, if the only potential problem is multiple realization. That should be uncontroversial—at least for geology.[11] We need, therefore, to consider the status of the anomalism of the mental doctrine.

Since the multiple physical realizability of the mental and the consequent extreme unlikelihood of psychophysical bridge laws is insufficient to distinguish psychology from other special sciences with respect to anomalism, great weight must be laid upon the main pillar of anomalism. This is what Davidson calls "the holistic character of the cognitive field" (1980, 231) and the "conceptually hermaphroditic character of its generalizations" (p. 240). I shall call it "Davidson's holism." Davidson's holism is brought to bear on the psychology-geology distinction as follows. Objects such as rivers are, *pace* Heraclitus, relatively sharp spatiotemporal particulars whose properties can be investigated by many means: they can be mapped, chemically analyzed, have their temperature gradients measured, and so on. On the other hand, the Davidsonian continues, mental states can only be imputed to agents on the basis of our interpretation of their actions in accordance with considerations of rationality, coherence, and the like. There is no independent empirical access to mental states to enable us to improve upon our folk psychological generalizations. In order to achieve a fully airtight, "exact" explanation[12] of an action one would, in the end, be forced to bring in an indefinitely long account of the interplay of factors in the agent's mental makeup. In the lawlike special sciences we can, however, always appeal "in principle" to other sciences for additional methods of observation and more exact means of measurement in order to improve and better confirm our generalizations.[13]

This reply to the appeal to an analogy between psychology and the other special sciences leads to a dilemma for the Davidsonian. I take the first horn of this dilemma directly from Fodor.[14] If the holistic and hermaphroditic character of the mental is supposed to be a consequence of the fact that folk psychological generalizations cannot be significantly improved and made more exact *within* psychology, then psychology has still not been distinguished from geology. Consider the geological generalization: Meandering rivers erode their outside banks. This is obviously grossly inexact; someone might build a concrete wall on the outside banks, the water might freeze, etc. Improving this generalization requires moving outside the taxonomy derived from the vocabulary of geology. "Concrete wall," for example, needs to be brought in. Making this geological generalization exact, even in principle, will require ceteris paribus clauses of complexity comparable to those require for action explanations because of the indefinite number of things that can go "wrong" with meandering rivers.

If, however, the Davidsonian relies on the fact that the mental cannot be measured and specified except by attending to actions and considerations of rationality, coherence, and the like, he must deal with the second horn of the dilemma. It is that Davidson's holism is not supported by the inexactness and *internal* unimprovability of psychological generalizations any more than an analogous and clearly false holism about geology is. Davidson's holism is an independent thesis, so the positive argument for anomalism only receives support from it that is proportional to the strength of the reasons for accepting it. And whether independent measurements, specifications, and individuations of the mental are available is partly a matter of how to interpret the *empirical* psychological evidence, and not a fully philosophical matter. (Compare debates about the epistemological status of Quine's Indeterminacy of Translation thesis).

Let us recall that Davidson's rescue of causal individual action explanations from neo-Wittgensteinian difficulties can be regarded as employing two ideas—token identity–multiple realizability and mental anomalism–holism. The first idea suffices to show how action explanations can make true, singular causal claims and to block heavy-handed materialism. That is the hard part. Davidson's anomalism–holism plays a relatively small role in the attack on the neo-Wittgensteinians. It has been thought to leave some room for free will that might not be available otherwise, and it makes plausible the inadequacies of psychological science in the 1950s and 1960s that the neo-Wittgensteinians had their eyes on. But problems about free will are tangential to our present concerns, and the present state of psychology puts anomalism in some danger of proving to be empirically unreasonable.[15] In this light, it seems safe to conclude that any threat to social science that depends on anomalism must appear shadowy.

A comprehensive evaluation of anomalism is not possible here and not fully germane to present purposes, so I propose to set aside the evaluation. In the rest of this essay, I shall write as though we can rely on it because my primary purpose is to examine anomalism's connection to the alleged anomalism of the social. In order better to enforce my main contention that philosophy of social science can be more separate from philosophy of mind I shall, in fact, consider the implications of an even stronger form of anomalism than Davidson's. While all of the philosophers being called neo-Wittgensteinian opposed taking reasons for actions also to be causes, they were not so monolithic, nor were they always so clear, about what reasons and the like *were*. They often wrote as though they were content to think of reasons as *mental states*, much as Davidson is. This is not to say that they thought that reasons were states of a mental substance, something that Davidson also repudiates; they were considered to be states of something else—a person perhaps. Insofar as the neo-Wittgensteinians were content with this way of thinking, they departed from the teachings of the Master. Wittgenstein himself, I believe, objected to characterizing reasons and the rest as *states* and especially as *mental states*. I cannot defend this view or its attribution to Wittgenstein here, but an example may serve to convey enough of the view to show how it results in a strengthened form of anomalism.

Recall Wittgenstein's (1958, 20) account of expectation, a typical "intentional mental state,"

> What happens if from 4 till 4:30 A expects B to come to his room? In one sense in which the phrase "to expect something from 4 to 4:30" is used it certainly does not refer to one process or state of mind going on throughout that interval, but to a great many different activities and states of mind. If for instance I expect B to come to tea, what happens *may* be this: At four o'clock I look at my diary and see the name "B" against today's date; I prepare tea for two; I think for a moment "does B smoke?" and put out cigarettes; towards 4:30 I begin to feel impatient; I imagine B as he will look when he comes into my room. All this is called "expecting B from 4 to 4:30." And there are endless variations to this process which we all describe by the same expression.

Wittgenstein goes on to characterize expecting as a family resemblance notion by which he meant, among other things, that there is no single thing that all instances of expectation have in common. The same points are made in the description of hope (1958 [1953], sec. 584).

> Now suppose I sit in my room and hope that N.N. will come and bring me some money, and suppose one minute of this state could be isolated, cut out of its context; would what happened in it then not be hope?—Think, for example, of the words which you perhaps utter in this space of time. They are no longer part of language. And in different surroundings the institution of money doesn't exist either.

He is saying, I think, that "expects" and "hopes" can be properly applied only when a complicated set of conditions is met, and these conditions are by no means exclusively mental. He is not making the highly implausible suggestion that mental occurrences are not involved, but instead that they are not the whole story. The same goes for believing, desiring, wanting, reading, understanding, and the rest. I am interested here in the ontological thesis that this suggests. When *A* purchases bonds expecting the interest rate to fall, his expectation is not, on this view, *spatiotemporally* confined to him or to his mind. It is only appropriate to attribute this expectation to him if a complex set of social institutions is in place and *A* is in this instance appropriately related to them. Moreover, the term "expectation" will usually (but not always) be used inappropriately when it is confined to the mental.[16] If this is right, it might still be correct to say with Davidson that *A*'s buying the bonds was partially *caused* by his expectation, but it is very clear on this view that psychology is in worse shape than he thinks and anomalism will rule. Psychology has had enough trouble with the purely mental. If it also has essentially to incorporate social and physical conditions that can be characterized only unsystematically by "family resemblance" relationships, the enterprise will indeed be hopeless.[17]

I do not, of course, allude to this iconoclastic and scientifically pessimistic position on the ontology of the intentional (which may or may not be Wittgenstein's) in order to convince the reader of its truth. My strategy is to argue that social science is not rendered impossible even on the assumption of an anomalism *stronger* than the one most neo-Wittgensteinians argued for.[18] We can reject the idea that individual actions performed for specific reasons such as desires, wishes, or fears, are explained primarily by some locally supervenient mental state while accepting the PITS injunction that social reasons need to be somehow analyzed into individual (though perhaps "nonmental") reasons. We can maintain sensible versions of MI and PITS, yet still give a nonmiraculous account of how social science might work. In the rest of the essay, I shall mobilize two main kinds of arguments for a weakened social-mental connection and then provide the appropriate formulation of PITS.

The first consideration is based on a familiar type of phenomenon. Every action has consequences whose occurrence does not form part of the reason for the action; these consequences are unintended. Most of the time unintended consequences are so insignificant or so remote from the action and its intended consequences that they are of no interest to either the agent or someone trying to explain the agent's behavior. When one types on a keyboard, one also produces clacking or beeping noises, disturbs microscopic flora, and so on. The author of a philosophy paper is usually unaware of these relatively insignificant consequences of typing. Typing a philosophy paper might also lead through some weird and unforeseen causal chains to a truly momentous policy decision by the president (the First Dog may be disturbed by the clacking typewriter keys, start barking, interrupt the president's sleep, etc.; or, probably less likely, the president might actually read the paper). Psychology will be mostly powerless to explain unintended effects, and the explanations that we can sometimes give are rarely of any interest at the individual level. At the social level, however, the cumulative effect of the consequences of many unintended acts can be of great importance, and explanations of such effects will be interesting.[19]

A famous example of a significant aggregative social effect of individually unintended actions is the pollution of natural resources. An example of how unintended social regularities might arise is the Smithian emergence of an ordered economy from the selfish strivings of individual agents. In both of these cases it is easy to see why many have thought that interesting social scientific laws are awaiting discovery. Perhaps effects like pollution put nomological limits on the sizes or levels of development that societies can normally attain. Similarly, there are no strong prima facie reasons for thinking there are no scientific laws governing aggregate production and exchange in societies of certain kinds. The possibility of these laws need not rest in any way on psychology because the consequences of human actions that are important to these cases can be unintended and, therefore, outside its scope. Consequently, the question of anomalousness is irrelevant here because even if there were laws, they would in no way underlie the social laws in question—psychology might as well be anomalous for all the social scientists will care. I conclude that these examples provide a prima facie case against psychological anomalism entailing social anomalism. Let us postpone a consideration of objections to this until I bring out the second consideration.[20]

A second way of seeing how to break the connection between mental and social anomalism relies on another idea that is familiar in a different social scientific context. Suppose that anomalism rules at the individual level because the imprecision and extreme context dependence of family resemblance relations among reasons for action precludes interesting generalizations. It might still be possible to remove this impediment at the social level because the information that is lost in aggregating descriptions of individual reasons and actions into social ones might reveal lawlike regularities in what remains.

The following fantasy illustrates the idea. Lots of people may decide to buy bonds because (as we ordinarily say) they expect the interest rate to fall. At the same time, we may suppose that plenty of people who expect the interest rate to fall do not buy bonds and that plenty buy bonds for other reasons. Putting this together with Wittgenstein's reminder that "expecting the rate to fall" is realized in different instances in highly diverse ways, we suppose that a nomological connection between expectation and purchase is not forthcoming. Nevertheless, in a large economy it may well happen that there is a socially significant sale of bonds and that a good nomological explanation of this is that people (in general) expect the interest rate to fall. None of this depends on denying that it is *true* that lots of individuals, a majority say, expect interests rates to fall—Wittgenstein does not object to saying *A* expected *B*. Nor does it conflict with PITS. It is individuals that buy bonds and it is presumably their reactions to social conditions that usually lead them to individually expect that the interest rate will fall. Here is another example, this one loosely adapted from the theory of Durkheim's mentioned above. "Social cohesion" might turn out to be a quantifiable feature of societies and figure in significant social laws even though (a) it ontologically consists in degrees of cohesion among individuals, and (b) cohesion is inoperative or anomalous at the individual level. I am only relying on the assumption that the individual-transmission-of-the-social is, in this case, anomalous and that this case can be generalized.

A salient potential objection to this line of reasoning is that the envisioned aggregative explanation amounts to a winnowing that separates inessential complications from the kernels of law-supporting psychological natural kinds. If that is right, then the Wittgensteinian view is wrong and I am not, after all, entitled to assume anomalism. Some of the plausibility of this line of objection comes from what appears to be a close parallel between the imagined social law and the corresponding law for individuals. "When people generally expect the interest rate to fall, bonds sell faster" seems naturally related to "When an agent expects the interest rate to fall, he buys bonds." In other words, the objection continues, nomological relations at the social level that *result* from aggregation are the manifestations of psychological laws that are obscured by irrelevant complexity or "interference" at the individual level. Perhaps if it were possible to experimentally control for this interference, we could discover the psychological laws. Another thought is that such laws could emerge after numerous observations thanks to the "Law of Averages," if the interference is "random." Moreover, my examples from economics may be thought to work in just this way. A common (albeit not universally accepted) interpretation of how economic explanations work runs thus: first assume that individual agents act nomologically, attempting to maximize utility subject to income constraints. Next, use these (and some additional constraints) and mathematics to derive lawlike descriptions of some social entity, a market or an entire economy. Finally, insofar as the social descriptions are successful, it is supposed to be because the assumptions made about the nomological behavior of individuals are justified, either straightforwardly or by facts about averages. Even though few actual individuals might exhibit "average behavior," it is sometimes thought that the assumed law about individual behavior causally underlies the rather erratic behavior most actual agents exhibit.[21]

My reply to this objection comes in two parts. We should remember that economists' empirical descriptions of individual agents do not, as a matter of fact, work particularly well. It is undeniable, for example, that economists are more successful in empirically determining plausible supply and demand schedules for entire markets than they are for individuals. The thesis that individual economic behavior *directly* reflects utility maximization is no longer very popular. It can be concluded that the example of bond buying cannot be used to support the objection that social laws flow directly from analogous individual laws. So we should not be too quick to worry about the apparently easy reducibility of aggregative economic explanations to individualistic ones (compare Nelson 1984).

The point about averages is more subtle. With regard to economics in particular, it should be noted that derivations of successful macrolaws from postulated microlaws do not by themselves support the claim that success at the macro-level comes from successfully describing "average" individual behavior. Real macro-level regularities can exist even when behavior at the micro-level is highly diverse or random, so successful economics at what I am calling the social level does not clearly depend on individual or "average individual" nomologicity. It may be a law of some modest kind that Dodgers team batting average is always near .270, but it can be true at the same time that no Dodger regularly hits near .270.[22]

We are now in a position to reexamine PITS. There are two main reasons for adopting it. One is ontological—we do not want social entities to be causally efficacious without all the real work being done by individuals. The second attraction of PITS depends on the intentional nature of social scientific explanations. It would seem mysterious if nomological concepts that social science borrows from intentional explanations of individual behavior were proved unscientific in psychology. For both of these reasons, it may still seem difficult to reconcile PITS with this essay's arguments for separating the fates of mental science and social science. I see two principal difficulties. The first step in resolving them is to dismantle unacceptably strong versions of PITS and state more precisely what should be required by the Principle.

It requires above all that there be a nonsocially described mechanism through which social causes produce their effects. We now can see, however, that there is no good reason to require that the mechanism be psychological. Davidson said that reasons could be causes because reason/action pairs instantiated mechanisms covered by *physical* laws. Why should PITS require more? We need *some* mechanism covered by some laws to transmit social causes in an ontologically acceptable way, but a sociobiological, neurological, or even physical mechanism could do that job. It is a mistake to state PITS in terms of psychological transmission of the social. Why then is it called the Principle of the *Individualistic* Transmission of the Social if there are no descriptions of individual people in the approved mechanism?

PITS should indeed require that there be an analysis of social cause-effects pairs into individual behavior; it is the nomologicity that can be excluded. The problem is that when contemplating such analyses, there is a temptation to adopt two harmful Dogmas of Analysis. The First Dogma has it that the analysans and analysandum must be nomologically related to each other. For example, if pains are to be analyzed into C-fiber stimulations, it requires that it be true by law that pains are C-fiber stimulations. Davidson, Putnam, and Fodor have taught us that multiple realizability makes this First Dogma false. The Second Dogma says that the elements of the analysans must be covered by laws.[23] This Dogma seems better than the First. There are supposed to be laws governing C-fibers. Water is analyzed into bonded hydrogen and oxygen and these elements are covered by laws.

The Dogma is dangerous nevertheless. It is often possible to give a very enlightening scientific explanation of something by showing how it is instantiated at a more basic level of description. Pretend that we could show how consciousness could be instantiated in a machine. That would be enlightening even if the machine was very erratic and did not behave nomologically from the Maytag repairperson's point of view. Why-questions have to stop somewhere; optimistic mechanists like Boyle explained many phenomena in terms of colliding corpuscles, but they could not, and did not need to, produce a law explaining *why* solid corpuscles are inpenetrable. Successful explanatory analysis must, in principle, be able to strike laws of some kind at *some* level, that is partly why PITS requires a non-social mechanism, but not *every* level of analysis needs to be covered by laws. PITS is not violated by an analysis of the social into anomalous descriptions of individuals; it is only violated if there is no analysis into individuals.

I shall now state the suitably weakened and flexible version of PITS:

> Social scientific laws must in principle be (a) analyzable into individualistic instantiations, and (b) compatible with a nomological mechanism at some more basic level.

It is worth noting that nothing in this statement of PITS or the arguments leading up to it precludes the possibility of developing ways of using fact about individuals to measure and specify social properties. Surveys of individual expectations, for instance, may be a valuable source of information about "national confidence." Likewise, information derived from social observations might assist in the specification and measurement of individual properties. Nor does PITS require that these procedures be discoverable. Social science might or might not be improvable in Davidson's sense by reference to other sciences and vice versa. It also emerges that PITS is primarily an ontological restriction on theorizing and not an explanatory one. The individualistically inclined are suspicious of talk of social facts and even more suspicious of social forces so they are pleased to see more tangible mechanisms. It is not so much that social explanations are inherently unsatisfying, it is the worry that they essentially employ objectionable ontology that drives most individualists. PITS ensures that both social facts and social forces are ontologically acceptable and thus constitutes the true core of MI.

I shall consider one final apparent difficulty for my attempt to put some distance between the social and the mental. It is very natural to assume that the intentional concepts employed at the social and individual levels are *identical.* So when a social scientist asserts, "Americans believe that pork rinds are good" and a news reporter asserts, "Bush believes that pork rinds are good," it is often assumed that exactly the same thing is being asserted of Americans and of Bush. If that were right, then it really would be hard to accept an explanation of the social that did not build on an explanation of the individual. But the assumption of identity is unwarranted. If, as I have suggested, instances of intentional concepts are related only by family resemblance, then beliefs attributed to social entities obviously share no essence with individual beliefs. If they are in the same family as the individual beliefs, they will probably be distant relatives. Even those who reject the family resemblance story should be very wary of identifying the social and individual concepts. Much could be said here, but I confine myself to the following. "Bush wants pork rinds" on traditional accounts relates some pork rinds to the individual Bush, but "Americans want pork rinds" does not, on most traditional accounts, relate pork rinds to any individual. "Americans" do not even constitute a neat abstract individual like the University of California. Similarly (and more convincingly), on the traditional view, "Bush wants pork rinds" attributes a mental state to Bush, but "Americans want pork rinds" cannot sensibly be said to attribute a mental state at all. So why assume that the *relations* "*a* wants *x*" (individual) and "*B* wants *Y*" (social) are the same at both levels when there is this big difference in their behaviors?

I conclude that sensible guidelines for social science do not conflict with problems arising in attempts scientifically to explain the mental. I have tried to show that the natural seeming connections between the mental and the social are not strong enough to link their scientific fates.[24] Social scientists searching for laws incorporating intentional concepts are likely to benefit from examining advances in psychology, supposing that there are any. They might even be aided by studying the philosophy of mind. These benefits are not to be counted upon, however. Naturalistic social science should not be methodologically constrained by the results of our struggle to understand the mental.

Notes

1. I shall not continue always to distinguish sharply between the philosophy of social science and the social sciences themselves. Philosophy and physics were fairly cleanly separated three hundred years ago, but the umbilical cord had not been cut at all in the social sciences until very recently and even now it has not been completely cut through.

   A. Rosenberg has written a detailed genealogy of how the social sciences are enmeshed with the mental (1988). The organization of the present essay has been influenced by Rosenberg's work. Also useful in this regard are Macdonald and Pettit (1981) and Papineau (1978).
2. It is plain that some social scientific theories are not so clearly intentional. I do not know whether there are any interesting theories that have been completely purged of intentional concepts. It is also plain, however, that most theories presented as nonintentional are modeled after behavioristic psychology or are otherwise motivated by suspicions about the scientific tractability or, indeed, the very existence of the mental. Though I shall not discuss such theories, I note that they have been developed in accordance with methodological strictures against the mental. My conclusions are, therefore, also relevant to them.
3. I make this assumption to simplify the argument, but the essay's conclusions do not count against the possibility that interpretive social science is or will be more important than the "scientistic," naturalistic variety.
4. Discussions of MI that I have found especially useful are Miller (1978), Williams (1985), Kincaid (1986), Macdonald (1986a), and Sensat (1988). Taken together, these pieces reliably summarize the older literature.
5. The principle is very nicely developed for the following example in more detail in Rosenberg (1980, chapter 3). My summary is based on his treatment. Rosenberg also provides a thematic application of PITS to some structuralist anthropological theses. The PITS idea also appears in Macdonald and Pettit (1981, 127).
6. For a discussion of the relevant notion of "complete explanation," see Kincaid (1986).
7. It is this last consequence that Durkheim seems to have missed in formulating his radically anti-individualistic stance. It is sometimes suggested that he adopted this posture to facilitate the prizing of research monies from well-connected psychologists.
8. I shall roughly follow the exposition in Davidson (1980, 3–19). He gives a representative, though not complete, list of sources (p. 3). All references to Davidson's essays will be to page numbers in this edition.
9. The ideas appear throughout Davidson (1980), but they are most explicit in pp. 3–19, 207–24, and 245–59.
10. Unless otherwise indicated, I shall use "anomalism" to refer to the thesis that there are no mental-mental laws.
11. Well, maybe not for geology. Geology (the example given in Fodor 1987, 4–6) in fact seems to be in the process of being devoured by geophysics whose vocabulary tends toward the basic: normal modes, wave propagation, forces, fields, etc. One is reminded of the way that "political economy" was mostly devoured by neoclassical economics in the 1950s and 1960s. Perhaps the only uncontroversially "physics-like" special sciences are chemistry and biochemistry. If that turns out to be true, it is not clear whether it would be good news or bad news for the special sciences. It could mean that they are all, natural and social, not "really scientific," or it could mean that they are all "scientific" but very different from physics and the few physics-like sciences. Time will tell; meanwhile let us pretend, if necessary, that geology has the same status as chemistry.
12. In this context, part of the meaning of "fully airtight explanation" is an explanation strong enough to convince us that it relied on a causal generalization, that is, it would conform to the deductive-nomological model that Davidson and his intended audience apply to scientific explanation.
13. For a clear, detailed, textual treatment of Davidson's view on this matter, see Rosenberg (1985).

    Perhaps the starkest case of the difficulty Davidson has in mind is the revealed preference interpretation of economic utility theory. According to that doctrine, an economic agent's preferences (i.e., desires for commodities) are nothing more than a formal representation of some of the agent's intentional actions. For arguments concluding that economics so interpreted is patently unscientific, see Nelson (1986; 1990).
14. (1987, 4–6). Fodor is perhaps the greatest philosophical exponent of the view that fully intentional psychology is on equal methodological footing with other special sciences.

15. A very challenging critique of the doctrines, shared by Davidson and Fodor, about scientific properties and predicates that issue in this supposed difference between special sciences and physics can be found in Wilson (1985). Nelson (1985) is an unfortunately inadequate attempt to come to grips with Wilson's presentation. An interesting perspective on aspects of the Davidson-Fodor dispute is in Macdonald (1986b).

    For a fresh defense of the optimistic view about psychology, see Cummins (1989). Two recent approaches to reconciling a fully intentional approach to mental states with developments in non-intentional (or at least less intentional) sciences can be found in Dretske (1988) and Cummins (1989, chaps. 10 and 11). I emphasize that optimism about cognitive science is consistent with the main conclusions drawn in this essay.
16. I find myself at odds here with the interesting reconstruction required by the interpretation of Wittgenstein in McGinn (1984, 95–110).
17. One might think that the considerations Wittgenstein draws our attention to can be accommodated by the distinction between broad and narrow contents of mental states. Tyler Burge has argued (1979, for example) that the contents of mental states, and therefore the states themselves, cannot be fully individuated by reference to what supervenes on a person's body; his environment needs to be considered as well. This might inspire someone to say that mental states have a narrow content that can be fully individuated by reference to what supervenes on the body and a broad content that is individuated by other facts as well (see Fodor 1987, chap. 2). One might continue to say (against Wittgenstein) that mental state terms refer to things that are entirely within the body, but (more or less with Wittgenstein) that these referents cannot be entirely understood without information about the context of use. So psychology might, after all, be able to proceed by attending only to the "purely mental." I acknowledge that psychology might be defended against the Wittgensteinian considerations adumbrated above by making use of the notion of broad (see Burge [1986] for a worked-out example) or narrow (Fodor, 1987) content. But this is not the view that I am entertaining. I want the extreme case: if we insist on making *A*'s expectation an entity (and we probably should not), it is a spatiotemporally scattered object.
18. It should be mentioned that some have thought that even spatiotemporally scattered objects can be causally related (Thomson 1977, for example). A more moderate suggestion that has been applied to psychology (Horgan and Woodward 1985, 217–19) is that such objects can causally interact in virtue of some of their parts interacting in the ordinary way. A skeptical account of these theories is in Hornsby (1985). I shall not consider these ways of attacking anomalism.
19. This account could be enriched with a description of the different ways in which consequences of actions (and actions themselves) can be unintended, and the different kinds of remote effects of actions.
20. The complement of unintended consequences—unrecognized reasons—is also of interest here. The aforementioned social sciences of individual behavior often attempt to make use of unrecognized social reasons for individual actions. One may sometimes act for unrecognized or unacknowledged reasons of class membership, for instance.
21. A detailed description and critique of this interpretation is in Nelson (1986).
22. This point, along with a development of the conditions under which macrolaws *do* support claims about the average behavior of individuals, is worked out in Nelson (1989).
23. The term "analysis" and the treatment of what I call the Second Dogma is fairly loosely adapted from the highly illuminating Cummins (1983, chap. 1).
24. I have argued that if psychology were a grand failure, that would not entail that the social sciences must fail. My primary purpose, however, is not to provide a philosophical defense of social science; it is instead to argue that the connection between the scientific explanation of the social and of the mental is not as secure as has often been thought. To reinforce this position, it should be pointed out that a kind of inverse thesis is reasonable. The lawlikeness of the mental would not entail the lawlikeness of the social. Some arguments for this are canvassed in Nelson (1990).

References

Burge, T. 1979. "Individualism and the Mental." *Midwest Studies in Philosophy* 4: 73–121.
Burge, T. 1986. "Individualism and Psychology." *Philosophical Review* 95: 3–45.
Cummins, R. 1983. *The Nature of Psychological Explanation*. Cambridge, Mass.
Cummins, R. 1989. *Meaning and Mental Representation*. Cambridge, Mass.

Davidson, D. 1980. *Essays on Actions and Events.* Oxford.
Dray, W. 1957. *Laws and Explanation in History.* London.
Dretske, F. 1988. *Explaining Behavior.* Cambridge, Mass.
Durkheim, E. 1951 [1897]. *Suicide.* Translated by J. Spaulding and G. Simpson. Glencoe, III.
Fodor, J. 1987. *Psychosemantics.* Cambridge, Mass.
Fodor, J. 1975. *The Language of Thought.* New York.
Horgan, T., and J. Woodward. 1985. "Folk Psychology is Here to Stay." *Philosophical Review* 94: 197–226.
Hornsby, J. 1985. "Physicalism, Events, and Part-Whole Relations." In *Actions and Events,* edited by E. LePore and B. McLaughlin, Oxford.
Kincaid, H. 1986. "Reduction, Explanation, and Individualism." *Philosophy of Science* 53: 492–513.
LePore, E., and B, McLaughlin, editors. 1985. *Actions and Events.* Oxford.
Macdonald, G. 1986a. "Modified Methodologial Individualism." *Proceedings of the Aristotelian Society* 86: 199–211.
Macdonald, G. 1986b. "The Possibility of the Disunity of Science." In *Fact, Science and Morality,* edited by G. Macdonald and C. Wright. Oxford.
Macdonald, G., and P. Pettit. 1981. *Semantics and Social Science.* London.
McGinn, C. 1984. *Wittgenstein on Meaning.* Oxford.
Miller, R. 1978. "Methodological Individualism and Social Explanation." *Philosophy of Science* 45: 387–414.
Nelson, A. 1984. "Some Issues Surrounding the Reduction of Macroeconomics to Microeconomics." *Philosophy of Science* 51: 573–94.
Nelson, A. 1985. "Physical Properties." *Pacific Philosophical Quarterly* 66: 268–82.
Nelson, A. 1986. "New Individualistic Foundations for Economics." *Nous* 20: 469–90.
Nelson, A. 1989. "Average Explanations." *Erkenntnis* 30: 23–42.
Nelson, A. 1990. "Are Economic Kinds Natural?" *Scientific Theories: Minnesota Studies in the Philosophy of Science,* vol. 14. Minneapolis.
Papineau, D. 1978. *For Science in the Social Sciences.* London.
Rosenberg, A. 1980. *Sociobiology and the Preemption of Social Science.* Baltimore, Md.
Rosenberg, A. 1985. "Davidson's Unintended Attack on Psychology." In *Actions and Events,* edited by E. LePore and B. McLaughlin. Oxford.
Rosenberg, A. 1988. *Philosophy of Social Science.* Boulder, Colo.
Sensat, J. 1988. "Methodological Individualism and Marxism." *Economics and Philosophy* 4: 189–220.
Thomson, J. 1977. *Acts and Other Events.* Ithaca, N.Y.
Williams, B. 1985. "Formal and Substantial Individualism." *Proceedings of the Aristotelian Society* 85: 119–32.
Wilson, M. 1985. "What Is This Thing Called "Pain"?—The Philosophy of Science Behind the Contemporary Debate." *Pacific Philosophical Quarterly* 66: 227–67.
Winch, P. 1958. *The Idea of a Social Science.* London.
Wittgenstein, L. 1958 [1953]. *Philosophical Investigations,* 3d ed., translated by G. Anscombe. New York.
Wittgenstein, L. 1958. *The Blue and Brown Books.* Oxford.

PART VII

Objectivity and Values

Introduction to Part VII

Critics of the social sciences have often maintained that they are subjective while the natural sciences are objective. By "subjective" and "objective" one could mean at least two things. First, the subjective-objective contrast might simply be the contrast between inner mental phenomena and other types of phenomena. The claim that social scientific findings are subjective would then mean that they are about inner mental phenomena. Given this understanding of "subjective," there is some truth to the claim. Some psychological research, for examples, in psychophysics and clinical psychology, does lead to findings about inner mental states, whereas natural scientific research is never about these. However, not all social scientific research is subjective in this sense. Much research in macroeconomics and demography, for example, has little to do with establishing facts about inner mental events and processes.

In any case, this psychological sense of "subjective" is not usually what people have in mind when they say that social science is subjective and natural science is not. What is usually meant is that the findings of the social sciences are biased and unreliable. Closely connected to the claims about the subjectivity of the social sciences in this sense is the thesis that values influence social scientific inquiry. Indeed, this is one reason often given for supposing that the social sciences are biased.

Several crucial questions therefore need to be asked about the objectivity of the social sciences and the influence of values on them. How do values influence the social sciences? Can social science be value neutral? If it can, should it be? Are the findings of the natural sciences less biased than those of the social sciences? In particular, do values have less influence on the findings of the natural than on those of the social sciences? If so, what impact will this have on social science explanation? Will erosion of objectivity render the study of society nonscientific? Should social science have as one of its goals exposing the false assumptions, value influences, and biases that shape our thinking and actions? If so, how can this be done?

These questions are addressed in the selections that follow. Max Weber and Charles Taylor provide reasons for supposing that the social sciences are not value free or objective, and Ernest Nagel and Michael Martin defend the objectivity of the social sciences. Naomi Weisstein, Alison Wylie, and Donald E. Comstock, in turn, show how by exposing the false assumptions and biases that dominate our minds the social sciences can be critical.

In a selection from *The Methodology of the Social Sciences* in chapter 34, Weber, perhaps the most influential social scientist of the twentieth century, presents a complex position on the objectivity of the social sciences. Arguing on the one hand that there cannot be an objective analysis of social phenomena, he maintains that our values influence which problems are selected for social scientific investigation and how general laws are applied in explaining concrete social reality. All knowledge of concrete social reality, Weber says, is from "particular points of view." Maintaining on the other hand that a social scientist's values influence the construction of the conceptual scheme that is used in an investigation but not its truth, Weber nevertheless holds that the results of social research are valid not just for one person. "For scientific truth is precisely what is *valid* for all who *seek* the truth."

Whereas Weber discusses social science in general, Charles Taylor, in "Neutrality in Political Science" (chapter 35) challenges the accepted view that political science is value neutral. Citing studies from political science such as S. M. Lipset's *Political Man*, Gabriel Almond's, *Politics of the Developing Areas*, and Harold Laswell's, *Power and Society*, he argues that the theoretical frameworks that are indispensable in political science implicitly support certain values. For example, Harold Laswell's theory of the "dimensions of variations of political society," such as the degree to which polities further equality and impartiality and the degree to which they are libertarian or authoritarian, tacitly supports the values of democracy. Taylor goes on to criticize a crucial philosophical assumption of the value neutrality position: the possibility of the separation of facts and values. Arguing against this separation, he maintains that to say of something that it fulfills human needs and wants "always constitutes a *prima facie* reason for calling it 'good.'"

In contrast Ernest Nagel, in a selection from *The Structure of Science* (chapter 36) in which he considers four ways that values might enter into science, rejects conclusions such as Taylor's. Like Weber, he points out that values can enter into the selection of problems. He also demonstrates that they enter into the determination of conclusions, the identification of facts, and the assessment of evidence. But Nagel argues that none of these influences distinguishes the social sciences from the natural sciences or shows that unbiased conclusions in the social sciences are impossible. He stresses the importance of distinguishing value judgments that express approval (appraising value judgments) from ones that estimate the degree to which some attribute is embodied in a given instance (characterizing value judgments). Once this distinction is drawn, he says, there is no reason to suppose "that an ethically neutral social science is inherently impossible." Ethical neutrality would result in being committed only to characterizing value judgments. One wonders if this distinction could be applied to Weber's and Taylor's arguments and, if it can, how these authors might reply to Nagel.

One type of bias Nagel discusses is that found in the conclusions of scientific investigations. A striking instance of this is experimenter bias, that is, the unconscious bias exerted by experimenters on the results of their experiments. In a number of studies Robert Rosenthal, a social psychologist, has shown dramatically that the expectations of experimenters can bias their experimental results. The so-called Rosenthal effect has been cited in support of the claim that the social sciences are intrinsically subjective. But is this the appropriate conclusion to draw from Rosenthal's studies? Michael Martin, in "The Philosophical Importance of the Rosenthal Effect" (chapter 37), suggests a more optimistic appraisal of their import. Arguing that Rosenthal proposes experimental techniques for avoiding the Rosenthal effect—for example, by minimizing contact between the experimenter and the subjects and by correcting bias by statistical methods—he maintains that Rosenthal's research indicates that objectivity is possible.[1]

Arguing that the very preconceptions of male and female differences that have been used to dominate and oppress women have tainted the conclusions of psychologists, Naomi Weisstein, in "Psychology Constructs the Female" (chapter 38), provides another example of how values can influence scientific conclusions.[2] She maintains that theories about the innate nature of women are suspect since the best evidence suggests that both women and men tend to behave in terms of social expectations rather than innate psychological dynamics. Do the findings of Weisstein and other feminist scholars who have documented the gender biases of their fields show the inherent bias of social scientific research? It might be argued that this work demonstrates that biases are detectable but not inevitable, that they can be brought to the light, and that once they are they can be corrected and eliminated. How can this be accomplished? Weisstein suggests that one must not only challenge the prevailing theories but transcend the perspectives of men and women whose minds these theories have dominated and oppressed.

Does feminist methodology provide the rationale for this transcendence? Alison Wylie, in "Reasoning About Ourselves: Feminist Methodology in the Social Sciences" (chapter 39), argues that some feminists have imposed unnecessary a priori limits on social science research by insisting that it remain within female experience. She maintains that the consciousness raising that is so crucial for women must go beyond the outlook of oppressed women. Thus, Wylie explicitly stresses what Weisstein merely implies: the importance of going beyond the view of the participants. In taking this position Weisstein and Wylie seem to challenge the claim of the interpretive approach that one must take the point of view of the participants in social practices. They also suggest a goal for the social sciences quite different from those posited by either naturalism (prediction and explanation) or the interpretive approach (hermeneutic understanding): freedom from false and oppressive ideas that enslave our minds and our lives.

This strategy of transcending the oppressive beliefs of the social actors in a situation in order to gain freedom and enlightenment has been developed by critical social science into a codified approach to social science research.[3] Adopting this position, Donald E. Comstock, in "A Method of Critical Research" (chapter 40), argues that social scientific investigation has the ability to liberate human beings from the false assumptions that are used to dominate them. He presents a general procedure for a critical social science that he contrasts with positive social science—what we have here called the naturalistic approach. One key question that he does not address concerns the relevance of the naturalistic approach to critical social science. Could the discovery of causal laws of society, such as, laws that relate ideology to economic factors, be useful in liberating human beings? If so, would not this discovery necessitate transcending the perspective of the social actors and, thus, going beyond the methodology of interpretive social science? Considerations of questions about objectivity thus link up with earlier issues concerning explanations.

Notes

1. For ways that evidence constrains theoretical bias and thus prevents subjectivity see chapter 48: Alison Wylie, "Evidential Constraints: Pragmatic Objectivism in Archaelogy."
2. For further evidence of the gender bias of social science see chapter 48: Wylie, "Evidential Constraints: Pragmatic Objectivism in Archaelogy."
3. See chapter 48, where Wylie speaks of "the enlightenment tradition of empiricism."

Chapter 34

"Objectivity" in Social Science and Social Policy

Max Weber

There is no absolutely "objective" scientific analysis of culture—or put perhaps more narrowly but certainly not essentially differently for our purposes—of "social phenomena" independent of special and "one-sided" viewpoints according to which—expressly or tacitly, consciously or unconsciously—they are selected, analyzed, and organized for expository purposes. The reasons for this lie in the character of the cognitive goal of all research in social science which seeks to transcend the purely *formal* treatment of the legal or conventional norms regulating social life.

The type of social science in which we are interested is an *empirical science* of concrete *reality* (*Wirklichkeitswissenschaft*). Our aim is the understanding of the characteristic uniqueness of the reality in which we move. We wish to understand on the one hand the relationships and the cultural significance of individual events in their contemporary manifestations and on the other the causes of their being historically *so* and not *otherwise*. Now, as soon as we attempt to reflect about the way in which life confronts us in immediate concrete situations, it presents an infinite multiplicity of successively and coexistently emerging and disappearing events, both "within" and "outside" ourselves. The absolute infinitude of this multiplicity is seen to remain undiminished even when our attention is focused on a single "object," for instance, a concrete act of exchange, as soon as we seriously attempt an exhaustive description of *all* the individual components of this "individual phenomena," to say nothing of explaining it causally. All the analysis of infinite reality which the finite human mind can conduct rests on the tacit assumption that only a finite portion of this reality constitutes the object of scientific investigation, and that only it is "important" in the sense of being "worthy of being known." But what are the criteria by which this segment is selected? It has often been thought that the decisive criterion in the cultural sciences, too, was in the last analysis, the "regular" recurrence of certain causal relationships. The "laws" which we are able to perceive in the infinitely manifold stream of events must—according to this conception—contain the scientifically "essential" aspect of reality. As soon as we have shown some causal relationship to be a "law," that is, if we have shown it to be universally valid by means of comprehensive historical induction or have made it immediately and tangibly plausible according to our subjective experience, a great number of similar cases order themselves under the formula thus attained. Those elements in each individual event which are left unaccounted for by the selection of their elements subsumable under the "law" are considered as scientifically unintegrated residues which will be taken care of in the further perfection of the system of "laws." Alternatively they will be viewed as "accidental" and therefore scientifically unimportant *because* they do not fit into the structure of the "law"; in other words, they are not typical of the event and hence can only be the objects of "idle curiosity." Accordingly, even among the followers of the Historical School we continually find the attitude which declares that the ideal which all the sciences, including the cultural sciences, serve and toward which they should strive even in the remote future is a system of propositions from which reality can be "deduced." As is well known, a leading natural scientist believed that he could designate the (factually unattainable) ideal goal of such a treatment of cultural reality as a sort of "*astronomical*" knowledge.

Let us not, for our part, spare ourselves the trouble of examining these matters more closely—however often they have already been discussed. The first thing that impresses one is that the "astronomical" knowledge which was referred to is not a system of laws at all. On the contrary, the laws which it presupposes have been taken from other disciplines like mechanics. But it too concerns itself with the question of the *individual* consequence which the working of these laws in a unique *configuration* produces, since it is these individual configurations which are *significant* for us. Every individual constellation which it "explains" or predicts is causally explicable only as the consequence of another equally individual constellation which has preceded it. As far back as we may go into the grey mist of the far-off past, the reality to which the laws apply always remains equally *individual*, equally *undeducible* from laws. A cosmic "primeval state" which had no individual character or less individual character than the cosmic reality of the present would naturally be a meaningless notion. But is there not some trace of similar ideas in our field in those propositions sometimes derived from natural law and sometimes verified by the observation of "primitives," concerning an economic-social "primeval state" free from historical "accidents," and characterized by phenomena such as "primitive agrarian communism," sexual "promiscuity," etc., from which individual historical development emerges by a sort of fall from grace into concreteness?

The social-scientific interest has its point of departure, of course, in the *real*, i.e., concrete, individually structured configuration of our cultural life in its universal relationships which are themselves no less individually structured, and in its development out of other social cultural conditions, which themselves are obviously likewise individually structured. It is clear here that the situation which we illustrated by reference to astronomy as a limiting case (which is regularly drawn on by logicians for the same purpose) appears in a more accentuated form. Whereas in astronomy, the heavenly bodies are of interest to us only in their *quantitative* and exact aspects, the *qualitative* aspect of phenomena concerns us in the social sciences. To this should be added that in the social sciences we are concerned with psychological and intellectual (*geistig*) phenomena the empathic understanding of which is naturally a problem of a specifically different type from those which the schemes of the exact natural sciences in general can or seek to solve. Despite that, this distinction in itself is not a distinction in principle, as it seems at first glance. Aside from pure mechanics, even the exact natural sciences do not proceed without qualitative categories. Furthermore, in our own field we encounter the idea (which is obviously distorted) that at least the phenomena characteristic of a money-economy—which are basic to our culture—are quantifiable and on that account subject to formulation as "laws." Finally it depends on the breadth or narrowness of one's definition of "law" as to whether one will also include regularities which because they are not quantifiable are not subject to numerical analysis. Especially insofar as the influence of psychological and intellectual (*gestige*) factors is concerned, it does not in any case exclude the establishment of *rules* governing rational conduct. Above all, the point of view still persists which claims that the task of psychology is to play a role comparable to mathematics for the

Geisteswissenschaften in the sense that it analyzes the complicated phenomena of social life into their psychic conditions and effects, reduces them to their most elementary possible psychic factors, and then analyzes their functional interdependences. Thereby, a sort of "chemistry" if not "mechanics" of the psychic foundations of social life would be created. Whether such investigations can produce valuable and—what is something else—useful results for the cultural sciences, we cannot decide here. But this would be irrelevant to the question as to whether the aim of social-economic knowledge in our sense, i.e., knowledge of *reality* with respect to its cultural *significance* and its causal relationships, can be attained through the quest for recurrent sequences. Let us assume that we have succeeded by means of psychology or otherwise in analyzing all the observed and imaginable relationships of social phenomena into some ultimate elementary "factors," that we have made an exhaustive analysis and classification of them and then formulated rigorously exact laws covering their behavior.—What would be the significance of these results for our knowledge of the *historically* given culture or any individual phase thereof, such as capitalism, in its development and cultural significance? As an analytical tool, it would be as useful as a textbook of organic chemical combinations would be for our knowledge of the biogenetic aspect of the animal and plant world. In each case, certainly an important and useful preliminary step would have been taken. In neither case can concrete reality be deduced from "laws" and "factors." This is not because some higher mysterious powers reside in living phenomena (such as "dominants," "entelechies," or whatever they might be called). This, however, is a problem in its own right. The real reason is that the analysis of reality is concerned with the *configuration* into which those (hypothetical!) "factors" are arranged to form a cultural phenomenon which is historically significant to us. Furthermore, if we wish to "explain" this individual configuration "causally" we must invoke other equally individual configurations on the basis of which we will explain it with the aid of those (hypothetical!) "laws."

The determination of those (hypothetical) "laws" and "factors" would in any case only be the first of the many operations which would lead us to the desired type of knowledge. The analysis of the historically given individual configuration of those "factors" and their *significant* concrete interaction, conditioned by their historical context and especially the *rendering intelligible* of the basis and type of this significance would be the next task to be achieved. This task must be achieved, it is true, by the utilization of the preliminary analysis but it is nonetheless an entirely new and *distinct* task. The tracing as far into the past as possible of the individual features of these historically evolved configurations which are *contemporaneously* significant, and their historical explanations by antecedent and equally individual configurations would be the third task. Finally the prediction of possible future constellations would be a conceivable fourth task.

For all these purposes, clear concepts and the knowledge of those (hypothetical) "laws" are obviously of great value as heuristic means—but only as such. Indeed they are quite indispensable for this purpose. But even in this function their limitations become evident at a decisive point. In stating this, we arrive at the decisive feature of the method of the cultural sciences. We have designated as "cultural sciences" those disciplines which analyze the phenomena of life in terms of their cultural significance. The *significance* of a configuration of cultural phenomena and the basis of this significance cannot however be derived and rendered intelligible by a system of analytical laws (*Gesetzesbegriffen*), however perfect it may be, since the significance of cultural events presupposes a *value-orientation* toward these events. The concept of culture is a *value-concept*. Empirical reality becomes "culture" to us because and insofar as we relate it to value ideas. It includes those segments and only those segments of reality which have become significant to us because of this value-relevance. Only a small portion of existing concrete reality is colored by our value-conditioned interest and it alone is significant to us. It is significant because it reveals relationships which are important to us due to their connection with our values. Only because and to the extent that this is the case is it worthwhile for us to know it in its individual features. We cannot discover, however, what is meaningful to us by means of a "presuppositionless" investigation of empirical data. Rather perception of its meaningfulness to us is the presupposition of its becoming an *object* of investigation. Meaningfulness naturally does not coincide with laws as such, and the more general the law the less the coincidence. For the specific meaning which a phenomenon has for us is naturally *not* to be found in those relationships which it shares with many other phenomena.

The focus of attention on reality under the guidance of values which lend it significance and the selection and ordering of the phenomena which are thus affected in the light of their cultural significance is entirely different from the analysis of reality in terms of laws and general concepts. Neither of these two types of the analysis of reality has any necessary logical relationship with the other. They can coincide in individual instances but it would be most disastrous if their occasional coincidence caused us to think that they were not distinct *in principle*. The *cultural significance* of a phenomenon, e.g., the significance of exchange in a money economy, can be the fact that it exists on a mass scale as a fundamental component of modern culture. But the historical fact that it plays this role must be causally expained in order to render its cultural significance understandable. The analysis of the *general* aspects of exchange and the technique of the market is a—highly important and indispensable—*preliminary task*. For not only does this type of analysis leave unanswered the question as to how exchange historically acquired its fundamental significance in the modern world; but above all else, the fact with which we are primarily concerned, namely, the *cultural significance* of the money-economy, for the sake of which we are interested in the description of exchange technique and for the sake of which alone a science exists which deals with that technique—is not derivable from any "law." The *generic features* of exchange, purchase, etc., interest the jurist—but we are concerned with the analysis of the *cultural significance* of the concrete *historical* fact that today exchange exists on a mass scale. When we require an explanation, when we wish to understand what distinguishes the social-economic aspects of our culture, for instance, from that of antiquity in which exchange showed precisely the same generic traits as it does today and when we raise the question as to where the significance of "money economy" lies, logical principles of quite heterogeneous derivation enter into the investigation. We will apply those concepts with which we are provided by the investigation of the general features of economic mass phenomena—indeed, insofar as they are relevant to the meaningful aspects of our culture, we shall use them as *means* of exposition. The *goal* of our investigation is not reached through the exposition of those laws and concepts, precise as it may be. The question as to what should be the object of universal conceptualization cannot be decided "presuppositionlessly" but only with reference to the *significance* which certain segments of that infinite multiplicity which we call "commerce" have for culture. We seek knowledge of a historical phenomenon, meaning by historical: significant in its individuality (*Eigenart*). And the decisive element in this is that only through the presupposition that a finite part alone of the infinite variety of phenomena is significant, does the knowledge of an individual phenomenon become meaningful. Even with the widest imaginable knowledge of "laws," we are helpless in the face of the question: how is the *causal explanation* of an *individual* fact possible—since a *description* of even the smallest slice of reality can never be exhaustive? The number and type of causes which have influenced any given event are always infinite and there is nothing in the things themselves to set some of them apart as alone meriting attention. A chaos of "existential judgments" about countless individual events would be the only result of a serious attempt to analyze reality "without presuppositions." And even this result is only seemingly possible, since every single perception discloses on closer examination an infinite number of constituent perceptions which can never be exhaustively expressed in a judgment. Order is brought into this chaos only on the condition that in every case only a *part* of concrete reality is interesting and *significant* to us, because only it is related to the *cultural values* with which we approach reality. Only certain sides of the infinitely complex concrete phenomenon, namely, those to which we attribute a general *cultural significance*—are therefore worthwhile knowing. They alone are objects of causal explanation. And even this causal explanation evinces the same character; an *exhaustive* causal investigation of any concrete phenomenon in its full reality is not only practically impossible—it is simply nonsense. We select only those causes to which are to be imputed in the individual case, the "essential" feature of an event. Where the *individuality* of a phenomenon is concerned, the question of causality is not a question of *laws* but of concrete causal *relationships*; it is not a question of the subsumption of the event under some general rubric as a representative case but of its imputation as a consequence of some constellation. It is in brief a *question of imputation*. Wherever the causal explanation of a "cultural" phenomenon—a "historical individual"[1] is under consideration, the knowledge of causal *laws* is not the *end* of the investigation but only a *means*. It facilitates and renders possible the causal imputation to their concrete causes of those components of the phenomenon the individuality of which is culturally significant. So far and only so far as it achieves this, is it valuable for our knowledge of concrete relationships. And the more "general," i.e., the more abstract the laws, the less they can contribute to the causal imputation of *individual* phenomena and, more indirectly, to the understanding of the significance of cultural events.

What is the consequence of all this?

Naturally, it does not imply that the knowledge of *universal* propositions, the construction of abstract concepts, the knowledge of regularities, and the attempt to formulate "*laws*" have no scientific justification in the cultural sciences. Quite the contrary, if the causal knowledge of the historians consists of the imputation of concrete effects to concrete causes, a *valid* imputation of any individual effect without the application of "*nomological*" *knowledge*—i.e., the knowledge of recurrent causal sequences—would in general be impossible. Whether a single individual component of a relationship is, in a concrete case, to be assigned causal responsibility for an effect, the causal explanation of which is at issue, can in doubtful cases be determined only by estimating the effects which we *generally* expect from it and from the other components of the same complex which are relevant to the explanation. In other words, the "*adequate*" effects of the causal elements involved must be considered in arriving at any such conclusion. The extent to which the historian (in the widest sense of the word) can perform this imputation in a reasonably certain manner with his imagination sharpened by personal experience and trained in analytic methods and the extent to which he must have recourse to the aid of special disciplines which make it possible varies with the individual case. Everywhere, however, and hence also in the sphere of complicated economic processes, the more certain and the more comprehensive our general knowledge the greater is the *certainty* of imputation. This proposition is not in the least affected by the fact that even in the case of all so-called economic laws without exception, we are concerned here not with "laws" in the narrower exact natural science sense, but with *adequate* causal relationships expressed in rules and with the application of the category of "objective possibility." The establishment of such regularities is not the *end* but rather the *means* of knowledge. It is entirely a question of expediency, to be settled separately for each individual case, whether a regularly recurrent causal relationship of everyday experience should be formulated into a "law." Laws are important and valuable in the exact natural sciences, in the measure that those sciences are *universally valid*. For the knowledge of historical phenomena in their concreteness, the most general laws, because they are most devoid of content are also the least valuable. The more comprehensive the validity—or scope—of a term, the more it leads us away from the richness of reality since in order to include the common elements of the largest possible number of phenomena, it must necessarily be as abstract as possible and hence *devoid* of content. In the cultural sciences, the knowledge of the universal or general is never valuable in itself.

The conclusion which follows from the above is that an "objective" analysis of cultural events, which proceeds according to the thesis that the ideal of science is the reduction of empirical reality to "laws," is meaningless. It is not meaningless, as is often maintained, because cultural or psychic events, for instance, are "objectively" less governed by laws. It is meaningless for a number of other reasons. First, because the knowledge of social laws is not knowledge of social reality but is rather one of the various aids used by our minds for attaining this end; second, because knowledge of *cultural* events is inconceivable except on a basis of the *significance* which the concrete constellations of reality have for us in certain *individual* concrete situations. In *which* sense and in *which* situations this is the case is not revealed to us by any law; it is decided according to the *value-ideas* in the light of which we view "culture" in each individual case. "Culture" is a finite segment of the meaningless infinity of the world process, a segment on which *human beings* confer meaning and significance. This is true even for the human being who views a *particular* culture as a mortal enemy and who seeks to "return to nature." He can attain this point of view only after viewing the culture in which he lives from the standpoint of his values, and finding it "too soft." This is the purely logical-formal fact which is involved when we speak of the logically necessary rootedness of all historical entities (*historische Individuen*) in "evaluative ideas." The transcendental presupposition of every *cultural science* lies not in our finding a certain culture or any "culture" in general to be *valuable* but rather in the fact that we are *cultural beings*, endowed with the capacity and the will to take a deliberate attitude toward the world and to lend it *significance*. Whatever this significance may be, it will lead us to judge certain phenomena of human existence in its light and to respond to them as being (positively or negatively) meaningful. Whatever may be the content of this attitude—these phenomena have cultural significance for us and on this significance alone rests its scientific interest. Thus when we speak here of the conditioning of cultural knowledge through *evaluative* ideas (*Wertideen*) (following the terminology of modern logic), it is done in the hope that we will not be subject to crude misunderstandings such as the opinion that cultural significance should be attributed only to *valuable* phenomena. Prostitution is a *cultural* phenomenon just as much as religion or money. All three are cultural phenomena *only* because and *only* insofar as their existence and the form which they historically assume touch directly or indirectly on our cultural *interests* and arouse our striving for knowledge concerning problems brought into focus by the evaluative ideas which give *significance* to the fragment of reality analyzed by those concepts.

All knowledge of cultural reality, as may be seen, is always knowledge from *particular points of view*. When we require from the historian and social research worker as an elementary presupposition that they distinguish the important from the trivial and that they should have the necessary "point of view" for this distinction, we mean that they must understand how to relate the events of the real world consciously or unconsciously to universal "cultural values" and to select out those relationships which are significant for us. If the notion that those standpoints can be derived from the "facts themselves" continually recurs, it is due to the naive self-deception of the specialist who is unaware that it is due to the evaluative ideas with which he unconsciously approaches his subject matter, that he has selected from an absolute infinity a tiny portion with the study of which he *concerns* himself. In connection with this selection of individual special "aspects" of the event which always and everywhere occurs, consciously or unconsciously, there also occurs that element of cultural-scientific work which is referred to by the often-heard assertion that the "personal" element of a scientific work is what is really valuable in it, and that personality must be expressed in every work if its existence is to be justified. To be sure, without the investigator's evaluative ideas, there would be no principle of selection of subject matter and no meaningful knowledge of the concrete reality. Just as without the investigator's conviction regarding the significance of particular cultural facts, every attempt to analyze concrete reality is absolutely meaningless, so the direction of his personal belief, the refraction of values in the prism of his mind, gives direction to his work. And the values to which the scientific genius relates the object of his inquiry may determine, i.e., decide the "conception" of a whole epoch, not only concerning what is regarded as "valuable" but also concerning what is significant or insignificant, "important" or "unimportant" in the phenomena.

Accordingly, cultural science in our sense involves "subjective" presuppositions insofar as it concerns itself only with those components of reality which have some relationship, however indirect, to events to which we attach cultural *significance*. Nonetheless, it is entirely *causal* knowledge exactly in the same sense as the knowledge of significant concrete (*individueller*) natural events which have a qualitative character. Among the many confusions which the overreaching tendency of a formal-juristic outlook has brought about in the cultural sciences, there has recently appeared the attempt to "refute" the "materialistic conception of history" by a series of clever but fallacious arguments which state that since all economic life must take place in legally or conventionally *regulated forms*, all economic "development" must take the form of striving for the creation of new *legal* forms. Hence, it is said to be intelligible only through ethical maxims and is on this account essentially different from every type of "natural" development. Accordingly the knowledge of economic development is said to be "teleological" in character. Without wishing to discuss the meaning of the ambiguous term "development," or the logically no less ambiguous term "teleology" in the social sciences, it should be stated that such knowledge need not be "teleological" in the sense assumed by this point of view. The cultural significance of normatively regulated legal *relations* and even norms themselves can undergo fundamental revolutionary changes even under conditions of the formal identity of the prevailing legal norms. Indeed, if one wishes to lose one's self for a moment in fantasies about the future, one might theoretically imagine, let us say, the "socialization of the means of production" unaccompanied by any conscious "striving" toward this result, and without even the disappearance or addition of a single paragraph of our legal code; the statistical frequency of certain legally regulated relationships might be changed fundamentally, and in many cases, even disappear entirely; a great number of legal norms might become *practically* meaningless and their whole cultural significance changed beyond identification. *De lege ferenda* discussions may be justifiably disregarded by the "materialistic conception of history" since its central proposition is the indeed inevitable change in the *significance* of legal institutions. Those who view the painstaking labor of causally understanding historical reality as of secondary importance can disregard it, but it is impossible to supplant it by any type of "teleology." From our viewpoint, "purpose" is the conception of an *effect* which becomes a cause of an action. Since we take into account every cause which produces or can produce a significant effect, we also consider this one. Its specific significance consists only in the fact that we not only *observe* human conduct but can and desire to understand it.

Undoubtedly, all evaluative ideas are "subjective." Between the "historical" interest in a family chronicle and that in the development of the greatest conceivable cultural phenomena which were and are common to a nation or to mankind over long epochs, there exists an infinite gradation of "significance" arranged into an order which differs for each of us. And they are, naturally, historically variable in accordance with the character of the culture and the ideas which rule men's minds. But it obviously does not follow from this that research in the cultural sciences can only have results which are "subjective" in the sense that they are *valid* for one person and not for others. Only the degree to which they interest different persons varies. In other words, the choice of the object of investigation and the extent or depth to which this investigation attempts to penetrate into the infinite causal web are determined by the evaluative ideas which dominate the investigator and his age. In the *method* of investigation, the guiding "point of view" is of great importance for the *construction* of the conceptual scheme which will be used in the investigation. In the mode of their *use*, however, the investigator is obviously bound by the norms of our thought just as much here as elsewhere. For scientific truth is precisely what is *valid* for all who *seek* the truth.

However, there emerges from this the meaninglessness of the idea which prevails occasionally even among historians, namely, that the goal of the cultural sciences, however far it may be from realization, is to construct a closed system of concepts, in which reality is synthesized in some sort of *permanently* and *universally* valid classification and from which it can again be deduced. The stream of immeasurable events flows unendingly toward eternity. The cultural problems which move men form themselves ever anew and in different colors, and the boundaries of that area in the infinite stream of concrete events which acquires meaning and significance for us, i.e., which becomes a "historical individual," are constantly subject to change. The intellectual contexts from which it is viewed and scientifically analyzed shift. The points of departure of the cultural sciences remain changeable throughout the limitless future as long as a Chinese ossification of intellectual life does not render mankind incapable of setting new questions to the eternally inexhaustible flow of life. A systematic science of culture, even only in the sense of a definitive, objectively valid, systematic fixation of the problems which it should treat, would be senseless in itself. Such an attempt could only produce a collection of numerous, specifically particularized, heterogeneous, and disparate viewpoints in the light of which reality becomes "culture" through being significant in its unique character....

Returning to our special case.[2] It may be asserted without the possibility of a doubt that as soon as one seeks to derive concrete directives from practical political (particularly economic and social-political) evaluations, (1) the indispensable means, and (2) the inevitable repercussions, and (3) the thus conditioned competition of numerous possible evaluations in their *practical* consequences, are all that an *empirical* discipline can demonstrate with the means at its disposal. Philosophical disciplines can go further and lay bare the "meaning" of evaluations, i.e., their ultimate meaningful structure and their meaningful consequences, in other words, they can indicate their "place" within the totality of all the possible "ultimate" evaluations and delimit their spheres of meaningful validity. Even such simple questions as the extent to which an end should sanction unavoidable means, or the extent to which undesired repercussions should be taken into consideration, or how conflicts between several concretely conflicting ends are to be arbitrated, are entirely matters of choice or compromise. There is no (rational or empirical) scientific procedure of any kind whatsoever which can provide us with a decision here. The social sciences, which are strictly empirical sciences, are the least fitted to presume to save the individual the difficulty of making a choice, and they should therefore not create the impression that they can do so.

Finally it should be explicitly noted that the recognition of the existence of this situation is, as far as our disciplines are concerned, completely independent of the attitude one takes toward the very brief remarks made above regarding the theory of value. For there is, in general, no logically tenable standpoint from which it could be denied except a hierarchical ordering of values unequivocally prescribed by *ecclesiastical* dogmas. I need not consider whether there really are persons who assert that such problems as (*a*) does a concrete event occur thus and so or otherwise, or (*b*) why do the concrete events in question occur thus and so and not otherwise, or (*c*) does a given event ordinarily succeed another one according to a certain law and with what degree of probability—are not basically different from the problems: (a^1) what should one do in a concrete situation, or (b^2) from which standpoints may those situations be satisfactory or unsatisfactory, or (c^3) whether they are—whatever their form—generally formulatable propositions (axioms) to which these standpoints can be reduced. There are many who insist further that there is no logical disjunction between such enquiries as, (*a*) in which direction will a concrete situation (or generally, a situation of a certain type) develop and with what greater degree of probability in which particular direction than in any other and (*b*) a problem which investigates whether one *should* attempt to influence the development of a certain situation in a given direction—regardless of whether it be the one in which it would also move if left alone, or the opposite direction or one which is different from either. There are those who assert that (*a*) the problem as to which attitudes toward any given problem specified persons or an unspecified number of persons under specified conditions will probably or even certainly take and (*b*) the problem as to whether the attitude which emerged in the situation referred to above is *right*—are in no way different from one another. The proponents of such views will resist any statement to the effect that the problems in the above-cited juxtapositions do not have even the slightest connection with one another and that they really are "to be separated from one another." These persons will insist furthermore that their position is not in contradiction with the requirements of scientific thinking. Such an attitude is by no means the same as that of an author who conceding the absolute heterogeneity of both types of problems, nevertheless, in one and the same book, on one and the same page, indeed in a principal and subordinate clause of one and the same sentence, makes statements bearing on each of the two heterogeneous problems referred to above. Such a procedure is strictly a matter of choice. All that can be demanded of him is that he does not unwittingly (or just to be clever) deceive his readers concerning the absolute heterogeneity of the problems. Personally I am of the opinion that nothing is too "pedantic" if it is useful for the avoidance of confusions.

Thus, the discussion of value-judgments can have only the following functions:

a. The elaboration and explication of the ultimate, internally "consistent" value-axioms, from which the divergent attitudes are derived. People are often in error, not only about their opponent's evaluations, but also about their own. This procedure is essentially an operation which begins with concrete particular evaluations and analyzes their meanings and then moves to the more general level of irreducible evaluations. It does not use the techniques of an empirical discipline and it produces no new knowledge of facts. Its "validity" is similar to that of logic.

b. The deduction of "implications" (for those accepting certain value-judgments) which follow from certain irreducible value-axioms, when the practical evaluation of factual situations is based on these axioms alone. This deduction depends on one hand, on logic, and on the other, on empirical observations for the completest possible casuistic analyses of all such empirical situations as are in principle subject to practical evaluation.

c. The determination of the factual consequences which the realization of a certain practical evaluation must have: (1) in consequence of being bound to certain indispensable means, (2) in consequence of the inevitability of certain, not directly desired repercussions. These purely empirical observations may lead us to the conclusion that (*a*) it is absolutely impossible to realize the object of the preference, even in a remotely approximate way, because no means of carrying it out can be discovered; (*b*) the more or less considerable improbability of its complete or even approximate realization, either for the same reason or because of the probable appearance of undesired repercussions which might directly or indirectly render the realization undesirable, (*c*) the necessity of taking into account such means or such repercussions as the proponent of the practical postulate in question did not consider, so that his evaluation of end, means, and repercussions becomes a new problem for him. Finally: (*d*) the uncovering of new axioms (and the postulates to be drawn from them) which the proponent of a practical postulate did not take into consideration. Since he was unaware of those axioms, he did not formulate an attitude toward them although the execution of his own postulate conflicts with the others either (1) in principle or (2) as a result of the practical consequences, (i.e., logically or actually). In (1) it is a matter in further discussion of problems of type (*a*); in (2), of type (*c*).

Far from being meaningless, value-discussions of this type can be of the greatest utility as long as their potentialities are correctly understood.

The utility of a discussion of practical evaluations at the right place and in the correct sense is, however, by no means exhausted with such direct "results." When correctly conducted, it can be extremely valuable for empirical research in the sense that it provides it with problems for investigation.

The problems of the empirical disciplines are, of course, to be solved "nonevaluatively." They are not problems of evaluation. But the problems of the social sciences are selected by the value-relevance of the phenomena treated. Concerning the significance of the expression "relevance to values" I refer to my earlier writings and above all to the works of Heinrich Rickert and will forbear to enter upon that question here. It should only be recalled that the expression "relevance to values" refers simply to the philosophical interpretation of that specifically scientific "interest" which determines the selection of a given subject-matter and the problems of an empirical analysis.

Notes

1. We will use the term which is already occasionally used in the methodology of our discipline and which is now becoming widespread in a more precise formulation in logic.
2. The concluding pages of this selection are reprinted from pp. 18–22. Max Weber, *Methodology of the Social Sciences.*

Chapter 35

Neutrality in Political Science

Charles Taylor

1

I

A few years ago one heard it frequently said that political philosophy was dead, that it had been killed by the growth of science, the growth of positivism, the end of ideology, or some combination of these forces, but that, whatever the cause, it was dead.

It is not my intention to rake over the coals of this old issue once more. I am simply using this as a starting point for a reflection on the relation between political science and political philosophy. For behind the view that political philosophy was dead, behind any view which holds that it *can* die, lies the belief that its fate can be separated from that of political science; for no one would claim that the science of politics is dead, however one might disapprove of this or that manner of carrying it on. It remains a perpetually possible, and indeed important enterprise.

The view was indeed that political science has come of age in freeing itself finally of the incubus of political philosophy. No more would its scope be narrowed and its work prejudiced by some value position which operated as an initial weight holding back the whole enterprise. The belief was that political science had freed itself from philosophy in becoming value free and in adopting the scientific method. These two moves were felt to be closely connected; indeed, the second contains the first. For scientific method is, if nothing else, a dispassionate study of the facts as they are, without metaphysical presuppositions, and without value biases.

As Vernon van Dyke puts it:

> *science* and *scientific,* then, are words that relate to only one kind of knowledge, i.e., to knowledge of what is observable, and not to any other kinds of knowledge that may exist. They do not relate to alleged knowledge of the normative—knowledge of what ought to be. Science concerns what has been, is, or will be, regardless of the "oughts" of the situation (*Political Science.* Stanford and London: Stanford University Press, 1960, p. 192).

Those who could hold that political philosophy was dead, therefore, were those who held to a conception of the social sciences as *wertfrei*; like natural science, political science must dispassionately study the facts. This position received support from the views of the logical empiricists who had, for philosophers, an extraordinarily wide influence among scientists in general, and among the sciences of man in particular. Emboldened by their teaching, some orthodox political scientists tended to claim that the business of normative theory, making recommendations and evaluating different courses of action, could be entirely separated from the study of the facts, from the theoretical attempt to account for them.

Many, of course, had doubts; and these doubts seem to be growing today among political scientists. But they do not touch the thesis of the logical separation between fact and value. They center rather around the possibility of setting one's values to one side when one undertakes the study of politics. The relation between factual study and normative beliefs is therefore thought of in the same traditional positivist way: that the relationship if any is from value to fact, not from fact to value. Thus, scientific findings are held to be neutral: that is, the facts as we discover them do not help to establish or give support to any set of values; we cannot move from fact to value. It is, however, often admitted that our values can influence our findings. This can be thought of as a vicious interference, as when we approach our work with bias which obscures the truth, or as something anodyne and inevitable, as when our values select for us the area of research on which we wish to embark. Or it can be thought of as a factor whose ill effects can be compensated by a clear consciousness of it: thus many theorists today recommend that one set out one's value position in detail a the beginning of a work so as to set the reader (and perhaps also the writer) on guard.

Value beliefs remain therefore as unfounded on scientific fact for the new generation of more cautious theorists as they were for the thinkers of the hey-day of "value-freedom." They arise, as it were, from outside factual study; they spring from deep choices which are independent of the facts. Thus David Easton, who goes on to attempt to show that "whatever effort is exerted, in undertaking research we cannot shed our values in the way we remove our coats" (*The Political System*, New York: Knopf, 1953, p. 225), nevertheless states his acceptance at the outset of the "working assumption" which is "generally adopted today in the social sciences," and which "holds that values can ultimately be reduced to emotional responses conditioned by the individual's total life-experiences" (p.221). Thus there is no question of founding values on scientific findings. Emotional responses can be explained by life-experience, but not justified or shown to be appropriate by the facts about society:

> The moral aspect of a proposition ... expresses only the emotional response of an individual to a state of real or presumed facts.... Although we can say that the aspect of a proposition referring to a fact can be true or false, it is meaningless to characterize the value aspect of a proposition in this way. (ibid.)

The import of these words is clear. For, if value positions could be supported or undermined by the findings of science, then they could not simply be characterized as emotional responses, and we could not say simply that it was *meaningless* (although it might be misleading) to speak of them as true or false.

Political philosophy, therefore, as reasoned argument about fundamental political values, can be entirely separated from political science, even on the mitigated positivist view which is now gaining ground among political scientists. "Values" steer, as it were, the process of discovery, but they do not gain or lose plausibility by it. Thus, although values may be somehow ineradicable from political science, reasoned argument concerning them would seem easily separable (though theorists may differ as to whether this is wise or not: cf. Easton, op. cit.). Indeed, it is hard to see in what such reasoned argument could consist. The findings of science will be relevant to our values, of course, in this sense, that they will tell us how to realize the goals we set ourselves. We can reconstruct political science in the mold of a "policy science," like engineering and medicine, which shows us how to attain our goals. But the goals and values still come from somewhere else; they are founded on choices whose basis remains obscure.

The aim of this chapter is to call into question this notion of the relation of factual findings in politics to value positions, and thus the implied relation between political science and political philosophy. In particular my aim is to call into question the view that the findings of political science leave us, as it were, as free as before, that they do not go some way to establishing particular sets of values and undermining others. If this view is shown to be mistaken, then we will have to recognize a convergence between science and normative theory in the field of politics.

It is usual for philosophers, when discussing this question, to leave the realms of the sciences of man and launch into a study of "good," or commending, or emotive meaning, and so on. I propose to follow another course here, and to discuss the question first in connection with the disciplines in terms of which I have raised it, namely, political philosophy and political science. When we have some understanding of the relations between these two on the ground, as it were, it will be time to see if these are considered possible in the heavens of philosophy.

II

The thesis that political science is value neutral has maximum plausibility when we look at some of its detailed findings. That French workers tend to vote Communist may be judged deplorable or encouraging, but it does not itself determine us to accept either of these judgments. It stands as a fact, neutral between them.

If this were all there is to political science, the debate would end here. But it is no more capable than any other science of proceeding by the random collection of facts. At one time it was believed that science was just concerned with the correlation of observable phenomena—the observables concerned being presumed to lie unproblematically before our gaze. But this position, the offshoot of a more primitive empiricism, is abandoned now by almost everyone, even those in the empiricist tradition.

For the number of features which any given range of phenomena may exhibit, and which can thus figure in correlations, is indefinite; and this because the phenomena themselves can be classified in an indefinite number of ways. Any physical object can be classified according to shape, color, size, function, aesthetic properties, relation to some process, etc.; when we come to realities as complex as political society, the case is no different. But among these features only a limited range will yield correlations which have some explanatory force.

Nor are these necessarily the most obtrusive. The crucial features, laws or correlations concerning which will explain or help to explain phenomena of the range in question, may at a given stage of the science concerned be only vaguely discerned if not frankly unsuspected. The conceptual resources necessary to pick them out may not yet have been elaborated. It is said, for instance, that the modern physical concept of mass was unknown to the ancients, and only slowly and painfully evolved through the searchings of the later Middle Ages. And yet it is an essential variable in the modern science. A number of more obtrusive features may be irrelevant; that is, they may not be such that they can be linked in functions explanatory of the phenomena. Obvious distinctions may be irrelevant, or have an entirely different relevance from that attributed to them, such as the distinction between Aristotle's "light" and "heavy" bodies.

Thus when we wish to go beyond certain immediate low-level correlations whose relevance to the political process is fairly evident, such as the one mentioned above;

when we want to explain why French workers vote Communist, or why McCarthyism arises in the United States in the late 1940s, or why the level of abstentionism varies from election to election, or why new African regimes are liable to military take-over, the features by reference to which we can explain these results are not immediately in evidence. Not only is there a wider difference of opinion about them, but we are not even sure that we have as yet the conceptual resources necessary to pick them out. We may easily argue that certain more obtrusive features, those pertaining, say, to the institutional structure, are not relevant, while others less obtrusive, say, the character structure prevalent in certain strata of the society, will yield the real explanation. We may, for instance, refuse to account for McCarthyism in terms of the struggle between executive and legislature, and look rather to the development of a certain personality structure among certain sections of the American population. Or else we may reject both these explanations and look to the role of a new status group in American society, newly rich but excluded from the eastern establishment. Or we may reject this, and see it as a result of the new position of the United States in the world.

The task of theory in political science, one which cannot be forgone if we are to elaborate any explanations worth the name, is to discover what are the kinds of features to which we should look for explanations of this kind. In which of the above dimensions are we to find an explanation for McCarthyism? Or rather, since all of these dimensions obviously have relevance, how are we to relate them in explaining the political phenomena? The task of theory is to delineate the relevant features in the different dimensions and their relation so that we have some idea of what can be the cause of what, of how character affects political process, or social structure affects character, or economic relations affect social structure, or political process affects economic relations, or vice versa; how ideological divisions affect party systems, or history affects ideological divisions, or culture affects history, or party systems affect culture, or vice versa. Before we have made some at least tentative steps in this direction we don't even have an idea where to look for our explanations; we don't know which facts to gather.

It is not surprising, then, that political science should be the field in which a great and growing number of "theoretical frameworks" compete to answer these questions. Besides the Marxist approach, and the interest-group theory associated with the name of Bentley, we have seen the recent growth of "structural-functional" approaches under the influence of systems theory; there have been approaches which have attempted to relate the psychological dimension to political behavior (e.g., Lasswell), different applications of sociological concepts and methods (e.g., Lipset and Almond), applications of game theory (e.g., Downs and Riker), and so on.

These different approaches are frequently rivals, since they offer different accounts of the features crucial for explanation and the causal relations which hold. We can speak of them, along with their analogues in other sciences, as "conceptual structures" or "theoretical frameworks," because they claim to delimit the area in which scientific inquiry will be fruitful. A framework does not give us at once all the variables which will be relevant and the laws which will be true, but it tells us what needs to be explained, and roughly by what kinds of factors. For instance, if we accept the principle of Inertia, certain ways of conceiving bodies and therefore certain questions are beyond the pale. To pursue them is fruitless, as was the search for what kept the cannon ball moving in pre-Galilean physics. Similarly an orthodox Marxist approach cannot allow that McCarthyism can be explained in terms of early upbringing and the resultant personality structure.

But we can also see a theoretical framework as setting the crucial dimensions through which the phenomena can vary. For it sets out the essential functional relations by which they can be explained, while at the same time ruling out other functional relations belonging to other, rival frameworks. But the given set of functional relations defines certain dimensions in which the phenomena can vary; a given framework therefore affirms some dimensions of variation and denies others. Thus, for a Marxist, capitalist societies do not vary as to who wields power, no matter what the constitution or the party in office; supposed variations in these dimensions, which are central to a great many theories, are sham; the crucial dimension is that concerning class structure.

In the more exact sciences theoretical discovery may be couched in the form of laws and be called principles, such as, e.g., of inertia, or the rectilinear propagation of light. But in the less exact, such as politics, it may consist simply of a general description of the phenomena couched in the crucial concepts. Or it may be implicit in a series of distinctions which a given theory makes (e.g., Aristotle's classification of the types of polity), or in a story of how the phenomena came to be (e.g., the myth of the social contract), or in a general statement of causal relations (e.g., Marx's Preface to *A Contribution to the Critique of Political Economy*).

But, however expressed, theoretical discovery can be seen as the delineating of the important dimensions of variation for the range of phenomena concerned.

III

Theoretical discovery of this kind is thus one of the concerns of modern political science, as we have seen. But it also is a traditional concern of what we call political philosophy, that is, normative political theory. It is not hard to see why. Normative theorists of the tradition have also been concerned with delineating crucial dimensions of variation—of course, they were looking for the dimensions which were significant for judging of the value of polities and policies rather than for explaining them. But the two types of research were in fact closely interwoven so that in pursuing the first they were also led to pursue the second.

Aristotle, for instance, is credited with a revision of Plato's threefold classification of political society which enhanced its explanatory value. He substituted for the number criterion a class criterion which gives a more revealing classification of the differences, and allows us to account for more: it made clear what was at stake between democracy and oligarchy; it opened up the whole range of explanations based on class composition, including the one for which Aristotle is known in history, the balancing role of the middle class.

But this revision was not unconnected with differences in the normative theory of the two thinkers. Plato attempted to achieve a society devoid of class struggle, either in the perfect harmony of the *Republic*, or in the single class state of the *Laws*. Aristotle is not above weaving the dream of the ideal state in one section of the *Politics*, but there is little connection between this and the political theory of the rest of the work. This latter is solidly based on the understanding that class differences, and hence divergence of interest and tension, are here to stay. In the light of this theory, Plato's idea in the *Republic* of overcoming class tension by discipline, education, a superior constitution, and so on, is so much pie-in-the-sky (not even very tasty pie in Aristotle's view, as he makes clear in Book II, but that is for other reasons).

Aristotle's insight in political science is incompatible with Plato's normative theory, at least in the *Republic*, and the *Politics* therefore takes a quite different line (for other reasons as well, of course). The difference on this score might perhaps be expressed in this way: both Plato and Aristotle held that social harmony was of crucial importance as a value. But Plato saw this harmony as achieved in the ending of all class conflict; Aristotle saw it as arising from the domestication of this conflict. But crucial to this dispute is the question of the causal relevance of class tension: is it an eradicable blot on social harmony, in the sense that one can say, for instance, that the violent forms of this conflict are? Or is it ineradicable and ever-present, only varying in its forms? In the first case one of the crucial dimensions of variation of our explanatory theory is that concerning the presence or absence of class conflict. In the second case, this dimension is not even recognized as having a basis in fact. If this is so, then the normative theory collapses, or rather is shifted from the realm of political philosophy to that we call utopia building. For the idea of a society without class conflict would be one to which we cannot even approach. Moreover, the attempt to approach it would have all the dangerous consequences attendant on large-scale political changes based on illusory hopes.

Thus Plato's theory of the *Republic*, considered as the thesis that a certain dimension of variation is normatively significant, contains claims concerning the dimensions of variation which are relevant for explanation, for it is only compatible with those frameworks which concede the reality of the normatively crucial dimension. It is incompatible with any view of politics as the striving of different classes, or interest groups, or individuals against one another.

It is clear that this is true of any normative theory, that it is linked with certain explanatory theory or theories, and incompatible with others. Aristotle's dimension whereby different constitutions were seen as expressing and molding different forms of life disappears in the atomistic conception of Hobbes. Rousseau's crucial dimension of the *Social Contract*, marking a sharp discontinuity between popular sovereignty and states of dependence of one form or another, could not survive the validation of the theories of Mosca, or Michels, or Pareto.

Traditional political philosophy was thus forced to engage in the theoretical function that we have seen to be essential to modern political science; and the more elaborate and comprehensive the normative theory, the more complete and defined the conceptual framework which accompanied it. That is why political science can learn something still from the works of Aristotle, Hobbes, Hegel, Marx, and so on. In the tradition one form of inquiry is virtually inseparable from the other.

2

I

This is not a surprising result. Everyone recognized that political philosophers of the tradition were engaged in elaborating on, at least embryonic, political science. But, one might say, that is just the trouble; that is why political science was so long in getting started. Its framework was always set in the interests of some normative theory. In order to progress science must be liberated from all *parti pris* and be value neutral. Thus, if normative theory requires political science and cannot be carried on without it, the reverse is not the case; political science can and should be separated from the older discipline. Let us examine some modern attempts to elaborate a science of politics to see if this is true.

Let us look first at S. M. Lipset's *Political Man* (New York: Doubleday, 1959). In this work Lipset sets out the conditions for modern democracy. He sees societies as existing in two dimensions—conflict and consensus. Both are equally necessary for democracy. They are not mere opposites as a simple-minded view might assume. Conflict here is not seen as a simple divergence of interest, or the existence of objective relations of exploitation, but as the actual working out of these through the struggle for power and over policy.

> Surprising as it may sound, a stable democracy requires the manifestation of conflict or cleavage so that there will be struggle over ruling positions, challenges to parties in power, and shifts of parties in office; but without consensus—a political system allowing the peaceful "play" of power, the adherence of the "outs" to decisions made by the "ins," and the recognition by the "ins" of the rights of the "outs"—there can be no democracy. The study of the conditions encouraging democracy must therefore focus on the sources of both cleavage and consensus. (*Political Man*, p. 21).

And again, "Cleavage—where it is legitimate—contributes to the integration of societies and organizations" (ibid.). The absence of such conflict, such as where a given group has taken over, or an all-powerful state can produce unanimity, or at least prevent diversity from expressing itself, is a sign that the society is not a free one. De Tocqueville feared (*Political Man*, p. 27) that the power of the state would produce apathy and thus do away even with consensus.

> Democracy in a complex society may be defined as a political system which supplies regular constitutional opportunities for changing the governing officials, and a social mechanism which permits the largest possible part of the population to influence major decisions by choosing among contenders for political office. (ibid., p. 45)

Such a society requires the organization of group interests to fight for their own goals—provided that this is done in a peaceful way, within the rules of the game, and with the acceptance of the arbiter in the form of elections by universal suffrage. If groups are not organized, they have no real part, their interests are neglected, and they cannot have their share of power; they become alienated from the system.

Now this view can at once be seen to conflict with a Rousseauian view which disapproves of the organization of "faction," and which sees consensus as arising out of isolated individuals. It also goes against the modern conservative view that to organize people on a class basis gratuitously divides the society. In face of Rousseau, Lipset holds that the absence of close agreement among all concerning the general will is not a sign that something has gone wrong. There are ineradicable basic divergences of interest; they have to be adjusted. If we get to some kind of conflictless state, this can only be because some of the parties have been somehow done down and prevented from competing. For Lipset, absence of conflict is a sure sign that some groups are being excluded from the public thing.

This difference closely parallels the one mentioned above between Plato and Aristotle. Indeed, Lipset points out on several occasions the similarity between his position and that of Aristotle. And it is clear that it is a difference of the same kind, one in which a normative theory is undermined because the reality of its crucial dimension of variation is challenged. A similar point can be made concerning the difference with conservatives who allow for divergence in the state, but resist class parties. Here the belief is that the divergence is gratuitous, that the real differences lie elsewhere, either in narrower or in broader interests, and that these are obfuscated and made more difficult of rational adjustment by class divisions. More, the state can be torn apart if these divisions are played up. Conservatives tend to feel about class in politics as liberals do about race in politics. Once again, Lipset's view would undermine the position, for he holds that class differences are at the center of politics, and cannot be removed except by reducing the number of players, as it were. They are therefore the very stuff of democratic politics, provided they are moderately and peacefully expressed. The struggle between rich and poor is ineradicable; it can take different forms, that's all.

Attempts to break outside of this range are thus irrational and dysfunctional. Irrational, because based on false premises; and dysfunctional, because the goal of conflictlessness or absence of class tension can only be achieved at the expense of features of the system which most will accept as valuable; by oppressing some segment of the population, or by its apathy and lack of organization. That is, of course, the usual fate of theories with a false base in politics; as was remarked above, they are not just erroneous, but positively dangerous.

It can be seen that the value consequences of Lipset's theory are fairly widespread even restricting ourselves to the alternatives which it negates or undermines. An examination of some of the factors which tend to strengthen democracy according to the theory will increase this list of rejected alternatives. Lipset holds that economic development is conducive to the health of democracy, in that, inter alia, it narrows gaps in wealth and living standards, tends to create a large middle class, and increases the "cross-pressures" working to damp down class conflict. For a society cannot function properly as a democracy unless, along with an articulation of class differences, there is some consensus which straddles them. Now Lipset's "cross-pressures"—typically exercised by religious affiliation, for instance, which cuts across class barriers—are the "opiates" of a strict Marxist. For they are integrators which prevent the system's coming apart at the social seam, and thus prevent the class war from coming to a head. But we are not dealing here simply with two value-judgments about the same facts understood in the same way. The crucial difference is that for Lipset the stage beyond the class struggle does not and cannot exist; the abolition of the conflict in unanimity is impossible; his view is "the rich ye have always with you." But in this case the integrating factors cease to be "opiates," breeding false consciousness and hiding the great revolutionary potentiality. There is nothing there to hide. Lipset's view therefore negates revolutionary Marxism in a direct way—in the same way as it negates the views above—by denying that the crucial dimensions of variation have reality.

But if we examine this last example a little more closely, we can see even wider normative consequences of Lipset's view. For if we rule out the transformation to the classless society, then we are left with the choice between different kinds of class conflict: a violent kind which so divides society that it can only survive under some form of tyranny, or one which can reach accommodations in peace. This choice, set out in these terms, virtually makes itself for us. We may point out that this does not cover the range of possibility, since there are also cases in which the class conflict is latent, owing to the relative absence of one party. But this is the result of underdevelopment, of a lack of education, or knowledge, or initiative on the part of the underprivileged. Moreover, it unfailingly leads to a worsening of their position relative to the privileged. As Lipset says in the statement of his political position which forms the introduction to the Anchor Edition of *Political Man*, "I believe with Marx that all privileged classes seek to maintain and *enhance* their advantages against the desire of the underprivileged to reduce them" (Anchor Edition, p. xxii, emphasis in original).

Thus, for Lipset, the important dimension of variation for political societies can be seen as L-shaped, as it were. On the one end lie societies where the divisions are articulated but are so deep that they cannot be contained without violence, suppression of liberty, and despotic rule; on the other end lie societies which are peaceful but oligarchic and which are therefore run to secure the good of a minority ruling group. At the angle are the societies whose differences are articulated but which are capable of accommodating them in a peaceful way, and which therefore are characterized by a high degree of individual liberty and political organization.

Faced with this choice, it is hard to opt for anywhere else but the angle. For to do so is either to choose violence and despotism and suppression over peace, rule by consent, and liberty, or to choose a society run more for the benefit of a minority over a society run more for the benefit of all, a society which exploits and/or manipulates over a society which tends to secure the common good as determined by the majority. Only in the angle can we have a society really run for the common good, for at one end is oligarchy based on an unorganized mass, at the other despotism.

Lipset himself makes this option explicit:

> A basic premise of this book is that democracy is not only or even primarily a means through which different groups can attain their ends or seek the good society; it is the good society itself in operation. Only the give-and-take of a free society's internal struggles offers some guarantee that the products of the society will not accumulate in the hands of a few power-holders, and that men may develop and bring up their children without fear of persecution. (p. 403)

This is a succinct statement of the value position implicit in *Political Man*, but it is wrongly characterized as a "premise." The use of this term shows the influence of the theory of value-neutrality, but it is misplaced. It would be less misleading to say "upshot," for the value position flows out of the analysis of the book. Once we accept Lipset's analysis concerning the fundamental role of class in politics, that it always operates even when division is not overt, and that it can never be surmounted in unanimity, then we have no choice but to accept democracy as he defines it, as a society in which most men are doers, take their fate in their own hands, or have a hand in determining it, and at least reduce the degree to which injustice is done to them, or their interests are unfavourably handled by others, as the good society.

II

But now we have gone far beyond the merely negative consequences noted above for Marxism, conservatism, or Rousseau's general will. We are saying that the crucial dimensions of variation of Lipset's theory not only negate dimensions crucial to other normative theories but support one of their own, which is implicit in the theory itself. But this conclusion, if true, goes against the supposed neutrality of scientific fact. Let us examine it a bit more closely.

We have said above that faced with the choice between a regime based on violence and suppression, and one based on consent, between regimes which serve the interests more or less of all versus regimes which serve the interests only of a minority, the choice is clear. Is this simply a rhetorical flourish, playing on generally accepted values among readers? Or is the connection more solid?

Granted that we wish to apply "better" and "worse" to regimes characterized along this dimension, can one conceive of reversing what seemed above to be the only possible judgment? Can one say: yes, a regime based on minority rule with violent suppression of the majority is better than one based on general consensus, where all have a chance to have their interests looked to? Certainly this is not a logically absurd position in itself. But if someone accepted the framework of Lipset and proceeded to make this judgment, surely we would expect him to go on and mention some other considerations which led him to this astounding conclusion. We might expect him to say that only minorities are creative, that violence is necessary to keep men from stagnating, or something of this kind. But supposing he said nothing of the sort? Supposing he just maintained that violence was better than its opposite, not qua stimulus to creativity, or essential element in progress, but just qua violence; that it was better that only the minority interest be served, not because the minority would be more creative but just because it was a minority? A position of this kind would be unintelligible. We could understand that the man was dedicating himself to the furtherance of such a society, but the use of the words "good" or "better" would be totally inappropriate here, for there would be no visible grounds for applying them. The question would remain open whether the man had understood these terms, whether, e.g., he had not confused "good" with "something which gives me a kick," or "aesthetically pleasing."

But, it might be argued, this is not a fair example. Supposing our unorthodox thinker did adduce other grounds for preferring violence and majority rule? Surely, then, he would be permitted to differ from us? Yes, but then it is very dubious whether he could still accept Lipset's framework. Suppose, for instance, that one believed (as Hegel did about war) that violence was morally necessary from time to time for the well-being of the state. This would not be without effect on one's conception of political science; the range of possible regimes would be different from that which Lipset gives us; for peaceful democratic regimes would suffer a process of stagnation which would render them less viable; they would not in fact be able to maintain themselves, and thus the spectrum of possible regimes would be different from the one Lipset presents us with; the most viable regime would be one which was able to ration violence and maintain it at a nondisruptive level without falling over into stagnation and decay.

But why need this change of values bring along with it a chance in explanatory framework? We seem to be assuming that the evils of internal peace must be such as to have a political effect, to undermine the viability of the political society. Is this assumption justified? Normally, of course, we would expect someone putting forward a theory of this kind to hold that inner violence is good because it contributes to the dynamism, or creativity of people, or progress of the society, or something of the kind which would make peaceful societies less viable. But supposing he chose some other benefits of violence which had nothing to do with the survival or health of political society? Let us say that he held that violence was good for art, that only in societies rent by internal violence could great literature, music, painting be produced? The position, for instance, of Harry Lime in *The Third Man?*

This certainly is a possible case. But let us examine it more closely. Our hypothetical objector has totally forsaken the ground of politics, and is making his judgment on extraneous (here aesthetic) grounds. He cannot deny that, setting these grounds aside, the normal order of preference is valid. He is saying in effect that, although it is better abstracting from aesthetic considerations that society be peaceful, nevertheless this must be overridden in the interests of art.

This distinction is important. We must distinguish between two kinds of objection to a given valuation. It may be that the valuation is accepted, but that its verdict for our actual choices is overridden, as it were, by other more important valuations. Thus we may think that freedom of speech is always a good, while reluctantly conceding that it must be curtailed in an emergency because of the great risks it would entail here. We are in this case self-consciously curtailing a good. The other kind of objection is the one which undermines the valuation itself, seeks to deprive the putative good of its status. This is what Lipset does, for instance, to spiritual followers of Rousseau in showing that their harmony can only be the silence of minority rule.[1] In one case we are conceding that the thing in question does really have the properties which its proponents attribute to it (e.g., that free speech does contribute to justice, progress, human development, or whatever), but we are adding that it also has other properties which force us to proceed against it (e.g., it is potentially disruptive) temporarily or permanently. In the other case, we are denying the condition in question the very properties by which it is judged good (e.g., that the legislation of the society without cleavage emanates from the free conscious will of all its citizens). Let us call these two objections respectively overriding and undermining.

Now what is being claimed here is that an objection which undermines the values which seem to arise out of a given framework must alter the framework; that in this sense the framework is inextricably connected to a certain set of values; and that if we can reverse the valuation without touching the framework, then we are dealing with an overriding.

To go back to the example above. In order to undermine the judgment against violence we would have to show that it does not have the property claimed for it. Now obviously violence has the property of killing and maiming which goes some way toward putting it in the list of undesirables, one might think irrevocably; so that it could only be overridden. But here we are not dealing with a judgment about violence per se, but rather with one concerning the alternative of peace and violence; and the judgment rests on the ground that violence has properties which peace has not, that the evils obviously attributed to violence are effectively avoided by peace. But if one can show that peace leads to stagnation, and thus to breakdown (and hence eventual chaos or violence) or foreign conquest, then the supposed gap between the two narrows. On the contrary, one is presented with a new alternative, that between more or less controlled violence and the destructive uncontrolled kind associated with internal breakdown or foreign conquest. What the undermining job has done is to destroy the alternative on which the original judgment was based, and thus deprive the previously preferred alternative of its differential property for which it was valued.

But any undermining of this kind is bound to alter the explanatory framework of which the original alternative was an essential part. If we cannot maintain a peaceful polity, then the gamut of possiblilities is very different, and Lipset is guilty of neglecting a whole host of factors, to do with the gamut tension-stagnation.

To take the other example, let our objector make a case for rule by the minority. Let him claim that only the minority are creative, that if they are not given preference, then they will not produce, and then everyone will suffer. Thus the supposed difference between rule for the minority and that for all, viz. that the ordinary bloke gets something out of the second that he does not out of the first, is set aside; rather the opposite turns out to be the case. The value is undermined. But so is the political framework altered, for now we have an elitist thesis about the importance of minority rule; another variable has entered the picture which was not present in the previous framework and which cuts across it, insofar as the previous framework presented the possibility of good progressive societies run for all.

Let us hold, however, that violence or elite rule is good for painting, and we have an overruling; for it remains the case that it would be better to have no violence and everybody getting a square deal, but alas....

Thus the framework does secrete a certain value position, albeit one that can be overridden. In general we can see this arising in the following way: the framework gives us as it were the geography of the range of phenomena in question, it tells us how they can vary, what are the major dimensions of variation. But since we are dealing with matters which are of great importance to human beings, a given map will have, as it were, its own built-in value slope. That is to say, a given dimension of variation will usually determine for itself how we are to judge of good and bad, because of its relation to obvious human wants and needs.

Now this may seem a somewhat startling result, since it is well known that there are wide differences over what human needs, desires, and purposes are. Not that there is not a wide area of agreement over basic things like life; but this clearly breaks down when one tries to extend the list. There can thus be great disagreement over the putative human need for self-expression or for autonomous development, both of which can and do play important parts in debates and conflicts over political theory.

Does this mean, therefore, that we can reject the previous result and imagine a state of affairs where we could accept the framework of explanation of a given theory, and yet refuse the value judgments it secretes, because we took a different view of the schedule of human needs?[2] Or, to put it another way, does this mean that the step between accepting a framework of explanation and accepting a certain notion of the political good is mediated by a premise concerning human needs, which may be widely enough held to go unnoticed, but which nevertheless can be challenged, thus breaking the connection?

The answer is no. For the connection between a given framework of explanation and a certain notion of the schedule of needs, wants, and purposes which seems to mediate the inference to value theory is not fortuitous. If one adopted a quite different view of human need, one would upset the framework. Thus to pursue another example from Lipset, stable democracies are judged better than stable oligarchies, since the latter can only exist where the majority is so uneducated and tradition bound or narrowed that it has not yet learned to demand its rights. But suppose we tried to upset this judgment by holding that underdevelopment is good for men, that they are happier when they are led by some unquestioned norms, do not have to think for themselves, and so on? One would then be reversing the value-judgment. But at the same time one would be changing the framework. For we are introducing a notion of anomie here, and we cannot suppose this factor to exist without having some important effect on the working of political society. If anomie is the result of the development of education and the breakdown of tradition, then it will affect the stability of the societies which promote this kind of development. They will be subject to constant danger of being undermined as their citizens, suffering from anomie, look for havens of certainty. If men are made unhappy by democracy, then undoubtedly it is not as good as its protagonists make out, but it is not so viable either.

The view above that we could accept the framework of explanation and reject the value conclusion by positing a different schedule of needs cannot be sustained. For a given framework is linked to a given conception of the schedule of human needs, wants, and purposes, such that, if the schedule turns out to have been mistaken in some significant way, the framework itself cannot be maintained. This is for the fairly obvious reason that human needs, wants, and purposes have an important bearing on the way people act, and that therefore one has to have a notion of the schedule which is not too wildly inaccurate if one is to establish the framework for any science of human behavior, that of politics not excepted. A conception of human needs thus enters into a given political theory, and cannot be considered something extraneous which we later add to the framework to yield a set of value judgments.

This is not to say that there cannot be needs or purposes which we might add to those implicit in any framework, and which would not alter the framework since their effect on political events might be marginal. But this would at most give us the ground of an overruling, not for an undermining. In order to undermine the valuation we would have to show that the putative need fulfilled was not a need, or that what looked like fulfilling a need, or a want, or a human purpose was really not so, or really did the opposite. Now even an overruling might destroy the framework, if a new need were introduced which was important enough motivationally to dictate quite different behavior. But certainly an undermining, which implies that one has misidentified the schedule of needs, would do so.

III

It would appear from the above example that the adoption of a framework of explanation carries with it the adoption of the "value slope" implicit in it, although the valuations can be overruled by considerations of an extra-political kind. But it might be objected that the study of one example is not a wide enough base for such a far-reaching conclusion. The example might even be thought to be peculiarly inappropriate because of Lipset's closeness to the tradition of political philosophy, and particularly his esteem for Aristotle.

If we wish, however, to extend the range of examples, we can see immediately that Lipset's theory is not exceptional. There is, for instance, a whole range of theories in which the connection between factual base and valuation is built in, as it were, to the conceptual structure. Such is the case of many theories which make use of the notion of function. To fulfill a function is to meet a requirement of some kind, and when the term is used in social theory, the requirement concerned is generally connected with human needs, wants, and purposes. The requirement or end concerned may be the maintenance of the political system which is seen as essential to man, or the securing of some of the benefits which political systems are in a position to attain for men—stability, security, peace, fulfilment of some wants, and so on. Since politics is largely made up of human purposeful activity a characterization of political societies in terms of function is not implausible. But insofar as we characterize societies in terms of their fulfilling in different ways and to different degrees the same set of functions, the crucial dimension of variation for explanatory purposes is also a normatively significant one. Those societies which fulfill the functions more completely are *pro tanto* better.

We can take as an example the "structural-functional" theory of Gabriel Almond as outlined in his *Politics of the Developing Areas* (Princeton: Princeton University Press, 1963). Among the functions Almond outlines that all polities must fulfill is that of "interest articulation." It is an essential part of the process by which the demands, interests, and claims of members of a society can be brought to bear on government and produce some result. Almond sees four main types of structures as involved in interest articulation.[3] Of three of these (institutional, nonassociational, and anomic interest groups), he says that a prominent role for them in interest articulation tends to indicate poor "boundary maintenance" between society and polity. Only the fourth (associational interest groups) can carry the main burden of interest articulation in such a way as to maintain a smooth-running system "by virtue of the regulatory role associational interest groups in processing raw claims or interest articulations occurring elsewhere in the society and the political system, and directing them in an orderly way and in aggregable form through the party system, legislature, and bureaucracy."[4]

The view here is of a flow of raw demands which have to be processed by the system before satisfaction can be meted out. If the processing is inefficient, then the satisfaction will be less, the system will increase frustration, uncertainty, and often as a consequence instability. In this context boundary maintenance between society and polity is important for clarity and efficiency. Speaking of the functions of articulation and aggregation together, Almond says:

> Thus, to attain a maximum flow of inputs of raw claims from the society, a low level of processing into a common language of claims is required which is performed by associated interest groups. To assimilate and transform these interests into a relatively small number of alternatives of policy and personnel, a middle range of processing is necessary. If these two functions are performed in substantial part before the authoritative governmental structures are reached, then the output functions of rule-making and rule application are facilitated, and the political and governmental processes become calculable and responsible. The outputs may be related to and controlled by the inputs, and thus circulation becomes relatively free by virtue of good boundary maintenance or division of labor.[5]

Thus in characterizing different institutions by the way they articulate or aggregate interests, Almond is also evaluating them. For obviously a society with the above characteristics is preferable to one without, where, that is, there is less free circulation, where "outputs" correspond less to "inputs" (what people want, claim, or demand), where government is less responsible, and so on. The characterization of the system in terms of function contains the criteria of "eufunction" and "dysfunction," as they are sometimes called. The dimension of variation leaves only one answer to the question, Which is better? because of the clear relation in which it stands to men's wants and needs.

Theories of this kind include not only those which make explicit use of "function" but also other derivatives of systems theory and frameworks which build on the analogy with organisms. This might be thought to include, for instance, David Easton (cf. *A Framework for Political Analysis*, Englewood Cliffs, N.J.: Prentice-Hall, 1965, and *A Systems Analysis of Political Life*, New York: Wiley, 1965) and Karl Deutsch (*The Nerves of Government*, Glencoe, Ill: The Free Press, 1963). For the requirements by which we will judge the performance of different political systems are explicit in the theory.

But what about theories which set out explicitly to separate fact from evaluations, to "state conditions" without in any way "justifying preferences"? What about a theory of the "behavioral" type, like that of Harold Lasswell?

IV

Harold Lasswell is clearly a believer in the neutrality of scientific findings. Lasswell is openly committed to certain values, notably those of the democratic society as he defines it, a society "in which human dignity is realized in theory and fact."[6] He believes that scientific findings can be brought to bear on the realization of these goals. A science so oriented is what he calls a "policy science." But this does not affect the neutrality of the findings: a policy science simply determines a certain grouping and selection of findings which help us to encompass the goal we have set. It follows that if there are policy sciences of democracy, "there can also be a 'policy science of tyranny.'"[7]

In Lasswell's "configurative analysis," then, both fact and valuation enter; but they remain entirely separable. The following passage from the introduction of *Power and Society* makes the point unambiguously:

> The present conception conforms ... to the philosophical tradition in which politics and ethics have always been closely associated. But it deviates from the tradition in giving full recognition to the existence of two distinct components in political theory—the empirical propositions of political science and the value judgments of political doctrine. Only statements of the first kind are formulated in the present work. (p. xiii)

Yet the implied separation between factual analysis and evaluation is belied by the text itself. In the sections dealing with different types of polity,[8] the authors introduce a number of dimensions of variation of political society. Polities vary (1) as to the allocation of power (between autocracy, oligarchy, republic), (2) as to the scope of power (society either undergoes greater regimentation or liberalization), (3) as to the concentration or dispersion of power (taking in questions concerning the separation of powers, or federalism), (4) as to the degree to which a rule is egalitarian (the degree of equality in power potential), (5) the degree to which it is libertarian or authoritarian, (6) the degree to which it is impartial, (7) and the degree to which it is juridical or tyrannical. Democracy is defined as a rule which is libertarian, juridical, and impartial.

It is not surprising to find one's sympathies growing toward democracy as one ploughs through this list of definitions. For they leave us little choice. Dimension (5) clearly determines our preference. Liberty is defined not just in terms of an absence of coercion, but of genuine responsibility to self. "A rule is libertarian where initiative, individuality and choice are widespread; authoritarian, if obedience, conformity and coercion are characteristic."[9] Quoting Spinoza with approval, Lasswell and Kaplan come down in favor of a notion of liberty as the capacity to "live by ... free reason." "On this conception, there is liberty in a state only where each individual has sufficient self-respect to respect others."[10]

Thus it is clear that liberty is preferable to its opposite. Many thinkers of the orthodox school, while agreeing with this verdict, might attribute it simply to careless wording on the author's part, to a temporary relaxation of that perpetual vigil which must be maintained against creeping value bias. It is important to point out therefore that the value force here is more than a question of wording. It lies in the type of alternative which is presented to us: on the one hand, a man can be manipulated by others, obeying a law and standards set up by others which he cannot judge; on the other hand, he is developed to the point where he can judge for himself, exercise reason, and apply his own standards; he comes to respect himself and is more capable of respecting others. If this is really the alternative before us, how can we fail to judge freedom better (whether or not we believe there are overriding considerations)?

Dimension (6) also determines our choice. "Impartiality" is said to "correspond in certain ways to the concepts of 'justice' in the classical tradition,"[11] and an impartial rule is called a "commonwealth," "enhancing the value position of all members of the society impartially, rather than that of some restricted class,"[12] Now if the choice is simply between a regime which works for the common good and a regime which works for the good of some smaller group, there is no doubt which is better in the absence of any overriding considerations.

Similarly dimension (7) is value-determinate. "Juridical" is opposed to "tyrannical" and is defined as a state of affairs where "decisions are made in accord with specified rules ... rather than arbitrarily"[13] or where a "decision is challenged by an appraisal of it in terms of ... conditions, which must be met by rulers as well as ruled." Since the alternative presented here is *arbitrary* decision, and one which cannot be checked by any due process, there is no question which is preferable. If we had wanted to present a justification of rule outside law (such as Plato did), we would never accept the adjective "arbitrary" in our description of the alternative to "juridical."

As far as the other dimensions are concerned, the authors relate them to these three key ones, so that they too cannot be seen as neutral, although their value relevance is derivative. Thus voluntarization is better for liberty than regimentation, and the dispersion of power can be seen as conducive to juridicalness. In short we come out with a full-dress justification of democracy, and this in a work which claims neutrality. The work, we are told in the introduction, "contains no elaborations of political doctrine, of what the state and society *ought* to be."[14] Even during the very exposition of the section on democracy, there are ritual disclaimers: for instance, when the term "justice" is mentioned, a parenthesis is inserted: "the present term, however, is to be understood altogether in a descriptive, non-normative sense";[15] and at the end of the chapter: "the formulations throughout are descriptive rather than normatively ambiguous."[16]

But neutral they are not, as we have seen: we cannot accept these descriptions and fail to agree that democracy is a better form of government than its opposite (a "tyrannical," "exploitative," "authoritarian" rule: you can take your choice). Only the hold of the neutrality myth can hide this truth from the authors.

Of course these sections do not represent adequately Lasswell's total work. Indeed, one of the problems in discussing Lasswell is that he has espoused a bewildering variety of conceptual frameworks of explanation. This is evident from a perusal of *Power and Society* alone, quite apart from his numerous other works. These may all cohere in some unified system, but if this is the case, it is far from obvious. Yet the link between factual analysis and evaluation reappears in each of the different approaches. There is not space to cover them all; one further example will have to suffice here.

In the later psychiatrically oriented works, such as *Power and Personality*, or "The Democratic Character,"[17] the goal explicitly set for policy science is democracy. But the implication that this is a goal chosen independently of what is discovered to be true about politics is belied all along the line. For the alternative to a society where people have a "self-system" which suits the democratic character is one in which various pathologies, often of a dangerous kind, are rampant. The problem of democracy is to create, among other things, a self-system which is "multivalued, rather than single-valued, and ... disposed to share rather than to hoard or to monopolize."[18] One might have some quarrel with this: perhaps single-minded people are an asset to society. But after seeing the alternative to multi-valuedness as set out in the "Democratic Character,"[19] one can understand why Lasswell holds this view. Lasswell lays out for us a series of what he describes frankly at one point as "character deformations."[20] In talking about the *homo politicus* who concentrates on the pursuit of power, he remarks, "The psychiatrist feels at home in the study of ardent seekers after power in the arena of politics because the physician recognizes the extreme egocentricity and sly ruthlessness of some of the paranoid patients with whom he has come in contact in the clinic" (p. 498).

The point here is not that Lasswell introduces valuation illegitimately by the use of subtly weighted language, or unnecessarily pejorative terms. Perhaps politicians do tend to approximate to unbalanced personalities seeking to make up deprivation by any means. The point is that, if this is true, then some important judgments follow about political psychiatry. And these are not, as it were, suspended on some independent value judgment, but arise from the fact themselves. There *could* be a policy science of tyranny, but then there could also be a medical science aimed at producing disease (as when nations do research into bacteriological warfare). But we could not say that the second was more worthy of pursuit than the first, unless we advanced some very powerful overriding reasons (which is what proponents of bacteriological warfare try—unsuccessfully—to do). The science of health, however, needs no such special justification.

3

I

The thesis we have been defending, however plausible it may appear in the context of a discussion of the different theories of political science, is unacceptable to an important school of philosophy today. Throughout the foregoing analysis, philosophers will have felt uneasy. For this conclusion tells against the well-entrenched doctrine according to which questions of value are independent of questions of fact: the view which holds that before any set of facts we are free to adopt an indefinite number of value positions. According to the view defended here, on the other hand, a given framework of explanation in political science tends to support an associated value position, secretes its own norms for the assessment of polities and policies.

It is, of course, this philosophical belief which, because of its immense influence among scientists in general and political scientists as well, has contributed to the cult of neutrality in political science, and the belief that genuine science gives no guidance as to right and wrong. It is time, therefore, to come to grips with this philosophical view.

There are two points about the use of "good" which are overlooked or negated by the standard "nonnaturalist" view: (1) to apply "good" may or may not be to commend, but it is always to claim that there are reasons for commending whatever it is applied to, (2) to say of something that it fulfills human needs, wants, or purposes always constitutes a prima facie reason for calling it "good," that is, for applying the term in the absence of overriding considerations.[21]

Now the nonnaturalist view, as expressed, for instance, by Hare of Stevenson, denies both these propositions. Its starting point is the casting of moral argument in deductive form—all the arguments against the so-called naturalistic fallacy have turned on the validity of deductive inference. The ordinary man may think that he is moving from a factual consideration about something to a judgment that it is good or bad, but in fact one cannot deduce a statement concerning the goodness or badness of something from a statement attributing some descriptive property to it. Thus the ordinary man's argument is really an enthymeme: he is assuming some major premise: when he moves from "X will make men happy" to "X is good," he is operating with the suppressed premise, "What makes men happy is good," for only by adding this can one derive the conclusion by valid inference.

To put the point in another way: the ordinary man sees "X will make men happy" as the reason for his favorable verdict on it. But on the nonnaturalist view, it is a reason only because he accepts the suppressed major premise. For one could, logically, reject this premise, and then the conclusion would not follow at all. Hence, that something is a reason for judging X good depends on what values the man who judges holds. Of course, one can find reasons for holding these values. That is, facts from which we could derive the major premise, but only by adopting a higher major which would allow us to derive our first major as a valid conclusion. Ultimately, we have to decide beyond all reasons, as it were, what our values are. For at each stage where we adduce a reason, we have already to have accepted some value (enshrined in a major premise) in virtue of which this reason is valid. But then our ultimate major premises stand without reasons; they are the fruit of a pure choice.

Proposition (1) above, then, is immediately denied by nonnaturalism. For in the highest major premises "good" is applied to commend without the claim that there are reasons for this commendation. And (2) also is rejected, for nothing can claim always to constitute a reason for calling something good. Whether it does or not depends on the decisions a man has made about his values, and it is not logically impossible that he should decide to consider human needs, wants, and purposes irrelevant to judgments about good and bad. A reason is always a reason-for-somebody, and has this status because of the values he has accepted.

The question at issue, then, is first whether "good" can be used where there are no reasons, either evident or which can be cited for its application.[22] Consider the following case:[23] There are two segregationists who disapprove of miscegenation. The first claims that mixing races will produce general unhappiness, a decline in the intellectual capacity and moral standards of the race, the abolition of a creative tension, and so on. The second, however, refuses to assent to any of these beliefs; the race will not deteriorate, men may even be happier; in any case they will be just as intelligent, moral, etc. But, he insists, miscegenation is bad. When challenged to produce some substitute reason for this judgment, he simply replies: "I have no reasons; everyone is entitled, indeed has to accept some higher major premise and stop the search for reasons somewhere. I have chosen to stop here, rather than seeking grounds in such fashionable quarters as human happiness, moral stature, etc." Or supposing he looked at us in puzzlement and said: "Reasons? why do you ask for reasons? Miscegenation is just bad."

Now no one would question that the first segregationist was making the judgment "miscegenation is bad." But in the case of the second, a difficulty arises. This can be seen as soon as we ask the question: how can we tell whether the man is really making a judgment about the badness of miscegenation and not just, say, giving vent to a strongly felt repulsion, or a neurotic phobia against sexual relations between people of different races? Now it is essential to the notions "good" and "bad" as we use them in judgments that there be a distinction of this kind between these judgments and expressions of horror, delight, liking, disliking, and so on. It is essential that we be able, e.g., to correct a speaker by saying: "What you want to say would be better put as 'miscegenation horrifies me' or 'miscegenation makes me go all creepy inside'" Because it is an essential part of the grammar of "good" and "bad" that they claim more than is claimed by expressions of delight, horror, etc. For we set aside someone's judgment that X is good when we say: "All you are saying is that you *like* X." To which the man can hotly reply: "I do not like X any more than you do, but I recognize that it is good."

There must therefore be criteria of distinction between these two cases if "good" and "bad" are to have the grammar that they have. But if we allow that our second segregationist is making the judgment "miscegenation is bad," then no such distinction can be made. A judgment that I like something does not need grounds. That is, the absence of grounds does not undermine the claim "I like X" (though other things, e.g., in my behavior, may undermine it). But unless we adduce reasons for it (and moreover reasons of a certain kind as we shall see below) we cannot show that our claim that X is good says more than "I like X." Thus a man can only defend himself against the charge that all he is saying is that he likes X by giving his grounds. If there are no grounds, then judgment becomes indistinguishable from expression; which means that there are no more judgments of good and bad, since the distinction is essential to them as we have seen.

Those who believe in the fact-value dichotomy have naturally tired to avoid this conclusion; they have tried to distinguish the two cases by fastening on the use made of judgments of good and bad in commending, prescribing, expressing approval, and so on. Thus, no matter what a man's grounds, if any, we could know that he was making a judgment of good and bad by the fact that he was commending, prescribing, or committing himself to pursue the thing in question, or something of the kind. But this begs the question, for we can raise the query: what constitutes commending, or prescribing, or committing myself, or expressing approval, or whatever? How does one tell whether a man is doing one of these things as against just giving vent to his feelings?

If we can say that we can tell by what the man accepts as following from his stand—whether he accepts that he should strive to realize the thing in question—then the same problem breaks out afresh: how do we distinguish his accepting the proposition that he should seek the end and his just being hell-bent on seeking this end? Presumably, both our segregationists would agree that they should fight miscegenation, but this would still leave us just as puzzled and uncertain about the position of the second. Perhaps we can tell by whether they are willing to universalize their prescription? But here again we have no touchstone, for both segregationists would assent that everyone should seek racial purity, but the question would remain open whether this had a different meaning in the two cases. Perhaps the second one just means that he cannot stand interracial mating, whether done by himself or by anyone else. Similarly, a compulsive may keep his hands scrupulously clean and feel disgust at the uncleanliness of others, even plead with them to follow his example; but we still want to distinguish his case from one who had judged that cleanliness was good.

Can we fall back on behavioral criteria, meaning by "behavior" what a man does in contrast to how he thinks about what he does? But there is no reason why a man with a neurotic phobia against X should not do all the things which the man who judges X is bad does, i.e., avoiding X himself, trying to stop others from doing it, and so on.

Thus the nonnaturalists would leave us with no criteria except what the man was willing to say. But then we would have no way of knowing whether the words were correctly applied or not, which is to say that they would have no meaning. All that we achieve by trying to mark the distinction by what follows from the judgment is that the same question which we raise about "X is bad" as against "X makes me shudder" can be raised about the complex "X is bad, I/you should not do X" as against the complex "X makes me shudder, please I/you do not do X." We simply appeal from what the man is willing to say on the first question to what he is willing to say on the second. The distinction can only be properly drawn if we look to the reasons for the judgment, and this is why a judgment without reasons cannot be allowed, for it can no longer be distinguished from an expression of feeling.[24]

II

This analysis may sound plausible for "miscegenation is bad," but how about, "anything conducive to human happiness is good"? What can we say here, if asked to give grounds for this affirmation? The answer is that we can say nothing, but also we need say nothing. For that something conduces to human happiness is already an adequate ground for judging it good—adequate, that is, in the absence of countervailing considerations. We come, then, to the second point at issue, the claim that to say of something that it fulfills human needs, wants, or purposes always constitutes a prima facie reason for calling it "good."

For in fact it is not just necessary that there be grounds for the affirmation if we are to take it at its face value as an attribution of good or bad; they must also be grounds of a certain kind. They must be grounds which relate in some intelligible way to what men need, desire, or seek after. This may become clearer if we look at another example. Suppose a man says: "To make medical care available to more people is good"; suppose, then, that another man wishes to deny this. We could, of course, imagine reasons for this: world population will grow too fast, there are other more urgent claims on scarce resources, the goal can only be obtained by objectionable social policies, such as socialized medicine, and so on. The espousal of any of these would make the opposition to the above judgment intelligible, even it not acceptable, and make it clear that it was *this* judgment that was being denied, and not just, say, an emotional reaction which was being countered with another. If, however, our objector said nothing, and claimed to have nothing to say, his position would be unintelligible, as we have seen; or else we would construe his words as expressing some feeling of distaste or horror or sadness at the thought.

But supposing he was willing to give grounds for his position, but none of the above or their like, saying instead, for instance, "There would be too many doctors," or "Too many people would be dressed in white"? We would remain in doubt as to how to take his opposition, for we would be led to ask of his opposition to the increase of doctors, say, whether he was making a judgment concerning good and bad or simply expressing a dislike. And we would decide this question by looking at the grounds he adduced for *this* position. And if he claimed to have nothing to say, his position would be unintelligible in exactly the same way as if he had decided to remain silent at the outset and leave his original statement unsupported. "What is this?" we would say, "You are against an increase in medical services, because it would increase the number of doctors? But are you just expressing the feelings of dislike that doctors evoke in you or are you really trying to tell us that the increase is bad?" In the absence of any defense on his part, we would take the first interpretation.

It is clear that the problem would remain unsolved, if our opponent grounded his opposition to doctors on the fact that they generally wore dark suits or washed their hands frequently. We might at this point suspect him of having us on. So that the length or elaboration of the reasoning has nothing to do with the question one way or another.

What would make his position intelligible, and intelligible as a judgment of good and bad, would be his telling some story about the evil influence doctors exercise on society, or the sinister plot they were hatching to take over and exploit the rest of mankind, or something of the kind. For this would relate the increase of doctors in an intelligible way to the interests, needs, or purposes of men. In the absence of such a relation, we remain in the dark, and are tempted to assume the worst.

What is meant by "intelligibility" here is that we can understand the judgment as a use of "good" and "bad." It is now widely agreed that a word gets its meaning from its place in the skein of discourse; we can give its meaning, for instance, by making clear its relations to other words. But this is not to say that we can give the meaning in a set of logical relations of equivalence, entailment, and so on, that an earlier positivism saw as the content of philosophical endeavor. For the relation to other terms may pass through a certain context. Thus, there is a relation between "good" and commending, expressing approval, and so on. But this is not to say that we can construe "X is good," for instance, as *meaning* "I commend X."[25] Rather, we can say that "good" can be used for commending, that to apply the word involves being ready to commend in certain circumstances, for if you are not then you are shown to have been unserious in your application of it, and so on.[26]

The relation between "good" and commending, expressing approval, persuading, and so on has been stressed by nonnaturalist theorists of ethics (though not always adequately understood, because of the narrow concentration on logical relations), but the term has another set of relations, to the grounds of its predication, as we have tried to show. These two aspects correspond respectively to what has often been called the evaluative, emotive, or prescriptive meaning on one hand (depending on the theory) and the "descriptive" meaning on the other. For half a century an immense barrage of dialectical artillery has been trained on the so-called naturalistic fallacy in an effort to pry "good" loose from any set range of descriptive meanings. But this immense effort has been beside the point, for it has concentrated on the nonexistence of logical relations between descriptive predicates and evaluative terms. But the fact that one cannot find equivalences, make valid deductive argument, and so on, may show nothing about the relation between a given concept and others.

Just as with the "evaluative" meaning above, so with the "descriptive" meaning: "good" does not *mean* "conducive to the fulfillment of human wants, needs, or purposes"; but its use is unintelligible outside of any relationship to wants, needs, and purposes, as we saw above. For if we abstract from this relation, then we cannot tell whether a man is using "good" to make a judgment, or simply express some feeling; and it is an essential part of the meaning of the term that such a distinction can be made. The "descriptive"[27] aspects of "good's" meaning can rather be shown in this way: "good" is used in evaluating, commending, persuading, and so on by a race of beings who are such that through their needs, desires, and so on, they are not indifferent to the various outcomes of the world process. A race of inactive, godless angels, as really disinterested spectators, would have no use for it, could not make use of it, except in the context of cultural anthropology, just as human anthropologists use "mana." It is because "good" has this use, and can only have meaning because there is this role to fill in human life, that it becomes unintelligible when abstracted from this role. Because its having a use arises from the fact that we are not indifferent, its use cannot be understood where we cannot see what there is to be not-indifferent about, as in the strange "grounds" quoted by our imaginary opponent above. Moreover, its role is such that it is supposed to be predicated on general grounds, and not just according to the likes and dislikes or feelings of individuals. This distinction is essential since (among other things) the race concerned spends a great deal of effort achieving and maintaining consensus within larger or smaller groups, without which it would not survive. But where we cannot see what the grounds could be, we are tempted to go on treating the use of "good" as an expression of partiality, only of the more trivial, individual kind.

We can thus see why, for instance, "anything conducive to human happiness is good" does not need any further grounds to be adduced on its behalf. In human happiness, which by definition men desire, we have an adequate ground. This does not mean that all argument is foreclosed. We can try to show that men degenerate in various ways if they seek only happiness, and that certain things which also make men unhappy are necessary for their development. Or we can try to show that there is a higher and a lower happiness, that most men seek under this title only pleasure, and that this turns them away from genuine fulfillment; and so on. But unless we can bring up some countervailing consideration, we cannot deny a thesis of this kind. The fact that we can always bring up such countervailing considerations means that we can never say that "good" *means* "conducive to human happiness," as Moore saw. But that something is conducive to human happiness, or in general to the fulfillment of human needs, wants, and purposes, is a prima facie reason for calling it good, which stands unless countered.

Thus the nonneutrality of the theoretical findings of political science need not surprise us. In setting out a given framework, a theorist is also setting out the gamut of possible polities and policies. But a *political* framework cannot fail to contain some, even implicit, conception of human needs, wants, and purposes. The context of this conception will determine the value slope of the gamut, unless we can introduce countervailing considerations. If these countervailing factors are motivationally marginal enough not to have too much relevance to political behavior, then we can speak of the original valuation as being only overridden. For that part of the gamut of possibilities which we originally valued still has the property we attributed to it and thus remains valuable for us in one aspect, even if we have to give it low marks in another. For instance, we still will believe that having a peaceful polity is good, even if it results in bad art. But if the countervailing factor is significant for political behavior, then it will lead us to revise our framework and hence our views about the gamut of possible polities and policies; this in turn will lead to new valuations. The basis of the old values will be undermined. Thus, if we believe that an absence of violence will lead to stagnation and foreign conquest or breakdown, then we change the gamut of possibility: the choice no longer lies between peace and violence, but between, say, controlled violence and greater uncontrolled violence. Peace ceases to figure on the register: it is not a good we can attain.

Of course, the countervailing factor may not revise our gamut of choices so dramatically. It may simply show that the values of our originally preferred regime cannot be integrally fulfilled or that they will be under threat from a previously unsuspected quarter, or that they will be attended with dangers or disadvantages or disvalues not previously taken into account, so that we have to make a choice as in the peace-versus-good-art case above. Thus not all alterations of the framework will undermine the original values. But we can see that the converse does hold, and all undermining will involve a change in the framework. For if we leave the original framework standing, then the values of its preferred regime will remain as fully realizable goods, even if they are attended with certain evils which force on us a difficult choice, such as that between peace and good art, or progress and psychic harmony, or whatever.

In this sense we can say that a given explanatory framework secretes a notion of good, and a set of valuations, which cannot be done away with—though they can be overridden—unless we do away with the framework. Of course, because the values can be overridden, we can only say that the framework tends to support them, not that it establishes their validity. But this is enough to show that the neutrality of the findings of political science is not what it was thought to be. For establishing a given framework restricts the range of value positions which can be defensibly adopted. For in the light of the framework certain goods can be accepted as such without further argument, whereas other rival ones cannot be adopted without adducing overriding considerations. The framework can be said to distribute the onus of argument in a certain way. It is thus not neutral.

The only way to avoid this while doing political science would be to stick to the narrow-gauge discoveries which, just because they are, taken alone, compatible with a great number of political frameworks, can bathe in an atmosphere of value neutrality. That Catholics in Detroit tend to vote Democrat and can consort with almost anyone's conceptual scheme, and thus with almost anyone's set of political values. But to the extent that political science cannot dispense with theory, with the search for a framework, to that extent it cannot stop developing normative theory.

Nor need this have the vicious results usually attributed to it. There is nothing to stop us making the greatest attempts to avoid bias and achieve objectivity. Of course, it is hard, almost impossible, and precisely because our values are also at stake. But it helps, rather than hinders, the cause to be aware of this.

Notes

1. Of course, Rousseau's general will may remain a value in the hypothetical world he casts for it, but that concerns utopia building, not political philosophy.
2. This could involve either an undermining or an overriding of the value judgment. For we can deny something, a condition or outcome, the property by which it is judged good not only by denying it a property by which it fulfills certain human needs, wants, or purposes, but also by denying that these needs, wants, or purposes exist. And we can override the judgment that it is good by pointing to other needs, wants, or purposes that it frustrates.
3. *Politics of the Developing Areas*, p. 33.
4. Ibid., pp. 35–6.
5. Ibid., p. 39.
6. "The Democratic Character," in *Political Writings* (Glencoe, Ill.: The Free Press, 1951), p. 473.
7. Ibid., p. 471n.

8. *Power and Society* (New Haven, Conn.: Yale University Press, 1952), Ch. 9, sections 3 and 4.
9. Ibid., p. 228.
10. Ibid., p. 229.
11. Ibid., p. 231.
12. Ibid.
13. Ibid., p. 232.
14. Ibid., p. xi.
15. Ibid., p. 231.
16. Ibid., p. 239.
17. *Political Writings.*
18. Ibid., pp. 497–8.
19. Ibid., pp. 497–502.
20. Ibid., p. 500.
21. We might also speak of "interests" here, but this can be seen as included in "wants" and "needs." Interest may deviate from want, but can only be explicated in terms of such concepts as "satisfaction," "happiness," "unhappiness," etc., the criteria for whose application are ultimately to be found in what we want.
22. In what follows I am indebted to the arguments of Mrs. P. Foot, e.g., to her "When Is a Principle a Moral Principle?" in *Aristotelian Society, Supplementary Vol.* xxviii (1954), and her "Moral Arguments" in *Mind*, A.S.S.V. lxvii (1958), although I do not know whether she would agree with the conclusions I draw from them.
23. Borrowed with changes from Hare's *Freedom and Reason* (Oxford: Clarendon Press, 1963).
24. We may use behavior, of course, to judge which of the two constructions to put on a man's words, but the two are not distinguished by behavioral criteria alone, but also by what a man thinks and feels. It is possible, of course, to challenge a man's even sincere belief that he is judging of good and bad, and to disvalue it on the grounds that one holds it to be based largely on irrational prejudice or unavowed ambitions or fears. Thus our first segregationist may be judged as not too different from our second. For there is some evidence that segregationist ideas can at least partly be assimilated to neurotic phobias in their psychological roots. But this is just why many people look on the judgments of segregationists as self-deception and unconscious sham. "Really," they are just expressions of horror. But this respects the logic of "good" as we have outlined it: for it concludes that if the rational base is mere show, then the judgment is mere show. Segregationists, for their part, rarely are of the second type, and pay homage to the logic of "good" by casting about for all sorts of specious reasons of the correct form.
25. Cf. John Searle's "Meaning and Speech Acts," *Philosophical Review*, lxxi (1962) 423–32.
26. Thus, if I say, "This is a good car," and then my friend comes along and says, "Help me choose a car," I have to eat my words if I am not willing to commend the car to him, *unless* I can adduce some other countervailing factor such as price, my friend's proclivity to dangerous driving, or whatever. But this complex relationship cannot be expressed in an equivalence, e.g., "This is a good car" entails, "If you are choosing a car, take this."
27. The terms "descriptive meaning" and "evaluative meaning"' can be seen to be seriously misleading, as is evident from the discussion. For they carry the implication that the meaning is "contained" in the word, and can be "unpacked" in statements of logical equivalence. There is rather a descriptive aspect and an evaluative aspect of its role or use, which are, moreover, connected, for we cannot see whether a use of the term carries the evaluation force of "good" unless we can also see whether it enters into the skein of relations which constitute the descriptive dimension of its meaning.

Chapter 36

The Value-Oriented Bias of Social Inquiry

Ernest Nagel

We turn, finally, to the difficulties said to confront the social sciences because the social values to which students of social phenomena are committed not only color the contents of their findings but also control their assessment of the evidence on which they base their conclusions. Since social scientists generally differ in their value commitments, the "value neutrality" that seems to be so pervasive in the natural sciences is therefore often held to be impossible in social inquiry. In the judgment of many thinkers, it is accordingly absurd to expect the social sciences to exhibit the unanimity so common among natural scientists concerning what are the established facts and satisfactory explanations for them. Let us examine some of the reasons that have been advanced for these contentions. It will be convenient to distinguish four groups of such reasons, so that our discussion will deal in turn with the alleged role of value judgments in (1) the selection of problems, (2) the determination of the contents of conclusions, (3) the identification of fact, and (4) the assessment of evidence.

1

The reasons perhaps most frequently cited make much of the fact that the things a social scientist selects for study are determined by his conception of what are the socially important values. According to one influential view, for example, the student of human affairs deals only with materials to which he attributes "cultural significance," so that a "value orientation" is inherent in his choice of material for investigation. Thus, although Max Weber was a vigorous proponent of a "value-free" social science—i.e., he maintained that social scientists must appreciate (or "understand") the values involved in the actions or institutions they are discussing but that it is not their business as objective scientists to approve or disapprove either those values or those actions and institutions—he nevertheless argued that

> The concept of culture is a *value-concept*. Empirical reality becomes "culture" to us because and insofar as we relate it to value ideas. It includes those segments and only those segments of reality which have become significant to us because of this value-relevance. Only a small portion of existing concrete reality is colored by our value-conditioned interest and it alone is significant to us. It is significant because it reveals relationships which are important to us due to their connection with our values. Only because and to the extent that this is the case is it worthwhile for us to know it in its individual features. We cannot discover, however, what is meaningful to us by means of a "presuppositionless" investigation of empirical data. Rather perception of its meaningfulness to us is the presupposition of its becoming an *object* of investigation.[1]

It is well-nigh truistic to say that students of human affairs, like students in any other area of inquiry, do not investigate everything, but direct their attention to certain selected portions of the inexhaustible content of concrete reality. Moreover, let us accept the claim, if only for the sake of the argument, that a social scientist addresses himself exclusively to matters which he believes are important because of their assumed relevance to his cultural values.[2] It is not clear, however, why the fact that an investigator selects the materials he studies in the light of problems which interest him and which seem to him to bear on matters he regards as important, is of greater moment for the logic of social inquiry than it is for the logic of any other branch of inquiry. For example, a social scientist may believe that a free economic market embodies a cardinal human value, and he may produce evidence to show that certain kinds of human activities are indispensable to the perpetuation of a free market. If he is concerned with processes which maintain this type of economy rather than some other type, how is this fact more pertinent to the question whether he has adequately evaluated the evidence for his conclusion than is the bearing upon the analogous question of the fact that a physiologist may be concerned with processes which maintain a constant internal temperature in the human body rather than with something else? The things a social scientist *selects for study* with a view to determining the conditions or consequences of their existence may indeed be dependent on the indisputable fact that he is a "cultural being." But similarly, were we not human beings though still capable of conducting scientific inquiry, we might conceivably have an interest neither in the conditions that maintain a free market, nor in the processes involved in the homeostasis of the internal temperature in human bodies, nor for that matter in the mechanisms that regulate the height of tides, the succession of seasons, or the motions of the planets.

In short, there is no difference between any of the sciences with respect to the fact that the interests of the scientist determine what he selects for investigation. But this fact, by itself, represents no obstacle to the successful pursuit of objectively controlled inquiry in any branch of study.

2

A more substantial reason commonly given for the value-oriented character of social inquiry is that, since the social scientist is himself affected by considerations of right and wrong, his own notions of what constitutes a satisfactory social order and his own standards of personal and social justice do enter, in point of fact, into his analyses of social phenomena. For example, according to one version of this argument, anthropologists must frequently judge whether the means adopted by some society achieves the intended aim (e.g., whether a religious ritual does produce the increased fertility for the sake of which the ritual is performed); and in many cases the adequacy of the means must be judged by admittedly "relative" standards, i.e., in terms of the ends sought or the standards employed by that society, rather than in terms of the anthropologist's own criteria. Nevertheless, so the argument proceeds, there are also situations in which

> we must apply absolute standards of adequacy, that is evaluate the end-results of behavior in terms of purposes we believe in or postulate. This occurs, first, when we speak of the satisfaction of psycho-physical "needs" offered by any culture; secondly, when we assess the bearing of social facts upon survival; and thirdly, when we pronounce upon social integration and stability. In each case our

> statements imply judgments as to the worth-whileness of actions, as to "good" or "bad" cultural solutions of the problems of life, and as to "normal" and "abnormal" states of affairs. These are basic judgments which we cannot do without in social enquiry and which clearly do not express a purely personal philosophy of the enquirer or values arbitrarily assumed. Rather do they grow out of the history of human thought, from which the anthropologist can seclude himself as little as can anyone else. Yet as the history of human thought has led not to one philosophy but to several, so the value attitudes implicit in our ways of thinking will differ and sometimes conflict.[3]

It has often been noted, moreover, that the study of social phenomena receives much of its impetus from a strong moral and reforming zeal, so that many ostensibly "objective" analyses in the social sciences are in fact disguised recommendations of social policy. As one typical but moderately expressed statement of the point puts it, a social scientist

> cannot wholly detach the unifying social structure that, as a scientist's theory, guides his detailed investigations of human behavior, from the unifying structure which, as a citizen's ideal, he thinks ought to prevail in human affairs and hopes may sometimes be more fully realized. His social theory is thus essentially a program of action along two lines which are kept in some measure of harmony with each other by that theory—action in assimilating social facts for purposes of systematic understanding, and action aiming at progressively molding the social pattern, so far as he can influence it, into what he thinks it ought to be.[4]

It is surely beyond serious dispute that social scientists do in fact often import their own values into their analyses of social phenomena. It is also undoubtedly true that even thinkers who believe human affairs can be studied with the ethical neutrality characterizing modern inquiries into geometrical or physical relations, and who often pride themselves on the absence of value judgments from their own analyses of social phenomena, do in fact sometimes make such judgments in their social inquiries.[5] Nor is it less evident that students of human affairs often hold conflicting values; that their disagreements on value questions are often the source of disagreements concerning ostensibly factual issues; and that, even if value predications are assumed to be inherently capable of proof or disproof by objective evidence, at least some of the differences between social scientists involving value judgments are not in fact resolved by the procedures of controlled inquiry.

In any event, it is not easy in most areas of inquiry to prevent our likes, aversions, hopes, and fears from coloring our conclusions. It has taken centuries of effort to develop habits and techniques of investigation which help safeguard inquiries in the natural sciences against the intrusion of irrelevant personal factors; and even in these disciplines the protection those procedures give is neither infallible nor complete. The problem is undoubtedly more acute in the study of human affairs, and the difficulties it creates for achieving reliable knowledge in the social sciences must be admitted.

However, the problem is intelligible only on the assumption that there is a relatively clear distinction between factual and value judgments, and that however difficult it may sometimes be to decide whether a given statement has a purely factual content, it is in principle possible to do so. Thus, the claim that social scientists are pursuing the twofold program mentioned in the above quotation makes sense, only if it is possible to distinguish between, on the one hand, contributions to theoretical understanding (whose factual validity presumably does not depend on the social ideal to which a social scientist may subscribe), and on the other hand contributions to the dissemination or realization of some social ideal (which may not be accepted by all social scientists). Accordingly, the undeniable difficulties that stand in the way of obtaining reliable knowledge of human affairs because of the fact that social scientists differ in their value orientations are practical difficulties. The difficulties are not necessarily insuperable, for since by hypothesis it is not impossible to distinguish between fact and value, steps can be taken to identify a value bias when it occurs, and to minimize if not to eliminate completely its perturbing effects.

One such countermeasure frequently recommended is that social scientists abandon the pretense that they are free from all bias, and that instead they state their value assumptions as explicitly and fully as they can.[6] The recommendation does not assume that social scientists will come to agree on their social ideals once these ideas are explicitly postulated, or that disagreements over values can be settled by scientific inquiry. Its point is that the question of how a given ideal is to be realized, or the question whether a certain institutional arrangement is an effective way of achieving the ideal, is on the face of it not a value question, but a factual problem—to be resolved by the objective methods of scientific inquiry—concerning the adequacy of proposed means for attaining stipulated ends. Thus, economists may permanently disagree on the desirability of a society in which its members have a guaranteed security against economic want, since the disagreement may have its source in inarbitrable preference for different social values. But when sufficient evidence is made available by economic inquiry, economists do presumably agree on the factual proposition that, *if* such a society is to be achieved, then a purely competitive economic system will not suffice.

Although the recommendation that social scientists make fully explicit their value commitments is undoubtedly salutary, and can produce excellent fruit, it verges on being a counsel of perfection. For the most part we are unaware of many assumptions that enter into our analyses and actions, so that despite resolute efforts to make our preconceptions explicit some decisive ones may not even occur to us. But in any event, the difficulties generated for scientific inquiry by unconscious bias and tacit value orientations are rarely overcome by devout resolutions to eliminate bias. They are usually overcome, often only gradually, through the self-corrective mechanisms of science as social enterprise. For modern science encourages the invention, the mutual exchange, and the free but responsible criticisms of ideas; it welcomes competition in the quest for knowledge between independent investigators, even when their intellectual orientations are different; and it progressively diminishes the effects of bias by retaining only those proposed conclusions of its inquiries that survive critical examination by an indefinitely large community of students, whatever be their value preferences or doctrinal commitments. It would be absurd to claim that this institutionalized mechanism for sifting warranted beliefs has operated or is likely to operate in social inquiry as effectively as it has in the natural sciences. But it would be no less absurd to conclude that reliable knowledge of human affairs is unattainable merely because social inquiry is frequently value oriented.

3

There is a more sophisticated argument for the view that the social sciences cannot be value free. It maintains that the distinction between fact and value assumed in the preceding discussion is untenable when purposive human behavior is being analyzed, since in this context value judgments enter inextricably into what appear to be "purely descriptive" (or factual) statements. Accordingly, those who subscribe to this thesis claim that an ethically neutral social science is in principle impossible, and not simply that it is difficult to attain. For if fact and value are indeed so fused that they cannot even be distinguished, value judgments cannot be eliminated from the social sciences unless all predications are also eliminated from them, and therefore unless these sciences completely disappear.

For example, it has been argued that the student of human affairs must distinguish between valuable and undesirable form of social activity, on pain of failing in his "plain duty" to present social phenomena truthfully and faithfully:

> Would one not laugh out of court a man who claimed to have written a sociology of art but who actually had written a sociology of trash? The sociologist of religion must distinguish between phenomena which have a religious character and phenomena which are a-religious. To be able to do this, he must understand what religion is.... Such understanding enables and forces him to distinguish between genuine and spurious religion, between higher and lower religions; these religions are higher in which the specifically religious motivations are effective to a higher degree.... The sociologist of religion cannot help noting the difference between those who try to gain it by a change of heart. Can he see this difference without seeing at the same time the difference between a mercenary and nonmercenary attitude?... The prohibition against value-judgments in social science would lead to the consequence that we are permitted to give a strictly factual description of the overt acts that can be observed in concentration camps, and perhaps an equally factual analysis of the motivations of the actors concerned: we would not be permitted to speak of cruelty. Every reader of such a description who is not completely stupid would, of course, see that the actions described are cruel. The factual description would, in truth, be a bitter satire. What claimed to be a straightforward report would be an unusually circumlocutory report.... Can one say anything relevant on public opinion polls ... without realizing the fact that many answers to the questionnaires are given by unintelligent, uninformed, deceitful, and irrational people, and that not a few questions are formulated by people of the same caliber—can one say anything relevant about public opinion polls without committing one value-judgment after another?[7]

Moreover, the assumption implicit in the recommendation discussed above for achieving ethical neutrality is often rejected as hopelessly naive—that is the assumption, it will be recalled, that relations of means to ends can be established without commitments to these ends, so that the conclusions of social inquiry concerning such relations are objective statements which make *conditional* rather than categorical assertions about values. This assumption is said by its critics to rest on the supposition that men attach value only to the ends they seek, and not to the means for realizing their aims. However, the supposition is alleged to be grossly mistaken. For the character of the means one employs to secure some goal affects the nature of the total outcome; and the choice men make between alternative means for obtaining a given end depends on the values they ascribe to those alternatives. In consequence, commitments to specific valuations are said to be involved even in what appear to be purely factual statements about means-ends relations.[8]

We shall not attempt a detailed assessment of this complex argument, for a discussion of the numerous issues it raises would take us far afield. However, three claims made in the course of the argument will be admitted without further comment as indisputably correct: that a large number of characterizations sometimes assumed to be purely factual descriptions of social phenomena do indeed formulate a type of value judgment; that it is often difficult, and in any case usually inconvenient in practice, to distinguish between the purely factual and the "evaluative" contents of many terms employed in the social sciences; and that values are commonly attached to means and not only to ends. However, these admissions do not entail the conclusion that, in a manner unique to the study of purposive human behavior, fact and value are fused beyond the possibility of distinguishing between them. On the contrary, as we shall try to show, the claim that there is such a fusion and that a value-free social science is therefore inherently absurd, confounds two quite different senses of the term "value judgment": the sense in which a value judgment expresses *approval or disapproval* either of some moral (or social) ideal, or of some action (or institution) because of a commitment to such an ideal; and the sense in which a value judgment expresses *an estimate* of the degree to which some commonly recognized (and more or less clearly defined) type of action, object, or institution is embodied in a given instance.

It will be helpful to illustrate these two senses of "value judgment" first with an example from biology. Animals with bloodstreams sometimes exhibit the condition known as "anemia." An anemic animal has a reduced number of red blood corpuscles, so that, among other things, it is less able to maintain a constant internal temperature than are members of its species with a "normal" supply of such blood cells. However, although the meaning of the term "anemia" can be made quite clear, it is not in fact defined with complete precision; for example, the notion of a "normal" number of red corpuscles that enters into the definition of the term is itself somewhat vague, since this number varies with the individual members of a species as well as with the state of a given individual at different times (such as its age or the altitude of its habitat). But in any case, to decide whether a given animal is anemic an investigator must judge whether the available evidence *warrants* the conclusion that the specimen is anemic.[9] He may perhaps think of anemia as being of several distinct kinds (as is done in actual medical practice), or he may think of anemia as a condition that is realizable with greater or lesser completeness (just as certain plane curves are sometimes described as better or worse approximations to a circle as defined in geometry); and, depending on which of these conceptions he adopts, he may decide either that his specimen has a certain kind of anemia or that it is anemic only to a certain degree. When the investigator reaches a conclusion, he can therefore be said to be making a "value judgment," in the sense that he has in mind some standardized type of physiological condition designated as "anemia" and that he *assesses* what he knows about his specimen with the measure provided by this assumed standard. For the sake of easy reference, let us call such evaluations of the evidence, which conclude that a given characteristic is in some degree present (or absent) in a given instance, "characterizing value judgments."

On the other hand, the student may also make a quite different sort of value judgment, which asserts that, since an anemic animal has diminished powers of maintaining itself, anemia is an undesirable condition. Moreover, he may apply this general judgment to a particular case, and so come to deplore the fact that a given animal is anemic. Let us label such evaluations, which conclude that some envisaged or actual state of affairs is worthy of approval or disapproval, "appraising value judgments."[10] It is clear, however, that an investigator making a characterizing value judgment is not thereby logically bound to affirm or deny a corresponding appraising evaluation. It is no less evident that he cannot consistently make an appraising value judgment about a given instance (e.g., that it is undesirable for a given animal to continue being anemic), unless he can affirm a characterizing judgment about that instance independently of the appraising one (e.g., that the animal is anemic). Accordingly, although characterizing judgments are necessarily entailed by many appraising judgments, making appraising judgments is not a necessary condition for making characterizing ones.

Let us now apply these distinctions to some of the contentions advanced in the argument quoted above. Consider first the claim that the sociologist of religion must recognize the difference between mercenary and nonmercenary attitudes, and that in consequence he is inevitably committing himself to certain values. It is certainly beyond dispute that these attitudes are commonly distinguished; and it can also be granted that a sociologist of religion needs to understand the difference between them. But the sociologist's obligation is in this respect quite like that of the student of animal physiology, who must also acquaint himself with certain distinctions—even though the physiologist's distinction between, say, anemic and nonanemic may be less familiar to the ordinary layman and is in any case much more precise than is the distinction between mercenary and nonmercenary attitudes. Indeed, because of the vagueness of these latter terms, the scrupulous sociologist may find it extremely difficult to decide whether or not the attitude of some community toward its acknowledged gods is to be characterized as mercenary; and if he should finally decide, he may base his conclusion on some inarticulated "total impression" of that community's manifest behavior, without being able to state exactly the detailed grounds for his decision. But however this may be, the sociologist who claims that a certain attitude manifested by a given religious group is mercenary, just as the physiologist who claims that a certain individual is anemic, is making what is primarily a characterizing value judgment. In making these judgments, neither the sociologist nor the physiologist is necessarily committing himself to any values other than the values of scientific probity; and in this respect, therefore, there appears to be no difference between social and biological (or for that matter, physical) inquiry.

On the other hand, it would be absurd to deny that in characterizing various actions as mercenary, cruel, or deceitful, sociologists are frequently (although perhaps not always wittingly) asserting appraising as well as characterizing value judgments. Terms like "mercenary," "cruel," or "deceitful" as commonly used have a widely recognized pejorative overtone. Accordingly, anyone who employs such terms to characterize human behavior can normally be assumed to be stating his disapprobation of that behavior (or his approbation, should he use terms like "nonmercenary," "kindly," or "truthful"), and not simply characterizing it.

However, although many (but certainly not all) ostensibly characterizing statements asserted by social scientists undoubtedly express commitments to various (not always compatible) values, a number of "purely descriptive" terms as used by natural scientists in certain contexts sometimes also have an unmistakably appraising value connotation. Thus, the claim that a social scientist is making appraising value judgments when he characterizes respondents to questionnaires as uninformed, deceitful, or irrational can be matched by the equally sound claim that a physicist is also making such judgments when he describes a particular chronometer as inaccurate, a pump as inefficient, or a supporting platform as unstable. Like the social scientist in this example, the physicist is characterizing certain objects in his field of research; but, also like the social scientist, he is in addition expressing his disapproval of the characteristics he is ascribing to those objects.

Nevertheless—and this is the main burden of the present discussion—there are no good reasons for thinking that it is inherently impossible to *distinguish* between the characterizing and the appraising judgments implicit in many statements, whether the statements are asserted by students of human affairs or by natural scientists. To be sure, it is not always easy to make the distinction formally explicit in the social sciences—in part because much of the language employed in them is very vague, in part because appraising judgments that may be implicit in a statement tend to be overlooked by us when they are judgments to which we are actually committed though without being aware of our commitments. Nor is it always useful or convenient to perform this task. For many statements implicitly containing both characterizing and appraising evaluations are sometimes sufficiently clear without being reformulated in the manner required by the task; and the reformulations would frequently be too unwieldy for effective communication between members of a large and unequally prepared group of students. But these are essentially practical rather than theoretical problems. The difficulties they raise provide no compelling reasons for the claim that an ethically neutral social science is inherently impossible.

Nor is there any force in the argument that, since values are commonly attached to means and not only to ends, statements about means-ends relations are not value free. Let us test the argument with a simple example. Suppose that a man with an urgent need for a car but without sufficient funds to buy one can achieve his aim by borrowing a sum either from a commercial bank or from friends who waive payment of any interest. Suppose further that he dislikes becoming beholden to his friends for financial favors, and prefers the impersonality of a commercial loan. Accordingly, the comparative values this individual places upon the alternative means available to him for realizing his aim obviously control the choice he makes between them. Now the *total* outcome that would result from his adoption of one of the alternatives is admittedly different from the *total* outcome that would result from his adoption of the other alternative means, each of them would achieve a result—namely, his purchase of the needed car—that is common to both the total outcomes. In consequence, the validity of the statement that he could buy the car by borrowing money from a bank, as well as of the statement that he could realize this aim by borrowing from friends, is unaffected by the valuations placed upon the means, so that neither statement involves any special appraising evaluations. In short, the statements about means-ends relations are value free.

4

There remains for consideration the claim that a value-free social science is impossible, because value commitments enter into the very *assessment of evidence* by social scientists, and not simply into the content of the conclusions they advance. This version of the claim itself has a large number of variant forms, but we shall examine only three of them.

The least radical form of the claim maintains that the conceptions held by a social scientist of what constitute cogent evidence or sound intellectual workmanship are the products of his education and his place in society, and are affected by the social values transmitted by this training and associated with this social position; accordingly, the values to which the social scientist is thereby committed determine which statements he *accepts* as well-grounded conclusions about human affairs. In this form, the claim is a *factual* thesis, and must be supported by detailed empirical evidence concerning the influences exerted by a man's moral and social values upon what he is ready to acknowledge as sound social analysis. In many instances such evidence is indeed available; and differences between social scientists in respect to what they accept as credible can sometimes be attributed to the influence of national, religious, economic, and other kinds of bias. However, this variant of the claim excludes neither the possibility of recognizing assessments of evidence that are prejudiced by special value commitments, nor the possibility of correcting for such prejudice. It therefore raises no issue that has not already been discussed when we examined the second reason for the alleged value-oriented character of social inquiry.

Another but different form of the claim is based on recent work in theoretical statistics dealing with the assessment of evidence for so-called statistical hypotheses—hypotheses concerning the probabilities of random events, such as the hypothesis that the probability of a male human birth is one-half. The central idea relevant to the present question that underlies these developments can be sketched in terms of an example. Suppose that, before a fresh batch of medicine is put on sale, tests are performed on experimental animals for its possible toxic effects because of impurities that have not been eliminated in its manufacture, for example, by introducing small quantities of the drug into the diet of one hundred guinea pigs. If no more than a few of the animals show serious after-effects, the medicine is to be regarded as safe, and will be marketed; but if a contrary result is obtained the drug will be destroyed. Suppose now that three of the animals do in fact become gravely ill. Is this outcome significant (i.e., does it indicate that the drug has toxic effects), or is it perhaps an "accident" that happened because of some peculiarity in the affected animals? To answer the question, the experimenter must *decide* on the basis of the evidence between the hypothesis H_1: the drug is toxic, and the hypothesis H_2: the drug is not toxic. But how is he to decide, if he aims to be "reasonable" rather than arbitrary? Current statistical theory offers him a rule for making a reasonable decision, and bases the rule on the following analysis.

Whatever decision the experimenter may make, he runs the risk of committing either one of two types of errors: he may reject a hypothesis though in fact it is true (i.e., despite the fact that H_1 is actually true, he mistakenly decides against it in the light of the evidence available to him); or he may accept a hypothesis though in fact it is false. His decision would therefore be eminently reasonable, were it based on a rule guaranteeing that no decision ever made in accordance with the rule would commit either type of error. Unhappily, there are no rules of this sort. The next suggestion is to find a rule such that, when decisions are made in accordance with it, the relative frequency of each type of error is quite small. But unfortunately, the risks of committing each type of error are not independent; for example, it is in general logically impossible to find a rule so that decisions based on it will commit each type of error with a relative frequency not greater than one in a thousand. In consequence, before a reasonable rule can be proposed, the experimenter must compare the relative importance to himself of the two types of error, and state what risk he is willing to take of committing the type of error he judges to be the more important one. Thus, were he to reject H_1 though it is true (i.e., were he to commit an error of the first type), all the medicine under consideration would be put on sale, and the lives of those using it would be endangered; on the other hand, were he to commit an error of the second type with respect to H_1, the entire batch of medicine would be scrapped, and the manufacturer would incur a financial loss. However, the preservation of human life may be of greater moment to the experimenter than financial gain; and he may perhaps stipulate that he is unwilling to base his decision on a rule for which the risk of committing an error of the first type is greater than one such error in a hundred decisions. If this is assumed, statistical theory can specify a rule satisfying the experimenter's requirement, though how this is done, and how the risk of committing an error of the second type is calculated, are technical questions of no concern to us. The main point to be noted in this analysis is that the rule presupposes certain appraising judgments of value. In short, if this result is generalized, statistical theory appears to support the thesis that value commitments enter decisively into the rules for assessing evidence for statistical hypotheses.[11]

However, the theoretical analysis upon which this thesis rests does not entail the conclusion that the rules actually employed in every social inquiry for assessing evidence necessarily involve some *special* commitments, i.e., commitments such as those mentioned in the above example, as distinct from those generally implicit in science as an enterprise aiming to achieve reliable knowledge. Indeed, the above example illustrating the reasoning in current statistical theory can be misleading, insofar as it suggests that alternative decisions between statistical hypothesis must invariably lead to alternative actions having immediate practical consequences upon which different special values are placed. For example, a theoretical physicist may have to decide between two statistical hypotheses concerning the probability of certain energy exchanges in atoms; and a theoretical sociologist may similarly have to choose between two statistical hypotheses concerning the relative frequency of childless marriages under certain social arrangements. But neither of these men may have any *special* values at stake associated with the alternatives between which he must decide, other than the values, to which he is committed as a member of a scientific community, to conduct his inquiries with probity and responsibility. Accordingly, the question whether any special value commitments enter into assessments of evidence in either the natural or social sciences is not settled one way or the other by theoretical statistics; and the question can be answered only by examining actual inquiries in the various scientific disciplines.

Moreover, nothing in the reasoning of theoretical statistics depends on what particular subject matter is under discussion when a decision between alternative statistical hypotheses is to be made. For the reasoning is entirely general; and reference to some special subject matter becomes relevant only when a definite numerical value is to be assigned to the risk some investigator is prepared to take of making an erroneous decision concerning a given hypothesis. Accordingly, if current statistical theory is used to support the claim that value commitments enter into the assessment of evidence for statistical hypotheses in social inquiry, statistical theory can be used with equal justification to support analogous claims for all other inquiries as well. In short, the claim we have been discussing establishes no difficulty that supposedly occurs in the search for reliable knowledge in the study of human affairs which is not also encountered in the natural sciences.

A third form of this claim is the most radical of all. It differs from the first variant mentioned above in maintaining that there is a necessary *logical* connection, and not merely a contingent or causal one, between the "social perspective" of a student of human affairs and his standards of competent social inquiry, and in consequence the influence of the special values to which he is committed because of his own social involvements is not eliminable. This version of the claim is implicit in Hegel's account of the "dialectical" nature of human history and is integral to much Marxist as well as non-Marxist philosophy that stresses the "historically relative" character of social thought. In any event, it is commonly based on the assumption that, since social institutions and their cultural products are constantly changing, the intellectual apparatus required for understanding them must also change; and every idea employed for this purpose is therefore adequate only for some particular stage in the development of human affairs. Accordingly, neither the substantive concepts adopted for classifying and interpreting social phenomena, nor the logical canons used for estimating the worth of such concepts, have a "timeless validity"; there is no analysis of social phenomena which is not the expression of some special social standpoint, or which does not reflect the interests and values dominant in some sector of the human scene at a certain stage is its history. In consequence, although a sound distinction can be made in the natural sciences between the origin of a man's views and their factual validity, such a distinction allegedly cannot be made in social inquiry; and prominent exponents of "historical relativism" have therefore challenged the universal adequacy of the thesis that "the genesis of a proposition is under all circumstances irrelevant to its truth." As one influential proponent of this position puts the matter.

> The historical and social genesis of an idea would only be irrelevant to its ultimate validity if the temporal and social conditions of its emergence had no effect on its content and form. If this were the case, any two periods in the history of human knowledge would only be distinguished from one another by the fact that in the earlier period certain things were still unknown and certain errors still existed which, through later knowledge were completely corrected. This simple relationship between an earlier incomplete and a later complete period of knowledge may to a large extent be appropriate for the exact sciences.... For the history of the cultural sciences, however, the earlier stages are not quite so simply superseded by the later stages, and it is not so easily demonstrable that early errors have subsequently been corrected. Every epoch has its fundamentally new approach and its characteristic point of view, and consequently sees the "same" object from a new perspective.... The very principles, in the light of which knowledge is to be criticized, are themselves found to be socially and historically conditioned. Hence their application appears to be limited to given historical periods and the particular types of knowledge then prevalent.[12]

Historical research into the influence of society upon the beliefs men hold is of undoubted importance for understanding the complex nature of the scientific enterprise; and the sociology of knowledge—as such investigations have come to be called—has produced many clarifying contributions to such an understanding. However, these admittedly valuable services of the sociology of knowledge do not establish the radical claim we have been stating. In the first place, there is no competent evidence to show that the principles employed in social inquiry for assessing the intellectual products are *necessarily* determined by the social perspective of the inquirer. On the contrary, the "facts" usually cited in support of this contention establish at best only a contingent causal relation between a man's social commitments and his canons of cognitive validity. For example, the once fashionable view that the "mentality" or logical operations of primitive societies differ from those typical in Western civilization—a discrepancy that was attributed to differences in the institutions of the societies under comparison—is now generally recognized to be erroneous, because it seriously misinterprets the intellectual processes of primitive peoples. Moreover, even extreme exponents of the sociology of knowledge admit that most conclusions asserted in mathematics and natural science are neutral to differences in social perspective of those asserting them, so that the genesis of these propositions is irrelevant to their validity. Why cannot propositions about human affairs exhibit a similar neutrality, at least in some cases? Sociologists of knowledge do not appear to doubt that the truth of the statement that two horses can in general pull a heavier load than can either horse alone, is logically independent of the social status of the individual who happens to affirm the statement. But they have not made clear just what are the inescapable considerations that allegedly make such independence inherently impossible for the analogous statement about human behavior, that two laborers can in general dig a ditch of given dimensions more quickly than can either laborer working alone.

In the second place, the claim faces a serious and frequently noted dialectical difficulty—that proponents of the claim have succeeded in meeting only by abandoning the substance of the claim. For let us ask what is the cognitive status of the thesis that a social perspective enters essentially into the content as well as the validation of every assertion about human affairs. Is this thesis meaningful and valid only for those who maintain it and who thus subscribe to certain values because of their distinctive social commitments? If so, no one with a different social perspective can properly understand it; its acceptance as valid is strictly limited to those who can do so, and social scientists who subscribe to a different set of social values ought therefore dismiss it as empty talk. Or is the thesis singularly exempt from the class of assertions to which it applies, so that its meaning and truth are not inherently related to the social perspectives of those who assert it? If so, it is not evident why the thesis is so exempt; but in any case, the thesis is then a conclusion of inquiry into human affairs that is presumably "objectively valid" in the usual sense of this phrase—and, if there is one such conclusion, it is not clear why there cannot be others as well.

To meet this difficulty, and to escape the self-defeating skeptical relativism to which the thesis is thus shown to lead, the thesis is sometimes interpreted to say that, though "absolutely objective" knowledge of human affairs is unattainable, a "relational" form of objectivity called "relationism" can nevertheless be achieved. On this interpretation, a social scientist can discover just what his social perspective is; and if he then formulates the conclusions of his inquiries "relationally," so as to indicate that his findings conform to the cannons of validity implicit in his perspective, his conclusions will have achieved a "relational" objectivity. Social scientists sharing the same perspective can be expected to agree in their answers to a given problem when the canons of validity characteristic of their common perspective are correctly applied. On the other hand, students of social phenomena who operate within different but incongruous social perspectives can also achieve objectivity, if in no other way than by a "relational" formulation of what must otherwise be incompatible results obtained in their several inquiries. However, they can also achieve it in "a more roundabout fashion," by undertaking "to find a formula for translating the results of one into those of the other and to discover a common denominator for these varying perspectivistic insights."[13]

But it is difficult to see in what way "relational objectivity" differs from "objectivity" without the qualifying adjective and in the customary sense of the word. For example, a physicist who terminates an investigation with the conclusion that the velocity of light in water has a certain numerical value when measured in terms of a stated system of units, by a stated procedure, and under stated experimental conditions, is formulating his conclusion in a manner that is "relational" in the sense intended; and his conclusion is marked by "objectivity," presumably because it mentions the "relational" factors upon which the assigned numerical value of the velocity depends. However, it is fairly standard practice in the natural sciences to formulate certain types of conclusions in this fashion. Accordingly, the proposal that the social sciences formulate their findings in an analogous manner carries with it the admission that it is not in principle impossible for these disciplines to establish conclusions having the objectivity of conclusions reached in other domains of inquiry. Moreover, if the difficulty we are considering is to be resolved by the suggested translation formulas for rendering the "common denominators" of conclusions stemming from divergent social perspectives, those formulas cannot in turn be "situationally determined" in the sense of this phrase under discussion. For if those formulas were so determined, the same difficulty would crop up anew in connection with them. On the other hand, a search for such formulas is a phase in the search for invariant relations in a subject matter, so that formulations of these relations are valid irrespective of the particular perspective one may select from some class of perspectives on that subject matter. In consequence, in acknowledging that the search for such invariants in the social sciences is not inherently bound to fail, proponents of the claim we have been considering abandon what at the outset was its most radical thesis.

In brief, the various reasons we have been examining for the intrinsic impossibility of securing objective (i.e., value free and unbiased) conclusions in the social sciences do not establish what they purport to establish, even though in some instances they direct attention to undoubtedly important practical difficulties frequently encountered in these disciplines.

Notes

1. Max Weber, *The Methodology of the Social Sciences,* The Free Press, New York, 1949, p. 76.
2. This question receives some attention below in the discussion of the fourth difficulty.
3. S. F. Nadel, *The Foundations of Social Anthropology,* The Free Press, New York, 1951, pp. 53–54. The claim is sometimes also made that the exclusion of value judgments from social science is undesirable as well as impossible. "We cannot disregard all questions of what is socially desirable without missing the significance of many social facts; for since the relation of means to ends is a special form of that between parts and wholes, the contemplation of social ends enables us to see the relations of whole groups of facts to each other and to larger systems of which they are parts." Morris R. Cohen, *Reason and Nature,* New York, 1931, p. 343.
4. Edwin A. Burtt, *Right Thinking,* New York, 1946, p. 522.
5. For a documented account, see Gunnar Myrdal, *Value in Social Theory,* London, 1958, pp. 134–52.
6. See, e.g., S. F. Nadel, op. cit., p. 54; also Gunnar Myrdal, op. cit., p. 120, as well as his *Political Element in the Development of Economic Theory,* Cambridge, Mass., 1954, esp. Chap. 8.

7. Leo Strauss, "The Social Science of Max Weber," *Measure*, Vol. 2 (1951), pp. 211–14. For a discussion of this issue as it bears upon problems in the philosophy of law, see Lon Fuller, "Human Purpose and Natural Law," *Natural Law Forum*, Vol. 3 (1958), pp. 68–76; Ernest Nagel, "On the Fusion of Fact and Value: A Reply to Professor Fuller," op. cit., pp. 77–82; Lon L. Fuller, "A Rejoinder to Professor Nagel," op. cit., pp. 83–104; Ernest Nagel, "Fact, Value, and Human Purpose," *Natural Law Forum*, Vol. 4 (1959). pp. 26–43.
8. Cf. Gunnar Myrdal, *Value in Social Theory*, London, 1958, pp. xxii, 211–13.
9. The evidence is usually a count of red cells in a sample from the animal's blood. However, it should be noted that "the red cell count gives only an estimate of the *number of cells per unit quantity of blood*," and does not indicate whether the body's total supply of red cells is increased or diminished. Charles H. Best and Norman B. Taylor, *The Physiological Basis of Medical Practice*, 6th ed., Baltimore, 1955, pp. 11, 17.
10. It is irrelevant to the present discussion what view is adopted concerning the ground upon which such judgments supposedly rest—whether those grounds are simply arbitrary preferences, alleged intuitions of "objective" values, categorical moral imperatives, or anything else that has been proposed in the history of value theory. For the distinction made in the text is independent of any particular assumption about the foundations of appraising value judgments, "ultimate" or otherwise.
11. The above example is borrowed from the discussion in J. Neymann, *First Course in Probability and Statistics*, New York, 1950, Chap. 5, where an elementary technical account of recent developments in statistical theory is presented. For a nontechnical account, see Irwin D. J. Bross, *Design for Decision*, New York, 1953, also R. B. Braithwaite, *Scientific Explanation*, Cambridge, Eng., 1953, Chap. 7.
12. Karl Mannheim, *Ideology and Utopia*, New York, 1959, pp. 271, 288, 292. The essay from which the above excerpts are quoted was first published in 1931, and Mannheim subsequently modified some of the views expressed in it. However, he reaffirmed the thesis stated in the quoted passages as late as 1946, the year before his death. See his letter to Kurt H. Wolff, dated April 15, 1946, quoted in the latter's "Sociology of Knowledge and Sociological Theory," in *Symposium on Sociological Theory* (ed. by Llewellyn Gross), Evanston, Ill., 1959, p. 571.
13. Karl Mannheim, op. cit., pp. 300–1.

Chapter 37

The Philosophical Importance of the Rosenthal Effect

Michael Martin

Robert Rosenthal and his colleagues have performed psychological experiments the philosophical importance of which has yet to be seriously considered. The purpose of this chapter will be to describe briefly Rosenthal's work and its importance for philosophical issues connected with social sciences.

Rosenthal's Experiments

Rosenthal's early experimental studies were connected with the effect an experimenter's expectancy about the results of an experiment have on the outcome of the experiment. He claimed to have shown in experiments with human and animal subjects that the expectancy an experimenter has about the outcome of an experiment unwittingly affects the outcome of the experiment in the direction of the expectancy. In later experiments Rosenthal claimed to have shown that the expectancies of teachers affect their students' behavior in the direction of the expectancies. This effect on behavior (of both experimenter and teacher) has come to be known in the literature as the Rosenthal effect. Let us consider three experiments.

One experiment was concerned with the effect that an experimenter's expectancy has on subjects' ratings of photographs.[1] Two groups of subjects, group G_1 and group G, were asked to rate photographs of people on a scale from -10 to $+10$ in terms of whether the people in the photographs had recently experienced success or failure. (The photographs actually had been chosen so that on the average the people should be seen as neither successful nor unsuccessful, but as neutral.) The experimenters who administered the test to the subjects in group G_1 were told that their group should average about $+5$, while the experimenters who administered the test to the subjects in group G_2 were told that their group should average -5. Aside from this difference the instructions to the experimenters for both groups were the same. The experimenters read exactly the same instructions to their subjects. Nevertheless, the results of the experiment for the two groups were different: group G_1 averaged $+0.40$ while group G_2 averaged -0.08 in their ratings. The experiment with minor variations was replicated several times.

Another experiment was concerned with the effect experimenter expectancy has on the performance of animals.[2] One group of experimenters was given rats which they were told were "maze bright" rats; a second group of experimenters was given rats which they were told were "maze dull" rats. (In reality the rats were randomly assigned to the two groups and were not bred for maze learning.) The two groups of rats were taught to run a T-maze by the two groups of experimenters. It turned out that the rats designated as maze bright learned to run the maze better than the rats designated as maze dull. In a similar experiment similar results were obtained with rats learning

Skinner-box problems.[3] Rats thought by experimenters to be "Skinner-box bright" did better in their learning tasks than rats thought by the experimenters to be "Skinner-box dull." According to Rosenthal, the combined probability that the results of these two experiments could have occurred by chance was 0.0007.[4]

In another experiment the experimenter was, as it were, replaced by a teacher.[5] The experiment tested the effect of a teacher's expectancy on students' behavior (Rosenthal refers to this experiment as the Pygmalion experiment). At the beginning of the year children in an elementary school were given a nonverbal test of intelligence by their teachers. The teachers were not told the test was an intelligence test and were made to believe that the test would predict academic "blooming." In each class one group of children was selected as children who "would show unusual academic development during the coming year." The teachers thought these groups were selected by the test results; actually the children were randomly assigned to the groups by Rosenthal. Thus the teacher in each class was given the expectancy that some children in his or her class were "late bloomers." The children were tested again at the end of the year. Children whose teachers thought they were late bloomers gained significantly more in IQ than the control group, that is, those children who were not designated as late bloomers. The Pygmalion experiment has been replicated may times.[6]

It is not known how widespread the Rosenthal effect is in the social sciences. Rosenthal suggests that one might assess the generalization of his findings by considering the similarity of the design of his experiments to the design of typical experiments in psychology.[7] Judged in these terms the generalizability of the experiments is impressive. To be sure most of the experimenters used by Rosenthal were graduate students. But this should not affect the generalizability of the results. First, there is good reason to suppose that professional, competent, and higher-status experimenters would be more likely to bias the results of an experiment than would more amateur experimenters. Second, most of the experimenters in psychological experiments today who come into direct contact with the subjects of the experiment are graduate students; the typical highly competent professional psychological experimenter has graduate students gather his data.

As far as the subjects of the Rosenthal experiments are concerned they were very much like typical subjects of psychological experiments: thus the generality of the results would not be affected by nonrepresentative subjects. Moreover, the situations in which the experiments were performed were varied, thereby increasing the generalizability of the results. The experiments were conducted in several universities, in different geographical areas, in different types of laboratories. The tasks included photograph rating, verbal learning, and taking standardized and projective psychological tests; for animal subjects the tasks included learning in T-mazes and Skinner boxes. One might suspect then that the Rosenthal effect is indeed widespread.

Exactly how the experimenters influence their human subjects to conform to their expectancies is not well understood, but it is generally agreed that the way the influence occurs is very complex and subtle. ESP was actually considered a possible means of communication between experimenter and subject at one point in Rosenthal's studies.[8] However, the ESP hypothesis was tested and dropped, and the hypothesis that subtle and complex auditory and visual cues are the means of influence is now favored.[9] How these cues work is unclear but there is reason to think that they can to a certain extent influence the results of an experiment at a very early stage, even before that subjects have responded to the tasks set by the experimenter.[10]

The way experimenters influence their animal subjects is even more difficult to understand. Rosenthal cites evidence suggesting that the quality and quantity of the handling of the rats "communicates" the experimenter's expectancy. He also suggests that the experimenters working with "bright" rats watch their rats more carefully, thus reinforcing the desired responses.[11]

Issues in the Philosophy of the Social Sciences

The Rosenthal effect raises several closely related philosophical issues connected with the social sciences. First of all, there is the issue of self-fulfilling prophecy: can the Rosenthal effect be considered a type of reflexive prediction—what has been commonly called a self-fulfilling prophecy? We shall argue that the Rosenthal effect cannot be so construed but that it, as well as a self-fulfilling prophecy, is a special case of a broader notion, one that will be called a reflexive truth vehicle. Second, the Rosenthal effect raises questions about the objectivity of the social sciences. In particular it can be asked how the Rosenthal studies relate to the recent discussions of objectivity by philosophers of science such as Scheffler and Popper. We shall argue that Scheffler's argument leaves unanswered an important question raised by Rosenthal's work and that Popper's strategy to achieve objectivity in science is not adequate to overcome the problem raised by Rosenthal's work. Nevertheless, it shall be maintained that Rosenthal does provide a way of achieving objectivity in the social sciences, a way of eliminating the Rosenthal effect. Third, there is the issue of the coherence of Rosenthal's thesis: Can Rosenthal consistently defend the widespread nature of the Rosenthal effect and the validity of his own work? Does not a Rosenthal effect affect Rosenthal's own studies, paradoxically calling them into question? We shall argue that the possibility of a meta-Rosenthal effect is indeed a problem but that there is a way of rationally defending Rosenthal's thesis against the charge of incoherence.

Reflexive Prediction

Rosenthal certainly thinks that the Rosenthal effect is a type of self-fulfilling prophecy and often speaks of the effect of the expectancies of experimenters and teachers in these terms. There does indeed seem to be an interesting analogy between the Rosenthal effect and the standard examples of self-fulfilling prophecies discussed in philosophical literature, e.g., the public prediction that a bank will fail, causing a run on the bank. On the other hand, there seem to be some disanalogies as well. Rosenthal's graduate student experimenters did not in any obvious sense make a prophecy or a prediction about the results of the experiment (although they might have made one if they had been asked). Clearly one must proceed with caution here lest crucial distinctions be overlooked.

After all what is the definition of a self-fulfilling prophecy? Although many people speak of self-fulfilling prophecies there have been very few actual analyses of this notion. To my knowledge the first extended treatment of the problem of the definition of reflexive prediction (which would include self-fulfilling prophecies as well as self-defeating prophecies) was Roger Buck's.

Buck suggests that the following four conditions define the concept of a reflexive prediction:[12]

1. Its truth value would have been different had its dissemination status been different.
2. The dissemination status it actually had was causally necessary for the social actors involved to hold relevant and causally efficacious beliefs.
3. The prediction was, or if disseminated would have been, believed and acted upon.
4. Something about the dissemination status or its causal consequences was abnormal, or at the very least unexpected by the predictor, by whoever calls it reflexive, or by those to whose attention its reflexive character is called.

Buck argues that his analysis accounts for typical cases of reflexive prediction in the social sciences. For example, an economist makes a public prediction that a certain bank may fail. The truth of the prediction would have been different had the dissemination status of the prediction been different, e.g., had the prediction not have been made public. Making the prediction public caused the patrons of the bank to hold certain beliefs which they acted on, causing a run on the bank; this in turn caused the bank to fail.

Construed in one way, Buck's analysis certainly applies in the Rosenthal experiments. Compare the case of the bank's failure and the experiments of rats learning the T-maze. One can say that Rosenthal and his associates are analogous to the economist and the graduate student experimenters are analogous to the patrons of the bank. Rosenthal and his associates disseminated certain information to the graduate student experimenters who affected the behavior of the rats. In this case it is not implausible to suppose that the experimenters (unwittingly) acted upon certain beliefs in affecting the rats' behavior.

But construed in another way, Buck's analysis fails. Suppose graduate student experimenters are analogous to the economist; the rats are analogous to the patrons of the bank. The graduate student experimenters somehow "disseminated the prediction" to the rats. Of course, it is at this point that the analogy breaks down, for it is surely implausible to suppose that the rats had certain beliefs and acted on these beliefs.

Which analogy should we appeal to in understanding the Rosenthal effect? No doubt both analogies are helpful in bringing out different aspects of the experiment. The first analogy is helpful in bringing out the effect of Rosenthal's statements on his experimenter subjects; the second analogy is helpful in bringing out the effect of the experimenter subjects' expectancy on the rats. However, the Rosenthal effect in this experiment, as Rosenthal conceives of it, is the effect of the experimenter subjects on the rats, not his effect on the experimenter subjects. Consequently when he says that the Rosenthal effect is a self-fulfilling prediction he has in mind the second analogy. But Buck's analysis, as we have seen, does not work in this case.

There is good reason, however, to suppose that Buck's analysis is inadequate. George Romanos has argued persuasively that Buck's argument attempting to justify his restriction to acting-on-beliefs is circular.[13] Romanos has suggested the following as an alternative analysis of reflexive prediction: "The formulation/dissemination style of the prediction must be a causal factor relative to the prediction's coming out true or false."

Is the Rosenthal effect a reflexive prediction given this new construal? To be able to tell we need to understand what a formulation/dissemination style is. According to Romanos the formulation style (F-style) of a prediction is the formal or syntactical properties of the prediction. Every prediction must be formulated in some way; there must be something like a sequence of sounds or inscriptions which are said to convey or express the prediction. This formulation need not be in some natural language which is spoken or written but "may be constructed out of such things as electric impulses, bodily movements, puffs of smoke, or anything else which can be interpreted as expressing a prediction."[14] According to Romanos, the dissemination style (D-style) of a prediction is the manner of reproduction and/or transmission of the prediction in a certain F-style.

It is not clear that the Rosenthal effect is a reflexive prediction in Romanos's view, either. First, there is a conceptual problem for it is not clear what should be counted as a syntax of a language. Even if Rosenthal is correct that the maze bright rats were handled differently by the experimenters from the maze dull rats, it is not clear that the experimenters' behavior can be considered to have expressed a prediction in, e.g., some syntax of bodily movement. One certainly does not want to say that *any* piece of behavior causally related to a person's expectance expresses a prediction. Suppose there was a stockbroker who, whenever he thought there was going to be a sharp rise in the market, broke out in hives. We would not want to say that the hives outbreak expressed the prediction that the market would rise sharply (although someone might *use* this information to make a prediction about a rise in the market). The behavior of the graduate student experimenters and the behavior of our hypothetical stockbroker may not be very much alike. But still the graduate student experiments' behavior is a far cry from the behavior of someone in the standard case of making a prediction. Hence our hesitation.

Second, there is an empirical problem. Experimental psychologists do not know with any certainty how the rats were affected by the experimenters. Rosenthal's suggestion that the two groups of rats performed differently because they were handled differently needs further testing before it can be accepted. One might suppose that the situation is different with the human subjects in the photograph-rating experiment. But again it is not well understood how the experimenters influenced the human subjects, and consequently it is not clear whether it is plausible to interpret the experimenters' behavior as expressing a prediction. Rosenthal may have this in mind when he says:

> We cannot be sure, however, that these changes in experimenter behavior are themselves conveyors of information to the subjects as to how they should respond. Possibly, those subjects who later go on to confirm or disconfirm the experimenter's hypothesis affect the experimenter differently early in the experiment. The experimenter then behaves differently towards these subjects but without necessarily conveying response-related information to the subject. In other words, differential treatment by the experimenter may be quite incidental to the question of whether a subject goes on to confirm or disconfirm the experimenter's hypothesis.[15]

Given Romanos's analysis, there is another reason why the Rosenthal effect should not be interpreted as a reflexive prediction. Insofar as one can legitimately speak of a prediction's being made, strictly speaking in some of the Rosenthal experiments the prediction comes out false. Recall that the one group of experimenters had the expectancy that their subjects would on the average rate the photographs about +5. The actual result was +0.4. The other group of experimenters had the expectancy that the subjects would on the average rate the photos as −5. The result was −0.08.

Consequently it hardly seems plausible to say that the F/D-style of the prediction was a causal factor relative to the truth or falsehood of the prediction. At best the F/D-style of the prediction was a causal factor affecting the degree of approximation of the prediction to the true value.

The above points should not be taken as criticism of Romanos's analysis of reflexive prediction. Rather they are difficulties in interpreting the Rosenthal effect as a reflexive prediction. Only if one is inclined preanalytically to include the Rosenthal effect as a reflexive prediction will these problems incline one to reject Romanos's analysis.

One thing is clear. Even if one is not inclined to include the Rosenthal effect under reflexive prediction, reflexive predictions and the Rosenthal effect seem to belong to the same species of methodological problem. They both have a reflexive nature: one is a prediction, the other is an expectancy. In a reflexive prediction the truth or falsity is affected; in a reflexive expectancy only the degree of approximation to the truth may be affected. In a prediction the reflexive aspect is a function of the F- and D-style of the prediction; in an expectancy the causal factors affecting the reflexive aspect are not well understood.

What is needed is some larger category in which both reflexive prediction and reflexive expectancy are included as special cases. I would suggest that such a category be that of a truth vehicle. Let us understand a truth vehicle as anything that can legitimately be said to be true or false in the relevant sense of true or false, e.g., beliefs, predictions, statements, and so on, but not friends. However, the terms "belief," "prediction," "statement" are ambiguous. One could be referring to what is believed, predicted, stated; one could also be referring to the act or state of believing, predicting, stating. Clearly it is the former and not the latter that are truth vehicles. One must also distinguish between the generic act or state of believing, predicting, or stating. Thus, one must distinguish between the generic act or state of predicting that a bank will fail and the act or state of predicting that a bank will fail by a particular person at a particular time. We shall refer to the particular act or state of believing, predicting, or stating as the instantiation of a truth vehicle.

Thus an instantiation of the belief that group G_1 will have an average score of $+5$ might be Bill Jones's belief (a particular psychological or neurological state) at noontime on July 14, 1970, that group G_1 will have an average score of $+5$. An instantiation of the prediction that the Bank of Douglas will fail might be Professor Smith's prediction (the act of uttering a sentence in a particular way) made to her class at the University of Arizona during the May 18, 1960, seminar on banking. An instantiation of the prediction that some particular missile will miss the target might be the prediction (the producing of a particular electrical impulse traveling along a certain wire) made by a particular IBM computer during training practice on April 1, 1955, at 9.00 A.M. (It should be clear that although truth vehicles can be spoken of as true or false, instantiations of truth vehicles cannot be spoken of as true or false. Thus, it is wrong to speak of a neurological state or the producing of an electrical impulse as true or false.)

Let us call the various instantiations of some truth vehicle V_1, instantiation $I_1 V_1$, instantiation $I_2 V_1$, instantiation $I_3 V_1, \ldots$, instantiation $I_n V_1$. One can, then, define the reflexivity of an arbitrary instantiation I_i of an arbitrary truth vehicle V_i as follows: $I_i V_i$ is reflexive if $I_i V_i$ is a causal factor either affecting the degree of approximation of V_i to the truth or affecting whether V_i comes out true or false. Given this analysis both the Rosenthal effect and reflexive prediction become special cases of the reflexivity of instantiated truth vehicles.

Reflexive instantiated truth vehicles are found in both natural and social science. Given a sufficiently liberal interpretation of reflexive prediction, reflexive predictions are found in the natural sciences. And reflexive instantiated truth vehicles that are not predictions are found in the natural sciences. It would seem possible that a physicist (because of an expectancy that his experiment would turn out in a certain way) would unwittingly set up experimental equipment in a way so as to speciously get the results expected. However, it may not be plausible to interpret the physicist's behavior as expressing a prediction.

Objectivity

The Rosenthal effect and its prima facie generality seems to have serious ramifications with respect to the objectivity of the behavioral sciences for, if the expectancy of the experimenter in the behavioral sciences influences the result of the experiment, doubt seems to be cast on the objectivity of the results of the experiment. The results of behavioral experiments may all be biased and may not constitute reliable knowledge.

Some recent philosophers of science have used psychological experiments to bring into question the objectivity of the natural sciences. It may be useful to relate their arguments to the Rosenthal effect. Thomas Kuhn, for example, has argued that a scientist's theoretical orientation influences what the scientist sees. He argues that scientists tend to be blinded to negative evidence because of the influence of their theoretical orientation on their observations; they tend to see what they believe and to fail to see what they do not believe. He used Bruner and Postal's experiment with playing cards as evidence for this influence.[16] In the Bruner and Postal experiment subjects were shown playing cards. Many of the cards were normal but some of them were anomalous, e.g., a red six of spades. The subjects tended to identify the anomalous cards as normal, e.g., a black four of hearts as a black four of spades.

Some defenders of the objectivity of science have argued against Kuhn as follows. First, it is important to distinguish between the influence of the theoretical premises held by a scientist on his observations and the influence of the theoretical categories held by the scientist on his observations. Thus a scientist may see the world in Newtonian categories, e.g., mass and force, without necessarily having all his observations conform to the predictions of Newtonian theory. Second, although the influence of a scientist's theoretical premises on observation may be strong, it is not inevitable. As Scheffler has put it: "Our expectations strongly structure what we see, but do not wholly eliminate unexpected sights.... There is no evidence for a general incapacity to learn from contrary observations, no proof of a preestablished harmony between what we believe and what we see."[17] Even in the Bruner and Postal experiment some subjects were finally able to recognize the anomalous cards as anomalous.

One naturally wonders whether the same sort of strategy can be used to defend the objectivity of behavioral research against the attack of the Rosenthal effect. Before such strategy is attempted, however, it is important to see exactly what has been accomplished by defenses like Scheffler's. I think it is fair to say that Scheffler has shown at most that a scientist's theoretical orientation does not necessarily make his observations confirm his theories; that specious confirmation is not inevitable. However, this establishes only that objectivity via observational testing is not logically impossible in science.

This result is, of course, extremely important since some philosophers have seemed to deny that objectivity is possible. However, it still might be true that as a matter of fact in the vast majority of cases the theoretical beliefs of scientists blind them to negative evidence. The extent of this influence may be so wide that the validity of most of our alleged scientific knowledge may be in question even though objectivity is possible.

I believe that a similar Schefflerian move can be made with respect to the Rosenthal effect and a similar reply can be given. Rosenthal has not shown that there is any preestablished harmony between experimenters' expectancies and the results of their experiments. After all, some experiments do seem to conflict with the experimenters' expectancies. But even granted this, it still may be true in the vast majority of experiments that the Rosenthal effect is very strong. Consequently the objectivity of the results of behavioral experiments can be called into question.

The Rosenthal effect also seems to call into question the way in which some philosophers of science have supposed objectivity can be achieved in science. Popper, for example, argues that the objectivity of science is not a function of the individual scientist but is a function of the essential social character of science. He argues that there are two important aspects to this social character of scientific objectivity: first, there is something approaching free criticism in that scientists put forth theories and other scientists criticize these theories; second, scientists avoid speaking at cross-purposes, for they state their conflicting theories in testable form and evaluate their theories by experience which is an impartial arbitrator. Popper says: "when speaking of 'experience' I have in mind experience of a 'public' character like observations, and experiments." He argues, "everyone who has learned the technique of understanding and testing scientific theories can always repeat the experiment and judge for himself."[18]

Popper's two aspects of objectivity are no doubt important. Indeed, free criticism and testable theories may be necessary conditions for the objectivity of science. But in the light of Rosenthal's results they do not seem to be a sufficient condition. The results of an experiment may be the results of the experimenter's expectancy. Consequently experience may not be an impartial arbitrator at all, as Popper supposes.

This does not mean that objectivity cannot be achieved in behavioral research. However, techniques will have to be used to meet the particular problems raised by Rosenthal's research. Mere appeal to intersubjective criticism in terms of repeatable experiments is not enough. Rosenthal himself has suggested a number of strategies that go a long way toward meeting the problem. Let us briefly consider a few of these strategies.[19]

One important strategy to control the Rosenthal effect is to use a number of experimenters (say drawn randomly from a population of experimenters) each of whom would deal with a small number of subjects. Such a technique would have the effect of tending to eliminate experimental bias because there is reason to suppose that experimenters unintentionally learn from their subjects how to influence them. With fewer subjects the chance of this learning taking place may decrease. Also, with a number of experimenters it is likely that there will be experimenters with different biases; these biases would statistically tend to cancel one another out.

It may also be possible to learn the bias of the experimenter before the experiment takes place. Once a bias is learned it can be correlated with the results of the experiment. If the correlation is large, the experimental effect can be corrected by such well-known statistical methods as partial correlations or analyses of co-variance. If the experimental effect is insignificant, it can be ignored.

Another strategy to diminish the Rosenthal effect is to minimize and standardize contact between the experimenter and the subjects. It should be recalled that the Rosenthal effect is produced by experimenters dealing with different subjects in subtly different ways, e.g., in their instructions to the the subject. One way of eliminating such differential treatment would be to have the instructions to a group of subjects prerecorded, e.g., on videotapes. Both experimental and control groups would be shown the same tapes. In order that the subjects would not feel that the experimental situation was unrealistic they could be led to believe that their instructions were coming "live" over closed-circuit TV. In order to further lessen the influence of the experimenter, the subjects might (when this was feasible) record their own responses.

If an experiment required that the experimenter have direct contact with the subjects, techniques could be introduced which would reduce the cues available to both the experimenter and the subjects. There is good reason to suppose that visual and auditory cues were very important in producing the Rosenthal effect. Imposing a screen between the experimenter and the subject would eliminate visual cues; auditory cues could be eliminated by having the subjects and experimenter communicate by writing.

Still another technique that could be used to correct the Rosenthal effect is the so-called expectancy control group method. Suppose we were interested in whether some psychological treatment T was effective. Standard experimental studies using an experimental and control group would be unreliable for it would be unclear if the results of the studies were due to treatment T or to the Rosenthal effect. However, this problem could be overcome in the following way. One experimenter would be chosen with an expectancy that treatment T would be effective and would use treatment T on experimental group E_1, and not on control group C_1. Another experimenter would be chosen with an expectancy that treatment T would not be effective and would use treatment T on experimental group E_2 and not on control group C_2. By a two-way analysis of variance a main effect attributable to treatment T, a main effect attributable to experimental effect, and an interaction to these two effects could be determined.

In conclusion, the impossibility of objectivity in the social sciences has not been established by Rosenthal's work, but it is possible that most social science research is biased as a result of experimental expectancy. However, the same sort of thing may well be true in the natural sciences. Although there is no reason to suppose that objectivity in the natural sciences is impossible, most observations used in testing theories may as a matter of fact be biased by the theoretical commitment of the observer.

Furthermore, there is no reason to think that the Rosenthal effect must be overcome by certain methods or approaches unique to the social sciences, e.g., *Verstehen*, or wholistic-phenomenological orientation. As we have seen, the Rosenthal effect can be controlled by particular experimental techniques, e.g., certain types of control groups, statistical analyses of the results of experiments. *Verstehen* and other such methods, whatever their value and whatever they may involve, do not seem to be needed.

The Meta-Rosenthal Effect

So far we have not considered a problem that has been lurking in the background in much of our discussion of the Rosenthal effect. Rosenthal claims to have shown that an experimenter's expectancy influences the results of experiments and he maintains that such biasing of the results of experiments is widespread. But if this is so it seems paradoxically to apply to the results of his own experiments. For Rosenthal's experiments are also performed by experimenters, e.g., Rosenthal. These experimenters have a certain expectancy, namely, that experimental expectancy will affect the result of the experiment in the direction of the expectancy. This expectancy may affect the result of *their* experiments in the direction of *their* expectancy. Let us call the Rosenthal effect as it affects Rosenthal's own research the meta-Rosenthal effect.

There are three basic questions that one can ask about the meta-Rosenthal effect. We might wonder what a meta-Rosenthal effect would show. We might wonder if it would be coherent to maintain that the Rosenthal effect is widespread and yet deny that the meta-Rosenthal effect exists. We might ask whether there is a meta-Rosenthal effect.

With respect to the first question one can say at least that if there is a meta-Rosenthal effect it would tend to undercut our trust in Rosenthal's studies. Furthermore, it would seem to give comfort to defenders of the objectivity of the social sciences. For if the Rosenthal studies are untrustworthy because of the meta-Rosenthal effect, one could not claim that the social sciences lacked objectivity on the basis of the Rosenthal experiments. A similar argument was used by Scheffler in defending the objectivity of the natural sciences against critics like Kuhn who used psychological studies and historical evidence to show that natural science was subjective. Scheffler argued:

> And indeed there is a striking self-contradictoriness in the effort to persuade others by argument that communication, and hence argument is impossible; in appeal to the facts about observation in order to deny that commonly observable facts exist, in arguing from the hard realities of the history of science to the conclusion that reality is not discovered but made by the scientist. To accept these claims is to deny all force to the arguments brought forward for them.[20]

In order to answer the second question one must distinguish two different positions. It would be contradictory to maintain the following theses:

(1) The meta-Rosenthal effect does not exist.

(2) The Rosenthal effect is found in every experiment.

Since Rosenthal's research is an experiment, by (2) the Rosenthal effect would influence his own experiment. But this is denied by (1). However, Rosenthal need not assert (1) and (2). Rosenthal can coherently maintain the following:

(1′) The meta-Rosenthal effect does not exist.

(2′) The Rosenthal effects is found in all experiments except the Rosenthal experiments.

Actually Rosenthal seems to believe something even weaker than (2′), namely:

(2″) The Rosenthal effects is likely to be found in many experiments but not in the Rosenthal experiment.

The question is whether Rosenthal and his colleagues have any justification for the last clause in (2′) and (2″). For it might be suggested that unless they do they escape incoherence only by an arbitrary restriction.

This brings us to the last question. Rosenthal would argue, I believe, that it is unlikely that the meta-Rosenthal effect exists because he and his colleagues (unlike most other experimenters) have taken the trouble to control for experimental expectancy in their research; they have instituted techniques that would tend to eliminate the bias that results from the Rosenthal effect. We have reviewed some of Rosenthal's techniques above. However, one technique we did not mention and the only technique explicitly mentioned[21] by Rosenthal in his own research is the technique of the total double-blind experiment.

In this experimental set-up, no one knows the experimental condition to which any subject is assigned. For example, in the photograph rating experiment ten experimenters were used. Each experimenter was assigned at random to ten different research rooms. Furthermore, ten sets of instructions were randomly assigned to the rooms (five inducing the $+5$ expectancy and five inducing the -5 expectancy). Subjects were also randomly assigned to the rooms. The experimenter read over the instructions when they arrived at the room, thus creating the experimental expectancy. It was not until the end of the experiment that anyone knew what instructions each experimenter had gotten.

Now although the total double-blind experiment may eliminate some aspects of the Rosenthal effect it is doubtful whether it could eliminate all. For just because Rosenthal and his colleagues did not know which graduate student experimenters would get which instructions, Rosenthal et al. still could have influenced the graduate student experimenters in a general way before they went to the research rooms. The graduate student might have been unwittingly influenced to influence their subjects' behavior in the direction of their expectancy rather than not at all or in the opposite direction. The total double-blind situation would not seem to eliminate this problem.

However, Rosenthal could have eliminated this problem by other techniques which he suggests but apparently does not use. For example, Rosenthal-type experiments can be conducted by various experimenters—some sympathetic to Rosenthal, some with an expectancy that there would be no Rosenthal effect. The result of different expectancies would tend to cancel each other out in the final result.

I believe one can conclude that although the problem of the meta-Rosenthal effect is not completely eliminated by Rosenthal's experimental design, the problem could be in principle eliminated. Consequently, it is possible that with more research Rosenthal could claim (2″) with some certainty.

Conclusion

The methodological importance of the Rosenthal effect should not lead us to draw unwarranted inferences concerning it. The Rosenthal effect, despite its similarity to phenomena of self-fulfilling prophecy, is not easily assimilated to this phenomena. Furthermore, it would not be warranted to dismiss Rosenthal's position as self-refuting because of the meta-Rosenthal effect. For although the possibility of the meta-Rosenthal effect is not completely eliminated in Rosenthal's work it is in principle completely eliminable. Thus there is no paradox or arbitrary restrictions in maintaining that the Rosenthal effect is widespread but absent from Rosenthal's own studies. Finally, although the Rosenthal effect raises serious questions about the objectivity of social science research it would be a mistake to conclude that Rosenthal has shown that objectivity is impossible in the social sciences. Indeed, Rosenthal shows how the Rosenthal effect can be overcome and how objectivity is still possible.

Notes

1. R. Rosenthal and K. L. Fode, "Three Experiments in Experimenter Bias," *Psychological Reports*, 1963, 12, pp. 491–511; see also R. Rosenthal, *Experimenter Effect in Behavioral Research* (New York: Appleton-Century-Crofts, 1966), pp. 145–57.
2. R. Rosenthal and K. L. Fode "The Effect of Experimental Bias on the Performance of the Albino Rat," *Behavioral Science*, 1963, 8, pp. 183–9; R. Rosenthal, *Experimenter Effect*, pp. 158–65.
3. R. Rosenthal and R. Lawson, "A Longitudinal Study of the Effect of Experimenter Bias on the Operant Learning of Laboratory Rats," *Journal of Psychiatric Research*, 1964, 2, pp. 61–72; R. Rosenthal, *Experimenter Effect*, pp. 165–76.
4. Ibid., p. 176.
5. R. Rosenthal and L. Jacobson, *Pygmalion in Classroom* (New York; Holt, Rinehart & Winston, 1968).
6. For a review of this literature see R. Rosenthal, "On the Social Psychology of the Self-fulfilling Prophecy: Further Evidence for Pygmalion Effects and Their Mediating Mechanisms," MSS Modular Publications, Inc., New York (1974), Module 53, pp. 1–28.
7. R. Rosenthal, *Experimenter Effect*, Ch. 17. Not all psychologists have agreed that the Rosenthal effect is widespread, and indeed some have challenged the statistical basis of Rosenthal's experiments. See Theodore X. Barber and Maurice J. Silver, "Fact, Fiction, and the Experimenter Bias Effect," *Psychological Bulletin Monograph Supplement*, 1968, 70, pp. 1–29. It will be assumed in this chapter that these statistical objections can be met. See R. Rosenthal, "Experimenter Expectancy and the Reassuring Nature of the Null Hypothesis Decision Procedure," *Psychological Bulletin Monograph Supplement*, 1968, 70, pp. 30–47; T. X. Barber and M. J. Silver, "Pitfalls in Data Analysis and Interpretation: A Reply to Robert Rosenthal," *Psycholgical Bulletin Monograph Supplement*, 1969, 70, pp. 48–62.
8. Ibid., p. 282.
9. Ibid., pp. 281–9. There is reason to think that expectancy can be communicated by auditory cues alone. See J. G. Adair and J. S. Epstein, "Verbal Cues in Mediation of Experimenter Bias," *Psychological Reports*, 1968, 22, pp. 1045–53.
10. R. Rosenthal, *Experimenter Effect*, pp. 289–93.
11. Ibid., p. 178; see also R. Rosenthal, "On the Social Psychology of the Self-fulfilling Prophecy," pp. 4–5.
12. R. C. Buck, "Reflexive Prediction," *Philosophy of Science*, 1964, 30, pp. 359–69; see also A. Grunbaum, "Comments on Professor Roger Buck's Paper 'Reflexive Prediction,'" *Philosophy of Science*, 1963, 30, pp. 370–2; R. C. Buck, "Rejoinder to Grunbaum," *Philosophy of Science*, 1963, 30, pp. 373–4.
13. George Romanos, "Reflexive Predictons," *Philosophy of Science*, 1973, 40, pp. 97–109.
14. Ibid., p. 105.
15. R. Rosenthal, *Experimenter Effect*, pp. 300–1.
16. Thomas Kuhn, *The Structure of Scientific Revolutions* (Chicago: University of Chicago Press, 1962), p. 63.
17. See I. Scheffler, *Science and Subjectivity* (Indianapolis, Ind.: Bobbs-Merrill, 1967), p. 44.
18. Karl Popper, "The Sociology of Knowledge," in (ed.) Philip P. Weiner, *Readings in the Philosophy of Science* (New York: Charles Scribner & Sons, 1953), p. 362.
19. Rosenthal, *Experimenter Effect*, Chs. 19–23.
20. Scheffler, op. cit., pp. 21–2.
21. Rosenthal, *Experimenter Effect*, p. 373. Elsewhere Rosenthal has argued that in 185 studies of experimenters' expectation 63 of these studies showed the effect of experimental expectancy at the 5 percent level of significance. He concludes from this that it is overwhelmingly likely that the Rosenthal effect exists. However, these results may be due to a meta-Rosenthal effect. See Rosenthal. "On the Psychology of the Self-fulfilling Prophecy," p. 5.

Chapter 38

Psychology Constructs the Female

Naomi Weisstein

It is an implicit assumption that the area of psychology which concerns itself with personality has the onerous but necessary task of describing the limits of human possibility. Thus when we are about to consider the liberation of women, we naturally look to psychology to tell us what "true" liberation would mean: what would give women the freedom to fulfill their own intrinsic natures. Psychologists have set about describing the true natures of women with a certainty and a sense of their own infallibility rarely found in the secular world. Bruno Bettelheim, of the University of Chicago, tells us (1965) that "we must start with the realization that, as much as women want to be good scientists or engineers, they want first and foremost to be womanly companions of men and to be mothers." Erik Erikson of Harvard University (1964), upon noting that young women often ask whether they can "have an identity before they know whom they will marry, and for whom they will make a home," explains somewhat elegiacally that "much of a young woman's identity is already defined in her kind of attractiveness and in the selectivity of her search for the man (or men) by whom she wishes to be sought." Mature womanly fulfillment, for Erikson, rests on the fact that a woman's "somatic design harbors an 'inner space' destined to bear the offspring of chosen men, and with it, a biological, psychological, and ethical commitment to take care of human infancy." Some psychiatrists even see the acceptance of woman's role by women as a solution to societal problems. "Woman is nurturance," writes Joseph Rheingold (1964), a psychiatrist at Harvard Medical School; "anatomy decrees the life of a woman ... when women grow up without dread of their biological functions and without subversion by feminist doctrine, and therefore enter upon motherhood with a sense of fulfillment and altruistic sentiment, we shall attain the goal of a good life and a secure world in which we live it" (p. 714).

These views from men who are assumed to be experts reflect, in a surprisingly transparent way, the cultural consensus. They not only assert that a woman is defined by her ability to attract men, they see no alternative definitions. They think that the definition of a woman in terms of a man is the way it should be; and they back it up with psychosexual incantation and biological ritual curses. A woman has an identity if she is attractive enough to obtain a man, and thus, a home; for this will allow her to set about her life's task of "joyful altruism and nurturance."

Business certainly does not disagree. If views such as Bettelheim's and Erikson's do indeed have something to do with real liberation for women, then seldom in human history has so much money and effort been spent on helping a group of people realize their true potential. Clothing, cosmetics, home furnishings, are multimillion dollar businesses: if you don't like investing in firms that make weaponry and flaming gasoline, then there's a lot of cash in "inner space." Sheet and pillowcase manufacturers are concerned to fill this inner space:

> Mother, for a while this morning, I thought I wasn't cut out for married life. Hank was late for work and forgot his apricot juice and walked out without kissing me, and when I was all alone I started crying. But then the postman came with the sheets and towels you sent, that look like big bandana handkerchiefs, and you know what I thought? That those big red and blue handkerchiefs are for girls like me to dry their tears on so they can get busy and do what a housewife has to do. Throw open the windows and start getting the house ready, and the dinner, maybe clean the silver and put new geraniums in the box. *Everything to be ready for him when he walks through that door.* (Fieldcrest 1966; emphasis added.)

Of course, it is not only the sheet and pillowcase manufacturers, the cosmetics industry, the home furnishings salesmen who profit from and make use of the cultural definitions of man and woman. The example above is blatantly and overtly pitched to a particular kind of sexist stereotype: the child nymph. But almost all aspects of the media are normative, that is, they have to do with the ways in which beautiful people, or just folks, or ordinary Americans, or extraordinary Americans should live their lives. They define the possible; and the possibilities are usually in terms of what is male and what is female. Men and women alike are waiting for Hank, the Silva Thins man, to walk back through that door.

It is interesting but limited exercise to show that psychologists and psychiatrists embrace these sexist norms of our culture, that they do not see beyond the most superficial and stultifying media conceptions of female nature, and that their ideas of female nature serve industry and commerce so well. Just because it is good for business doesn't mean it's wrong. What I will show is that it *is wrong*; that there isn't the tiniest shred of evidence that these fantasies of servitude and childish dependence have anything to do with women's true potential; that the idea of the nature of human possibility which rests on the accidents of individual development or genitalia, on what is possible today because of what happened yesterday, on the fundamentalist myth of sex organ causality, has strangled and deflected psychology so that it is relatively useless in describing, explaining, or predicting humans and their behavior. It then goes without saying that present psychology is less than worthless in contributing to a vision which could truly liberate—men as well as women.

The central argument of my chapter, then, is this. Psychology has nothing to say about what women are really like, what they need and what they want, essentially because psychology does not know. I want to stress that this failure is not limited to women; rather, the kind of psychology which has addressed itself to how people act and who they are has failed to understand, in the first place, why people act the way they do, and certainly failed to understand what might make them act differently.

The kind of psychology which has addressed itself to these questions divides into two professional areas: academic personality research, and clinical psychology and psychiatry. The basic reason for failure is the same in both these areas: the central assumption for most psychologists of human personality has been that human behavior rests on an individual and inner dynamic, perhaps fixed in infancy, perhaps fixed by genitalia, perhaps simply arranged in a rather immovable cognitive network. But this assumption is rapidly losing ground as personality psychologists fail again and again to get consistency in the assumed personalities of their subjects (Block 1968). Meanwhile, the evidence is collecting that what a person does, and who he believes himself to be, will in general be a function of what people around him expect him to be, and what the overall situation in which he is acting implies that he is. Compared to the influence of the social context within which a person lives, his or her history and "traits," as well as biological makeup, may simply be random variations, "noise" superimposed on the true signal which can predict behavior.

Some academic personality psychologists are at least looking at the counterevidence and questioning their theories; no such corrective is occurring in clinical psychology and psychiatry. Freudians and neo-Freudians, Adlerians and neo-Adlerians, classicists and swingers, clinicians and psychiatrists, simply refuse to look at the evidence against their theory and practice. And they support their theory and their practice with stuff so transparently biased as to have absolutely no standing as empirical evidence.

To summarize: the first reason for psychology's failure to understand what people are and how they act is that psychology has looked for inner traits when it should have been looking for social context; the second reason for psychology's failure is that the theoreticians of personality have generally been clinicians and psychiatrists, and they have never considered it necessary to have evidence in support of their theories.

Theory without Evidence

Let us turn to this latter cause of failure first: the acceptance by psychiatrists and clinical psychologists of theory without evidence. If we inspect the literature of personality, it is immediately obvious that the bulk of it is written by clinicians and psychiatrists, and that the major support for their theories is "years of intensive clinical experience." This is a tradition started by Freud. His "insights" occurred during the course of his work with his patients. Now there is nothing wrong with such an approach to theory *formulation*; a person is free to make up theories with any inspiration which works: divine revelation, intensive clinical practice, a random numbers table. But he is not free to claim any validity for his theory until it has been tested and confirmed. But theories are treated in no such tentative way in ordinary clinical practice. Consider Freud. What he thought constituted evidence violated the most minimal conditions of scientific rigor. In *The Sexual Enlightenment of Children* (1963), the classic document which is supposed to demonstrate empirically the existence of a castration complex and its connection to a phobia, Freud based his analysis not on the little boy who had the phobia, but on the reports of the father of the little boy, himself in therapy, and a devotee of Freudian theory. I really don't have to comment further on the contamination in this kind of evidence. It is remarkable that only recently has Freud's classic theory on the sexuality of women—the notion of the double orgasm—been actually tested physiologically and found just plain wrong. Now those who claim that fifty years of psychoanalytic experience constitute evidence enough of the essential truths of Freud's theory should ponder the robust health of the double orgasm. Did women, until Masters and Johnson (1966), believe they were having two different kinds of orgasm? Did their psychiatrists intimidate them into reporting something that was not true? If so, were there other things they reported that were also not true? Did psychiatrists ever learn anything different than their theories had led them to believe? If clinical experience means anything at all, surely we should have been done with the double orgasm myth long before the Masters and Johnson studies.

But certainly, you may object, "years of intensive clinical experience" is the only reliable measure in a discipline which rests for its findings on insight, sensitivity, and intuition. The problem with insight, sensitivity, and intuition is that they can confirm for all time the biases that one started out with. People used to be absolutely convinced of their ability to tell which of their number were engaging in witchcraft. All it required was some sensitivity to the workings of the devil.

Years of intensive clinical experience is not the same thing as empirical evidence. The first thing an experimenter learns in any kind of experiment which involves humans is the concept of the "double blind." The term is taken from medical experiments, where one group is given a drug which is presumably supposed to change behavior in a certain way, and a control group is given a placebo. If the observers or the subjects know which group took which drug, the result invariably comes out on the positive side for the new drug. Only when it is not known which subject took which pill is validity remotely approximated. In addition, with judgments of human behavior, it is so difficult to precisely tie down just what behavior is going on, let alone what behavior should be expected, that one must test again and again the reliability of judgments. How many judges, blind, will agree in their observations? Can they replicate their own judgments at some later time? When, in actual practice, these judgment criteria are tested for clinical judgments, then we find that the judges cannot judge reliably, nor can they judge consistently: they do no better than chance in identifying which of a certain set of stories were written by men and which by women; which of a whole battery of clinical test results are the products of homosexuals and which are the products of heterosexuals (Hooker 1957); and which of a battery of clinical test results *and* interviews (where questions are asked such as "Do you have delusions?"—Little and Schneidman 1959) are products of psychotics, neurotics, psychosomatics, or normals. Lest this summary escape your notice, let me stress the implications of these findings. The ability of judges, chosen for their clinical expertise, to distinguish male heterosexuals from male homosexuals on the basis of three widely used clinical projective tests—the Rorschach, the TAT, and the MAP—was *no better than chance*. The reason this is such devastating news, of course, is that sexuality is supposed to be of fundamental importance in the deep dynamic of personality; if what is considered gross sexual deviance cannot be caught, then what are psychologists talking about when they, for example, claim that at the basis of paranoid psychosis is "latent homosexual panic"? They can't even identify what homosexual anything is, let alone "latent homosexual panic."[1] More frightening, expert clinicians cannot be consistent on what diagnostic category to assign to a person, again on the basis of both tests and interviews; a number of normals in the Little and Schneidman study were described as psychotic, in such categories as "schizophrenic with homosexual tendencies" or "schizoid character with depressive trends." But most disheartening, when the judges were asked to rejudge the test protocols some weeks later, their diagnoses of the same subjects on the basis of the same protocol differed markedly from their initial judgments. It is obvious that even simple descriptive conventions in clinical psychology cannot be consistently applied; that these descriptive conventions have any explanatory significance is therefore, of course, out of the question.

As a graduate student at Harvard some years ago, I was a member of a seminar which was asked to identify which of two piles of a clinical test, the TAT, had been written by males and which by females. Only four students out of twenty identified the piles correctly, and this was after one and a half months of intensively studying the differences between men and women. Since this result is below chance—that is, this result would occur by chance about four out of a thousand times—we may conclude that there is finally a consistency here; students are judging knowledgeably within the context of psychological teaching about the differences between men and women; the teachings themselves are simply erroneous.

You may argue that the theory may be scientifically "unsound" but at least it cures people. There is no evidence that it does. In 1952, Eysenck reported the results of what is called an "outcome of therapy" study of neurotics which showed that, of the patients who received psychoanalysis the improvement rate was 44 percent; of the patients who received psychotherapy the improvement rate was 64 percent; and of the patients who received no treatment at all the improvement rate was 72 percent. These findings have never been refuted; subsequently, later studies have confirmed the negative results of the Eysenck study (Barron and Leary 1955; Bergin 1963; Cartwright and Vogel 1960; Truax 1963; Powers and Witmer 1951). How can clinicians and psychiatrists, then, in all good conscience, continue to practice? Largely by ignoring these results and being careful not to do outcome-of-therapy studies. The attitude is nicely summarized by Rotter (1960) (quoted in Astin 1961): "Research studies in psychotherapy tend to be concerned with psychotherapeutic procedure and less with outcome ... to some extent, it reflects an interest in the psychotherapy situation as a kind of personality laboratory." Some laboratory.

The Social Context

Thus, since clinical experience and tools can be shown to be worse than useless when tested for consistency, efficacy, agreement, and reliability, we can safely conclude that theories of a clinical nature advanced about women are also worse than useless. I want to turn now to the second major point in my chapter, which is that, even when psychological theory is constructed so that it may be tested, and rigorous standards of evidence are used, it has become increasingly clear that in order to understand why people do what they do, and certainly in order to change what people do, psychologists must turn away from the theory of the causal nature of the inner dynamic and look to the social context within which individuals live.

Before examining the relevance of this approach for the question of women, let me first sketch the groundwork for this assertion.

In the first place, it is clear (Block 1968) that personality tests never yield consistent predictions; a rigid authoritarian on one measure will be an unauthoritarian on the next. But the reason for this inconsistency is only now becoming clear, and it seems overwhelmingly to have much more to do with the social situation in which the subject finds himself than with the subject himself.

In a series of brilliant experiments, Rosenthal and his co-workers (Rosenthal and Jacobson 1968; Rosenthal 1966) have shown that if one group of experimenters has one hypothesis about what it expects to find, and another group of experimenters has the opposite hypothesis, both groups will obtain results in accord with their hypotheses. The results obtained are not due to mishandling of data by biased experimenters; rather, somehow, the bias of the experimenter creates a changed environment in which subjects actually act differently. For instance, in one experiment, subjects were to assign numbers to pictures of men's faces, with high numbers representing the subject's judgment that the man in the picture was a successful person, and low numbers representing the subject's judgment that the man in the picture was an unsuccessful person. One group of experimenters was told that the subjects tended to rate the faces high; another group of experimenters was told that the subjects tended to rate the faces low. Each group of experimenters was instructed to follow precisely the same procedure: they were required to read to subjects a set of instructions, and to *say nothing else*. For the 375 subjects run, the results showed clearly that those subjects who performed the task with experimenters who expected high ratings gave high ratings, and those subjects who performed the task with experimenters who expected low ratings gave low ratings. How did this happen? The experimenters all used the same words; it was something in their conduct which made one group of subjects do one thing, and another group of subjects do another thing.[2]

The concreteness of the changed conditions produced by expectation is a fact, a reality: even with animal subjects, in two separate studies (Rosenthal and Fode 1960; Rosenthal and Lawson 1961), those experimenters who were told that rats learning mazes had been especially bred for brightness obtained better learning from their rats than did experimenters believing their rats to have been bred for dullness. In a very recent study, Rosenthal and Jacobson (1968) extended their analysis to the natural classroom situation. Here, they tested a group of students and reported to the teachers that some among the students tested "showed great promise." Actually, the students so named had been selected on a random basis. Some time later, the experimenters retested the group of students: those students whose teachers had been told that they were "promising" showed real and dramatic increments in their IQs as compared to the rest of the students. Something in the conduct of the teachers toward those whom the teachers believed to be the "bright" students made those students brighter.

Thus, even in carefully controlled experiments, and with no outward or conscious difference in behavior, the hypotheses we start with will influence enormously the behavior of another organism. These studies are extremely important when assessing the validity of psychological studies of women. Since it is beyond doubt that most of us start with notions as to the nature of men and women, the validity of a number of observations of sex differences is questionable, even when these observations have been made under carefully controlled conditions. Second, and more important, the Rosenthal experiments point quite clearly to the influence of social expectation. In some extremely important ways, people are what you expect them to be or at least they behave as you expect them to behave. Thus, if women, according to Bettelheim, want first and foremost to be good wives and mothers, it is extremely likely that this is what Bruno Bettelheim, and the rest of society, want them to be.

There is another series of brilliant social psychological experiments which point to the overwhelming effect of social context. These are the obedience experiments of Stanley Milgram (1965) in which subjects are asked to obey the orders of unknown experimenters, orders which carry with them the distinct possibility that the subject is killing somebody.

In Milgram's experiments, a subjects is told that he is administering a learning experiment, and that he is to deal out shocks each time the other "subject" (in reality, a confederate of the experimenter) answers incorrectly. The equipment appears to provide graduated shocks ranging upward from 15 volts through 450 volts; for each of four consecutive voltages there are verbal descriptions such as "mild shock," "danger, severe shock," and, finally, for the 435 and 450 volt switches, a red XXX marked over the switches. Each time the stooge answers incorrectly the subject is supposed to increase the voltage. As the voltage increases, the stooge begins to cry in pain; he demands that the experiment stop; finally, he refuses to answer at all. When he stops responding, the experimenter instructs the subject to continue increasing the voltage;

for each shock administered the stooge shrieks in agony. Under these conditions, about 62.5 percent of the subjects administered shock that they believed to be possibly lethal.

No tested individual differences between subjects predicted how many would continue to obey, and which would break off the experiment. When forty psychiatrists predicted how many of a group of 100 subjects would go on to give the lethal shock, their predictions were orders of magnitude below the actual percentages; most expected only one-tenth of 1 percent of the subjects to obey to the end.

But even though *psychiatrists* have no idea how people will behave in this situation, and even though individual differences do not predict which subjects will obey and which will not, it is easy to predict when subjects will be obedient and when they will be defiant. All the experimenter has to do is change the social situation. In a variant of the experiment, Milgram had two stooges present in addition to the "victim"; these worked along with the subject in administering electric shocks. When these two stooges refused to go on with the experiment, only 10 percent of the subjects continued to the maximum voltage. This is critical for personality theory. It says that behavior is predicted from the social situation, not from the individual history.

Finally, an ingenious experiment by Schachter and Singer (1962) showed that subjects injected with adrenalin, which produces a state of physiological arousal in all but minor respects identical to that which occurs when subjects are extremely afraid, became euphoric when they were in a room with a stooge who was acting euphoric, and became extremely angry when they were placed in a room with a stooge who was acting extremely angry.

To summarize: If subjects under quite innocuous and noncoercive social conditions can be made to kill other subjects and under other types of social conditions will positively refuse to do so; if subjects can react to a state of physiological fear by becoming euphoric because there is somebody else around who is euphoric or angry because there is somebody else around who is angry; if students become intelligent because teachers expect them to be intelligent, and rats run mazes better because experimenters are told the rats are bright, then it is obvious that a study of human behavior requires, first and foremost, a study of the social contexts within which people move, the expectations as to how they will behave, and the authority which tells them who they are and what they are supposed to do.

Biologically Based Theories

Biologists also have at times assumed they could describe the limits of human potential from their observations of animal rather than human behavior. Here, as in psychology, there has been no end of theorizing about the sexes, again with a sense of absolute certainty. These theories fall into two major categories.

One biological theory of differences in nature argues that since females and males differ in their sex hormones, and sex hormones enter the brain (Hamburg and Lunde in Maccoby 1966), there must be innate behavioral differences. But the only thing this argument tells us is that there are differences in physiological state. The problem is whether these differences are at all relevant to behavior.

Consider, for example, differences in testosterone levels. A man who calls himself Tiger[3] has recently argued (1970) that the greater quantities of testosterone found in human males as compared with human females (of a certain age group) determines innate differences in aggressiveness, competitiveness, dominance, ability to hunt, ability to hold public office, and so forth. But Tiger demonstrates in this argument the same manly and courageous refusal to be intimidated by evidence which we have already seen in our consideration of the clinical and psychiatric tradition. The evidence does not support his argument, and in some cases, directly contradicts it. Testosterone level co-varies neither with hunting ability, nor with dominance, nor with aggression, nor with competitiveness. As Storch has pointed out (1970), all normal male mammals in the reproductive age group produce much greater quantities of testosterone than females; yet many of these males are neither hunters nor are they aggressive. Among some hunting mammals, such as the large cats, it turns out that more hunting is done by the female than the male. And there exist primate species where the female is clearly more aggressive, competitive, and dominant than the male (Mitchell 1969; and see below). Thus, for some species, being female, and therefore, having less testosterone than the male of that species, means hunting more, or being more aggressive, or being more dominant. Nor does having *more* testosterone preclude behavior commonly thought of as "female": there exist primate species where females do not touch infants except to feed them; the males care for the infants (Mitchell 1969; see fuller discussion below). So it is not clear what testosterone or any other sex-hormonal difference means for differences in nature of sex-role behavior.

In other words, one can observe identical sex-role behavior (e.g., "mothering") in males and females despite known differences in physiological state, i.e., sex hormones. What about the converse to this? That is, can one obtain differences in behavior given a single physiological state? The answer is overwhelmingly yes, not only as regards non-sex-specific hormones (as in the Schachter and Singer 1962 experiment cited above), but also as regards gender itself. Studies of hermaphrodites with the same diagnosis (the genetic, gonadal, hormonal sex, the internal reproductive organs, and the ambiguous appearances of the external genitalia were identical) have shown that one will consider oneself male or female depending simply on whether one was defined and raised as male or female (Money 1970; Hampton and Hampton 1961): "There is no more convincing evidence of the power of social interaction on gender-identity differentiation than in the case of congenital hermaphrodites who are of the same diagnosis and similar degree of hermaphroditism but are differently assigned and with a different postnatal medical and life history" (Money 1970, p. 432).

Thus, for example, if out of two individuals diagnosed as having the adrenogenital syndrome of female hermaphroditism, one is raised as a girl and one as a boy, each will act and identify her/himself accordingly. The one raised as a girl will consider herself a girl; the one raised as a boy will consider himself a boy; and each will conduct her/himself successfully in accord with that self-definition.

So, identical behavior occurs given different physiological states; and different behavior occurs given an identical physiological starting point. So it is not clear that differences in sex hormones are at all relevant to behavior.

There is a second category of theory based on biology, a reductionist theory. It goes like this. Sex-role behavior in some primate species is described, and it is concluded that this is the "natural" behavior for humans. Putting aside the not insignificant problem of observer bias (for instance, Harlow 1962, of the University of Wisconsin, after observing differences between male and female rhesus monkeys, quotes Lawrence Sterne to the effect that women are silly and trivial, and concludes that "men and women have differed in the past and they will differ in the future"), there are a number of problems with this approach.

The most general and serious problem is that there are no grounds to assume that anything primates do is necessary, natural, or desirable in humans, for the simple reason that humans are not nonhumans. For instance, it is found that male chimpanzees placed alone with infants will not "mother" them. Jumping from hard data to ideological speculation researchers conclude from this information that *human* females are necessary for the safe growth of human infants. It would be as reasonable to conclude, following this logic, that it is quite useless to teach human infants to speak, since it has been tried with chimpanzees and it does not work.

One strategy that has been used is to extrapolate from primate behavior to "innate" human preference by noticing certain trends in primate behavior as one moves phylogenetically closer to humans. But there are great difficulties with this approach. When behaviors from lower primates are directly opposite to those of higher primates, or to those one expects of humans, they can be dismissed on evolutionary grounds—higher primates and/or humans grew out of that kid stuff. On the other hand, if the behavior of higher primates is counter to the behavior considered natural for humans, while the behavior of some lower primate is considered the natural one for humans, the higher primate behavior can be dismissed also, on the grounds that it has diverged from an older, prototypical pattern. So either way, one can select those behaviors one wants to prove as innate for humans. In addition, one does not know whether the sex-role behavior exhibited is dependent on the phylogenetic rank, or on the environmental conditions (both physical and social) under which different species live.

Is there then any value at all in primate observations as they relate to human females and males? There is a value but it is limited: its function can be no more than to show some extant examples of diverse sex-role behavior. It must be stressed, however, that this is an extremely limited function. The extant behavior does not begin to suggest all the possibilities, either for nonhuman primates or for humans. Bearing these caveats in mind, it is nonetheless interesting that if one inspects the limited set of existing nonhuman primate sex-role behaviors, one finds, in fact, a much larger range of sex-role behavior than is commonly believed to exist. "Biology" appears to limit very little; the fact that a female gives birth does not mean, even in nonhumans, that she necessarily cares for the infant (in marmosets, for instance, the male carries the infant at all times except when the infant is feeding [Mitchell 1969]); "natural" female and male behavior varies all the way from females who are much more aggressive and competitive than males (e.g., Tamarins; see Mitchell 1969) and male "mothers" (e.g., Titi monkeys, night monkeys, and marmosets; see Mitchell 1969)[4] to submissive and passive females and male antagonists (e.g., rhesus monkeys).

But even for the limited function that primate arguments serve, the evidence has been misused. Invariably, only those primates have been cited which exhibit exactly the kind of behavior that the proponents of the biological basis of human female behavior wish were true for humans. Thus, baboons and rhesus monkeys are generally cited: males in these groups exhibit some of the most irritable and aggressive behavior found in primates, and if one wishes to argue that females are naturally passive and submissive, these groups provide vivid examples. There are abundant counterexamples, such as those mentioned above (Mitchell 1969); in fact, in general, a counterexample can be found for every sex-role behavior cited, including, as mentioned in the case of marmosets, male "mothers."

But the presence of counterexamples has not stopped florid and overarching theories of the natural or biological basis of male privilege from proliferating. For instance, there have been a number of theories dealing with the innate incapacity in human males for monogamy. Here, as in most of this type of theorizing, baboons are a favorite example, probably because of their fantasy value: the family unit of the hamadryas baboon, for instance, consists of a highly constant pattern of one male and a number of females and their young. And again, the counterexamples, such as the invariably monogamous gibbon, are ignored.

An extreme example of this maiming and selective truncation of the evidence in the service of a plea for the maintenance of male privilege is a recent book, *Men in Groups* (1969) by Tiger (see above and note 3). The central claim of this book is that females are incapable of honorable collective action because they are incapable of "bonding" as in "male bonding." What is "male bonding"? Its surface definition is simple: "a particular relationship between two or more males such that they react differently to members of their bonding units as compared to individuals outside of it" (pp. 19–20). If one deletes the word "male," the definition, on its face, would seem to include all organisms that have any kind of social organization. But this is not what Tiger means. For instance, Tiger asserts that females are incapable of bonding; and this alleged incapacity indicates to Tiger that females should be restricted from public life. Why is bonding an exclusively male behavior? Because, says Tiger, it is seen in male primates. All male primates? No, very few male primates. Tiger cites two examples where male bonding is seen: rhesus monkeys and baboons. Surprise, surprise. But not even all baboons: as mentioned above, the hamadryas social organization consists of one-male units; so does that of the Gelada baboon (Mitchell 1969). And the great apes do not go in for male bonding much either. The "male bond" is hardly a serious contribution to scholarship one reviewer for *Science* has observed that the book "shows basically more resemblance to a partisan political tract than to a work of objective social science," with male bonding being "some kind of behavioral phlogiston" (Fried 1969, p. 884).

In short, primate arguments have generally misused the evidence; primate studies themselves have, in any case, only the very limited function of describing some possible sex-role behavior; and at present, primate observations have been sufficiently limited so that even the range of possible sex-role behavior of nonhuman primates is not known. This range is not known since there is only minimal observation of what happens to behavior if the physical or social environment is changed. In one study (Itani 1963), different troops of Japanese macaques were observed. Here, there appeared to be cultural differences: males in three out of the eighteen troops observed differed in their amount of aggressiveness and infant-caring behavior. There could be no possibility of differential evolution here; the differences seemed largely transmitted by infant socialization. Thus, the very limited evidence points to some plasticity in the sex-role behavior of nonhuman primates; if we can figure out experiments which massively change the social organization of primate groups, it is possible that we might observe great changes in behavior. At present, however, we must conclude that, since given a constant physical environment nonhuman primates do not seem to change their social conditions very much by themselves, the "innateness" and fixedness of their behavior is simply not known. Thus, even if there were some way, which there isn't, to settle on the behavior of a particular primate species as being the "natural" way for humans, we would not know whether or not this were simply some function of the present social organization of that species. And finally, once again it must be stressed that even if nonhuman primate behavior turned out to be relatively fixed, this would say little about our behavior. More immediate and relevant evidence, i.e., the evidence from social psychology, points to the enormous plasticity in human behavior, not only from one culture to the next, but from one experimental group to the next. One of the most salient features of human social organization is its variety; there are a number of cultures where there is at least a rough equality between men and women (Mead 1949). In summary, primate arguments can tell us very little about our "innate" sex-role behavior; if they tell us anything at all, they tell us that there is no one biologically "natural" female or male behavior, and that sex-role behavior in nonhuman primates is much more varied than has previously been thought.

Conclusion

In brief, the uselessness of present psychology (and biology) with regard to women is simply a special case of the general conclusion: one must understand the social conditions under which women live if one is going to attempt to explain the behavior of women. And to understand the social conditions under which women live, one must be cognizant of the social expectations about women.

How are women characterized in our culture, and in psychology? They are inconsistent, emotionally unstable, lacking in a strong conscience or superego, weaker, "nurturant" rather than productive, "intuitive" rather than intelligent, and, if they are at all "normal," suited to the home and the family. In short, the list adds up to a typical minority group stereotype of inferiority (Hacker 1951): if they know their place, which is in the home, they are really quite lovable, happy, childlike, loving creatures. In a review of the intellectual differences between little boys and little girls, Eleanor Maccoby (1966) has shown that there are no intellectual differences until about high school, or, if there are, girls are slightly ahead of boys. At high school, girls begin to do worse on a few intellectual tasks, such as arithmetic reasoning, and beyond high school, the achievement of women now measured in terms of productivity and accomplishment drops off even more rapidly. There are a number of other, nonintellectual tests which show sex differences; I chose the intellectual differences since it is seen clearly that women start becoming inferior. It is no use to talk about women being different but equal; all of the tests I can think of have a "good" outcome and a "bad" outcome. Women usually end up at the "bad" outcome. In light of social expectations about women, what is surprising is not that women end up where society expects they will; what is surprising is that little girls don't get the message that they are supposed to be stupid until high school; and what is even more remarkable is that some women resist this message even after high school, college, and graduate school.

My chapter began with remarks on the task of the discovery of the limits of human potential. Psychologists must realize that it is they who are limiting discovery of human potential. They refuse to accept evidence, if they are clinical psychologists, or, if they are rigorous, they assume that people move in a context-free ether, with only their innate dispositions and their individual traits determining what they will do. Until psychologists begin to respect evidence, and until they begin looking at the social contexts within which people move, psychology will have nothing of substance to offer in this task of discovery. I don't know what immutable differences exist between men and women apart from differences in their genitals; perhaps there are some other unchangeable differences; probably there are a number of irrelevant differences. But it is clear that until social expectations for men and women are equal, until we provide equal respect for both men and women, our answers to this question will simply reflect our prejudices.

Notes

1. It should be noted that psychologists have been as quick to assert absolute truths about the nature of homosexuality as they have about the nature of women. The arguments presented in this chapter apply equally to the nature of homosexuality; psychologists know nothing about it; there is no more evidence for the "naturalness" of heterosexuality than for the "naturalness" of homosexuality. Psychology has functioned as a pseudo-scientific buttress for our cultural sex-role notions, that is, as a buttress for patriarchal ideology and patriarchal social organization: women's liberation and gay liberation fight against a common victimization.
2. I am indebted to Jesse Lemisch for his valuable suggestions in the interpretation of these studies.
3. Schwarz-Belkin (1914) claims that the name was originally *Mouse*, but this may be a reference to an earlier L. Tiger (putative).
4. All these are lower-order primates, which makes their behavior with reference to humans unnatural, or more natural; take your choice.

References

Astin, A. W., "The Functional Autonomy of Psychotherapy." *American Psychologist*, 1961, 16, 75–78.

Barron, F., and Leary, T., "Changes in Psychoneurotic Patients with and without Psychotherapy." *Journal of Consulting Psychology*, 1955, 19, 239–245.

Bergin, A. E., "The Effects of Psychotherapy: Negative Results Revisited." *Journal of Consulting Psychology*, 1963, 10, 244–250.

Bettelheim, B., "The Commitment Required of a Woman Entering a Scientific Profession in Present-Day American Society." *Woman and the Scientific Professions*, the MIT Symposium on American Women in Science and Engineering, 1965.

Block, J., "Some Reasons for the Apparent Inconsistency of Personality." *Psychological Bulletin*, 1968, 70, 210–212.

Cartwright, R. D., and Vogel, J. L., "A Comparison of Changes in Psychoneurotic Patients during Matched Periods of Therapy and No-therapy." *Journal of Consulting Psychology*, 1960, 24, 121–127.

Erikson. E., "Inner and Outer Space: Reflections on Womanhood." *Daedalus*, 1964, 93, 582–606.

Eysenck, H. J., "The Effects of Psychotherapy: An Evaluation." *Journal of Consulting Psychology*, 1952, 16, 319–324.

Fieldcrest—Advertisement in the *New Yorker*, 1965.

Fried, M. H., "Mankind Excluding Woman," review of Tiger's *Men in Groups*. *Science*, 165, 1969, 883–884.

Freud, S., *The Sexual Enlightenment of Children*. Collier Books Edition, 1963.

Goldstein, A. P., and Dean, S. J., *The Investigation of Psychotherapy: Commentaries and Readings*. New York: John Wiley & Sons, 1966.

Hamburg, D. A., and Lunde, D. T., "Sex Hormones in the Development of Sex Differences in Human Behavior." In Maccoby (ed.), *The Development of Sex Differences*. Stanford University Press, 1966, 1–24.

Hacker, H. M., "Women as a Minority Group." *Social Forces*, 1951, 30, 60–69.

Hampton, J. L., and Hampton, J. C., "The Ontogenesis of Sexual Behavior in Man." In W. C. Young (ed.), *Sex and Internal Secretions*, 1961, 1401–1432.

Harlow, H. F., "The Heterosexual Affectional System in Monkeys." *American Psychologist*, 1962, 17, 1–9.

Hooker, E., "Male Homosexuality in the Rorschach." *Journal of Projective Techniques*, 1957, 21, 18–31.

Itani, J. "Paternal Care in the Wild Japanese Monkeys, *Macaca fuscata*." In C. H. Southwick (ed.), *Primate Social Behavior*. Princeton: Van Nostrand, 1963.

Little, K. B., and Schneidman, E. S., "Congruences among Interpretations of Psychological and Anamnestic Data." *Psychological Monographs*, 1959, 73, 1–42.

Maccoby, Eleanor E., "Sex Differences in Intellectual Functioning." In Maccoby (ed.), *The Development of Sex Differences*. Stanford University Press, 1966, 25–55.

Masters, W. H., and Johnson, V. E., *Human Sexual Response*. Boston: Little, Brown, 1966.

Mead, M., *Male and Female: A Study of the Sexes in a Changing World*. New York: William Morrow, 1949.

Milgram, S., "Some Conditions of Obedience and Disobedience to Authority." *Human Relations*, 1965a, 18, 57–76.
Milgram, S., "Liberating Effects of Group Pressure." *Journal of Personality and Social Psychology*, 1965b, 1, 127–134.
Mitchell, G. D., "Paternalistic Behavior in Primates." *Psychological Bulletin*, 1969, 71, 339–417.
Money, J., "Sexual Dimorphism and Homosexual Gender Identity." *Psychological Bulletin*, 1970, 74, 6, 425–440.
Powers, E., and Witmer, H., *An Experiment in the Prevention of Delinquency*. New York: Columbia University Press, 1951.
Rheingold, J., *The Fear of Being a Woman*. New York: Grune & Stratton, 1964.
Rosenthal, R., "On the Social Psychology of the Psychological Experiment: The Experimenter's Hypothesis as Unintended Determinant of Experimental Results." *American Scientist*, 1963, 51, 268–283.
Rosenthal, R., *Experimenter Effects in Behavioral Research*. New York: Appleton-Century-Crofts, 1966.
Rosenthal, R., and Fode, K. L., "The Effect of Experimenter Bias on the Performance of the Albino Rat." Unpublished Manuscript, Harvard University, 1960.
Rosenthal, R., and Jacobson, L., *Pygmalion in the Classroom: Teacher Expectation and Pupils' Intellectual Development*. New York: Holt, Rinehart & Winston, 1968.
Rosenthal, R., and Lawson, R., "A Longitudinal Study of the Effects of Experimenter Bias on the Operant Learning of Laboratory Rats." Unpublished manuscript, Harvard University, 1961.
Rotter, J. B., "Psychotherapy." *Annual Review of Psychology*, 1960, 11, 381–414.
Schachter, S., and Singer, J. E., "Cognitive, Social and Physiological Determinants of Emotional State." *Psychological Review*, 1962, 69, 379–399.
Schwarz-Belkin, M. "Les Fleurs de Mal." In *Festschrift for Gordon Piltdown*. New York: Ponzi Press, 1914.
Storch, M., "Reply to Tiger," 1970. Unpublished manuscript.
Tiger, L., *Men in Groups*. New York: Random House, 1969.
Tiger, L., "Male Dominance? Yes. Alas. A Sexist Plot? No." *New York Times Magazine*, Section N, Oct. 25, 1970.
Truax, C. B., "Effective Ingredients in Psychotherapy: An Approach to Unraveling the Patient-Therapist Interaction. *Journal of Counseling Psychology*, 1963, 10, 256–263.

Chapter 39

Reasoning about Ourselves: Feminist Methodology in the Social Sciences

Alison Wylie

Feminist revaluations of science are now at a critical juncture. Having exposed sexist bias in an enormous range of disciplines we face a new question: how is it possible to do better? What would a nonsexist or, indeed, a feminist science look like?

Guidelines for nonsexist research have been developed in a number of contexts, as components of funding guidelines and disciplinary codes of practice, as well as in connection with feminist critiques of science and general methodological discussions (for an example of the former, see the Social Sciences and Humanities Research Council of Canada handbook, "On the Treatment of the Sexes in Research"; Eichler and Lapointe 1985). Perhaps the most detailed is Eichler's "Practical Guide" to *Nonsexist Research Methods* (Eichler 1988): as an antidote to various "primary problems" that she identifies with extant practice, she urges that researchers guard against "gender insensitivity," that they not treat (analyze, label) otherwise identical phenomena differently because of their gender association, and that they not generalize the attributes of one gender to the whole of a population or treat them as an analytic or evaluative norm (Eichler 1988, 6).

But more challenging and controversial are a range of directives that have been articulated for self-consciously *feminist* research. From the outset, those who have suspected that extant methods are part of the problem—that they replicate, perhaps even generate, sexist bias—have asked whether there might not be distinctively feminist methods for inquiry. Perhaps these would be phenomenological and qualitative methods, rather than quantitative, or some feminist transformation of historical materialism (O'Brien 1981), or a "political hermeneutics" (MacKinnon 1982). Increasingly, however, this seems the wrong question to ask. As Harding puts it, the "preoccupation with method mystifies what have been the most interesting aspects of feminist research process" (Harding 1987, 19); it is the "fetishization of method itself" (1987, 20) that feminists should resist. This is, in fact, a recurrent theme that incorporates a number of different considerations. Vickers warns that if a "uniform reconstruction of method" were to emerge (Vickers 1982, 43), it would certainly bring with it "the dangers of a new orthodoxy."[1] Numerous others have rejected the quest for a "correct" approach, a method that is proof against sexist bias, on the ground that this puts misplaced trust in technical fixes for conceptual problems; any method can be used in a sexist manner (Eichler 1987, 32). Longino observes, in this connection, that the sciences are so diverse methodologically, it is implausible that they "might be equally transformed by [a feminist] framework" (Eichler 1987, 53).

Nevertheless, Longino, and most of those who mistrust the quest for a feminist method, maintain that there is something or, perhaps, a range of things distinctive about feminist research qua feminist. Typically this is identified with a "general 'orientation'" derived from feminist political theory and practice (Stanley and Wise 1979,

273). Feminist research, on this account, should be a matter of "doing feminism in another context" (Stanley and Wise 1983a, 195) or, as Longino puts it, of "doing science as a feminist," rather than of doing "feminist science" (Longino 1987, 53). What I propose to do here is examine various proposals that have been made for "doing science as a feminist" with the aim of identifying what they share, as feminist, and of articulating, if not resolving, some problems raised by their divergence on a number of crucial issues.

Two Models for Feminist Research

Whatever else feminist research may be, it addresses a set of problems that bear at least a family resemblance to one another in being of concern to feminists. Most widely, it is conceived as research that provides a basis for critically reassessing extant ideology and theory where this leaves out women altogether or significantly distorts or devalues their activities and lives as women. In this, its encompassing aim is to empower women by recovering the details of their experience and activities, by "piecing together a way of understanding the world from the point of view of women" on this basis (Brunsdon 1978, 26), by delineating "pervasive patterns of subordination" that have "marked the fortunes of women" (Keohane and Gelpi 1982, x), and by providing an explanatory understanding of the nature and sources of the patriarchal oppression revealed in these emerging patterns. Some add the requirement that problems be addressed that are of concern to particular groups of women and that research results be presented in a form that is both accessible and useful to them (e.g., Acker, Barry, and Esseveld 1984, 425).

Often, commitment to these general aims is associated with the more specific requirement that a central feature of feminist research must be a consideration of women's "lived reality," their concrete, particular, personal experience. This is advocated not just as a necessary basis for addressing questions of concern to women—for getting at the gendered dimensions of social life that conventional categories of analysis obscure, and for "test[ing] thinking against experience, making sure that it remains rooted in the real lives of women" (Keohane and Gelpi 1982, ix)—but as a consequence of commitment to feminist principles (Eichler 1987, 33); the articulation and "validation" of women's experience is an end to be valued in itself whatever other ends it may support. It is in this connection that consciousness raising, conceived as a process of recovering and analyzing "the politics of the personal," is often invoked as a model for research practice (e.g., in MacKinnon 1982).[2] Where the "first principle" of consciousness raising, as articulated in the early 1970s (e.g., in *Ms.* 1972, 1), is never to challenge or question experiential accounts, sometimes the commitment to focus on women's experience is aligned with a requirement to treat it as veridical.

While most feminist researchers would subscribe to the general aim of empowering women in various of the senses I have described, differences do arise when they consider the question of what this commitment entails for practice. This is especially clear when they discuss the practical implications of making experience central and ask whose experience is at issue, how it is to ground the articulation of general explanatory accounts of women's situation, and whether the commitment to take women's experience seriously really means that they can never question experience, or "go beyond" it in pursuit of these explanatory aims. I discern two models of research practice—two distinct "general orientations"—emerging in response to these questions: a "collectivist" model of inquiry and a "self-study" model.

The Collectivist Model

The first model[3] focuses on the experience of women as research subjects and takes the requirement of fidelity to this experience to mean that women should be asked rather than told what they are experiencing and why; "as researchers, we must not impose our definitions of reality on those researched" (Acker, Barry, and Esseveld 1983, 425). At a descriptive level, this means that researchers should be particularly careful to design research so that discrepancies between their constructs and the understanding of subjects can be identified (e.g., Acker, Barry, and Esseveld 1983, 430). On the consciousness-raising model, research is to be conducted as a dialectical process in which the feminist investigator "continually test[s] the plausibility of [her] work against her own experience as well as against the experience of other women" (Vickers 1982, 36).[4] At an explanatory level, the commitment to take subjects' experience seriously requires that women be acknowledged to have a credible theoretical, explanatory grasp of what goes on in their lives. Indeed, those sympathetic to ethnomethodology argue that the explanatory constructs a researcher generates are simply another "version," an alternative with no particular claim to privilege over the accounts formulated by (other) participants in the subject context.

This attention to the experience and self-understanding of women is aligned with various ethical requirements. Minimally, research must not be exploitative; "research that aims to be liberating should not in the process become only another mode of oppression" (Acker, Barry, and Esseveld 1983, 425).[5] Beyond this it is routinely argued that subjects should not be "objectified," turned "into objects of scrutiny and manipulation" (Acker, Barry, and Esseveld 1983, 425), that the "illusion" of a sharp separation of researcher (qua "knower") from the objects of study be abandoned, and, most important, that the power differences constituting the standard hierarchical relationship between researcher and subject be eliminated as far as possible (Oakley 1981, 41). These recommendations have implications for the roles of both researcher and researched in feminist inquiry. Researchers, it is argued, must be acknowledged to enter the research relationship as concretely situated (social) individuals whose subjective experience and social engagement with the subjects of study inevitably affect what they come to understand. The virtue in this is that the processes by which knowledge is constructed, specifically, their partiality, are made "visible," a subject of the research process itself and a basis for reflexive critique.[6]

This revaluation of the status of the researcher is secondary, however, to the emphasis put on the importance of transforming the role of research subjects. Again and again it is insisted that subjects must be actively involved in the research process as coparticipants at various levels, sometimes in determining the direction of research, more often as collaborators in the description and interpretation of their experience, and sometimes in formulating and assessing explanatory, theoretical constructs. Oakley's critique of standard guidelines for interviewing that demand both detachment and rapport—the "mythology of 'hygienic' research" (Oakley 1981, 58)—represents one strategy for implementing these directives. She argues, on moral and methodological grounds, for a collaborative, nonhierarchical relationship with interviewees governed by the principle, "no intimacy without reciprocity" (Oakley 1981, 49).[7] Acker, Barry, and Esseveld push further this commitment to active involvement. They made a concerted effort not just to engage subjects at a personal level in the process of data collection, but to involve them directly in analysis and interpretation. This proved difficult in a number of respects, however. They worked with a group of women who were by no means all feminists and frequently "had to confront discrepancies between our ideas and interpretations and those of the women we interviewed" (Acker, Barry, and Esseveld 1983, 427). They were especially concerned that the results of their analysis—an identification of common themes in the experiences of their subjects as women reentering the work force—would be upsetting to at least some of those they had interviewed, and decided, in the end, "not to include them as active participants in the analysis." (Acker, Barry, and Esseveld 1983, 429).

Acker, Barry, and Esseveld thus confront a dilemma. It seems that the ideal of involving subjects in research on a fully active, egalitarian basis—as coparticipants in a potentially emancipatory process of inquiry—can only be realized when dealing with subjects who are "very much like us," and yet this carries with it the threat of "eliminating most women from our view and limiting the usefulness of our projects" (Acker, Barry, and Esseveld 1983, 434). In the end, Acker, Barry, and Esseveld find this latter consequence unacceptable, more unacceptable, that is, than qualifying the commitment to fully engage (all) subjects in research. So it is the ideal of conducting research as a collective enterprise that they compromise.

A further problem arises even when it is possible to involve subjects in analysis. Acker, Barry, and Esseveld observe that, when it came to analysis, "we had to assume the role of the people with the power to define" (Acker, Barry, and Esseveld 1983, 429); it was they who determined the categories and dimensions of analysis along which commonalities and differences might be assessed. What they sought from subjects was, primarily, a critical check on the adequacy of their constructs. In this capacity, they found they could not avoid some degree of objectification of their subjects: "If we were to fulfil the emancipatory aim for the people we were studying [indeed, as demanded by them],[8] we had to go beyond the faithful representation of their experience, beyond 'letting them talk for themselves' and put those experiences into the theoretical framework with which we started the study, a framework that links women's oppression to the structure of Western capitalist society" (Acker, Barry, and Esseveld 1983, 429–30). In fact, this sometimes meant not just taking their subjects' experiences as an "object," in the sense of requiring explanation in terms other than those constitutive of the experience, but also, on occasion, treating them in terms that countered the self-understanding of the research subjects themselves. Consequently, despite a deep commitment to "understand reality from the perspective of the people experiencing it" (1983, 431)—to "grant them full subjectivity," to avoid "violating their reality" (429)—Acker, Barry, and Esseveld found that they could not always take the experiential perspective of subjects at face value if they were to pursue the further aims of understanding how it arose, what it has in common with other women's experiences, and how it is, in this sense, "political." They see their main challenge as that of maintaining a "difficult balance between granting respect to the other's interpretation of her reality, while going beyond that interpretation to comprehend its underlying relations" (Acker, Barry, and Esseveld 1983, 429).

This appreciation of conflict between the animating commitments of feminist research, and the conclusion that an interest in understanding the patriarchal conditions of experience requires sometimes that its integrity be questioned, is by no means unique to the three researchers whose work I have been describing. In an early paper, Dorothy Smith declares at one point that, contra the administrative impulse of sociology to control and appropriate the experience of others, in feminist research "their reality, their varieties of experience must be an unconditional datum" (Smith 1987, 93).

But she then goes on to insist that "no amount of observation of face-to-face relations, no amount of analysis of commonsense knowledge of everyday life, will take us beyond our essential ignorance of how it is put together" (94); the emancipatory potential of feminist research requires us to "posit ... a total socioeconomic order 'in back' of [any given experiential moment]," to seek the "determinants" of experience outside experience (94).[9]

More recently, there has been a spate of literature directly criticizing "experientialism." In a 1987 article, Grant rejects outright the notion that direct experience is, or should be taken as, a "necessary precondition for knowledge" (Grant 1987, 105) and inveighs against the threat of relativism that she sees looming in an uncritical privileging of "unmediated experience" (Grant 1987, 110–13). Brunsdon argues, in a similar vein, that "the dependence on direct experience [and common sense] is one of the aspects of our oppression" (Brunsdon 1978, 25); it precludes "the construction of a coherent oppositional world.... It produces an understanding of events as episodic and random" (Brunsdon 1978, 25). The effect is often to confirm subordination as inevitable, and to limit the scope and effectiveness of political action. In the view of these critics, the feminist revaluation of women's experience should be construed as a commitment to take their subjects' experience seriously *as a point of departure,* not as immune to challenge and criticism.

The Self-study Model

However compelling these critiques of experientialism may be, there remain proponents of the feminist commitment to recover and revalue the experience of women who reject, out of hand, the balancing of compromises associated with the collectivist model of practice. Stanley and Wise, for example, hold that the essential feature of feminist practice must be its "insistence on the validity of women's experience" (Stanley and Wise 1983b, 135), and they will countenance no qualification of this commitment. In particular, they reject any form of argument for "going beyond" the personal, for treating it as a point of departure, a "resource," to be used in constructing explanatory theories that postulate structures, processes, or, indeed, a "reality" beyond women's experience (Stanley and Wise 1983b, 81). Feminist researchers who are prepared to question the integrity or veracity of experiential reports, in the course of treating them as a "springboard" to theory, risk slipping back into precisely the "'expert' analytical and theoretical approaches" that have so effectively silenced and excluded women in the past (Stanley and Wise 1983b, 56).

Where the commitment to privilege women's experience comes into conflict with a demand for broader theoretical understanding, Stanley and Wise are prepared to repudiate any mode of theorizing that might involve reassessment or displacement of the theories and perceptions of subjects, "what women and men have to tell us about their lives" (Stanley and Wise 1983b, 83–84). This includes any quest after structural explanations (Stanley and Wise 1983b, 83, 85), any attempt to "generaliz[e] from [subjective experiences]" (83), and sometimes any postulation of "'real' conditions of oppression outside experience and understanding" (81). Realities are, they argue, negotiated constructs and any theorizing that suggests otherwise just constructs a myth, usually an oppressive myth. Their constructive proposal is that feminists should adhere to the model of consciousness-raising practice that, they argue, requires that they "go back into the personal," rather than beyond it, in order to "explicate ... to examine in detail exactly what this experience is" (84).

On some occasions they seem to argue this approach on pragmatic, methodological grounds. Feminists *need not* go beyond "the personal" because any conditions relevant for understanding women's experience and oppression can be revealed within (or, through direct analysis of) experience, contra the arguments of Dorothy Smith cited earlier. They insist, in this connection, that "power and its use can be examined within personal life" and that, in fact, "the political *must* be examined in this way" because "the system [the array of institutions that oppress women] is experienced *in* everyday life, and isn't separate from it" (Stanley and Wise 1983b, 54).[10]

On other occasions, however, their argument turns on considerations of (feminist) principle. Stanley and Wise insist that, above all else, feminists have a moral and political responsibility to resist any imposition of the reality or standpoint of one person on others or any privileging of one perspective over another (1983b, 112).[11] When they confront the possibility that the experience of diverse others may be contradictory, or reported interpretations of experience implausible or incoherent, they reiterate that, because all realities are constructed, there can never be grounds for judging a subject "wrong" in what she claims about *her* reality. In one instance they observe that "if a housebound, depressed, battered mother of six with an errant spouse says she's *not* oppressed, there's little point in us telling her she's got it wrong because of the objective reality of her situation" (Stanley and Wise 1983b, 112); her account is "*truth* for her," and there can be no justification for "attempting to impose our reality on [such subjects]" (113).[12] Thus, they seem prepared to embrace precisely the relativist implications that led others to qualify feminist proscriptions against judgment.

When they consider the implications of taking seriously the perceptions of deeply sexist men (rapists, batterers, obscene phone callers) they are prepared to conclude that, although people must not be treated as objects, it is not incumbent on the feminist researcher to involve subjects as active participants in research; "we do not want people, 'the researched,' to have more involvement in designing questionnaires, interpreting statistical or other results" (Stanley and Wise 1983b, 170). In particular, they see no justification for allowing the input of sexist subjects to shape feminist findings. When they discuss "the place of the personal within research," they turn immediately to questions about the "presence of the researcher" (150), a presence that cannot be avoided and that they say is, in any case, the primary source and grounding of all theoretical understanding.[13] In fact, despite references to the experiences of subjects as the proper object of inquiry, these "others" continually slip out of view; it is the experience of the researcher that draws Stanley and Wise's attention.

Two principles reinforce what I identify as a persistent tendency, in Stanley and Wise's discussion, to shift the focus of feminist research inward. One is epistemic: implicit in their proscriptions against imposition is a deep skepticism about whether we can ever grasp the experience of others. Perhaps we inevitably "impose," or project, our own reality when we purport to describe that of another; perhaps we can never do more than describe *our* experience of their actions, descriptions, testimonials, or interactions with us. But more fundamentally, Stanley and Wise embrace what amounts to an ethical principle to the effect that not only should we avoid imposing our realities on others, we should avoid speaking for them. I detect in this something of the feminist stance recently articulated by Trebilcot when she says, "I speak only for myself ... in the sense that an account I give reports only my understanding of the world or, more accurately, only my *with*standing in the world" (Trebilcot 1988, 4),[14] and adds to this a stringent "principle of nonpersuasion" (5). Taken together, the implication of commitment to feminist principles seems to be that the experience of the researcher becomes not just the (only) source and ground of understanding, but also its primary object.

This approach to doing research as a feminist is embodied, at least in part, in the early work of the Vancouver Women's Research Center; this is, in fact, one example of feminist research that Stanley and Wise identify as going some distance toward realizing their ideals (Stanley and Wise 1983b). One member of the Center describes feminist research as a matter of "study[ing] your own community" (Jacobson 1977) that requires, first, that you, the researcher, "locate yourself" in your community, and then that you proceed by analyzing your own "everyday world" as a member of this community, gradually "broadening the boundaries" of your exploration so you begin to formulate an account of this community "as you know it" (Jacobson 1977, 11–13). There is no discussion of how the perceptions of others might be incorporated; the report closes with the observation that such research is never "complete" inasmuch as there are always alternative accounts that could be given. Similarly, Stanley and Wise's work on obscene phone calls is largely an analysis of the impact of these calls on their own lives and understanding of patriarchy (Stanley and Wise 1979). They give a brief thematic analysis of the content of obscene calls that they received while running a lesbian feminist help line, but do not return to them either to develop the content analysis or to consider factors that might be responsible for the unrelenting hostility of those responsible for the calls. In short, they do not "go beyond" their own experience of the calls; they present, in general terms, the conception of hostile patriarchy that they formed as a result of suffering the harassment of these calls over an extended period.

In this second model of what it is to do research as a feminist, then, the commitment to respect the integrity of "the personal"—meaning each individual's "subjectivity" or "reality"—takes precedence over all other concerns. In the process, it displaces both the ideal of conducting research as an egalitarian, collective effort at consciousness raising, and the objective of achieving any general, explanatory understanding of women's situation. Followed through consistently as proposed by Stanley and Wise, these principles dramatically narrow the scope of feminist research—it becomes, in their hands, virtually autobiographical—hence, my reference to this approach as one of "self-study," rather than of collective study.[15]

"Back Into" Consciousness Raising

Faced with this divergence of thinking about what it is to do research (specifically, social scientific research) "as a feminist," one response might be to encourage all the various alternatives and see what they contribute. This is attractive, especially given that each may have different ranges of application and given the diversity of problems that concern feminists. Certainly it is important to feminists to understand something about rapists and obscene phone callers, as well as about women's lives (their own and those of others), and these subjects pose rather different methodological problems. But beyond complementary differences in approach, there are some quite deep problems inherent in the way interpretations of core feminist commitments have diverged, even when the same commitments are recognized as the source of inspiration and direction for feminist research.

The difficulty with the self-study model is, most simply, that it endorses a relativism that compromises the emancipatory potential of feminism and feminist research. As Acker, Barry, and Esseveld argue, "When all accounts are equally valid, the search for 'how it actually works' becomes meaningless" (Acker, Barry, and Esseveld 1983, 429), and yet the search for just this sort of understanding has been a central goal of feminism both in its critical dimension, when it means to oppose dominant theories that deny or trivialize women's experience of oppression, and in its constructive dimension, when it seeks an understanding of how patriarchal structures work as a necessary basis for effective political action. Ironically, even Stanley and Wise claim, at one point, that they do not mean to deny "the existence of an 'objective reality,'" adding that "social facts" are "as real and constraining as tables and chairs" (Stanley and Wise 1983b, 111). Presumably they would agree, given this, that the emancipatory potential of feminism depends on establishing just exactly what these real and constraining "facts" come to. It is perhaps significant, in this connection, that Stanley and Wise do allow, in the case where the battered woman's "objective reality" seemed to belie her experiential tesimony, that it may sometimes be appropriate to ask "how and why people construct realities in the way that they do" (Stanley and Wise 1983b, 112). This is the only point at which they make this proposal, not surprisingly given that it suggests that "reality disjunctures" may require one to "go beyond" the experience in question and inquire into conditions that shape it of which the subject herself is not conscious. But it is a significant break with their uncompromising position on the integrity of experience. It suggests a perfectly good sense in which you can discover (or conclude) that a subjects' assessment of her own experience *is* "wrong," namely, when this assessment is internally incoherent or inconsistent with the experience that she herself reports. And it suggests an appreciation that the possibility of critically rethinking your experience is an essential part of any emancipatory program of research or action.

Where the practice of consciousness raising is cited on all sides as a source of guidelines for research practice, it is perhaps relevant to ask what veracity is accorded experience in these context. It is certainly the case that the widely cited "Guide to Consciousness Raising" published in *Ms.* (1972), identifies as the "most important [rule]" that of "never challenging another woman's experience": "in describing personal feelings and experience, there is no right or wrong," therefore it is never appropriate to "give advice or judge a woman's testimony" (*Ms.* 1972, 22); "what a sister says may seem inaccurate to you but it is true for her at that moment" (23). Participants are, moreover, encouraged to "speak personally, subjectively and specifically"; generalizations and abstractions inevitably "misrepresent" precisely the details of diverse personal experience that consciousness raising means to bring into view (22). Nevertheless, the purpose of giving testimony in this way is described as that of learning "who we are," of discerning "common elements in our experience" and determining how these relate to the role and status of women "as a group" (23). To this end, it is recommended that consciousness-raising sessions conclude with a "summing up" discussion in which commonalities are noted and further questions raised, and, in this context, critical reflection seems crucial. The "Guide" suggests, for example, that questions be raised about language use as a "key," one assumes to assumptions about the subordinate status of women that are so pervasive they shape even the reflective discourse of women engaged in consciousness raising (23).

Accounts of consciousness-raising practice in succeeding years return again and again to these latter, often proscribed, critical, and generalizing dimensions of the process. A *Ms.* editor, reporting on her experiences the next year, describes her consciousness-raising group as "tough and candid" and seems to consider these attributes the crucial catalyst for changes in her self-understanding and political appreciation of the status of women (Pogrebin 1973, 104). Indeed, the process of *raising* one's consciousness is routinely described as transformative. It is a matter of coming to consciousness of things "not formerly perceive[d]" (Cassell 1977, 16), of recognizing the extent to which experiences thought to be idiosyncratic are shared by women, as women (33), of developing an awareness that this "larger group is degraded and oppressed" (50), and, finally, of coming to a critical understanding of how patriarchal institutions—the family, marriage—systematically oppress women. In short, consciousness raising is a process by which women come to see that the personal is political and thereby "reconsider ... their political, economic, and social position in relation to that of men" (Jenkins and Kramer 1978, 72). It is a process by which they redefine themselves.[16]

In all of this, experience and personal testimony is to be respected. Often it is observed that consciousness-raising groups were, for many women, the first and perhaps the only context in which they could articulate their perceptions without being corrected, spoken for, or silenced. But it does not follow from this that such testimony must be treated as inviolate, as an "unconditional" datum. Even in the *Ms.* "Guide," the assertion that a woman's account of her own experience is to be treated as "true for her at that moment" is followed by the observation that if she is to discover it wrong, she must do this herself (*Ms.* 1972, 23). The emphasis is on *change*, on critical reassessment of experience and of the standard construals of experience that shape it and inform social action. Most important, the process of change in question is not one that can be undertaken in isolation; it is crucially a matter of recovering the *socio-political* dimension of experience—of learning in what sense you are a (gendered) social being—and this can only be accomplished collectively. Indeed, it is properly a collective accomplishment. The reason you "speak your experience" *to and with others* is precisely so that, with their critical input and their experience as a basis for comparison, you can begin to see where your "constructions of reality" are suspect, what you may have forgotten or suppressed, and where you have internalized general schemas for interpreting experience that are neither consistent with your own experience nor with that of others on whom you had projected it. Taken as a whole, the process is deeply and essentially critical—not just self-critical but mutually critical—and it is centrally one of "going beyond" your own experience. It is a process of learning to "make 'she' speak for more than 'herself'" that certainly takes as its point of departure a commitment to "first learn[ing] to speak *for and of herself*" (Brunsdon 1978, 28),[17] but by no means remains at the level of speaking experience.

The Place of Criticism

The limitations of consciousness raising that leaves experience where it finds it in direct testimony are notorious: "All the sympathy and empathy and sexual solidarity in the world cannot together substitute for a clear-headed [ideological] understanding of the causes of oppression and the psychological reflex within ourselves" (Brunsdon 1978, 73). Strikingly similar concerns have been raised by Geiger and Zita about feminist pedagogy when it is seen to encourage personal disclosure without critical analysis: "The group begins to uncritically validate each and every reaction ... [often] leading to the vacuous belief that personal descriptive truth to which the individual allegedly has privileged epistemic access is all that needs to be said" (Geiger and Zita 1985, 112–13). The difficulty here is not just that the quality of classroom dialogue is compromised but that, in fact, this sort of practice systematically obscures the extent to which experience is itself concretely situated, and the speaking of it a socially constructed event: "Well-intentioned liberal acceptance of 'every woman's point of view' [results in] the responses [being] perceived as idiosyncrasies of the individual, as testimonials pure and simple, rather than as particular points of view that reflect specific social positions" (Geiger and Zita 1985, 144). Although they are not specifically concerned with feminist research practice, I suggest that Geiger and Zita address the same destructive paradox in feminist principles as I have found at the heart of the self-study model of feminist research. The commitment to comprehend/disclose the *political* nature of the personal is defeated by an impulse to support and validate formerly suppressed points of view at all costs. Experience is merely revalued rather than reevaluated.

I propose that it is as important in feminist research as in feminist teaching to cultivate what Geiger and Zita describe as "a willingness to practice thoughtful self-criticism and self-censorship and to accept criticism as well as support from others" (Geiger and Zita 1985, 117). Without the development of critical discourse and critical inquiry, the potential for transformative change is irretrievably lost, both for individuals and for women as a collectivity. In this spirit, I argue that feminist research should, indeed, take consciousness raising as a model but should draw inspiration specifically from the probing, political analyses women have developed together in these contexts. The most compelling reason to embrace the collectivist model of feminist research is that it builds on these (often suppressed) aspects of consciousness raising; it urges us to "go beyond experience" in a search for explanatory understanding of its conditions and genesis that requires precisely the forms of critical (collective) inquiry that have proven essential to the emancipatory potential of feminist practice in a wide range of contexts.

I would add two further considerations that tell in favor of the collectivist model. The first is that, where feminist theory has been centrally concerned to bring into view the socially constructed nature of our identities and consciousness, it is at least incongruous, if not inconsistent, to insist that feminist researchers should treat experience as an "unconditional datum." The central impetus for and paradox of feminist inquiry is awareness that

> feelings are ... both access to truth ... and an artifact of politics. There is both suspicion of feelings and an affirmation of their health. They become simultaneously an inner expression of outer lies and a less contaminated resource for verification....
>
> Taking situated feelings and common detail (common here meaning both ordinary and shared) as the matter of political analysis, it explores the terrain that is most damaged, most contaminated, yet therefore most women's own, most intimately known, most open to reclamation. (MacKinnon 1982, 21, 22)

Consequently, the integrity and efficacy of "personal" experience should stand as a primary object of critical investigation, not as a presupposition of inquiry. The second consideration is that, where a degree of objectification of ourselves—of going beyond our own experience—seems an essential condition of *self-understanding*, there is no good reason to rule it out a priori where others are concerned on grounds that it is an inappropriate way of treating intentional "subjects."

Given this, I conclude that the most pressing task for feminist research is to develop the analytic tools necessary for investigating the relationship between personal experience and the context of experience, structured as it is by factors of race, class, age, ethnicity, nationality, and sexual orientation, as well as by gender. The process of inquiry, in which these conceptual tools are hammered out, is necessarily dialectical, one of "political hermeneutics," to borrow MacKinnon's term, given that, where feminist research is concerned, we have no option but to seek an understanding of "women's condition from within the perspective of [our] experience, not outside it" (MacKinnon 1982, 22).[18] The crucial task will be to give an account of criticism that differentiates the forms necessary for feminist inquiry—forms that are constructive, respectful of and responsive to women's experience—from the oppressive, and destructive forms all too common in dominant disciplinary traditions of research.

Notes

1. Stanley and Wise object that "the idea that there is only 'one road' to the feminist revolution, only one type of 'truly feminist' research, is as limiting and as offensive as male-biased accounts of research that have gone before" (Stanley and Wise 1983b, 26). And McCormack argues that methodology should be seen as a means, not an end, something that should be responsive to change in the content of science and that need not (indeed, probably cannot) be determined in advance of inquiry (McCormack 1987, 20).
2. From the outset feminist theorists have argued that understanding women's experience and oppression as women is a matter of coming to see that the "*personal* is political." Brunsdon insists, in this connection, that "the essence of feminism" is its "revaluation of women's experience" in just this sense (1978, 24). Likewise, Stanley and Wise maintain that the "exploration and analysis of consciousness is the key to everything else about feminism" (1983b, 149), while MacKinnon insists that "feminism is the theory of women's point of view," and that "consciousness raising," conceived as a process of recovering and articulating this point of view through the collective "speaking of experience," is its "quintessential expression" (1982, 21). It is in this spirit that Keohane and Gelpi advocate feminist research that is conceived as an extension of the "experiential thinking" they find central to consciousness raising (1982, xi).
3. The collectivist model might alternatively be described as a communalistic model of research, following Carlson's distinction between communal and agentic research (Bernard 1973; Eichler 1987; Reuben 1978).
4. Frequently it is concern with the inadequacies (specifically, the gender bias) of standard quantitative instruments and methodologies that gives rise to discussions about the ways in which researchers' presuppositions may systematically distort (or, indeed, simply exclude) the experience and perceptions of women. Those who do not want to see feminist researchers abandon quantitative methodologies altogether recommend that they be supplemented by qualitative approaches on the principle that "we need information from women, about women's lives and work, before we can adequately generate the framework [i.e., define variables, frame questions] that will allow us to carry out survey research that fully includes women" (Jacobson 1977, 6). In analysis of the limitations of Canadian census data, Armstrong and Armstrong argue that qualitative approaches can serve both as a check on the theoretical assumptions that inform quantitative research design—a way of determining whether the range of census options for reporting household organization or employment status capture the full range of women's activities and roles—and as a means of filling in the details of "the broad outline" provided by statistical data (Armstrong and Armstrong 1978, 58, 76). Oakley and Oakley (1979) make similar points about the British census.

5. In fact, this is not so "minimal" a requirement, as Stacey describes in reflection on her misgivings about ethnographic methods. Although these are widely seen as more compatible with feminist commitments than less engaged (more "remote") methodologies, she comes to the conclusion that, in fact, "the appearance of greater respect for and equality with research subjects ... masks a deeper, more dangerous form of exploitation" (Stacey 1988, 22).
6. Smith argues that "an alternative sociology [one appropriate to feminism] must be reflexive.... [It is] one that preserves in it the presence, concerns, and experience of the sociologist as knower and discoverer" (Smith 1987, 92). Likewise, Stanley and Wise insist that unless research analysis and reports include an account/consideration of the research process itself, "the sources of the researcher's knowledge are hidden from scrutiny" (1983a, 195).
7. See, however, Stacey's discussion of Oakley's recommendations for further thoughts on the complexities and dangers inherent in these sorts of intimate and reciprocal research relationships (Stacey 1988, 22).
8. Acker, Barry, and Esseveld note that, in fact, their research subjects would now allow them to abdicate this role: "They were hesitant about being negative but were clearly critical [of our initial efforts]. What they wanted, they said, was more of our own sociological analysis. They wanted us, the researchers, to interpret their experience to them" (1983, 429–30).
9. Smith adds that "it is not possible to account for one's directly experienced world or how it is related to the worlds which others directly experience who are differently placed by remaining within the boundaries of the former" (1987, 94).
10. Stanley and Wise conclude, on the basis of the discussion elaborating these points, that all the institutions through which women are oppressed "can best be examined and understood through an exploration of relationships and experiences within everyday life" (1983b, 54, 58).
11. An earlier statement of this principle is particularly clear: "Feminism directly confronts the idea that one person or set of people have the right to impose definitions of reality on others" (Stanley and Wise 1979, 373).
12. Stanley and Wise conclude this statement with the intriguing qualification that interference or imposition is unjustifiable "when they [the subjects] don't want us to" (1983b, 113), suggesting that one might be justified in imposing their "reality" on another if given some indication that this other did want them to do this. What would this amount to in the case of an individual whose oppression so undermines her sense of self she is inclined to accept or invite the imposition of any strong will and associated definition of reality?
13. They reject "positivist" methodologies that require disengagement (Stanley and Wise 1983b, 157), and insist that "there is no way we can avoid deriving theoretical constructs from experience," in the sense that the process of research is centrally one in which the researchers' theoretical perceptions are changed as their experiences in research contexts unfold (1983b, 161). They go on to say that it is the "presence of the personal *within research experience*" that must be reclaimed on feminist principles, and the associated discussion suggests that it is specifically the experiences of the researcher that are at issue (1983b, 158).
14. Trebilcot elaborates: "I use 'withstanding' to mean both that I am standing with wimmin and that I am withstanding patriarchy" (1988, 4).
15. In a sense, this introspective focus may also broaden the scope of feminist research. Where the researcher is under no compunction to involve subjects in the research process, it is possible to consider any group of people, activities, or context, as they impinge on the researcher and her experience, without compromising the feminist character of the research (as in the case of Stanley and Wise's study of obscene telephone calls). Stanley and Wise are, in fact, explicit on the point that the reason they consider the researcher under no constraint to involve subjects in the design, direction, or interpretation of her research is because they "reject a feminist research which is concerned with women" (1983b, 170).
16. In a retrospective assessment, MacKinnon argues that the slogan "the personal is political" should not be treated as a metaphor or analogy; it is a literal, descriptive claim that to know the politics of women's situation is "to know their personal lives" and that the "authentic politics of women's personal lives" as revealed to themselves and others in the process of giving testimony, is "powerlessness to men" (1982, 21).
17. In this quotation, Brunsdon is alluding to Rowbotham's observation that "the present inability of 'she' to speak for more than herself is a representation of reality," which she cited a few pages earlier (Brunsdon 1978, 26).

18. MacKinnon's full statement is: "Through consciousness raising, women grasp the collective reality of women's condition from within the perspective of that experience, not outside it" (1982, 22). She goes on to say that this is a process that aims at "the collective reconstitution of the meaning of women's social experience as women live through it.... Its method stands inside its own determinations to uncover them, just as it criticizes them in order to value them in its own terms, in order to have its own terms at all" (29).

References

Acker, Joan, and Kate Barry. 1984. "Comments on MacKinnon's 'Feminism, Marxism, Method, and the State.'" *Signs* 10, no. 1: 175–79.

Acker, Joan, Kate Barry, and Joke Esseveld. 1983. "Objectivity and Truth: Problems in Doing Feminist Research." *Women's Studies International Forum* 6, no. 4: 423–35.

Armstrong, Pat, and Hugh Armstrong. 1987. "Beyond Numbers: Problems with Quantitative Data." In *Women and Men: Interdisciplinary Readings on Gender*, ed. Greta Hofmann Nemiroff, 54–79. Montreal: Fitzhenry and Whiteside.

Bernard, Jessie. 1973. "My Four Revolutions: An Autobiographical History of the ASA." In *Changing Women in a Changing Society*, ed. Joan Huber, 11–39. Chicago: University of Chicago Press.

Brunsdon, Charlotte. 1978. "'It is Well Known That by Nature Women are Inclined to Be Rather Personal.'" In *Women Take Issue: Aspects of Women's Subordination*, ed. Women's Studies Group, Center for Cultural Studies, University of Birmingham, 18–34. London: Hutchinson.

Carlson, Rae. 1972. "Understanding Women: Implications for Personality Theory and Research." *Journal of Social Issues 28*, no. 2: 17–32.

Cassell, Joan. 1977. *A Group Called Women: Sisterhood and Symbolism in the Feminist Movement.* New York: David McKay.

Eichler, Margrit. 1987. "The Relationship between Sexist, Nonsexist, Woman-centered and Feminist Research in the Social Sciences." In *Women and Men: Interdisciplinary Readings on Gender*, ed. G. Hofmann Nemiroff, 21–53. Montreal: Fitzhenry and Whiteside.

Eichler, Margrit. 1988. *Nonsexist Research Methods: A Practical Guide.* Boston: Allen and Unwin.

Eichler, Margrit, and Jeanne Lapointe. 1985. *On the Treatment of the Sexes in Research.* Ottawa: Social Sciences and Humanities Research Council of Canada.

Geiger, Susan, and Jacqueline N. Zita. 1985. "White Traders: The Caveat Emptor of Women's Studies." *Journal of Thought* 20: 106–21.

Grady, Kathleen E. 1981. "Sex Bias in Research Design." *Psychology of Women Quarterly* 5, no. 4: 628–36.

Grant, Judith. 1987. "I Feel Therefore I Am: A Critique of Female Experience as the Basis for a Feminist Epistemology." *Women and Politics* 7, no. 3: 99–114.

Harding, Sandra. 1986. *The Science Question in Feminism.* Ithaca, N.Y.: Cornell University Press.

Harding, Sandra. 1987. "The Method Question." *Hypatia* 2, 3: 19–36.

Jacobson, Helga E. 1977. *How to Study Your Own Community: Research from the Perspective of Women.* Vancouver: Vancouver Women's Research Center.

Jenkins, Lee, and Cheris Kramer. 1978. "Small Group Process: Learning from Women." *Women's Studies International Quarterly* 1: 67–84.

Kelly, Alison. 1978. "Feminism and Research." *Women's Studies International Quarterly* 1: 225–32.

Keohane, Nannerl O., and Barbara C. Gelpi. 1982. "Foreword." In *Feminist Theory: A Critique of Ideology*, ed. N. O. Keohane, M. Rosaldo, and B. C. Gelpi, vii–xii. Chicago: University of Chicago Press.

Longino, Helen. 1987. "Can There Be a Feminist Science?" *Hypatia* 2, no. 3: 51–64.

McCormack, Thelma. 1987. "Feminism and the New Crisis in Methodology." Paper presented at the conference "The Effects of Feminist Approaches on Research Methodologies," University of Calgary.

MacKinnon, Catharine A. 1982. "Feminism, Marxism, Method, and the State: An Agenda for Theory." In *Feminist Theory: A Critique of Ideology*, ed. N. O. Keohane, M. Rosaldo, and B. C. Gelpi, 1–30. Chicago: University of Chicago Press.

Ms. 1972. Editorial. "Women's Body, Women's Mind: A Guide to Consciousness Raising." *Ms.* 1 (July): 18–23.

The Nebraska Feminist Collective. 1983. "A Feminist Ethic for Social Science Research." *Women's Studies International Forum* 6, no. 5: 535–43.

Oakley, Ann. 1981. "Interviewing Women: A Contradiction in Terms." In *Doing Feminist Research*, ed. Helen Roberts, 30–61. London: Routledge and Kegan Paul.

Oakley, Ann, and Robin Oakley. 1979. "Sexism in Official Statistics." In *Demystifying Social Statistics,* ed. John Irvine, Ian Miles, and Jeff Evans, 172–89. London: Pluto Press.

O'Brien, Mary. 1981. *The Politics of Reproduction.* Boston: Routledge and Kegan Paul.

Pogrebin, Letty Cotlin. 1973. "Rap Groups: The Feminist Connection." *Ms.* 1 (March): 80–83, 98–100.

Reuben, Elaine. 1978. "In Defiance of the Evidence: Notes on Feminist Scholarship." *Women's Studies International Quarterly* 1: 215–18.

Smith, Dorothy E. 1987. "A Women's Perspective as a Radical Critique of Sociology." In *Feminism and Methodology: Social Science Issues,* ed. Sandra Harding, 84–96. Bloomington: Indiana University Press.

Stacey, Judity. 1988. "Can There Be a Feminist Ethnography?" *Women's Studies International Forum* 11: 21–27.

Stanley, L., and S. Wise. 1979. "Feminist Research, Feminist Consciousness, and Experiences of Sexism." *Women's Studies International Quarterly* 2: 359–74.

Stanley, L., and S. Wise. 1983a. "'Back into the Personal' or: Our Attempt to Construct 'Feminist Research.'" In *Theories of Women's Studies,* ed. Gloria Bowles and Renate Duelli Klein, 192–209. London: Routledge and Kegan Paul.

Stanley, L., and S. Wise. 1983b. *Breaking Out: Feminist Consciousness and Feminist Research.* London: Routledge and Kegan Paul.

Trebilcot, Joyce. 1988. "Dyke Methods." *Hypatia* 3, no. 2: 1–14.

Vickers, Jill McCalla. 1982. "Memoirs of an Ontological Exile: The Methodological Rebellions of Feminist Research." In *Feminism in Canada,* ed. A. Miles and G. Finn, 27–46. Montreal: Blackrose Books.

Chapter 40

A Method for Critical Research

Donald E. Comstock

Take the ideas of the masses (scattered and unsystematic ideas) and concentrate them (through study turn them into concentrated and systematic ideas), then go to the masses and propagate and explain these ideas until the masses embrace them as their own, hold fast to them and translate them into action, and test the correctness of these ideas in such action. Then once again concentrate ideas from the masses and once again go to the masses.... And so on, over and over again in an endless spiral, with the ideas becoming more correct, more vital and richer each time. Such is the Marxist theory of knowledge.—Mao Tse-tung[1]

In the last decade there has been a remarkable number of books and articles intended to familiarize American sociologists with the critical theories of society associated with the Frankfurt Institute for Social Research. However, partly because of considerable ambivalence by the members of the "Frankfurt school" toward empirical research, and partly because, since the 1930s, they have been isolated from political practice, limited attention has been given to research methods appropriate to a critical social science.[2] In this chapter I outline such a method with the hope of stimulating a reconsideration of research activities which are appropriate to a social science of *praxis*. Following a brief discussion of the need for such a method, I will point out several distinctions between positive social science and critical social science which are relevant to research methods. In the final section of the chapter I will present a sequence of steps which outline a critical method of research. In this context "method" is not to be confused with specific research techniques involved in data collection and analysis. Rather, I use the term to refer to the general procedure by which we go about studying society, including selecting research problems, constructing and evaluating theories, and disseminating our findings.

The Necessity for a Critical Method of Research

The central argument of this chapter is that the development of critical theories of contemporary social institutions requires a critical research method. We cannot simply apply the investigative logic developed by the positive social sciences to new topics and expect to develop a truly critical social science. Moreover, most critical theory, and neo-Marxist analysis in general, is presently far removed from the people and class it purports to enlighten and is of very little use to those engaged in concrete struggles for progressive change. Despite the best intentions of its practitioners radical theory remains entrapped in the academic practice of theory building. The questions it addresses are theoretical questions and the answers it supplies are meant to advance theoretical knowledge, not political practice. This is because both mainstream social science and radical analysis is practiced almost entirely in academic settings and because the first has consciously and the second has unconsciously adopted the epistemology of positive science.[3]

The function of a critical social science is to increase the awareness of social actors of the contradictory conditions of action which are distorted or hidden by everyday understandings. It is founded on the principle that all men and women are potentially active agents in the construction of their social world and their personal lives: that they can be the subjects, rather than the objects, of sociohistorical processes. Its aim is self-conscious practice which liberates humans from ideologically frozen conceptions of the actual and the possible. The method of research appropriate to this endeavor cannot be adopted from positive social or natural sciences. The method of positive social science reflects the empiricist assumption that society is a neutral datum for a systematic observation. This is both a reflection of, and a contribution to, the reification of monopoly and state capitalist societies. This method objectifies the human subjects of an investigation by treating their behavior as raw data which is external to the scientific enterprise. The positive research method reifies social processes by naturalizing social phenomena, addressing them as eternal to our understandings, and denying their sociohistorical constructedness. The consequence is to reinforce the alienaton of the subjects of social science research from their social, political, and economic institutions.

A consistent critical method which treats society as a human construction and people as the active subjects of that construction would be based on a *dialogue* with its subjects rather than the observation or experimental manipulation of people. A critical social science must directly contribute to the revitalization of moral discourse and revolutionary action by engaging its subjects in a process of active self-understanding and collective self-formation. In this way, science becomes a method for self-conscious action rather than an ideology for the technocratic domination of a passive populace.

Some Differences between Critical Social Science and Positive Social Science

The roots of scientific thought lie in the Enlightenment belief that genuine knowledge is the most effective means for the liberation of humans. Yet modern positive science has excluded, on epistemological and methodological grounds, any such interest. Instead, science is viewed as a neutral means equally utilizable for liberation or oppression. The specific use to which knowledge will be put is excluded from the province of scientific understanding itself. The self-image of contemporary social science is dominated by the epistemology and methodology of the natural sciences which assumes a fundamental separation of the knower from the known, of the subject from the object.

Critical theory is the historical contradiction to this scientistic self-image of modern society. Critical theory expressly denies the view that humans and society can be understood via the assumptions and methods of the natural sciences that deny the self-creative character of human thought and action. A brief comparison of positive and critical social sciences will cover four points of divergence: (1) their different images of society and human nature, (2) the constitution of our knowledge of social processes, (3) the form that scientific explanations of social processes must take, and (4) the role of the social scientist.[4]

Images of Society and Human Nature
Positive social science sees society as an objective phenomenon which can be described as a set of ahistoric forces which constrain human behavior. Critical social science, on the other hand, views society as a human construction which is altered through people's progressive understanding of historically specific processes and structures. Positive social science takes human nature as a constant datum while the critical perspective argues that humans change themselves by reconstituting their society. By attributing a "thing-like" quality to society, positive social sciences endow social processes with what Barry Smart calls pseudo-concreteness.[5] Reflecting the commonsense objectifications of the alienated individual in capitalist society, positive social science fails to analyze society as a human construction. This approach may have predictive validity in the short run so long as the reificatory power of ideology remains unchallenged. However, since it fails to see the social processes and structures it studies as historically specific, positive social science can neither account for nor serve as a guide for fundamental social change.[6]

In contrast, a critical social science, which self-consciously participates in the reconstruction of social structures and collective meanings, is an immediately historical project. Rather than an exclusive concern with naive descriptions of what exists, critical social science directs attention to the *possibilities* immanent in the historical development of social processes created by human understanding and action. As Horkheimer wrote, "the critical theory of society ... has for its objective men as producers of their own historical way of life in its totality."[7] This quite different view of society and human nature has profound implications for the constitution of social science knowledge.

The Constitution of Social Science Knowledge
For a positive social science, knowledge is constituted by the theoretical ordering of empirical observations of an objective reality. The data are descriptions of social behaviors and the subjective beliefs, attitudes, and values of individuals.[8] These are assumed to exhaust the field of inquiry and to provide an external or objective datum for testing theoretical models. Positive social science attempts to model parts of this reality through the use of value-neutral concepts. According to Habermas, "in this way the naive idea that knowledge describes reality itself becomes prevalent."[9] Objectivism and the value neutrality of concepts become guiding principles of positive social science such that facts are sharply distinguished from values and theory from practice. Although unacknowledged, the intention that has justified and legitimated positive science has been the prediction and instrumental control of nature.[10] Thus, as Fay shows, when this interest is applied to the study of humans, the result is the manipulation of social relationships, the enthronement of technical means over moral ends, the impoverishment of political discussion, and the support of dominant classes.[11]

From the perspective of a critical social science, knowledge may fulfill two human interests, in addition to technical control. Since human life is social life and is governed by normative principles and moral precepts, a second interest consists of the practical requirement for intersubjective understanding and agreement on norms, values, and meanings. The social sciences are therefore ineluctably hermeneutic: they rest on an interpretation of the intersubjective meanings and practices which order everyday life. In turn, they contribute to the maintenance or change of those meanings and practices. For this reason, it becomes necessary to recognize a third human interest and a third type of knowledge: that which frees participants from outmoded and reified conceptions of reality. This emancipatory interest is the basis for a critique of ideology and for knowledge that informs fundamental social change.

A critical social science refuses to accept current social practices as the final context of validation. Both meanings and actions at any historical moment are transitory manifestations of changing social structures. A critical social science must analyze these in the light of their arrested and denied possibilities.[12] Under conditions of domination, actors' understandings are historically frozen by ideologies which legitimate and attempt to perpetuate existing relations of power. Yet social action continues to change the social structure in ways that contradict these meanings. Possibilities for different understandings and actions remain hidden from view only so long as the powerful and their agents can maintain distorting ideas.

The intention which constitutes a critical social science is the emancipation of its subjects from ideologically frozen conceptions and the development of self-sustaining processes of enlightenment and self-conscious political action. Critical knowledge is never neutral; it is always for some particular subject. The validity of its concepts, data, and theory is related to the historical aims and purposes of that subject.[13] Thus instead of simply describing a particular historical formation, a critical social science attempts to elucidate the possible courses of action and meanings which are immanent in the development of the present order. To do this it must combine a theory of structural change with a critique of its subjects' ideologically distorted understandings. In this way a critical understanding of society mediates theory and practice.

The Form of Scientific Knowledge

Positive social science assumes that a single form of explanation is appropriate to all sciences. The deductive nomological or covering-law is paradigmatic, even if rarely achieved in the social sciences. The heart of this form of explanation lies in the appeal to ahistoric laws or nomological statements of deterministic or probablistic relations between phenomena. Explanation is symmetrical with prediction: the deduction of an event post factum from laws and empirical conditions is identical in form to the prediction of future events. The validity of these laws rests on their capacity to conditionally predict future events. All positive scientific activities—collecting and refining data, discovering correlations, and formulating empirical generalizations, hypotheses, and models—are directed toward the growth of testable and well-confirmed laws which are ordered by logically coherent theories.

A critical social science challenges both the possibility of this model of explanation and the advisability of its function as a paradigm for the social sciences. While positive social science studies human *behavior*, critical social science studies human *action* and seeks to make manifest the processes by which social structures are constructed by human action and ordered by intersubjective meanings. Critical accounts relate social conditions to the subjects' actions, not directly or mechanically, but as they are interpreted and ordered by their understandings and motives. Since all human actions are consequences of socially interpreted sets of conditions, we cannot predict behavior directly from social conditions. Instead, critical explanations must recognize the mediation of meanings by which members make sense of their own and others' acts. Once we assume that social processes are constituted through meaningful action, the prediction of social action is possible only so long as constitutive meanings do not change. Since humans are self-defining agents who reflect on and interpret their actions, and since all fundamental social change is the product of innovation in both meanings and social practices, future conditions and regularities are not predetermined. Moreover, as Taylor notes, even the categories and concepts we use today are likely to be superseded in the future.[14] (Curiously, positivists assume this in the development of their own scientific disciplines but exclude it from their considerations on other social processes.)

If all social processes must be understood as the product of meaningful human actions, then all critical accounts must begin from the intersubjective meanings, values, and motives of historically specific groups of actors. Interpretive accounts of action, in contrast to causal explanations, take the form of practical inferences whereby acts are deduced from a knowledge of the actors' intentions and conceptions of what must be done to accomplish those intentions.[15] Moreover, these meanings, values, and motives must be related to social processes in a way which clearly demonstrates how they have been constructed by human action and reflection.

Critical explanations include a theory of basic structural change and an account of the genesis of meanings, values, and motives which arise in consequence of structural changes. They must also reveal the contradictory consequences in social structure which result from acting in accordance with dominant meanings—consequences which render actors' intentions unachievable in the context of changed circumstances. For example, the idea of social mobility may have been supported by the personal experiences of a visible and vocal minority during the early development of capitalist social relations—especially in the United States. However, since that time continued interpersonal competition has solidified the position of an elite and severely reduced the possibility of significant personal mobility. A critical social science would endeavor to show both how the intention of individual mobility is unrealizable under present social conditions and how the ideology of mobility serves to reinforce the existing inequalities of opportunity and achievement which are experienced by actors as a frustration of personal values.

Critical explanations explicate the historical dialectic between subjective and objective factors whereby historically preformed social structures and processes set the conditions for human action and these conditions are, in turn, negated by social actions. Actors interpret social and natural conditions, they act to achieve certain intersubjectively valid intentions, and they thereby alter the conditions of action. Under conditions of ideologically frozen understandings—the tendency in any hierarchical social formation—the objective conditions of action come to contradict the intersubjective meanings attributed to them. In this way, the unanticipated consequences of action bring about changes in social conditions which render ideologies less and less able to account for actors' experiences.

Radical changes occur because of crises that result from emerging contradictions in social processes. These are revealed in the divergence of social conditions from ideologies. A fundamental contradiction will exist when the changed social conditions give rise to contradictory action imperatives which cannot be integrated and legitimated by the old ideology (for example, state regulation and protection of monopolies versus laissez-faire competition). Eventually a crisis threatens when actors experience these changes as both critical for their continued material existence and threatening to their social identity.[16] One or more new meaning systems arise and contend with the old and some of these new meanings will express progressive possibilities immanent in the contradictory social conditions. A group or class that is dominated and frustrated by current conditions will organize for political struggle and social change.

e outcome of these political struggles depends upon the degree to which social conditions have changed as well as the relative effectiveness of progressive interests vis-à-vis the old power holders. If fundamental contradictions have not yet matured, adjustments in the ideology or power structure may suffice to temporize dissatisfactions while competing leaders are co-opted, isolated, or expelled. At some point a crisis arises which cannot be resolved through adjustment and co-optation and a radical change must occur in both the social structure and the system of meanings. The timing of such radical changes cannot be predicted from knowledge of the contradictions immanent in a social structure because human action is not the mechanical result of natural forces but the product of conscious choice based on varying degrees of social awareness by the participants. Thus, the aim of a critical theory is not to predict social change, but to explicate the immanent tendencies in the historical development of a social formation so that the participants may create social change. This implies a radically different role for the social scientist than the one idealized by positive social science.

The Role of the Social Scientist

Horkheimer noted that traditional, that is, positive and interpretive, social sciences, in an attempt to deny the political aspects of knowledge, separate the role of the social scientist *qua* scientist from her or his role as a political actor.[17] In a contemporary manifestation of Marx's distinction between *bourgeois* and *citoyen*, the scientist is divided into two beings: a nonpolitical, value-free observer and theorizer on the one hand, and a political person who expresses values and interests on the other. The positivist injunction is to always keep these roles separate: to cultivate a disinterested attitude when investigating social phenomena and only as a private citizen to bring one's tentative knowledge to bear on vital issues of the day. According to this, the task of the social scientist is to describe and explain the facts, not to make prescriptive statements about what ought to be.[18]

Critical social science recognizes that social scientists are participants in the sociohistorical develpment of human action and understanding. As such, they must decide the interests they will serve. Since all knowledge is rooted in social practices and social practices are informed and ordered, in part, by social science explanations, theory and action are inseparable and all facts and theories are warranted as valid from a particular framework of social practices. The only legitimate activity of a critical social scientist is to engage in the collective enterprise of progressive enlightenment with the aim of showing how his or her accounts are valid in the light of the subjects' oppressive social position and the specific values and actions possible in that position.

In summary, a critical social science sees society as humanly constructed and, in turn, human nature as a collective self-construction. This emphasizes the historicity of social structure, processes, and meanings and directs attention to the possible world immanent in present formations. Social science knowledge is, like all reflexive understanding of social practice, constituted by the understandings, values, and goals of particular individuals, groups, or classes. Critical social science is distinguished by its interest in the emancipation of those groups and classes that are presently dominated. This makes it a science of *praxis* in which action serves as both the source and the validation of its theories. It rejects the criterion of prediction as reifying social practices and alienating humans from their social products. In aim and method it must be self-consciously dealienating in order to enable humans, as subjects, to reappropriate the world they have constructed. The form of its explanations is historical. It traces the dialectic between preformed structural conditions, human understandings, and social action while endeavoring to reveal how ideological distortions have masked contradictions between intentions and structural possibilities. It tries to show its subjects how they can emancipate themselves by conceiving and acting upon the social order in new ways.

The criterion for the truth of critical theorems is the response of the theory's subjects. Because it begins from the meanings its subjects attribute to social processes and attempts to rectify ideologically distorted meanings and values, its method of investigation and validation is based on *dialogue* with its subjects. It does not simply observe humans as objects but endeavors to *engage* them in self-conscious action. In the next section I will outline a method of research which rests on the paradigm of dialogue and participation rather than observation and manipulation.

The Method of Critical Research

Critical social research begins from the life problems of definite and particular social agents who may be individuals, groups, or classes that are oppressed by and alienated from social processes they maintain or create but do not control. Beginning from the practical problems of everyday existence it returns to that life with the aim of enlightening its subjects about unrecognized social constraints and possible courses of action by which they may liberate themselves. Its aim is enlightened self-knowledge and effective political action. Its method is dialogue, and its effect is to heighten its subjects' self-awareness of their collective potential as the active agents of history. Practically, this requires the critical investigator to begin from the intersubjective understandings of the participants of a social setting and to return to these participants with a program of education and action designed to change their understandings and their social conditions. Analytically, critical research must, first, provide an account of the dynamics of the social situation of its subjects, that is, a theory of the genesis and maintenance of both social conditions and intersubjective understandings, and, second, must offer a critique of ideologies based on a comparison of the social structure with participants' understandings of it. Critical research links depersonalized social processes to its subjects' choices and actions with the goal of eliminating unrecognized and contradictory consequences of collective action.[19]

A similar logic of investigation applies at any level of analysis from the political struggles of a ghetto neighborhood to the class struggles of world capitalism. The method outlined in this section is phrased in terms of the struggles of local groups and movements—the arena of most progressive action today. However, it is intended, at least in broad outline, as a logic of investigation equally applicable at the societal or world-system levels. Its repeated movement through four phases—interpretive, empirical-analytic, critical-dialectical, and practical—is intended as a fair reading of the method Marx used in his critique of liberal capitalism. Critiques of advanced capitalism must similarly combine structural analyses with critiques of the ideologies that command contemporary thought. Only in this way will radical analysis encourage self-conscious revolutionary action.

Research for local struggles must begin from a dialectical view of the totality of advanced capitalism, however limited our understanding may be at the present. This view of the totality will guide the selection of progressive issues and movements, the interpretation of ideologies, and the selection and analysis of empirical data. In turn, microanalyses of particular struggles will serve to modify and elaborate macrotheories of advanced capitalism. Critical micro- and macroanalyses thus proceed in dialectical tension ... and unity.

In the following paragraphs I will present a critical research method in seven steps. While this may seem a bit programmatic and mechanical, my intention in setting out this method so explicitly is to clearly display the necessary differences between the critical method and the method of positive social research. This critical method is summarized and contrasted with the positive method in table 40.1.

Step 1: Identify Movements or Social Groups Whose Interests Are Progressive

Critical research is not *about* a social process but rather is *for* particular social groups—groups that represent progressive tendencies currently obscured and dominated. Abstract categories such as mankind, the people, the working class, women, or minorities cannot be agents of social change. Instead, we must identify organizations, parties, and movements that represent these categories and that are not only able but also willing to put the research findings into practice.[20]

The identification of such groups is not a simple matter and ultimately relies on critical theories of the totality of advanced capitalism. Contemporary commentators do

Table 40.1
Steps in the research methods of positive and critical social sciences

Positive social science	Critical social science
1. Identify a scientific problem by studying the results of past empirical and theoretical work.	1. Identify social groups or movements whose interests are progressive.
2. Develop empirically testable hypotheses which promise to improve the theory's explanatory and predictive power.	2. Develop an interpretive understanding of the intersubjective meanings, values, and motives held by all groups in the settings.
3. Select a setting (community, group, organization, etc.) which is suitable to the scientific problem.	3. Study the historical development of the social conditions and the current soical structures that constrain actions and shape understandings.
4. Develop measures and data-gathering strategies based on: Previous research Observations and interviews in the setting The investigator's own common sense Knowledge of social processes	4. Construct models of the relations between social conditions, intersubjective interpretations of those conditions, and participants' actions.
5. Gather data through: Experiments Existing documents and texts Surveys and interviews Observations	5. Elucidate the fundamental contradictions which are developing as a result of actions based on ideologically frozen understandings: Compare conditions with understandings Critique the ideology Discover immanent possibilities for action
6. Analyze data to test hypotheses.	6. Participate in a program of education with the subjects that gives them new ways of seeing their situation.
7. Alter laws and theory in light of findings and restate scientific problem to be addressed by subsequent research.	7. Participate in a theoretically grounded program of action which will change social conditions and will also engender new, less alienated, understandings and needs.
Return to Step 1.	Return to Step 2.

not agree on this totality. For example, Sandberg suggest these will be primarily organizations of the working class while Piccone goes so far as to argue that the Marxist notion of class is no longer useful for identifying progressive movements.[21] I wish to propose that groups are progressive insofar as they express interests, purposes, or human needs which cannot be satisfied within the context of a social order characterized by material and ideological domination. In other words, their interests require more or less fundamental changes in the direction of increased participation in the collective and self-conscious control of social institutions.

Appropriate subjects for critical research would include organizations of the trade union movement including rank and file movements to restore democracy, but it would also include groups not always identified as working class movements such as associations of neighborhood residents, environmental groups, organizations representing the needs of women, minorities, poor people, and other such unpopular peoples' movements. In each case the investigator must determine if the group is both willing and able to participate in the investigation and put the findings into practice.

Step 2: Develop an Interpretive Understanding of the Intersubjective Meanings, Values, and Motives Held by All Groups of Actors in the Subjects' Milieu

Critical research begins with a study of the subjects' own life world—the constitutive meanings, social rules, values, and typical motives which govern action in their particular setting. Social action is dominated by models of what the world is like so that what people are and what they do is determined by the understandings that penetrate their thought and action.[22] This requires that critical accounts be grounded in the meanings and values held by its subjects and those with whom they interact. The second step of critical research is thus hermeneutic: the investigator seeks, through dialogue with the participants, to construct a coherent account of the understandings they have of their world.[23] The resulting accounts will present the practical inferences the actors use to warrant their own actions and the conduct of others.[24] Thus both typical motives and intentions as well as conceptions of conditions will be explicated. These accounts will serve both as a basis for empirical-analytic studies and as a corrective to the investigator's preconceptions regarding the subjects' life-world and experiences.

The critical investigator must determine if meanings are differentiated, that is, if the understandings of some subjects differ from the dominant ideology. Such a differentiation of meaning will provide clues to the contradictions immanent in the dominant system of action and indicate the maturity of the resistance by the dominated group. Moreover, a differentiation of meaning *within* the dominated group may indicate emerging contradictions within the movement itself which must be addressed by the members. The possibility of ideologically distorted self-understandings within progressive movements must not be discounted.

The accuracy of these interpretive accounts is to be judged by the degree to which the investigator and the subjects come to talk about the latter's actions and beliefs in the same way.[25] Only in dialogue with the subjects can the investigator determine if his or her understanding of their world is adequate for further analysis and critique. In this dialogue the investigator must learn and use the subjects' language and must avoid esoteric mainstream or Marxist social science terms. Only later, in the pedagogical step of the investigation, should the researcher introduce critical concepts and theories.

Meanings, values, and motives are not reducible to individual psychological attributes. The failure of positive social science to consider subjectivity as *inter*subjectivity, that is, as a sociohistorical construct and a collective reality, results in an inability to recognize the historical specificity of the particular understandings held by actors.[26] Further, it also leads to the inability to account for crises and fundamental changes which result from attempts to redefine the world and hence reconstitute social practices.[27] Human actions and interpretations are historical; "they take place within a context preconditioned by the sedimentations of the past."[28] The critical study of intersubjective meanings makes sense of what the subjects do by reference to the structural bases of present meanings, values, and motives.

Step 3: Study the Historical Development of the Social Conditions and the Current Social Structures That Constrain the Participants' Actions and Shape Their Understandings
Social reality is not limited to the intersubjective understandings of historically specific groups or classes. Under conditions of domination, these are, to a greater or lesser extent, mystifying and distorting ideologies. In order to critically engage the participants in dialogue about their world, the researcher must also carry out empirical studies of social structures and processes. These studies will elucidate the specific determinants of the participants' beliefs and the existing constraints on social practices.[29] According to Adorno the task for a critical social science is "to confront all its statements on the subjective experience ... of human beings and human groups, with the objective factors determining their existence." This means, "it will emphasize ... those determining factors that are connected with the subjective thoughts, feelings and behaviors of those whom it is investigating."[30] These determining factors will be found in historically constructed social structures.

While eschewing the epistemological assumptions of positivism, the critical researcher may fruitfully use empirical findings of past research as well as conduct his or her own empirical studies. Especially important will be studies of the macrostructures of advanced capitalism—structures that form the context for the dominated group's actions and interpretations. For example, in working with progressive labor union movements, it is crucial to confront their anticommunism with accounts of the innumerable red-scare tactics used by capitalists to purge labor of its best organizers. Similarly, the critical researcher working with an antinuclear power group will find it necessary to use or develop analyses of the economic and political structure of the energy industry, while a group opposing banks that red-line neighborhoods will need empirical studies of the banking industry. What is important here is to present such empirical findings and analytic theories in ways that clearly show the historicity and constructedness of social conditions. Conditions must be shown not to be the consequences of immutable laws but to be structures and processes constructed by elites with specific interests and intentions. Only in this way can oppressed groups see social structures both as constraints and as processes subject to conscious direction.

These empirical analyses of social structures must be referred to the specific experiences of the study's subjects. They must appear to the group to recount real events, issues, and processes which the participants or their predecessors have experienced—even though the subjects may differ from the investigator in the significance or importance they attribute to certain events and processes. In these analyses the origins of specific ideologies must be distinguished from the social processes which serve to maintain them. It is in the latter that we will find immanent contradictions which will lead to crises. It is also in the processes by which ideologies are maintained that we will find the specific social conditions which must be targeted for change.

These empirical and historical analyses, along with the interpretive accounts developed in step two, are the basic components of the dialectical and critical steps which follow.

Step 4: Construct Models of the Determinate Relations between Social Conditions, Intersubjective Interpretations of Those Conditions, and Participants' Actions

All critical accounts are based on an understanding of the historical dialectic by which social processes and intersubjective meanings have developed. The aim is to show how the meanings are the product of specific historical conditions and, further, to show how social conditions have changed in such a way as to render these meanings partially or wholly invalid. We must keep in mind that neither social conditions nor intersubjective meanings alone constitute the whole of social reality. Rather, in contrast to both positive and phenomenological approaches, a critical social science focuses on the dialectical tension between the historically created conditions of action and the actors' understandings of these conditions. This requires that the intersubjective meanings, values, and intentions, which may be false or distorting ideologies, be linked to the social processes and structures that create and maintain them. In this way, the critical investigator prepares the way for the comparison of objective social conditions, with intersubjective understandings in the way Adorno suggested in the passage quoted above.

This phase of the investigation takes the form of describing the social processes and structures that gave rise to particular understandings and that presently serve to reinforce or maintain meanings, values, and motives. For example, the meaning of being a university professor today includes the necessity of carrying out well-funded research programs. This is maintained by the historically developed university dependence on government and foundation support. At least in part, the ideology of research productivity and graduate education has developed to legitimate the university's dependence upon this support. Without an understanding of the historical development of both the structures of dependence and the legitimating meaning systems, it would be difficult to account for the frenzied search for grants and contracts which characterizes most faculty activity today. Such a dialectical account lays bare particular relationships between social conditions, ideologies, and actions and provides the foundation for the critique which follows.

Step 5: Elucidate the Fundamental Contradictions Which Are Developing as a Result of Current Actions Based on Ideologically Frozen Understandings

Under conditions of domination and ideologically frozen understandings, many actions are the result of social conditions over which actors have no conscious control.[31] As Giddens notes, "the production or constitution of society is a skilled accomplishment of its members, but one that does not take place under conditions that are wholly intended or wholly comprehended by them."[32] The critical investigator studies the historical consequences of actions in order to uncover the unanticipated and contradictory social conditions which result from ideologically determined actions. Thus, in the university large numbers of graduate students are recruited to work on funded research projects. Many of these students eventually graduate and, in increasing numbers, join private research organizations which compete with university faculty for grants and contracts. The result is a reduction in support for funded research at universities—a consequence surely not intended and one which contradicts present efforts in the universities to maintain research activity. Thus, actions to increase funding for research create the conditions which contradict both the ideology of research productivity and the administrative structure and social practices of the university which have been built up over the past few decades. The ensuing crisis may result in a minor adjustment, such as university research centers adopting organizational forms of private organizations and greatly reducing graduate education, or it may result in more radical changes both in social conditions and in the meanings, values, and motives for research. The outcome of such crises depends upon the critical awareness of groups of scientists and administrators with varying degrees of power and differing interests in either maintaining or changing present relations of power.

The search for fundamental contradictions follows what Adorno called immanent analysis, or the "analysis of the internal consistency or inconsistency of the opinion itself and of its relationship to its object ... the social structure that undergirds it."[33] This analysis sets the stage for a critique of the dominating ideology which prevents the participants from recognizing the possibilities immanent in the present. Such a critique must proceed from the progressive elements of current understandings (e.g., the desirability and possibility of socially useful research at universities) to show how such possibilities are frustrated by other intentions and practices. In this way the elements of truth in an ideology are extracted from their context of falsity.

Against the background of a historical account and an analysis of the relations between social conditions, ideology, and actions, the critical researcher helps the participants to see why past social conditions cannot be recaptured. The researcher must show how present intentions are unrealizable in the context of changed circumstances (e.g., the return to days of large and numerous research grants or, in other struggles, the return to laissez-faire capitalist economy) or, alternatively, must show how the ideology was internally contradictory and therefore never did accurately interpret social conditions (for example, relations of equal exchange between workers and capitalists).

At this point, having initiated the interpretive, empirical, and dialectical phases of critical research, investigators must participate in a program of education and provide assistance in developing strategies for change. The final two steps in this research method demonstrate the *practical intent* of critical research.

Step 6: Participate in a Program of Education with the Subjects That Gives Them New Ways of Seeing Their Situation

Critical social science is distinguished fundamentally from positive social science by its goal of involving the actors it studies. According to Fay the aim is to enlighten subjects so that, coming to see themselves and their social situation in a new way, they *themselves* can decide to alter the conditions which they find repressive.[34] A program of education links social conditions to the subjects' actions by explicating the contradictions that have developed historically. Moreover, it shows how conditions have changed and how they can be changed in the future given increased understanding and self-awareness. Finally, a program of education encourages the participants to develop new understandings and actions which are appropriate and possible under changed conditions.

The model of education which is appropriate is not the familiar one of formal schooling but, rather, a model of dialogue in which the critical researcher attempts to either problematize certain meanings, motives, or values accepted by his or her subjects or to respond to issues which are already perceived by them as problematic. The impetus to search for new understandings and new actions must come from the subjects, not from the researcher. In response, the researcher must express his or her critical accounts in the subjects' own language and in terms of acting *agents* rather than in terms of impersonal social processes.[35] These accounts must show them how they can *act* to change the situation that frustrates them rather than merely showing them how that situation came to exist. It must point to theoretically and empirically warrantable *possibilities* immanent in the development of social conditions. The pegagogy of Paulo Freire provides a model for such an education.[36] Freire's image of critical education has been characterized as "the process in which people, not as recipients, but as knowing subjects, achieve a deepening awareness both of the sociohistorical reality which shapes their lives and of their capacity to transform that reality."[37]

A critical education program provides one criterion of the validity of critical accounts: the response which its subjects make to its claims. This is one reason why education, as well as political action, is not simply "tacked on" at the end of the critical research program but is central to testing and improving its accounts. Dialogic education is integral to every research program which treats subjects as active agents instead of objectifying them and reifying their social conditions. The aim of critical research, however, is not simply to enlighten but also to inform and initiate political action.

Step 7: Participate in a Theoretically Grounded Program of Action Which Will Change Social Conditions and, in Addition, Will Engender New Less Alienated Understandings and Needs

The purpose of critical research and theory is to initiate action by providing an adequate knowledge of the historical development of social conditions and meanings and a vision of a desirable and possible future.[38] This final step of political action links the subjects' actions back to social conditions in order to reduce or eliminate the irrational construction of contradictory social conditions. Action becomes conscious and reflective through critical education while social conditions become intentional constructions through action informed by a critical analysis. What is objective—social structure and process—becomes subjectively or meaningfully comprehended and what is subjectively comprehended becomes objectified in social process and structure. The subjects' existence and their self-understandings are brought into theoretical and practical unity and critical thought becomes an active social force. Through critical action the subjective and objective factors of revolutionary change are united. Despite the dangers to theoretical independence which inhere in too great an identification with a progressive or revolutionary movement (for example, subordination to the party), the validity of critical theorems can only be tested in practice and this requires the involvement of the researchers in the subjects' political activity.

Another major reason for continuing involvement in political action carried on by the subjects of a critical investigation is the necessity to combat tendencies for groups to become reformist. If the researcher merely responds to the felt needs and frustrations of dominated groups as Fay suggests, she or he may only help to satisfy immediate needs and intentions within the context of a fundamentally unchanged social structure.[39] Since the needs of members of society are generated by the concretely existing and alienating social conditions, the critical researcher must encourage political action which constructs new social conditions conducive to nonalienated needs such as participation, love, creativity, and collective control. The challenge for the researcher as participant is to continually broaden and deepen the participants' awareness of the meaning and probable outcome of their political action. The aim is the subject's progressive movement toward an understanding of the totality of historical circumstances that affect them. This requires participation in a continuous cycle of critical analysis, education, and action.

Critical researchers do not, therefore, enter progressive groups on an episodic basis to solve clearly defined problems. Since their aim is to stimulate a self-sustaining process of critical analysis and enlightened action, it becomes necessary for critical researchers to ally themselves with progressive groups and work with them for considerable periods of time. This increases the problems of selecting a group willing and able to participate in critical research for it requires that they become progressively more self-critical and willing to analyze their own values, motives, and understandings as well as critically evaluate the results of their political actions. The role of the critical researcher is, through dialogue and analysis, to return the group's attention to their understandings and intentions (Step 2) and to reinitiate the critical analysis of actions (Step 4) in the light of the further development of historical forces (Step 3). The aim here is to develop the group's own capacity to discover the contradictions which may be developing as a result of their actions (Step 5) and to educate themselves for more enlightened and self-conscious political action (Steps 6 and 7).

Conclusion

A look at table 40.1 will highlight the differences between positive and critical social research. Positive social research begins with identification of a scientific problem, proceeds through data gathering and hypothesis testing, and returns to confirm or revise theoretical understandings. Critical research begins from the practical problems and ideologically distorted understandings of groups that are dominated and frustrated by present social conditions. It proceeds through interpretive, empirical, and dialectical phases of analysis with the intent to inform the emancipatory practices of these groups. It is a method of *praxis* for it combines disciplined analysis with practical action. It is aimed not merely at understanding the world, but at changing it. Instead of objectifying people and society, it enables its subjects to reappropriate their life-world and become self-conscious agents of sociohistorical progress. It is democratic rather than elitist, and it is enlightening instead of mystifying. Such a critical research method is the basis for critical theories which have practical utility in the political struggle for freedom.

Notes

I would like to thank Richard Appelbaum, Walter Feinberg, and T. R. Young for their comments on earlier drafts of this chapter. I hope the present version fairly reflects many of their suggestions while maintaining enough of our original differences to prompt further dialogue.

1. Mao Tsetung, "Some Questions Concerning Methods of Leadership," *Selected Readings from the Works of Mao Tsetung* (Peking: Foreign Language Press, 1943), pp. 287–94.
2. Martin Jay, *The Dialectical Imagination* (Boston: Little, Brown, 1973); Perry Anderson, *Considerations on Western Marxism* (London: New Left Books, 1976).
3. Anderson, *Considerations on Western Marxism.*
4. Cf. John H. Sewart, "Critical Theory and the Critique of Conservative Method," *American Sociologist* 13 (1978): 15–22.
5. Barry Smart, *Sociology, Phenomenology and Marxist Analysis* (Boston: Routledge and Kegan Paul, 1976).

6. Max Horkheimer, *Critical Theory: Selected Essays* (New York: Herder and Herder, 1972).
7. Ibid., p. 244.
8. Charles Taylor, "Interpretation and the Sciences of Man," *Review of Metaphysics* 25, no. 3 (1971): 1–51.
9. Jürgen Habermas, *Knowledge and Human Interests* (Boston: Beacon, 1971), p. 69.
10. Ibid.: Richard J. Bernstein, *The Restructuring of Social and Political Theory* (New York: Harcourt, Brace and Jovanovich, 1976).
11. Brian Fay, *Social Theory and Political Practice* (New York: Holmes and Meier, 1976), p. 57.
12. James Farganis, "A Preface to Critical Theory," *Theory and Society* 2 (1975): 467–482.
13. Paul Piccone, "Phenomenological Marxism," in *Towards a New Marxism*, ed. B. Grahl and P. Piccone (St. Louis, Mo.: Telos Press, 1973).
14. Taylor, "Interpretation."
15. Georg Henrik von Wright, *Explanation and Understanding* (Ithaca, N.Y.: Cornell University Press, 1971).
16. Jürgen Habermas, *Legitimation Crisis* (Boston: Beacon, 1975).
17. Horkheimer, *Critical Theory.*
18. Bernstein, *Restructuring of Social and Political Theory*, p. 44.
19. Ake Sandberg, *The Limits to Democratic Planning* (Stockholm: Liber Förlag, 1976), p. 45.
20. Ibid., p. 227.
21. Piccone, "Phenomenological Marxism," p. 157.
22. Bernstein, *Restructuring of Social and Political Theory*, p. 63.
23. Hans-Georg Gadamer, "The Historicity of Understanding," in *Critical Sociology*, ed. P. Connerton (New York: Penguin, 1976), pp. 117–133.
24. Von Wright, *Explanation and Understanding.*
25. Fay, *Social Theory and Political Practice*, p. 82.
26. Friedrich Pollock, "Empirical Research into Public Opinion," in *Critical Sociology*, ed. Connerton, pp. 225–236.
27. Taylor, "Interpretation."
28. Piccone, "Phenomenological Marxism," p. 141.
29. Richard P. Appelbaum, "Marxist Method: Structural Constraints and Social Praxis," *American Sociologist* 13 (1978): 73–81.
30. Theodor W. Adorno, "Sociology and Empirical Research," in *Critical Sociology*, ed. Connerton, pp. 237–257.
31. Fay, *Social Theory and Political Practice*, p. 95.
32. Anthony Giddens, *New Rules of Sociological Method* (New York: Basic Books, 1976), p. 102.
33. Adorno, "Sociology and Empirical Research," p. 256.
34. Fay, *Social Theory and Political Practice*, p. 102.
35. Sandberg, *Limits to Democratic Planning*, p. 227.
36. Paulo Freire, *Pegagogy of the Oppressed* (New York: Seabury, 1970); Paulo Freire, *Cultural Action for Freedom* (Cambridge, Mass.: Center for the Study of Development and Social Change, 1970).
37. Freire, *Cultural Action*, p. 27n.
38. Fay, *Social Theory and Political Practice.*
39. Agnes Heller, "Theory and Practice: Their Relation to Human Needs," *Social Praxis* 1 (1973): 359–373.

PART VIII

Problems of the Special Sciences

Introduction to Part VIII

It is appropriate at the end of such a book, devoted primarily to philosophical issues that affect the method of the social sciences as a whole, to realize that many important philosophical issues also have been played out within each of the social sciences and that there are problems unique to each discipline. Critics sometimes accuse the special sciences of reinventing the wheel, as they consider problems that already have been better developed in the general literature of the philosophy of science. But if the philosophy of social science is to have an impact on the practice of working social scientists, it must address the issues specific to each discipline, in addition to resolving more general philosophical disputes. In this part, we have chosen selections by both philosophers and social scientists that speak to the problems found in the philosophy of economics, psychology, history, and archaeology.

Economics has long been regarded as the most "scientific" of the social sciences, leading one to expect that its method would be closest to that used in the physical sciences. Milton Friedman makes the most of this analogy in "The Methodology of Positive Economics" (chapter 41), in which he argues that "positive economics is, or can be, an 'objective' science, in precisely the same sense as any of the physical sciences." Friedman develops this theme by advocating the nomological and predictive character of economics and comparing the objections it faces about the realism of its assumptions with the role of simplifying assumptions in physics. As in physics, Friedman maintains, one cannot assess the merit of a theory by the "realism" of its assumptions but only by its predictive success. But how successful has economics been in fulfilling this desideratum?

Alexander Rosenberg, in a carefully reasoned essay that asks, "If Economics Isn't Science, What Is It?" (chapter 42), reconsiders the scientific status of economics, given its empirical failure to meet the self-imposed standards for predictive and explanatory success and its unwillingness to surrender or compromise the methods by which it is practiced. In the end, Rosenberg argues that economics is not a science at all but is more akin to a branch of mathematics, thus providing a powerful tonic to Friedman's optimistic analogy with physics.

Psychologists too have been concerned with the scientific status of their discipline. Yet the philosophy of psychology has gone well beyond this more general question recently, to ask about the specifics of attempts to formulate psychological explanations and to question what the appropriate level of inquiry should be in offering such accounts. Donald Davidson's "Actions, Reasons, and Causes" (chapter 43) has become a modern classic in this area, framing the debate about whether reasons can be considered as causes for human action. Davidson maintains that generalizations connecting reasons to actions cannot be codified into lawlike explanations that afford accurate predictions of human behavior. This chapter offers further insight into Davidson's famous argument against social scientific laws, set out in "Philosophy as Psychology" in chapter 6 of this book.

Jerry Fodor's "Special Sciences (or: The Disunity of Science as a Working Hypothesis)" (chapter 44) is one of the most important chapters in this final part and provides a touchstone for thinking about the credibility of the existence of the special sciences—including psychology—at all. He too is preoccupied with the level of explanation appropriate to psychology, primarily as a result of the reductionist program advocated by scientifically minded psychologists who would like to get rid of psychological explanations altogether. Such arguments have led to far-reaching assertions that psychology must be reduced to neuroscience, because the laws that are behind human behavior are at base physiological ones, and not social scientific at all. Yet even if the latter is true, Fodor contends, there is room for inquiry at the secondary level. Thus, Fodor defends psychology along with the rest of the special sciences against those who would accommodate it as nonnomological inquiry (like Davidson) or would wish to abandon it totally as an explanatory enterprise.

The status of historical explanation has been no less controversial than psychology. Are the ways historians attempt to understand the past different in principle from the ways scientists attempt to understand the world? If so, how does historical understanding differ from scientific understanding? Paul Roth and Louis O. Mink address these questions.

Historians often attempt to understand events in terms of narratives; that is, they tell stories about past events that allegedly provide understanding of the past. However, a number of objections have been raised to narratives as ways of understanding. In chapter 45, "Narrative Explanations: The Case of History," Roth critically considers two reasons to suppose that in the light of these objections, historians are wrong to use narrative explanations. According to the methodological objection, narrative explanation do not have the proper scientific form. In particular, they do not appeal to general laws and thus are too far removed from scientific explanations to provide understanding. According to the metaphysical objection, narrative explanations neglect the requirement that historical explanations must correspond to the historical facts and are too close to fictional narratives. Roth rejects both of these objections and, in so doing, defends historians' use of narrative explanations.

Rejecting the view that history is a proto-science, Mink, in "The Autonomy of Historical Understanding" (chapter 46), argues that historical understanding is autonomous. By elucidating six characteristics of historical understanding, he attempts to show that history has a distinctively different mode of understanding from science. According to Mink, the idea of historical synthesis constitutes a common theme running through these characteristics. History, he argues, cultivates synoptic judgments that "convert congeries of events into cancatenations." Whether Mink's thesis that historical understanding is autonomous is acceptable depends not only on his portrayal of history, therefore, but also on his contrasting characterization of science.

The field of archaeology is often neglected by philosophers of social science. One finds them illustrating their theories with examples drawn from economics, cultural anthropology, psychology, and history but almost never from archaeology. Archaeologists themselves, however, have not only been interested in the philosophy of science but have been greatly influenced by it. In particular, a recent movement known as the new archaeology has explicitly borrowed the covering-law model of explanation, as well as other ideas from the philosophy of science literature.[1] The question of whether such ideas are appropriate to archaeology has sparked a lively debate that is virtually unknown to many philosophers. The chapters by Merrilee Salmon and Alison Wylie set out many of the important issues in the philosophy of archaeology.

In "On the Possibility of Lawful Explanation in Archaeology" (chapter 47), Salmon addresses three questions about functional explanations: Has the presence of an item in the archaelogical record, for example, a pottery bowl, been explained when its function has been identified? If so, are such explanations dependent upon laws? And if they are so dependent, what can we say about such laws? These questions are, of course, closely connected to ones taken up earlier in this book. And, as Salmon makes clear, they reach to the very foundations of archaeology.

In "Evidential Constraints: Pragmatic Objectivism in Archaeology," Wylie (chapter 48) reviews a variety of claims about nationalistic, racist, and gender bias in archaeology that seem to threaten the possibility of an objective science of archaeology. Acknowledging the validity of these claims, she attempts to show how they are in fact underwritten by the possibility of objectivity. Wylie's argument here relates back to part VII, Objectivity and Values, in this book in three ways: it complements Naomi Weisstein's account of sexist bias in psychology, it bolsters the thesis of those, like Ernest Nagel, who argue that social science can be objective, and it provides an empirical rationale for one element of Donald Comstock's program of critical social science. Thus, the case of archaeology once again demonstrates how specific debates within the special sciences lead us back to more general disputes in the philosophy of social science.

Note

1. Merrilee H. Salmon, *Philosophy and Archaeology* (New York: Academic Press, 1982), pp. 1–3.

Chapter 41
The Methodology of Positive Economics
Milton Friedman

In his admirable book on *The Scope and Method of Political Economy* John Neville Keynes distinguishes among "a *positive science* ... [,] a body of systematized knowledge concerning what is; a *normative* or *regulative* science ... [,] a body of systematized knowledge discussing criteria of what ought to be ... ; an *art* ... [,] a system of rules for the attainment of a given end"; comments that "confusion between them is common and has been the source of many mischievous errors"; and urges the importance of "recognizing a distinct positive science of political economy."[1]

This chapter is concerned primarily with certain methodological problems that arise in constructing the "distinct positive science" Keynes called for—in particular, the problem how to decide whether a suggested hypothesis or theory should be tentatively accepted as part of the "body of systematized knowledge concerning what is." But the confusion Keynes laments is still so rife and so much of a hindrance to the recognition that economics can be, and in part is, a positive science that it seems well to preface the main body of the chapter with a few remarks about the relation between positive and normative economics.

The Relation between Positive and Normative Economics

Confusion between positive and normative economics is to some extent inevitable. The subject matter of economics is regarded by almost everyone as vitally important to himself and within the range of his own experience and competence; it is the source of continuous and extensive controversy and the occasion for frequent legislation. Self-proclaimed "experts" speak with many voices and can hardly all be regarded as disinterested; in any event, on questions that matter so much, "expert" opinion could hardly be accepted solely on faith even if the "experts" were nearly unanimous and clearly disinterested.[2] The conclusions of positive economics seem to be, and are, immediately relevant to important normative problems, to questions of what ought to be done and how any given goal can be attained. Laymen and experts alike are inevitably tempted to shape positive conclusions to fit strongly held normative preconceptions and to reject positive conclusions if their normative implications—or what are said to be their normative implications—are unpalatable.

Positive economics is in principle independent of any particular ethical position or normative judgments. As Keynes says, it deals with "what is," not with "what ought to be." Its task is to provide a system of generalizations that can be used to make correct predictions about the consequences of any change in circumstances. Its performance is to be judged by the precision, scope, and conformity with experience of the predictions it yields. In short, positive economics is, or can be, an "objective" science, in precisely the same sense as any of the physical sciences. Of course, the fact that economics deals with the interrelations of human beings, and that the investigator is himself part of the subject matter being investigated in a more intimate sense than in the physical sciences, raises special difficulties in achieving objectivity at the same time that it provides the social scientist with a class of data not available to the physical scientist. But neither the one nor the other is, in my view, a fundamental distinction between the two groups of sciences.[3]

Normative economics and the art of economics, one the other hand, cannot be independent of positive economics. Any policy conclusion necessarily rests on a prediction about the consequences of doing one thing rather than another, a prediction that must be based—implicitly or explicitly—on positive economics. There is not, of course, a one-to-one relation between policy conclusions and the conclusions of positive economics; if there were, there would be no separate normative science. Two individuals may agree on the consequences of a particular piece of legislation. One may regard them as desirable on balance and so favor the legislation; the other, as undesirable and so oppose the legislation.

I venture the judgment, however, that currently in the Western world, and especially in the United States, differences about economic policy among disinterested citizens derive predominantly from different predictions about the economic consequences of taking action—differences that in principle can be eliminated by the progress of positive economics—rather than from fundamental differences in basic values, differences about which men can ultimately only fight. An obvious and not unimportant example is minimum-wage legislation. Underneath the welter of arguments offered for and against such legislation there is an underlying consensus on the objective of achieving a "living wage" for all, to use the ambiguous phrase so common in such discussions. The difference of opinion is largely grounded on an implicit or explicit difference in predictions about the efficacy of this particular means in furthering the agreed-on end. Proponents believe (predict) that legal minimum wages diminish poverty by raising the wages of those receiving less than the minimum wage as well as of some receiving more than the minimum wage without any counterbalancing increase in the number of people entirely unemployed or employed less advantageously than they otherwise would be. Opponents believe (predict) that legal minimum wages increase poverty by increasing the number of people who are unemployed or employed less advantageously and that this more than offsets any favorable effect on the wages of those who remain employed. Agreement about the economic consequences of the legislation might not produce complete agreement about its desirability, for differences might still remain about its political or social consequences; but, given agreement on objectives, it would certainly go a long way toward producing consensus.

Closely related differences in positive analysis underlie divergent views about the appropriate role and place of trade unions and the desirability of direct price and wage controls and of tariffs. Different predictions about the importance of so-called economies of scale account very largely for divergent views about the desirability or necessity of detailed government regulation of industry and even of socialism rather than private enterprise. And this list could be extended indefinitely.[4] Of course, my judgment that the major differences about economic policy in the Western world are of this kind is itself a "positive" statement to be accepted or rejected on the basis of empirical evidence.

If this judgment is valid, it means that a consensus on "correct" economic policy depends much less on the progress of normative economics proper than on the progress of a positive economics yielding conclusions that are, and deserve to be, widely accepted. It means also that a major reason for distinguishing positive economics sharply from normative economics is precisely the contribution that can thereby be made to agreement about policy.

Positive Economics

The ultimate goal of a positive science is the development of a "theory" or "hypothesis" that yields valid and meaningful (i.e., not truistic) predictions about phenomena not yet observed. Such a theory is, in general, a complex intermixture of two elements. In part, it is a "language" designed to promote "systematic and organized methods of reasoning."[5] In part, it is a body of substantive hypotheses designed to abstract essential features of complex reality.

Viewed as a language, theory has no substantive content; it is a set of tautologies. Its function is to serve as a filing system for organizing empirical material and facilitating our understanding of it; and the criteria by which it is to be judged are those appropriate to a filing system. Are the categories clearly and precisely defined? Are they exhaustive? Do we know where to file each individual item, or is there considerable ambiguity? Is the system of headings and subheadings so designed that we can quickly find an item we want, or must we hunt from place to place? Are the items we shall want to consider jointly filed together? Does the filing system avoid elaborate cross-references?

The answers to these questions depend partly on logical, partly on factual, considerations. The canons of formal logic alone can show whether a particular language is complete and consistent, that is, whether propositions in the language are "right" or "wrong." Factual evidence alone can show whether the categories of the "analytical filing system" have a meaningful empirical counterpart, that is, whether they are useful in analyzing a particular class of concrete problems.[6] The simple example of "supply" and "demand" illustrates both this point and the preceding list of analogical questions. Viewed as elements of the language of economic theory, these are the two major categories into which factors affecting the relative prices of products or factors of production are classified. The usefulness of the dichotomy depends on the "empirical generalization that an enumeration of the forces affecting demand in any problem and of the forces affecting supply will yield two lists that contain few items in common."[7] Now this generalization is valid for markets like the final market for a consumer good. In such a market there is a clear and sharp distinction between the economic units that can be regarded as demanding the product and those that can be regarded as supplying it. There is seldom much doubt whether a particular factor should be classified as affecting supply, on the one hand, or demand, on the other; and there is seldom much necessity for considering cross-effects (cross-references) between the two categories. In these cases the simple and even obvious step of filing the relevant factors under the headings of "supply" and "demand" effects a great simplification of the problem and is an effective safeguard against fallacies that otherwise tend to occur. But the generalization is not always valid. For example, it is not valid for the day-to-day fluctuations of prices in a primarily speculative market. Is a rumor of an increased excess-profits tax, for example, to be regarded as a factor operating primarily on today's supply of corporate equities in the stock market or on today's demand for them? In similar fashion, almost every factor can with about as much justification be classified under the heading "supply" as under the heading "demand." These concepts can still be used and may not be entirely pointless; they are still "right" but clearly less useful than in the first example because they have no meaningful empirical counterpart.

Viewed as a body of substantive hypotheses, theory is to be judged by its predictive power for the class of phenomena which it is intended to "explain." Only factual evidence can show whether it is "right" or "wrong" or, better, tentatively "accepted" as valid or "rejected." As I shall argue at greater length below, the only relevant test of the *validity* of a hypothesis is comparison of its predictions with experience. The hypothesis is rejected if its predictions are contradicted ("frequently" or more often than predictions from an alternative hypothesis); it is accepted if its predictions are not contradicted; great confidence is attached to it if it has survived many opportunities for contradiction. Factual evidence can never "prove" a hypothesis; it can only fail to disprove it, which is what we generally mean when we say, somewhat inexactly, that the hypothesis has been "confirmed" by experience.

To avoid confusion, it should perhaps be noted explicitly that the "predictions" by which the validity of a hypothesis is tested need not be about phenomena that have not yet occurred, that is, need not be forecasts of future events; they may be about phenomena that have occurred but observations on which have not yet been made or are not known to the person making the prediction. For example, a hypothesis may imply that such and such must have happened in 1906, given some other known circumstances. If a search of the records reveals that such and such did happen, the prediction is confirmed; if it reveals that such and such did not happen, the prediction is contradicted.

The validity of a hypothesis in this sense is not by itself a sufficient criterion for choosing among alternative hypotheses. Observed facts are necessarily finite in number; possible hypotheses, infinite. If there is one hypothesis that is consistent with the available evidence, there are always an infinite number that are.[8] For example, suppose a specific excise tax on a particular commodity produces a rise in price equal to the amount of the tax. This is consistent with competitive conditions, a stable demand curve, and a horizontal and stable supply curve. But it is also consistent with competitive conditions and a positively or negatively sloping supply curve with the required compensating shift in the demand curve or the supply curve; with monopolistic conditions, constant marginal costs, and stable demand curve, of the particular shape required to produce this result; and so on indefinitely. Additional evidence with which the hypothesis is to be consistent may rule out some of these possibilities; it can never reduce them to a single possibility alone capable of being consistent with the finite evidence. The choice among alternative hypotheses equally consistent with the available evidence must to some extent be arbitrary, though there is general agreement that relevant considerations are suggested by the criteria "simplicity" and "fruitfulness," themselves notions that defy completely objective specification. A theory is "simpler" the less the initial knowledge needed to make a prediction within a given field of phenomena; it is more "fruitful" the more precise the resulting prediction, the wider the area within which the theory yields predictions, and the more additional lines for further research it suggests. Logical completeness and consistency are relevant but play a subsidiary role; their function is to ensure that the hypothesis says what it is intended to say and does so alike for all users—they play the same role here as checks for arithmetical accuracy do in statistical computations.

Unfortunately, we can seldom test particular predictions in the social sciences by experiments explicitly designed to eliminate what are judged to be the most important disturbing influences. Generally, we must rely on evidence cast up by the "experiments" that happen to occur. The inability to conduct so-called controlled experiments does not, in my view, reflect a basic difference between the social and physical sciences both because it is not peculiar to the social sciences—witness astronomy—and because the distinction between a controlled experiment and uncontrolled experience is at best one of degree. No experiment can be completely controlled, and every experience is partly controlled, in the sense that some disturbing influences are relatively constant in the course of it.

Evidence cast up by experience is abundant and frequently as conclusive as that from contrived experiments; thus the inability to conduct experiments is not a fundamental obstacle to testing hypotheses by the success of their predictions. But such evidence is far more difficult to interpret. It is frequently complex and always indirect and incomplete. Its collection is often arduous, and its interpretation generally requires subtle analysis and involved chains of reasoning, which seldom carry real conviction. The denial to economics of the dramatic and direct evidence of the "crucial" experiment does hinder the adequate testing of hypotheses; but this is much less significant than the difficulty it places in the way of achieving a reasonably prompt and wide consensus on the conclusions justified by the available evidence. It renders the weeding out of unsuccessful hypotheses slow and difficult. They are seldom downed for good and are always cropping up again.

There is, of course, considerable variation in these respects. Occasionally, experience casts up evidence that is about as direct, dramatic, and convincing as any that could be provided by controlled experiments. Perhaps the most obviously important example is the evidence from inflations on the hypothesis that a substantial increase in the quantity of money within a relatively short period is accompanied by a substantial increase in prices. Here the evidence is dramatic, and the chain of reasoning required to interpret it is relatively short. Yet, despite numerous instances of substantial rises in prices, their essentially one-to-one correspondence with substantial rises in the stock of money, and the wide variation in other circumstances that might appear to be relevant, each new experience of inflation brings forth vigorous contentions, and not only by the lay public, that the rise in the stock of money is either an incidental effect of a rise in prices produced by other factors or a purely fortuitous and unnecessary concomitant of the price rise.

One effect of the difficulty of testing substantive economic hypotheses has been to foster a retreat into purely formal or tautological analysis.[9] As already noted, tautologies have an extremely important place in economics and other sciences as a specialized language or "analytical filing system." Beyond this, formal logic and mathematics, which are both tautologies, are essential aids in checking the correctness of reasoning, discovering the implications of hypotheses, and determining whether supposedly different hypotheses may not really be equivalent or wherein the differences lie.

But economic theory must be more than a structure of tautologies if it is to be able to predict and not merely describe the consequences of action; if it is to be something different from disguised mathematics.[10] And the usefulness of the tautologies themselves ultimately depends, as noted above, on the acceptability of the substantive hypotheses that suggest the particular categories into which they organize the refractory empirical phenomena.

A more serious effect of the difficulty of testing economic hypotheses by their predictions is to foster misunderstanding of the role of empirical evidence in theoretical work. Empirical evidence is vital at two different, though closely related, stages: in constructing hypotheses and in testing their validity. Full and comprehensive evidence on the phenomena to be generalized or "explained" by a hypothesis, besides its obvious value in suggesting new hypotheses, is needed to ensure that a hypothesis explains what it sets out to explain—that its implications for such phenomena are not contradicted in advance by experience that has already been observed.[11] Given that the hypothesis is consistent with the evidence at hand, its further testing involves deducing from it new facts capable of being observed but not previously known and checking these deduced facts against additional empirical evidence. For this test to be relevant, the deduced facts must be about the class of phenomena the hypothesis is designed to explain; and they must be well enough defined so that observation can show them to be wrong.

The two stages of constructing hypotheses and testing their validity are related in two different respects. In the first place, the particular facts that enter at each stage are partly an accident of the collection of data and the knowledge of the particular investigator. The facts that serve as a test of the implications of a hypothesis might equally well have been among the raw material used to construct it, and conversely. In the second place, the process never begins from scratch; the so-called initial stage itself always involves comparison of the implications of an earlier set of hypotheses with observation; the contradiction of these implications is the stimulus to the construction of new hypotheses or revision of old ones. So the two methodologically distinct stages are always proceeding jointly.

Misunderstanding about this apparently straightforward process centers on the phrase "the class of phenomena the hypothesis is designed to explain." The difficulty in the social sciences of getting new evidence for this class of phenomena and of judging its conformity with the implications of the hypothesis makes it tempting to suppose that other, more readily available, evidence is equally relevant to the validity of the hypothesis—to suppose that hypotheses have not only "implications" but also "assumptions" and that the conformity of these "assumptions" to "reality" is a test of the validity of the hypothesis *different from or additional to* the test by implications. This widely held view is fundamentally wrong and productive of much mischief. Far from providing an easier means for sifting valid from invalid hypotheses, it only confuses the issue, promotes misunderstanding about the significance of empirical evidence for economic theory, produces a misdirection of much intellectual effect devoted to the development of positive economics, and impedes the attainment of consensus on tentative hypotheses in positive economics.

Insofar as a theory can be said to have "assumptions" at all, and insofar as their "realism" can be judged independently of the validity of predictions, the relation between the significance of a theory and the "realism" of its "assumptions" is almost the opposite of that suggested by the view under criticism. Truly important and significant hypotheses will be found to have "assumptions" that are widely inaccurate descriptive representations of reality, and, in general, the more significant the theory, the more unrealistic the assumptions (in this sense).[12] The reason is simple. A hypothesis is important if it "explains" much by little, that is, if it abstracts the common and crucial elements from the mass of complex and detailed circumstances surrounding the phenomena to be explained and permits valid predictions on the basis of them alone.

To be important, therefore, a hypothesis must be descriptively false in its assumptions; it takes account of, and accounts for, none of the many other attendant circumstances, since its very success shows them to be irrelevant for the phenomena to be explained.

To put this point less paradoxically, the relevant question to ask about the "assumptions" of a theory is not whether they are descriptively "realistic," for they never are, but whether they are sufficiently good approximations for the purpose in hand. And this question can be answered only by seeing whether the theory works, which means whether it yields sufficiently accurate predictions. The two supposedly independent tests thus reduce to one test.

The theory of monopolistic and imperfect competition is one example of the neglect in economic theory of these propositions. The development of this analysis was explicitly motivated, and its wide acceptance and approval largely explained, by the belief that the assumptions of "perfect competition" or "perfect monopoly" said to underlie neoclassical economic theory are a false image of reality. And this belief was itself based almost entirely on the directly perceived descriptive inaccuracy of the assumptions rather than on any recognized contradiction of predictions derived from neoclassical economic theory. The lengthy discussion on marginal analysis in the *American Economic Review* some years ago is an even clearer, though much less important, example. The articles on both sides of the controversy largely neglect what seems to me clearly the main issue—the conformity to experience of the implications of the marginal analysis—and concentrate on the largely irrelevant question whether businessmen do or do not in fact reach their decisions by consulting schedules, or curves, or multivariable functions showing marginal cost and marginal revenue.[13] Perhaps these two examples, and the many others they readily suggest, will serve to justify a more extensive discussion of the methodological principles involved than might otherwise seem appropriate.

Can a Hypothesis Be Tested by the Realism of Its Assumptions?

We may start with a simple physical example, the law of falling bodies. It is an accepted hypothesis that the acceleration of a body dropped in a vacuum is a constant—g, or approximately 32 feet per second per second on the earth—and is independent of the shape of the body, the manner of dropping it, etc. This implies that the distance traveled by a falling body in any specified time is given by the formula $s = \frac{1}{2}gt^2$, where s is the distance traveled in feet and t is time in seconds. The application of this formula to a compact ball dropped from the roof of a building is equivalent to saying that a ball so dropped behaves *as if* it were falling in a vacuum. Testing this hypothesis by its assumptions presumably means measuring the actual air pressure and deciding whether it is close enough to zero. At sea level the air pressure is about 15 pounds per square inch. Is 15 sufficiently close to zero for the difference to be judged insignificant? Apparently it is, since the actual time taken by a compact ball to fall from the roof of a building to the ground is very close to the time given by the formula. Suppose, however, that a feather is dropped instead of a compact ball. The formula then gives wildly inaccurate results. Apparently, 15 pounds per square inch is significantly different from zero for a feather but not for a ball. Or, again, suppose the formula is applied to a ball dropped from an airplane at an altitude of 30,000 feet. The air pressure at this altitude is decidedly less than 15 pounds per square inch. Yet, the actual time of fall from 30,000 feet to 20,000 feet, at which point the air pressure is still much less than at sea level, will differ noticeably from the time predicted by the formula—much more noticeably than the time taken by a compact ball to fall from the roof of a building to the ground. According to the formula, the velocity of the ball should be gt and should therefore increase steadily. In fact, a ball dropped at 30,000 feet will reach its top velocity well before it hits the ground. And similarly with other implications of the formula.

The initial question whether 15 is sufficiently close to zero for the difference to be judged insignificant is clearly a foolish question by itself. Fifteen pounds per square inch is 2,160 pounds per square foot, or 0.0075 ton per square inch. There is no possible basis for calling these numbers "small" or "large" without some external standard of comparison. And the only relevant standard of comparison is the air pressure for which the formula does or does not work under a given set of circumstances. But this raises the same problem at a second level. What is the meaning of "does or does not work"? Even if we could eliminate errors of measurement, the measured time of fall would seldom if ever be precisely equal to the computed time of fall. How large must the difference between the two be to justify saying that the theory "does not work"? Here there are two important external standards of comparison. One is the accuracy achievable by an alternative theory with which this theory is being compared and which is equally acceptable on all other grounds. The other arises when there exists a theory that is known to yield better predictions but only at a greater cost. The gains from greater accuracy, which depend on the purpose in mind, must then be balanced against the costs of achieving it.

This example illustrates both the impossibility of testing a theory by its assumptions and also the ambiguity of the concept "the assumptions of a theory." The formula $s = \frac{1}{2}gt^2$ is valid for bodies falling in a vacuum and can be derived by analysing the behavior of such bodies. It can therefore be stated: under a wide range of circumstances, bodies that fall in the actual atmosphere behave *as if* they were falling in a vacuum. In the language so common in economics this would be rapidly translated into: the formula assumes a vacuum. Yet it clearly does no such thing. What it does say is that in many cases the existence of air pressure, the shape of the body, the name of the person dropping the body, the kind of mechanism used to drop the body, and a host of other attendant circumstances have no appreciable effect on the distance the body falls in a specified time. The hypothesis can readily be rephrased to omit all mention of a vacuum: under a wide range of circumstances, the distance a body falls in a specified time is given by the formula $s = \frac{1}{2}gt^2$. The history of this formula and its associated physical theory aside, is it meaningful to say that it assumes a vacuum? For all know there may be other sets of assumptions that would yield the same formula. The formula is accepted because it works, not because we live in an approximate vacuum—whatever that means.

The important problem in connection with the hypothesis is to specify the circumstances under which the formula works or, more precisely, the general magnitude of the error in its predictions under various circumstances. Indeed, as is implicit in the above rephrasing of the hypothesis, such a specification is not one thing and the hypothesis another. The specification is itself an essential part of the hypothesis, and it is a part that is peculiarly likely to be revised and extended as experience accumulates.

In the particular case of falling bodies a more general, though still incomplete, theory is available, largely as a result of attempts to explain the errors of the simple theory, from which the influence of some of the possible disturbing factors can be calculated and of which the simple theory is a special case. However, it does not always pay to use the more general theory because the extra accuracy it yields may not justify the extra cost of using it, so the question under what circumstances the simpler theory works "well enough" remains important. Air pressure is one, but only one, of the variables that define these circumstances; the shape of the body, the velocity attained, and still other variables are relevant as well. One way of interpreting the variables other than air pressure is to regard them as determining whether a particular departure from the "assumption" of a vacuum is or is not significant. For example, the difference in shape of the body can be said to make 15 pounds per square inch significantly different from zero for a feather but not for a compact ball dropped a moderate distance. Such a statement must, however, be sharply distinguished from the very different statement that the theory does not work for a feather because its assumptions are false. The relevant relation runs the other way: the assumptions are false for a feather because the theory does not work. This point needs emphasis, because the entirely valid use of "assumptions" in *specifying* the circumstances for which a theory holds is frequently, and erroneously, interpreted to mean that the assumptions can be used to *determine* the circumstances for which a theory holds, and has, in this way, been an important source of the belief that a theory can be tested by its assumptions.

Let us turn now to another example, this time a constructed one designed to be an analogue of many hypotheses in the social sciences. Consider the density of leaves around a tree. I suggest the hypothesis that the leaves are positioned as if each leaf deliberately sought to maximize the amount of sunlight it receives, given the position of its neighbors, as if it knew the physical laws determining the amount of sunlight that would be received in various positions and could move rapidly or instantaneously from any one position to any other desired and unoccupied position.[14] Now some of the more obvious implications of this hypothesis are clearly consistent with experience: for example, leaves are in general denser on the south than on the north side of trees but, as the hypothesis implies, less so or not at all on the northern slope of a hill or when the south side of the trees is shaded in some other way. Is the hypothesis rendered unacceptable or invalid because, so far as we know, leaves do not "deliberate" or consciously "seek," have not been to school and learned the relevant laws of science or the mathematics required to calculate the "optimum" position, and cannot move from position to position? Clearly, none of these contradictions of the hypothesis is vitally relevant; the phenomena involved are not within the "class of phenomena the hypothesis is designed to explain"; the hypothesis does not assert that leaves do these things but only that their density is the same *as if* they did. Despite the apparent falsity of the "assumptions" of the hypothesis, it has great plausibility because of the conformity of its implications with observation. We are inclined to "explain" its validity on the ground that sunlight contributes to the growth of leaves and that hence leaves will grow denser or more putative leaves survive where there is more sun, so the result achieved by purely passive adaptation to external circumstances is the same as the result that would be achieved by deliberate accommodation to them. This alternative hypothesis is more attractive than the constructed hypothesis not because its "assumptions" are more "realistic" but rather because it is part of a more general theory that applies to a wider variety of phenomena, of which the position of leaves around a tree is a special case, has more implications capable of being contradicted, and has failed to be contradicted under a wider variety of circumstances. The direct evidence for the growth of leaves is in this way strengthened by the indirect evidence from the other phenomena to which the more general theory applies.

The constructed hypothesis is presumably valid, that is, yields "sufficiently" accurate predictions about the density of leaves, only for a particular class of circumstances. I do not know what these circumstances are or how to define them. It seems obvious, however, that in this example the "assumptions" of the theory will play no part in specifying them: the kind of tree, the character of the soil, etc., are the types of variables that are likely to define its range of validity, not the ability of the leaves to do complicated mathematics or to move from place to place.

A largely parallel example involving human behavior has been used elsewhere by Savage and me.[15] Consider the problem of predicting the shots made by an expert billiard player. It seems not at all unreasonable that excellent predictions would be yielded by the hypothesis that the billiard player made his shots *as if* he knew the complicated mathematical formulas that would give the optimum directions of travel, could estimate accurately by eye the angles, etc., describing the location of the balls, could make lightning calculations from the formulas, and could then make the balls travel in the direction indicated by the formulas. Our confidence in this hypothesis is not based on the belief that billiard players, even expert ones, can or do go through the process described; it derives rather from the belief that, unless in some way or other they were capable of reaching essentially the same result, they would not in fact be *expert* billiard players.

It is only a short step from these examples to the economic hypothesis that under a wide range of circumstances individual firms behave *as if* they were seeking rationally to maximize their expected returns (generally if misleadingly called "profits")[16] and had full knowledge of the data needed to succeed in this attempt; *as if*, that is, they knew the relevant cost and demand functions, calculated marginal cost and marginal revenue from all actions open to them, and pushed each line of action to the point at which the relevant marginal cost and marginal revenue were equal. Now, of course, businessmen do not actually and literally solve the system of simultaneous equations in terms of which the mathematical economist finds it convenient to express this hypothesis, any more than leaves or billiard players explicitly go through complicated mathematical calculations or falling bodies decide to create a vacuum. The billiard player, if asked how he decides where to hit the ball, may say that he "just figures it out" but then also rubs a rabbit's foot just to make sure; and the businessman may well say that he prices at average cost, with of course some minor deviations when the market makes it necessary. The one statement is about as helpful as the other, and neither is a relevant test of the associated hypothesis.

Confidence in the maximization-of-returns hypothesis is justified by evidence of a very different character. This evidence is in part similar to that adduced on behalf of the billiard-player hypothesis—unless the behavior of businessmen in some way or other approximated behavior consistent with the maximization of returns, it seems unlikely that they would remain in business for long. Let the apparent immediate determinant of business behavior be anything at all—habitual reaction, random chance, or whatnot. Whenever this determinant happens to lead to behavior consistent with rational and informed maximization of returns, the business will prosper and acquire resources with which to expand; whenever it does not, the business will tend to lose resources and can be kept in existence only by the addition of resources from outside. The process of "natural selection" thus helps to validate the hypothesis—or, rather, given natural selection, acceptance of the hypothesis can be based largely on the judgment that it summarizes appropriately the conditions for survival.

An even more important body of evidence for the maximization-of-returns hypothesis is experience from countless applications of the hypothesis to specific problems and the repeated failure of its implications to be contradicted. This evidence is extremely hard to document; it is scattered in numerous memorandums, articles, and monographs concerned primarily with specific concrete problems rather than with submitting the hypothesis to test. Yet the continued use and acceptance of the hypothesis over a long period, and the failure of any coherent, self-consistent alternative to be developed and be widely accepted, is strong indirect testimony to its worth. The evidence *for* a hypothesis always consists of its repeated failure to be contradicted, continues to accumulate so long as the hypothesis is used, and by its very nature is difficult to document at all comprehensively. It tends to become part of the tradition and folklore of a science revealed in the tenacity with which hypotheses are held rather than in any textbook list of instances in which the hypothesis has failed to be contradicted.

Conclusion

Economics as a positive science is a body of tentatively accepted generalizations about economic phenomena that can used to predict the consequences of changes in circumstances. Progress in expanding this body of generalizations, strengthening our confidence in their validity, and improving the accuracy of the predictions they yield is hindered not only by the limitations of human ability that impede all search for knowledge but also by obstacles that are especially important for the social sciences in general and economics in particular, though by no means peculiar to them. Familiarity with the subject matter of economics breeds contempt for special knowledge about it. The importance of its subject matter to everyday life and to major issues of public policy impedes objectivity and promotes confusion between scientific analysis and normative judgment. The necessity of relying on uncontrolled experience rather than on controlled experiment makes it difficult to produce dramatic and clear-cut evidence to justify the acceptance of tentative hypotheses. Reliance on uncontrolled experience does not affect the fundamental methodological principle that a hypothesis can be tested only by the conformity of its implications or predictions with observable phenomena; but it does render the task of testing hypotheses more difficult and gives greater scope for confusion about the methodological principles involved. More than other scientists, social scientists need to be self-conscious about their methodology.

One confusion that has been particularly rife and has done much damage is confusion about the role of "assumptions" in economic analysis. A meaningful scientific hypothesis or theory typically asserts that certain forces are, and other forces are not, important in understanding a particular class of phenomena. It is frequently convenient to present such a hypothesis by stating that the phenomena it is desired to predict behave in the world of observation *as if* they occurred in a hypothetical and highly simplified world containing only the forces that the hypothesis asserts to be important. In general, there is more than one way to formulate such a description—more than one set of "assumptions" in terms of which the theory can be presented. The choice among such alternative assumptions is made on the grounds of the resulting economy, clarity, and precision in presenting the hypothesis; their capacity to bring indirect evidence to bear on the validity of the hypothesis by suggesting some of its implications that can be readily checked with observation or by bringing out its connection with other hypotheses dealing with related phenomena; and similar considerations.

Such a theory cannot be tested by comparing its "assumptions" directly with "reality." Indeed, there is no meaningful way in which this can be done. Complete "realism" is clearly unattainable, and the question whether a theory is realistic "enough" can be settled only by seeing whether it yields predictions that are good enough for the purpose in hand or that are better than predictions from alternative theories. Yet the belief that a theory can be tested by the realism of its assumptions independently of the accuracy of its predictions is widespread and the source of much of the perennial criticism of economic theory as unrealistic. Such criticism is largely irrelevant, and, in consequence, most attempts to reform economic theory that it has stimulated have been unsuccessful.

The irrelevance of so much criticism of economic theory does not, of course, imply that existing economic theory deserves any high degree of confidence. These criticisms may miss the target, yet there may be a target for criticism. In a trivial sense, of course, there obviously is. Any theory is necessarily provisional and subject to change with the advance of knowledge. To go beyond this platitude, it is necessary to be more specific about the content of "existing economic theory" and to distinguish among its different branches; some parts of economic theory clearly deserve more confidence than others. A comprehensive evaluation of the present state of positive economics, summary of the evidence bearing on its validity, and assessment of the relative confidence that each part deserves is clearly a task for a treatise or a set of treatises, if it be possible at all, not for a brief paper on methodology.

About all that is possible here is the cursory expression of a personal view. Existing relative price theory, which is designed to explain the allocation of resources among alternative ends and the division of the product among the cooperating resources and which reached almost its present form in Marshall's *Principles of Economics*, seems to me both extremely fruitful and deserving of much confidence for the kind of economic system that characterizes Western nations. Despite the appearance of considerable controversy, this is true equally of existing static monetary theory, which is designed to explain the structural or secular level of absolute prices, aggregate output, and other variables for the economy as a whole and which has had a form of the quantity theory of money as its basic core in all of its major variants from David Hume to the Cambridge School to Irving Fisher to John Maynard Keynes. The weakest and least satisfactory part of current economic theory seems to me to be in the field of monetary dynamics, which is concerned with the process of adaptation of the economy as a whole to changes in conditions and so with short-period fluctuations in aggregate activity. In this field we do not even have a theory that can appropriately be called "the" existing theory of monetary dynamics.

Of course, even in relative price and static monetary theory there is enormous room for extending the scope and improving the accuracy of existing theory. In particular, undue emphasis on the descriptive realism of "assumptions" has contributed to neglect of the critical problem of determining the limits of validity of the various hypotheses that together constitute the existing economic theory in these areas. The abstract models corresponding to these hypotheses have been elaborated in considerable detail and greatly improved in rigor and precision. Descriptive material on the characteristics of our economic system and its operations has been amassed on an unprecedented scale. This is all to the good. But, if we are to use effectively these abstract models and this descriptive material, we must have a comparable exploration of the criteria for determining what abstract model it is best to use for particular kinds of problems, what entities in the abstract model are to be identified with what observable entities, and what features of the problem or of the circumstances have the greatest effect on the accuracy of the predictions yielded by a particular model or theory.

Progress in positive economics will require not only the testing and elaboration of existing hypotheses but also the construction of new hypotheses. On this problem there is little to say on a formal level. The construction of hypotheses is a creative act of inspiration, intuition, invention; its essence is the vision of something new in familiar material. The process must be discussed in psychological, not logical, cateories; studied in autobiographies and biographies, not treatises on scientific method; and promoted by maxim and example, not syllogism or theorem.

Notes

I have incorporated bodily in this chapter without special reference most of my brief "Comment" in *A Survey of Contemporary Economics*, vol. II, ed. B. F. Haley (Chicago: Richard D. Irwin, Inc., 1952), pp. 455–7. I am indebted to Dorothy S. Brady, Arthur F. Burns, and George J. Stigler for helpful comments and criticism.

1. (London: Macmillan & Co., 1891), pp. 34–5 and 46.
2. Social science or economics is by no means peculiar in this respect—witness the importance of personal beliefs and of "home" remedies in medicine wherever obviously convincing evidence for "expert" opinion is lacking. The current prestige and acceptance of the views of physical scientists in their fields of specialization—and, all to often, in other fields as well—derives, not from faith alone, but from the evidence of their works, the success of their predictions, and the dramatic achievements from applying their results. When economics seemed to provide such evidence of its worth, in Great Britain in the first half of the nineteenth century, the prestige and acceptance of "scientific economics" rivaled the current prestige of the physical sciences.
3. The interaction between the observer and the process observed that is so prominent a feature of the social sciences, besides its more obvious parallel in the physical sciences, has a more subtle counterpart in the indeterminacy principle arising out of the interaction between the process of measurement and the phenomena being measured. And both have a counterpart in pure logic in Gödel's theorem, asserting the impossibility of a comprehensive self-contained logic. It is an open question whether all three can be regarded as different formulations of an even more general principle.
4. One rather more complex example is stabilization policy. Superficially, divergent views on this question seem to reflect differences in objectives; but I believe that this impression is misleading and that at bottom the different views reflect primarily different judgments about the source of fluctuations in economic activity and the effect of alternative countercyclical action. For one major positive consideration that accounts for much of the divergence see "The Effects of a Full-Employment Policy on Economic Stability: A Formal Analysis," Milton Friedman, *Essays in Positive Economics* (Chicago: University of Chicago Press, 1953), pp. 117–32. For a summary of the present state of professional views on this question see "The Problem of Economic Instability," a report of a subcommittee of the Committee on Public Issues of the American Economic Association, *American Economic Review*, vol. xl (September 1950), pp. 501–38.
5. Final quoted phrase from Alfred Marshall, "The Present Position of Economics" (1885), reprinted in *Memorials of Alfred Marshall*, ed. A. C. Pigou (London: Macmillan & Co., 1925), p. 164. See also "The Marshallian Demand Curve," pp. 56–7, 90–1 [of Friedman 1953].
6. See "Lange on Price Flexibility and Employment: A Methodological Criticism," pp. 282–9 [of Friedman 1953].
7. "The Marshallian Demand Curve," p. 57 [of Friedman 1953].
8. The qualification is necessary because the "evidence" may be internally contradictory, so there may be no hypothesis consistent with it. See also "Lange on Price Flexibility and Employment," pp. 282–3 [of Friedman 1953].
9. See "Lange on Price Flexibility and Employment" [in Friedman 1953], passim.

10. See also Milton Friedman and L. J. Savage, "The Expected-Utility Hypothesis and the Measurability of Utility," *Journal of Political Economy*, vol. lx (December 1953), pp. 463–74, esp. pp. 465–7.
11. [We have omitted a long, technical footnote discussing a possible distinction between a model and a structure—Eds.]
12. The converse of the proposition does not, of course, hold: assumptions that are unrealistic (in this sense) do not guarantee a significant theory.
13. See R. A. Lester, "Shortcomings of Marginal Analysis for Wage-Employment Problems," *American Economic Review*, vol. xxxvi (March 1946), pp. 62–82; Fritz Machlup, "Marginal Analysis and Empirical Research," *American Economic Review*, vol. xxxvi (September 1946), pp. 519–54; R. A. Lester, "Marginalism, Minimum Wages, and Labor Markets," *American Economic Review*, vol. xxxvii (March 1947), pp. 135–48; Fritz Machlup, "Rejoinder to an Antimarginalist," *American Economic Review*, vol. xxxvii (March 1947), pp. 148–54; G. J. Stigler, "Professor Lester and the Marginalists," *American Economic Review*, vol. xxxvii (March 1947), pp. 154–7; H. M. Oliver, Jr., "Marginal Theory and Business Behavior," *American Economic Review*, vol. xxxvii (June 1947), pp. 375–83; R. A. Gordon, "Short-Period Price Determination in Theory and Practice," *American Economic Review*, vol. xxxviii (June 1948), pp. 265–88.

    It should be noted that, along with much material purportedly bearing on the validity of the "assumptions" of marginal theory, Lester does refer to evidence on the conformity of experience with the implications of the theory, citing the reactions of employment in Germany to the Papen plan and in the United States to changes in minimum-wage legislation as examples of lack of conformity. However, Stigler's brief comment is the only one of the other papers that refers to this evidence. It should also be noted that Machlup's thorough and careful exposition of the logical structure and meaning of marginal analysis is called for by the misunderstandings on this score that mar Lester's paper and almost conceal the evidence he presents that is relevant to the key issue he raises. But, in Machlup's emphasis on the logical structure, he comes perilously close to presenting the theory as a pure tautology, though it is evident at a number of points that he is aware of this danger and anxious to avoid it. The papers by Oliver and Gordon are the most extreme in the exclusive concentration on the conformity of the behavior of businessmen with the "assumptions" of the theory.
14. This example, and some of the subsequent discussion, though independent in origin, is similar to and in much the same spirit as an example and the approach in an important paper by Armen A. Alchian, "Uncertainty, Evolution, and Economic Theory," *Journal of Political Economy*, vol. lviii (June 1950), pp. 211–21.
15. Milton Friedman and L. J. Savage, "The Utility Analysis of Choices Involving Risk," *Journal of Political Economy*, vol. lvi (August 1948), p. 298. Reprinted in American Economic Association, *Readings in Price Theory* (Chicago: Richard D. Irwin, Inc., 1952), pp. 57–96.
16. It seems better to use the term "profits" to refer to the difference between actual and "expected" results, between ex post and ex ante receipts. "Profits" are then a result of uncertainty and, as Alchian (op. cit., p. 212), following Tintner, points out, cannot be deliberately maximized in advance. Given uncertainty, individuals or firms choose among alternative anticipated probability distributions of receipts or incomes. The specific content of a theory of choice among such distributions depends on the criteria by which they are supposed to be ranked. One hypothesis supposes them to be ranked by the mathematical expectation of utility corresponding to them (see Friedman and Savage, "The Expected-Utility Hypothesis and the Measurability of Utility," op. cit.). A special case of this hypothesis or an alternative to it ranks probability distributions by the mathematical expectation of the money receipts corresponding to them. The latter is perhaps more applicable, and more frequently applied, to firms than to individuals. The term "expected returns" is intended to be sufficiently broad to apply to any of these alternatives.

    The issues alluded to in this note are not basic to the methodological issues being discussed, and so are largely bypassed in the discussion that follows.

Chapter 42

If Economics Isn't Science, What Is It?

Alexander Rosenberg

In a number of papers, and in *Microeconomic Laws*,[1] I argued that economic theory is a conceptually coherent body of causal general claims that stand a chance of being laws. My arguments elicited no great sigh of relief among economists, for they are not anxious about the scientific respectability of their discipline.[2] But others eager to adopt or adapt microeconomic theory to their own uses have appealed to these and other arguments which attempt to defend economic theory from a litany of charges that are as old as the theory itself.[3] Among these charges, the perennial ones were those that denied to economic theory the status of a contingent empirical discipline because it failed to meet one or another fashionable positivist or Popperian criterion of scientific respectability. With the waning of positivism these charges have seemed less and less serious to philosophers, although they have retained their force for the few economists still distracted by methodology.[4] But among philosophers charges that economics does not measure up to standards for being a science have run afoul of the general consensus that we have no notion of science good enough to measure candidates against. This makes it difficult to raise the question of whether economics is a science, and tends to leave economists, and their erstwhile apologists like me, satisfied with the conclusion that since there is nothing logically or conceptually incoherent about economics, it must be a respectable empirical theory of human behavior and/or its aggregate consequences.

The trouble with this attitude is that it is unwarrantably complacent. It is all well and good to say that economics is conceptually coherent, and that there are no uncontroversial standards against which economics may be found wanting, but this attitude will not make the serious anomalies and puzzles about economic theory go away. These puzzles surround its thorough-going predictive weakness. The ability to predict and control may be neither necessary nor sufficient criteria for cognitively respectable scientific theories. But the fact is that microeconomic theory has made no advances in the management of economic processes since its current formalism was first elaborated in the nineteenth century. And this surely undermines a complacent conviction that the credentials of economics as a science are entirely in order. For a long time after 1945 it might confidently have been said that Keynesian macroeconomics was a theory moving in the right direction: although a macro theory, it would ultimately provide the sort of explanatory and predictive satisfaction characteristic of science. But the simultaneous inflation and unemployment levels of the last decade and the economy's imperviousness to fiscal policy have eroded the layman's and the economist's confidence in the theory. Moreover the profession's reaction to the failures of Keynesian theory is even more disquieting to those who view economic theory as unimpeachably a scientific enterprise. For a large part of the response to its failures has been a return to the microeconomic theories which it was sometimes claimed to supersede. The diagnosis offered for the failure of the Keynesian theory has been that it does not accord individual agents the kind of rationality in the use of information and the satisfaction of preferences that neoclassical microeconomic theory accords them.[5] The alternative offered to Keynesian theory in the light of this result is nothing more nor less than a return to the status quo ante, to the neoclassical theory of Walras, Marshall, and the early Hicks, that Keynesianism had preempted.[6] This cycle brings economic theory right back to where it was before 1937, and it should seriously undermine the confidence of anyone's beliefs that economics is an empirical science, with aims and standards roughly identical to other empirical sciences. For the twentieth-century history of economic theory certainly does not appear to be that of an empirical science.

Of course, eighty years is not a long time in the life of a science, or even a theory, so the fact that economics has not substantially changed, either in its form or in its degree of confirmation, since Walras, or arguably since Adam Smith, is no reason to deny it scientific respectability. But it is reason to ask why economics has not moved away from the theoretical strategies that have characterized it at least since 1874, in spite of their practical inapplicability to crucial matters like the business cycle, economic development, or stagflation. On some views of proper scientific method, of course, economists have been doing just what they should be doing. Since the nineteenth century they have been pursuing a single research strategy, acting in accordance with a ubiquitous and powerful paradigm. For, economists have been steadily elaborating a theory whose *form* is identical to that of the great theoretical breakthroughs in science since the sixteenth century. Accordingly it may be argued that it would be irrational for economists to surrender this strategy short of a conclusive demonstration that it is inappropriate to the explanation of economic activity. The strategy is that of viewing the behavior economists seek to explain as reflecting forces which always move toward stable equilibria that maximize or minimize some theoretically crucial variable. In the case of microeconomics, this crucial variable is utility (or its latter day surrogates), and the equilibrium is given by a level of price in all markets that maximizes this variable. This strategy is most impressively exemplified in Newtonian mechanics and in the Darwinian theory of natural selection. It is no surprise that a strategy which serves so well in these two signal accomplishments of science should have as strong a grip in other domains to which it seems applicable. Moreover, the constraints on theoretical and empirical developments that this strategy imposes can explain many of the greatest successes of Newtonian and Darwinian science, and much of the puzzling character of developments in economic theory.

I call this strategy the extremal strategy, because it is especially apparent in Newtonian mechanics when that theory is expressed in so-called extremal principles, according to which a system's behavior always minimizes or maximizes variables reflecting the mechanically possible states of the system. In the theory of natural selection this strategy assumes that the environment acts so as to maximize fitness. This strategy is crucial to the success of these theories because of the way it directs and shapes the research motivated by them. Thus, if we believe that a system always acts to maximize the value of a mechanical variable, for example, total energy, and our measurements of the observable value of that variable diverge from the predictions of the theory and the initial conditions, we do not infer that the system described is failing to maximize the value of the variable in question. We do not falsify the theory. We assume that we have incompletely described the constraints under which the system is actually operating. In Newtonian mechanics attempts to more completely describe the systems under study resulted in the discovery of new planets, Neptune, Uranus, and Pluto, the invention of new instruments, and eventually in the discovery of new laws, like those of thermodynamics. Similarly in biology, assuming that fitness is maximized led to the discovery of forces not previously recognized to effect genetic variation within a population, and more important, led to the discovery of genetic laws that explain the persistence in a population of apparently maladaptive traits, like sickle-cell anemia, for instance. Because these theories are "extremal" ones, differential calculus may be employed to express and interrelate their leading ideas. Microeconomics is an avowedly extremal theory, asserting that the systems it describes maximize utility (or some surrogate). That is why it can be couched in the language of differential calculus. It is the extremal character of the theory, and not the fact that it deals with "quantifiable" variables, like money, that makes microeconomics a quantitively expressed theory.

More important than the fact that they all employ differential calculus, these theories are all committed *to explain everything in their domains* because of their extremal character. In virtue of the claim that systems in their domains always behave in a way which maximizes or minimizes some quantity, the theories ipso facto provide the explanation of all of their subjects' behavior by citing the determinants of all their subjects' relevant states. An extremal theory cannot be treated as only a partial account of the behavior of objects in its domain, or as enumerating just *some* of the many determinants of its subject's states; for any behavior that actually fails to maximize or minimize the value of the privileged variable simply refutes the theory *tout court*. In fact, the pervasive character of extremal theories insulates them from falsification to a degree absent from nonextremal theories. All theories are strictly unfalsifiable, simply because testing them involves auxiliary hypotheses. But extremal theories are not only insulated against strict falsification, they are also insulated against the sort of actual falsification that usually overthrows theories, instead of auxiliary hypotheses. In the case of a nonextremal theory falsification may lead us either to revise the auxiliary assumptions about test conditions, or to revise the theory by adding new antecedent clauses to its generalizations, or new qualifications to its ceteris paribus clauses. But this is not possible in the case of extremal theories. The axioms of theories like Newton's, or Darwin's, or Walras's do not embody even implicit ceteris paribus clauses. Microeconomics does not, for example, assume that agents maximize utility, ceteris paribus. With these theories the choice is always between rejecting the auxiliary hypotheses—the description of test conditions—or rejecting the theory altogether. For the only change that can be made to the theory is to deny that its subjects invariably maximize or minimize its chosen variable. This is why high-level extremal theories like Newtonian mechanics are left untouched by apparent counter-instances; why they are not simply improved by qualifications and caveats in their antecedent conditions; why they are superseded only by utterly new theories, in which the values of very different variables are maximized, or minimized.

Extremal theories are an important methodological strategy because they are so well insulated from falsification. This has enabled them to function at the core of research programs, turning what otherwise might be anomalies and counterinstances into new predictions and new opportunities for extending their domains and deepening their precision. Accordingly, it may be argued, economists' attachment to their extremal theory represents not complacency, but a well-grounded methodological conservatism. Given the fantastic successes of this approach in such diverse areas as mechanics and biology, it would be unreasonable to forgo similar strategies in the attempt to explain human behavior. So viewed, the history of attempts to make recalcitrant facts about human behavior, and the economic systems humans have constructed, fit the extremal theory of microeconomics reflects a commitment that is on a par with astronomers' attempts to make recalcitrant facts about planetary perihelions fit the demands of Newtonian mechanics; it is on a par with biologists' attempts to make the persistent genetic predisposition to malaria fit the facts of adaptation demanded by the theory of natural selection. Since these attempts do not discredit their theories as empty or unfalsifiable, it should not be inferred that there is anything improper in the economists' attempts to do the same thing. Or so it may be argued.

It is certainly correct that much of the commitment to microeconomic strategies does in fact reflect this sort of reasoning. After all, it is not just the intellectual prestige associated with the scope for differential calculus, topology, and differential geometry attending any extremal theory that explains the reluctance of economists to forgo the strategy. But this conservative rationale for the attachment of economists to extremal theories is vitiated by a crucial disanalogy between microeconomics and mechanics or evolution. Economists would indeed be well advised not to surrender their extremal research program, if only they could boast even a small part of the startling successes that other extremal research programs have achieved. But two hundred years of work in the same direction have produced nothing comparable to the physicists' discovery of new planets, or of new technologies by which to control the mechanical phenomena that Newton's laws systemized. Economists have attained no independently substantiated insight into their domain to rival the biologists' understanding of macroevolution and its underlying mechanism of adaptation and heredity. There has been no signal success of economic theory akin to these advances of extremal theory. This is a disanalogy important enough to bear explaining. Failing a satisfactory explanation, the difference is significant enough to make economists question the merits of their extremal approach, and to make us query the scientific credentials of economic theory. There is, of course, a vast literature on why economics has so little in the way of predictive content, and on how a theory so dependent on idealizations and factually false assumptions as microeconomics can nevertheless constitute a respectable scientific enterprise. This literature goes back to John Stuart Mill[7] and forward to, for example, Hal Varian.[8] The only two things clear about this literature are that economists have found it almost universally satisfying and legitimating, and noneconomists have consistently been left unsatisfied, insisting that methodological excuses are no substitute for attempting to do what economics has hitherto not done: improve its predictive content.

Having shaken free from the complacent attitude toward economic theory evinced in *Microeconomic Laws*, I have come to think that the failure of economics is not methodological, or conceptual, but very broadly empirical. Despite its conceptual integrity, microeconomics, together with all the sciences of human action and its aggregation, rests on a false but central conviction that vitiates its axioms and so bedevils the theorems deduced from them. Economic theory assumes that the categories of preference and expectation are the classes in which economic causes are to be systematized, and that the events to be explained are properly classified as actions like buying, selling, and the movements of markets, industries, and economies that these actions aggregate to. The theory has made this assumption, because of course it is an assumption we all make about human behavior; our behavior constitutes action and is caused by the joint operation of our desires and beliefs. Marginalists of the late nineteenth century like Wicksteed saw clearly that microeconomics is but the formalization of this commonsense notion, and the history of the theory of consumer behavior is the search for laws that will express the relations between desire, belief, and action, first in terms of cardinal utility and certainty, later in terms of ordinal utility, revealed preference and expected utility under varying conditions of uncertainty and risk. The failure to find such a law or any approximation to it that actually improves our ability to predict consumer behavior any better than Adam Smith could have resulted on the one hand in a reinterpretation of the aims of economic theory away from explaining individual human action, and on the other in the tissue of apologetics with which the consumer of economic methodology is familiar.[9]

The real trouble with economics, the real source of its failure to find *improvable* laws of economic behavior, is something that has only become clear in philosophy's recent attempts to understand and improve the foundations of another science in trouble: psychology, and particularly behavioral and cognitive psychology. Philosophers have shown that the terms in which ordinary thought and the behavioral sciences describe the causes and effects of human action do not describe "natural kinds," they do not divide nature at the joints. They do not label categories of states that share the same manageably small set of causes and effects, and so cannot be brought together in causal generalizations that improve on our ordinary level of prediction and control of human actions, let alone attain the sort of continuing improvement characteristic of science. I cannot hope to do more in the present compass than identify the conclusions to which thirty-five years of work in the philosophy of psychology has arrived, and to show their bearing on economic theory. This work has been devoted to understanding "intentional" terms like "belief," "desire," "action," and their vast hoard of cognates. It is worth noting that the term "intentional" does not mean simply "purposive"; rather, it is employed by philosophers to note that the mental states like beliefs and desires are identified by their "contents," by the propositions they "contain" or the ends and objects to which they are "directed." The trouble with beliefs and desires is that when people have them the propositions they "contain" need not be true or false and the objects they are directed at need not exist or even be possible objects. So we can't decide whether a person is in a given mental state by determining either the truth or falsity of any statement open to our confirmation, or the existence or attainment of any object or end of human action. In psychology and philosophy the two leading attempts to solve this problem of other minds have been behaviorism and the so-called identity theory, the claim that each *type* of mental state ordinary language distinguishes is identical to, or is correlated with, an identifiable type of brain state. Behaviorism is false, however, just because no intentional state can be identified without making assumptions about other mental states. I can't identify your beliefs by observing your actions unless I know what your desires are; indeed I can't identify your movements as actions at all, as opposed to mere reflexes, unless I assume they are caused by the joint operation of your beliefs and desires. There is an intentional circle into which we cannot break by discoveries about mere movement, about behavior. The identity theory seems equally fruitless an approach, because the brain states neuroscience identifies just don't seem to line up in lockstep with mental states that introspection reports. This means that the intentional vocabulary of common sense, (intentional) psychology, and the social sciences are fated to remain isolated from any other conceptual scheme that identifies and systematizes the mental states and human behavior they describe and explain.[10] This isolation is an obstacle to improving the predictive power of these disciplines, because improvement in any explanatory theory requires the variables of the theory to be measured with increasing accuracy. This sort of precision is plainly impossible when the theory's variables are restricted to an interdefinable circle alone, and are not independently identifiable. But the failure of behaviorism and neurological reduction forecloses all such identifications. Accordingly we can't expect to improve our intentional explanations of action beyond their present levels of predictive power. But this level of predictive power is no higher than Plato's. The predictive weakness of theories couched in intentional vocabulary reflects the fact that the terms of this vocabulary do not correlate in a manageable way the vocabulary of other successful scientific theories; they don't divide nature at the joints, insofar as its joints are revealed in already successful theories like those of neuroscience. Some philosophers hold that we must reconcile ourselves to this state of affairs, and reduce our expectations about the possibilities of a predictively powerful theory of human behavior. Others are more optimistic, but insist that we must jettison "folk psychology" and its intentional idiom if we are to hit upon an improvable theory in the science of psychology.[11] This choice extends, of course, beyond psychology to all the other intentional sciences, of which economics, with its reliance on expectation and preference, is certainly one.[12]

Applying the new orthodoxy in the philosophy of psychology, it becomes clear that economics' predictive weakness hinges on the intentional typology of the phenomena it explains and the causes it identifies. Its failure to uncover laws of human behavior is due to its wrongly assuming that these laws will trade in desires, beliefs, or their cognates. And the system of propositions about markets and economies that economists have constructed on the basis of its assumptions about human behavior is deprived of improving explanatory and predictive power because its assumptions can't be improved in a way that transmits improved precision to their consequences. Thus the failure of economics is traced not to a conceptual mistake, or to the inappropriateness of extremal theories and their elegant mathematical apparatus to human action, but to a false assumption economists share with all other social scientists, indeed with everyone who has ever explained their own or others' behavior by appeal to the operation of desires and beliefs.

Just as economists have been given no pause by previous attacks on their discipline, they are unlikely to down tools in the fact of this diagnosis either. Indeed, the persistence of economists in pursuing the extremal and intentional approach that has been conventional for well over a century suggests that nothing could make them give it up, or at any rate that nothing which would make empirical scientists give up a theory will make economists give up their theoretical strategy. But this conviction leads to the conclusion that economics is not empirical science at all. Despite its appearances, and the interest of some economists in applying their formalism to practical matters, this formalism does not any longer have the aims, nor does it make the claims, of an unequivocally empirical theory.

My diagnosis and conclusion face many potential rejoinders and several serious questions. In the remainder of this chapter I shall address one of these rejoinders and attempt to answer one of these questions. The rejoinder takes the following form:

> Surely the fact that the fundamental axioms of economic theory fail to divide nature at the joints does not vitiate the entire enterprise. After all, your explanation would have us forgo not just the abstract claims about preference orders and individual choice under uncertainty, but also the laws of supply and demand. Yet surely these are useful approximations, regularities roughly and frequently enough instanced to reflect some underlying truth about economic behavior, enough at any rate to make economics a worthwhile pursuit even if all your claims are correct. Though you may have provided an abstract explanation of why we cannot improve the current theory indefinitely, you have not given enough reason to deprive it of its usefulness, nor deny it scientific standing.

The question I want to address is closely related to this rejoinder:

> If the extremal intentional research program of economics is an empirical failure, and if economists are not about to surrender either the intentional stance or the extremal method, what is it on your view that economists are doing? What after all is economics? On your view, it does not constitute a scientific discipline. What then is it? This query is not just a rhetorical demand that I attach a label to the subject, but the reasonable request that my diagnosis must come complete with an explanation of what is really going on in economics, if as I have claimed what is going on cannot be viewed as empirical science.

The strength of the rejoinder rests on two undeniable considerations. One is that for all its infirmities, economic theory does at least sometimes *seem* to be insightful. Occasionally, qualitative predictions are borne out, and even more frequently, retrospective economic explanations of events that were unexpected, like a 15 per cent reduction in the consumption of gasoline, can be given. The second consideration is more abstract, but quite telling as a point in the philosophy of science: There are several scientific theories which to varying degrees fail to divide nature at the joints, and yet they are useful approximations, even if they cannot be reduced to more fundamental theories that do divide nature at the joints (or that are currently believed to do so). For instance, the Mendelian unit of inheritance cannot be reduced to the molecular gene and so does not divide its phenomena at the joints. Yet Mendel's laws are useful approximations that we would be silly to forgo. Mutatis mutandis for economic theory. Thus, even if my claims about the intentional vocabulary of economics are correct, they are not sufficient reason to surrender the theory.

It is quite correct that the problems I have noted for economics have parallels in other successful scientific theories. On the other hand these theories have proved successful, on standards of *improving* technological and predictive success that economics has not met. The fact that such theories do not carve nature at the joints, that they are to a degree incommensurable with their successors, or with more fundamental theories, is a problem in the philosophy of science. But it is a different problem from that which faces economics. For economics has not met with anything like the success they have met with, and what is required by this fact is an explanation of why it has not. I claim that part of the explanation is that the descriptive vocabulary fails to divide nature at the joints, so that no improvement of current economic theory can provide laws governing intentional economic activities. A better comparison for economic theory than Mendelian genetics is phlogiston theory, whose failure is traceable to its *incommensurability* with the oxygen theory that superseded it. Phlogiston theory is a scientific dead end, because there is no such thing as phlogiston, because the notion of phlogiston does not divide nature at any joint. Phlogiston is not a natural kind. This is an empirical fact about nature, and the claim that intentional notions are not natural kinds is equally a contingent claim. Thus, economics and phlogiston theory are not methodologically defective. They are simply false.

But, the rejoinder continues, what of the successes of economic theory? How can I square my arcane philosopher's argument with the evident applicability of such staples as the laws of supply and demand? After all, it is a fact about markets in all commodities that *eventually* price will influence demand and supply in the directions that microeconomic theories of economic action dictate. Surely this is an economic regularity and surely it is a consequence of individual choices, preferences, and beliefs.

Bringing what little utility there is in economic theory into harmony with my diagnosis of its ills is a task for which there are many solutions. The first thing to note is this: although the laws of supply and demand and other market-level general statements are deduced from claims about the intentional determinants of individual actions, they are logically separable from such claims, and, more important, they can be shown to follow from assumptions which are the direct denial of these general claims about rational action. From the assumption that individuals behave in purely habitual ways, always purchasing the same or the most nearly similar bundle of commodities available, no matter what the price, the law of downward sloping demand follows, as it does from the assumption that their purchases are all impulsively random.[13] The same can be shown for the choices of entrepreneurs. So surrendering the extremal intentional approach to human behavior does not logically or even theoretically oblige us to surrender these "laws." On the other hand, we cannot sharpen their applicability beyond the most qualitative or generic levels, or quantify the values of their parameters like elasticity, or improve our foresight or hindsight in the employment of these principles. Now the fact that we can usefully employ false or vacuous general statements, up to certain limits, is no mystery in the philosophy of science at all. The clearest instance of such restrictedly useful though false or vacuous general statements is Euclidean geometry. For millennia this axiomatic system was viewed as the science of space, and the great mystery which surrounded it was how we can have the apparently a priori knowledge of the nature of the world that the science of space, Euclidean geometry, gave us. Since Poincaré and Einstein this problem has been largely resolved. Prior to the twentieth century Euclidean geometry was equivocally interpreted as both a pure axiomatic system about abstract objects, one that constituted the implicit definitions of its terms, and was therefore a priori true, and as a body of claims about actual spatial relations among real objects in the world. The equivocation between these two interpretations in part caused Kant's problem of how synthetic propositions could be known a priori. Once distinguished, we came to discover that, interpreted as a theory of actual spatial relations, Euclidean geometry is false, and interpreted as a body of a priori truths implicitly defining the terms that figure in it, Euclidean geometry is vacuous. More important for our purposes, it was shown to be useless and inapplicable as a body of conventions, beyond certain values of distance and mass in space. In retrospect, we can explain why no one ever noticed these facts about geometry, and why before 1919 it proved entirely satisfactory for settling empirical questions of geography, surveying, engineering, mechanics, and astronomy. The reason, of course, is that for these questions we neither needed nor had the means to make measurements fine enough to reveal the inadequacies of Euclidean geometry.

When we need to improve our measurements beyond this level of fineness, in contemporary cosmology, for instance, we must forgo Euclidean geometry in favor of one or another of its non-Euclidean alternatives. One way to describe the twentieth-century fate of Euclidean geometry is to say that its kind terms proved not to name *natural kinds*: nature diverges from the predictions of an applied Euclidean geometry, because it does not contain examples, realizations, and instances of the kind terms of that theory. There are no Euclidean triangles. This is something we came to learn only with the advent of another theory, the general theory of relativity, which not only revealed this fact but also explained the degree of success Euclidean geometry does in fact attain when applied to small regions of space.

Of course, economic theory has attained nothing like the success of Euclidean geometry. But the apparent applicability of some of its claims is to be explained by appeal to the same factors which explain why we can employ, e.g., the Pythagorean theorem, even though there are no Euclidean triangles and no Euclidean straight lines. We can employ the laws of supply and demand, even though human beings are not economically rational agents; that is, we can employ these "laws" even though individuals do not make choices reflecting any empirical regularity governing their expectations and their intentions. We can employ them all right, but the laws of supply and demand cannot be applied with the usefulness and exactitude of the Pythagorean theorem, just because the kind terms of economic theory are different from the real kinds in which human behavior is correctly classified. And this difference is comparatively much greater than the difference between the kind terms of applied geometry and those of physics. There are no Euclidean triangles, but we know why, and we can calculate the amount of the divergence between any physical triangle and the Euclidean claims about it, because we have a physical theory to make these corrections, the very one which showed Euclidean geometry to be factually false. We can make no such improvements in the application of the laws of supply and demand; we can never do any better than apply them retrospectively or generically; we cannot specify their parameters, or their exceptions, because the axiomatic system in which they figure diverges from the facts very greatly, and because we have no associated theory that enables us to measure this divergence and make appropriate corrections for it. This is a difference in kind between Euclidean geometry and economic theory. They differ in applicability only by degree, the predicates of neither pick out natural kinds; but they differ in kind because for Euclidean geometry there is a theory, physics, that enables us to correct and improve the applicability of its implications. There is no such theory that enables us to improve on the applicability of economic theory.

Such a theory is of course logically possible, says a version of cognitive psychology, that provides bridges from economic variables like preference and expectation to independently identifiable psychological states. Such a theory might enable us to actually predict individual economic choices and to correct our microeconomic predictions of them, when these predictions go wrong. It would either enable us to improve microeconomics beyond the level at which it has been stuck for a hundred years, or it would show that the determinants of human behavior are so orthogonal to the theory's assumptions about them that microeconomics is best given up altogether. In other words, such a theory would show either that economic theory is like Euclidean geometry or that it is like phlogiston theory, or perhaps that it is somewhere in between. But the fact is that no such theory is in the offing, or on the horizon. What is worse, even if it were available, it is not likely to actually deflect practicing economists from their intentional extremal research program. And the reason is not that they are satisfied with the level of success their theory has attained, a level much closer to that of phlogiston theory than to that of Euclidean geometry. Rather, the reason is that they are not really much interested in questions of empirical applicability at all. Otherwise some of the attractive nonintentional and/or nonextremal approaches to economic behavior that are available would long ago have elicited more interest from economists than they have.[14]

My explanation for the failures of economics does economists more credit than several possible alternatives. It does not, for example, simply write off economic theory as the ideological rationalization of bourgeois capitalism; it renders the immense amount of sheer genius bestowed on the development of this theory its due. The explanation does not stigmatize the methods of economists as conceptually confused or misdirected. It isolates the failings of the theory in an empirical supposition about the determinants of human behavior, one that economists share with all of us. But this supposition is false, and so economics rests on a purely contingent, though nevertheless central, mistaken belief; just as the conviction that Euclidean geometry was the science of space rested on a purely contingent, almost equally central mistaken belief that the paths of light rays are Euclidean straight lines.

But, as I have said, the history of economic theory shows that economists cannot be expected to surrender their commitment to an extremal intentional research program, no matter what its empirical inadequacies. And this raises the question that goes together with the rejoinder: If economics does not behave the way a science does, what sort of an activity is it? If economists have not in fact been elaborating a contingent, empirical theory that successively improves our explanatory and predictive understanding of economic behavior, what has this notable intellectual achievement been aimed at? The parallel I have drawn with Euclidean geometry can help answer this question.

Euclidean geometry was once styled the science of space, but calling it a science did not make it one, and we have come to view advances in the axiomatization and extension of geometry as events not in science, but in mathematics. Economics is often defined as the science of the distribution of scarce resources, but calling it a science does not make it one. For much of their histories, since 1800, advances in both these disciplines have consisted in improvements of deductive rigor, economy, and elegance of expression, in better axiomatizations, and in the proofs of more and more general results, without much concern as to the usefulness of these results. In geometry, the fifth axiom, the postulate of the parallels, came increasingly to be the focus of attention, not because it was in doubt, but because it seemed so much more ampliative than the others. The crisis of nineteenth-century geometry was provoked by the discovery that denying the postulate of the parallels did not generate a logically inconsistent axiomatic system. Thus the question of the cognitive status of geometry became acute. Some, following Plato, held it to be an intuitively certain body of abstract truths. Some, following Mill, held it to be a body of empirical generalizations. Others, following Kant, viewed it as a body of synthetic a priori truths. Matters were settled by distinguishing between geometry as a pure axiomatic system, composed of analytic truths about abstract objects with or without real physical instances; and geometry as an applied theory about the path of light rays, which was shown to be false for reasons given in the general theory of relativity. Moreover, the abstract and apparently pointless exercises of nineteenth-century geometers in developing non-Euclidean geometries turned out to have an altogether unexpected and important empirical role to play in helping us understand the structure of space after all. For they apparently describe the real structure of space in the large. Of course pure geometry, both Euclidean and non-Euclidean, has continued to be a subject of sustained mathematical interest, and both have had applications undreamed of eighty years ago.

Compare the history of economic theory during the same time. Unlike physical theory, or for that matter the other social sciences, economics has been subject to exactly the same conceptual pigeonholing as geometry. Some have viewed it, with Lionel Robbins, as a Platonic body of intuitively obvious, idealized but nonetheless correct descriptions of human behavior. Others, following Ludwig Von Mises, have insisted it is a Kantian body of synthetic a priori truths about rationality. Others, like the geometrical conventionalists and following T. W. Hutchison, have derided it as a body of tautologies, as a pure system of implicit definitions without any grip on the real world. Still others, following Mill, have held it to be a body of idealizations of rough empirical regularities. Finally some, following Friedman, have treated it as an uninterpreted calculus in the way positivists treated geometry.[15] But most economists, like most geometers, have gone about their business proving theorems, and deriving results, without giving much thought at all to the question of economic theory's cognitive status. For them the really important question, the one which parallels the geometer's concern about the postulate of the parallels, was whether Walras's theorem that a general market clearing equilibrium exists, that it is stable and unique, follows from the axioms of microeconomic theory. Walras offered this result in 1874, as a formalization of Adam Smith's conviction about decentralized economies, but he was unable to give more than intuitive arguments for the theorem. It was only in 1934 that Abraham Wald provided an arduous and intricate satisfactory proof, and much work since his time has been devoted to producing more elegant, more intuitive, and more powerful proofs of new wrinkles on the theorem.[16] Just as geometers in the nineteenth century explored the ramifications of varying the strongest assumptions of Euclidean geometry, economists have devoted great energies to varying equally crucial assumptions about the number of agents, their expectations, returns to scale and divisibilities, and determining whether a consistent economy—a market clearing equilibrium—will still result, will be stable, and will be unique. Their interests in this formal result are quite independent of, indeed are in spite of, the fact that its assumptions about production, distribution, and information are manifestly false. The proof of general equilibrium is the crowning achievement of mathematical economics. But just as geometry as a science faced a crisis in the 1919 observations that confirmed the general theory of relativity, so too economic theory faced a crisis in the evident fact of the Great Depression. For a long time after 1929 the economists lost the *conviction* that the Walrasian general equilibrium was at least a state toward which markets must, in the long run, move. The main reaction to this crisis was, of course, Keynesianism. Insofar as this extremal theory rests on a denial of the fundamental microeconomic assumption that economic agents' expectations are rational, that they do not suffer from money illusions, that they will tailor their actions to current and future economic environments, Keynesian theory represents as much of a conceptual revolution as non-Euclidean geometry did. Keynes, of course, did not entirely win the field, even during the period when his theory appeared to explain why the market clearing general equilibrium might never be approached, let alone realized. One reason for this is that many economists continued to be interested in the purely formal questions of the conventional theory, quite regardless of its irrelevance to understanding the actual world. These economists were implicitly treating microeconomics as a pure axiomatic system, whose terms may not be instantiated in the real world, but which is of great interest, like Euclidean geometry, whether its objects actually exist. More crucially for the history of economics, there never was and is not yet a theory which can play a role for economics like the role played for geometry by physical theory. Physics enables us to choose between alternative applied geometries, and to explain the deviations from actual observation of the ones we reject. There is no such theory to serve as an auxiliary in any choice between an applied neoclassical equilibrium theory and a Keynesian equilibrium theory. When, in the 1970s, Keynesian theory foundered on empirical facts of joint unemployment and inflation, as unremitting as was the fact of the apparent nonmarket clearing equilibrium of the 1930s, the result was an eager return to the traditional theory. Economists have not forgotten the Great Depression, but their interest in it seems limited to showing that, after all, the Walrasian approach is at least logically consistent with it, something Keynes's earliest opponents could have vouchsafed them. In short, the theory and its development have been as insulated from empirical influences as geometry ever was before Einstein. All this suggests that, like geometry, economics is best viewed as a branch of mathematics somewhere on the intersection between pure and applied axiomatic systems.

Much of the mystery surrounding the actual development of economic theory—its shifts in formalism, its insulation from empirical assessment, its interest in proving purely formal, abstract possibilities, its unchanged character over a period of centuries, the controversies about its cognitive status—can be comprehended and properly appreciated if we give up the notion that economics any longer has the aims or makes the claims of an empirical science of human behavior. Rather we should view it as a branch of mathematics, one devoted to examining the formal properties of a set of assumptions about the transitivity of abstract relations: axioms that implicitly define a technical notion of "rationality," just as geometry examines the formal properties of abstract points and lines. This abstract term "rationality" may have far more potential interpretations than economists themselves realize,[17] but rather less bearing on human behavior and its consequences than we have unreasonably demanded economists reveal.

There are some important practical consequences of this answer to the question of what economics really is. If it is best viewed as more akin to a branch of mathematics on the intersection between pure axiomatization and applied geometry, then not only are several cognitive mysteries about economics solved, but more important our perspective on the bearing of economic theory must be fundamentally altered. For if this view is correct we cannot demand that it provide the reliable guide to the behavior of economic agents and the performance of economies as a whole for which the formulation of public policy looks to economics. We should neither attach much confidence to predictions made on its basis nor condemn it severely when these predictions fail. For it can no more be relied on or faulted than Euclidean geometry should be in the context of astrophysics. Admittedly this attitude leaves a vacuum in the foundations of public policy. For without economics we lose even the illusion that we understand the probable, or potential, long-term or merely possible consequences of choices that policy makers are forced to make. Of course, the caution that loss of illusions may foster is certain to be salubrious. On the other hand the vacuum may attract a really useful foundation for decisions about the economy and its improvement.

Notes

For helpful comments on earlier drafts of this chapter I owe thanks to Jeffrey Straussman, R. J. Wolfson, and participants in a colloquium at the University of Colorado, Boulder. Research was supported by the American Council of Learned Societies and the John Simon Guggenheim Memorial Foundation.

1. *Microeconomic Laws: A Philosophical Analysis* (Pittsburgh: University of Pittsburgh Press, 1976).
2. Cf. Scott Gordon, "Should Economists Pay Attention to Philosophers?" *Journal of Political Economy*, 86 (1978). His answer is no.
3. Cf., for instance, R. A. Posner, "Some Uses and Abuses of Economics in Law," *University of Chicago Law Review*, 46 (1979), 281–306.
4. Cf. L. Boland, "A Critique of Friedman's Critics," *Journal of Economic Literature*, 17 (1979), 502–22 and M. Willes, "Rational Expectations as a Counterrevolution," *Public Interest* (1980), 81–96 in which Milton Friedman's classic, "The Methodology of Positive Economics," *Essays in Positive Economics* (Chicago: University of Chicago Press, 1953), is rehashed.
5. Exponents of "rational expectation theory" have shown that, on the basis of attributions of microeconomic rationality to individuals operating in an economy regulated in accordance with Keynesian policies, we can expect just what has happened: increasing the deficit or the money supply does nothing but increase inflation, without lowering unemployment. Cf. B. Kantor, "Rational Expectations and Economic Thought," *Journal of Economic Literature*, 17 (1979), 1427–1441.
6. Willes, op. cit., note 4, reprinted in *The Crisis in Economic Theory* (Basic Books, 1982).
7. *The System of Logic* (London, 1867). Mill's views are ably expounded in D. Hausman, "John Stuart Mill's Philosophy of Economics," *Philosophy of Science*, 48 (1981), 362–85.
8. Hal Varian and Alan Gibbard, "Economic Models," *Journal of Philosophy*, 75 (1978), 669–77.
9. Cf. A. Rosenberg, "Obstacles to Nomological Connection of Reasons and Actions," *Philosophy of Social Science*, 10 (1980), 79–91.
10. Cf. D. Dennett, *Content and Consciousness* (London: Routledge and Kegan Paul, 1966) and D. Davidson, "Mental Events," in Foster, et al., *Experience and Theory* (Amherst, MA: University of Massachusetts Press and Duckworth, 1970), pp. 79–101.
11. Cf. Paul Churchland, "Eliminative Materialism and the Propositional Attitudes," *Journal of Philosophy*, 78 (1981), 67–90.
12. Cf. A. Rosenberg, *Sociobiology and the Preemption of Social Science* (Baltimore: Johns Hopkins Press, 1980), where I expound these implications in detail.
13. Gary Becker, "Irrational Behavior and Economic Theory," *Journal of Political Economy*, 70 (1962), 1–13.
14. In this space I can mention only three such approaches, but all three undercut the rejoinder's suggestion that we cannot give up the intentional approach to economic behavior. There is first of all the behavioral approach to economic behavior associated with Herbert Simon or Richard Cyert, which forgoes both the generality and the elegance of extremal theories in favor of an attempt to uncover at least rough empirical generalizations about the actual behavior of economic units like the household and the firm. This approach involves treating the economic units as feedback systems, though not intentional ones, which respond to environmental forces as adaptive mechanisms and in turn generate outputs that affect other economic units. Still another approach associated with the work of S. G. Winter involves applying a natural selection model to attempt to account for the behavior of economic agents. Finally, N. Georgescu-Roegen's insistence on the relevance of thermodynamic approaches to economic processes represents a clear alternative to the conventional attitude. Adopting or adapting any of these alternatives will result in a theory very different from the classical one, and they all show that the extremal intentional research program is by no means unavoidable and inevitable as an approach to economic phenomena. Moreover, so far as preserving whatever might be useful in conventional theory is concerned, these alternatives are explicitly designed to have no less applicability to market phenomena, and even to individual behavior, than neoclassical theory now has. If there is anything in the laws of supply and demand, in the possibility of stable or unique partial or general equilibria, at least some of these approaches are designed to capture it without burdening themselves by commitments to the causal force of preference and expectation operating uniformly and invariably on choice. Cf. R. Cyert and J. A. March, *A Behavioral Theory of the Firm* (Englewood Cliffs, N. J.: Prentice-Hall, 1963) and R. Cyert and G. Pottinger, "Towards a Better Microeconomic Theory," *Philosophy of Science*, 46 (1979), 204–22, and H. Simon, "A Behavioral Model of Rational Choice," *Quarterly Journal of Economics*, 69 (1955), 99–118; S. J. Winter, "Economic Natural Selection and the Theory of the Firm," *Yale Economic Essay*, 4 (1967), 224–72; N. Georgescu-Roegen, *The Entropy Law and the Economic Process* (Cambridge, MA: Harvard University Press, 1971); and R. Cyert and C. L. Hendrick, "Theory of the Firm," *Journal of Economic Literature*, 10 (1972), 398–414.

15. Lionel Robbins, *An Essay on the Nature and Significance of Economic Science* (London: St. Martins, 1932). T. W. Hutchison, *The Significance and Basic Postulates of Economics* (London: McMillan, 1938); L. Von Mises, *Human Action* (1949); and Milton Friedman, op. cit., note 4.
16. Leonn Walras, *Elements of Pure Economics*, trans. W. Jaffe (Homewood, IL; Irwin, 1954). A. Wald, "On Some Systems of Equations for Mathematical Economics," *Econometrica*, 19, 368–403. For a contemporary version of the proof, cf. Gerard Debreu, *Theory of Value* (New York: Wiley, 1959).
17. Indeed, like apparently useless arcana of nineteenth-century non-Euclidean geometry, the formalism and results of general equilibrium theory are turning out to have applications undreamed of by the economists who have proved the impressive theorems in this system. For their results are being taken over and reinterpreted by mathematical ecology: stripped of their intentional interpretation, they provide proofs and stability conditions for unique stable equilibria, that modern evolutionary biology requires in the development of its own extremal theory of balance and competition in the evolution of the biosphere. Cf., for instance, Oster and Wilson, *The Social Insects* (Cambridge, MA: Harvard University Press, 1978), and R. May, *Stability and Complexity in Model Eco-System* (Princeton, N.J.: Princeton University Press, 1973).

Chapter 43
Actions, Reasons, and Causes
Donald Davidson

What is the relation between a reason and an action when the reason explains the action by giving the agent's reason for doing what he did? We may call such explanations *rationalizations*, and say that the reason *rationalizes* the action.

In this chapter I want to defend the ancient—and commonsense—position that rationalization is a species of causal explanation. The defense no doubt requires some redeployment, but it does not seem necessary to abandon the position, as has been urged by many recent writers.[1]

I

A reason rationalizes an action only if it leads us to see something the agent saw, or thought he saw, in his action—some feature, consequence, or aspect of the action the agent wanted, desired, prized, held dear, thought dutiful, beneficial, obligatory, or agreeable. We cannot explain why someone did what he did simply by saying the particular action appealed to him; we must indicate what it was about the action that appealed. Whenever someone does something for a reason, therefore, he can be characterized as (a) having some sort of pro attitude toward actions of a certain kind, and (b) believing (or knowing, perceiving, noticing, remembering) that his action is of that kind. Under (a) are to be included desires, wantings, urges, promptings; and a great variety of moral views, aesthetic principles, economic prejudices, social conventions, and public and private goals and values insofar as these can be interpreted as attitudes of an agent directed toward actions of a certain kind. The word "attitude" does yeoman service here, for it must cover not only permanent character traits that show themselves in a lifetime of behavior, like love of children or a taste for loud company, but also the most passing fancy that prompts a unique action, like a sudden desire to touch a woman's elbow. In general, pro attitudes must not be taken for convictions, however temporary, that every action of a certain kind ought to be performed, is worth performing, or is, all things considered, desirable. On the contrary, a man may all his life have a yen, say, to drink a can of paint, without ever, even at the moment he yields, believing it would be worth doing.

Giving the reason why an agent did something is often a matter of naming the pro attitude (a) or the related belief (b) or both; let me call this pair the *primary reason* why the agent performed the action. Now it is possible both to reformulate the claim that rationalizations are causal explanations and to give structure to the argument by stating two theses about primary reasons:

1. In order to understand how a reason of any kind rationalizes an action it is necessary and sufficient that we see, at least in essential outline, how to construct a primary reason.
2. The primary reason for an action is its cause.

I shall argue for these points in turn.

II

I flip the switch, turn on the light, and illuminate the room. Unbeknown to me I also alert a prowler to the fact that I am home. Here I need not have done four things, but only one, of which four descriptions have been given.[2] I flipped the switch because I wanted to turn on the light and by saying I wanted to turn on the light I explain (give my reason for, rationalize) the flipping. But I do not, by giving this reason, rationalize my alerting of the prowler or my illuminating of the room. Since reasons may rationalize what someone does when it is described in one way and not when it is described in another, we cannot treat what was done simply as a term in sentences like, "My reason for flipping the switch was that I wanted to turn on the light"; otherwise we would be forced to conclude, from the fact that flipping the switch was identical with alerting the prowler, that my reason for alerting the prowler was that I wanted to turn on the light. Let us mark this quasi-intensional[3] character of action descriptions in rationalizations by stating a bit more precisely a necessary condition for primary reasons:

> C1. *R* is a primary reason why an agent performed the action *A* under the description *d* only if *R* consists of a pro attitude of the agent toward actions with a certain property, and a belief of the agent that *A*, under the description *d*, has that property.

How can my wanting to turn on the light be (part of) a primary reason, since it appears to lack the required element of generality? We may be taken in by the verbal parallel between "I turned on the light" and "I wanted to turn on the light." The first clearly refers to a particular event, so we conclude that the second has this same event as its object. Of course it is obvious that the event of my turning on the light can't be referred to in the same way by both sentences since the existence of the event is required by the truth of "I turned on the light" but not by the truth of "I wanted to turn on the light." If the reference were the same in both cases, the second sentence would entail the first; but in fact the sentences are logically independent. What is less obvious, at least until we attend to it, is that the event whose occurrence makes "I turned on the light" true cannot be called the object, however intentional, of "I wanted to turn on the light." If I turned on the light, then I must have done it at a precise moment, in a particular way—every detail is fixed. But it makes no sense to demand that my want be directed to an action performed at any one moment or done in some unique manner. Any one of an indefinitely large number of actions would satisfy the want and can be considered equally eligible as its object. Wants and desires often are trained on physical objects. However, "I want that gold watch in the window" is not a primary reason and explains why I went into the store only because it suggests a primary reason—for example, that I wanted to buy the watch.

Because "I wanted to turn on the light" and "I turned on the light" are logically independent, the first can be used to give a reason why the second is true. Such a reason gives minimal information: it implies that the action was intentional, and wanting tends to exclude some other pro attitudes, such as a sense of duty or obligation. But the exclusion depends very much on the action and the context of explanation. Wanting seems pallid beside lusting, but it would be odd to deny that someone who lusted after a woman or a cup of coffee wanted her or it. It is not unnatural, in fact, to treat wanting as a genus including all pro attitudes as species. When we do this and when we know some action is intentional, it is easy to answer the question, "Why did you do it?" with, "For no reason," meaning not that there is no reason but that there is no *further* reason, no reason that cannot be inferred from the fact that the action was done intentionally; no reason, in other words, besides wanting to do it. This last point is not essential to the present argument, but it is of interest because it defends the possibility of defining an intentional action as one done for a reason.

A primary reason consists of a belief and an attitude, but it is generally otiose to mention both. If you tell me you are easing the jib because you think that will stop the main from backing, I don't need to be told that you want to stop the main from backing; and if you say you are biting your thumb at me because you want to insult me, there is no point in adding that you think that by biting your thumb at me you will insult me. Similarly, many explanations of actions in terms of reasons that are not primary do not require mention of the primary reason to complete the story. If I say I am pulling weeds because I want a beautiful lawn, it would be fatuous to eke out the account with, "And so I see something desirable in any action that does, or has a good chance of, making the lawn beautiful." Why insist that there is any *step*, logical or psychological, in the transfer of desire from an end that is not an action to the actions one conceives as means? It serves the argument as well that the desired end explains the action only if what are believed by the agent to be means are desired.

Fortunately, it is not necessary to classify and analyze the many varieties of emotions, sentiments, moods, motives, passions, and hungers whose mention may answer the question, "Why did you do it?" in order to see how, when such mention rationalizes the action, a primary reason is involved. Claustrophobia gives a man's reason for leaving a cocktail party because we know people want to avoid, escape from, be safe from, put distance between themselves and what they fear. Jealousy is the motive in a poisoning because, among other things, the poisoner believes his action will harm his rival, remove the cause of his agony, or redress an injustice, and these are the sorts of things a jealous man wants to do. When we learn that a man cheated his son out of greed, we do not necessarily know what the primary reason was, but we know there was one, and its general nature. Ryle analyzes "he boasted from vanity" into "he boasted on meeting the stranger and his doing so satisfies the lawlike proposition that whenever he finds a chance of securing the admiration and envy of others, he does whatever he thinks will produce this admiration and envy" (89). This analysis is often, and perhaps justly, criticized on the ground that a man may boast from vanity just once. But if Ryle's boaster did what he did from vanity, then something entailed by Ryle's analysis is true: the boaster wanted to secure the admiration and envy of others, and he believed that his action would produce this admiration and envy; true or false, Ryle's analysis does not dispense with primary reasons, but depends upon them.

To know a primary reason why someone acted as he did is to know an intention with which the action was done. If I turn left at the fork because I want to get to

Katmandu, my intention in turning left is to get to Katmandu. But to know the intention is not necessarily to know the primary reason in full detail. If James goes to church with the intention of pleasing his mother, then he must have some pro attitude toward pleasing his mother, but it needs more information to tell whether his reason is that he enjoys pleasing his mother, or thinks it right, his duty, or an obligation. The expression "the intention with which James went to church" has the outward form of a description, but in fact its syncategorematic and cannot be taken to refer to an entity, state, disposition, or event. Its function in context is to generate new descriptions of actions in terms of their reasons; thus, "James went to church with the intention of pleasing his mother" yields a new, and fuller, description of the action described in "James went to church." Essentially the same process goes on when I answer the question, "Why are you bobbing around that way?" with, "I'm knitting, weaving, exercising, sculling, cuddling, training fleas."

Straight description of an intended result often explains an action better than stating that the result was intended or desired. "It will soothe your nerves" explains why I pour you a shot as efficiently as "I want to do something to soothe your nerves," since the first in the context of explanation implies the second; but the first does better, because, if it is true, the facts will justify my choice of action. Because justifying and explaining an action so often go hand in hand, we frequently indicate the primary reason for an action by making a claim which, if true, would also verify, vindicate, or support the relevant belief or attitude of the agent. "I knew I ought to return it," "The paper said it was going to snow," "You stepped on *my* toes," all, in appropriate reason-giving contexts, perform this familiar dual function.

The justifying role of a reason, given this interpretation, depends upon the explanatory role, but the converse does not hold. Your stepping on my toes neither explains nor justifies my stepping on your toes unless I believe you stepped on my toes, but the belief alone, true or false, explains my action.

III

In the light of a primary reason, an action is revealed as coherent with certain traits, long- or short-termed, characteristic or not, of the agent, and the agent is shown in his role of rational animal. Corresponding to the belief and attitude of a primary reason for an action, we can always construct (with a little ingenuity) the premises of a syllogism from which it follows that the action has some (as Anscombe calls it) "desirability characteristic."[4] Thus there is a certain irreducible—though somewhat anemic—sense in which every rationalization justifies: from the agent's point of view there was, when he acted, something to be said for the action.

Noting that nonteleological causal explanations do not display the element of justification provided by reasons, some philosophers have concluded that the concept of cause that applies elsewhere cannot apply to the relation between reasons and actions, and that the pattern of justification provides, in the cause of reasons, the required explanation. But suppose we grant that reasons alone justify actions in the course of explaining them; it does not follow that the explanation is not also—and necessarily—causal. Indeed our first condition for primary reasons (C1) is designed to help set rationalizations apart from other sorts of explanation. If rationalization is, as I want to argue, a species of causal explanation, then justification, in the sense given by C1, is at least one differentiating property. How about the other claim: that justifying is a kind of explaining, so that the ordinary notion of cause need not be brought in? Here it is necessary to decide what is being included under justification. It could be taken to cover only what is called for by C1: that the agent have certain beliefs and attitudes in the light of which the action is reasonable. But then something essential has certainly been left out, for a person can have a reason for an action, and perform the action, and yet this reason not be the reason why he did it. Central to the relation between a reason and an action it explains is the idea that the agent performed the action *because* he had the reason. Of course, we can include this idea too in justification; but then the notion of justification becomes as dark as the notion of reason until we can account for the force of that "because."

When we ask why someone acted as he did, we want to be provided with an interpretation. His behavior seems strange, alien, outré, pointless, out of character, disconnected; or perhaps we cannot even recognize an action in it. When we learn his reason, we have an interpretation, a new description of what he did, which fits it into a familiar picture. The picture includes some of the agent's beliefs and attitudes; perhaps also goals, ends, principles, general character traits, virtues, or vices. Beyond this, the redescription of an action afforded by a reason may place the action in a wider social, economic, linguistic, or evaluative context. To learn, through learning the reason, that the agent conceived his action as a lie, a repayment of a debt, an insult, the fulfillment of an avuncular obligation, or a knight's gambit is to grasp the point of the action in its setting of rules, practices, conventions, and expectations.

Remarks like these, inspired by the later Wittgenstein, have been elaborated with subtlety and insight by a number of philosophers. And there is no denying that this is true: when we explain an action, by giving the reason, we do redescribe the action; redescribing the action gives the action a place in a pattern, and in this way the action is explained. Here it is tempting to draw two conclusions that do not follow. First, we can't infer, from the fact that giving reasons merely redescribes the action and that causes are separate from effects, that therefore reasons are not causes. Reasons, being beliefs and attitudes, are certainly not identical with actions; but, more important, events are often redescribed in terms of their causes. (Suppose someone was injured. We could redescribe this event "in terms of a cause" by saying he was burned.) Second, it is an error to think that, because placing the action in a larger pattern explains it, therefore we now understand the sort of explanation involved. Talk of patterns and contexts does not answer the question of how reasons explain actions, since the relevant pattern or context contains both reason and action. One way we can explain an event is by placing it in the context of its cause; cause and effect form the sort of pattern that explains the effect, in a sense of "explain" that we understand as well as any. If reason and action illustrate a different pattern of explanation, that pattern must be identified.

Let me urge the point in connection with an example of Melden's. A man driving an automobile raises his arm in order to signal. His intention, to signal, explains his action, raising his arm, by redescribing it as signaling. What is the pattern that explains the action? Is it the familiar pattern of an action done for a reason? Then it does indeed explain the action, but only because it assumes the relation of reason and action that we want to analyze. Or is the pattern rather this: the man is driving, he is approaching a turn; he knows he ought to signal; he knows how to signal, by raising his arm. And now, in this context, he raises his arm. Perhaps, as Melden suggests, if all this happens, he does signal. And the explanation would then be this; if, under these conditions, a man raises his arm, then he signals. The difficulty is, of course, that this explanation does not touch the question of why he raised his arm. He had a reason to raise his arm, but this has not been shown to be the reason why he did it. If the description "signaling" explains his action by giving his reason, then the signaling must be intentional; but, on the account just given, it may not be.

If, as Melden claims, causal explanations are "wholly irrelevant to the understanding we seek" of human action (184) then we are without an analysis of the "because" in "He did it because ..." where we go on to name a reason. Hampshire remarks, of the relation between reasons and action, "In philosophy one ought surely to find this ... connection altogether mysterious" (166). Hampshire rejects Aristotle's attempt to solve the mystery by introducing the concept of wanting as a causal factor, on the grounds that the resulting theory is too clear and definite to fit all cases and that, "There is still no compelling ground for insisting that the word 'want' *must* enter into every full statement of reasons for acting" (168). I agree that the concept of wanting is too narrow, but I have argued that, at least in a vast number of typical cases, some pro attitude must be assumed to be present if a statement of an agent's reasons in acting is to be intelligible. Hampshire does not see how Aristotle's scheme can be appraised as true or false, "for it is not clear what could be the basis of assessment, or what kind of evidence could be decisive" (167). But I would urge that, failing a satisfactory alternative, the best argument for a scheme like Aristotle's is that it alone promises to give an account of the "mysterious connection" between reasons and actions.

IV

In order to turn the first "and" to "because" in "He exercised *and* he wanted to reduce and thought exercise would do it," we must, as the basic move,[5] augment condition C1 with:

> C2. A primary reason for an action is its cause.

The considerations in favor of C2 are by now, I hope, obvious; in the remainder of this chapter I wish to defend C2 against various lines of attack and, in the process, to clarify the notion of causal explanation involved.

A

The first line of attack is this. Primary reasons consist of attitudes and beliefs, which are states or dispositions, not events; therefore they cannot be causes.

It is easy to reply that states, dispositions, and conditions are frequently named as the causes of events: the bridge collapsed because of a structural defect; the plane crashed on takeoff because the air temperature was abnormally high; the plate broke because it had a crack. This reply does not, however, meet a closely related point. Mention of a causal condition for an event gives a cause only on the assumption that there was also a preceding event. But what is the preceding event that causes an action?

In many cases it is not difficult at all to find events very closely associated with the primary reason. States and dispositions are not events, but the onslaught of a state or disposition is. A desire to hurt your feelings may spring up at the moment you anger me; I may start wanting to eat a melon just when I see one; and beliefs may begin at the moment we notice, perceive, learn, or remember something. Those who have argued that there are no mental events to qualify as causes of actions have often missed the obvious because they have insisted that a mental event be observed or noticed (rather than an observing or a noticing) or that it be like a stab, a qualm, a prick or a quiver, a mysterious prod of conscience or act of the will. Melden, in discussing the driver who signals a turn by raising his arm, challenges those who want to explain actions causally to identify "an event which is common and peculiar to all such cases" (87), perhaps a motive or an intention, anyway "some particular feeling or experience" (95). But of course there is a mental event; at some moment the driver noticed (or thought he noticed) his turn coming up, and that is the moment he signaled. During any continuing activity, like driving, or elaborate performance, like swimming the Hellespont, there are more or less fixed purposes, standards, desires, and habits that give direction and form to the entire enterprise, and there is the continuing input of information about what we are doing, about changes in the environment, in terms of which we regulate and adjust our actions. To dignify a driver's awareness that his turn has come by calling it an experience, or even a feeling, is no doubt exaggerated, but whether it deserves a name or not, it had better be the reason why he raises his arm. In this case, and typically, there may not be anything we would call a motive, but if we mention such a general purpose as wanting to get to one's destination safely, it is clear that the motive is not an event. The intention with which the driver raises his arm is also not an event, for it is no thing at all, neither event, attitude, disposition, nor object. Finally, Melden asks the causal theorist to find an event that is common and peculiar to all cases where a man intentionally raises his arm, and this, it must be admitted, cannot be produced. But then neither can a common and unique cause of bridge failures, plane crashes, or plate breakings be produced.

The signaling driver can answer the question, "Why did you raise your arm when you did?" and from the answer we learn the event that caused the action. But can an actor always answer such a question? Sometimes the answer will mention a mental event that does not give a reason: "Finally I made up my mind." However, there also seem to be cases of intentional action where we cannot explain at all why we acted when we did. In such cases, explanation in terms of primary reasons parallels the explanation of the collapse of the bridge from a structural defect: we are ignorant of the event or sequence of events that led up to (caused) the collapse, but we are sure there was such an event or sequence of events.

B

According to Melden, a cause must be "logically distinct from the alleged effect" (52); but a reason for an action is not logically distinct from the action; therefore, reasons are not causes of actions.[6]

One possible form of this argument has already been suggested. Since a reason makes an action intelligible by redescribing it, we do not have two events, but only one under different descriptions. Causal relations, however, demand distinct events.

Someone might be tempted into the mistake of thinking that my flipping of the switch caused my turning on of the light (in fact it caused the light to go on). But it does not follow that it is a mistake to take "My reason for flipping the switch was that I wanted to turn on the light" as entailing, in part, "I flipped the switch, and this action is further describable as having been caused by wanting to turn on the light." To describe an event in terms of its cause is not to confuse the event with its cause, nor does explanation by redescription exclude causal explanation.

The example serves also to refute the claim that we cannot describe the action without using words that link it to the alleged cause. Here the action is to be explained under the description: "my flipping the switch," and the alleged cause is "my wanting to turn on the light." What relevant logical relation is supposed to hold between these phrases? it seems more plausible to urge a logical link between "my turning on the light" and "my wanting to turn on the light," but even here the link turns out, on inspection, to be grammatical rather than logical.

In any case there is something very odd in the idea that causal relations are empirical rather than logical. What can this mean? Surely not that every true causal statement is empirical. For suppose "A caused B" is true. Then the cause of $B = A$; so substituting, we have "The cause of B caused B," which is analytic. The truth of a causal statement depends on *what* events are described; its status as analytic or synthetic depends on *how* the events are described. Still, it may be maintained that a reason rationalizes an action only when the descriptions are appropriately fixed, and the appropriate descriptions are not logically independent.

Suppose that to say a man wanted to turn on the light *meant* that he would perform any action he believed would accomplish his end. Then the statement of his primary reason for flipping the switch would entail that he flipped the switch—"straightway he acts," as Aristotle says. In this case there would certainly be a logical connection between reason and action, the same sort of connection as that between, "It's water soluble and was placed in water" and "It dissolved." Since the implication runs from description of cause to description of effect but not conversely, naming the cause still gives information. And, though the point is often overlooked, "Placing it in water caused it to dissolve" does not entail "It's water soluble"; so the latter has additional explanatory force. Nevertheless, the explanation would be far more interesting if, in place of solubility, with its obvious definitional connection with the event to be explained, we could refer to some property, say a particular crystalline structure, whose connection with dissolution in water was known only through experiment. Now it is clear why primary reasons like desires and wants do not explain actions in the relatively trivial way solubility explains dissolvings. Solubility, we are assuming, is a pure disposition property: it is defined in terms of a single test. But desires cannot be defined in terms of the actions they may rationalize, even though the relation between desire and action is not simply empirical; there are other, equally essential criteria for desires—their expression in feelings and in actions that they do not rationalize, for example. The person who has a desire (or want or belief) does not normally need criteria at all—he generally knows, even in the absence of any clues available to others, what he wants, desires, and believes. These logical features of primary reasons show that it is not just lack of ingenuity that keeps us from defining them as dispositions to act for these reasons.

C

According to Hume, "we may define a cause to be an object, followed by another, and where all the objects similar to the first are followed by objects similar to the second." But, Hart and Honoré claim, "the statement that one person did something because, for example, another threatened him, carries no implication or covert assertion that if the circumstances were repeated the same action would follow" (52). Hart and Honoré allow that Hume is right in saying that ordinary singular causal statements imply generalizations, but wrong for this very reason in supposing that motives and desires are ordinary causes of actions. In brief, laws are involved essentially in ordinary causal explanations, but not in rationalizations.

It is common to try to meet this argument by suggesting that we do have rough laws connecting reasons and actions, and these can, in theory, be improved. True, threatened people do not always respond in the same way; but we may distinguish between threats and also between agents, in terms of their beliefs and attitudes.

The suggestion is delusive, however, because generalizations connecting reasons and actions are not—and cannot be sharpened into—the kind of law on the basis of which accurate predictions can reliably be made. If we reflect on the way in which reasons determine choice, decision, and behavior, it is easy to see why this is so. What emerges, in the ex post facto atmosphere of explanation and justification, as *the* reason frequently was, to the agent at the time of action, one consideration among many, *a* reason. Any serious theory for predicting action on the basis of reasons must find a way of evaluating the relative force of various desires and beliefs in the matrix of decision; it cannot take as its starting point the refinement of what is to be expected from a single desire. The practical syllogism exhausts its role in displaying an action as falling under one reason; so it cannot be subtilized into a reconstruction of practical reasoning, which involves the weighing of competing reasons. The practical syllogism provides a model neither for a predictive science of action nor for a normative account of evaluative reasoning.

Ignorance of competent predictive laws does not inhibit valid causal explanation, or few causal explanations could be made. I am certain the window broke because it was struck by a rock—I saw it all happen; but I am not (is anyone?) in command of laws on the basis of which I can predict what blows will break which windows. A generalization like "Windows are fragile, and fragile things tend to break when struck hard enough, other conditions being right" is not a predictive law in the rough—the predictive law, if we had it, would be quantitative and would use very different concepts. The generalization, like our generalizations about behavior, serves a different function: it provides evidence for the existence of a causal law covering the case at hand.

We are usually far more certain of a singular causal connection than we are of any causal law governing the case; does this show that Hume was wrong in claiming that singular causal statements entail laws? Not necessarily, for Hume's claim, as quoted above, is ambiguous. It may mean that "*A* caused *B*" entails some particular law involving the predicates used in the descriptions *A* and *B*, or it may mean that "*A* caused *B*" entails that there exists a causal law instantiated by some true descriptions of *A* and *B*.[7] Obviously, both versions of Hume's doctrine give a sense to the claim that singular causal statements entail laws, and both sustain the view that causal explanations "involve laws." But the second version is far weaker, in that no particular law is entailed by a singular causal claim, and a singular causal claim can be defended, if it needs defense, without defending any law. Only the second version of Hume's doctrine can be made to fit with must causal explanations; it suits rationalizations equally well.

The most primitive explanation of an event gives its cause; more elaborate explanations may tell more of the story, or defend the singular causal claim by producing a relevant law or by giving reasons for believing such exists. But it is an error to think no explanation has been given until a law has been produced. Linked with these errors is the idea that singular causal statements necessarily indicate, by the concepts they employ, the concepts that will occur in the entailed law. Suppose a hurricane, which is reported on page 5 of Tuesday's *Times*, causes a catastrophe, which is reported on page 13 of Wednesday's *Tribune*. Then the event reported on page 5 of Tuesday's *Times* caused the event reported on page 13 of Wednesday's *Tribune*. Should we look for a law relating events of these *kinds*? It is only slightly less ridiculous to look for a law relating hurricane and catastrophe. The laws needed to predict the catastrophe with precision would, of course, have no use for concepts like hurricane and catastrophe. The trouble with predicting the weather is that the descriptions under which events interest us—"a cool, cloudy day with rain in the afternoon"—have only remote connections with the concepts employed by the more precise known laws.

The laws whose existence is required if reasons are causes of actions do not, we may be sure, deal in the concepts in which rationalizations must deal. If the causes of a class of events (actions) fall in a certain class (reasons) and there is a law to back each singular causal statement, it does not follow that there is any law connecting events classified as reasons with events classified as actions—the classifications may even be neurological, chemical, or physical.

D

It is said that the kind of knowledge one has of one's own reasons in acting is not compatible with the existence of a causal relation between reasons and actions: a person knows his own intentions in acting infallibly, without induction or observation, and no ordinary causal relation can be known in this way. No doubt our knowledge of our own intentions in acting will show many of the oddities peculiar to first-person knowledge of one's own pains, beliefs, desires, and so on; the only question is whether these oddities prove that reasons do not cause, in any ordinary sense at least, the actions that they rationalize.

You may easily be wrong about the truth of a statement of the form, "I am poisoning Charles because I want to save him pain," because you may be wrong about whether you are poisoning Charles—you may yourself be drinking the poisoned cup by mistake. But it also seems that you may err about your reasons, particularly when you have two reasons for an action, one of which pleases you and one which does not. For example, you do want to save Charles pain; you also want him out of the way. You may be wrong about which motive made you do it.

The fact that you may be wrong does not show that in general it makes sense to ask you how you know what your reasons were or to ask for your evidence. Though you may, on rare occasions, accept public or private evidence as showing you are wrong about your reasons, you usually have no evidence and make no observations. Then your knowledge of your own reasons for your actions is not generally inductive, for where there is induction, there is evidence. Does this show the knowledge is not causal? I cannot see that it does.

Causal laws differ from true but nonlawlike generalizations in that their instances confirm them; induction is, therefore, certainly a good way to learn the truth of a law. It does not follow that it is the only way to learn the truth of a law. In any case, in order to know that a singular causal statement is true, it is not necessary to know the truth of a law; it is necessary only to know that some law covering the events at hand exists. And it is far from evident that induction, and induction alone, yields the knowledge that a causal law satisfying certain conditions exists. Or, to put it differently, one case is often enough, as Hume admitted, to persuade us that a law exists, and this amounts to saying that we are persuaded without direct inductive evidence that a causal relation exists.

E

Finally I should like to say something about a certain uneasiness some philosophers feel in speaking of causes of actions at all. Melden, for example, says that actions are often identical with bodily movements, and that bodily movements have causes; yet he denies that the causes are causes of the actions. This is, I think, a contradiction. He is led to it by the following sort of consideration: "It is futile to attempt to explain conduct through the causal efficacy of desire—all *that* can explain is further happenings, not actions performed by agents. The agent confronting the causal nexus in which such happenings occur is a helpless victim of all that occurs in and to him" (128, 129). Unless I am mistaken, this argument, if it were valid, would show that actions cannot have causes at all. I shall not point out the obvious difficulties in removing actions from the realm of causality entirely. But perhaps it is worth trying to uncover the source of the trouble. Why on earth should a cause turn an action into a mere happening and a person into a helpless victim? Is it because we tend to assume, at least in the arena of action, that a cause demands a causer, agency an agent? So we press the question; if my action is caused, what caused it? If I did, then there is the absurdity of infinite regress; if I did not, I am a victim. But of course the alternatives are not exhaustive. Some causes have no agents. Among these agentless causes are the states and changes of state in persons which, because they are reasons as well as causes, constitute certain events as free and intentional actions.

Notes

1. Some examples: Gilbert Ryle, *the Concept of Mind,* G. E. M. Anscombe, *Intention,* Stuart Hampshire, *Thought and Action,* H. L. A. Hart and A. M. Honoré, *Causation in the Law,* William Dray, *Laws and Explanation in History,* and most of the books in the series edited by R. F. Holland, *Studies in Philosophical Psychology,* including Anthony Kenny, *Action, Emotion and Will,* and A. I. Melden, *Free Action.* Page references in parentheses are to these works.
2. We might not call my unintentional alerting of the prowler an action, but it should not be inferred from this that alerting the prowler is therefore something different from flipping the switch, say just its consequence. Actions, performances, and events not involving intention are alike in that they are often referred to or defined partly in terms of some terminal stage, outcome, or consequence.
3. The word "action" does not very often occur in ordinary speech, and when it does it is usually reserved for fairly portentous occasions. I follow a useful philosophical practice in calling anything an agent does intentionally an action, including intentional omissions. What is really needed is some suitably generic term to bridge the following gap: suppose *A* is a description of an action, *B* is a description of something done voluntarily, though not intentionally, and *C* is a description of something done involuntarily and unintentionally; finally, suppose $A = B = C$. Then *A, B,* and *C* are the same—what? "Action," "event," "thing done," each has at least in some contexts, a strange ring when coupled with the wrong sort of description. Only the question, "Why did you (he) do *A*?" has the true generality required. Obviously, the problem is greatly aggravated if we assume, as Melden does, that an action ("raising one's arm") can be identical with a bodily movement ("one's arm going up"). "Quasi-intentional" because, besides its intentional aspect, the description of the action must also refer in rationalizations; otherwise it could be true that an action was done for a certain reason and yet the action not have been performed. Compare "the author of *Waverley*" in "George IV knew the author of Waverley wrote *Waverley.*"

4. Anscombe denies that the practical syllogism is deductive. This she does partly because she thinks of the practical syllogism, as Aristotle does, as corresponding to a piece of practical reasoning (whereas for me it is only part of the analysis of the concept of a reason with which someone acted), and therefore she is bound, again following Aristotle, to think of the conclusion of a practical syllogism as corresponding to a judgment, not merely that the action has a desirable characteristic, but that the action is desirable (reasonable, worth doing, etc.).
5. I say "as the basic move" to cancel any suggestion that C1 and C2 are jointly *sufficient* to define the relation of reasons to the actions they explain.
6. This argument can be found in one or more versions, in Kenny, Hampshire, and Melden, as well as in P. Winch, *The Idea of a Social Science*, and R. S. Peters, *The Concept of Motivation*. In one of its forms, the argument was, of course, inspired by Ryle's treatment of motives in *the Concept of Mind*.
7. We could roughly characterize the analysis of singular causal statements hinted at here as follows: "*A* caused *B*" is true if and only if there are descriptions of *A* and *B* such that the sentence obtained by putting these descriptions for *A* and *B* in "*A* caused *B*" follows from a true causal law. This analysis is saved from triviality by the fact that not all true generalizations are causal laws; causal laws are distinguished (though of course this is no analysis) by the fact that they are inductively confirmed by their instances and by the fact that they support counterfactual and subjunctive singular causal statements.

Chapter 44

Special Sciences (or: The Disunity of Science as a Working Hypothesis)

Jerry Fodor

A typical thesis of positivistic philosophy of science is that all true theories in the special sciences should reduce to physical theories in the long run. This is intended to be an empirical thesis, and part of the evidence which supports it is provided by such scientific successes as the molecular theory of heat and the physical explanation of the chemical bond. But the philosophical popularity of the reductivist program cannot be explained by reference to these achievements alone. The development of science has witnessed the proliferation of specialized disciplines at least as often as it has witnessed their reduction to physics, so the widespread enthusiasm for reduction can hardly be a mere induction over its past successes.

I think that many philosophers who accept reductivism do so primarily because they wish to endorse the generality of physics vis à vis the special sciences: roughly, the view that all events which fall under the laws of any science are physical events and hence fall under the laws of physics.[1] For such philosophers, saying that physics is basic science and saying that theories in the special sciences must reduce to physical theories have seemed to be two ways of saying the same thing, so that the latter doctrine has come to be a standard construal of the former.

In what follows, I shall argue that this is a considerable confusion. What has traditionally been called "the unity of science" is a much stronger, and much less plausible, thesis than the generality of physics. If this is true it is important. Though reductionism is an empirical doctrine, it is intended to play a regulative role in scientific practice. Reducibility to physics is taken to be a *constraint* upon the acceptability of theories in the special sciences, with the curious consequence that the more the special sciences succeed, the more they ought to disappear. Methodological problems about psychology, in particular, arise in just this way: the assumption that the subject matter of psychology is part of the subject matter of physics is taken to imply that psychological theories must reduce to physical theories, and it is this latter principle that makes the trouble. I want to avoid the trouble by challenging the inference.

I

Reductivism is the view that all the special sciences reduce to physics. The sense of "reduce to" is, however, proprietary. It can be characterized as follows.[2]

Let

$$S_1 x \rightarrow S_2 x \qquad (1)$$

be a law of the special science S. ((1) is intended to be read as something like "all S_1 situations bring about S_2 situations." I assume that a science is individuated largely by reference to its typical predicates, hence that if S is a special science S_1 and S_2 are not predicates of basic physics. I also assume that the "all" which quantifies laws of the special sciences needs to be taken with a grain of salt; such laws are typically *not* exceptionless. This is a point to which I shall return at length.) A necessary and sufficient condition of the reduction of (1) to a law of physics is that the formulas (2) and (3) be laws, and a necessary and sufficient condition of the reduction of S to physics is that all its laws be so reducible[3]

$$S_1 x \leftrightarrows P_1 x \tag{2a}$$

$$S_2 x \leftrightarrows P_2 x \tag{2b}$$

$$P_1 x \rightarrow P_2 x. \tag{3}$$

P_1 and P_2 are supposed to be predicates of physics, and (3) is supposed to be a physical law. Formulas like (2) are often called "bridge" laws. Their characteristic feature is that they contain predicates of both the reduced and the reducing science. Bridge laws like (2) are thus contrasted with "proper" laws like (1) and (3). The upshot of the remarks so far is that the reduction of a science requires that any formula which appears as the antecedent or consequent of one of its proper laws must appear as the reduced formula in some bridge law or other.[4]

Several points about the connective $\rightarrow$ are in order. First, whatever other properties that connective may have, it is universally agreed that it must be transitive. This is important because it is usually assumed that the reduction of some of the special sciences proceeds via bridge laws which connect their predicates with those of intermediate reducing theories. Thus, psychology is presumed to reduce to physics via, say, neurology, biochemistry, and other local stops. The present point is that this makes no difference to the logic of the situation so long as the transitivity of $\rightarrow$ is assumed. Bridge laws which connect the predicates of S to those of S^* will satisfy the constraints upon the reduction of S to physics so long as there are other bridge laws which, directly or indirectly, connect the predicates of S^* to physical predicates.

There are, however, quite serious open questions about the interpretations of $\rightarrow$ in bridge laws. What turns on these questions is the respect in which reductivism is taken to be a physicalist thesis.

To begin with, if we read $\rightarrow$ as "brings about" or "causes" in proper laws, we will have to have some other connective for bridge laws, since bringing about and causing are presumably *a*symmetric, while bridge laws express symmetric relations. Moreover, if $\rightarrow$ in bridge laws is interpreted as any relation other than identity, the truth of reductivism will only guarantee the truth of a weak version of physicalism, and this would fail to express the underlying ontological bias of the reductivist program.

If bridge laws are not identity statements, then formulas like (2) claim at most that, by law, *x*'s satisfaction of a P predicate and *x*'s satisfaction of an S predicate are causally correlated. It follows from this that it is nomologically necessary that S and P predicates apply to the same things (i.e., that S predicates apply to a subset of the things that P predicates apply to). But, of course, this is compatible with a nonphysicalist ontology since it is compatible with the possibility that *x*'s satisfying S should not itself *be* a physical event. On this interpretation, the truth of reductivism does *not* guarantee the generality of physics vis à vis the special sciences since there are some events (satisfactions of S predicates) which fall in the domains of a special science (S) but not in the domain of physics, (One could imagine, for example, a doctrine according to which physical and psychological predicates are both held to apply to organisms, but where it is denied that the event which consists of an organism's satisfying a psychological predicate is, in any sense, a physical event. The upshot would be a kind of psychophysical dualism of a non-Cartesian variety, a dualism of event and/or properties rather than substances.)

Given these sorts of considerations, many philosophers have held that bridge laws like (2) ought to be taken to express contingent event identities, so that one would read (2a) in some such fashion as "every event which consists of x's satisfying S_1 is identical to some event which consists of x's satisfying P_1 and vice versa." On this reading, the truth of reductivism would entail that every event that falls under any scientific law is a physical event, thereby simultaneously expressing the ontological bias of reductivism and guaranteeing the generality of physics vis à vis the special sciences.

If the bridge laws express event identities, and if every event that falls under the proper laws of a special science falls under a bridge law, we get the truth of a doctrine that I shall call "token physicalism." Token physicalism is simply the claim that all the events that the sciences talk about are physical events. There are three things to notice about token physicalism.

First, it is weaker than what is usually called "materialism." Materialism claims *both* that token physicalism is true *and* that every event falls under the laws of some science or other. One could therefore be a token physicalist without being a materialist, though I don't see why anyone would bother.

Second, token physicalism is weaker than what might be called "type physicalism," the doctrine, roughly, that every *property* mentioned in the laws of any science is a physical property. Token physicalism does not entail type physicalism because the contingent identity of a pair of events presumably does not guarantee the identity of the properties whose instantiation constitutes the events, not even where the event identity is nomologically necessary. On the other hand, if every event is the instantiation of a property, then type physicalism does entail token physicalism: two events will be identical when they consist of the instantiation of the same property by the same individual at the same time.

Third, token physicalism is weaker than reductivism. Since this point is, in a certain sense, the burden of the argument to follow, I shan't labor it here. But, as a first approximation, reductivism is the conjunction of token physicalism with the assumption that there are natural kind predicates in an ideally completed physics which correspond to each natural kind predicate in any ideally completed special science. It will be one of my morals that the truth of reductivism cannot be inferred from the assumption that token physicalism is true. Reductivism is a sufficient, but not a necessary, condition for token physicalism.

In what follows, I shall assume a reading of reductivism which entails token physicalism. Bridge laws thus state nomologically necessary contingent event identities and a reduction of psychology to neurology would entail that any event which consists of the instantiation of a psychological property is identical with some event which consists of the instantiation of some neurological property.

Where we have got to is this: reductivism entails the generality of physics in at least the sense that any event which falls within the universe of discourse of a special science will also fall within the universe of discourse of physics. Moreover, any prediction which follows from the laws of a special science and a statement of initial conditions will also follow from a theory which consists of physics and the bridge laws, together with the statement of initial conditions. Finally, since "reduces to" is supposed to be an asymmetric relation, it will also turn out that physics is *the* basic science; that is, if reductivism is true, physics is the only science that is general in the sense just specified. I now want to argue that reductivism is too strong a constraint upon the unity of science, but that the relevantly weaker doctrine will preserve the desired consequences of reductivism: token physicalism, the generality of physics, and its basic position among the sciences.

II

Every science implies a taxonomy of the events in its universe of discourse. In particular, every science employs a descriptive vocabulary of theoretical and observation predicates such that events fall under the laws of the science by virtue of satisfying those predicates. Patently, not every true description of an event is a description in such a vocabulary. For example, there are a large number of events which consist of things having been transported to a distance of less than three miles from the Eiffel Tower. I take it, however, that there is no science which contains "is transported to a distance of less than three miles form the Eiffel Tower" as part of its descriptive vocabulary. Equivalently, I take it that there is no natural law which applies to events in virtue of their being instantiations of the property *is transported to a distance of less than three miles from the Eiffel Tower* (though I suppose it is conceivable that there is some law that applies to events in virtue of their being instantiations of some distinct but coextensive property). By way of abbreviating these facts, I shall say that the property *is transported* ... does not determine a *natural kind,* and that predicates which express that property are not natural kind predicates.

If I knew what a law is, and if I believed that scientific theories consist just of bodies of laws, then I could say that P is a natural kind predicate relative to S iff S contains proper laws of the form $P_x \rightarrow \alpha_x$ or $\alpha_x \rightarrow P_x$; roughly, the natural kind predicates of a science are the ones whose terms are the bound variables in its proper laws. I am inclined to say this even in my present state of ignorance, accepting the consequence that it makes the murky notion of a natural kind viciously dependent on the equally murky notions *law* and *theory*. There is no firm footing here. If we disagree about what is a natural kind, we will probably also disagree about what is a law, and for the same reasons. I don't know how to break out of this circle, but I think that there are interesting things to say about which circle we are in.

For example, we can now characterize the respect in which reductivism is too strong a construal of the doctrine of the unity of science. If reductivism is true, then *every* natural kind is, or is coextensive with, a physical natural kind. (Every natural kind *is* a physical natural kind if bridge laws express property identities, and every natural kind is coextensive with a physical natural kind if bridge laws express event identities.) This follows immediately from the reductivist premise that every predicate which appears as the antecedent or consequent of a law of the special sciences must appear as one of the reduced predicates in some bridge, together with the assumption that the natural kind predicates are the ones whose terms are the bound variables in proper laws. If, in short, some physical law is related to each law of a special science in the way that (3) is related to (1), then every natural kind predicate of a special science is related to a natural kind predicate of physics in the way that (2) relates S_1 and S_2 to P_1 and P_2.

I now want to suggest some reasons for believing that this consequence of reductivism is intolerable. These are not supposed to be knock-down reasons; they couldn't be, given that the question whether reductivism is too strong is finally an *empirical* question. (The world could turn out to be such that every natural kind corresponds to a physical natural kind, just as it could turn out to be such that the property *is transported to a distance of less than three miles from the Eiffel Tower* determines a natural kind in, say, hydrodynamics. It's just that, as things stand, it seems very unlikely that the world *will* turn out to be either of these ways.)

The reason it is unlikely that every natural kind corresponds to a physical natural kind is just that (a) interesting generalizations (e.g., counterfactual supporting generalizations) can often be made about events whose physical descriptions have nothing in common, (b) it is often the case that *whether* the physical descriptions of the events subsumed by these generalizations have anything in common is, in an obvious sense, entirely irrelevant to the truth of the generalizations, or to their interestingness, or to their degree of confirmation of, indeed, to any of their epistemologically important properties, and (c) the special sciences are very much in the business of making generalizations of this kind.

I take it that these remarks are obvious to the point of self-certification; they leap to the eye as soon as one makes the (apparently radical) move of taking the special sciences at all seriously. Suppose, for example, that Gresham's "law" really is true. (If one doesn't like Gresham's law, then any true generalization of any conceivable future economics will probably do as well.) Gresham's law says something about what will happen in monetary exchanges under certain conditions. I am willing to believe that physics is general *in the sense that it implies that any event which consists of a monetary exchange* (hence any event which falls under Gresham's law) *has a true description in the vocabulary of physics and in virtue of which it falls under the laws of physics.* But banal considerations suggest that a description which covers all such events must be wildly disjunctive. Some monetary exchanges involve strings of wampum. Some involve dollar bills. And some involve signing one's name to a check. What are the chances that a disjunction of physical predicates which covers all these events (i.e., a disjunctive predicate which can form the right hand side of a bridge law of the form "x is a monetary exchange $\leftrightarrows$ . . .") expresses a physical natural kind? In particular, what are the chances that such a predicate forms that antecedent or consequent of some proper law of physics? The point is that monetary exchanges have interesting thing in common; Gresham's law, if true, says what one of these interesting things is. But what is interesting about monetary exchanges is surely not their commonalities under *physical* description. A natural kind like a monetary exchange *could* turn out to be coextensive with a physical natural kind; but if it did, that would be an accident on a cosmic scale.

In fact, the situation for reductivism is still worse than the discussion thus far suggests. For, reductivism claims not only that all natural kinds are coextensive with physical natural kinds, but that the coextensions are nomologically necessary: bridge laws are *laws*. So, if Gresham's law is true, it follows that there is a (bridge) law of nature such that "x is a monetary exchange $\leftrightarrows$ x is P," where P is a term for a physical natural kind. But, surely, there is no such law. If there were, then P would have to cover not only all the systems of monetary exchange that there *are*, but also all the systems of monetary exchange that there *could be*; a law must succeed with the counterfactuals. What physical predicate is a candidate for P in "x is a nomologically possible monetary exchange iff P_x"?

To summarize: an immortal econophysicist might, when the whole show is over, find a predicate in physics that was, in brute fact, coextensive with "is a monetary exchange." If physics is general—if the ontological biases of reductivism are true—then there must *be* such a predicate. But (a) to paraphrase a remark Donald Davidson made in a slightly different context, nothing but brute enumeration could convince us of this brute coextensivity, and (b) there would seem to be no chance at all that the physical predicate employed in stating the coextensivity is a natural kind term, and (c) there is still less chance that the coextension would be lawful (i.e., that it would hold not only for the nomologically possible world that turned out to be real, but for any nomologically possible world at all).

I take it that the preceding discussion strongly suggests that economics is not reducible to physics in the proprietary sense of reduction involved in claims for the unity of science. There is, I suspect, nothing special about economics in this respect; the reasons why economics is unlikely to reduce to physics are paralleled by those which suggest that psychology is unlikely to reduce to neurology.

If psychology is reducible to neurology, then for every psychological natural kind predicate there is a coextensive neurological natural kind predicate, and the generalization which states this coextension is a law. Clearly, many psychologists believe something of the sort. There are departments of "psychobiology" or "psychology and brain science" in universities throughout the world whose very existence is an institutionalized gamble that such lawful coextensions can be found. Yet, as has been frequently remarked in recent discussions of materialism, there are good grounds for hedging these bets. There are no firm data for any but the grossest correspondence between types of psychological states and types of neurological states, and it is entirely possible that the nervous system of higher organisms characteristically achieves a given psychological end by a wide variety of neurological means. If so, then the attempt to pair neurological structures with psychological functions is foredoomed. Physiological psychologists of the stature of Karl Lashley have held precisely this view.

The present point is that the reductivist program in psychology is, in any event, *not* to be defended on ontological grounds. Even if (token) psychological events are (token) neurological events, it does not follow that the natural kind predicates of psychology are coextensive with the natural kind predicates of any other discipline (including physics). That is, the assumption that every psychological event is a physical event does not guarantee that physics (or, a fortiori, any other discipline more general than psychology) can provide an appropriate vocabulary for psychological theories. I emphasize this point because I am convinced that the make-or-break commitment of many physiological psychologists to the reductivist program stems precisely from having confused that program with (token) physicalism.

What I have been doubting is that there are neurological natural kinds coextensive with psychological natural kinds. What seems increasingly clear is that, even if there is such a coextension, it cannot be lawlike. For, it seems increasingly likely that there are nomologically possible systems other than organisms (namely, automata) which satisfy natural kind predicates in psychology, and which satisfy no neurological predicates at all. Now, as Putnam has emphasized, if there are any such systems, then there are probably vast numbers, since equivalent automata can be made out of practically anything. If this observation is correct, then there can be no serious hope that the class of automata whose psychology is effectively identical to that of some organism can be described by *physical* natural kind predicates (though, of course, if token physicalism is true, that class can be picked out by some physical predicate or other). The upshot is that the classical formulation of the unity of science is at the mercy of progress in the field of computer simulation. This is, of course, simply to say that formulation was too strong. The unity of science was intended to be an empirical hypothesis, defeasible by possible scientific findings. But no one had it in mind that it should be defeated by Newell, Shaw, and Simon.

I have thus far argued that psychological reductivism (the doctrine that every psychological natural kind is, or is coextensive with, a neurological natural kind) is not equivalent to, and cannot be inferred from, token physicalism (the doctrine that every psychological event is a neurological event). It may, however, be argued that one might as well take the doctrines to be equivalent since the only possible *evidence* one could have for token physicalism would also be evidence for reductivism: namely, the discovery of type-to-type psychophysical correlations.

A moment's consideration shows, however, that this argument is not well taken. If type-to-type psychophysical correlations would be evidence for token physicalism, so would correlations of other specifiable kinds.

We have type-to-type correlations where, for every *n*-tuple of events that are of the same psychological kind, there is a correlated *n*-tuple of events that are of the same neurological kind. Imagine a world in which such correlations are *not* forthcoming. What is found, instead, is that for every *n*-tuple of type identical psychological events, there is a spatiotemporally correlated *n*-tuple of type *distinct* neurological events. That is, every psychological event is paired with some neurological event or other, but psychological events of the same kind may be paired with neurological events of different kinds. My present point is that such pairings would provide as much support for token physicalism as type-to-type pairings do *so long as we are able to show that the type distinct neurological events paired with a given kind of psychological event are identical in respect of whatever properties are relevant to type-identification in psychology*. Suppose, for purposes of explication, that psychological events are type identified by reference to their behavioral consequences.[5] Then what is required of all the neurological events paired with a class of type homogeneous psychological events is only that they be identical in respect of their behavioral consequences. To put it briefly, type identical events do not, of course, have *all* their properties in common, and type distinct events must nevertheless be identical in *some* of their properties. The empirical confirmation of token physicalism does not depend on showing that the neurological counterparts of type identical psychological events are themselves type identical. What needs to be shown is only that they are identical in respect of those properties which determine which kind of *psychological* event a given event is.

Could we have evidence that an otherwise heterogeneous set of neurological events have these kinds of properties in common? Of course, we could. The neurological theory might itself explain why an *n*-tuple of neurologically type distinct events are identical in their behavioral consequences, or, indeed, in respect of any of indefinitely many other such relational properties. And, if the neurological theory failed to do so, some science more basic than neurology might succeed.

My point in all this is, once again, not that correlations between type homogeneous psychological states and type heterogeneous neurological states would prove that token physicalism is true. It is only that such correlations might give us as much reason to be token physicalists as type-to-type correlations would. If this is correct, then the epistemological arguments from token physicalism to reductivism must be wrong.

It seems to me (to put the point quite generally) that the classical construal of the unity of science has really misconstrued the *goal* of scientific reduction. The point of reduction is *not* primarilty to find some natural kind predicate of physics coextensive with each natural kind predicate of a reduced science. It is, rather, to explicate the physical mechanisms whereby events conform to the laws of the special sciences. I have been arguing that there is no logical or epistemological reason why success in the second of these projects should require success in the first, and that the two are likely to come apart *in fact* wherever the physical mechanisms whereby events conform to a law of the special sciences are heterogeneous.

III

I take it that the discussion thus far shows that reductivism is probably too strong a construal of the unity of science; on the one hand, it is incompatible with probable results in the special sciences, and, on the other, it is more than we need to assume if what we primarily want is just to be good token physicalists. In what follows, I shall try to sketch a liberalization of reductivism which seems to me to be just strong enough in these respects. I shall then give a couple of independent reasons for supposing that the revised doctrine may be the right one.

The problem all along has been that there is an open empirical possibility that what corresponds to the natural kind predicates of a reduced science may be a heterogeneous and unsystematic disjunction of predicates in the reducing science, and we do not want the unity of science to be prejudiced by this possibility. Suppose, then, that we allow that bridge statements may be of the form

$$S_x \leftrightarrows P_1x \vee P_2x \vee \ldots \vee P_nx, \qquad (4)$$

where $P_1 \vee P_2 \vee \ldots \vee P_n$ is *not* a natural kind predicate in the reducing science. I take it that this is tantamount to allowing that at least some "bridge laws" may, in fact, not turn out to be laws, since I take it that a necessary condition on a universal generalization being lawlike is that the predicates which consitute its antecedent and consequent should pick out natural kinds. I am thus supposing that it is enough, for purposes of the unity of science, that every law of the special sciences should be reducible to physics by bridge statements which express true empirical generalizations. Bearing in mind that bridge statements are to be construed as a species of identity statements, (4) will be read as something like "every event which consists of x's satisfying S is identical with some event which consists of x's satisfying some or other predicate belonging to the disjunction $P_1 \vee P_2 \vee \ldots \vee P_n$."

Now, in cases of reduction where what corresponds to (2) is not a law, what corresponds to (3) will not be either, and for the same reason. Namely, the predicates appearing in the antecedent or consequent will, by hypothesis, not be natural kind predicates. Rather, what we will have is something that looks like (5).

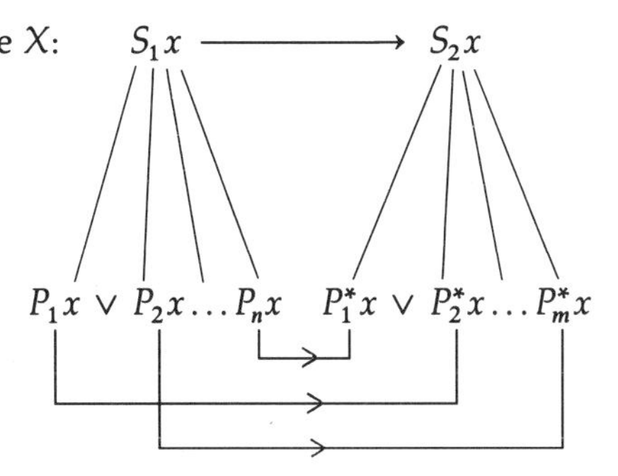

That is, the antecedent and consequent of the reduced law will each be connected with a disjunction of predicates in the reducing science, and, if the reduced law is exceptionless, there will be laws of the reducing science which connect the satisfaction of each member of the disjunction associated with the antecedent to the satisfaction of some member of the disjunction associated with the consequent. That is, if $S_1x \rightarrow S_2x$ is exceptionless, then there must be some proper law of the reducing science which either states or entails that $P_1x \rightarrow P^*$ for some P^*, and similarly for P_2x through P_nx. Since there must be such laws, it follows that each disjunct of $P_1 \vee P_2 \vee \ldots \vee P_n$ is a natural kind predicate, as is each disjunct of $P_1^* \vee P_2^* \vee \ldots \vee P_n^*$.

This, however, is where push comes to shove. For, it might be argued that if each disjunct of the P disjunction is lawfully connected to some disjunct of the P^* disjunction, it follows that (6) is itself a law.

$$P_1x \vee P_2x \vee \ldots \vee P_nx \rightarrow P_1^*x \vee P_2^*x \vee \ldots \vee P_n^*x. \quad (6)$$

The point would be that (5) gives us $P_1x \rightarrow P_2^*x$, $P_2x \rightarrow P_m^*x$, etc., and the argument from a premise of the form $(P \supset R)$ and $(Q \supset S)$ to a conclusion of the form $(P \vee Q) \supset (R \vee S)$ is valid.

What I am inclined to say about this is that it just shows that "it's law that ———" defines a nontruth functional context (or, equivalently for these purposes, that not all truth functions of natural kind predicates are themselves natural kind predicates). In particular, that one may not argue from "it's a law that P brings about R" and "it's a law that Q brings about S" to "it's a law that P or Q brings about R or S." (Though, of course, the argument from those premises to "P or Q brings about R or S" *simpliciter* is fine.) I think, for example, that it is a law that the irradiation of green plants by sunlight causes carbohydrate synthesis, and I think that it is a law that friction causes heat, but I do not think that it is a law that (either the irradiation of green plants by sunlight or friction) causes (either carbohydrate synthesis or heat). Correspondingly, I doubt that "is either carbohydrate synthesis or heat" is plausibly taken to be natural kind predicate.

It is not strictly mandatory that one should agree with all this, but one denies it at a price. In particular, if one allows the full range of truth functional arguments inside the context "it's a law that ———," then one gives up the possibility of identifying the natural kind predicates of a science with those predicates which appear as the antecedents or the consequents of its proper laws. (Thus (6) would be a proper law of physics which fails to satisfy that condition.) One thus inherits the need for an alternative construal of the notion of a natural kind, and I don't know what that alternative might be like.

The upshot seems to be this. If we do not require that bridge statements must be laws, then either some of the generalizations to which the laws of special sciences reduce are not themselves lawlike, or some laws are not formulable in terms of natural kinds. Whichever way one takes (5), the important point is that it is weaker than standard reductivism: it does not require correspondences between the natural kinds of the reduced and the reducing science. Yet it is physicalistic on the same assumption that makes standard reductivism physicalistic (namely, that the bridge statements express true token identies). But these are precisely the properties that we wanted a revised account of the unity of science to exhibit.

I now want to give two reasons for thinking that this construal of the unity of science is right. First, it allows us to see how the laws of the special sciences could reasonably have exceptions, and, second, it allows us to see why there are special sciences at all. These points in turn.

Consider, again, the model of reduction implicit in (2) and (3). I assume that the laws of basic science are strictly exceptionless, and I assume that it is common knowledge that the laws of the special sciences are not. But now we have a painful dilemma. Since $\rightarrow$ expresses a relation (or relations) which must be transitive, (1) can have exceptions only if the bridge laws do. But if the bridge laws have exceptions, reductivism loses its ontological bite, since we can no longer say that every event which consists of the instantiation of an S predicate is identical with some event which consists of the instantiation of a P predicate. In short, given the reductionist model, we cannot consistently assume that the bridge laws and the basic laws are exceptionless while assuming that the special laws are not. But we cannot accept the violation of the bridge laws unless we are willing to vitiate the ontological claim that is the main point of the reductivist program.

We can get out of this (*salve* the model) in one of two ways. We can give up the claim that the special laws have exceptions or we can give up the claim that the basic laws are exceptionless. I suggest that both alternatives are undesirable. The first because it flies in the face of fact. There is just no chance at all that the true, counterfactual supporting generalizations of, say, psychology, will turn out to hold in strictly each and every condition where their antecedents are satisfied. Even where the spirit is willing, the flesh is often weak. There are always going to be behavioral lapses which are physiologically explicable but which are uninteresting from the point of view of psychological theory. The second alternative is only slightly better. It may, after all, turn out that the laws of basic science have exceptions. But the question arises whether one wants the unity of science to depend upon the assumption that they do.

On the account summarized in (5), however, everything works out satisfactorily. A nomologically sufficient condition for an exception to $S_1x \rightarrow S_2x$ is that the bridge statements should identify some occurrence of the satisfaction of S_1 with an occurrence of the satisfaction of a P predicate which is not itself lawfully connected to the satisfaction of any P^* predicate. (Suppose S_1 is connected to P' such that there is no law which connects P' to any predicate which bridge statements associate with S_2. Then any instantiation of S_1 which is contingently identical to an instantiation of P' will be an event which constitutes an exception to $S_1x \rightarrow S_2x$.) Notice that, in this case, we need assume no exceptions to the laws of the *reducing* science since, by hypothesis, (6) *is not a law.*

In fact, strictly speaking, (6) has no status in the reduction at all. It is simply what one gets when one universally quantifies a formula whose antecedent is the physical disjunction corresponding to S_1 and whose consequent is the physical disjunction corresponding to S_2. As such, it will be true when $S_1 \rightarrow S_2$ is exceptionless and false otherwise. What does the work of expressing the physical mechanisms whereby n-tuples of events conform, or fail to conform, to $S_1 \rightarrow S_2$ is not (6) but the laws which severally relate elements of the disjunction $P_1 \vee P_2 \vee \cdots \vee P_n$ to elements of the disjunction $P_1^* \vee P_2^* \vee \cdots \vee P_n^*$. When there *is* a law which relates an event that satisfies one of the P disjuncts to an event which satisfies one of the P^* disjuncts, the pair of events so related conforms to $S_1 \rightarrow S_2$. When an event which satisfies a p predicate is *not* related by law to an event which satisfies a P^* predicate, that event will constitute an exception to $S_1 \rightarrow S_2$. The point is that none of the laws which effect these several connections need themselves have exceptions in order that $S_1 \rightarrow S_2$ should do so.

To put this discussion less technically: we could, if we liked, *require* the taxonomies of the special sciences to correspond to the taxonomy of physics by insisting upon distinctions between the natural kinds postulated by the former wherever they turn out to correspond to distinct natural kinds in the latter. This would *make* the laws of the special sciences exceptionless if the laws of basic science are. But it would also lose us precisely the generalizations which we want the special sciences to express. (If economics were to posit as many *kinds* of monetary systems as there are kinds of physical realizations of monetary systems, then the generalizations of economics *would* be exceptionless, but, presumably, only vacuously so, since there would be no generalizations left to state. Graham's law, for example, would have to be formulated as a vast, open disjunction about what happens in monetary system$_1$ or monetary system$_n$ under conditions which would themselves defy uniform characterization. We would not be able to say what happens in monetary systems *tourt court* since, by hypothesis, "is a monetary system" corresponds to no natural kind predicate of physics.)

In fact, what we do is precisely the reverse. We allow the generalizations of the special sciences to *have* exceptions, thus preserving the natural kinds to which the generalizations apply. But since we know that the *physical* descriptions of the natural kinds may be quite heterogeneous, and since we know that the physical mechanisms which connect the satisfaction of the antecedents of such generalizations to the satisfaction of their consequents may be equally diverse, we expect both that there will be exceptions to the generalizations and that these exceptions will be "explained away" at the level of the reducing science. This is one of the respects in which physics really is assumed to be bedrock science; exceptions to *its* generalizations (if there are any) had better be random, because there is nowhere "further down" to go in explaining the mechanism whereby the exceptions occur.

This brings us to why there are special sciences at all. Reductivism as we remarked at the outset, flies in the face of the facts about the scientific institution: the existence of a vast and interleaved conglomerate of special scientific disciplines which often appear to proceed with only the most token acknowledgment of the constraint that their theories must turn out to be physics "in the long run." I mean that the acceptance of this constraint, *in practice*, often plays little or no role in the validation of theories. Why is this so? Presumably, the reductivist answer must be *entirely* epistemological. If only physical particles weren't so small (if only brains were on the *out*side, where one can get a look at them), *then* we would do physics instead of paleontology (neurology instead of psychology; psychology instead of economics; and so on down). There is an epistemological reply; namely, that even if brains were out where they can be looked *at*, as things now stand, we wouldn't know what to look *for*: we lack the appropriate theoretical apparatus for psychological taxonomy of neurological events.

If it turns out that the functional decomposition of the nervous system corresponds to its neurological (anatomical, biochemical, physical) decomposition, then there are only epistemological reasons for studying the former instead of the latter. But suppose there is no such correspondence? Suppose the functional organization of the nervous system cross-cuts its neurological organization (so that quite different neurological structures can subserve identical psychological functions across times or across organisms). Then the existence of psychology depends not on the fact that neurons are so sadly small, but rather on the fact that neurology does not posit the natural kinds that psychology requires.

I am suggesting, roughly, that there are special sciences not because of the nature of our epistemic relation to the world, but because of the way the world is put together: not all natural kinds (not all the classes of things and events about which there are important, counterfactual supporting generalizations to make) are, or correspond to, physical natural kinds. A way of stating the classical reductionist view is that things which belong to different physical kinds ipso facto can have no projectible descriptions in common; that if x and y differ in those descriptions by virtue of which they fall under the proper laws of physics, they must differ in those descriptions by virtue of which they fall under any laws at all. But why should we believe that this is so? Any pair of entities, however different their physical structure, must nevertheless converge in indefinitely many of their properties. Why should there not be, among those convergent properties, some whose lawful interrelations support the generalizations of the special sciences? Why, in short, should not the natural kind predicates of the special sciences *cross-classify* the physical natural kinds?[6]

Physics develops the taxonomy of its subject matter which best suits its purposes: the formulation of exceptionless laws which are basic in the several senses discussed above. But this is not the only taxonomy which may be required if the purposes of science in general are to be served: e.g., if we are to state such true, counterfactual supporting generalizations as there are to state. So, there are special sciences, with their specialized taxonomies, in the business of stating some of these generalizations. If science is to be unified, then all such taxonomies must apply *to the same things*. If physics is to be basic science, then each of these things had better be a physical thing. But it is not further required that the taxonomies which the special sciences employ must themselves reduce to the taxonomy of physics. It is not required, and it is probably not true.

Notes

I wish to express my gratitude to Ned Block for having read a version of this chapter and for the very useful comments he made.

1. I shall usually assume that sciences are about events, in at least the sense that it is the occurrence of events that makes the laws of a science true. But I shall be pretty free with the relation between events, states, things, and properties. I shall even permit myself some latitude in construing the relation between properties and predicates. I realize that all these relations are problems, but they aren't my problem in this chapter. Explanation has to *start* somewhere, too.
2. The version of reductionism I shall be concerned with is a stronger one than many philosophers of science hold, a point worth emphasizing since my argument will be precisely that it is too strong to get away with. Still, I think that what I shall be attacking is what many people have in mind when they

refer to the unity of science, and I suspect (though I shan't try to prove it) that many of the liberalized versions suffer from the same basic defect as what I take to be the classical form of the doctrine.

3. There is an implicit assumption that a science simply *is* a formulation of a set of laws. I think this assumption is implausible, but it is usually made when the unity of science is discussed, and it is neutral so far as the main argument of this chapter is concerned.
4. I shall sometimes refer to "the predicate which constitutes the antecedent or consequent of a law." This is shorthand for "the predicate such that the antecedent or consequent of a law consists of that predicate, together with its bound variables and the quantifiers which bind them." (Truth functions of elementary predicates are, of course, themselves predicates in this usage.)
5. I don't think there is any chance at all that this is true. What is more likely is that type-identification for psychological states can be carried out in terms of the "total states" of an abstract automaton which models the organism. For discussion, see Block and Fodor (1972).
6. As, by the way, the predicates of natural languages quite certainly do. For discussion, see Chomsky (1965).

Bibliography

Block, N., and Fodor, J., "What Psychological States Are Not," *Philosophical Review* 81 (1972), 159–181.
Chomsky, N., *Aspects of the Theory of Syntacs*, MIT Press, Cambridge, 1965.

Chapter 45

Narrative Explanations: The Case of History

Paul A. Roth

Narratives are stories, a telling that something happened. A narrative explanation, presumably, presents an account of the linkages among events as a process leading to the outcome one seeks to explain. Examples of explanations in a storylike format are readily found in history books, certain anthropological accounts, case histories in psychoanalytic writings, and the sort of stories one hears daily from students and colleagues as to why this paper was not done or that committee meeting was not attended. The use of narratives to explain is unquestioned; what is subject to philosophical dispute is whether this habit is to be tolerated or condemned.

An important focus of this dispute is not the fact that much is obscure with regard to the notion of narrative. Rather, objections arise because the notion of explanation is deemed by some clear enough to rule out any category of narrative explanation, no matter how "narrative" is to be understood.

Indeed, the very idea of a narrative explanation invites two objections. The first I term methodological. It runs as follows. Explanations have a characteristic logical form. And while the precise constituents of narrative form are a subject of much study and debate in literary theory, there exists a *prima facie* distinction between narratives and the standard form of a proper scientific explanation. Specifically, narratives relate discrete events; they do not invoke laws. The methodological complaint, in other words, is that narrative structure is too far from the form of a scientific explanation to count as an explanation. There cannot be narrative explanations, then, because such a category runs afoul of a received explication of "explanation."

This objection is closely associated, of course, with positivism. Although my purpose in this chapter is not to review the too familiar debate inspired by positivist models of historical explanation, I sketch reasons for believing that much of the debate —both pro and con—on *the* form of historical explanation is misguided.

The second objection I call metaphysical. This objection may be formulated in the following way. The academic division of labor is such that while, for example, historians work to construct true accounts of the past, philosophers toil to understand by what marks the truth may be known. Any satisfactory analysis of the notion of explanation, and so of historical explanation, should reveal the conditions which must be satisfied if that explanation is to be counted true. Attention to narrative form, however, slights this critical point. Since analyses of narrative structure underline the parallels between history and fiction, the study of narrative is not going to illuminate the relevant differentia of historical explanations. The complaint, in brief, is that emphasis on narrative structure situates historical practice too close to the writing of fiction. So the category of narrative explanation is rejected given the nature of narrative and its contrast to the purpose of historical inquiry.

Notice that the objections require only the assumption that history is a nonfiction discipline. This hardly seems disputable. Yet, if nonfiction, history either is a science or it is not. If it is, then narrative explanations will not do for formal or methodological reasons. But suppose, if you wish, that history is not science-like. Perhaps the nature of historical inquiry is only to provide an *understanding* of events. To invoke a traditional distinction, history is an *idiographic* and not a *nomothetic* discipline. Historians, on this account, study unique and nonrepeating occurrences, or, at least, what is unique about events.[1]

Yet even on this conception of history, a question remains concerning how to verify a narrative. And the issue of verification does not intersect, in any obvious or interesting way, with the issue of narrative form. The extent to which history respects canons of narrative construction might influence the literary merit of that history. But it hardly seems relevant to determining the conditions under which that history is true. Thus, whether the emphasis of a historian's task is taken to be explanation or is defined as understanding, verificationist concerns seem to rule out the relevance of narrative form.

Both of these objections, I argue, are illfounded. The reasons in each case are quite different. The methodological objection and the dispute regarding the status of historical explanation can be disposed of by undercutting the view of knowledge which motivates it. The metaphysical objection is more subtle and stubborn. It is with this objection that I am primarily concerned. What is metaphysical about the objection is that it assumes a correspondence theory of historical knowledge. This assumption, I argue, is incoherent.

A consequence of rejecting this correspondence view is that it no longer makes sense to speak of historical narratives as true or false. At first blush, this sounds troubling. I suggest why, properly understood, it is not. Concluding considerations related to the suggested logic of narrative explanation are meant to illuminate why the failure of narrative form as such to be true or false engenders no special problem for assessing the objectivity or explanatory utility of narratives qua explanations.

Why insist on the Procrustean exercise of rendering histories into a format dictated by the current favorite model of scientific explanation? The problem is what it means to do science. A remark by Hempel offers a glimpse of what lies at the heart of this issue. "The necessity, in historical inquiry, to make extensive use of universal hypotheses of which at least the overwhelming majority come from fields of research traditionally distinguished from history is just one of the aspects of what may be called the methodological unity of empirical science."[2] The methodological objection, this suggests, is not tied to the viability of some particular model of scientific explanation, such as the covering-law model; the issue is what disciplines yield knowledge. Hempel's remark points to the fact that behind the old debate on the applicability of the covering-law model to history is the unity-of-method thesis.

Positivism attempted to legislate to the republic of letters a general criterion of what could count as knowledge. Is there still a basis for mandating that some one form or other is, for example, *the* form of explanation? The failures of positivism remain a source of important and instructive lessons. Perhaps the most instructive failure can be seen in the history of the efforts, beginning with Carnap's *Aufbau* and continuing to Hempel's "Empiricist Criteria of Cognitive Significance: Problems and Changes," to provide a reconstruction of scientific knowledge by their own standards. Positivism was done in by its own best advocates. It ceased to be a viable research program not for reasons tangential to its concerns, such as an inability to provide a plausible reconstruction of historical explanation. The failure took place at the heart, in the discovery that its methods were inadequate and inappropriate to characterize scientific explanation. The broader epistemological objections later developed by Quine and by Sellars argued convincingly that the problems are irremediable.

The question of what to count as an explanation becomes, in part, a question of the use of this term. The methodological objection assumes that a proper subset of disciplines ought to serve to define for the rest what this standard is. This debate on explanation has interesting parallels to the problem I have elsewhere termed the *Rationalitätstreit*.[3] This problem concerns whether standards of rationality vary radically or whether one may insist, following Martin Hollis, on the "epistemological unity of mankind." Each side of this debate, I maintain, is committed to a view I dubbed "methodological exclusivism."[4] Exclusivists (of whatever stripe) presume that there is exactly one correct methodological approach to a subject matter. Yet, once the philosophical presumptions of methodological exclusivism are exposed, exclusivism loses its appeal.

As to explanation, it is worth reminding ourselves there is no good reason to believe that there is just one correct explication of the notion of explanation. Such claims to explication come to have a purely stipulative or legislative force in the absence of some notion of analyticity.

My suggestion has been that the methodological objection presupposes the plausibility of some exclusivist explication of explanation. These explications appeal, in the case at hand, either to the unity-of-method thesis or some implicit notion of analytic equivalence. Only by presupposing such problematic philosophical doctrines does one justify demands either for countenancing or failing to countenance narrative as a form of explanation. Indeed, there is no clear candidate for the title of *the* logic of explanation.

Yet, without some sense of what the logical form is, determination of truth conditions—however those are to be spelled out—and of implication remain obscure. And to the extent they remain obscure, the rational evaluation of issues is frustrated or precluded. I challenge the view that precisely one logical form is appropriate to explicating the notion of explanation. A positive case for a category of narrative explanation would require, inter alia, exposing enough formal properties of narrative accounts to establish how such explanations are viable candidates for objective evaluation. Resolving general objections is a mere prolegomenon to that undertaking.

There is, I suggested, a second general objection to the possibility of narrative explanations which bears examination. I termed this the metaphysical objection. The type of objection I have in mind here is made by Maurice Mandelbaum in *The Anatomy of Historical Knowledge*. The impulse to link history and narrative is one which Mandelbaum deems "unfortunate," because it emphasizes what is, strictly speaking, a purely incidental aspect of historical inquiry. Narrative structure is, on his view, a mere stylistic device. Whereas the methodological objection centered on the issue of adherence to certain formal constraints, the metaphysical objection emphasizes the relation of what is written to what is being written about. Indeed, Mandelbaum invokes an almost Rankean image of the historian recounting the past "as it actually was":

> Describing history as narrative suggests—and I assume is meant to suggest—that historiography is to be compared with telling a tale or story. This is misleading even when applied to the most traditional histories. A historian dealing with any subject matter must first attempt to discover what occurred in some segment of the past, and establish how these occurrences were related to one another. Once this research has been carried forward to a partial conclusion, he must, of course, think about how he will best present his findings, and this ... may be regarded as "constructing a narrative." Such a narrative, however, is not independent of his antecedent research, nor is that research merely incidental to it; the historian's "story"—if one chooses to view it merely as a story—must emerge from his research and must be assumed to be at every point dependent on it. It is therefore misleading to describe what historians do as if this were comparable to what is most characteristic of the storyteller's art.[5]

Mandelbaum's artless Baconian conception of historical research stops just short of endorsing what might be called a correspondence theory of historical truth and objectivity. The reluctance to endorse directly a correspondence theory is a consequence of contrasting the complexity of the "full" historical picture and any historian's necessarily limited depiction of it. His version of the sort of metaphysical picture I ultimately want to reject has it that events enter into processes by some natural *historical* dynamic inherent in the events and processes of which they are parts. He argues:

> From what has been said it can be seen that the events with which a historian deals in tracing a process may belong together either because they are, quite simply, constitutive parts within that process, or because they have entered it through influencing one or more of these parts. In speaking of the constitutive parts of a series of events, I refer to the fact that when a historian seeks to understand the nature of and changes in a society ... he is dealing with a complex whole, some of whose parts he already knows. It is these parts—and any others whose existence he uncovers—that are parts of the whole.... Thus, one can see that whenever a historian correctly analyzes the structures present in a society, or whenever he gives correct information as to the sequence of changes that it ... has undergone, he has dealt with events that belong together because they are the parts of the continuing whole.
>
> Such a whole is not formed merely because the historian has defined his subject matter in a certain way and has confined the scope of his inquiry to what occurred with respect to that particular subject matter.... Rather, the events that he includes as belonging within the series of occurrences with which he is to deal are those between which he finds inherent connections because they have influenced one another.[6]

Mandelbaum's guiding analogy is likening history to mapmaking.[7] Both maps and histories may differ in terms of scale, scope, detail. Both may be subject to change over time. However, histories, like maps, are guides over existing terrains:

> One may hold that a basic structure is imposed on a historical account by the evidence on which it rests; the existence of lacunae in that evidence, and the new questions that are present in it, direct the historian's attention to the need for further evidence of a specific kind.... Thus, whatever evidence is originally available to a historian will not be an inchoate mass, and the more evidence there

> is, the less choice he will have as to the alternative ways in which he may reasonably structure his account....
>
> It is on the basis of the connections inherent in the evidence with which historians work that they can propose concrete causal analyses of the events with which they deal.[8]

Historical pictures are successively filled in by collecting more evidence concerning the events of interest. The picture is always partial; but what history provides is an ever clearer picture of things as they actually were. The past exists in itself; in Louis Mink's phrase, it exists as an "untold story."[9] A history is, of course, more than a mere chronicle. But the work of a historian, in Mandelbaum's conception, is more like that of a scribe than an author.

The sort of metaphysical assumption which underwrites Mandelbaum's rejection of narrative, however philosophically tenuous Mandelbaum's own exposition of it, has deep intuitive roots. It is anchored in an intuition that, as Mink puts it, "the story of the past needs only to be communicated, not constructed."[10] What needs to be rejected is the picture of a past that is simply there waiting for a historian to come along. Construing history on the model of narrative appears inappropriate so long as the historian's art is assumed to consist in chipping off the excrescences of time so that the past can stand revealed.

The assumption on which the metaphysical objection is predicated is difficult to attack because it is most commonly implicitly assumed rather than articulated. As Mink notes, "But that past actuality is an untold-story is a presupposition, not a proposition which is often consciously asserted or argued. I do not know a single historian, or indeed anyone, who would subscribe to it as a consciously held belief; yet if I am right, it is implicitly presupposed as widely as it would be explicitly rejected."[11] No sophisticated person, I presume, doubts that stories about the past can be constructed in many ways. But this belief is consistent with an assumption "that everything that has happened belongs to a single and determinate realm of unchanging actuality."[12] The past is a *Ding-an-sich* at a temporal remove.

The metaphysical objection to narrative explanations in history presupposes the cogency of conceiving of an objectively desirable past. What I propose to do is to give this metaphysical assumption of the objective past the most plausible form that I can, and then show that the assumption is untenable.

The metaphysical presupposition is made compelling in a device made famous by Arthur Danto in his excellent *Analytical Philosophy of History*.[13] In the context of his seminal discussion of what he terms "narrative sentences," Danto introduces as expository devices the notions of an Ideal Chronicle and, correlatively, an Ideal Chronicler. The purpose of these devices is to suggest a case in which the factual record of the past is as complete as can be imagined *at the moment at which events occur*.

> We can imagine a description which really is a full description, which tells everything and is perfectly isomorphic with an event. Such a description then will be *definitive*: it shows the event *wie es eigentlich gewesen ist*.... I now want to insert an Ideal Chronicler into my picture. He knows whatever happens the moment it happens, even in other minds. He is also to have the gift of instantaneous transcription: everything that happens across the whole forward rim of the Past is set down by him, as it happens, the *way* it happens. The resultant running account I shall term the Ideal Chronicle.[14]

Having assumed for the sake of argument that such a complete record exists, Danto then convincingly shows that there are statements true of some time *t* in the past which cannot have been known to be true at that time. These statements will not appear even in an Ideal Chronicle. Examples are easy to generate. Simply formulate descriptions known to be true of persons at a time later than *t* and use them to refer to those persons at *t*. The result—what Danto calls "narrative sentences"—is sentences true at *t* but which could not have been known at *t*, and so escape even the ideal Chronicler.

Consider, for example, someone who viewed *Bedtime for Bonzo* when it was first released (1951). That person could not say truly, at that moment, that he had just seen a movie starring a future president of the United States. But we can describe the matter in that way; we can give a true description of what happened at time *t* which is missing from the Ideal Chronicle. An example of Danto's is: "The Thirty Years War begins in 1618." Assuming that that war is so named because of its length, the sentence is true of some event (or series of events) in 1618, but could not appear in even an Ideal Chronicle of events for 1618.

Danto's device vividly illustrates that what is interesting and important about events, what is of historical interest, is characteristically known only after the fact. A perfect witness to the past does not pick out or observe all there is to be known about the past. Danto's narrative sentences are sentences true of the past but not knowable in the past. They "belong to stories which historians alone can tell."[15] Danto nicely summarizes his own point as follows:

> For there is a class of descriptions of any event under which the event cannot be witnessed, and these descriptions are necessarily and systematically excluded from the I. C. The whole truth concerning an event can only be known after, and sometimes only *long* after the event itself has taken place, and this part of the story historians alone can tell. It is something even the best sort of witness cannot know.[16]

Danto's characterization of narrative sentences is ingenious and, I believe, correct. But how does any of this bear on the metaphysical objection with which I began? It is relevant in the following way. Recall that I claimed that this objection to narratives as a form of explanation takes its force not from the sort of flat-footed exposition which one finds in Mandelbaum, but from the intuition behind that exposition, the sort of intuition captured in Nietzsche's remark that the past is a rock you cannot move. The past is *there*. But if the fixity of the past is a coherent notion, as it seems to be, then this implies that there could be an Ideal Chronicle. Danto, for one, explicitly draws this conclusion in a passage I cited above. And even Danto betrays more allegiance to this notion of a fixed past than he otherwise claims to have by suggesting, as noted above, that the whole truth of an event might be known.

No matter that an Ideal Chronicle lacks narrative sentences; that is not the issue which now concerns us. If the past is fixed, if it is a story waiting to be told, then it must be logically possible to have some chronicle of it of the sort Danto imagines. What I argue is that the notion of an Ideal Chronicle is *not* coherent, and so we must reject as well the metaphysical picture which implies it.

The critical difficulty with the notion of an Ideal Chronicle is hinted at in the following passage from Mink.

> I refer to the Ideal Chronicle ... to point out, merely, that we *understand* the idea of it perfectly clearly. And we could not conceive or imagine an Ideal Chronicle at all *unless* we already had the concept of a totality of "what really happened." We reject the possibility of a historiographical representation of this totality, but the very rejection presupposes the concept of the totality itself. It is in that presupposition that the idea of Universal History lives on.[17]

Mink is, I suggest, right in sensing a difficulty, but he does not develop an argument. In order, then, to make the problem explicit, imagine the I.C. at work. What does the I.C. record? Danto's suggestion is everything, and at once at that. But in agreeing that the I.C. can write anything at all we have, in a Wittgensteinian sense, been tricked; the very first step is the fatal one. The conjuring trick is complete once one concedes that there is *anything* for an Ideal Chronicler to record.

What is the basic unit of the posited perfect record? They are events of every sort: visits home, heartbeats, a first kiss, the jump of an electron from one orbital position to another. But, as we know, events may be sliced thick or thin; a glance may be identified as an isolated event or as an instance in an event. What the unit-event is depends on the telling of it. Given the instructions to record "everything that happens, as it happens," the problem is not that there is too much for an Ideal Chronicler to record; the irony is that there are *no things* in the abstract to be recorded. An Ideal Chronicler never gets started because there are no ideal events to chronicle.

What sort of things are events? On one standard account, events are identified only under a description. A reason for worrying whether events exist in some philosophically relevant sense of that term—that is, whether they count as legitimate objects of discourse—is that assuming their existence proves a convenience for the purposes of explicating the logical form of sentences about actions. Countenancing events facilitates the ability to draw permissible inferences which otherwise cannot be readily managed if events are ruled out as individuated objects.

To show, then, that my claim of a paragraph back does not simply beg the question against events as objects, consider someone such as Davidson who has argued for tolerating such an ontology.[18] But a Davidsonian ontology does not help the ideal Chronicler with her task. Without some description or other, there are no specific events; with an identifying description, we still do not know if the event is of the requisite ideal sort—that is, not primarily of our making.

The specification of identity conditions does not solve the problem of underdetermination which has bedeviled philosophers of science. There is no unique physical theory entailed by the available evidence; incompatible theories can be formulated compatible with whatever data are at hand. My point about putative "ideal events"—those recounted in some Ideal Chronicle—is that treating such events as objects independent of our object (and event) positing scheme of things runs afoul of what we know about the relation of evidence to theory. The very possibility of an Ideal Chronicle presumes not just identity conditions for events, but their existence apart from our theoretical specification of them. But it is precisely this realist inference which is unjustified by any set of identity conditions for events and which, given the problem of the underdetermination of theories, is patently unjustifiable.

The problem is, of course, not ameliorated by shifting to some set of identity conditions for events other than Davidson's. Let events be as well individuated as you please; as noted, I grant we might even be allowed to quantify them (meaning, in nonphilosophic jargon, that events are treated on a par with individual objects). This does not change the problem. The objection arose not because of some inability to identify events, but due to a question about the status of these events apart from some object- (or event-) positing scheme or other. The issue is their metaphysical status, whether or not we may presume some correspondence between our talk of events and events-in-themselves. To assume that logically adequate identity conditions for events is tantamount to proving that this is how things must be with the world is, of course, to beg the question at issue.

Events *simpliciter* cannot be shown to exist; they are not known to be of nature's making rather than of ours. Events exist only by proxy. This is why one cannot presume that there are any *ideal* events for our erstwhile chronicler to chronicle; knowledge of events is restricted to happenings isolated under descriptions provided by interested parties.

Can this problem be solved by augmenting the Ideal Chronicler with a complete set of descriptions? (I continue to exempt, for the sake of argument, the type of descriptions used in Danto's narrative sentences.) Does the notion of completeness make sense here? The metaphysical assumption requires that the completeness be of the past *wie es eigentlich gewesen*. The I.C. is an objective record; a transcribing of all that has come to pass.

All statements of events appearing in the imagined chronicle are true. Therefore, they must be consistent with one another. But if all descriptions are allowed in, inconsistencies will cloud the chronicle. Consider the events depicted in Kurosawa's film *Rashomon*. The story of what happened in the forest is successively retold from the perspectives of the husband, the wife, and the robber. None tells the same story; indeed, their stories are inconsistent. One tells a tale of rape and humiliation and a husband's cowardice; in other tales, one or more of these descriptions is contradicted. The audience sees what happens each time through the narrator's eyes; it is just that the narrators see different things. Events off the screen certainly have this quality as well. If the conjunction of all possible descriptions is included in the chronicle, the imagined purpose of the chronicle is defeated. Since it contains inconsistencies, it is no longer the hoped-for objective record of what actually happened. But if some descriptions of events are excluded, then the chronicle is incomplete, contrary to its intended purpose. Hence, if complete, then inconsistent, and if consistent, then incomplete. There can be no Ideal Chronicle.

My argument, so far, assumes the premise that events are not natural entities; they exist only under a description. I then argued for the premise that the varying ways of individuating events are not always mutually consistent. Granted these premises, Roth's incompleteness theorem for Ideal Chronicles follows.

But let us make another pass at attempting to fill out the notion of an Ideal Chronicle. Perhaps what I have shown is that it is futile to imagine that there could be an Ideal Chronicle if such a chronicle requires a summing of all descriptions of events as various individuals perceive these matters. But an Ideal Chronicler need not proceed in this way. The charge of the Ideal Chronicler is not, after all, to be faithful to this or that perspective; the task is to record what happened, individual perspectives be damned.

This way of putting the matter is tantamount to denying my first premise—the claim that events are not natural entities and exist only under a description. The problem, as I originally developed it, does not assume that some fact is left out; the problem is a failure of people to agree on what counts as the event to be described. Is there a way to include all events and exclude the descriptions of human agents?

Boethius imagined that God saw everything at once; all actions at all times stood revealed simultaneously to God. Certainly this is a way of capturing all that happens. Moreover, the advantage of a Boethian Chronicler is that this person need not rely, or so I shall assume, on potentially conflicting descriptions. This account, however, still will not do, not even if we cut it down so that at time *t*, everything up to and including what is happening at *t* stands so revealed to the Ideal Chronicler. The problem is that the Boethian vision, though comprehensive, still does not contain events, or, alternatively, it contains just one event, the total picture at *t*.

The past so pictured presents not a chronicle, moreover, but a Jamesian buzzing, booming confusion. Put another way, the identification of events from the Boethian tapestry of the past requires separating the simultaneous presentation of happenings which Boethius imagined into particular strands, the ones that interest us. God may see everything at once; an Ideal Chronicler, within a temporal limit, may do the same, or so I asked you to imagine. But this chronicle gives us less than we have even now. It is not just that there might be a need to factor in cultural conditioning and personal quirks when discussing what we see; seeing is not perceiving, not in any simple sense. The basic problem is more elementary than that. When we view a snapshot or read a page of a book, if the object is not at the proper distance from our eyes, in appropriate light, and so on, we cannot see what we want to see. If someone pushes the book or picture up so it touches our nose, we see something—but not, for example, the picture of the picnic or the story of the latest Reagan gaffe. Given the Boethian view, the Ideal Chronicler is in just this position, or leaves us in this position when consulting the resulting tapestry of happenings. The Boethian Chronicler has no natural point of focus. But without a focus, either nothing appear—the booming, buzzing confusion—or God-knows-what looms before us, like the photo pressed too close for one to view. Total information gives us less than we need to know.

Given the Boethian picture, it does not follow that human beings could say anything about it at all. Chronicles presuppose categorizations of time and events, and there is no reason to believe the Boethian account could be a chronicle. Nothing in that account, filled though it is with every conceivable happening, entails that there are humanly identifiable events arranged in recognizable order. If events are picked out by human agents, the chronicle is not ideal; if the world is viewed from the eye of God, there is no chronicle. A Boethian chronicle cuts things too coarsely to solve the problem of identifying events in an objective way.

The point at which the discussion has arrived is this: if events are individuated by some favored set of identity conditions, the notion of there being an ideal chronicle self-destructs; such a chronicle is logically impossible. If we imagine the chronicle along Boethian lines, the notion still cannot be made cogent, for the Boethian image cannot be translated into the form of a chronicle.

But perhaps the Boethian picture is a start. It is, at least, complete. The problem is to find a finer grained description of matters uninfected by conflicting descriptions; this would preserve the metaphysical assumption that the past exists objectively as an untold story.

Problems arise, we just noted, if there is total information and no categories by which to organize and focus viewing. Perhaps a solution to this problem is a Carnapian Chronicler. The C.C., let us imagine, defines a language—ideal-in-L—which contains rules and definitions such that, given certain state descriptions, Ideal-in-L permits the derivation of the event which took place. Consistency is thus ensured and no ambiguity threatens. But this is no Ideal Chronicle in the desired respect. The question of which events exist has now been relegated to the status of an internal question; the existence of events is explicitly relativized to a particular set of rules. This preserves consistency, but it defeats the purpose of positing the chronicle. The purpose is to explicate how to construct a complete and objective record of the past. The correspondence theory of historical truth remains unvindicated by appeal to a Carnapian Chronicler.

The only refinement on the matter I have left to suggest would be to limit the Ideal Chronicler's task. Do the problems abate if we imagine an Ideal Boswell? The task is cut down by giving I.B. the more modest task of compiling a complete record for a single individual. If history is, as Carlyle claimed, but so many biographies, then the I.B. would preserve the metaphysical assumption. But, alas, the Ideal Boswell too produces only a blur. The root of the problem is not in the scope of the enterprise but in its completeness. Unless we equip I.B. with our categories, there are no recognizable events. But if equipped with our categories, he ceases to be ideal. He is just one of us, albeit a tad more compulsive. I conclude that the notions of an Ideal Chronicle and an Ideal Chronicler cannot be coherently fleshed out, and so the metaphysical objection fails.

Viewing the world *sub specie aeternitatis*, an ideal chronicler is imagined to see events bare, shorn of the misperceptions and oversights to which mere mortals are prone. In particular, historical events are conceived as having their own pristine ontological integrity. Caesar crosses the Rubicon in 49 B.C. or he does not; if true, the chronicler notes it and if not, not. (The dating here makes this a narrative sentence, but this complication can be ignored.) A disinterested chronicle seems impossible. The core of my complaint has been that it is the pretense to disinterestedness and completeness which makes Danto's fiction ultimately incoherent. Given the lofty God's-eye perspective, no events appear. A less lofty perspective defeats the purpose of the literary conceit. The philosophical moral is one pressed by philosophers from Kant to Davidson. We may query the world and learn a great deal, but it is a confusion to think that the categories in which the questions are posed and the answers framed constitute, to paraphrase Rorty, History's Own Vocabulary.

My primary concern throughout has been with prima facie objections to the notion of a narrative explanation. My handling of these objections, even if convincing, is as yet no delineation, however, of what counts as a proper narrative explanation. My concluding remarks are, in this regard, only programmatic.

Histories ought to contain only true statements. What remains problematic is the narrative structure which presents the verifiable statements as steps in a process which effects change. The facts with which a historian works may be, in Hayden White's term, emplotted in various ways.[19] White, as is well known, claims that there are basic narrative strategies—fundamental tropes—for emplotting events, and that these incompatible forms of emplotment are products of the historian's art in telling about the events. There is no truth-value, for example, to the statement that such and such a happening is tragic; there is only a telling which so presents it.[20] Insofar as events and processes are artifacts of different strategies of emplotment, the narrative is neither true nor false in any sense congruent with the correspondence theory.

Narratives are constrained by the facts, since they are constructed from verifiable statements. They are subject to objective evaluation because, as both White and Gene Wise argue, narrative forms in history must function as methodological paradigms. Paradigms, in the methodological sense, provide problem-solving models and, as a consequence, function to direct research.[21] Narrative forms can, then, be judged relative to their fruitfulness in guiding research and their resources for solving problems.

Finally, the narrative patterns which are candidates for explanation forms are, White suggests, themselves artifacts of our cultural heritage. What counts as an explanation may, then, be a historically contingent phenomenon. And insofar as methodological paradigms serve as basis for historical explanations, as Wise claims, an account of explanation as pattern finding and problem solving is suggested. The analysis of the notion of explanation, in short, is quite possibly a question which belongs as much to cognitive psychology and cultural history as it does to the logic and philosophy of science.

Narrative explanations, as is to be expected, are underdetermined by their evidence; agreement on the evidence still allows for the construction of logically incompatible histories. But a twist arises in the historical case that further complicates the epistemological picture. Settling the scientific world picture does not settle which macroscopic events there are; each person's story in *Rashomon* involves, *ex hypothesi*, no alteration of the facts, and so no change in any scientific inventory of what there is. Historical events, this suggests, escape specification even within a completed theory of the physical universe. Surrendering a belief in a God's-eye chronicle, and so a metaphysical commitment to the past as an untold story, does not impugn the tie of historical inquiry to the world. My penultimate remark suggests a reason as well for believing that historical inquiry is not identical with natural science.

Notes

1. Ernest Nagel traces this terminology to Windelband. See Ernest Nagel, *The Structure of Science* (New York, 1961). 547–548.
2. Carl Hempel, *Aspects of Scientific Explanation* (New York, 1965), 243.
3. See Paul A. Roth, "Resolving the Rationalitätstreit," *Archives Européennes de Sociologie* 26 (1985), 142–157.
4. I discuss this issue further in *Meaning and Method in Social Science: A Case for Pluralism* (Ithaca, 1987).
5. Maurice Mandelbaum, *The Anatomy of Historical Knowledge* (Baltimore, 1977), 25.
6. Ibid., 126–127.
7. Ibid., 15–17.
8. Ibid., 192–193.
9. Louis O. Mink, "Narrative Form as a Cognitive Instrument," in *The Writing of History: Literary Form and Historical Understanding*, ed. R. Canary and H. Kozicki (Madison, Wisc., 1978), 140.
10. Ibid., 135.
11. Idem.
12. Ibid., 141.
13. First published in 1968; revised edition, *Narration and Knowledge* (New York, 1985). All references are to the later edition.
14. Ibid., 148–149.
15. Louis O. Mink, "Philosophical Analysis and Historical Understanding," *Review of Metaphysics* 21 (1968), 690.
16. Danto, 151.
17. Mink, "Narrative Form as a Cognitive Instrument," 141.

18. See, for example, essays 1, 6–10 in Donald Davidson, *Essays on Actions and Events* (Oxford, 1980).
19. Arguments that differences in historical explanations are *not* necessarily differences over matters of fact but disagreements concerning modes of emplotment are found in Hayden White, *Tropics of Discourse* (Baltimore, 1978), and *Metahistory* (Baltimore, 1973), and Gene Wise, *American Historical Explanations*, 2d ed., rev. (Minneapolis, 1980).
20. See particularly "Interpretation in History" and "The Historical Text as Literary Artifact," in White, *Tropics of Discourse.*
21. See Wise, chapter 5, for an especially interesting discussion of paradigms as "explanation forms."

Chapter 46

The Autonomy of Historical Understanding

Louis O. Mink

I

Classical philosophy of history, which claimed to disclose the secret of human progress or to discover the overarching meaning of universal history, was consumed in the holocaust of two world wars, but it has latterly arisen from its own ashes in the guise of the theory of historical knowledge. It would be misleading, however, to refer to a renaissance of interest in the philosophy of history. It is plain that there are two revivals, not one, and that they suffer from a lack of mutual communication even more remarkable—because prima facie less justified—than the general breakdown of a community of discourse in a world of increasing intellectual specialization. Yet although each of these revivals can be fairly clearly identified and characterized, and although each is represented by a large and growing literature, neither seems to take account of the other or even to be curious about its existence and direction. Moreover, one is represented entirely by professional historians and the other entirely by professional philosophers. It would be surprising if there were not differences of problems and points of view between two groups, each with much training in and constant exposure to the subtler nuances of guildsmanship. But even so it is surprising to find no community of interest among subgroups at least; and the fact seems to be that there is an absence of either agreement or controversy between philosophers and historians who devote some thought to problems of historical knowledge.

One symptom of the absence of discourse is the extraordinary difference, in writing by representatives of the philosophical revival and of the historical revival, between footnotes and bibliographies. The latter invariably cover both revivals with at least quantitative fairness; the former infallibly reveal the hermetic limits of each. For example, the latest report of the Social Science Research Council's Committee on Historiography[1] (*Generalization in the Writing of History*) contains an extensive bibliography, with many pages devoted to the category of "Philosophical Discussions"; and most, although not all, of the books and articles comprising the recent debate among philosophers on the logic of historical explanation are listed under this category. Yet—and despite the fact that Hans Meyerhoff, as philosophical consultant to the committee, explained that to philosophers "generalization, explanation and causation form a syndrome from a logical point of view" (vii, n.)—in the text of the twelve articles comprising the report there is but *one* passing reference (a single sentence, in fact) to any of the problems or arguments which appear in the philosophical literature.

On the other hand, this latest report is the third in a series that has been, among historians, the center of discussion and controversy on the nature of historical knowledge and inquiry. Previous reports were: *Theory and Practice in Historical Study*, published in 1946 as Bulletin 54 of the Social Science Research Council; and *The Social*

Sciences in Historical Study, published in 1954 by the same Council as Bulletin 64. While these two earlier reports can hardly be said to be the Old Testament and the New Testament of contemporary historiography, they have been the agencies by which theoretical and methodological questions have been most forcibly brought to the attention of the historical guild. Yet, to choose only one example, while both reports are dutifully listed in the bibliography of Gardiner's well-known anthology,[2] there is just *one* passing reference to either of them in the text of the articles by contemporary philosophers printed or reprinted in Part II of that anthology. And this is not an isolated instance. While there are numerous references by the philosophers represented to historical works (usually as sources of examples for analysis), there are almost no references to what historians have written on historiography—although, again, many such items are included in the bibliography.

That the work of guild-historians and guild-philosophers is cited in each other's bibliographies indicates that neither group is ignorant of the other's efforts; but the absence of each in the discussions of the other suggests that they do not speak to the same questions. It is understandable that this should be so, although the reason for it is seldom remarked: what the philosophers are really discussing is logical theory, and historical inference interests them not because it is historical but because it is *inference*. Historians suspect this, and to them it seems, as one has said, that "what philosophers seem to be interested in are the remains rather than the views of historians."[3] On the other hand, what the historians are really discussing is whether history is a discipline or just an aggregation of parts of a number of other disciplines, and they are interested not in the logic of argument in general but in the *differentiae of historical* arguments. Philosophers are not unaware of this, but they have no intellectual or existential stake in the self-definition of the historical guild.

Yet although these differences arise because the organization of research increasingly reflects the fragmentation of the university into academic departments by lines as sharp as they are arbitrary, it does not follow that the problems of history and the problems of philosophy are not alternative ways of formulating common intellectual problems whose importance transcends the provincial problems of academic guilds. And this, I believe, is demonstrably the case and makes the lack of communication between historians and philosophers a serious problem rather than an incidentally curious one. Both historians and philosophers are too sophisticated to reveal—even when they are sufficiently sophisticated to recognize—the deepest concerns and creeds which inform their work. Yet these, suggested rather than stated and obliquely expressed rather than explicitly formulated, are involved in and account for the fact that despite the lack of communication between historians and philosophers, *some* philosophers are closer to *some* historians than they are to other philosophers, and *some* historians are closer to *some* philosophers than they are to other historians.

The differences are clearest among philosophers, since their main business is to sharpen them. Prima facie, almost all of the philosophical literature on philosophy of history in the last decade has dealt with the logic of explanation, and specifically has consisted of defense and criticism of an increasingly sophisticated version of the "covering-law" model of explanation and a fortiori of historical explanation. Yet the gravamen of the issue has rarely been fully evident. Historians, generally but not unanimously, have felt that the covering-law analysis bears little relation to what they understand as their own mode of inquiry, and they have applauded the sustained criticism of it by other philosophers, especially by Dray. Other historians have sided with the Hempelians. But only recently have the elective affinities been made fully visible.

One of the latest and most comprehensive defenses of the covering-law model is May Brodbeck's "Explanation, Prediction, and 'Imperfect Knowledge.'"[4] The criticism of deductive explanation, Miss Brodbeck charges, is not only logically defective but obscurantist, because it is in effect an attack on the "scientific study of man and society" which seeks to discover "quantified relations among relevant variables" (p. 232). "There is no such thing as 'historical' explanation," she maintains, "only the [scientific] explanation of historical events" (p. 254); and the claim that there is a uniquely historical mode of explanation results in, if it is not motivated by, the rejection of any "science of man" (p. 272).

On the other hand, a leading critic of the covering-law model has lately made the other side of the underlying issue explicit: "What drives us to the study of history, as much as anything else," Dray says, "is a humane curiosity: an interest in discovering and imaginatively reconstructing the life of people at other times and places.... My chief complaint against acceptance of the covering law doctrine in history is not the difficulty of operating it, in either fully deductive or mutilated form. It is rather that it sets up a kind of *conceptual barrier* to a humanistically oriented historiography."[5]

Juxtaposed, these asides to the reader make it clear that although it is to questions of logical theory that the recent philosophical literature has been addressed, the underlying issue remains the antagonistic confrontation of the scientific culture and the humanistic culture (I do not say: "of science and the humanities"), and the claim of each to privileged inquiry into at least some areas closed in principle to the other. Each side, moreover, charges the other with being the aggressor. Brodbeck presumably will not deny historians the right to indulge in "imaginative reconstruction" if it pleases them, but is outraged by the defense of such activity as yielding *historical* explanation because such a defense precludes *scientific* investigation of the same events. Dray does not claim that any events are closed in principle to "covering-law" explanations but argues against the imperialistic claim that no alternative modes of explanation are logically defensible.

Now the distinction between universally quantified and statistical hypotheses, to mention only one point which has emerged in the refinement of the covering-law model, may well be immediately relevant only to logical theory; but the underlying issue is one of general intellectual importance, because it concerns man's understanding of his own past and of himself in the light of that past. And the attempt of historians to place themselves in relation to contemporary social science (identified primarily in terms of its commitment to a scientific methodology to which the "covering-law" doctrine belongs) reflects an increasing intellectual and practical awareness of the issue. Historians and philosophers, in fact, seem to have been unwittingly converging on a common point from opposite poles.

Philosophers have always been logicians, and the main problems of modern philosophy since the seventeenth century have arisen in the analysis of the logic of scientific theory, especially physics. Historians, on the other hand, have been researchers and stylists, and the main problems of modern history since the early nineteenth century have been the discovery and synthesis of "facts." Philosophers and historians have, in general, been content to identify themselves with the roles outlined for them by Aristotle: philosophy as the concern with the universal, history as the concern with the particular. But in recent years there have been revolutions in both fields against the received doctrines: philosophers (or some of them) have been attempting to break the spell of the idea of a *single* logical model of knowledge with formal properties applicable to inquiry into any subject matter. Historians, on the other hand, have been trying to break the spell of the "school of the unique." Bulletin 64 of the Social Science Research Council attempted with no notable success to convince historians that they should *generalize* more, e.g., by formulating testable general hypotheses. The succeeding Committee on Historiography adopted the subtler strategy of convincing historians that they cannot and in fact do not avoid generalization in any case, for example, because they necessarily use concepts, like "revolution," which are classificatory or "labeling" generalizations. The conclusion of this volume of essays by historians of otherwise divergent views is that, as the editor summarizes the consensus, "the historian willy-nilly uses generalizations at different levels and of different kinds" (p. 208).

The upshot of all this is that any distinction resembling Windelband's classification of "nomothetic," or theoretical-explanatory, and "idiographic," or particularizing, sciences has been rejected from both sides as untenable in the special case of historical inquiry. Some philosophers have advanced reasons for denying the philosophical doctrine that history must be nomothetic if it is to be a science; and most historians have proved susceptible to the argument that it cannot be merely idiographic if it is to be a discipline. Philosophers and historians, so to speak, have backed into each other, although they are still facing in different directions. The question at issue is whether historians will back all the way into the received philosophical doctrine which has increasingly become credal in the social sciences as it has increasingly inspired apostasy among philosophers themselves.

II

The gist of the received philosophical doctrine is that history is not yet a science but by the explicit adoption of scientific methods and criteria may become one. Bemused by the contradictions and ambiguities of "common sense" and ordinary language, history as we know it is in the chrysalid stage of *proto-science*; and the apparent differences between historical and scientific methods can be reconciled theoretically by analysis and practically by increased methodological sophistication on the part of historians.

Common to all arguments that there is no irreducible difference between historical and scientific method is an acceptance of something like these propositions: (1) No valid distinction can be made between *Naturwissenschaften* and *Geisteswissenschaften* in terms of a difference in the logical and conceptual structure of the explanations which they provide; there is a single logical mode of explanation. (This is the principle of the methodological unity of science.) (2) The explanation of any phenomenon involves its subsumption under general principles which are not ad hoc and for which disconfirming instances are not known. Specifically, the explanation of a phenomenon requires showing the statement describing it to be a deductive consequence of a set of general laws ("covering laws") together with particular statements describing initial conditions. (This is the principle of explanation by hypothetico-deductive theory.) (3) The hypothetico-deductive method is not a method of discovery, and creative imagination finds its place in the formulation of new concepts and hypotheses; nevertheless, the *test* of such original formulations is whether they can be ordered in theory, deductively formulated, and empirically confirmed. (This is the principle of empirical correspondence.) (4) Hypotheses cannot be confirmed as isolated statements; experimental or observational confirmation is always of a set of hypotheses, many of which may not be explicitly stated, just as the results of an experiment in the behavior of neutrons may not only confirm a hypothesis in nuclear physics but also laws of electromagnetism involved in the construction of the experimental apparatus. (This is the principle of theoretical coherence.) The latter principle makes possible the division of labor among scientists: a single scientist may (and must) use as auxiliary hypotheses the results of the research of many other investigators without repeating their work.

These four principles jointly indicate an emphasis on conceptual theory at some variance with the traditional concern of historians with interpretative narrative; and each principle denies some feature of ordinary historical practice. But it is clearly not obligatory on a logical analysis of knowledge to justify everything which *claims* to be knowledge: its purpose would be defeated if it did not have normative efficacy, that is, the capacity to distinguish between valid and invalid claims. Moreover, it would indeed block the path of inquiry to argue a priori that there could be or should be no attempt to develop a methodology of history in accordance with these principles—an attempt which, if successful, would convert history into a true general sociology. But the question is not whether there can be a scientific explanation of what are called "historical" phenomena; it is rather whether historical knowledge can be analyzed *without remainder* on such principles.

Supporters of the proto-science view are of course quite aware that most historians have, to speak mildly, something less than the shock of self-recognition when confronted with a rigorous exposition of scientific method. But, it may be argued, at every point at which a doubt is raised, it can be shown that the doubting historian in fact tacitly agrees in principle without knowing it, just as a scientist might have difficulty in giving an exact description of a method which he nevertheless *uses* correctly and easily. It is not necessarily an embarrassment to the proto-science view when historians refer to "historical insight" or when they deny that they seek historical laws or even make use of general non-historical laws. If history be proto-science, astute analysis can reveal to historians that their habits of work and processes of inference are more complex than they realize, so that greater logical sophistication will enable them to enjoy and extend the fruits of the scientific habit while avoiding the lapses which tempt the uninitiated.

Now the primary assumption of the proto-science view is that there can be no legitimate mode of understanding not analyzable by an explicit methodology. If it can be shown that history is autonomous, and not proto-science, it must be, I think, not by showing that there is some fact or set of facts which can be explained "historically" but not "scientifically," nor even by providing alternative models of scientific explanation, but by a critique of this assumption. One may accept the thesis that there is no fact incapable of being scientifically explained and yet hold without inconsistency that there are other ways of understanding the same facts (although I would not concede, except for purposes of argument, that there is no problem about the meaning of "same facts," since what will *count as a fact* is in part prescribed by the adoption of an explicit methodology). Whether historical understanding is a first approximation to scientific explanation, or whether it can support the claim to autonomy, I propose to examine by reference to six features of historical practice and description which are generally (although of course not universally) accepted by historians. The proto-science view can to a considerable degree account for these features as evidences of methodological immaturity. The question is whether they can be *justified* by any alternative account incompatible with the proto-science view.

III

1. *Historians generally claim that they can give at least partial explanations of past events; but they do not ordinarily undertake to predict the future, even at the level of incompleteness and generality at which they "explain" the past.* Nor does it, by and large, seem to them an oddity that the past should be explainable but the future unpredictable. The reasons for dissociating explanation and prediction are not often articulated by historians, but probably in any given case include one or more of the following: (a) explanation and prediction are *logically* different, and success in one is not intrinsically connected with success in the other; (b) the past consists of settled fact while the future—at least the future of human decisions and their consequences—is as yet indeterminate; (c) historical explanation requires a certain "perspective," which the historian can achieve only after a connected sequence of events (such as a revolution or a reign) has been completed; (d) too much contemporary evidence (classified data, memoirs, diaries, letters, confidential reports, etc.) is not available to the historian at the time when it is relevant for prediction, although most such material will be available to future historians at a time when it is useful for explanation.

To the proto-science view of history none of these can be expanded into cogent reasons, because there is, in this view, neither a logical nor a practical (so far as inquiry alone is concerned) difference between explanation and prediction. Either to explain or to predict an event is to show that the statement describing its occurrence follows logically from a set of general laws together with (true) statements of initial conditions. If the event is past, this process yields an explanation; if the event is future, it yields a prediction. And every prediction which conforms to this logical pattern and is confirmed in experience becomes an explanation as soon as the event has occurred. The difference between explanation and prediction may be important for one's practical interests (since one may wish to prepare for the event or to avoid the prediction by changing the conditions), but the *logical* structure is identical in both. No one would argue that we can predict eclipses, but that we cannot really "explain" them until after they have occurred. Why should historical explanations be different?

It therefore seems (still runs the proto-science view) that a historical explanation of an event is unwarranted precisely to the extent that the historian *would have been unable* to predict the even before its occurrence, assuming his present professional equipment and knowledge of antecedent facts. Nothing in the lapse of time will convert a logically inadequate prediction into a logically satisfactory explanation. But even though historical explanations are poor creatures indeed by the standards of the physical sciences, the historian need not despair to the extent of seeking refuge in appeals to "perspective" or "insight." He has in fact, without realizing it, been using this logical model of explanation to provide what Hempel has called an "explanation sketch." Compared with an ideal explanation, his statements of initial conditions are factually incomplete, imperfectly analyzed, and imprecise. His general laws, probably not explicitly acknowledged, tend to be commonsense generalizations about human behavior not yet clarified and confirmed by reconstruction in psychological, sociological, or economic theory. But these lacks point the directions which historical inquiry should take: the development of analytical techniques, the quantification of data, and the search for causal and statistical laws. Even with cumulative success in these enterprises, history may fall short of completely accurate specific predictions—but, of course, so does much of science. History *may* be able, like the natural sciences, given specific conditions to predict within a reasonable margin of error the changes in a relatively isolated system.

The force of this proto-science view, however, lies not so much in the proposal the historians should be satisfied with nothing less than accurate predictions as in the argument that they should be dissatisfied with present explanations. Successful prediction is a *test* of theory (although it is not even that in such sciences as geology and cosmogony), but the *condition* of theory is its logical structure which makes prediction possible as more than an informed guess. With a logical model of explanation at hand, the reasons for dissociating explanation and prediction disappear (and all other reasons are in principle disposed of): (a) the belief that explanation and prediction are "logically" different is clearly refuted in the case of scientific knowledge, and one cannot argue that they differ in historical even though not in scientific knowledge without begging the question at issue: (b) the argument that human actions are explainable but not predictable is either ruled out in any particular inquiry (as in psychology) or else outflanked by the admission of statistical macro-laws (as in demography); (c) the appeal to "historical perspective" can be explained in terms of the ex post facto selection by historians of commonsense generalizations about human behavior which plausibly fit the completed data. (Men sometimes sacrifice themselves for a cause, and sometimes betray a cause in their own interests; so a historian may "explain" a given action by either of these incomplete generalizations *after* the action is completed and its circumstances known); (d) and the unavailability of contemporary evidence is at best a practical difficulty but not a theoretical objection; the same difficulty obtains, say, in meteorology.

Now to the extent that historians seek to explain events as instances of a general theory, the proto-science view is, I believe, unobjectionable. But it must clearly beg the question if it undertakes to convince them that they *should* accept only theoretical explanations as satisfying the desire to understand. And if one entertains the possibility that historians mean *something* by the claim that the past is explainable but the future unpredictable, to accept the proto-scientific view rules out the possibility of discovering what this meaning is. It may be that historians use other modes of explanation than the theoretical mode; and only an a priori argument could rule out this possibility.

It is to describe one such mode that W. H. Walsh has revived Whewell's term "colligation," by which Walsh intends to describe "the procedure of explaining an event by tracing its intrinsic relations to other events and locating it in its historical context."[6] Explanation by colligation, he suggests, is appropriate in cases where a purpose or policy has found expression in a series of actions each "intrinsically" related to the others in the series. This does not exclude the possibility that many historical explanations, or parts of every historical explanation, also make use of "explanation of a quasi-scientific type involving the application of general principles to particular cases." But even so it is questionable whether the facts of historical practice correctly described as "colligative" pose an insuperable difficulty for the proto-scientific view, unless the latter is committed additionally to a mechanistic or behavioristic theory of human behavior which cannot admit purposes as causes. If it is possible in principle to formulate generalizations to the effect that men seeking certain goals under certain conditions will act in specific and predictable ways, then colligation would turn out to be not a mode of explanation but merely a description of very incomplete approximations to the ideal model of explanation proposed in the proto-science interpretation.

Nevertheless, the historian's insistence on "perspective" seems to me more than a mere recommendation of the attitude of objectivity or an excuse to avoid making predictions. It is at least in part a claim that for the historical understanding of an event one must know its consequences as well as its antecedents; that the historian must look before *and* after (and not pine for what is not); that in *some* sense we may understand a particular event by locating it correctly in a narrative sequence as well as by classifying it as an instance of a law.

Not infrequently we ask, "And then what happened?" not merely out of curiosity but in order to understand what we have already been told. Suppose that one were watching a pantomime of an exceptionally tall man preparing for the night in the upper berth of a Pullman—but without knowing the subject of the pantomime. The mimickry of gesture, at first utterly cryptic, may gradually become intelligible, but a necessary condition of its intelligibility is the significant *order* of represented action. Whether the downward motion of a hand represents the closing of a shade or the drawing of a vertical line depends on its position in a sequence of actions, and a sequence in which the subsequent actions are as significant as those antecedent. It would seem gratuitous to deny that understanding may result from a narrative answer to the question, "What happened then?" And one may call such an answer a *sequential* explanation without implying that it is the *only* explanation possible of a specific fact or set of facts, or that it is a satisfactory answer to the different question, "Why did it happen?" Moreover, the state of understanding produced by a sequential explanation is clearly complicated, and may include, perhaps in all occurrences, commonsense generalizations and implicit theories. But I am concerned only to argue that the latter in some cases presuppose and do not in any reconstruction *replace* the limited understanding afforded by a sequential explanation. If this is so, it indicates that "historical perspective" is not merely an honorific name for an explanation sketch but refers to a kind of understanding which can account for the difference between "hindsight" and "foresight" on grounds which do not necessarily deny the logical identity of scientific explanation and prediction. And it preserves, one might add, the possibility of genetic explanation.

2. *Historians may often prove false a "hypothesis" about a historical event or period without concluding that it is false in any other case or as such.* Thus it may be found that the distribution of votes in a parliamentary division on a specific issue is not explicable in terms of the economic classes and interests of its members, and yet no inference may be drawn that the hypothesis of economic determinism, however stated, is thereby disconfirmed.

Now the proto-scientific view can of course account for this fact as follows: although it is true for logical reasons that a hypothesis is disconfirmed by a single negative instance (assuming that it is not one of the auxiliary hypotheses which has really been falsified), the case is no different in history from that in science, because at least two senses of "hypothesis" must be distinguished. In a trivial sense, any particular statement whose truth has not been established can be called a hypothesis. A chemist may entertain of an unknown the "hypothesis" that it is an acid, and quickly test it with a piece of litmus paper. This would be analogous to a historian's guess: "This revolution failed because it lacked leadership." In neither case would the falsification of the hypothesis preclude the employment of a similar hypothesis in another case. But such a particular hypothesis must be distinguished from generalized hypotheses such as "Acids turn litmus paper red," or "Revolutions without strong leadership invariably fail," either of which *would* be falsified by a negative instance. There is no reason to believe, therefore, that the role of hypotheses or the logic of confirmation is subject to special qualifications when applied to historical explanation.

The difficulty with this account, however, is that even generalized hypotheses are not abandoned by historians when disconfirmed, nor can the survival be dismissed as unfortunately ideological. It is highly doubtful whether any historians regard the "economic interpretation of history" as a hypothesis, or theory, to be accepted only so long as some single bristlingly irrefragable fact (perhaps from the Fourth Ming Dynasty) remains undiscovered. The reply to this by the proto-science view, of course, is that the vagueness of terms and the absence of measurability in historical hypotheses make it impossible to decide empirically whether or not they are confirmed by any given data. (How shall we define "leadership" and measure relative strengths of leadership when investigating revolution?) Moreover, historians are often rough-and-ready in their use of terms. A leading defender of Frederick Jackson Turner's "frontier hypothesis," for example, observes of Turner's critics that they refused to recognize "that Turner was advancing an hypothesis rather than attempting to prove a theory."[7] To a logician this is puzzling: a theory *is* a set of hypotheses, and it is logically impossible to entertain a hypothesis except in connection with a theory to be proved or disproved.

Yet such a curious usage may reveal that historians mean something quite different by the term "hypothesis" or that hypotheses function quite differently in historical inquiry than in scientific inquiry—including social science. It may not be a logical gaffe but a significant symptom that Turner's "frontier hypothesis" has never been precisely formulated and yet has been a supremely fertile source of suggestions of specific inquiries to be undertaken.

One might say that for the proto-science view of history, a hypothesis is in the first instance a *candidate*, regarded with interest and some hopeful expectation but without approval until it passes its examinations and is admitted (on condition of continued good behavior) as a law. But historians seem generally to regard generalized hypotheses not as potential laws but as *guides* whose services they employ. This Virgilian function of hypotheses is of course not absent in natural science; witness the commonly accepted criterion of "fertility" for good hypotheses. But the historian seems to use hypotheses as *suggestively* rather than *deductively* fertile. For him, a hypothesis is not a tentative law but a rule for asking questions, a rule for delimiting the scope of inquiry, and a rule for determining the relevance of evidence. So it is in science too, but in an incidental rather than in an essential way. For the scientist, the hypothesis is the target; for the historian, a signpost. In history, the informal rule function of hypotheses is not, as it is in science, preserved in the results obtained; hypotheses are left behind, so to speak, like the wives of some great men. In Kantian language, one might say that hypotheses as rules for inquiry are constitutive in science, regulative in history.

This distinction would be otiose, however, if it served merely as an indirect restatement of the familiar view that science seeks the universal, history the particular. The latter distinction is at best one of emphasis and degree; and so would be the distinction between the scientific use of data to confirm hypotheses and the historical use of hypotheses to lead to data *unless the distinction points to different kinds of understanding* rather than merely to different stages of the same pattern of inquiry. Nor would it be enlightening to point out that the use even of false hypotheses (as, for example, racist theories) may lead historians to stumble onto areas of evidence which accidentally turn out to be significant in other contexts; for the history of natural science is not wanting either in examples of false theories which led to happy discoveries. The question is whether the historian seeks a distinctively different *mode of understanding* in the attainment of which generalized hypotheses are means but not ends. Most of the recent literature on historical explanation has sought to justify the ways in which general statements which are not universally quantified statements enter into historical thought and writing; hence do not have the logical form of laws, and hence cannot serve as premises for rigorously deductive explanation.[8]

But suggestions that models of historical explanation weaker than the classic Hempelian model are more adequately descriptive of actual historical practice still share with the proto-science view the assumption that historical understanding can be analyzed into explicit methodology. They call attention to the ubiquity in history (although not *only* in history) of "limited" or "guarded" generalizations, but they do not even claim to account for *interpretative* hypotheses such as the "frontier hypothesis" or the Pirenne thesis. And it is these which historians generally believe in some way distinguish history, as *interpretative narrative,* from chronology on the one hand and "science" on the other. It is perhaps for this reason that historians have on the whole been unabashed by the notion of *empathy* which philosophers have been so eager to relegate to that limbo of philosophical fictions already populated by final causes, entelechies, dormitive properties, and mental substances. Nevertheless:

3. *Historians very often testify that they find it useful or necessary to "relive" or "recreate" in imagination the events which they investigate*; and they have not been reluctant to call this activity "insight" or "intuition." Dilthey, Croce, and Collingwood, among others, seized on this fact as the defining difference of historical inquiry, arguing that human actions are always the expressions of states of mind and that the distinctive act of historical understanding is the reexperiencing by the historian (and by his readers if he is as adept at writing as he is in research) of these states of mind.

The proto-science interpretation of this claim is quite explicit. Of course, its supporters say, *Nacherleben* or "psychological empathy" may be a useful *heuristic* device—as the use of pictorial models by natural scientists has heuristic value—but like all heuristic devices it must be regarded as an initial stage in investigation, not as its aim or as a test of its conclusions. Thus, according to Edgar Zilsel, "when a city is bombed it is plausible that intimidation and defeatism of the population result. But it is plausible as well that the determination to resist increases.... Which process takes place cannot be decided by psychological empathy but by statistical observation only.... The method of 'understanding' ('insight') ... is not sufficient when investigating historical laws."[9]

Now it should not be necessary to observe that a historian trying to understand what took place in Guernica or Rotterdam is hardly likely to consider himself to be "investigating historical laws." Certainly the "method of understanding" has, to my knowledge, *never* been recommended as a substitute for scientific method in the discovery of laws but has always been intended to describe a kind of reflective activity in the understanding of events. A working historian may well object to Zilsel's example,

> Of course, the bombing of a city may lead in general to defeatism or to stiffened morale. But I am interested in *this* city, at *this* time, in *these* circumstances, and the more I learn about this case, the more clearly can I understand why one result rather than the other obtained. What I come to understand is the way in which all of the events—the contamination of the water supply, the mayor's speech, the death of a child—are interrelated as constituents of the total event.

Such a reply is relevant at least in its denial that the aim of inquiry is to discover what variables are required for the formulation of a general law connecting "bombing" with "intimidation" and "determination."

Criticisms of the "method of understanding" seem invariably based on the unwarranted assumption that it is intended to verify general laws or to establish particular facts for which there is no independent evidence; and the instances of "understanding" considered in such criticisms are invariably anonymous examples wrenched loose from the rich and detailed context with which the historian works. No doubt this is an inescapable error, since one can hardly quote a written history in full to illustrate a point. But the error is also in part due to the fact that the proto-science rejection of empathy is based on the assumption that to understand an event is equivalent to seeing it as an instance of a law, and this is the very question at issue.

What is required, evidently, is an elucidation of the concept of "understanding," and one which clarifies its connection with other related concepts. Now the proto-science view provides an admirably clear conceptual map, in which the concept of *understanding* is identified with the concept of *explanation,* and the concept of explanation is identified, so far as its formal properties are concerned, with the concept of *prediction,* through a logical model which specifies the relation of both to other concepts such as "law," "deducibility," etc. Most philosophical criticism of the proto-science view has been devoted to prying loose the concept of explanation form that of prediction; but we might also try to see to what degree, and by what arguments, we can loosen the tie that binds the concept of *understanding* to that of *explanation.*

The key to an alternate account of understanding is perhaps in the term "context." The minimal description of historical practice is that the historian deals with complex events in terms of the interrelationship of their constituent events (leaving open entirely the question whether there are "unit events" in history). Even supposing that all of the facts of the case are established, there is still the problem of comprehending them in an act of judgment which manages to hold them together rather than reviewing them *seriatim.* This is something like, in fact, the sense in which one can *think* of a family as a group of related persons rather than as a set of persons plus their individual relations of kinship. It is in the latter way alone that one can *describe* a family, but this is an accidental consequence of the fact that language is discursive and that one must name one after the other the constituents of a group which one can think of (as one can see them) simultaneously. It is this elementary fact, I suspect, which has led critics of the "method of understanding" to center attention on the accidental features of the way in which it is described and illustrated rather than on the act of judgment to which it refers. Nevertheless, it is important to recognize that it is not in any proper sense a "method"; it is neither a technique of proof nor an organon of discovery but a type of reflective judgment. And it is a kind of judgment, if I am not mistaken, for which the proto-science view cannot account—and does not wish to.

4. *Historians generally do not adopt one another's significant conclusions unless convinced by their own thorough inspection of the argument*; unlike scientists in general, they must read one another's books instead of merely noting their results. Except for monographic studies, historical "abstracts" (like philosophical abstracts) will tell a historian only whether he should study the argument abstracted. To the proto-scientific view, this merely confirms the belief that in the absence of conceptual theory no two historians (perhaps no historian at two different times) mean the same thing by such central terms as "feudalism." At best the lack of method can lead to irresolvable disagreements about the weight and adequacy of evidence; at worst it is an invitation to mistake arbitrary and a priori "interpretations" for evidenced conclusions. It is no wonder, then, that history is not a cumulative science in which an investigator can use the results of others' research without repeating for himself the entire process of their research (always excepting monographic results). Thus the second Committee on Historiography of the Social Science Research Council, having learned from discussions with other social scientists that "the most portentous development in modern social science is pioneering work in providing a more coherent basis for cumulative theory," urged that "to give impetus to the process of cumulative analysis is, in the opinion of the Committee, one of the most important ends to be achieved by the use of social science methods in history."[10]

Now cumulative knowledge rests on the possibility of what might be called "detachable conclusions," and there is no question that the use of detachable conclusions affords enormous economy and efficiency in the administration of scientific research. A scientist will normally repeat another scientist's work only if he suspects the possibility of error; a historian may repeat another's work if he is convinced that it has afforded significant results. And despite the lack of rigorously defined method, there may be very considerable agreement among historians who have examined the same materials. The proto-scientific view, of course, interprets such agreement as indicating that tacit hypotheses and principles of inference are shared by the parties to the agreement; and it proposes that these can be articulated in an explicitly formulated methodology. But if the account of historical judgment so far adumbrated is tenable, there are other reasons for this agreement in the absence of such a method.

One might plausibly argue that the historian is dealing with types of evidence to which quantitative weights cannot be assigned; or that, even if such assignment were possible, the total weight of evidence is not a function of the weights of individual items of evidence taken separately (there is, I think, at least a question whether in history or in law there can be an explicit logic of corroborative as distinguished from demonstrative evidence); or that there is a different sort of relation between evidence and hypothesis in historical and some other arguments from that which one finds in science. But to expand any of these possibilities would be in effect to accept the gambit of the proto-science view: that the problem is one of the logic of confirmation. The major point of difficulty in attempting to transform history into a cumulative science, I believe, is not one of the *logic of evidence* but one of the *meaning of conclusions.*

Detachable conclusions are possible in science because—and only because—of its theoretical structure. The division of labor in research requires that concepts have a uniformity of meaning, and the methodological problem of definition therefore becomes central. But despite the fact that a historian may "summarize" conclusions in his final chapter, it seems clear that these are seldom or never detachable conclusions; not merely their validity but their meaning refers backward to the ordering of evidence in the total argument. The significant conclusions, one might say, are ingredient in the argument itself, not merely in the sense that they are scattered through the text but in the sense that they are *represented by the narrative order itself*. As ingredient conclusions they are *exhibited* rather than *demonstrated*. Articulated as separate statements in a grand finale, they are not conclusions but reminders to the reader (and to the historian himself) of the topography of events to which the entire narrative has given order. In this one respect, at least, history is akin to poetry in its reliance on ingredient conclusions rather than detachable ones. It was suggested above that the conclusions of historical inquiry dispense with the hypotheses which served as rules to guide inquiry, whereas the conclusions of scientific inquiry contain its hypotheses; we might expand this suggestion and say that scientific conclusions dispense with their evidence while historical conclusions contain it. This sounds odd if the comparison is made only in terms of the logic of evidence; it is less odd if we are thinking of the rules of meaning.

Is a historian cheating because he never defines the term "capitalism"? Yes and no. Yes, if the aim of his study is to discover socioeconomic laws about the development of economic institutions; no, if his aim is to understand the interrelationships of particular events, ideas, and institutions in their complicated development. It is not uncommon to find historians, bemused by methodology, defining terms and then using these terms in senses not allowed by the definitions. But most often it is no matter unless one is looking for illustrations for a logic text rather than for historical understanding. The actual meanings are provided by the total context; they may be consistently or inconsistently used, but one's judgment about this has exactly the same grounds as the judgment whether the argument exhibits significant conclusions.

Even historians who have been attracted to the proto-science view claim that history has a distinctive aim. The second Committee on Historiography reminded itself that the "analysis of interrelations goes on in all social science, but the attempt to make a general synthesis of all major factors at work in a given conjuncture of events is peculiar to historical studies.... The conscientious historian, even when engaged upon monographic research, never permits himself to forget the final goal, namely, comprehensive synthesis."[11] In view of the total impression left by this report, it seems necessary to observe that although the "analysis of interrelations" by the use of hypothetico-deductive method may *enable* synthesis, it will not of itself *produce* it, and it may possibly distract the historian from his habit of seeing things together to a program of explaining things separately. Because there is no logical method of testing the synoptic judgment which is his real aim, he may take refuge in a piecemeal series of "testable" statements.

What the historian does in aiming at synoptic judgment, I suggest, is much like what everyone does in interpreting the meaning of a statement. The meaning of an actual utterance can be analyzed as a function of the meaning of its individual terms plus its syntax plus emphasis, all interpreted in a particular context of discourse. Analogously, the historian tries to understand a complex process as a function of its component events plus their interrelationships (including causal relationships) plus their importance, all interpreted in a larger context of change. It is primarily the syntax of events, of course, in which he is interested. One might even say that the aim of historical knowledge is to discover the *grammar* of events, whereas in the proto-science view it is, so to speak, the *logic* of events alone that could deserve the name of knowledge. The figure can be extended even further: as the meaning of a statement may be questionable because of semantic ambiguity, syntactical ambiguity, amphiboly, or because it has been taken out of context, so there may be historical disagreement about the correct description of individual events, of their interrelations, of their relative importance, or of the significance of the process as a part of its larger history. Those who hold the proto-science view of history tend as a principle of methodology to reject the possibility of formulating exact rules for the unequivocal determination of meaning of all statements under all conditions. The desire to approach this goal has led in fact to the proposal that an artificially constructed "language of science" be substituted for the hopeless complexities, ambiguities, and nuances of living language. But unfortunately the historian cannot solve his problem by constructing an artificial past, as the logician solves his by constructing an ideal language. The danger is rather that the historian may become obsessed with compulsive neatness and try to do so.

5. *Historians generally agree that there are resemblances among complex events (e.g., revolutions), but also insist that no two such events are identical*; and they often add that their aim is to understand an event as "unique" rather than as typical. This is in the first instance a description of historical practice, but has also been the basis of attempts, notably by Windelband and Rickert, to *define* history as knowledge of the concrete and the particular rather than the abstract and general. But the proto-science view can cogently argue that this will not serve to distinguish history from science, because no two physical events are exactly identical either, and, moreover, a physical fact is just as concrete and particular as a historical fact. There is no reason, according to the proto-science view, why the historian's impressionistic awareness of resemblance cannot be replaced by precise analytic techniques, and no more ultimate significance in the fact that no two revolutions are exactly alike than in the fact that the spectrographs of no two stars are identical. But here again, I suspect, historians have been right in what they have denied, although mistaken in what they have affirmed. No doubt too great emphasis has been given to the irreducible uniqueness of events; but while this is an error if taken as a theory about events, it can be defended as obliquely revealing a distinctive characteristic of the historical *judgment*. There is reason to believe that the recognition of resemblances is a kind of *terminal* judgment, which is not replaceable by an analysis of factors, as an explicit methodology would require. When one observes that two people "look alike," one does not ordinarily compare a series of physiognomic details and *then* infer a similarity; the recognition of resemblance is immediate and total. Of course, it is possible to carry out an analysis of the resemblance noted, comparing cephalic indices, lengths of noses, and so on. However, the question is not whether the immediate recognition of resemblance can in every case be *supported* by something like a graphic chart of similar characteristics, but whether the latter can *substitute* for the former as a description of the fact of resemblance. I think that we should no more accept a set of comparative indices as the *fact* of resemblance than we should fall in love with a Bertillon identification. One could never predict by looking at sets of physical measurements whether the persons they describe resemble each other if one were not antecedently capable of recognizing resemblance without such analytical tools.

This is not merely an observation that the whole is greater than the sum of its parts, or a rejection in general of the possibility of exhaustive analysis. But analytic comparisons do seem plainly inadequate as a *substitute* for the immediate recognition of resemblances, and this inadequacy is at least a clue to the nature of historical understanding. The suggestion is that the distinctive characteristic of historical understanding consists of comprehending a complex event by "seeing things together" in a total and synoptic judgment which cannot be replaced by any analytic technique.

Now it is in the effort to attain this judgment and not as a method of "confirmation" that empathy or the so-called method of understanding is serviceable. It is misleading to say that in order to understand Caesar's decision to cross the Rubicon I must somehow "become" Caesar or "relive" his decision. Obviously, I can imagine myself as Caesar no matter how little I know about him, and such imagination is worthless from the standpoint of knowledge, although it has produced some interestingly imaginative dramas. But the more I know about the facts of the case, the more necessary it becomes to use something like empathy in order to convert an indigestible heap of data into a synoptic judgment by which I can "see together" all these facts in a single act of understanding. Otherwise, if I am asked what I have learned, I can only point mutely to my filing cabinet. On the other hand, it would be misleading to suggest that historical inquiry consists of the indefatigable collection of facts and *then* a grand swoop of synthesis. The complicated connections between facts and inference and the ways in which inference informs the process of "factual research" have often enough been pointed out. What is here called "synoptic judgment" is, I think, both a characterization of the type of historical thought in the process of research and also a description of its final aim.

The proto-science view also fails to recognize that practical experience is necessary both to attain and to understand such synoptic judgment. It is often observed that Caucasians have great difficulty in telling apart strangers of another race: "All Chinese look alike." Similarly those with little knowledge of history are likely to see resemblances which the working historian considers insignificant, and to fail to see resemblances which he finds important—as Mussolini saw rather too much resemblance between the Roman Empire and the Fascist State, and not enough between himself and Renaissance *condottiere*.

Why then, if synoptic judgment is ingredient in historical skill and an aim of historical understanding, has it been so little noted and so much misunderstood? First, I suspect, because the historian must set forth *in sequence* a narrative which, if I am correct, he understands or tries to understand *as a whole*. Thus the substance of essentially nondiscursive understanding may be obscured for his reader, or even to himself, by the compelling problems of discursive style. Second, a historian is apt to "summarize his conclusions," thus giving the impression that the latter, like the detachable conclusions of science, are inferred from the evidence, rather than being indicators which point to the way in which the evidence has been ordered. Hence if a historian lists characteristics in which the presidencies of Jackson and the second Roosevelt were alike, the proto-scientific interpreter will at once ask whether functional dependencies can be discovered between these characteristics *in general*: what, for example, is the connection, if any, between the fact that a statesman's (*any* statesman's) power rests on mass support and the character of his fiscal policies? Such a question is, of course, perfectly legitimate in its own right, but a generalization about popular democracy and economic policy is not necessarily the aim of historical inquiry. the synoptic judgment may be the satisfaction of its own aim, not a way of authenticating data in the service of sociology, economics, or political theory. Finally, the primary reason for misunderstanding is the tendency to divert attention from historical *judgment* as a reflective act to historical *explanation* as a series of statements, some of which are offered as reason for others. The latter is of course a partially correct description of a written history, as of any argumentative discourse. But it is clear that if there *were* such a thing as synoptic judgment, achieved by some historians and communicated in some histories, no analysis of a history as a series of statements would be adequate to rendering the sense of such judgment. Nor would an explicit methodology serve to produce such judgment. The logic of confirmation is appropriate to the testing of detachable conclusions, but ingredient meanings require a theory of judgment. Otherwise one is in the position of denying that a chair will stand alone because four one-legged stools will not stand unsupported.

But two surprising and connected consequences result from accepting the view that historical knowledge involves a distinctive kind of judgment rather than dealing with an autonomous subject matter or employing a unique "method." One is that temporal order is not of the essence of historical judgment; the other is that "historical" or synoptic judgment is not limited to the understanding of past events. Philosophical idealists such as Collingwood and Oakeshott have sensed the importance in historical inquiry of what here has been called "synoptic judgment"; but they seem to have concluded that because a synoptic judgment is a single and self-contained act of understanding which does not *contain* temporal sequence, it therefore cannot significantly *refer* to such sequence. As a result, such apparently paradoxical conclusions as "All history is contemporary history" are misleading, but not entirely wrong: misleading because they obscure the distinction between knowing history and history-as-known, but not wrong insofar as they direct attention to the act of "seeing things together." That events occur sequentially in time means not that the historian must "relive" them—by reproducing a determinate serial order in his own thought—to understand them, but that he must in an act of judgment hold together in thought events which, by the destructiveness of time, no one could experience together.

But this is the same type of synoptic judgment by which a critic "sees together" the complex of metaphor in a poem, by which the clinical psychologist "sees together" the responses and history of a patient, or by which the leader of a group "sees together" the mutually involved abilities, interests, and purposes of its members. Without aesthetic theory the critic may become bizarre; without psychological theory the clinician would be irresponsible; without political theory the leader is likely to become merely the foremost member of a clique. But it is not theory which elucidates the poem, cures the patient, or finds effective expression for the will of the group. Success in any of these enterprises depends at least as much on the ability to make synoptic judgments as on the correctness of theory. The proto-scientific view is quite correct in pointing out that judgments about concrete cases are tests of a theory; its failure is its inability to account for the fact that theory may lead to judgment and yet not be a part of it.

6. *Historians generally assume that they have a potentially universal audience, especially for the "comprehensive syntheses" at which they aim.* With special exceptions, such as economic history and history of science, written history has not ordinarily supposed special information or training on the part of its readers. Of course, historians write for one another in the sense that they seek to meet professional standards of competence; but few historians have abandoned the hope of educating a general audience *directly* in the knowledge produced by their inquiries rather than of serving the lot of us indirectly by applying the results of their labors—as does, say, a biologist or an econometrician. Traditionally, historians have believed that whatever utility may be ascribed to historical knowledge accrues from its intellectual *possession*, not from its

conversion into techniques, as mathematical principles can be converted into slide rules. The phrase "applied history" is at best a metaphor. But historians have not without reason believed that they are in a preeminent way the stewards of the funded wisdom of the race, with a responsibility to communicate it widely. The very name of their discipline meant originally "to inquire, to know, to *tell what one has learned.*"

Yet the historian has a peculiar problem of organizing (not merely "selecting," as is commonly said) what he has to tell, and not merely because he cannot reproduce exactly in the order of narrative the sequences and simultaneities of the events described. His problem becomes intelligible, however, if it is seen as an attempt to communicate his experience of seeing-things-together in the necessarily narrative style of one-thing-after-another. (It hardly needs saying that even if we possessed a time machine which enabled us to sit before a screen and directly review the past in its minutest details, we should still need the act of historical understanding to make intelligible this bewildering panorama. Under such circumstances, historical inquiry would no longer be necessary to search out and authenticate data of use to social scientists; but for exactly the same reason social science would still be unable to dispense with imaginative schemata of historical change which it could not produce form its own resources.) Given the historian's problem, it is not surprising that he finds himself attracted to the possibility of theoretical understanding, which is also a way of seeing-things-together—things, that is, which are instances of the same law or realizations of the same model. But this is a different way, and to adopt it as a logic of inquiry is not a solution of the historian's peculiar problem but a denial that the latter exists at all.

Thus it is proper for the historian to "tell what he has learned," not what he has demonstrated. His readers (including most of his colleagues) may be at his mercy so far as the authenticity of his "facts" is concerned, but he stands before the bar of his audience so far as his judgment is concerned. His judgment is, so to speak, perfectly transparent; one looks through his ostensible conclusions to the total argument which is their meaning. The proto-scientific view correctly objects that "plausibility" is no criterion of truth. The feeling of plausibility may "authenticate," as in better historical novels, narratives of events which never occurred; and it may very well rule out as implausible (like the Irishman who exclaimed, "It's a lie!" when he saw his first swordfish) events of whose occurrence the evidence leaves no reasonable doubt. But plausibility is subject to degrees, and it becomes a *better* criterion as we learn more of the circumstances of the event. It is perhaps implausible that an emperor should abdicate and enter a monastery; it is more plausible when we know more, not about the statistical behavior of emperors in general, but about Charles V. It is plausible that Hitler should have launched an invasion of England in 1940; it is less plausible when we learn more about Hitler's strategic plans, his own estimate of German and British capabilities, and his beliefs about British morale.

In the way it functions, historical judgment seems very like Aristotle's *phronesis*, a kind of sagacity or "practical wisdom." Like the latter, it is concerned with ultimate particulars and is an "intellectual" virtue, "possessing truth by affirmation or denial"; and like the latter, it can be a way of grasping a practical rule or principle which cannot be explicitly exhibited as the basis for a mechanical method. "Therefore we ought to attend to the undemonstrated sayings and opinions of ... people of practical wisdom not less than to demonstrations; for because experience has given them an eye they see aright."[12]

So historians may be wiser than they can say, but only if we hear what they have to tell, and not insist that they hand out piecemeal, like the slips in fortune cookies, tested hypotheses as "what history teaches." Moreover, there is some danger that out of respect for "cumulative analysis" and wistfulness for the rewards of group research historians may try to imitate the success of the sciences in finding within the division of labor useful jobs for second-rate as well as first-rate abilities. So long as the historian seeks synoptic judgment, difficult to attain and more difficult to communicate but distinctive of his inquiry, there is no substitute for the alembic of his mind, and no efficiency in detachable conclusions.

IV

A common theme runs through all six of the characteristics of historiography on which I have tried at least to focus attention: the idea of "historical synthesis" or "interpretative history"; and it is the special character of this as a mode of understanding which a theory of historical knowledge must recognize if the methodological autonomy of history is to be justified or preserved. Both the proto-science interpretation and the abundant criticisms of that interpretation have discreetly skirted the difficulties of this idea by limiting their attention to the explanation of discrete events, and the result has been to polarize the available accounts of explanation. On the one hand, the proto-science view finds historical explanations highly imperfect but implicitly deductive: the fall of a kingdom is more complex than a track in a Wilson cloud chamber, and the theory available for its explanation much thinner, but they differ in degree only, not in kind. On the other hand, critics of the proto-science view tend to regard historical explanations as commonsense explanations which just happen to be about interesting facts more remote than yesterday. As Scriven says of historical explanations: "There they are, simple and unadorned, logically no different from those [explanations] in commonsense talk about people and physical objects."[13] The assumption is that historical thinking is really nothing but common sense and a lot of factual research; one "explains" Caesar's actions as one would those of one's friends, and is historical only in finding out what Caesar's actions in fact were.

The proto-science view does not claim to provide an explicit model for everything which might be advanced in some context as an "explanation." Its power is prescriptive rather than descriptive, and it can legitimately argue that the concept of explanation can no more be clarified by regarding every putative explanation as in order as it stands than the concept of force can be made usefully precise by a definition which accounts for every appearance of the term in ordinary usage as well as for its theoretical meaning in physical inquiry. Moreover, there is no a priori reason to deny that *every* event is subject in principle to hypothetico-deductive explanation, *if* that model of explanation has the prescriptive right to determine what shall *count* as an event. Historians should realize that this is the nerve of the issue between the philosophical proponents of the proto-science view and its philosophical critics, and that it is essentially identical with the issue over behavioral psychology and sociology: can an instance of intentional *action* be analyzed without remainder into an instance of operationally describable *behavior*? This is not an empirical question but a conceptual issue: the dispute is over what *counts* as an event. Both parties to the dispute, however, agree that no events are characteristically and peculiarly *historical* in any sense other than that they are past.

I have tried, therefore, to raise a different problem, and to ask whether "history" differs from "science" *not* because it deals with different kinds of events and *not* because it uses models of explanation which differ from—or may include but go beyond—the received model of explanation in the natural sciences, but because it cultivates the specialized habit of understanding which converts congeries of events into concatenations, and emphasizes and increases the scope of synoptic judgment in our reflection on experience.

Now synoptic judgment is not a *substitute* for a methodology, any more than "empathy" is a substitute for evidence; and acknowledging it leaves open the questions whether "interpretative syntheses" can be logically compared, whether there are general grounds for preferring one to another, and whether there are criteria of historical objectivity and truth. So far it is only an attempt to identify what distinguishes sophisticated historical thinking from both the everyday explanations of common sense and the theoretical explanations of natural science. But the problems which remain in the elucidation of interpretative syntheses are problems of moment, because it is in such synthesis that history can stand apart, if at all, from the lure of the abstract and the tyranny of the particular, from both the computer and the data-retrieval system. And it may be in the understanding of synthesis that historians and philosophers can reach a synthesis of understanding.

Notes

1. *Generalization in the Writing of History*, ed. Louis Gottschalk (University of Chicago Press, Chicago, 1963).
2. P. Gardiner, ed., *Theories of History* (Free Press, Glencoe, 1959).
3. Leonard Krieger, "Comments on Historical Explanation," in *Philosophy and History*, ed. Sidney Hook (New York University Press, New York, 1936), p. 136.
4. *Minnesota Studies in the Philosophy of Science*, Vol. III (University of Minnesota Press, Minneapolis, 1963).
5. William H. Dray, "The Historical Explanation of Actions Reconsidered," in *Philosophy and History*, op. cit., pp. 132–133.
6. W. H. Walsh, *Introduction to Philosophy of History* (Hutchinson, London, 1951), pp. 59–64.
7. Ray Allen Billington, *The American Frontier* (American Historical Association, Washington, D.C., 1958) p. 4.
8. There is a comprehensive review of such suggestions in William Dray's "The Historical Explanation of Actions Reconsidered," in *Philosophy and History*, op. cit., pp. 105–133.
9. Edgar Zilsel, "Physics and the Problems of Historico-Sociological Laws," *Philosophy of Science*, VIII (1941), pp. 567–579.
10. Bulletin 64, Social Science Research Council, pp. 136, 138.
11. Bulletin 64, Social Science Research Council, p. 87.
12. Aristotle, *Nicomachean Ethics*, tr. W. D. Ross (Oxford University Press, London, 1915), 1143^b, 11–13.
13. Michael Scriven, "Truisms as the Grounds for Historical Explanations," in *Theories of History*, ed. P. Gardiner, p. 458.

Chapter 47

On the Possibility of Lawful Explanation in Archaeology

Merrilee H. Salmon

Serious concern with lawful explanation in archaeology is a relatively recent phenomenon. It coincides with a philosophical surge of interest in explanation during the past three or four decades. "New" archaeologists of the 1960s and 1970s expressed respect for the detailed taxonomies and cultural sequences constructed by their predecessors, but they were more ambitious. They wanted scientific explanations of such matters as why prehistoric peoples shifted from hunting and gathering to agriculture, why population centers developed and were abandoned just when they were, and why whole civilizations rose and fell. Ethnocentrically biased analogies or plausibility considerations rather than hard evidence supported earlier archaeologists' accounts of these phenomena. New archaeologists insisted on the possibility of lawful explanation.

In considering lawful explanations in archaeology, I will be concerned primarily with explanation of why particular features occur at an archaeological site or why archaeological discoveries have the features they do. It is at this level that archaeological theory is literally brought down to earth and tested in practice. Obviously, such small-scale phenomena must be accounted for if archaeology is to go on to provide large-scale explanations of social, behavioral, and cultural change.

Although the new archaeologists proclaim and are noted for their commitment to the aims and practices of science, this commitment does not adequately characterize the novelty of their position. As Alison Wylie (1990) points out, archaeologists at the turn of the century had proclaimed the need to seek scientific proof for their claims and to follow standard scientific practices in gathering and recording archaeological data. New archaeologists in the latter half of the twentieth century, however, focused on problems of explanation, turning to philosophy of science to acquire understanding of scientific explanation and confirmation. (Binford 1972, pp. 7–8; Watson, LeBlanc, and Redman 1971, 1984).

New archaeologists believed that modern technologies would continue to improve their data base, which was already adequate for many of their explanatory questions. Extracting knowledge from those data was the problem. The key to acquiring knowledge, they believed, was the proper scientific method, as set forth in the writings of contemporary philosophers of science, especially Hempel and Oppenheim's classic paper (1948) and Hempel's introductory text, *Philosophy of Natural Science* (1966).

New archaeologists urged their colleagues to adopt as ideal standards of scientific reasoning in archaeology the deductive-nomological model of explanation and the hypothetico-deductive model of confirmation. Not surprisingly, however, difficulties arose when they tried to construct covering-law explanations. For a time, discussion concentrated on establishing suitable explanatory laws. Archaeologists argued about how to distinguish laws from mere generalizations, about whether there were any genuine archaeological laws, and whether laws really are essential to scientific explanations. In response to these problems, some archaeologists professed an unwavering faith that laws would be found and offered examples of admittedly low-level archaeological laws, such as Schiffer's, "Loss probability [the probability that an object will be lost] varies inversely with an object's mass" (1976, p. 32). Others derided this approach, characterizing the so-called laws as trivial—"Mickey Mouse laws," Kent Flannery called them. Still other archaeologists looked for alternatives to Hempel's models of explanation that did not require laws. For example, Tuggle and his coauthors (1972) presented a "systems approach" to explanation, based on E. Meehan's *Explanation in Social Science—A Systems Paradigm* (1968). In this work, Meehan, a political scientist, argued that because of the problematic status of laws in the social sciences, adopting the standard covering-law models in those disciplines was inappropriate. In his proposed systems model, regularities, but no laws, are required. Archaeologists who believed that their discipline should be more closely aligned with the humanities than the social sciences felt their own position strengthened by problems about discovering laws of archaeology (Trigger 1978). Philosophers of science entered the fray, pointing out archaeologists' failure to comprehend Hempel's models adequately (Morgan 1973). They also noted problems with Hempel's models, as well as the lack of agreement among philosophers about how best to model scientific explanations (Salmon and Salmon 1979).

Robert Dunnell (1989) says that archaeologists lost interest in the problem of explanation when they saw that philosophers of science disagreed about the correct model of science. Some support for Dunnell's view comes from noting that in the 1980s, discussions of archaeological explanation shifted away from standard philosophical questions about the nature of laws and the formal properties of explanation to their content. This change in interest is reflected in the new title (*Archaeological Explanations*) as well as different focus and content of what was to have been simply a second edition of Watson, LeBlanc, and Redman's *Explanation in Archaeology*, a manifesto of new archaeology. Nevertheless, I will argue that although the focus of discussion has shifted, questions concerning lawful explanation remain important and unresolved issues in archaeological theory. While some archaeologists turn away from philosophical debates, others continue to press for a philosophically satisfactory account of archaeological explanation.

In a limited sense, archaeologists' commitment to lawful explanation is indisputable. Working archaeologists, whatever their theoretical leanings, use laws that were discovered or formulated in other disciplines to interpret archaeological remains. For more than a century, they have depended on geological principles of stratification and superposition to assign relative dates to archaeological materials. Today they rely on physical laws of radioactive decay and on biological laws governing tree ring growth to assign absolute dates. Using sensitive microscopes capable of detecting minute differences in the texture of lithic materials, archaeologists invoke physical laws of abrasion to analyze patterns of wear on prehistoric implements. Thus, the microscopically detectable surface features of the edge of a worked flint tool might be explained by saying these features result from pressure contact with animal hides. Archaeologists recognize that poor preservation or contamination of materials and their contexts can lead to mistaken interpretations and that refinement of techniques for application of the laws is possible, but the use of such laws in archaeological practice is generally regarded as uncontroversial.

Archaeologists, moreover, do not question whether physical sciences employ lawful explanation, nor do they doubt the existence of laws of nature. Generally, they are quite happy to avail themselves of any physical, chemical, or biological law that can help solve problems that interest them. So far at least, the archaeological literature is unaffected by the sorts of doubts van Fraassen raises in *Laws and Symmetry* (1989) about the existence of laws and their importance for science. Many archaeologists, however, are skeptical about whether it is possible to find explanatory laws that connect particular instances or patterns of human behavior with material remains. Thus, it is in the context of archaeology as a social or behavioral science that the controversial questions about lawful explanation typically arise.

One common type of question which archaeologists try to answer is why a particular kind of object or feature is found in a given archaeological context. Answers often take the form of stating the function of the object or feature. Why, for example, are the large water storage jars (*ollas*) found in Sonoran desert sites made of slightly porous clay? Because the porosity, which is achieved by using an organic temper that burns away in firing, permits evaporative cooling of the water. Functional ascriptions such as this are often based on arguments from analogy. Specifically, archaeologists infer that the function of an artifact (or feature) found in an archaeological site is the same as that of some ethnographically or historically known object when the forms of the two are analogous. This is true of the account of the *ollas* just given. Papago Indians living in the Sonoran desert were using *ollas* for storing drinking water at the time of European contact. They showed the newcomers how it was done, and both groups used *ollas* to store water until the advent of modern refrigerators. Implements, identified as corn scrapers, which Flannery and Winter (1976) found in archaeological sites in Oaxacan caves provide another example. The form of these implements, including such detailed features as patterns of edge wear, was identical to that of implements used by contemporary Oaxacan farmers for scraping kernels from corncobs.

Attributions of function provide a fundamental link between material remains and the behavior of people known only through their remains. We would hardly know how to describe archaeological findings without a functional-ascription vocabulary. "Bowl," "jar," "roof," "scraper and blade," for example, are all functional terms. Their archaeological use is grounded in implicit arguments from similar form to similar function. Although many functional ascriptions seem unproblematic, not all inference from form to function are reliable. Objects of similar form can be destined for different functions, and similar functions can be served by objects of diverse forms. Questions obviously arise about how to assign functions to archaeologically found objects when no ethnographic or historical analogues are known. Assigning all such objects to the remainder category of "ritual function" is an archaeological joke. Moreover, since perceived similarities of form depend to a large extent on the interests of the perceiver, mistakes can easily occur.

New archaeologists have been particularly critical of their predecessors (usually labeled "culture historians") for their unsubstantiated and sometimes whimsical assignments of functions. New archaeologists in turn are criticized by postprocessual archaeologists for ascribing only utilitarian functions to objects and refusing to pay attention to symbolic and meaningful aspects of human behavior. Strong political overtones color the current debate, with new archaeologists now being cast in the role of arch-conservatives who enforce their view, stereotyped by modern conceptions of male-female roles and political arrangements, of humans grubbing out a living by reacting to a usually hostile physical environment. The postprocessualists who paint this picture regard themselves as innovative, open-minded investigators of material evidence for the symbolic means by which humans actively create and define their environments. Whereas Hempel's naturalistic view of a science of human behavior is taken as a model by the new archaeologists, the postprocessualists turn to interpretivist accounts, such as that of R. G. Collingwood (Hodder 1985, 1987a; Salmon 1989). An extreme version of postprocessualism expresses views similar to those of critical theorists associated with the Frankfurt school. It regards any attempt to model a science of human behavior on the physical sciences as a ploy of dominant Western political ideology which is used to manipulate humans and to justify the maintenance of an unfair and discriminatory set of social practices (Shanks and Tilley 1987).

Problems about justifying the ascription of functions are related to the question of lawful explanation and will be taken up later. First, however, I will try to answer three questions: Has the presence of an item in the archaeological record been explained when its function is identified? If so, are such explanations dependent on laws? Finally, if a positive answer is given to our second question, what can we say about such laws?

To answer the first question, we need to look at what it means to ascribe a function to an object. In the archaeological literature "function" is rarely defined precisely but is usually characterized as that which an item is good for or used for with respect to some specific task.[1] Larry Wright, in his careful analysis of functional ascriptions (1976), insists, however, that attribution of function involves more than merely saying that the object is good for or used for some task at hand. He says that when we ascribe a function to a human artifact, we imply that the object was selected, designed, created, or modified to accomplish the desired outcome. This in turn says something about why the object came to be, or came to be where it is, or has the form it has. In this way functional ascriptions say something about the causal history of the item with the function. Insofar as functional ascriptions provide a causal etiology, they have, Wright insists, explanatory force.

In contrast to Hempel (1965) and Nagel (1977), who distinguish functional ascriptions from functional explanations, Wright argues that "the simple attribution of a function ipso facto *provides* that explanation (ascription-explanation) [of why the item is there], just as does the simple attribution of a goal to behavior" (1976, p. 80).

Wright (1976) offers the following standard formulation of such ascription-explanations:

> The function of X is Z iff:
>
> Z is a consequence (result) of X's being there;
>
> X is there because it does (results in) Z.

His use of the expression "is there" in this formulation is deliberately unspecific (pp. 81–82). In one context, the expression might mean "exists (at all)" (*Ollas* are there to store water); in another, "objects have them" (*Ollas* have slightly porous bodies to promote evaporative cooling).

The explanatory force of functional ascription depends heavily on distinguishing between the function(s) of an item and any other incidental or accidental purposes it might be good for or used for. For example, sewing shears might be used as a murder weapon, but that is not their function. Serving as murder weapons is not part of the causal story of how sewing shears came to be designed, manufactured, or found in sewing baskets. The exquisite decoration on Mayan pottery bowls has resulted in enriching the pockets of dealers who trade illegally in antiquities, but that is not the function of the decoration. In other words, the decoration is not there because it results in enriching traders, though it has that result. Only the first clause of the formulation that characterizes a function is fulfilled. In ordinary language, we often use the expression "function as" to acknowledge the distinction between the function(s) of an object and other things it might be good for—for example, "The screwdriver functions as a window opener, but that is not its function."

Whereas a sharp distinction must be drawn between an object's function (or functions) and its other "accidental" uses, the term should not be restricted to utilitarian or practical functions. Although in some contexts, it is appropriate to describe a feature as "nonfunctional, merely stylistic," the categories of function and style are best understood as coextensive rather than mutually exclusive. So-called stylistic features of objects frequently serve nonutilitarian expressive functions and occasionally even utilitarian ones. Wright's account of function can accommodate symbolic and expressive functions as well as practical functions. A brief discussion of some relationships between function and style will clarify this point.

Art historians, archaeologists, literary critics, and others have offered various analyses of style. One of the most useful analyses for archaeology, I believe, is that of Gombrich (1968) and Sackett (1982). They argue for a usage in which style represents the availability of alternatives or choices among various ways to accomplish some task. The task might be utilitarian (such as storing food) or symbolic (such as displaying status). The choices referred to can include craft traditions in a given culture as well as conscious individual choices. For example, consider a bowl made to store seed for next year's corn crop. The bowl's utilitarian function constrains its shape, size, and material composition to some degree. But even when operating under the functional imperative of making a seed storage bowl, an artisan is free to exercise some choices. She can, for example, paint on the exterior of the bowl or leave it unpainted. In the situation as described, a painted exterior is a stylistic feature of the bowl. Alternatives that are functionally equivalent with respect to seed storage are available to the artisan. Consider a different scenario, however, in which the artisan is a member of the rabbit clan in a culture which requires representing the totem animal on seed storage bowls. The painted design has the function of symbolically expressing or affirming cultural beliefs. The alternative of an unpainted seed storage bowl is not an available choice from the perspective of behaving in the culturally approved way. Of course, some stylistic variation is possible in the way in which the rabbit is painted, just as some degree of stylistic variation is possible with respect to size, shape, and material of the bowl. Style and function on this view are relative and complementary notions. Separating them analytically requires detailed contextual knowledge, for unless we can discern both the function of the artifact and the range of alternatives available to its maker, we cannot judge with any accuracy whether a given feature represents stylistic choice. Moreover, the judgment that a feature is stylistic does not preclude attributing a function to that feature.

Despite conflicting intuitions about the concept of style and complexities in the accounts offered by various authors, stylistic variation has been immensely important to archaeologists because the choices that style represents allow them to identify ethnic groups and to track their interactions. Wright's account of ascription-explanation accommodates expressive or symbolic functions of features that could be judged stylistic from a utilitarian perspective. Thus it accords well with postprocessual archaeologists' interest in the functional roles of stylistic variation. Stylistic features such as decorative painting on food containers, for example, not only can express cultural values but also can convey important information about the environment. A recently discovered, beautifully painted Mimbres bowl, for example, apparently depicts the important astronomical event that occurred in 1054 A.D. The account of functional ascription offered here should forestall the criticism that postprocessual archaeologists direct at new archaeologists for focusing too closely on "functional" or "passively adaptive" aspects of human behavior and not attending carefully enough to its symbolic or creative aspects.

Wright's account of functional ascription-explanations is, to my mind, the most detailed and satisfactory available and is particularly appropriate to archaeological theory, since artifacts designed by humans serve as paradigmatic cases for his analysis. With his account of the nature of functional ascription in hand and with an understanding that "function" includes expressive as well as utilitarian functions, we are ready to turn to our second question: Are functional ascription-explanations dependent on laws?

One way to try to answer this question is to see whether functional-ascription explanations conform to any of Hempel's covering-law models. These are, after all, the standard examples of lawful explanations. Hempel's covering-law models emphasize structural or logical features of scientific explanation. Explanations of individual events (including such events as the presence of a particular feature on an artifact), according to these models, have the structure of arguments in which the explanandum follows (deductively or with high probability) from the explanans. Laws are necessary components of the explanans because they provide the appropriate logical link between the particular explanatory facts cited in the explanans and the explanandum. The laws in these models might be, but need not be, causal.

In Hempel's account, the statement that a feature has a particular function (a functional ascription) is one of the initial conditions in a functional explanation. Typically, what is to be explained is the presence of some feature in a system (for example, an organ in a human body or a custom in a society). The explanans includes a law to the effect that systems of that type require for their maintenance the satisfaction of specific requirements. Initial conditions mentioned in the explanans include a statement that the system is operating more or less satisfactorily and that the feature mentioned in the explanandum is capable of fulfilling the requirement mentioned in the law (1965). The anthropologist Radcliffe-Brown (1952, chap. 4) offers many functional explanations of customs in primitive societies. For example, he says that the pattern of somewhat abusive conduct of a man toward his mother-in-law (joking relationship) in some relatively isolated small societies has the function of preventing conflict that would undermine social stability. The joking relationship provides a harmless outlet for tension between potentially antagonistic pairs of relatives. Hempel regards this sort of explanation as incomplete since the explanandum cannot be derived from the explanans. Different features, "functionally equivalent" to that mentioned in the explanandum, could satisfy the specific requirement mentioned in the explanans. The "explanation" explains why there is some mechanism for avoiding conflict but not why that particular custom serves the function. Anthropologists are well aware of functional equivalents. In another society studied by Radcliffe-Brown, for example, an avoidance relationship between son-in-law and his mother-in-law serves the same function as the joking relationship.

Nagel (1977), whose views are close to Hempel's, has tried to solve the problem of functional equivalents by arguing that functional explanations can often be supplemented by historical and other considerations to rule out functional equivalents and to bring the explanations into accord with deductive-nomological or inductive statistical models. Even when this can be accomplished, however, functional explanations cannot be causal explanations because the laws involved lack the proper temporal component. A law of the form, "Societies of type *A* maintain stability only when a joking relationship exists between men and their mothers-in-law," permits the joking relationship to be cotemporaneous with rather than antecedent to stability in the society.

Wright's account of functional ascription-explanation does not fit well with these models. In contrast to Hempel and Nagel, Wright discounts concern with structural features of explanation. Relying instead on the principle that a phenomenon is explained when its cause has been identified, Wright says that functional ascriptions are explanatory since they have etiological force—the item to which a function is ascribed exists, is where it is, because it has that function. Explanations, for Wright, need not be deductive or inductive arguments. Explanation consists in identifying the causes of phenomena. In Wright's analysis, the troubling question of functional equivalents loses force. Although different features could have been designed and employed to accomplish the same task, a functional ascription-explanation of a particular feature requires only that the required consequence etiology can be established for that feature.

Although the identification of causes is Wright's explanatory sine qua non, he provides no guidelines for recognizing causes and offers no analysis of the causal relationship itself. He maintains that causality is a primitive notion which cannot be elucidated in terms of any more fundamental relationship, and he says that at least in paradigm cases, causal connections are immediately perceived. In ascribing functions to human artifacts, the inherent causal etiologies normally refer to human intentions.[2] Like Donald Davidson (1980), Wright rejects the antimechanistic arguments of interpretivists who claim that because statements about the relationship between reasons for acting and the actions that arise from such reasons are analytic, there can be no causal mechanism connecting reasons with behavior (Wright 1976, ch. 4). Wright regards neurophysiological reductionism as a "plausible form of mechanism" which could provide "some law or rule governing the intention" (132). Explanations, for Wright, however, do not require the statement of such laws, for the causal relationship between reason and action can be identified even in the absence of knowledge of the mechanism.

Nickles (1977) argues for the independence of singular causal explanation in the social sciences, with special reference to archaeology. He bases his case on our ability to detect singular causal relationships even when we are at a loss to identify appropriate causal laws to ground such judgments. Other authors, of course, disagree with this position and have argued extensively that causal laws must underlie any singular attribution of causes (Hempel 1966; W. Salmon 1984). Davidson (1980), in his discussion of reasons and actions, argues that we are sometimes justified in asserting that a particular reason is the cause of an action, although we are unable to frame any appropriate underlying causal laws relating reasons to actions. Despite this, unlike Nickles, he insists that there must be some causal law to ground such claims. Davidson's position reveals the tension between the intuitive attraction of grounding singular causal claims in underlying causal laws and the special problems regarding the formulation of psychophysical laws. These issues are complex, but it at least is clear that the answer to our second question of whether functional ascription-explanations require laws depends on how the issue of singular causation vis-a-vis laws is resolved.

Pursuing our third question we ask, What would the laws that ascription-explanations depend on look like? This question is closely related to the previous one. Many authors are uncomfortable with insisting that causal laws are required to explain human behavior when these laws have proven so elusive. A look at Wright's (1976) explanation of why Cadillacs used to have fins will help focus on the relevant issues.

Wright says that the fins were there because of their popular appeal. Nevertheless, he notes that while the popular appeal of fins was the reason why (in the sense of causal etiology) Cadillacs had fins, popular appeal was neither a necessary nor a sufficient condition for the fins being there. The fins might have been there for aerodynamic reasons, or they might not have been there despite popular appeal because they were too expensive to tool up for. He makes this point to demonstrate that the causal relationship cannot be understood in terms of necessary and sufficient conditions. Thus, "Whenever a feature has popular appeal, it will be included in the design of automobiles," cannot be the proper statement of the law that underlies Wright's singular causal claim. If there is an underlying law, what is it?

Several responses are possible. We could qualify the antecedent of the proposed law in various ways ("unless the feature is too expensive to tool up for" or "unless the design staff is unaware of the public's taste"). The number of such modifications, however, seems to be limited only by the fertility of our imaginations. We could add an "all things being equal" or ceteris paribus clause, but normally all things are not equal.

We could point out that causal laws need not be deterministic and try to frame a probablistic version of the law. We probably lack statistical evidence that would enable us to frame a law of the form, "In *z* percent of cases ..." A vague probabilistic law, such as, "Usually, when a feature has popular appeal, it is incorporated into the design of a mass-produced item," however, is not noticeably distinct in meaning from the universal form with ceteris paribus clause. Even if we had statistical data, we would not be comfortable with using them to frame a law unless we were persuaded that the statistics captured something "real." But how or by what evidence would we be persuaded? Merely changing the form to that of a probabilistic rather than deterministic law does not avoid the problem of how to make the law precise, general, and strong enough to be worthy of the name.

Critics of laws in the social sciences (e.g., MacIntyre 1982) point to these problems and contrast the vague ceteris paribus clauses of so-called laws in the social sciences with the high degree of specification of boundary conditions on physical laws. For example, although the ideal gas laws are subject to the qualifying clause "under moderate temperature and pressure," this constraint can be spelled out precisely, whereas in the social sciences ceteris paribus clauses are vague and open ended.

A closer look at physical laws, however, shows that they are not strikingly different from laws in the social sciences. Physical laws similarly fail to specify all of the conditions under which the law would not apply. Whereas it is true that statements of physical laws do not explicitly invoke ceteris paribus clauses, Hempel (1988) points out that such clauses are nevertheless implicit. He argues, for example, that physical laws neither state precisely how all of the various surrounding circumstances could interfere with their operation nor that such conditions will not occur.

In the history of the physical sciences, an important focus of empirical investigations has been to specify increasingly precise boundary conditions for many physical laws. Ceteris paribus clauses in the statement of laws of social science can and sometimes do suggest where research efforts should be directed to attain more precise qualifying conditions. Market analysis or psychological or anthropological research might enable us to frame boundary conditions for a causal law relating popular appeal and features of expensive mass-produced objects, such as automobiles, that serve both utilitarian and expressive functions.

This optimistic approach must be tempered, however, by the acknowledged failure of social scientists to come up with many (some would say any) interesting *laws*, as opposed to mere summaries of practices within a given culture of historical period. Although the latter summaries might have predictive value and can even afford some insights into individual and social behavior, they somehow lack the force of explanatory laws. We feel keenly their limitations to time, place, and other circumstances, and when we try to generalize by stripping those limits away, we are left with truisms.

Some archaeologists have tried to avoid problems with behavioral laws by relying only on well-supported physical laws (Dunnell 1985, 1989). As noted earlier, some of the attributions of function made by archaeologists are strongly grounded in physical properties of the features that show traces or marks of interaction with some aspect of the physical environment. Striations made by brush strokes can be detected on pottery surfaces long after traces of paint have faded. Physical laws of abrasion and minute traces of substances provide strong support for claims that tools and containers were used in specific ways. In view of these considerations, we can ask whether at least some functional ascription-explanations are completely grounded in physical laws. The answer to this question is negative if we adopt Wright's analysis of functional ascription. Physical laws can support the claim that a piece of stone shaped by a human functioned as a hide scraper; but a further inference, not grounded in physical laws, is required to support the claim that hide scraping was its function (that the reason that the stone "is there" or has the form it has is its usefulness for scraping hides). Insofar as archaeologists are interested in ascriptions of functions, they must take into account purposive behavior or actions. I make this point not to cast doubt on functional ascriptions but only to indicate that one cannot completely bypass intentional considerations in offering ascription-explanations of human artifacts.

Philosophical criticism of attempts to formulate laws that connect intentions with behavior has been sustained on several fronts. Although Davidson believes that reasons are causes of actions, he says that numerous counterexamples show that the causal laws that underlie human behavior cannot be formed by generalizing such causal claims. Davidson believes that eventually laws that ground singular causal statements about human intentional behavior may be found but that they will not be psychophysical laws of any sort. Apparently, in his view, the discovery of the appropriate underlying laws waits on the successful development of our understanding of neurophysiology. Alexander Rosenberg (1988) also points to the failure of social science to go beyond folk psychology in framing laws of the form, "Given any person x, if x wants d and x believes that a is a means to attain d, under the circumstances, then x does a." He argues that this formulation "turns out not even to be of limited employment as a causal regularity, for the elements it connects cannot even in principle be shown to bear contingent relations to one another" (p. 49). These difficulties about laws pose serious problems for archaeological theory insofar as it is concerned with explaining material remains in terms of human intentional behavior.

That archaeology is no worse off than any other social science in this respect may not be particularly comforting to those committed to a view of archaeology as a social science, but foundational problems with laws of human behavior need not paralyze attempts to construct explanations in archaeology. After all, foundational problems concerning the nature of physical and biological laws have not halted explanatory efforts in physics or biology.

In that spirit, let us assume that despite the foundational problem discussed above, the project of explaining archaeological materials in behavioral terms is not totally misguided.[3] This assumption is shared by new archaeologists and their postprocessual critics. New archaeologists, however, have been extremely wary of trying to ascribe functions to archaeological materials in the absence of any close historical or ethnographic analogues and detailed physical evidence. As mentioned earlier, they have been more reluctant than postprocessualists to ascribe symbolic or expressive functions, that is to say, "to assign meanings," to features that do not bear traces of specific interactions with the physical environment. Many of the debates that exercise contemporary archaeologists focus on this point. The disagreement is sometimes characterized as a struggle between the scientific new archaeologists who seek (lawful) *explanations* and the humanistic postprocessualists who instead seek *understanding*, or *meanings*, while either ignoring or denying laws. More accurately, the two groups disagree about the following points:

1. The importance of symbolic or expressive aspects of human behavior for acquiring adequate understanding (or explanations) of people known through their archaeological remains.
2. The availability and quality of evidence to support the generalizations (*laws*) required to understand or explain archaeological findings in terms of symbolic behavior. Both groups recognize the central role of functional ascription, and, insofar as our analysis of ascription-explanation is applicable, both are concerned with intentional behavior of humans and with explanatory laws.

Having noted the common concerns, I now want to focus on the differences. New archaeologists have dismissed some attempts to reconstruct symbolic behavior of people known only through their archaeological remains as futile "paleopsychology" because, they say, neither analogical nor physical evidence is strong enough to support the required inferences. Ian Hodder (1987b), in contrast, claims that rigorous procedures can be developed for ascribing detailed expressive or symbolic functions, which he also calls "subjective meanings" and "symbolic meanings" (p. vii). Hodder develops his account of contextual archaeology to overcome criticisms that ascriptions of symbolic functions in his earlier work (1982) were scientifically unacceptable. An examination of this work, however, will show that insofar as the procedures suggested by Hodder have any rigor, they are identical with procedures already accepted and used by new archaeologists. Hodder's contribution in his (1987b) is a new set of lawlike principles rather than new procedures.

Hodder correctly notes that new archaeologists themselves assign meanings to archaeological remains. As we have seen, intentional (i.e., meaningful) behavior is attributed to the producers or users of human artifacts even when utilitarian functions are ascribed to them. We can go further than Hodder and point out that new archaeologists also frequently assign symbolic functions to archaeologically known features. For example, new archaeologists routinely explain the presence of valuable grave goods as expressions of the high status of the deceased. Although symbolic functions rarely impart physical traces that permit the identification of those functions, relevant historical and ethnographic analogies can often support the claim that an item was used for or good for a particular purpose and that the item "is there" because it served that purpose. The point of disagreement between new archaeologists and postprocessual archaeologists is not, therefore, whether or not meanings (intentions) are involved in the functional ascriptions or even whether symbolic behavior can ever be inferred but rather the nature of the evidence to support such ascriptions.

Questions about the quality of evidence can be asked separately about each of Wright's two conditions for ascribing functions:

1. Identifying what an item was good for or used for.
2. Determining whether it "is there" for that reason (has the proper causal etiology).

For features with utilitarian functions, either traces of physical interactions or ethnographic or historical analogies can be used to infer what an item was good for or used for. Analogies provide a major source of evidence for symbolic uses of items, as well as evidence for the requisite causal etiologies for both symbolic and utilitarian functions. Since the range of forms for features with symbolic functions is less restricted than for features with primarily utilitarian functions, historical and ethnographic analogies, even when available, are less compelling in symbolic cases. Obviously, arguments for ascribing functions depend on the context of the feature as well as its form. Ascription arguments are inductive and vary in strength according to available evidence.

Postprocessual archaeologists, following the lead of Ian Hodder, insist that despite difficulties, archaeologists can and must discern symbolic or expressive functions of the materials they study if they are to understand adequately the behavior of archaeologically known peoples.

Hodder believes symbolic meanings can be approached by first performing statistical analyses of similarities and differences along traditional archaeological dimensions such as space and time and then assuming some "universal 'language' in which similarities and differences are meaningful" (1987b, p. 5). The existence of this universal language, according to Hodder, allows us to frame some general principles (laws) such as, "Similarities and differences are constructed by making boundaries between things ... and repeating and correlating the same categories along different dimensions" (p. 7). These principles, along with detailed knowledge of archaeological contexts, are invoked to support ascription-explanations of archaeological materials. Hodder admits that multiple interpretations of symbolic functions are always possible. In fact, Hodder acknowledges, but offers no answer to, most of the standard criticisms of attempts to assign meanings to phenomena in cultures spatially or temporally distant from our own. Despite this, however, he maintains that "other historical contexts with their unique frameworks of meaning can be understood through an examination of material culture" (p. 10), and he proposes to demonstrate his claim through the case studies that comprise the book. Hodder's principles and his claim that a universal language can be read to discern the meanings of similarities and differences constitute a novel approach to archaeological explanation which would not be accepted by most new archaeologists.

To exemplify his new "rigorous procedures" for assigning meanings, Hodder discusses an unpublished study of prehistoric iron-smelting furnaces in one region of East Africa (Collett 1985). The study begins with measurements along several spatial dimensions of excavated furnaces. On the basis of various measures, the investigator divided the furnaces into two categories: deep and shallow. "Significant correlations" were also found to support an association between deep furnaces, decorated furnace bricks, and high iron slag. A corresponding association connected shallow furnaces, undecorated bricks, and low iron slag. Metallurgical analysis of the two forms of slag and "analogies with other pre-industrial smelting processes" (Hodder 1987b, p. 5) indicate that the deep furnaces were used to smelt ore, whereas the shallow furnaces were used for resmelting to refine the iron. Assuming that Collett's statistical correlations are in order, most archaeologists would regard these functional ascriptions to the two types of furnace as unproblematic even though there is no explicit concern with establishing the claim that the different types of furnaces "are there" *because* they are "good for" the two different types of smelting. That is, the procedures used thus far would be recognized as "rigorous" by new archaeologists, but these procedures constitute no novelty in Hodder's position, as he would himself admit.

Cooking pots with decorations similar to those on the decorated furnace bricks are also found near those bricks and the deep furnaces. Hodder claims that it is possible to explain also the similarity between the decorations on the bricks and the cooking pots. He believes that the "context" which associates cooking pots with one type of furnace can provide the explanatory link that will allow the archaeologists to "read" the universal language of relevant similarities and differences. At this point we expect his new methods to be revealed.

Hodder's approach requires the archaeologist "to imagine a dimension of variation that ma[kes] sense of the link between deep furnaces and cooking pots" (p. 6). Since scientific reasoning has traditionally benefited from imagination, his suggestion is hardly novel. Little imagination is required in any case to connect cooking and smelting on the grounds that both involve using heat to transform material. Hodder also cites ethnographic and historic studies conducted by Collett (1985) to support his claim that the decoration on the furnaces has a symbolic function. Collett has shown that furnaces in a number of African cultures have forms or decoration which serve expressive functions (for example, fertility motifs) that are explicitly acknowledged by the people who manufacture and use the furnaces. New archaeologists would readily agree that Collett's analogies are good evidence that the decorated bricks associated with the archaeologically known furnaces served some expressive function.

Hodder, however, goes beyond this general conclusion to state that the similar decorations on bricks associated with the deep furnace and the decorations on the cooking pots can be interpreted in terms of the "general anthropological understanding of nature/culture, raw/cooked dichotomies" (p. 6). In addition, Hodder explains the lack of decoration on the shallow furnaces, which would appear to involve transformation by heat as well, by saying that a second smelting involves only refinement and not transformation.

Hodder's symbolic ascription-explanation of the similar decorations on cooking pots and furnace bricks depends ultimately on his principles or "laws" for reading similarities and differences, since the other evidence that Hodder cites cannot support his specific interpretation. Although his principle would be judged too vague to support such an interpretation by new archaeologists, who would deny that Hodder has succeeded in providing a "rigorous procedure" for ascribing symbolic functions, Hodder's work does bring out the central role that laws play in archaeological explanations offered by postprocessualists.

Hodder's novel contribution to archaeological interpretation or explanation in this work resides in his proposal of new explanatory laws. In Hodder's introductory example, as in many other case studies presented in *The Archaeology of Contextual Meanings*, these novel explanatory laws play a crucial though often unacknowledged role. Thus, Dunnell's claim that "the immediate impact of the revelation that no single model of science was unchallenged in philosophy was a cessation of archaeological interest in the kinds of philosophical discussion that had characterized archaeological journals in the early and middle 1970's" (1989, p. 7) should not be interpreted as an indication that archaeologists are no longer interested in the possibility of lawful explanation in archaeology. In fact, we can describe the gulf that separates the new archaeologists from their postprocessual critics as a strong disagreement about the sorts of laws that can be used to explain archaeological phenomena.

Notes

The author is indebted to the other participants in the symposium on Explanation and Laws, sponsored by the Instituto de Investigaciones Filosóficas (UNAM), for helpful discussion and to Jeremy Sabloff, who read and commented on an earlier version of this chapter.

1. R. Dunnell (1978) does offer a definition: "Function is manifest as those forms that directly affect the Darwinian fitness of the populations in which they occur." This nonstandard definition has not received wide acceptance.
2. Some qualification is necessary because specifying the intention that gives rise to an artifact's function is complicated. For example, from the perspective of the artisan's cultural tradition, the rabbit design on seed storage bowls might be intended to ensure a good harvest. But ensuring a good harvest is not why the design "is there" in Wright's sense, for the design may not be efficacious. Following Merton's (and Radcliffe-Brown's) advice of seeking "latent" (or social) functions in such cases, we could say instead that the function of the design is to promote cultural or social solidarity. To justify this functional ascription, we try to establish that the design does promote solidarity and to show that it "is there" as a result of helping to maintain a society in which painting rabbits on seed storage bowls is culturally prescribed. Wright does not discuss latent functions, but his account can accommodate them. This note was prompted by a comment made by Raúl Orayen
3. Dunnell, who considers the social science model entirely inappropriate, would reject this assumption.

References

Binford, L. R. 1972. *An Archaeological Perspective*. New York: Harcourt.

Collett, D. 1985. "The Spread of Early Iron-Producing Communities in Eastern and Southern Africa." Ph.D. dissertation, University of Cambridge.

Davidson, D. 1980. *Essays on Actions and Events*. Oxford: Clarendon Press.

Dunnell, R. 1978. "Style and Function: A Fundamental Dichotomy." *American Antiquity* 43: 192–202.

———. 1985. "Methodological Issues in Contemporary Americanist Archaeology." In P. Asquith and P. Kitcher, eds., *PSA 1984*, vol. 2, *Proceedings of the 1984 Biennial Meeting of the Philosophy of Science Association*, East Lansing, Mich., pp. 719–744.

———. 1989. "Philosophy of Science and Archaeology." In *Critical Traditions in Contemporary Archaeology*, Edited by V. Pinsky and A. Wylie. Cambridge: Cambridge University Press.

Flannery, K., and M. C. Winter. 1976. "Analyzing Household Activities" In *The Early Mesoamerican Village*. Edited by K. Flannery. New York: Academic Press.

Gombrich, E. 1968. "Style." In *International Encyclopedia of the Social Sciences*, 15: 352–361.

Hempel, C. G., 1965. *Aspects of Scientific Explanation*. New York: Free Press.

———. 1966. *Philosophy of Natural Science*, Englewood Cliffs, N.J.: Prentice-Hall.

———. 1988. "Provisoes: A Problem Concerning the Inferential Function of Scientific Theories." *Erkenntnis* 28: 147–164.

Hempel, C. G., and P. Oppenheim. 1948. "Studies in the Logic of Explanation." *Philosophy of Science* 15: 135–175.
Hodder, I. 1985. "Digging for Symbols in Science and History: A Reply." *Proceedings of the Prehistoric Society* 51: 352–356.
Hodder, I., ed. 1982. *Symbolic and Structural Archaeology*. Cambridge: Cambridge University Press.
———. 1987a. *Archaeology as Long-Term History*. Cambridge: Cambridge University Press.
———. 1987b. *The Archaeology of Contextual Meanings*. Cambridge: Cambridge University Press.
MacIntyre, A. 1984. *After Virtue*. Notre Dame: University of Notre Dame Press.
Meehan, E. 1968. *Explanation in Social Science—A Systems Paradigm*. Homewood, Ill.: Dorsey Press.
Morgan, C. 1973. "Archaeology and Explanation." *World Archaeology* 4: 259–276.
Nagel, E. 1977. "Teleology Revisited: The Dewey Lectures." *Journal of Philosophy* 74: 261–301.
Nickles, T. 1977. "On the Independence of Singular Causal Explanation in the Social Sciences: Archaeology." *Philosophy of the Social Sciences* 7: 163–187.
Radcliffe-Brown, A. 1952. *Structure and Function in Primitive Society*. Glencoe: Free Press.
Rosenberg, A. 1988. *Philosophy of Social Science*, Boulder, Colo.: Westview Press.
Sackett, J. 1982. "Approaches to Style in Lithic Archaeology." *Journal of Anthropological Archaeology* 1: 59–112.
Salmon M. 1989. "Post-Processual Explanation in Archaeology: Whose Child Is This?" Paper presented at the First Joint Archaeological Congress, Baltimore, January 7.
Salmon, M., and W. Salmon. 1979. "Alternative Models of Scientific Explanation." *American Anthropologist* 81: 61–74.
Salmon, W. 1984. *Scientific Explanation and the Causal Structure of the World*. Princeton: Princeton University Press.
Schiffer, M. 1976. *Behavioral Archaeology*. New York: Academic Press.
Shanks, M., and C. Tilley. 1987. *Social Theory and Archaeology*. Albuquerque: University of New Mexico Press.
Trigger, B. 1978. *Time and Traditions*. New York: Columbia University Press.
Tuggle, D., A. H. Townsend, and T. Riley. 1972. "Laws, Systems, and Research Designs." *American Antiquity* 37: 3–12.
van Fraassen, B. 1989. *Laws and Symmetry*. Oxford: Oxford University Press.
Watson, P. J., LeBlanc, S., and Redman, C. 1971. *Explanation in Archaeology; An Explicitly Scientific Approach*. New York: Columbia University Press.
———. 1984. *Archaeological Explanation*. New York: Columbia University Press.
Wright, Larry. 1976. *Teleological Explanations*. Berkeley and Los Angeles: University of California Press.
Wylie, A. 1990. "A Proliferation of New Archaeologies: Scepticism, Processualism, and Post-Processualism." Paper presented to the Department of Archaeology, Boston University.

Chapter 48

Evidential Constraints: Pragmatic Objectivism in Archaeology

Alison Wylie

I begin with a digression that will situate my discussion of archaeological uses of evidence in the wider context of debate about the objectivity and value neutrality of archaeological understanding. My aim is to argue that although archaeology is a thoroughly social and political enterprise, evidential constraints are not entirely reducible to the interests of individual archaeologists or to the macro- and micropolitical dynamics of the contexts in which they operate. In fact, they are in some respects constitutive of political interests.

Archaeology as Politics by Other Means

This is a particularly appropriate time to take up questions about the salience of a certain kind of modified objectivism in a social science like archaeology, this being the year (and almost the date) of the Columbian quincentennial. The nature and significance of the historical facts defining the events marked by this anniversary have become the focus of deeply acrimonious public and professional debate, and archaeologists have been implicated at every turn in the construction and contestation of these facts. Archaeologists find their evidence and interpretations of precontact cultures and of the dynamics of contact appropriated as support or legitimation for positions lying across the whole length and breadth of the political spectrum, and just as often as their authority is invoked (sometimes with dizzying incongruity), they have found themselves the target of stinging critiques of Eurocentrism and racism from both within and outside the discipline. What this has brought home, in especially concrete and public terms, is the fact that archaeology is a profoundly political enterprise. However esoteric and inaccessible its practitioners may strive to make it, archaeology does do work in the world; more to the point, however pervasive and influential the rhetoric of objectivity may be among professional archaeologists, the practice and products of archaeology do reflect the standpoint and interests of its makers.

This is by no means a new insight within the discipline of archaeology, although it is one that a great many archaeologists regard with suspicion, if not outright hostility. It constitutes a profound challenge to the conviction—a central and defining tenet of North American archaeology since its founding early this century as a "profession"/discipline—that the social and political contexts of inquiry are properly external to the process of inquiry and to its products. In general terms, as Rouse describes these ideals, it is assumed that,

> Knowledge acquires its epistemological status independent of the operations of power....[1]

> Power can influence our motivation to achieve knowledge [in specific areas] and can deflect us from such achievement, but it can play no constructive role in determining what knowledge is. (Rouse 1987, pp. 13, 14)[2]

In archaeological contexts, especially in the 1960s and 1970s, the hope was that, if properly scientific modes of inquiry were adopted, archaeologists would secure a body of data that is autonomous of, and could provide a decisive check on, interpretive claims about the past, an evidential resource that could eliminate dependence on "mere speculation" and preserve them from the pernicious influence of standpoint specific interests either as they operate within the discipline (the micropolitics of the discipline or the interests of individual practitioners) or as they impinge on it from outside (external, sociopolitical factors operating in the encompassing society). As in many other social sciences, there has been enormous store set in establishing the scientific credibility and authority of archaeology. In North American archaeology, specifically, this took the form of widespread commitment to the pro-science goals of the "new" archaeology, a widely influential research program in which the ambition of instituting more scientific modes of practice in the field was characterized in explicitly positivist terms. Reconstructive hypotheses were to be treated as the starting point, not the end point, of research, and any investigation of the archaeological record was to be designed (on a hypothetico-deductive model of confirmation) as an empirical test of these hypotheses; whatever their sources, they were to be confronted with evidence from the surviving record of the pasts they purport to describe and accepted or rejected on this basis.

Despite the continuing influence of these ideals, there has been no shortage of critical analyses that demonstrate (with hindsight) how pervasively some of the best, most empirically sophisticated archaeological practice has reproduced manifestly nationalist, racist, and, on the most recent analyses, sexist understandings of the cultural past; confronting test hypotheses with evidence seems not, in itself, proof against intrusive bias. These critiques take a number of forms. By way of a short and selective summary, let me distinguish five types, or levels, of critique that have appeared in recent years. At the end of the chapter I will return to a detailed analysis of several examples of critique that expose specifically sexist bias, but for now let me indicate how broadly the critiques of value neutrality and objectivity have been framed in recent years.

1. First there are critiques that expose straightforward erasure, where the choice of research problem or the determination of "significant" sites or periods or cultural complexes systematically directs attention away from certain kinds of subjects, namely, those that might challenge the tenets of a dominant ideology or might be particularly relevant to the self-understanding of subordinate and oppressed groups. These include the critiques of colonial period archaeology in North America that have given rise to vigorous new areas of research focusing on African-American sites and material culture—for example, plantation archaeology and the archaeology of slavery (Singleton 1985). The motivating critique that, in part, gave rise to these new fields of interest was concern that, where archaeologists had failed to consider the material record of slavery, they had helped to ensure that silence on this aspect of U.S. history would be enforced by a lack of relevant data. Critiques from Latin America and from various parts of Africa make it clear that the typical preoccupations of first world and neocolonial research programs—such as discovery of the most primitive human and

hominid remains (e.g., paleoanthropology in the Rift Valley) and documentation of the now-eclipsed glories of "ancient civilizations" (in Mesoamerica and South America)—systematically direct attention away from the history of oppression and colonization that is crucially relevant to contemporary indigenous and mestizo populations in these areas (Schmidt 1992; Vargas 1992; Patterson 1992; Vargas 1990; Irele 1991, as cited by Vargas 1992 and Schmidt 1992). And most recently (since the late 1980s), a rapidly expanding body of feminist critiques documents how systematically women and gender have been left out of account even when they are a crucial part of the story to be told (see below; Conkey and Spector 1984; Spector and Whelan 1989).

2. Even when such "marginal" subjects are directly acknowledged and investigated as part of the subject domain of archaeology, they are often, notoriously, characterized in terms that legitimate their displacement or colonization. Thus, a common (second) type of critique identifies not erasure, but systematic, and manifestly "interested" (standpoint specific) distortions in how various archaeological subjects are understood. Some critics have argued that this is evident even in the new work on African-American sites and heritage (Potter 1992). Most often such critiques challenge the presuppositions of long-established research programs. One of the earliest, Trigger's discussion, "Archaeology and the Image of the American Indian" (1980), traces out the continuing legacy of nineteenth-century evolutionary thinking that has systematically obscured the complexity and diversity of native American cultures. He has recently extended this analysis to the presuppositions that lie behind a pervasively romantic view of early native American responses to contact with Europeans, a view that was intended to be a corrective to earlier accounts but represents native Americans as essentially tradition and culture bound. This fails to credit them with any significant capacity for rational self-determination, obscuring considerable diversity in their responses to Europeans that, Trigger argues, is evident in a number of aspects of the archaeological record of the period (Trigger 1991).

Trigger's original critique has been quite radically extended, for example, by Handsman (1989, 1990) and by Handsman and Richmond (Handsman and Richmond 1992), who decry the dependence of North American archaeologists on Eurocentric models of community and settlement. In the Northeast, Handsman argues (1990), the assumption that human presence must be marked by nucleated, permanent townsites has systematically obscured the presence of native Americans in their traditional homelands where, in many cases, they continued to live in dispersed homesteads throughout the periods of intense European occupation. Handsman and Richmond (1992) document how this failure to recognize native presence in anything but European-style settlements was crucial in establishing the rhetoric of absence that has been used, throughout the long history of native dispossession, to justify the appropriation of native lands. In a similar vein, Hall documents the inherent racism of "archaeolog[ies] of the colonized ... mostly practiced by the descendants of the colonizers" (1984, p. 455) in southern Africa, where presumptions of indigenous absence and/or primitiveness, and an erasure of class conflict, have been perpetuated by the dependence of archaeological analysis on orienting concepts of tribal identity. The feminist critiques of androcentrism in archaeological research often operate at this second level of analysis; they routinely object not just to the absence of any consideration of women and gender but to the unquestioned projection onto prehistory of profoundly presentist and ethnocentric (if not overtly sexist) assumptions about sexual divisions of labor and the status and roles of women in prehistory (see, for example, discussions in Spector and Whelan 1989 and in Conkey and Spector 1984 of the explicit androcentric bias in "man the hunter" models, and the critique, in Gero 1991a, of standard patterns of functional ascription to stone tools).

In all of these cases the imposition of prejudgments about what must have been (or most plausibly was) the case in the cultural past determines not just what range of reconstructive models will be considered but what sorts of data will be recovered and how they will be interpreted as evidence. At their most radical, and pessimistic, these critics insist that the stereotypes, evaluative commitments, and "mythologies" (to use a term from Thomas 1991) informing archaeological research have necessarily been self-perpetuating; they have foreclosed the collection or serious consideration of evidence that might run counter to or that have unseated them.

3. At a more general level, a number of synthetic critiques that have been advanced delineate broad patterns of congruence or "resonance" (Patterson 1986a, b) between the interests of large-scale geopolitical elites and entrenched archaeological research programs. For example, Trigger (1989) has published a compendious history of archaeological thought in which he documents the pervasive entanglement of archaeology, in every context in which it has flourished, with nationalist programs of territorial expansion and cultural legitimation. At a less global scale, Patterson (1986a) has argued that one can discern in the training and interpretive practices of North American archaeology—in the discourse, the "content and form, level of exposition, and the chosen vehicles for publication" in this field (p. 21; see also 1986b)—two distinct communities whose views of the past "resonate with" the interests of the eastern establishment (that is, international capital and its allies) on one hand, and with the "core culture" (midwestern, national capital and its power base) on the other.

4. At an even more general level are critiques of the enterprise of archaeology as a whole that indict its methodological and epistemic stance—its commitment to scientific ideals of objectivity—as effectively reinforcing, rather than countering, the partiality of its makers. The British critics of the new archaeology and its positivism have been among the most outspoken in this connection. For example, Tilley (1989) has argued that where "living in Western society of the 1980s is to be involved with and, in part, responsible for prevailing [grossly inequitable] social conditions" (p. 105), the attempts by archaeologists to establish their political neutrality, to retain a stance of disengagement, in itself sustains and legitimates this order.

5. While the foregoing types of critique reveal, at various levels of analysis, systematic gaps, biases, and distortions in the results of archaeological inquiry that we should be prepared to question, for the most part they provide no detailed explanation of how these compromising effects are produced or why they persist. That is, there is little account given of the conditions under which, or mechanisms by which, both local and global political interests come to shape the content of archaeological understanding, generating the sorts of resonances and congruences—the systematic silences and replication of stereotypes—that have been widely noted at a number of different levels of analysis. A fifth form of critique, perhaps the least developed but one that is crucially important in its potential to provide these missing explanatory links, consists of analyses of how the internal conditions of archaeological practice, the micropolitics of archaeology conceived as a community and a discipline articulated with or through a range of institutions, shape the direction and results of inquiry.

Several studies along these lines were reported in a landmark collection of essays, *The Socio-politics of Archaeology*, which appeared in 1983 (Gero, Lacy, and Blakey). In one contribution to this volume, Wobst and Keene (1983) made the argument that a number of active areas of research may attract attention because, structurally, they provide practitioners particularly good opportunities to establish their centrality to a field, not necessarily because they are inherently more challenging or intellectually significant than other areas. The construction of regional syntheses, for example, affords researchers the opportunity to subsume the work of a great many of their colleagues and ensures them wide recognition (which translates into extensive citations) if they can establish their organizational scheme as the framework for comparative analysis in a given region or period. Wobst and Keene suggest that the fascination with "origins" research must be attributed, at least in part, to the fact that an individual who controls the understanding of originary events or cultural formations must be acknowledged in various ways by all who work on later, linked periods and developments; they are, in a sense, the eye of the needle through which all else must pass. This line of argument has been extended and reframed in feminist terms by Conkey in collaboration with Williams (1991).

Although feminist scrutiny of the discipline has produced a number of critical sociological analyses, documenting pervasive and all too familiar patterns of differential support, training, and advancement of women in the field, as well as strong patterns of gender segregation in the areas in which women typically work (Gero 1985; Kramer and Stark 1988; Levine 1991; and other contributions to Walde and Willows 1991), much of this research remains disconnected from questions about androcentric or sexist bias in the content of archaeological accounts. One study that does make the link is Gero's (1991b) recent work on palaeoindian research (that is, research on the earliest human populations in the Americas). Gero notes a strong pattern of gender segregation whereby the predominantly male community of palaeoindian researchers focuses almost exclusively on stereotypically male activities, specifically, large-scale mammoth and bison kill sites, technologically sophisticated hunting tool assemblages and the replication of these tools, and the hunting and butchering practices they are thought to have facilitated. She finds that the women in this field have largely been displaced from these core research areas; they work on expedient blades, flake tools, and so-called domestic sites, and have focused on edge-wear analysis. This pattern of segregation in the workplace is reinforced by gender bias in citation patterns. In the field of lithics analysis generally, Gero argues, women are much less frequently cited than their male colleagues, even when they do research that is more typical for their male colleagues, except when they co-author with men. Moreover, their work on expedient blades and edge-wear patterns is almost completely ignored, despite the fact that it reveals patterns of exploitation of a wide range of plant materials, presumably foraged as a complement to the diet of Pleistocene mammals.

At the very least, these disciplinary dynamics—these "social *relations* of paleo research practice" (Gero 1991b p. 6)—would seem to enforce an unfortunate incompleteness in entrenched accounts of palaeoindian culture. But more seriously, Gero charges, they substantially derail (or impose a radically limiting "en-railment" of) the research program as a whole; "women's exclusion from pleistocene lithic and faunal analysis ... is intrinsic to, and necessary for, the bison-mammoth knowledge construct" (p. 7). The central problematics of palaeoindian research are created by the fact that the technology, subsistence activities, social organization, mobility, and patterns of occupation of the landscape by palaeoindians are all characterized exclusively in terms of male-associated hunting activities. It is this that generates the puzzles that dominate paleoindian research: how to explain or reconstruct what happened to the mammoth hunters when the mammoths went extinct. Did palaeoindians disappear or die out, to be replaced by the small-game and plant-foraging groups that appeared subsequently? Did they effect a miraculous transformation of their entire form of life as the subsistence base changed? These questions arise only, Gero argues, if you ignore the evidence that palaeoindians depended on a much more diversified set of subsistence strategies than allowed by standard "man the (mammoth/bison) hunter" models, precisely the evidence produced (largely) by women working (largely) on microblades and use-wear patterns. To set this incompleteness right requires not just that practitioners take into account female-associated tools but, in addition, that they revalue women's work—the work of women researchers and of women in paleoindian contexts—and undertake a profound rethinking of how palaeoindian culture has been conceived as a whole.

Taken together, these various critiques are seen by many to demonstrate not just that archaeology is partial—interest and standpoint specific—in the sense that external interests and power relations may determine what questions will be taken up and what uses will be made of the results of inquiry. This would leave disciplinary practice and the content of its products (as authoritative knowledge) uncompromised by values, interests, and the sociopolitical and material conditions of its operation. Rather, they are often understood to reveal sociopolitical entanglements—power relations and dynamics—that are intrinsic to disciplinary practice and, most important, constitutive of its results at all levels. They document how external factors determine what data will be collected and how they will be construed as evidence, what interpretive and explanatory hypotheses will be taken seriously and/or accepted (sometimes evidence notwithstanding), and what range of revisions or corrections will be considered when evidence resists appropriation in familiar terms. As Rouse (1987) has put this point, with reference to general challenges to objectivism, critiques such as these make it clear that "power does not merely impinge on science and scientific knowledge from without. Power relations permeate the most ordinary activities in scientific research. Scientific knowledge arises out of these power relations rather than in opposition to them" (p. 24).[3]

The Implications of Constructivism

What exactly follows from the argument that archaeological knowledge, including its evidential claims, is local, defeasible, and much more profoundly shaped by contemporary interests and conditions of practice than we might have realized? In some cases, the sorts of counterexamples to the externality thesis already described are parlayed into a completely general rejection of all concepts or ideals of objectivity through appeal to arguments that are familiar in philosophical contexts: underdetermination, theory-ladenness, various forms of the Duhem/Quine thesis (holism). Given the ladenness of evidence by theory, broadly construed—its ladenness by framework assumptions, background and collateral knowledge, a sense of the plausible that embodies the specific location and conceptual resources an investigator (or community of investigators)—there seems to be no basis, on these arguments, for any faith that evidence can provide a check on the interpretive claims, prejudices, or preferred constructions of a past that are congruent with, or in some sense required by, specific present interests. The production of knowledge that is tailored, intentionally or not, to fit the interests of its makers includes the production of appropriate evidence.

Within an archaeological context, something approaching this degree of epistemological cynicism or pessimism has been articulated, in some contexts, by proponents of the fourth level of critique; it is perhaps clearest in challenges brought against the new archaeology by British critics in the mid- to late 1980s, as when Hodder declared that archaeologists simply "create facts" (1983, p. 6) and that evidential claims depend on "an edifice of auxiliary theories and assumptions" that are essentially "unverifiable" and archaeologists (therefore) accept on purely conventional grounds (1984, p. 26). Hodder, and later his students, Shanks and Tilley (1987), concluded on the basis of these arguments that any attempt to use archaeological data to test reconstructive hypotheses about the past can "only result in tautology" (p. 111). On their account, archaeology is, quite literally, politics by other means; one should tell the stories that need to be told (see Hodder 1983; Shanks and Tilley 1987; for a more detailed account of these positions and the debates they have generated, see Wylie forthcoming).[4]

The difficulty is, however, that no matter how irresistible these sorts of arguments may seem in the abstract, they are often systematically subverted by the very contingencies that they mean to bring into view. However much a construct archaeological evidence may be, however inextricable from the power relations that constitute the thoroughly cultural enterprise of its production, it does routinely resist appropriation; it can, and regularly does, undermine our confidence even in presuppositions that are taken to be the necessary conditions for specific programs of scientific/disciplinary inquiry and for the forms of life they reflect, influence, or constitute. In fact, this is a feature of archaeological practice that the critics of ethnocentric, androcentric, and nationalistic bias in archeology routinely exploit. Time and again they make good use of recalcitrant evidence—evidence that resists appropriation in any of the terms compatible with dominant views or assumptions about the past—to expose the deeply political nature of the discipline. Of those critics cited, perhaps only Tilley and Gero would endorse an uncompromising constructivism, and even they depend on the very evidential constraints and empirical "routines" that they critique.

This points to a residual problem, where a rethinking of the relations between power and knowledge is concerned, that is largely sidestepped, even by a pragmatist like Rouse for whom "the real [just] is what we manipulate, what resists us, what we notice, and what we take account of without ever taking explicit notice" (1987, p. 156). It is the problem of determining how the thoroughly constructed materials of science—whatever counts (in a given context) as the evidential basis for a particular set of knowledge claims—can, in fact, "resist" theoretical appropriation, sometimes in quite unexpected and decisive ways.[5] This is a problem that is crucial to the possibility of libratory practice in science.[6] In taking it up, my larger aim is to recover and to make sense of the emancipatory capacity science may have, where systematic empirical inquiry continues to be an enormously powerful tool in contesting the prejudice and dogma that underwrite oppressive forms of life.

Theory Ladenness Reconsidered

The key to understanding how archaeological evidence can (sometimes) offer resistance or function, to some degree, as an autonomous constraint on claims about the cultural past is to look more closely at the various kinds of background knowledge—

the linking principles, "middle-range theory"—that mediate the interpretation of archaeological data as evidence and establish a connection between surviving archaeological traces and specific events or conditions in the past that are thought to have produced them. It is a well-worn truism that archaeological data only stand as evidence under interpretation, that they are laden by "theory." What the foregoing cases demonstrate is that, all too often, the crucial interpretive inferences are underwritten by "theories"—presuppositions or taken-for-granted beliefs—that derive from, reflect, or are rendered plausibly by standpoint-specific political interests. Hodder's charge of arbitrary conventionalism rings true in these cases. I will argue that the constitution of archaeological data as evidence is mediated by an enormous range of background assumptions, that the "theory" ladening archaeological data, establishing its evidential significance, is never monolithic or pervasive. It is, in a sense, in the anarchy and disunity of science that the potential for deploying empirical constraints arises.

I first sketch a model, based on both archaeological and philosophical discussions of evidence, that captures the central features of the process by which archaeological data are constituted as (potentially constraining) evidence and then show how these are deployed by recent feminist critics in archaeology.

The first and most obvious point is that, in some sense, the inferential move from data to evidence is only as strong as the mediating interpretive principles that establish a link between the surviving archaeological trace and specific antecedents in the cultural past; the challenge is to strengthen these linking principles. In this case it is the security of the ladening "theory" that matters. More specifically, I suggest, it is security understood in at least three senses that is at issue. On one hand, what counts is security in the sense described by Shapere (1985) in connection with his analysis of what physicists describe as observations of events at the center of the sun; it is the degree to which the linking assumptions can be regarded as "free from doubt" (29; see also Shapere 1982), given the results of inquiry in the contexts from which they derive, that determines the credibility of the evidential claims based on them. On the other hand, however, what matters is something like the additional considerations captured by Longino and Doell's (1983) spatial metaphor of distance (pp. 209–210; see also Longino 1990) and by Kosso's (1988) account of the directness, immediacy, and amount of interpretation required to establish a link between accessible "effects" and a transmitting "cause." In archaeological cases this is mainly a matter of judging the length and complexity of the causal chain by which archaeological remains have been produced (that is, considerations of the number of interactions and of different kinds of factors involved); this is discussed directly by archaeologists (e.g., by Schiffer 1983) in terms of the various "transforms" (cultural, natural, depositional, and so on) responsible for the production of an archaeological record. More often, however, the sort of security that concerns archaeologists is not so much the length or interpretive complexity of the linkage between a surviving record and the subject past as the nature of these linkage(s). The ideal of security in this third sense is realized when background knowledge supports (or provides) a biconditional linking principle to the effect that a surviving archaeological trace could have been produced by only one kind of antecedent condition, event, or behavior.[7]

While methodological discussions among the most recent advocates of scientific archaeology (heirs to the new archaeology of the 1960s and 1970s) have concentrated on considerations of security, especially security in the third sense, occasional references are made to the importance of making "ascriptions of meaning" (evidential significance) to archaeological data on the basis of principles that presuppose background knowledge about "processes that are *in no sense dependent* for their characteristics or patterns of interaction upon interactions [that constitute the subject of the reconstructive hypothesis under evaluation]" (Binford 1983, p. 135; see also Binford and Sabloff 1982). I read this as an appeal to exactly the sense of independence that Hacking (1983, pp. 183–185) and Kosso (1988, p. 456) find crucial in determining whether an observation can stand as evidence for or against a given test hypothesis—independence between the constituents and the conclusions of an inference that runs on what amounts to a vertical axis from some element of a given data base to claims about its linkage to (and significance as evidence of) some aspect of the cultural past.[8]

There is, a second sort of independence that plays an important role in stabilizing evidential claims in archaeology but is never commented upon in the archaeological literature and is discussed only briefly in philosophical contexts (Kosso 1989, p. 183): the independence of linking hypotheses from one another that arises when a number of such hypotheses are used to establish evidential claims about different aspects of a particular past event or set of conditions. It is analogous to the independence Hacking (1983, chap. 11) finds exploited by the makers and users of microscopes, where different physical processes (different interaction chains, different bodies of ladening "theory") are used to detect the same microscopic bodies or structural features of bodies. Certainly triangulation on a single aspect of an archaeological subject is sometimes possible and important. But more often, diverse (independent) resources are used to constitute evidence of quite distinct aspects of a past context. On the assumption that these are interconnected, the requirement of congruence among lines of evidence (that they yield a coherent model of the past context as a whole) sets up a system of mutual constraints on a horizontal dimension that can be as important in determining the credibility of any given bit of evidence as independence on the vertical dimension (the independence of a test hypothesis from the auxiliaries used to establish evidence for or against it).

There are, then, at least three sorts of security at issue in archaeological assessments of evidential claims: security as a function of the entrenchment or freedom from doubt of the background knowledge about the linkages between archaeological data and the antecedents that produced them; security that arises because of the overall length and complexity of the linkages; and security due to the nature of the linkages, specifically, the degree to which they are unique or deterministic. These are cross-cut by considerations of independence: vertical independence between linking assumptions and test hypotheses and horizontal independence between linking assumptions and the ascriptions of evidential significance that they support.

Archaeological Analyses of Gender

These sorts of evidential constraints are deployed to good effect by the critics of sexism, racism, nationalism, and other sorts of systematic bias in archaeological reconstructions of the past. They exploit the fact that the dependence of evidence on "theory," on linking principles and background or collateral knowledge, is by no means necessarily or pervasively circular. The systems of belief comprising a standpoint-specific view of the past rarely have the resources to constitute any wide range of data as evidence bearing on any very rich set of reconstructive hypotheses about this past. Even when they do—even when the framing presuppositions of a research program do directly inform the interpretation of data as evidence, as well as providing a suite of favored interpretive hypotheses—the circularity that results is not always irretrievably vicious.

Consider a feminist critique of research on the emergence of horticulture in the Eastern Woodlands developed by Watson and Kennedy (1991).[9] This is a critique that fits the first two of the five categories described at the outset. As they summarize extant thinking about emergence of horticulture, it depends on a set of underlying assumptions, uncritically appropriated from popular culture and traditional anthropology, according to which women are presumed to have been primarily responsible for gathering plants under earlier foraging adaptations and for the cultivation of domesticates when a horticultural regime was later established but are strikingly absent from accounts of the transition from foraging to horticulture. Whatever the specific mechanisms or processes postulated, the main contenders all assume that women could not have been actively responsible for the development of cultigens.

One model turns on the blatantly ad hoc proposal that shamans, consistently identified as male, were the instigators of this culture-transforming development. It was the knowledge they developed of plants for ritual purposes that led to production of the cultigens on which Eastern Woodlands horticultural practices were based. In effect, women passively "followed plants around" when foraging and then passively tended them when introduced as cultigens by men. The dominant alternative postulates a process of "coevolution" in which horticulture emerged as an adaptive response to a transformation of the plant resources that occurred without the benefit of any deliberate human intervention. At most, human patterns of refuse disposal in "domestilocalities" unintentionally introduced artificial selection pressures that generated the varieties of indigenous plants that became cultigens. On this account the plants effectively "domesticate themselves" and women are, once again, represented as passively adapting to imposed change.

Watson and Kennedy (1991) make much of the artificiality of both models. Why assume that dabbling for ritual purposes would be more likely to produce the knowledge and transformations of the resource base necessary for horticulture than the systematic exploitation of these resources as a primary means of subsistence? (Indeed, why assume that shamans were men?) Why deny human agency altogether and represent the emergence of horticulture as an "automatic process" (p. 266) when it seems that the most plausible ascription of agency (if any is to be made) must be to women (pp. 262–264)? They are "leery," they say, "of explanations that remove women from the one realm that is traditionally granted them, as soon as innovation or invention enters the picture" (p. 264). What they find wrong with these accounts—what this artificality reveals—is, in essence, a vicious circularity of interdependence between the assumptions that inform the interpretation of archaeological data as evidence (for example, of function and gender association) and the explanatory/reconstructive hypotheses on which this evidence is brought to bear, which postulate automatic transition or the mediation of shamans. The common ground of reconstructive inference at both levels is, they argue, a stock set of androcentric or, indeed, sexist assumptions to the effect that women could not have been responsible for any major culture-transforming exercise of human agency. In this case there is, then, an erasure of gender relations and of the role and presence of women (an instance of the first type of critique), predicated on underlying gender stereotypes (the second type of critique).

This would seem to be a worst-case scenario—exactly the sort of "nepotism" or "self-accounting" (in Kosso's terms) that is assumed, by the most outspoken critics of "objectivism" and empiricism in archaeology, to be typical of archaeological practice. But even here it is possible to establish grounds for questioning the assumptions that frame both the favored hypothesis and the constitution of data as the evidence from which it derives support.

Two strategies of critique are evident in Watson and Kennedy's analysis. The first exploits nonarchaeological resources, both conceptual and empirical, in an independent assessment of the framing assumptions that they find worrisome. In this connection, Watson and Kennedy draw attention to a straightforward contradiction inherent in current theorizing about the emergence of horticulture in the Eastern Woodlands. On their analysis, the models they discuss assume that women can be equated with plants and that they are essentially passive: they are identified as the tenders of plants, whether wild or under cultivation, and yet are systematically denied any role in the transformation of plants required to move from foraging to horticultural practices, whatever the cost in terms of theoretical elegance, plausibility, or explanatory power. To indicate just how high this cost may be, they draw on background (botanical) knowledge about the range and environmental requirements of the plant varieties that became domesticates to establish that they routinely appear in prehistoric contexts that were far from optimal (p. 266). In short, it is most implausible that they could have arisen "automatically" under conditions of neglect, as suggested by the coevolution model.

Where the shaman hypothesis is concerned, their analysis is informed by an appreciation that, over the past three decades, feminist anthropologists have documented enormous variability in the roles played by women, in the degrees to which they are active rather than passive, mobile rather than bound to a home base, and politically powerful rather than stereotypically dispossessed and victimized. This work seriously undermines any presupposition that women are inherently less capable of innovation, self-determination, and strategic manipulation of resources than their male counterparts and renders suspect any interpretation that depends on such an assumption, regardless of its archaeological implications. In this way, they undercut an unreflective confidence in the plausibility of the framework assumptions shared by the main lines of interpretive/explanatory theorizing about the transition, and they challenge the security of the linking principles that are necessary to establish evidential support for them.

In addition, however, even when the problem is "self-accounting"—vicious circularity—the archaeological data under interpretation can sometimes function as a second locus of evidential constraint. The archaeological record may be interpreted in sexist or androcentric terms—specific categories of material may be ascribed significance as evidence of activities whose gender associations are presumed obvious/unproblematic—but this does not necessarily ensure (indeed, as Watson and Kennedy point out, it has not ensured) that the record will obligingly provide evidence that activities identified as male, on these assumptions, mediated the transition from a foraging to a horticultural way of life, however strong the expectation that they must have. Indeed, where the coevolution model is concerned, most of the activities responsible for the creation of the "domestilocalities" in which cultigens emerged were women's activities, if the archaeological record of such sites is interpreted in the light of the traditional assumptions about gender relations that Watson and Kennedy find presupposed by this account (p. 262). If the interpretive assumptions in question constituted a more closely specified theory, the outlines of Glymour-type bootstrapping inference might emerge in cases like these, complete with internal-to-theory independence between linking and test hypotheses (Glymour 1980; Wylie 1986). In practice, then, even cases in which vicious circularity is a present danger can (sometimes) be quite closely constrained by both external and internal evidence. In this connection, Watson and Kennedy exploit considerations of security in the first sense and of independence in the first sense.

Straightforward circularity is generally not the central problem in archaeological interpretation, however. The "foreknowledge"—the theoretical presuppositions and political or evaluative commitments—that generates a set of favored reconstructive hypotheses typically lacks the conceptual resources to supply the linking principles necessary to interpret archaeological data as evidence that bears on them. Usually the basis for ascribing evidential significance to archaeological data is some form of analogical inference that draws on diverse sources, most of which are understood in terms of highly localized theory. And here the worry is not overdetermination by an all-encompassing conceptual framework but underdetermination due to a lack of generalizable knowledge about the conditions under which observed linkages between (archaeological) "statics" and (cultural, behavioral) "dynamics" may be projected onto past (or otherwise unobserved) contexts.

Although this means that the inferences involved are always tenuous, they are subject to at least two sets of evidential constraints that parallel those deployed by Watson and Kennedy: background knowledge that determines what can be claimed about the analogue based on knowledge of source contexts and the presence or absence, or structure, of material in the archaeological record to which the analogues might apply. These are at work in a second feminist critique: Brumfiel's (1991) argument for a thoroughgoing reconceptualization of the state formation processes responsible for the emergence of the Aztec state in the Valley of Mexico through the period when the Aztec state was establishing a tribute system in the region. Her analysis depends on a preliminary, and explicitly analogical, argument to the effect that spindle whorls and cooking pots and griddles were associated with women—that food and cloth production were organized in the same way—in archaeological as in ethnohistoric contexts in Mesoamerica. This is supported by a comparison of archaeological evidence from various periods with ethnohistoric records that indicate that this aspect of the material culture changed very little over time and provides strong prima facie support for assuming continuity in its gender associations.

Given this analogical "ascription of meaning" to these classes of artifacts, Brumfiel makes a series of intriguing observations about changing patterns in their density and distribution relative to the urban centers from which political control and demands for tribute emanated. She finds that the presence of spindle whorls, hence of fabric production by women, increased dramatically in outlying areas but, surprisingly, decreased in the vicinity of the urban centers as the Aztec practice of extracting tribute payments in cloth developed. At the same time, the density of griddles (associated with the production of labor-intensive and transportable cooked food based on tortillas) increased near the urban areas and decreased in the outlying areas where stew pots (indicative of less demanding pot-cooked food preparation) continued to predominate. Given these quite striking and formerly unnoticed patterns in the structure of these assemblages, she suggests that cloth may have been exacted directly as tribute in the hinterland, while populations living closer to the city centers intensified their production of transportable food so that they could participate in the markets and "extradomestic institutions" (p. 241) that were then emerging in the Valley of Mexico and required a mobile labor force. Presumably this emerging market system allowed them to meet tribute demands by other means than direct production of cloth. In either case, Brumfiel points out, the primary burden of meeting the tribute demands for cloth imposed by Aztec rule was shouldered by women and met by strategic realignments of their domestic labor. Where the Aztec state depended on tribute to maintain its political and economic hegemony, its emergence must be understood to have transformed, and to have been dependent on a transformation of, the way predominantly female domestic labor was organized and deployed—something that (evidently) does not figure at all in standard accounts of this state system.

The power of Brumfiel's challenge to extant models of the economic base of the Aztec empire—her argument for taking gender relations seriously in studies of state formation processes, for countering the absence or erasure of women and gender in existing accounts—depends fundamentally on the strength of the link she establishes between specific classes of artifact, an associated set of activities, and gendered divisions of labor and crucially on the fact that this link depends on background knowledge that is wholly independent (in content, source, and security) of both the hypotheses Brumfiel means to challenge and those she promotes. Thus, the "evidence" she cites stands, in this case and for these purposes, as a provisionally stable foundation for testing the limitations of extant models of state formation.

The strongest arguments of this sort depend on evidential constructs mediated by linking principles that are drawn from completely independent, nonethnographic sources, especially sources that specify unique causal antecedents for elements of the surviving record (that is, that establish strong security in the third sense). In a study that parallels Brumfiel's but focuses on pre-Hispanic sites in the central Andes, Hastorf (1991) draws on several lines of evidence to establish that gendered divisions of labor and participation in the public, political life of these highland communities were profoundly altered through the period when the Inka extended their control in the region. In short, the household structure and gender roles encountered in historic periods cannot be treated as a stable, traditional feature of Andean life that predates and underlies state formation as typically assumed (139). One source of support for this argument is the outcome of a comparison between the sexes of skeletal remains recovered from these sites; a stable isotope analysis of bone composition has established the proportions of types of food consumed by these individuals over their lifetimes. Although the dietary profiles of males and females are undifferentiated through the period preceding the advent of Inka control in the valley, Hastorf finds evidence of sharp divergence in the period when evidence of an Inka presence appears. Specifically, males show higher rates of consumption of foods that have the isotope values associated with maize than do females. If the background knowledge deployed in stable isotope analysis is considered secure in the first sense described and if the same can be said of the techniques for dating the skeletons (these assumptions are always open to critical reassessment), then this analysis establishes, in chemical terms, that substantial sex-linked changes in dietary intake had to have occurred at exactly the time when the Inka state expanded into the region. This constitutes as close to a biconditional linkage between antecedent conditions and surviving material traces as archaeologists are likely to get.[10]

It is important to note, however, that Hastorf cannot do much with this one line of evidence taken on its own. She relies on a number of collateral lines of evidence to establish what, exactly, the change in dietary intake involved and to argue its significance as evidence that standard models of Andean state formation should be revised to take changing gender relations into account. Paleobotanical analyses of plant remains found in household compounds occupied before and after the advent of Inka control establish what range of available plant resources were being used by the local population. Chemical and ethnohistorical evidence establishes that the maize component of their diet was being consumed, at least in part, in the form of maize beer (*chicha*). A comparative analysis of the density and distribution of maize remains on these sites indicates that while there was clear intensification of maize production over time, the activities of grinding and otherwise processing maize were increasingly restricted to specific locations within the domestic space of sites. And ethnohistoric records suggest a strong association of women with maize processing and testify to the gendered nature of Inka political structures; specifically, men were drawn out of local communities to act as heads of households in ritualized negotiations based on the consumption of maize beer (*chicha*) and to serve out a labor tax away from their villages compensated with maize and *chicha*. Drawing these quite distinct lines of evidence together, Hastorf concludes that, through this transitional period, the newly imposed political structures of the Inka empire forced a realignment of gender roles. Women "became the focus of [internal, social, and economic] tensions as they produced more beer while at the same time they were more restricted in their participation in the society" (p. 152).

While each line of evidence, taken on its own, enjoys some degree of credibility —although the security of the linkages varies considerably—the real strength of Hastorf's argument lies in the independence of these lines of inference from one another. The bone marrow—chemical evidence is especially telling not just because it relies on particularly secure and autonomous linking principles but because it could not have been expected to converge on, to complement, the other (mutually independent) lines of evidence based on ethnohistoric and bioecological assumptions. To put the point in general terms, the convergence of multiple, mutually independent lines of evidence on a given hypothesis is compelling because it is so implausible that it should be an artifact, the result of compensatory error in all the inferences establishing its evidential support (for philosophical discussion of these considerations, see Kosso 1988, p. 456; Hacking 1983, pp. 183–185).

But perhaps more significant than cases where convergence strengthens credibility are those where dissonance emerges among independently constituted lines of interpretation, making it clear that error lies somewhere in the system of background assumptions and linking inferences, however secure these may seem to be, or else in the range of favored interpretative or explanatory hypotheses that have been invoked —or constructed—to account for this evidence, however plausible and exhaustive they may seem to be. Often the real difficulty is that of finding one account, one reconstructive or explanatory hypothesis, that is consistent with all lines of evidence bearing on a particular set of past events or contexts—that tells a coherent story about how such an array of material remains could have been produced—rather than that of adjudicating between a number of different (plausible, coherent) explanatory options.

It is exactly this negative constraint that motivates many of the critiques described at the outset and many of the feminist analyses now emerging in archaeology. To take one final example, even with something as ephemeral as the interpretation of paleolithic art, feminist critics like Conkey (with the collaboration of Williams 1991, p. 121) argue that there are strong grounds for rejecting the range of manifestly sexist and presentist interpretations that are routinely projected onto the past, given how sharply dissonant these are with evidence bearing on virtually any other aspect of paleolithic life. All indications are that the paleolithic cultures in question must have been so vastly different from any with which we are familiar—in their technological resources, subsistence patterns, social organization—that the images comprising their artistic record cannot be assumed to have any transculturally stable meaning continous with that which it might have in any contemporary contexts. Whatever their import, it is most unlikely that these images and objects were instances of either commodified pornography or high art, as produced in contemporary contexts. Thus, the strategy of working with multiple lines of evidence can sometimes provide strong empirical grounds for resisting the imposition of favored interpretations and, in this, undermine formerly plausible claims about the past, even if it cannot underwrite a particular alternative conclusion or family of conclusions. It can make it clear what we cannot claim in connection with a particular past, and sometimes it can force a reconsideration of quite fundamental orienting assumptions about the nature of the subject domain and the limits for our understanding of it. It is, paradoxically, the fragmentary nature of the archaeological record that is its strength in setting up evidential constraints of these sorts, even establishing the limits of inquiry.

Conclusion

If archaeologists are to get at past forms of life, events, and trajectories of cultural development, much less the animating processes that might lie behind them, there is no avoiding the difficult task of constructing data as evidence; that is, they must laden it with "theory" that establishes a presumptive connection between the surviving material record and the past events and conditions that produced it. This is an undeniably tenuous undertaking and one that is open at every level to the play of interests, favored presuppositions, values, and prejudices that constitute the angle of vision of those engaged in the enterprise and those who support or otherwise control it. The politics of contemporary archaeology thrown into relief by the quincentennial debates and related critiques are symptoms of a much deeper entanglement of power and knowledge than has been thinkable for most archaeological practitioners or, indeed, for many philosophers of science. In this sense, entrenched assumptions about the externality of power to knowledge must be reconsidered; in particular, it must be recognized that the process of recovering and constituting data as evidence is a thoroughly social and political enterprise.

Nonetheless, a seamless holism—whether conceptual or sociological—seems untenable when considering, in any detail, the vagaries of practice in a field like archaeology. It is precisely in the complexity of the inferences and the diversity of the resources required to construct archaeological data as evidence that the potential lies for being rigorously empirical, perhaps even realising a degree of objectivity (in a suitably amended sense). Given the interplay of evidential resources described here, I argue that it is sometimes plausible to say we have "discovered" a fact about the world, or have shown a formerly plausible claim to be "just false," although such claims are clearly never unrevisable. And it is in this that the possibility resides for challenging taken-forgranteds, exclusions, and prejudgments about how the past must have been, given cherished views about the present and the future. Ironically, I take most of the critiques that demonstrate the political nature of archaeology to provide an argument for, rather than against, recognizing evidential constraints as a central feature of knowledge-producing enterprises that must figure in any reconsideration of the relations between power and knowledge, albeit not in the way idealized in many scientific disciplines or standard philosophical models of science. Indeed, they are one element—perhaps the central element—of the now widely repudiated enlightenment tradition of empiricism that I believe we should not give up.

Notes

I gratefully acknowledge the support of the Social Sciences and Humanities Council of Canada that provided for the research resulting in this chapter. I also thank Lee McIntyre for providing me an opportunity to present the orginal paper in the symposium on "The Revival of Empiricist Philosophy of Social Science" sponsored by the Boston Colloquium for the Philosophy of Science.

1. Rouse continues in his characterization of these ideals of value and standpoint neutrality: "Power can influence what de facto is known, but its being known, and what it is for it to be known, cannot be subject to the influence of power. That is, power can influence what we believe, but considerations of power are entirely irrelevant to which of our beliefs are true, which of these are known to be true, and what justifies their status as knowledge.... [Given this] it is generally believed that knowledge is best achieved within an inquiry freed from political pressure" (Rose 1987, p. 13).
2. Moreover, on these "received views" of science and knowledge as (ideally) value neutral, power is conceived in relatively narrow terms, consistent with the traditional accounts outlined by Rouse: as possessed and exercised by specific agents (individuals or corporate/collective entities); as operating on our representations but not on what is represented; as repressive, sometimes enabling, and never productive. A "conceptual separation" is maintained "between science viewed as a field of knowledge and science viewed as a field of power" (Rouse 1987, p. 17): internal (cognitive) and external (social-political) factors are sharply distinguished, and their effects on knowledge/inquiry are compartmentalized as sometimes (regrettably) influencing but still essentially external to the enterprise itself and its products.
3. By contrast with the conception of power presupposed by objectivist accounts (as that which is external to science and knowledge production), the power relations and dynamics revealed in the fifth sort of critique described above are very much like the "mundane," diffuse, Foucauldian complex of "capillary power effects" in which Rouse locates the political nature and import of science. Socio-political factors—the dynamics of power—are not restricted to the operation on scientific archaeology of intrusive (external) political interests, through the exercise of "repressive juridical power" ("power over" held by specific agents with definable interests; pp. 24, 210), but rather pervade and constitute the enterprise as a whole.
4. This strong a social constructivism is endorsed by those who advocate critiques at the fourth level described, and is also articulated in connection with critiques of the fifth sort; it is evident in Gero's discussion, for example (1991G). It is not widely embraced, however, by those responsible for critiques in the other categories described.
5. In this connection, Rouse (1987) suggests that, as a pragmatist, he finds it redundant to assert that the objects of inquiry are "real" or that claims about them are "true." The worry that what theories describe may be "no more than an artifact of our research activities" is unproblematic because, on his account, we deal with and know nothing but artifacts; it is in them that reality resides. He goes on to say that "these conditions [laboratory arrangements] themselves are a human artifact. Indeed, the construction of those conditions, like every other practical engagement with the world, is responsive to the capacities of the things we deal with. But these capacities emerge as intelligible only in relation to other things and the practices through which they are related." (p. 157).

   Even given this deflationary treatment of realist theses and claims about "truth," the problem remains of explaining how, or on what grounds, judgments are made that some claims are more reliable or accurate than others, or that we were mistaken in what we believed about a given subject domain (whether it is understood to consist of artifacts or to be, in some sense, an independent reality).

6. Rouse (1987) acknowledges the "ambivalence lurking within almost all liberationist philosophies of science" where these both critique and deploy the tools of science, recognizing them to be powerful, "a ... force to be reckoned with ... [satisfying] some important needs and desires, both intellectual and material" (p. 256). The philosophical problem that this raises is "to do justice to the discovery of forms of bias, exclusion, and oppression it [feminist and other liberationist critiques of science] identifies within many scientific practices and claims without having to reject wholesale the scientific enterprise and its achievements" (p. 257). The examples of feminist practice that I consider below are particularly intriguing in this connection; they make it clear that the power relations that constitute both androcentrism and the critique of androcentrism in archaeology do not go through unchecked, without resistance. In all cases, the process of assembling a data base, constructing evidence, and manipulating technical, material, and conceptual resources to effect a conjunction or convergence—a fit—between these resources is indeed a matter of work in and on an emerging world. But it is work, labor, that is continuously subverted and canalized by the materials on, or with, which it operates, materials that are not entirely or plausibly of our own making and are not wholly products of our constructive enterprise.
7. Longino and Doell (1983) treat these latter two considerations of security together when they characterize the concept of "distance" between data, evidence, and hypotheses as a "logical notion of being more or less directly consequential"; they also extend this metaphor to some quite distinct considerations of independence (pp. 209–210).
8. Longino and Doell's (1983) concept of distance also covers considerations of independence in this sense: "the less a description of fact is a direct consequence of the hypothesis for which it is taken to be evidence, the more *distant* that hypothesis is from its evidence" (p. 210, emphasis added).
9. This is the first of several examples of critiques of androcentrism and sexism in archaeology I will consider here; all but the last are drawn from the first collection of feminist critiques and research on gender to have appeared in archaeology (Gero and Conkey 1991). A more detailed discussion of these cases, and an account of how they have arisen in the last few years, is provided in Wylie (1992).
10. The independence and the security of linking arguments based on background knowledge of this physical, chemical, bioecological sort is routinely exploited by archaeologists: in morphological analyses of skeletal remains that provide evidence of pathologies and physical stress; in radiocarbon, archeomagnetic, and related methods of dating; and in reconstructions of prehistoric technology and paleoecology, to name a few such examples.

References

Binford, Lewis, R. 1983. *Working at Archaeology*. New York: Academic Press.

Binford, Lewis R., and Jeremy A. Sabloff. 1982. "Paradigms, Systematics, and Archaeology." *Journal of Anthropological Research* 38: 137–153.

Brumfiel, Elizabeth M. 1991. "Weaving and Cooking Women's Production in Aztec Mexico." In *Engendering Archaeology Women and Prehistory*, pp. 224–253. Edited by Joan M. Gero and Margaret W. Conkey. Cambridge: Basil Blackwell Press.

Conkey, Margaret W., and Janet D. Spector. 1984. "Archaeology and the Study of Gender" In *Advances in Archaeological Method and Theory*, 7: 1–38. Edited by Michael B. Schiffer. New York: Academic Press.

Conkey, Margaret W., with Sarah H. Williams. 1991. "Original Narratives: The Political Economy of Gender in Archaeology." In *Gender, Culture, and Political Economy Feminist Anthropology in the Post-Modern Era*, pp. 102–139. Edited by Micaela di Leonardo. Berkeley: University of California Press.

Gero, Joan M., 1991a. "Genderlithics: Women's Roles in Stone Tool Production." In *Engendering Archaeology: Women and Prehistory*, pp. 163–193. Edited by Joan M. Gero and Margaret W. Conkey. Cambridge: Basil Blackwell Press.

Gero, Joan M. 1991b. "The Social World of Prehistoric Facts: Gender and Power in Prehistoric Research." Working Papers Series #4. London: Centre for Women's Studies and Feminist Research, University of Western Ontario.

Gero, Joan M. 1985. "Socio-Politics and the Woman-at-Home Ideology," *American Antiquity* 50 (2): 342–350.

Gero, Joan M., and Margaret W. Conkey, eds. 1991. *Engendering Archaeology: Women and Prehistory*. Cambridge: Basil Blackwell Press.

Gero, Joan M., David M. Lacy, and Michael L. Blakey, eds. 1983. *The Socio-politics of Archaeology.* Research Report no. 23. Amherst: Department of Anthropology, University of Massachusetts.
Glymour, Clark. 1980. *Theory and Evidence.* Princeton: Princeton University Press.
Hacking, Ian. 1983. *Representing and Intervening: Introductory Topics in the Philosophy of Natural Science.* Cambridge: Cambridge University Press.
Hall, Martin. 1984. "The Burden of Tribalism: The Social Context of Southern African Iron Age Studies." *American Antiquity* 49(3): 455–467.
Handsman, Russell. 1990. "The Weantinock Indian Homeland Was Not a 'Desert.'" *Artifacts: The American Indian Archaeological Institute* 18(2): 3–7.
Handsman, Russell. 1989. "Native Americans and an Archaeology of Living Traditions." *Artifacts: The American Indian Archaeological Institute* 17(2): 3–5.
Handsman, Russell, and Trudie Lamb Richmond. 1992. "Confronting Colonialism: The Mahican and Schaghticoke Peoples and Us." Contribution to "Making Alternative Histories," School of American Research Seminar. Santa Fe, April.
Hastorf, Christine A. 1991. "Gender, Space, and Food in Prehistory." In *Engendering Archaeology: Women and Prehistory,* pp. 132–159. Edited by Joan M. Gero and Margaret W. Conkey. Cambridge: Basil Blackwell Press.
Hodder, Ian. 1984. "Archaeology in 1984." *Antiquity* 58: 25–32.
Hodder, Ian. 1983. "Archaeology, Ideology and Contemporary Society." *Royal Anthropological Institute News* 56: 6–7.
Irele, Abiola. 1991. "The African Scholar: Is Black Africa Entering the Dark Ages of Scholarship?" *Transition* 51: 56–69.
Kosso, Peter. 1989. "Science and Objectivity." *Journal of Philosophy* 86: 245–257.
Kosso, Peter. 1988. "Dimensions of Observability." *British Journal of Philosophy of Science* 39: 449–467.
Kramer, Carol, and Miriam Stark. 1988. "The Status of Women in Archaeology." *Anthropology Newsletter* 29(9): 1, 11–12.
Levine, Mary Ann. 1991. "An Historical Overview of Research on Women in Anthropology." In *The Archaeology of Gender: Proceedings of the 22nd Annual Chacmool Conference,* pp. 177–186. Edited by Dale Walde and Noreen D. Willows. Calgary: Archaeological Association of the University of Calgary.
Longino, Helen E. 1990. *Science as Social Knowledge: Values and Objectivity in Scientific Inquiry.* Princeton: Princeton University Press.
Longino, Helen, and Ruth Doell. 1983. "Body, Bias, and Behavior: A Comparative Analysis of Reasoning in Two Areas of Biological Science." *Signs* 9: 206–227.
Patterson, Thomas C. 1992. "*Indigenismo,* Indigenous Movements, and the State: Archaeology and History in Mexico and Peru." Contribution to "Making Alternative Histories," School of American Research Seminar. Santa Fe, April.
Patterson, Thomas C. 1986a. "The Last Sixty Years: Toward a Social History of Americanist Archaeology in the United States." *American Anthropologist* 88: 7–22.
Patterson, Thomas C. 1986b. "Some Postwar Theoretical Trends in U.S. Archaeology." *Culture* 11(1): 43–54.
Potter, Parker B. 1991. "What Is the Use of Plantation Archaeology?" *Historical Archaeology* 25(3): 94–107.
Rouse, Joseph. 1987. *Knowledge and Power: Toward a Political Philosophy of Science.* Ithaca N.Y.: Cornell University Press.
Schiffer, Michael B. 1983. "Toward the Identification of Formation Processes." *American Antiquity* 48: 675–706.
Schmidt, P. R. 1992. "Remaking History in Africa." Contribution to "Making Alternative Histories," School of American Research Seminar. Santa Fe, April.
Shapere, Dudley. 1985. "Observation and the Scientific Enterprise." In *Observation, Experiment, and Hypothesis in Modern Physical Science,* pp. 22–45. Edited by P. Achinstein and O. Hannaway. Cambridge, Mass.: MIT Press.
Shapere, Dudley. 1982. "The Concept of Observation in Science and Philosophy." *Philosophy of Science* 49: 485–525.
Shanks, Michael, and Christopher Tilley. 1987. *Re-constructing Archaeology.* Cambridge: Cambridge University Press.
Singleton, Theresa A., ed. 1985. *The Archaeology of Slavery and Plantation Life.* New York: Academic Press.

Spector, Janet D., and Mary K. Whelan. 1989. "Incorporating Gender into Archaeology Courses." In *Gender and Anthropology: Critical Reviews for Research and Teaching*, pp. 65–94. Edited by Sandra Morgen. Washington, D.C.: American Anthropological Association.

Tilley, Christopher. 1989. "Archaeology as Socio-political Action in the Present." In *Critical Traditions in Archaeology: Essays in the Philosophy, History and Socio-politics of Archaeology*, pp. 117–135. Cambridge: Cambridge University Press.

Thomas, David Hurst. 1991. "Cubist Perspective on the Spanish Borderlands: Past, Present, and Future." In *Columbian Consequences: The Spanish Borderlands in Pan-American Perspective*, 3: xiii–xix. Edited by David Hurst Thomas. Washington, D.C.: Smithsonian Institution Press.

Trigger, Bruce G. 1991. "Early Native North American Responses to European Contact: Romantic versus Rationalistic Interpretations." *Journal of American History* 4: 1195–1215.

Trigger, Bruce G. 1989. *A History of Archaeological Thought*. Cambridge: Cambridge University Press.

Trigger, Bruce G. 1980. "Archaeology and the Image of the American Indian." *American Antiquity* 45: 662–676.

Vargas Arenas, Iraida. 1992. "The Perception of History and Archaeology in Latin America: A Theoretical Approach." Contribution to "Making Alternative Histories," School of American Research Seminar. Santa Fe, April.

Vargas Arenas, Iraida, and Mario Sanoja. 1990. "Education and the Political Manipulation of History in Venezuela." In *The Excluded Past*, pp. 50–60. Edited by Peter Stone and R. McKenzie. London: Unwin Hyman.

Walde, Dale, and Noreen D. Willows, (eds.) 1991. *The Archaeology of Gender: Proceedings of the 22nd Annual Chacmool Conference*. Calgary: Archaeological Association of the University of Calgary.

Watson, Patty Jo, and Mary C. Kennedy. 1991. "The Development of Horticulture in the Eastern Woodlands of North America: Women's Role." In *Engendering Archaeology: Women and Prehistory*, pp. 255–275. Edited by Joan M. Gero and Margaret W. Conkey. Cambridge: Basil Blackwell Press.

Wobst, Martin H., and Arthur S. Keene. 1983. "Archaeological Explanation as Political Economy." In *The Socio-politics of Archaeology*, pp. 79–90. Edited by Joan M. Gero, David M. Lacy, and Michael L. Blakey. Research Report no. 23. Amherst: Department of Anthropology, University of Massachusetts.

Wylie, Alison. 1992. "On 'Heavily Decomposing Red Herrings' Scientific Method in Archaeology and the Ladening of Evidence with Theory." In *Metaarchaeology*, pp. 269–288. Edited by Lester Embree. Boston Studies in the Philosophy of Science. Boston: Kluwer.

———. 1992. "The Interplay of Evidential Constraints and Political Interests: Recent Archaeological Research on Gender." *American Antiquity* 57: 15–35.

———. 1986. "Bootstrapping in Un-Natural Sciences: An Archaeological Case." *PSA 1986*, 1: 314–322. Edited by A. Fine and P. Machamer. East Lansing, Mich.: Philosophy of Science Association.

Bibliography

Although we have tried to make this bibliography as complete as space limitations would allow, the philosophy of social science is so diverse that any attempt to compile a fully satisfactory, comprehensive bibliography would be virtually impossible. The reader should also note that we have omitted here citations for works reprinted in this book.

Achinstein, Peter. *Law and Explanation: An Essay in the Philosophy of Science*. Oxford: Clarendon Press, 1971.

Adorno, Theodor, et al. *The Positivist Dispute in German Sociology*. New York: Harper Row, 1969.

Apel, Karl Otto. "Types of Social Science in the Light of Human Interests of Knowledge." *Social Research* #44 (1977), 425–470.

———. *Understanding and Explanation: A Transcendental-Pragmatic Perspective*. Translated by Georgia Warnke. Cambridge: MIT Press, 1984.

Armstrong, D. M. *What Is a Law of Nature?* Cambridge: Cambridge University Press, 1983.

Beck, Lewis White. "the 'Natural Science Ideal' in the Social Sciences." *Scientific Monthly*, 68 (June 1949), 386–394.

Bell, Daniel. *The Social Sciences since the Second World War*. New Brunswick, N.J.: Transaction Books, 1982.

Berelson, Bernard, and Steiner, Gary. *Human Behavior: An Inventory of Scientific Findings*. New York: Harcourt, Brace, and World, 1964.

Bergner, Jeffrey. *The Origins of Formalism in Social Science*. Chicago: University of Chicago Press, 1981.

Bernstein, Richard. *The Restructuring of Social and Political Theory*. New York: Harcourt Brace Jovanovich, 1976.

———. *Praxis and Action: Contemporary Philosophies of Human Action*. Philadelphia: University of Pennsylvania Press, 1971.

———. *Beyond Objectivism and Relativism*. Philadelphia: University of Pennsylvania Press, 1983.

Blalock, Hubert M. *Causal Inferences in Nonexperimental Research*. Chapel Hill: University of North Carolina Press, 1964.

———. *Basic Dilemmas in the Social Sciences*. Beverly Hills: Sage Publications, 1984.

Bleicher, Josef. *Contemporary Hermeneutics*. London: Routledge and Kegan Paul, 1980.

Block, Ned (ed.). *Readings in the Philosophy of Psychology*. Cambridge: Harvard University Press, 1980.

Bohman, James. *New Philosophy of Social Science: Problems of Indeterminacy*. Cambridge: MIT Press, 1991.

Borger, Robert, and Cioffi, Frank (eds.). *Explanations in the Behavioral Sciences*. Cambridge: Cambridge University Press, 1970.

Boulding, Kenneth. *Beyond Economics: Essays on Society, Religion, and Ethics*. Ann Arbor: University of Michigan Press, 1968.

Braithwaite, Richard. *Scientific Explanation: A Study of the Function of Theory, Probability and Law in Science*. Cambridge: Cambridge University Press, 1968.

Braybrooke, David. *Philosophy of Social Science*. Englewood Cliffs, N.J.: Prentice-Hall, 1987.

———. *Philosophical Problems of the Social Sciences*. London: Macmillan, 1965.

Brodbeck, May. "On the Philosophy of the Social Sciences." In E. C. Harwood (ed.), *Reconstruction of Economics*. Great Barrington, Mass.: American Institute for Economic Research, 1955.

———. "Methodological Individualisms: Definition and Reduction." In May Brodbeck (ed.), *Readings in the Philosophy of the Social Sciences*, pp. 280–303. New York: Macmillan, 1968.

———. (ed.). *Readings in the Philosophy of the Social Sciences*. New York: Macmillan, 1968.

Bronowski, J. *The Identity of Man*. London: Heinemann, 1965.

Brown, Robert. *Explanation in Social Science*. London: Routledge and Kegan Paul, 1963.

———. *The Nature of Social Laws*. Cambridge: Cambridge University Press, 1984.

———. "Explanation by Laws in Social Science." *Philosophy of Science* 21 (1954), 25–32.

Brown, S. C. (ed.). *Philosophical Disputes in the Social Sciences.* Sussex: Harvester Press, 1979.

Bryson, Lyman. *Science and Freedom.* New York: Columbia University Press, 1947.

Buck, Roger C. "Reflexive Predictions." In May Brodbeck (ed.), *Readings in the Philosophy of the Social Sciences,* pp. 436–447. New York: Macmillan, 1968.

Cartwright, Nancy. *How the Laws of Physics Lie.* Oxford: Clarendon Press, 1983.

Chase, Stuart. *The Proper Study of Mankind: An Inquiry into the Science of Human Relations.* New York: Harper and Bros., 1948.

Cohen, Morris R. *Reason and Nature: An Essay on the Meaning of Scientific Method.* Glencoe, Ill.: Free Press, 1931.

Collingwood, R. G. *The Idea of History.* Oxford: Oxford University Press, 1946.

Cunningham, Frank. *Objectivity and Social Science.* Toronto: University of Toronto Press, 1973.

Dallmayr, Fred, and McCarthy, Thomas (eds.). *Understanding and Social Inquiry.* Notre Dame: University of Notre Dame Press, 1977.

Davidson, Donald. "Mental Events." In *Essays on Actions and Events.* Oxford: Oxford University Press, 1980.

———. *Essays on Actions and Events.* Oxford: Clarendon Press, 1980.

Dennett, Daniel. *Brainstorms.* Cambridge: MIT Press, 1978.

Donagan, Alan. "Are the Social Sciences Really Historical?" In Bernard Baumrin (ed.), *Philosophy of Science: The Delaware Seminar, vol. 1: 1961–1962.* New York: Interscience Publishers, 1963.

Doyal, Len, and Harris, Roger. *Empiricism, Explanation, and Rationality: An Introduction to the Philosophy of the Social Sciences.* London: Routledge and Kegan Paul, 1986.

Dray, William. *Laws and Explanation in History.* Oxford: Clarendon Press, 1957.

Dretske, Fred. *Explaining Behavior.* Cambridge: MIT Press, 1988.

Dreyfus, Hubert. "Why Current Studies of Human Capacities Can Never Be Scientific." *Berkeley Cognitive Science Report* #11 (January 1984).

Durkheim, Emile. *The Rules of Sociological Method.* Glencoe, Ill.: Free Press, 1965.

Dyke, Charles. *The Evolutionary Dynamics of Complex Systems: A Study in Biosocial Complexity.* Oxford: Oxford University Press, 1988.

Earman, John. *A Primer on Determinism.* Dordrecht: D. Reidel Publishing Co., 1986.

Elster, Jon. *Ulysses and the Sirens.* Cambridge: Cambridge University Press, 1979.

Emmet, Dorothy, and MacIntyre, Alasdair (eds.). *Sociological Theory and Philosophical Analysis.* New York: Macmillan, 1970.

Fay, Brian. *Critical Social Science: Liberation and Its Limits.* Ithaca: Cornell University Press, 1987.

———. *Social Theory and Political Practice.* London: George Allen and Unwin, 1975.

———. "Naturalism as a Philosophy of Social Science." *Philosophy of Social Science* 14 (1984), 529–542.

Feyerabend, Paul. *Problems of Empiricism: Philosophical Papers.* Vol. 2. Cambridge: Cambridge University Press, 1981.

Flew, Anthony. *Thinking about Social Thinking: The Philosophy of the Social Sciences.* Oxford: Basil Blackwell, 1985.

French, P., Vahling, T., and Wettstein, H. (eds.). *Midwest Studies in Philosophy, vol. 15: The Philosophy of the Human Sciences.* Notre Dame: University of Notre Dame Press, 1990.

Friedman, Milton. *Essays in Positive Economics.* Chicago: University of Chicago Press, 1953.

Gaffron, Hans. *Resistance to Knowledge.* San Diego: Salk Institute for Biological Studies, 1970.

Garfinkel, Alan. *Forms of Explanation.* New Haven: Yale University Press, 1981.

Geertz, Clifford. *The Interpretation of Cultures.* New York: Basic Books, 1973.

Gergen, Kenneth J. "Social Psychology as History." *Journal of Personality and Social Psychology,* 26, no. 2 (1973), 309–320.

Gewirth, Alan. "Can Men Change Laws of Social Science?" *Philosophy of Science* 21 (1954), 229–241.

Gibbard, Allan, and Varian, Hal R. "Economic Models." *Journal of Philosophy* 75 (1978), 664–677.

Gibson, Quentin. *The Logic of Social Enquiry.* London: Routledge and Kegan Paul, 1960.

Goldman, Alvin. "Actions, Predictions, and Books of Life." *American Philosophical Quarterly* 5, no. 3 (July 1968), 135–151.

Gordon, Scott. *Social Science and Modern Man.* Toronto: University of Toronto Press, 1970.

Grunberg, Emile. "'Complexity' and 'Open Systems' in Economic Discourse." *Journal of Economic Issues* 12, no. 3, (September 1978), 541–560.

Handy, Rollo. *Methodology of the Behavioral Sciences: Problems and Controversies.* Springfield, Ill.: Charles C. Thomas, 1964.

Handy, Rollo, and Kurtz, Paul. *A Current Appraisal of the Behavioral Sciences*. Great Barrington, Mass.: Behavioral Research Council, 1964.
Hanson, Norwood R. *Observation and Explanation: A Guide to the Philosophy of Science*. London: George Allen and Unwin, 1972.
Harding, Sandra (ed.). *Feminism and Methodology*. Bloomington: Indiana University Press, 1987.
Harrod, Roy. *Sociology, Morals, and Mystery*. London: Macmillan, 1971.
Harwood, E. C. (ed.) *Reconstruction in Economics*. Great Barrington, Mass.: American Institute for Economic Research, 1955.
Hausman, Daniel (ed.). *The Philosophy of Economics*. Cambridge: Cambridge University Press, 1984.
Hayek, F. A. *The Counter-Revolution of Science: Studies on the Abuse of Reason*. Indianapolis: Liberty Press, 1979.
———. "The Pretence of Knowledge." *Swedish Journal of Economics* 77 (December 1975), 433–442.
———. "Degrees of Explanation." *In Studies in Philosophy, Politics and Economics*. Chicago: University of Chicago Press, 1967.
———. "The Use of Knowledge in Society." *American Economic Review* 35, no. 4 (September 1945), 519–530.
———. Studies in Philosophy, Politics and Economics. Chicago: University of Chicago Press, 1967.
———. "The Results of Human Action But Not of Human Design." *Studies in Philosophy, Politics and Economics*. Chicago: University of Chicago Press, 1967.
Hempel, Carl G. "Logical Positivism and the Social Sciences." In P. Achinstein and S. Barker (eds.), *The Legacy of Logical Positivism*, pp. 163–194. Baltimore: Johns Hopkins Press, 1969.
———. *Philosophy of Natural Science*. Englewood Cliffs, N.J.: Prentice-Hall, 1966.
———. "Studies in the Logic of Explanation." *In Aspects of Scientific Explanation and Other Essays in the Philosophy of Science*. New York: Free Press, 1965.
———. *Aspects of Scientific Explanation and Other Essays in the Philosophy of Science*. New York: Free Press, 1965.
———. "Explanation in Science and in History." In R. G. Colodny (ed.), *Frontiers of Science and Philosophy*, pp. 9–33. Pittsburgh: University of Pittsburgh Press, 1962.
———. "Explanation and Prediction by Covering Laws." In Bernard Baumrin (ed.), *Philosophy of Science: The Delaware Seminar, vol. 1: 1961–1962*. New York: Interscience Publishers, 1963.
Hendrick, Clyde. "Social Psychology as Historical and as Traditional Science: An Appraisal." *PSPB* 2 (1976), 392–403.
Hindess, Barry. *Philosophy and Methodology in the Social Sciences*. Sussex: Harvester Press, 1977.
Hollis, Martin, and Lukes, Steven (eds.). *Rationality and Relativism*. Cambridge: MIT Press, 1982.
Homans, George C. *Social Behavior: Its Elementary Forms*. New York: Harcourt, Brace, and World, 1981.
———. *The Nature of Social Science*. New York: Harcourt, Brace, and World, 1967.
Hookway, C., and Pettit, P. (eds.). *Action and Interpretation: Studies in the Philosophy of the Social Sciences*. Cambridge: Cambridge University Press, 1978.
Hoy, David (ed.). *The Critical Circle*. Los Angeles: University of California Press, 1982.
Hughes, John. *The Philosophy of Social Research*. London: Longman Group, 1980.
Kaplan, Abraham. *The Conduct of Inquiry: Methodology for Behavioral Science*. San Francisco: Chandler Publishing, 1964.
Kaufmann, Felix. *Methodology of the Social Sciences*. New York: Humanities Press, 1958.
Keat, Russell. "Positivism, Naturalism, and Anti-Naturalism in the Social Sciences." *Journal for the Theory of Social Behaviour* 1 (1971) 3–17.
Kim, Jaegwon. "Concepts of Supervenience." *Philosophy and Phenomenological Research* 45, no. 2, (December 1989), 153–176.
Kincaid, Harold. "Confirmation, Complexity, and Social Laws." *PSA 1988* 2 (1988), 299–307.
Kitcher, Phillip, and Salmon, Wesley (eds.). *Scientific Explanation*. Minnesota Studies in the Philosophy of Science, vol. 13. Minneapolis: University of Minnesota Press, 1989.
Krimerman, Leonard I. (ed.). *The Nature and Scope of Social Science: A Critical Anthology*. New York: Appleton-Century-Crofts, 1969.
Kuhn, Thomas. *The Structure of Scientific Revolutions*. Chicago: University of Chicago Press, 1962.
Lennor, Kathleen. *Explaining Human Action*. Lasalle, Ill.: Open Court, 1990.
Lessnoff, Michael. *The Structure of Social Science: A Philosophical Introduction*. New York: International Publications Service, 1975.
Little, Daniel. *Varieties of Social Explanation*. Boulder, Colo.: Westview Press, 1990.
Louch, A. R. *Explanation and Human Action*. Berkeley: University of California Press, 1966.

Lundberg, George A. *Can Science Save Us?* New York: Longmans, Green, and Co., 1961.
———. "Alleged Obstacles to Social Science." *Scientific Monthly* (May 1950), 229–305.
Lynd, Robert S. *Knowledge for What? The Place of Social Science in American Culture.* Princeton: Princeton University Press, 1939.
Machlup, Fritz. "Friedrich Hayek on Scientific and Scientistic Attitudes." In *Methodology of Economics and Other Social Sciences.* New York: Academic Press, 1978.
———. "The Inferiority Complex of the Social Sciences." In *Methodology of Economics and Other Social Sciences*, pp. 334–344. New York: Academic Press, 1978.
———. "If Matter Could Talk." In *Methodology of Economics and Other Social Sciences*, pp. 309–332. New York: Academic Press, 1978.
———. *Methodology of Economics and Other Social Sciences.* New York: Academic Press, 1978.
MacIntyre, Alasdair. *Against the Self-Images of the Age: Essays on Ideology and Philosophy.* New York: Schocken Books, 1971.
———. *After Virtue: A Study in Moral Theory.* Notre Dame: University of Notre Dame Press, 1981.
MacIver, A. M. "Levels of Explanation in History." In May Brodbeck (ed.), *Readings in the Philosophy of the Social Sciences*, pp. 304–316. New York: Macmillan, 1968.
Manicas, Peter. *A History and Philosophy of the Social Sciences.* New York: Basil Blackwell, 1987.
Martin, Jane. *Explaining, Understanding and Teaching.* New York: McGraw-Hill. 1970.
Martin, Michael. "Explanation in Social Science: Some Recent Work." *Philosophy of the Social Sciences* 2 (1972), 66–81.
———. *Social Science and Philosophical Analysis: Essays in Philosophy of the Social Sciences.* Washington, D.C.: University Press of America, 1978.
Meehan, Eugene. *Explanation in Social Science: A System Paradigm.* Homewood, Ill.: Dorsey Press, 1968.
Michalos, Alex. "Philosophy of Social Science." In P. D. Asquith and H. E. Kyburg, Jr. (eds.), *Current Research in Philosophy of Science*, pp. 463–502. East Lansing: PSA, 1979.
Mill, J. S. *A System of Logic.* London: Longmans, 1961.
Miller, Richard W. "Fact and Method in the Social Sciences." In Daniel R. Sabia, Jr., and Jerald Wallulis (eds.), *Changing Social Science*, pp. 73–101. Albany: SUNY Press, 1983.
———. *Fact and Method.* Princeton: Princeton University Press, 1978.
Moon, J. Donald. "The Logic of Political Inquiry: A Synthesis of Opposed Perspectives." In F. Greenstein (ed.), *Handbook of Political Science.* Vol. 1. Reading, Mass.: Addison-Wesley, 1974.
Mukerjee, Radhakamal. *The Philosophy of Social Science.* London: Macmillan, 1960.
Nagel, Ernest. *The Structure of Science: Problems in the Logic of Scientific Explanation.* New York: Harcourt, Brace, and World, 1961.
Natanson, Maurice (ed.). *Philosophy of the Social Sciences: A Reader.* New York: Random House, 1963.
Neilsen, Joyce McCarl (ed.). *Feminist Research Methods.* Boulder, Colo.: Westview Press, 1990.
Neurathe, Otto. *Foundations of the Social Sciences.* Chicago: University of Chicago Press, 1944.
Nicholson, Michael. *The Scientific Analysis of Social Behaviour: A Defence of Empiricism in Social Science.* London: Frances Pinter Publishers, 1983.
Nisbett, R., and Wilson, T. "Telling More Than We Can Know: Verbal Reports on Mental Processes." *Psychological Review* 84 (1977), 231–259.
Outhwaite, William. *Understanding Social Life.* London: George Allen and Unwin, 1975.
Papineau, David. *For Science in the Social Sciences.* London: Macmillan, 1978.
Phillips, D. C. *Philosophy, Science and Social Inquiry: Contemporary Methodological Controversies in Social Science and Related Applied Fields of Research.* Oxford: Permagon Press, 1987.
Popper, Karl R. "Prediction and Prophecy in the Social Sciences." In *Conjectures and Refutations: The Growth of Scientific Knowledge*, pp. 336–346. New York: Harper Torchbooks, 1965.
———. *The Open Universe: An Argument for Indeterminism.* Totowa, N.J.: Rowman and Littlefield, 1982.
———. *The Logic of Scientific Discovery.* New York: Harper Torchbooks, 1959.
———. "The Logic of the Social Sciences." In *The Positivist Dispute in German Sociology.* London: Heinemann Educational Books, 1976.
———. *The Open Society and Its Enemies. Vols. 1, 2.* Princeton: Princeton University Press, 1962.
———. *The Poverty of Historicism.* New York: Harper Torchbooks, 1961.
Porpora, Douglas. "On the Prospects for a Nomothetic Theory of Social Structure." *Journal for the Theory of Social Behaviour* 13, (1983), 243–264.
Pratt, Vernon. *The Philosophy of the Social Sciences.* London: Methuen and Co., 1978.

Rabinow, Paul, and Sullivan, William (eds.). *Interpretive Social Science: A Reader.* Berkeley: University of California Press, 1979.

———. (eds.). *Interpretive Social Science: A Second Look.* Berkeley: University of California Press, 1987.

Radcliffe-Brown, A. R. *A Natural Science of Society.* Glencoe, Ill.: Free Press, 1957.

Railton, Peter. "A Deductive-Nomological Model of Probabilistic Explanation." *Philosophy of Science* 45 (1978), 206–226.

———. "Explanation and Metaphysical Controversy." In P. Kitcher and W. Salmon (eds.), *Scientific Explanation,* pp. 220–252. Minnesota Studies in the Philosophy of Science, vol. 13. Minneapolis: University of Minnesota Press, 1989.

———. "Probability, Explanation, and Information." *Synthese* 48 (1981), 233–256.

Rickman, H. P. *Understanding and the Human Studies.* London: Heinemann, 1967.

Riley, Gresham (ed.), *Values, Objectivity and the Social Sciences.* Reading, Mass.: Addison-Wesley, 1974.

Robinson, James Harvey. *The Humanizing of Knowledge.* New York: George H. Doran, 1923.

———. *The Mind in the Making: The Relation of Intelligence to Social Reform.* New York: Harper and Bros., 1921.

Rose, Arnold M. *Theory and Method in the Social Sciences.* Minneapolis: University of Minnesota Press, 1954.

Rosenberg, Alexander. *Microeconomic Laws: A Philosophical Analysis.* Pittsburgh: University of Pittsburgh Press, 1976.

———. *The Philosophy of Social Science.* Boulder, Colo.: Westview Press, 1988.

———. *Sociobiology and the Preemption of Social Science.* Baltimore: Johns Hopkins University Press, 1980.

Roth, Paul A. *Meaning and Method in the Social Sciences: A Case for Methodological Pluralism.* Ithaca: Cornell University Press, 1987.

Rothbard, Murray. *Individualism and the Philosophy of the Social Sciences.* San Francisco: Cato Institute, 1979.

Rudner, Richard. *Philosophy of Social Science.* Englewood Cliffs, N.J.: Prentice-Hall, 1966.

———. "Philosophy and Social Science." In E. C. Harwood (ed.), *Reconstruction of Economics.* Great Barrington, Mass.: American Institute for Economic Research, 1955.

Runciman, W. G. *Social Science and Political Theory.* Cambridge: Cambridge University Press, 1963.

Ryan, Alan. *The Philosophy of the Social Sciences.* New York: Pantheon Books, 1970.

——— (ed.). *The Philosophy of Social Explanation.* Oxford: Oxford University Press, 1973.

Sabia, Daniel R. and Wallulis, Jerald (eds.). *Changing Social Science.* Albany: SUNY Press, 1983.

Salmon, Merrilee. *Philosophy and Archaeology.* New York: Academic Press, 1982.

———. "Explanation in the Social Sciences." In P. Kitcher and W. Salmon (eds.), *Scientific Explanation,* pp. 384–409. Minnesota Studies in the Philosophy of Science, vol. 13. Minneapolis: University of Minnesota Press, 1989.

Salmon, Wesley. "Four Decades of Scientific Explanation." In P. Kitcher and W. Salmon (eds.), *Scientific Explanation,* pp. 3–219. Minnesota Studies in the Philosophy of Science, vol. 13. Minneapolis: University of Minnesota Press, 1989.

Sayer, Andrew. *Method in Social Science: A Realist Approach.* London: Hutchinson, 1984.

Schlenker, Barry R. "Social Psychology and Science." *Journal of Personality and Social Psychology* 27, no. 1 (1974), 1–15.

Scriven, Michael. "Explanations, Predictions, and Laws." In H. Feigl and G. Maxwell (eds.), Minnesota Studies in the Philosophy of Science, vol. 3. Minneapolis: University of Minnesota Press, 1962.

———. "Truisms as the Grounds for Historical Explanations." In P. Gardiner (ed.), *Theories of History,* pp. 443–471. Glencoe, Ill.: Free Press, 1959.

———. "Views of Human Nature." In T. W. Wann (ed.), *Behaviorism and Phenomenology: Contrasting Bases for Modern Psychology.* Chicago: University of Chicago Press, 1964.

———. "The Temporal Asymmetry of Explanations and Predictions." In Bernard Baumrin (ed.), *Philosophy of Science: The Delaware Seminar, vol. 1: 1961–1962.* New York: Interscience Publishers, 1963.

Simon, Herbert. "Bandwagon and Underdog Effects of Election Predictions." *Public Opinion Quarterly* 18 (1954), 245–253.

Simon, Michael. *Understanding Human Action: Social Explanation and the Vision of Social Science.* Albany: SUNY Press, 1982.

Smart, J. J. C. *Philosophy and Scientific Realism.* London: Routledge and Kegan Paul, 1963.

Sperber, Daniel. "Apparently Irrational Beliefs." In M. Hollis and S. Lukes (eds.), *Rationality and Relativism,* pp. 149–180. Cambridge: MIT Press, 1982.

Studdert-Kennedy, Gerald. *Evidence and Explanation in Social Science: An Interdisciplinary Approach.* London: Routledge and Kegan Paul, 1975.

Taylor, Charles. *Philosophy and the Human Sciences:* Philosophical Paper 2. Cambridge: Cambridge University Press, 1985.

Taylor, Charles. *The Explanation of Behaviour*. London: Routledge and Kegan Paul, 1964.

Thomas, David. *Naturalism and Social Science: A Post-Empiricist Philosophy of Social Science*. Cambridge: Cambridge University Press, 1979.

Trigg, Roger. *Understanding Social Science*. London: Blackwell, 1985.

Truzzi, Marcelo (ed.). *Verstehen*. Reading, Mass.: Addison-Wesley, 1974.

Van Parijs, Philippe. *Evolutionary Explanation in the Social Sciences*. Totowa, N.J.: Rowman and Littlefield, 1981.

Von Wright, George H. *Explanation and Understanding*. Ithaca: Cornell University Press, 1971.

Wallace, Walter. *The Logic of Science in Sociology*. New York: Aldine Publishing, 1971.

Wann, T. W. (ed.) *Behaviorism and Phenomenology: Contrasting Bases for Modern Psychology*. Chicago: University of Chicago Press, 1964.

Weber, Max. *The Methodology of the Social Sciences*. New York: Macmillan, 1950.

White, Morton. *Foundations of Historical Knowledge*. New York: Harper & Row, 1965.

Wilson, Bryan (ed.). *Rationality*. Oxford: Basil Blackwell, 1979.

Wimsatt, William C. "Reductionism, Levels of Organization, and the Mind-Body Problem." In G. G. Globus, G. Maxwell, and I. Savodnik (eds.), *Consciousness and the Brain: A Scientific and Philosophical Inquiry*, pp. 199–267. New York: Plenum Press, 1976.

———. "Complexity and Organization." In Kenneth F. Schaffner and Robert S. Cohen (eds.), *PSA 1972*, pp. 67–86. East Lansing, Michigan: The Philosophy of Science Association.

Winch, Peter. *The Idea of a Social Science and Its Relation to Philosophy*. London: Routledge and Kegan Paul, 1958.

Wisdom, J.O. *Philosophy of the Social Sciences I: A Metascientific Introduction*. London: Averbury, 1987.

———. *Philosophy of the Social Sciences II: Schemata*. London: Averbury, 1987.

Wootton, Barbara. *Testament for Social Science: An Essay in the Application of Scientific Method to Human Problems*. New York: W. W. Norton and Co., 1950.

Index